Computer and Machine Vision: Theory, Algorithms, Practicalities

This book is dedicated to my family.
To my late mother, Mary Davies, to record her never-failing
love and devotion.
To my late father, Arthur Granville Davies, who passed on to me
his appreciation of the beauties of mathematics and science.
To my wife, Joan, for love, patience, support, and inspiration.
To my children, Elizabeth, Sarah, and Marion, the music in my life.
To my grandson, Jasper, for reminding me of the carefree
joys of youth.

Computer and Machine Vision: Theory, Algorithms, Practicalities

Fourth Edition

E. R. DAVIES

Department of Physics
Royal Holloway,
University of London,
Egham, Surrey, UK

AMSTERDAM • BOSTON • HEIDELBERG • LONDON
NEW YORK • OXFORD • PARIS • SAN DIEGO
SAN FRANCISCO • SINGAPORE • SYDNEY • TOKYO
Academic Press is an imprint of Elsevier

Academic Press is an imprint of Elsevier
225 Wyman Street, Waltham, 02451, USA
The Boulevard, Langford Lane, Kidlington, Oxford OX5 1GB, UK

First edition 1990
Second edition 1997
Third edition 2005
Fourth edition 2012

Library of Congress Cataloging-in-Publication Data
A catalog record for this book is available from the Library of Congress

British Library Cataloguing-in-Publication Data
A catalogue record for this book is available from the British Library

ISBN: 978-0-12-386908-1

For information on all Elsevier publications
visit our website at elsevierdirect.com

Typeset by MPS Limited, a Macmillan Company, Chennai, India
www.macmillansolutions.com

Printed and bound by CPI Group (UK) Ltd, Croydon, CR0 4YY

Transferred to digital print 2012

Working together to grow
libraries in developing countries

www.elsevier.com | www.bookaid.org | www.sabre.org

ELSEVIER   BOOK AID International   Sabre Foundation

Contents

Topics Covered in Application Case Studies

	Tracking laparoscopic tools	Food inspection	Locating human irises and faces	Art, photography, image stitching	Locating contaminants in cereals	High speed cereal grain location	Surveillance	Monitoring of traffic flow	Model-based tracking of animals	Driver assistance systems	Road lane and sign detection	Locating vehicles and pedestrians	Vehicle guidance in agriculture	Designing inspection hardware
Affine motion models							√	√		√				
Belief networks							√	√						
Chamfer matching											√	√		
Circle and ellipse detection	√	√				√					√			√
Cost-speed tradeoffs														√
Decoupling shape and intensity									√					
Hough transform		√	√										√	√
Hysteresis thresholding													√	
Kalman filter							√	√		√				
Linear feature detection				√										
Median-filter based analysis				√										
Morphological processing	√			√										
Occlusion reasoning							√	√		√				
Optimal system design		√												√
Pattern recognition		√	√	√						√		√		
Perspective and vanishing points	√		√				√			√	√			
Principal components analysis									√					
RANSAC	√									√	√			
Real-time processing		√					√	√		√				√
Selection of hardware modules		√												√
Shape distortions		√		√										
Snake approximations and splines								√	√					
Speedup by sampling		√				√								√
Symmetrical object detection			√							√		√		
Temporal filtering							√	√	√	√				
Tracking and particle filters							√			√				
Two-stage matching				√										√

Influences Impinging upon Integrated Vision System Design

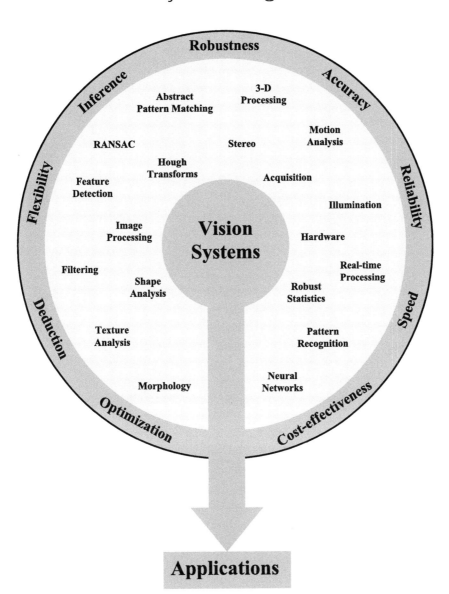

Influences Impinging upon Integrated Vision,
System Design

Robustness

Accuracy

Tolerance

3-D
Processing

Abstract
Pattern Matching

Motion
Analysis

Stereo

RANSAC

Reliability

Hough
Transforms

Feature
Detection

Flexibility

Acquisition

Illumination

Image
Processing

Hardware

Vision
Systems

Filtering

Real-time
Processing

Robust
Statistics

Shape
Analysis

Speed

Pattern
Recognition

Texture
Analysis

Discreteness

Neural
Networks

Morphology

Cost-effectiveness

Optimization

Applications

Foreword

Although computer vision is such a relatively young field of study, it has matured immensely over the last 25 years or so—from well-constrained, targeted applications to systems that learn automatically from examples.

Such progress over these 25 years has been spurred not least by mind-boggling advances in vision and computational hardware, making possible simple tasks that could take minutes on small images, now integrated as part of real-time systems that do far more in a fraction of a second on much larger images in a video stream.

This all means that the focus of research has been in a perpetual state of change, marked by near-exponential advances and achievements, and witnessed by the quality, and often quantity, of outstanding contributions to the field published in key conferences and journals such as ICCV and PAMI. These advances are most clearly reflected by the growing importance of the application areas in which the novel and real-time developments in computer vision have been applied to or developed for. Twenty-five years ago, industrial quality inspection and simple military applications ruled the waves, but the emphasis has since spread its wings, some slowly and some like wildfire, to many more areas, for example, from medical imaging and analysis to surveillance and, inevitably, complex military and space applications.

So how does Roy's book reflect this shift? Naturally, there are many fundamental techniques that remain the same, and this book is a wonderful treasure chest of tools that provides the fundamentals for any researcher and teacher. More modern and state-of-the-art methodologies are also covered in the book, most of them pertinent to the topical application areas currently driving not only the research agenda, but also the market forces. In short, the book is a direct reflection of the progress and key methodologies developed in computer vision over the last 25 years and more.

Indeed, while the third edition of this book was already an excellent, successful, and internationally popular work, this fourth edition is greatly enhanced and updated. All its chapters have been substantially revised and brought up to date by the inclusion of many new references covering advances in the subject made even in the past year. There are now also two entirely new chapters (to reflect the great strides that have been made in the area of video analytics) on surveillance and in-vehicle vision systems. The latter is highly relevant to the coming era of advanced driver assistance systems, and the former's importance and role requires no emphasis in this day and age where so many resources are dedicated to criminal and terrorist activity monitoring and prevention.

The material in the book is written in a way that is both approachable and didactic. It is littered with examples and algorithms. I am sure that this volume will be welcomed by a great many students and workers in computer and machine vision, including practitioners in academia and industry—from beginners who are starting out in the subject to advanced researchers and workers who need to gain insight into video analytics. I will also welcome it personally, for use by my own undergraduate and postgraduate students, and will value its presence on my bookshelf as an up-to-date reference on this important subject.

Finally, I am very happy to go on record as saying that Roy is the right person to have produced this substantial work. His long experience in the field of computer and machine vision surpasses even the "big bang" in computer vision around 25 years ago in the mid-80s when the Alvey Vision Conference (UK) and CVPR (USA) were only inchoates of what they have become today and reaches back to when ICPR and IAPR began to be dominated by image processing in the late 70s.

Majid Mirmehdi
University of Bristol, UK
September 2011

Preface

PREFACE TO THE FOURTH EDITION

The first edition came out in 1990, and was welcomed by many researchers and practitioners. However, in the subsequent two decades, the subject moved on at a rapidly accelerating rate, and many topics that hardly deserved a mention in the first edition had to be solidly incorporated in subsequent editions. It seemed particularly important to bring in significant amounts of new material on mathematical morphology, 3-D vision, invariance, motion analysis, object tracking, artificial neural networks, texture analysis, X-ray inspection, foreign object detection, and robust statistics. There are thus new chapters or appendices on these topics, and they have been carefully integrated with the existing material. The greater proportion of the new material has been included in Parts 3 and 4. So great has been the growth in work on 3-D vision and its applications that the original single chapter on 3-D vision had to be expanded into the set of *five* chapters on 3-D vision and motion forming Part 3, together with a further *two* chapters on surveillance and in-vehicle vision systems in Part 4. Indeed, these changes have been so radical that the title of the book has had to be modified to reflect them. At this stage, Part 4 encompasses such a range of chapters—covering applications and the components needed for constructing real-time visual pattern recognition systems— that it is difficult to produce a logical ordering for them: notably, the topics interact with each other at a variety of different levels—theory, algorithms, methodologies, practicalities, design constraints, and so on. However, this should not matter in practice, as the reader will be exposed to the essential richness of the subject, and his/her studies should be amply rewarded by increased understanding and capability.

It is worth remarking that, at this point in time, computer vision has attained a level of maturity that has made it substantially more rigorous, reliable, generic, and—in the light of the improved hardware facilities now available for its implementation (not least, FPGA and GPU types of solution)—capable of real-time performance. This means that workers are more than ever before using it in serious applications, and with fewer practical difficulties. It is intended that this edition of the book will reflect this radically new and exciting state of affairs at a fundamental level.

A typical final-year undergraduate course on vision for electronic engineering or computer science students might include much of the work of Chapters 1−10 and 14, 15, plus a selection of sections from other chapters, according to requirements. For MSc or PhD research students, a suitable lecture course might go on to cover Part 3 in depth, including several of the chapters in Part 4,[1] with many practical exercises being undertaken on an image analysis system. Here, much will depend on the research program being undertaken by each individual student. At this stage, the text will have to be used more as a handbook for research, and indeed, one of the prime aims of the volume is to act as a handbook for the researcher and practitioner in this important area.

As mentioned in the original Preface, this book leans heavily on experience I have gained from working with postgraduate students: in particular, I would like to express my gratitude to Mark Edmonds, Simon Barker, Daniel Celano, Darrel Greenhill, Derek Charles, Mark Sugrue, and Georgios Mastorakis, all of whom have in their own ways helped to shape my view of the subject. In addition, it is a special pleasure to recall very many rewarding discussions with my colleagues Barry Cook, Zahid Hussain, Ian Hannah, Dev Patel, David Mason, Mark Bateman, Tieying Lu, Adrian Johnstone, and Piers Plummer, the last two named having been particularly prolific in generating hardware systems for implementing my research group's vision algorithms. Next, I am immensely grateful to Majid Mirmehdi for reading much of the manuscript and making insightful comments and valuable suggestions. Finally, I am indebted to Tim Pitts of Elsevier Science for his help and encouragement, without which this fourth edition might never have been completed.

SUPPORTING MATERIALS

Elsevier's website for the book contains resources to help students and other readers using this text. For further information, go to the publisher's website:

http://www.elsevierdirect.com/companion.jsp?ISBN = 9780123869081

E. R. DAVIES
Royal Holloway,
University of London

PREFACE TO THE FIRST EDITION (1990)

Over the past 30 years or so, machine vision has evolved into a mature subject embracing many topics and applications: these range from automatic (robot) assembly to automatic vehicle guidance, from automatic interpretation of documents to verification of signatures, and from analysis of remotely sensed images to checking of fingerprints and human blood cells; currently, automated visual inspection is undergoing very substantial growth, necessary improvements in

---

[1]The importance of the appendix on robust statistics should not be underestimated once one gets onto serious work, although this will probably be outside the restrictive environment of an undergraduate syllabus.

quality, safety and cost-effectiveness being the stimulating factors. With so much ongoing activity, it has become a difficult business for the professional to keep up with the subject and with relevant methodologies: in particular, it is difficult to distinguish accidental developments from genuine advances. It is the purpose of this book to provide background in this area.

The book was shaped over a period of 10−12 years, through material I have given on undergraduate and postgraduate courses at London University, and contributions to various industrial courses and seminars. At the same time, my own investigations coupled with experience gained while supervising PhD and post-doctoral researchers helped to form the state of mind and knowledge that is now set out here. Certainly it is true to say that if I had had this book 8, 6, 4, or even 2 years ago, it would have been of inestimable value to myself for solving practical problems in machine vision. It is therefore my hope that it will now be of use to others in the same way. Of course, it has tended to follow an emphasis that is my own—and in particular one view of one path toward solving automated visual inspection and other problems associated with the application of vision in industry. At the same time, although there is a specialism here, great care has been taken to bring out general principles—including many applying throughout the field of image analysis. The reader will note the universality of topics such as noise suppression, edge detection, principles of illumination, feature recognition, Bayes' theory, and (nowadays) Hough transforms. However, the generalities lie deeper than this. The book has aimed to make some general observations and messages about the limitations, constraints, and tradeoffs to which vision algorithms are subject. Thus, there are themes about the effects of noise, occlusion, distortion and the need for built-in forms of robustness (as distinct from less successful *ad hoc* varieties and those added on as an afterthought); there are also themes about accuracy, systematic design, and the matching of algorithms and architectures. Finally, there are the problems of setting up lighting schemes which must be addressed in complete systems, yet which receive scant attention in most books on image processing and analysis. These remarks will indicate that the text is intended to be read at various levels—a factor that should make it of more lasting value than might initially be supposed from a quick perusal of the Contents.

Of course, writing a text such as this presents a great difficulty in that it is necessary to be highly selective: space simply does not allow everything in a subject of this nature and maturity to be dealt with adequately between two covers. One solution might be to dash rapidly through the whole area mentioning everything that comes to mind, but leaving the reader unable to understand anything in detail or to *achieve* anything having read the book. However, in a practical subject of this nature, this seemed to me a rather worthless extreme. It is just possible that the emphasis has now veered too much in the opposite direction, by coming down to practicalities (detailed algorithms, details of lighting schemes, and so on): individual readers will have to judge this for themselves. On the other hand, an author has to be true to himself and my view is that it is better for a reader or student to have mastered a coherent series of topics than to have a mish-mash of information that he is later unable to recall with any accuracy. This, then, is my justification for presenting this particular material in this particular way and for reluctantly omitting from detailed discussion such important topics as texture analysis, relaxation methods, motion, and optical flow.

As for the organization of the material, I have tried to make the early part of the book lead into the subject gently, giving enough detailed algorithms (especially in Chapters 2 and 6) to provide a sound feel for the subject—including especially vital, and in their own way quite intricate, topics such as connectedness in binary images. Hence, Part 1 provides the lead-in, although it is not always trivial material and indeed some of the latest research ideas have been brought in (e.g., on thresholding techniques and edge detection). Part 2 gives much of the meat of the book. Indeed, the (book) literature of the subject currently has a significant gap in the area of intermediate-level vision; while high-level vision (AI) topics have long caught the researcher's imagination, intermediate-level vision has its own difficulties which are currently being solved with great success (note that the Hough transform, originally developed in 1962, and by many thought to be a very specialist topic of rather esoteric interest, is arguably only now coming into its own). Part 2 and the early chapters of Part 3 aim to make this clear, while Part 4 gives reasons why this particular transform has become so useful. As a whole, Part 3 aims to demonstrate some of the practical applications of the basic work covered earlier in the book, and to discuss some of the principles underlying implementation: it is here that chapters on lighting and hardware systems will be found. As there is a limit to what can be covered in the space available, there is a corresponding emphasis on the theory underpinning practicalities. Probably, this is a vital feature, since there are many applications of vision both in industry and elsewhere, yet listing them and their intricacies risks dwelling on interminable detail, which some might find insipid; furthermore, detail has a tendency to date rather rapidly. Although the book could not cover 3-D vision in full (this topic would easily consume a whole volume in its own right), a careful overview of this complex mathematical and highly important subject seemed vital. It is therefore no accident that Chapter 16 is the longest in the book. Finally, Part 4 asks questions about the limitations and constraints of vision algorithms and answers them by drawing on information and experience from earlier chapters. It is tempting to call the last chapter the Conclusion. However, in such a dynamic subject area, any such temptation has to be resisted, although it has still been possible to draw a good number of lessons on the nature and current state of the subject. Clearly, this chapter presents a personal view but I hope it is one that readers will find interesting and useful.

About the Author

Roy Davies is Emeritus Professor of Machine Vision at Royal Holloway, University of London. He has worked on many aspects of vision, from feature detection and noise suppression to robust pattern matching and real-time implementations of practical vision tasks. His interests include automated visual inspection, surveillance, vehicle guidance, and crime detection. He has published more than 200 papers and three books—*Machine Vision: Theory, Algorithms, Practicalities* (1990), *Electronics, Noise and Signal Recovery* (1993), and *Image Processing for the Food Industry* (2000); the first of these has been widely used internationally for more than 20 years, and is now out in this much enhanced fourth edition. Roy is a Fellow of the IoP and the IET, and a Senior Member of the IEEE. He is on the Editorial Boards of *Real-Time Image Processing, Pattern Recognition Letters, Imaging Science,* and *IET Image Processing*. He holds a DSc at the University of London: he was awarded BMVA Distinguished Fellow in 2005 and Fellow of the International Association of Pattern Recognition in 2008.

Acknowledgements

The author would like to credit the following sources for permission to reproduce tables, figures and extracts of text from earlier publications:

Elsevier

For permission to reprint portions of the following papers from *Image and Vision Computing* as text in Chapters 5 and 14; as Tables 5.1–5.5; and as Figures 3.29, 5.2, 14.1, 14.2, 14.6:

> Davies (1984b, 1987c)
> Davies, E.R. (1991). *Image and Vision Computing* **9**, 252–261

For permission to reprint portions of the following paper from *Pattern Recognition* as text in Chapter 9; and as Figure 9.11:

> Davies and Plummer (1981)

For permission to reprint portions of the following papers from *Pattern Recognition Letters* as text in Chapters 3, 5, 11–14, 21, 24; as Tables 3.2; 12.3; 13.1; and as Figures 3.6, 3.8, 3.10; 5.1, 5.3; 11.1, 11.2a, 11.3b; 12.4, 12.5, 12.6, 12.7–12.10; 13.1, 13.3–13.11; 21.3, 21.6:

> Davies (1986a,b; 1987a,e,f; 1988b,c,e,f; 1989a)
> Davies et al. (2003a)

For permission to reprint portions of the following paper from *Signal Processing* as text in Chapter 3; and as Figures 3.15–3.20:

> Davies (1989b)

For permission to reprint portions of the following paper from *Advances in Imaging and Electron Physics* as text in Chapter 3:

> Davies (2003c)

For permission to reprint portions of the following article from *Encyclopedia of Physical Science and Technology* as Figures 9.9, 9.12, 10.1, 10.4:

> Davies, E.R. (1987). Visual inspection, automatic (robotics). In: Meyers, R.A. (ed.) *Encyclopedia of Physical Science and Technology, Vol. 14*. Academic Press, San Diego, pp. 360–377

The Committee of the Alvey Vision Club

For permission to reprint portions of the following paper as text in Chapter 14; and as Figures 14.1, 14.2, 14.6:

> Davies, E.R. (1988). An alternative to graph matching for locating objects from their salient features. *Proc. 4th Alvey Vision Conf., Manchester (31 August–2 September)*, pp. 281–286

CEP Consultants Ltd (Edinburgh)
For permission to reprint portions of the following paper as text in Chapter 20:

> Davies, E.R. (1987). Methods for the rapid inspection of food products and small parts. In: McGeough, J.A. (ed.) *Proc. 2nd Int. Conf. on Computer-Aided Production Engineering, Edinburgh (13−15 April)*, pp. 105−110

EURASIP
For permission to reprint portions of the following papers as text in Chapter 21; and as Figures 21.5, 21.7−21.11:

> Davies (1998)
> Davies et al. (1998b)

These papers were first published in the *Proceedings of the 9th European Signal Processing Conference (EUSIPCO-1998)* in *1998*, published by EURASIP.

IEEE
For permission to reprint portions of the following paper as text in Chapter 3; and as Figures 3.4, 3.5, 3.7, 3.11:

> Davies (1984a)

IET
For permission to reprint portions of the following papers from the *IET Proceedings and Colloquium Digests* as text in Chapters 3, 4, 6, 7, 13, 20−23, 25, 26; as Tables 3.3; 26.4, 26.5; and as Figures 3.21, 3.22, 3.24; 4.8−4.12; 6.5−6.9, 6.12; 7.6−7.8; 13.12; 20.2−20.4; 21.1, 21.2, 21.12; 22.16−22.18; 23.1, 23.3, 23.4; 25.4−25.8:

> Davies (1985; 1988a; 1997b; 1999f; 2000b,c; 2005; 2008b)
> Davies, E.R. (1997). Algorithms for inspection: constraints, tradeoffs and the design process. IEE Digest no. 1997/041, Colloquium on *Industrial Inspection*, IEE (10 Feb.), pp. 6/1−5
> Sugrue and Davies (2007)
> Mastorakis and Davies (2011)
> Davies et al. (1998a)
> Davies and Johnstone (1989)

IFS Publications Ltd
For permission to reprint portions of the following paper as text in Chapters 12, 20; and as Figures 12.1, 12.2, 20.5:

> Davies (1984c)

The Council of the Institution of Mechanical Engineers
For permission to reprint portions of the following paper as text in Chapter 26; and as Tables 26.1, 26.2:

Davies and Johnstone (1986)

MCB University Press (Emerald Group)
For permission to reprint portions of the following paper as Figure 20.6:

Patel et al. (1995)

The Royal Photographic Society
For permission to reprint portions of the following papers[1] as text in Chapter 3; as Table 3.4; and as Figures 3.12, 3.13, 3.25–3.28:

Davies (2000f)
Charles and Davies (2004)

Springer-Verlag
For permission to reprint portions of the following papers as text in Chapters 6, 21; and as Figures 6.2, 6.4:

Davies (1988d), Figs. 1–3
Davies, E.R. (2003). Design of object location algorithms and their use for food and cereals inspection. Chapter 15 in Graves, M. and Batchelor, B.G. (eds.). *Machine Vision Techniques for Inspecting Natural Products*. Springer-Verlag, pp. 393–420

Peter Stevens Photography
For permission to reprint a photograph as Figure 3.12(a).

F.H. Sumner
For permission to reprint portions of the following article from *State of the Art Report: Supercomputer Systems Technology* as text in Chapter 9; and as Figure 9.4:

Davies, E.R. (1982). Image processing. In: Sumner, F.H. (ed.) *State of the Art Report: Supercomputer Systems Technology*. Pergamon Infotech, Maidenhead, pp. 223–244

---

[1]See also the Maney website: www.maney.co.uk/journals/ims.

Unicom Seminars Ltd

For permission to reprint portions of the following paper as text in Chapter 20; and as Figures 11.2b, 11.3a, 12.3, 12.13, 13.12c, 14.8:

> Davies, E.R. (1988). Efficient image analysis techniques for automated visual inspection. *Proc. Unicom Seminar on Computer Vision and Image Processing, London (29 November–1 December)*, pp. 1–19

World Scientific

For permission to reprint portions of the following book as text in Chapters 7, 21, 22, 23, 26; and as Figures 7.1–7.4, 21.4, 22.20, 23.15, 23.16, 26.3:

> Davies (2000a)

Royal Holloway, University of London

For permission to reprint extracts from the following examination questions, originally written by E.R. Davies:

> EL385/97/2; EL333/98/2; EL333/99/2, 3, 5, 6; EL333/01/2, 4–6; PH5330/98/3, 5; PH5330/03/1–5; PH4760/04/1–5

University of London

For permission to reprint extracts from the following examination questions, originally written by E.R. Davies:

> PH385/92/2, 3; PH385/93/1–3; PH385/94/1–4; PH385/95/4; PH385/96/3, 6; PH433/94/3, 5; PH433/96/2, 5

Glossary of Acronyms and Abbreviations

1-D	one dimension/one-dimensional
2-D	two dimensions/two-dimensional
3-D	three dimensions/three-dimensional
3DPO	3-D part orientation system
ACM	Association for Computing Machinery (USA)
ADAS	advanced driver assistance system
ADC	analog to digital converter
AI	artificial intelligence
ANN	artificial neural network
APF	auxiliary particle filter
ASCII	American Standard Code for Information Interchange
ASIC	application-specific integrated circuit
ATM	automated teller machine
AUC	area under curve
AVI	audio video interleave
BCVM	between-class variance method
BetaSAC	beta [distribution] sampling consensus
BMVA	British Machine Vision Association
BRAM	block of RAM
BRDF	bidirectional reflectance distribution function
CAD	computed-aided design
CAM	computer-aided manufacture
CCD	charge-coupled device
CCTV	closed-circuit television
CDF	cumulative distribution function
CIM	computer integrated manufacture
CLIP	cellular logic image processor
CPU	central processing unit
DCSM	distinct class based splitting measure
DET	Beaudet determinant operator
DEXA	dual-emission X-ray absorptiometry
DG	differential gradient
DN	Dreschler–Nagel corner detector
DoF	degree of freedom
DoG	difference of Gaussians
DSP	digital signal processor
EM	expectation maximization
EURASIP	European Association for Signal Processing
FAST	features from accelerated segment test
FFT	fast Fourier transform

FN	false negative
fnr	false negative rate
FoE	focus of expansion
FoV	field of view
FP	false positive
FPGA	field programmable gate array
FPP	full perspective projection
fpr	false positive rate
GHT	generalized Hough transform
GLOH	gradient location and orientation histogram
GMM	Gaussian mixture model
GPS	global positioning system
GPU	graphics processing unit
GroupSAC	group sampling consensus
GVM	global valley method
HOG	histogram of orientated gradients
HSI	hue, saturation, intensity
HT	Hough transform
IBR	intensity extrema-based region detector
IDD	integrated directional derivative
IEE	Institution of Electrical Engineers (UK)
IEEE	Institute of Electrical and Electronics Engineers (USA)
IET	Institution of Engineering and Technology (UK)
ILW	iterated likelihood weighting
IMechE	Institution of Mechanical Engineers (UK)
IMPSAC	importance sampling consensus
ISODATA	iterative self-organizing data analysis
JPEG/JPG	Joint Photographic Experts Group
k-NN	*k*-nearest neighbor
KR	Kitchen–Rosenfeld corner detector
LED	light emitting diode
LFF	local-feature-focus method
LIDAR	light detection and ranging
LMedS	least median of squares
LoG	Laplacian of Gaussian
LS	least squares
LUT	lookup table
MAP	maximum a posteriori
MDL	minimum description length
MIMD	multiple instruction stream, multiple data stream
MIPS	millions of instructions per second
MISD	multiple instruction stream, single data stream
MLP	multi-layer perceptron

MoG	mixture of Gaussians
MP	microprocessor
MSER	maximally stable extremal region
NAPSAC	*n* adjacent points sample consensus
NIR	near infra-red
NN	nearest neighbor
OCR	optical character recognition
PC	personal computer
PCA	principal components analysis
PCB	printed circuit board
PE	processing element
P*n*P	perspective *n*-point
PR	pattern recognition
PROSAC	progressive sample consensus
PSF	point spread function
RAM	random access memory
RANSAC	random sample consensus
RGB	red, green, blue
RHT	randomized Hough transform
RKHS	reproducible kernel Hilbert space
RMS	root mean square
ROC	receiver operating characteristic
RoI	region of interest
RPS	Royal Photographic Society (UK)
SFOP	scale-invariant feature operator
SIAM	Society of Industrial and Applicative Mathematics
SIFT	scale-invariant feature transform
SIMD	single instruction stream, multiple data stream
SIR	sampling importance resampling
SIS	sequential importance sampling
SISD	single instruction stream, single data stream
SOC	sorting optimization curve
SOM	self-organizing map
SPIE	Society of Photo-optical Instrumentation Engineers
SPR	statistical pattern recognition
STA	spatiotemporal attention [neural network]
SURF	speeded-up robust features
SUSAN	smallest univalue segment assimilating nucleus
SVM	support vector machine
TM	template matching
TMF	truncated median filter
TN	true negative
tnr	true negative rate

TP	true positive
tpr	true positive rate
TV	television
ULUT	universal lookup table
USEF	unit step edge function
VLSI	very large scale integration
VMF	vector median filter
VP	vanishing point
WPP	weak perspective projection
ZH	Zuniga—Haralick corner detector

Vision, the Challenge

1.1 INTRODUCTION—MAN AND HIS SENSES

Of the five senses—vision, hearing, smell, taste, and touch—vision is undoubtedly the one that man has come to depend upon above all others, and indeed the one that provides most of the data he receives. Not only do the input pathways from the eyes provide megabits of information at each glance but the data rates for continuous viewing probably exceed 10 megabits per second (mbit/s). However, much of this information is redundant and is compressed by the various layers of the visual cortex, so that the higher centers of the brain have to interpret abstractly only a small fraction of the data. Nonetheless, the amount of information the higher centers receive from the eyes must be at least two orders of magnitude greater than all the information they obtain from the other senses.

Another feature of the human visual system is the ease with which interpretation is carried out. We see a scene as it is—trees in a landscape, books on a desk, widgets in a factory. No obvious deductions are needed and no overt effort is required to interpret each scene: in addition, answers are effectively immediate and are normally available within a tenth of a second. Just now and again some doubt arises—e.g. a wire cube might be "seen" correctly or inside out. This and a host of other optical illusions are well known, although for the most part we can regard them as curiosities—irrelevant freaks of nature. Somewhat surprisingly, illusions are quite important, since they reflect hidden assumptions that the brain is making in its struggle with the huge amounts of complex visual data it is receiving. We have to pass by this story here (although it resurfaces now and again in various parts of this book). However, the important point is that we are for the most part unaware of the complexities of vision. Seeing is not a simple process: it is just that vision has evolved over millions of years, and there was no particular advantage in evolution giving us any indication of the difficulties of the task (if anything, to have done so would have cluttered our minds with irrelevant information and slowed our reaction times).

In the present day and age, man is trying to get machines to do much of his work for him. For simple mechanistic tasks this is not particularly difficult, but for more complex tasks the machine must be given the sense of vision. Efforts have been made to achieve this, sometimes in modest ways, for well over 30 years. At first, schemes were devised for reading, for interpreting chromosome images, and so on, but when such schemes were confronted with rigorous practical tests, the problems often turned out to be more difficult. Generally, researchers react to finding that apparent "trivia" are getting in the way by intensifying their efforts and applying great ingenuity, and this was certainly so with early efforts at vision algorithm design. Hence, it soon became evident that the task really is a complex one, in which numerous fundamental problems confront the researcher, and the ease with which the eye can interpret scenes turned out to be highly deceptive.

Of course, one of the ways in which the human visual system gains over the machine is that the brain possesses more than 10^{10} cells (or neurons), some of which have well over 10,000 contacts (or synapses) with other neurons. If each neuron acts as a type of microprocessor, then we have an immense computer in which all the processing elements can operate concurrently. Taking the largest single man-made computer to contain several hundred million rather modest processing elements, the majority of the visual and mental processing tasks that the eye−brain system can perform in a flash have no chance of being performed by present-day man-made systems. Added to these problems of scale, there is the problem of how to organize such a large processing system, and also how to program it. Clearly, the eye−brain system is partly hard-wired by evolution but there is also an interesting capability to program it dynamically by training during active use. This need for a large parallel processing system with the attendant complex control problems shows that machine vision must indeed be one of the most difficult intellectual problems to tackle.

So what are the problems involved in vision that make it apparently so easy for the eye, yet so difficult for the machine? In the next few sections an attempt is made to answer this question.

1.2 THE NATURE OF VISION

1.2.1 The Process of Recognition

This section illustrates the intrinsic difficulties of implementing machine vision, starting with an extremely simple example—that of character recognition. Consider the set of patterns shown in Fig. 1.1(a). Each pattern can be considered as a set of 25 bits of information, together with an associated class indicating its interpretation. In each case imagine a computer learning the patterns and their classes by rote. Then any new pattern may be classified (or "recognized") by comparing it with

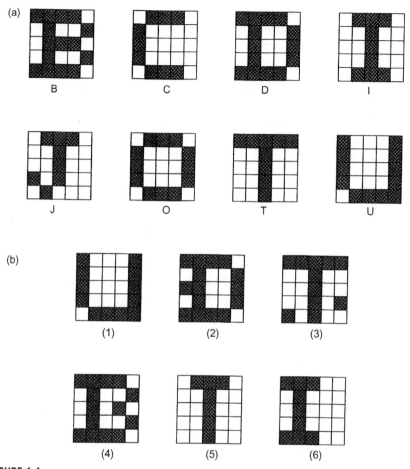

FIGURE 1.1

Some simple 25-bit patterns and their recognition classes used to illustrate some of the basic problems of recognition: (a) training set patterns (for which the known classes are indicated); (b) test patterns.

this previously learnt "training set," and assigning it to the class of the nearest pattern in the training set. Clearly, test pattern (1) (Fig. 1.1(b)) will be allotted to class U on this basis. Chapter 24 shows that this method is a simple form of the nearest-neighbor approach to pattern recognition.

The scheme outlined above seems straightforward and is indeed highly effective, even being able to cope with situations where distortions of the test patterns occur or where noise is present: this is illustrated by test patterns (2) and (3). However, this approach is not always foolproof. First, there are situations where distortions or noise are excessive, so errors of interpretation arise. Second, there are situations where patterns are not badly distorted or subject to obvious noise, yet are misinterpreted: this seems much more serious, since it indicates an unexpected limitation of the technique rather than a reasonable result of noise or distortion. In particular, these problems arise where the test pattern is displaced or misorientated relative to the appropriate training set pattern, as with test pattern (6).

As will be seen in Chapter 24, there is a powerful principle that indicates why the unlikely limitation given above can arise: it is simply that there are *insufficient training set patterns*, and that those that are present are *insufficiently representative* of what will arise in practical situations. Unfortunately, this presents a major difficulty, since providing enough training set patterns incurs a serious storage problem, and an even more serious search problem when patterns are tested. Furthermore, it is easy to see that these problems are exacerbated as patterns become larger and more real (obviously, the examples of Fig. 1.1 are far from having enough resolution even to display normal type-fonts). In fact, a combinatorial explosion[1] takes place. Forgetting for the moment that the patterns of Fig. 1.1 have familiar shapes, let us temporarily regard them as random bit patterns. Now the number of bits in these $N \times N$ patterns is N^2, and the number of possible patterns of this size is 2^{N^2}: even in a case where $N = 20$, remembering all these patterns and their interpretations would be impossible on any practical machine, and searching systematically through them would take impracticably long (involving times of the order of the age of the universe). Thus, it is not only impracticable to consider such brute-force means of solving the recognition problem, it is effectively also impossible theoretically. These considerations show that other means are required to tackle the problem.

1.2.2 Tackling the Recognition Problem

An obvious means of tackling the recognition problem is to standardize the images in some way. Clearly, normalizing the position and orientation of any 2-D picture object would help considerably: indeed this would reduce the number of degrees of freedom by three. Methods for achieving this involve centralizing the objects— arranging their centroids at the center of the normalized image—and making their major axes (deduced by moment calculations, for example) vertical or horizontal. Next, we can make use of the order that is known to be present in the image—and here it may be noted that very few patterns of real interest are indistinguishable from random dot patterns. This approach can be taken further: if patterns are to be nonrandom, isolated noise points may be eliminated. Ultimately, all these methods help by making the test pattern closer to a restricted set of training set patterns (although care must also be taken to process the training set patterns initially so that they are representative of the processed test patterns).

It is useful to consider character recognition further. Here, we can make additional use of what is known about the structure of characters—namely, that they consist of

---

[1]This is normally taken to mean that one or more parameters produce fast-varying (often exponential) effects, which "explode" as the parameters increase by modest amounts.

FIGURE 1.2

Use of thinning to regularize character shapes. Here, character shapes of different limb widths—or even varying limb widths—are reduced to stick figures or skeletons. Thus, irrelevant information is removed and at the same time recognition is facilitated.

limbs of roughly constant width. In that case the width carries no useful information, so the patterns can be thinned to stick figures (called skeletons—see Chapter 9); then, hopefully, there is an even greater chance that the test patterns will be similar to appropriate training set patterns (Fig. 1.2). This process can be regarded as another instance of reducing the number of degrees of freedom in the image, and hence of helping to minimize the combinatorial explosion—or, from a practical point of view, to minimize the size of the training set necessary for effective recognition.

Next, consider a rather different way of looking at the problem. Recognition is necessarily a problem of discrimination—i.e. of discriminating between patterns of different classes. However, in practice, considering the natural variation of patterns, including the effects of noise and distortions (or even the effects of breakages or occlusions), there is also a problem of *generalizing* over patterns of the same class. In practical problems there is a tension between the need to discriminate and the need to generalize. Nor is this a fixed situation. Even for the character recognition task, some classes are so close to others (*n*'s and *h*'s will be similar) that less generalization is possible than in other cases. On the other hand, extreme forms of generalization arise when, e.g., an *A* is to be recognized as an *A* whether it is a capital or small letter, or in italic, bold, suffix or other form of font—even if it is handwritten. The variability is determined largely by the training set initially provided. What we emphasize here, however, is that generalization is as necessary a prerequisite to successful recognition as is discrimination.

At this point it is worth considering more carefully the means whereby generalization was achieved in the examples cited above. First, objects were positioned and orientated appropriately; second, they were cleaned of noise spots; and third, they were thinned to skeleton figures (although the latter process is relevant only for certain tasks such as character recognition). In the last case we are generalizing over characters drawn with all possible limb widths, width being an irrelevant degree of freedom for this type of recognition task. Note that we could have generalized the characters further by normalizing their size and saving another degree of freedom. The common feature of all these processes is that they aim to give the characters a high level of standardization against known types of variability before finally attempting to recognize them.

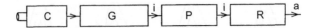

FIGURE 1.3

The two-stage recognition paradigm: C, input from camera; G, grab image (digitize and store); P, preprocess; R, recognize (i, image data; a, abstract data). The classical paradigm for object recognition is that of (i) preprocessing (image processing) to suppress noise or other artifacts and to regularize the image data, and (ii) applying a process of abstract (often statistical) pattern recognition to extract the very few bits required to classify the object.

The standardization (or generalization) processes outlined above are all realized by image processing, i.e. the conversion of one image into another by suitable means. The result is a two-stage recognition scheme: first, images are converted into more amenable forms containing the same numbers of bits of data; and second, they are classified, with the result that their data content is reduced to very few bits (Fig. 1.3). In fact, recognition is a process of data abstraction, the final data being abstract and totally unlike the original data. Thus, we must imagine a letter A starting as an array of perhaps 20 bits $\times$ 20 bits arranged in the form of an A, and then ending as the 7 bits in an ASCII representation of an A, namely 1000001 (which is essentially a random bit pattern bearing no resemblance to an A).

The last paragraph reflects to a large extent the history of image analysis. Early on, a good proportion of the image analysis problems being tackled were envisaged as consisting of an image "preprocessing" task carried out by image processing techniques, followed by a recognition task undertaken by statistical pattern recognition methods (Chapter 24). These two topics—image processing and statistical pattern recognition—consumed much research effort and effectively dominated the subject of image analysis, while "intermediate-level" approaches such as the Hough transform were, for the time, slower to develop. One of the aims of this book is to ensure that such intermediate-level processing techniques are given due emphasis, and indeed the best range of techniques is applied to any machine vision task.

1.2.3 Object Location

The problem that was tackled in the previous section—that of character recognition—is a highly constrained one. In a great many practical applications it is necessary to search pictures for objects of various types, rather than just interpreting a small area of a picture.

Search is a task that can involve prodigious amounts of computation and which is also subject to a combinatorial explosion. Imagine the task of searching for a letter E in a page of text. An obvious way of achieving this is to move a suitable "template" of size $n \times n$ over the whole image, of size $N \times N$, and to find where a match occurs (Fig. 1.4). A match can be defined as a position where

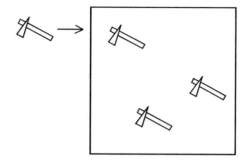

FIGURE 1.4

Template matching, the process of moving a suitable template over an image to determine the precise positions at which a match occurs—hence revealing the presence of objects of a particular type.

there is exact agreement between the template and the local portion of the image but, in keeping with the ideas of Section 1.2.1, it will evidently be more relevant to look for a best local match (i.e. a position where the match is locally better than in adjacent regions) and where the match is also good in some more absolute sense, indicating that an E is present.

One of the most natural ways of checking for a match is to measure the Hamming distance between the template and the local $n \times n$ region of the image, i.e. to sum the number of differences between corresponding bits. This is essentially the process described in Section 1.2.1. The places with a low Hamming distance are places where the match is good. These template matching ideas can be extended to cases where the corresponding bit positions in the template and the image do not just have binary values but may have intensity values over a range 0–255. In that case the sums obtained are no longer Hamming distances but may be generalized to the form:

$$\mathscr{D} = \sum_t |I_i - I_t| \tag{1.1}$$

where I_t is the local template value, I_i is the local image value, and the sum is taken over the area of the template. This makes template matching practicable in many situations: the possibilities are examined in more detail in subsequent chapters.

We referred above to a combinatorial explosion in this search problem too. The reason this arises is as follows. First, when a 5×5 template is moved over an $N \times N$ image in order to look for a match, the number of operations required is of the order of $5^2 N^2$, totaling some 1 million operations for a 256×256 image. The problem is that when larger objects are being sought in an image, the number of operations increases as the square of the size of the object, the total number of operations being N^2n^2 when an $n \times n$ template is used.[2] For a 30×30 template and a 256×256 image, the number of operations required rises to ~ 60 million.

Next, recall that in general, objects may appear in many orientations in an image (E's on a printed page are exceptional). If we imagine 360 possible orientations (i.e. one per degree of rotation), then a corresponding number of templates will in principle have to be applied in order to locate the object. This additional degree of freedom pushes the search effort and time to enormous levels, so far away from the possibility of real-time[3] implementation that new approaches must be found for tackling the task. Fortunately, many researchers have applied their minds to this problem and there are a good many ideas for tackling it. Perhaps the most important general means for saving effort on this sort of scale is that of two-stage (or multistage) template matching. The principle is to search for objects via their features. For example, we might consider searching for E's by looking for characters that have horizontal line segments within them. Similarly, we might search for hinges on a manufacturer's conveyor by looking first for the screw holes they possess. In general, it is useful to look for small features, since they require smaller templates and hence involve significantly less computation, as demonstrated above. This means that it may be better to search for E's by looking for corners instead of horizontal line segments.

Unfortunately, noise and distortions give rise to problems if we search for objects via small features—there is a risk of missing the object altogether. Hence, it is necessary to collate the information from a number of such features. This is the point where the many available methods start to differ from each other. How many features should be collated? Is it better to take a few larger features than a lot of smaller ones? And so on. Also, we have not answered in full the question of what types of feature are the best to employ. These and other questions are considered in the following chapters.

Indeed, in a sense, these questions are the subject of this book. Search is one of the fundamental problems of vision, yet the details and the application of the basic idea of two-stage template matching give the subject much of its richness: to solve the recognition problem the dataset needs to be explored carefully. Clearly, any answers will tend to be data-dependent but it is worth exploring to what extent there are generalized solutions to the problem.

1.2.4 Scene Analysis

The last subsection considered what is involved in searching an image for objects of a certain type: the result of such a search is likely to be a list of centroid

---

[2]Note that, in general, a template will be larger than the object it is used to search for, because some background will have to be included to help demarcate the object.

[3]A commonly used phrase meaning that the information has to be processed as it becomes available: this contrasts with the many situations (such as the processing of images from space-probes) where the information may be stored and processed at leisure.

coordinates for these objects, although an accompanying list of orientations might also be obtained. This subsection considers what is involved in scene analysis—the activity we are continually engaged in as we walk around, negotiating obstacles, finding food, and so on. Scenes contain a multitude of objects, and it is their interrelationships and relative positions that matter as much as identifying what they are. There is often no need for a search *per se* and we could in principle passively take in what is in the scene. However, there is much evidence (e.g. from analysis of eye movements) that the eye—brain system interprets scenes by continually asking questions about what is there. For example, we might ask the following questions: Is this a lamp-post? How far away is it? Do I know this person? Is it safe to cross the road? And so on. It is not the purpose here to dwell on these human activities or introspection about them but merely to observe that scene analysis involves enormous amounts of input data, complex relationships between objects within scenes and, ultimately, descriptions of these complex relationships. The latter no longer take the form of simple classification labels, or lists of object coordinates, but have a much richer information content: indeed, a scene will, to a first approximation, be better described in English than as a list of numbers. It seems likely that a much greater combinatorial explosion is involved in determining relationships between objects than in merely identifying and locating them. Hence, all sorts of props must be used to aid visual interpretation: there is a considerable evidence of this in the human visual system, where contextual information and the availability of immense databases of possibilities clearly help the eye to a considerable degree.

Note also that scene descriptions may initially be at the level of factual content but will eventually be at a deeper level—that of meaning, significance, and relevance. However, we shall not be able to delve further into these areas in this book.

1.2.5 Vision as Inverse Graphics

It has often been said that vision is "merely" inverse graphics. There is a certain amount of truth in this. Computer graphics is the generation of images by computer, starting from abstract descriptions of scenes and a knowledge of the laws of image formation. Clearly, it is difficult to quarrel with the idea that vision is the process of obtaining descriptions of sets of objects, starting from sets of images and a knowledge of the laws of image formation (indeed, it is good to see a definition that explicitly brings in the need to know the laws of image formation, since it is all too easy to forget that this is a prerequisite when building descriptions incorporating heuristics that aid interpretation).

However, this similarity in formulation of the two processes hides some fundamental points. First, graphics is a "feedforward" activity, i.e. images can be produced straightforwardly once sufficient specification about the viewpoint and the objects, and knowledge of the laws of image formation, has been obtained. True, considerable computation may be required but the process is entirely determined and predictable. The situation is not so straightforward for vision because search is involved and there is an accompanying combinatorial explosion. Indeed, some vision packages incorporate graphics (or CAD) packages (Tabandeh and Fallside, 1986) that are inserted into feedback loops for interpretation: the graphics package is then guided iteratively until it produces an acceptable approximation to the input image, when its input parameters embody the correct interpretation (there is a close parallel here with the problem of designing analog-to-digital converters by making use of digital-to-analog converters). Hence, it seems inescapable that vision is intrinsically more complex than graphics.

We can clarify the situation somewhat by noting that, as a scene is observed, a 3-D environment is compressed into a 2-D image and a considerable amount of depth and other information is lost. This can lead to ambiguity of interpretation of the image (both a helix viewed end-on and a circle project into a circle), so the 3-D to 2-D transformation is many-to-one. Conversely, the interpretation must be one-to-many, meaning that there are many possible interpretations, yet we know that only one can be correct: vision involves not merely providing a list of all possible interpretations but providing the most likely one. Hence, some additional rules or constraints must be involved in order to determine the single most likely interpretation. Graphics, in contrast, does not have these problems, as the above ideas show it to be a many-to-one process.

1.3 FROM AUTOMATED VISUAL INSPECTION TO SURVEILLANCE

So far we have considered the nature of vision but not what man-made vision systems may be used for. There is in fact a great variety of applications for artificial vision systems—including, of course, all of those for which man employs his visual senses. Of particular interest in this book are surveillance, automated inspection, robot assembly, vehicle guidance, traffic monitoring and control, biometric measurement, and analysis of remotely sensed images. By way of example, fingerprint analysis and recognition have long been important applications of computer vision, as have the counting of red blood cells, signature verification and character recognition, and aeroplane identification (both from aerial silhouettes and from ground surveillance pictures taken from satellites). Face recognition and even iris recognition have become practical possibilities, and vehicle guidance by vision will, in principle, soon be sufficiently reliable for urban use.[4]

Among the main applications of vision considered in this book are those of manufacturing industry—particularly, automated visual inspection and vision for

---

[4]Whether the public will accept this, with all its legal implications, is another matter, but note that radar blind-landing aids for aircraft have been in wide use for some years. In fact, last-minute automatic action to prevent accidents is a good compromise (see Chapter 23 for a related discussion on driver assistance schemes).

automated assembly. In these cases, much the same manufactured components are viewed by cameras: the difference lies in how the resulting information is used. In assembly, components must be located and orientated so that a robot can pick them up and assemble them. For example, the various parts of a motor or brake system need to be taken in turn and put into correct positions, or a coil may have to be mounted on a TV tube, an integrated circuit placed on a printed circuit board, or a chocolate placed into a box. At the time of inspection, objects may pass the inspection station on a moving conveyor at rates typically between 10 and 30 items per second and it has to be ascertained whether they have any defects. If any defects are detected, the offending parts will usually have to be rejected, i.e. the feedforward solution. In addition, a feedback solution may be instigated—i.e. some parameter may have to be adjusted to control plant further back down the production line (this is especially true for parameters that control dimensional characteristics such as product diameter). Inspection also has the potential for amassing a wealth of information that is useful for management, on the state of the parts coming down the line: the total number of products per day, the number of defective products per day, the distribution of sizes of products, and so on. The important feature of artificial vision is that it is tireless and that *all* products can be scrutinized and measured; thus, quality control can be maintained to a very high standard. In automated assembly too, a considerable amount of on-the-spot inspection can be performed and this may help to avoid the problem of complex assemblies being rejected, or having to be subjected to expensive repairs, because (for example) a proportion of screws was threadless and could not be inserted properly.

An important feature of most industrial tasks is that they take place in real time: if it is used, machine vision must be able to keep up with the manufacturing process. For assembly, this may not be too exacting a problem, since a robot may not be able to pick up and place more than one item per second—leaving the vision system a similar time to do its processing. For inspection, this supposition is rarely valid: even a single automated line (e.g. one for stoppering bottles) is able to keep up a rate of 10 items per second (and, of course, parallel lines are able to keep up much higher rates). Hence, visual inspection tends to press computer hardware very hard. Note in addition that many manufacturing processes operate under severe financial constraints so it is not possible to employ expensive multiprocessing systems or supercomputers. Hence, great care must be taken while designing of hardware accelerators for inspection applications. Chapter 26 aims to give some insight into these hardware problems.

Finally, we return to the starting discussion, about the huge variety of applications of machine vision—and it is interesting to consider surveillance tasks as the outdoor analogs of automated inspection (indeed, it is amusing to imagine that cars speeding along a road are just as subject to inspection as products speeding along a product line!). In fact, they have recently been acquiring close to exponentially increasing application. Thus, the techniques used for inspection have acquired an injection of vitality, and many more techniques have been developed. Naturally, this has meant the introduction of whole tranches of new subject matter, such as motion analysis and perspective invariants (see Part 3). It is also interesting that such techniques add a new richness to such old topics as face recognition (Section 17.10.1).

1.4 WHAT THIS BOOK IS ABOUT

The foregoing sections have examined something of the nature of machine vision and have briefly considered its applications and implementation. It is already clear that implementing machine vision involves considerable practical difficulties but, more important, these practical difficulties embody substantial fundamental problems: these include various factors giving rise to excessive processing load and time. Practical problems may be overcome by ingenuity and care; however, by definition, truly fundamental limitations cannot be overcome by *any* means—the best that we can hope for is that we will be able to minimize their effects following a complete understanding of their nature.

Understanding is thus a cornerstone for success in machine vision. It is often difficult to achieve, since the dataset (i.e. all pictures that could reasonably be expected to arise) is highly variegated. Indeed, much investigation is required to determine the nature of a given dataset, including not only the objects being observed but also the noise levels, degrees of occlusion, breakage, defect, and distortion that are to be expected, and the quality and nature of the lighting. Ultimately, sufficient knowledge might be obtained in a useful set of cases so that a good understanding of the milieu can be attained. Then it remains to compare and contrast the various methods of image analysis that are available. Some methods will turn out to be quite unsatisfactory for reasons of robustness, accuracy, or cost of implementation, or other relevant variables: and who is to say in advance what a relevant set of variables is? This, too, needs to be ascertained and defined. Finally, among the methods that could reasonably be used, there will be competition: tradeoffs between parameters such as accuracy, speed, robustness, and cost will have to be worked out first theoretically and then in numerical detail to find an optimal solution. This is a complex and long process in a situation where workers have in the past aimed to find solutions for their own particular (often short-term) needs. Clearly, there is a need to ensure that practical machine vision advances from an art to a science. Fortunately, this process has been developing for some years, and it is one of the aims of this book to throw additional light on the problem.

Before proceeding further, there are one or two more pieces to fit into the jigsaw. First, there is an important guiding principle: *if the eye can do it, so can the machine*. Thus, if an object is fairly well hidden in an image, yet the eye can see it and track it, then it should be possible to devise a vision algorithm that can do the same. Next, although we can expect to meet this challenge, should we set our sights even higher and aim to devise algorithms that can beat the eye? There seems no reason to suppose that the eye is the ultimate vision machine: it has been built through the vagaries of evolution, so it may be well adapted for finding berries or nuts, or for recognizing faces, but ill-suited for certain other tasks. One such task is that of measurement. The eye probably does not need to measure the sizes of objects, at a glance, to better than a few percent accuracy. However, it could be distinctly useful if the robot eye could achieve remote size measurement, at a glance, and with an accuracy of say 0.001%. Clearly, the robot eye could acquire capabilities superior to those of biological systems. Again, this book aims to point out such possibilities where they exist.

Finally, it will be useful to clarify the terms "Machine Vision" and "Computer Vision." In fact, these arose a good many years ago when the situation was quite different from what it is today. Over time, computer technology has advanced hugely and at the same time knowledge about the whole area of vision has been radically developed. In the early days, *Computer Vision* meant the study of the science of vision and the *possible* design of the software—and to a lesser extent with what goes into an integrated vision system, whereas *Machine Vision* meant the study not only of the software but also of the hardware environment and of the image acquisition techniques needed for real applications—so it was a much more engineering-orientated subject. At the present point in time, computer technology has advanced so far that a sizeable proportion of real-world and real-time applications can be realized on unaided PCs. This and many other developments in knowledge in this area have led to significant convergence between the terms, with the result that they are often used more or less interchangeably. But bear in mind that certain implementation tools such as those of embedded systems (Chapter 26) require expertise that only a limited number of vision practitioners possess. Overall, the present volume aims to embody a comprehensive view of the whole subject, which explains the combined title *Computer and Machine Vision*.

> Broadly, Computer Vision is the science of vision and Machine Vision is the study of methods, techniques and hardware whereby artificial vision systems can be constructed for practical applications. Recently, the advent of faster computers has led to significant convergence in the use of these terms.

1.5 THE FOLLOWING CHAPTERS

On the whole, the early chapters of the book (Chapters 2–4) cover rather simple concepts, such as the nature of image processing, how image processing algorithms may be devised, and the restrictions on intensity thresholding techniques. The next three chapters (Chapters 5–8) discuss edge detection and some classic binary image analysis techniques (although topics such as graph cuts and affine invariant feature detectors are quite recent and are still developing). Then, Chapters 9–14 move on to intermediate-level processing that has developed significantly in the past two decades, particularly in the use of transform techniques to deduce the presence of objects. Intermediate-level processing is important for the inference of complex objects, both in 2-D (Chapter 14) and subsequently in 3-D (Chapter 15). It also enables automated inspection to be undertaken efficiently (Chapter 20). Chapter 24 enlarges on the process of recognition that is fundamental to many inspection and other processes—as outlined earlier in this chapter. Chapters 25 and 26, respectively, outline the enabling technologies of image acquisition and vision hardware design; finally, Chapter 27 reiterates and highlights some of the lessons and topics dealt within the book, while Appendix A develops the subject of Robust Statistics that relates to a large proportion of the methods that are covered here.

To help give the reader more perspective on the 27 chapters, the main text has been divided into four parts: Part 1 (Chapters 2−8) is entitled "Low-Level Vision," Part 2 (Chapters 9−14) is called "Intermediate-Level Vision," Part 3 (Chapters 15−19) is entitled "3-D Vision and Motion," and Part 4 (Chapters 20−27) is headed "Towards Real-Time Pattern Recognition Systems." This last heading is used to emphasize real-world applications with immutable data flow rates, and the need to integrate all the necessary recognition processes into reliable working systems.

Although the sequence of chapters follows the somewhat logical order just described, the ideas outlined in the previous section—understanding of the visual process, constraints imposed by realities such as noise and occlusion, tradeoffs between relevant parameters, and so on—are mixed into the text at relevant junctures, as they reflect all-pervasive issues.

Finally, there are many topics that would ideally have been included in the book, yet space did not permit this. The chapter bibliographies, the main list of references and the indexes are intended to make good some of these deficiencies.

1.6 BIBLIOGRAPHICAL NOTES

The purpose of this chapter is to introduce the reader to some of the problems of machine vision, showing the intrinsic difficulties but not getting into details at this stage. For detailed references the reader should consult the later chapters. Meanwhile, some background on the world of pattern recognition can be obtained from Duda et al. (2001). In addition, some insight into human vision can be obtained from the fascinating monograph by Hubel (1995).

Low-Level Vision

This part of the book introduces images and image processing, and then proceeds to show how image processing may be developed in order to start the process of image analysis. By the end of Chapter 8, image analysis has been taken, *via* this "traditional" route, far enough to achieve a good many useful practical aims. The main topics to be developed are noise suppression, feature detection, object segmentation, and region analysis with the aid of morphology—itself a development of basic procedures defined and elaborated in the first two chapters of Part 1.

Low-Level Vision

This part of the book introduces images and image processing, and then proceeds to show how image processing may be developed in order to start the process of image analysis. By the end of Chapter 8, image analysis has been taken, via this traditional route, far enough to achieve a good many useful practical aims. The main topics to be developed are noise suppression, feature detection, object segmentation, and region analysis, with the bit of morphology—itself a development of basic procedures defined and elaborated in the first two chapters of Part 1.

Images and Imaging Operations

2

pĭ x'ĕllātèd, a. picture broken into a regular tiling
pĭ x' ĭlātèd, a. pixie-like, crazy, deranged

Images are at the core of vision, and there are many ways—from simple to sophisticated—for processing and analyzing them. This chapter concentrates on simple algorithms, which nevertheless need to be treated carefully as there are important subtleties to be learnt. Above all, the chapter aims to show that quite a lot can be achieved with such algorithms, which can readily be programmed and tested by the reader.

Look out for:

- the different types of image—binary, grayscale, and color.
- a compact notation for presenting image processing operations.
- basic pixel operations—clearing, copying, inverting, thresholding.
- basic window operations—shifting, shrinking, expanding.
- grayscale brightening and contrast-stretching operations.
- binary edge location and noise removal operations.
- multi-image and convolution operations.
- the distinction between sequential and parallel operations, and complications that can arise in the sequential case.
- problems that arise around the edge of the image.

Although elementary, this chapter actually provides basic methodology for the whole of Part 1 and much of Part 2 of the book, and its importance should not be underestimated: nor should the subtleties be ignored. Full understanding at this stage will save many complications later, while programming more sophisticated algorithms.

2.1 INTRODUCTION

This chapter is concerned with images and simple image processing operations. It is intended to lead on to more advanced image analysis operations that are of use for machine vision in an industrial environment. Perhaps the main purpose of the chapter is to introduce the reader to some basic techniques and notations that will be of use throughout the book. However, the image processing algorithms introduced here are of value in their own right in disciplines ranging from remote sensing to medicine, and from forensic to military and scientific applications.

This chapter deals with images that have already been obtained from suitable sensors: sensors are covered in Chapter 25. Typical of such images is that shown in Fig. 2.1(a). This is a gray-tone image that at first sight appears to be a normal "black and white" photograph. However, closer inspection shows that it is composed of a large number of individual picture cells, or "pixels." In fact, the image is a 128×128 array of pixels. To get a better feel for the limitations of such a digitized image, Fig. 2.1(b) shows a 42×42 section that has been subjected to a three-fold magnification so that the pixels can be examined individually.

It is not easy to see that these gray-tone images are digitized into a gray scale containing just 64 gray levels. To some extent high spatial resolution compensates for lack of grayscale resolution, and as a result it is difficult to see the difference between an individual shade of gray and the shade it would have had in an ideal picture. In addition, when we look at the magnified section of image in Fig. 2.1(b), it is difficult to understand the significance of the individual pixel intensities—the whole is becoming lost in a mass of small parts. Early TV cameras typically gave a grayscale resolution that was accurate only to about 1 part in 50, corresponding to about 6 bits of useful information per pixel. Modern solid-state cameras commonly give less noise and may allow 8 or even 9 bits of information per pixel. However, there are many occasions when it is not worthwhile to aim for such high grayscale resolutions, particularly when the result will not be visible to the human eye, or when there is an enormous amount of other data that a robot can use to locate objects within the field of view. Note that if the human eye can see an object in a digitized image of particular spatial and grayscale resolution, it is in principle possible to devise a computer algorithm to do the same thing.

Nevertheless, there is a range of applications for which it is valuable to retain good grayscale resolution, so that highly accurate measurements can be made from a digital image. This is the case in many robotic applications, where high-accuracy checking of components is critical. More will be said about this later. In addition, it will be seen in Part 2 that certain techniques for locating components efficiently require local edge orientation to be estimated to better than $1°$, and this can be achieved only if at least 6 bits of grayscale information are available per pixel.

(a)

(b)

FIGURE 2.1

Typical grayscale image: (a) grayscale image digitized into a 128×128 array of pixels; (b) section of image shown in (a) subjected to three-fold linear magnification: the individual pixels are now clearly visible.

2.1.1 Gray Scale Versus Color

Returning now to the image of Fig. 2.1(a), we might reasonably ask whether it would be better to replace the gray scale with color, using an RGB color camera and three digitizers for the three main colors. There are two aspects of color that are important for the present discussion. One is the *intrinsic value* of color in machine vision and the other is the *additional storage and processing penalty* it might bring. It is tempting to say that the latter aspect is of no great importance given the cheapness of modern computers which have both high storage and high speed. On the other hand, high-resolution images can arrive from a collection of CCTV cameras at huge data rates, and it will be many years before it will be possible to analyze *all* the data arriving from such sources as they come in. Hence, if color adds substantially to the storage and processing load, this will need to be justified.

Against this, the *potential* of color for helping with many aspects of inspection, surveillance, control, and a wide variety of other applications including medicine (color playing a crucial role in images taken during surgery) is enormous. This is illustrated with regard to robot navigation and driving in Figs. 2.2 and 2.3; and

FIGURE 2.2

Value of color for segmentation and recognition. In natural outdoor scenes such as this, color helps with segmentation and with recognition. While it may have been important to the early human when discerning sources of food in the wild, robot drones may benefit by using color to aid navigation (see also color Plate 2).

for food inspection in Figs. 2.4 and 2.5; for color filtering see Figs. 3.12 and 3.13. Note that some of these images almost have color for color's sake (especially in Figs. 2.4 and 2.5), although none of them are artificially generated. In others the color is more subdued (Fig. 2.3), and in Fig. 2.5 (excluding the tomatoes), it is quite subtle. The point to be made here is that for color to be useful it need not be garish, but can be subtle as long as it brings the right sort of information to bear on the task in hand. Suffice it to say that in some of the simpler inspection applications, where mechanical components are scrutinized on a conveyor or workbench, it is quite likely to be the shape that is in question rather than the color of the object or its parts. On the other hand, if an automatic fruit picker is to be devised, it will probably be more crucial to check color than specific shape. We leave it to the reader to imagine when and where color is particularly useful, or merely an unnecessary luxury.

FIGURE 2.3

Value of color in the built environment. Color plays an important role for the human in managing the built environment. In a vehicle, a plethora of bright lights, road signs, and markings (such as yellow lines) are coded to help the driver: they may likewise help a robot to drive more safely by the provision of crucial information (see also color Plate 3).

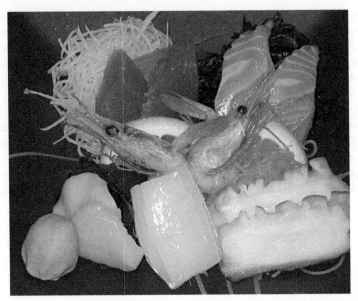

FIGURE 2.4

Value of color for food inspection. Much food is brightly colored, as with this Japanese meal. While this may be attractive to the human, it could also help the robot to check quickly for foreign bodies or toxic substances (see also color Plate 4).

Next, it is useful to consider the processing aspect of color. In many cases, good color discrimination is required to separate and segment two types of object from each other. Typically this will mean not using one or other specific color channel,[1] but subtracting two, or combining three in such a way as to foster discrimination. In the worst case of combining three color channels by simple arithmetic processing in which each pixel is treated identically, the processing load will be very light. In contrast, the amount of processing required to *determine* the optimal means of combining the data from the color channels, and to carry out different operations dynamically on different parts of the image, may be far from negligible, and some care will be needed in the analysis. These problems arise because color signals are inhomogeneous: this contrasts with the situation for grayscale images, where the bits representing the gray scale are all of the same type and take the form of a number representing the pixel intensity: they can thus be processed as a single entity on a digital computer.

---

[1]Here we use the term "channel" not just to refer to the red, green, or blue channel, but any derived channel obtained by combining the colors into a single-color dimension.

FIGURE 2.5

Subtle shades of color in food inspection. While much food is brightly colored, as for the tomatoes in this picture, green salad leaves show much more subtle combinations of color and may indeed provide the only reliable means of identification. This could be important for inspection both of the raw product and its state when it reaches the warehouse or the supermarket.

2.2 IMAGE PROCESSING OPERATIONS

In what follows, the images of Figs. 2.1(a) and 2.7(a) are considered in some detail, examining some of the many image processing operations that can be performed on them. The resolution of these images reveals a considerable amount of detail and at the same time shows how it relates to the more "meaningful" global information. This should help to make it clear how simple imaging operations contribute to image interpretation.

When performing image processing operations, we start with an image in one storage area and generate a new processed image in another storage area. In practice, these storage areas may either be in a special hardware unit called a frame store that is interfaced to the computer or they may be in the main memory of the computer or on one of its disks. In the past a special frame store was required to store images, since each image contains a good fraction of a megabyte of information and this amount of space was not available for normal users in the computer main memory. Nowadays this is less of a problem, but for image acquisition and display a frame store is still required. However, we shall not worry about such details here. Instead it will be assumed that all images are inherently visible and that they are stored in various image "spaces" P, Q, R, etc. Thus, we might start with an image in space P and copy it to space Q, for example.

2.2.1 Some Basic Operations on Grayscale Images

Perhaps the simplest of imaging operations is that of clearing an image or setting the contents of a given image space to a constant level. We need some way of arranging this, and accordingly the following C++ routine may be written for implementing it:[2]

```
for (j = 0;  j <= 127;  j++)
    for (i = 0;  i <= 127;  i++)
      P[j][i] = alpha;
```
(2.1)

In this routine the local pixel intensity value is expressed as P[j][i], since P-space is taken to be a two-dimensional array of intensity values (Table 2.1). In what follows, it will be advantageous to rewrite such routines in the more succinct form:

```
for all pixels in image do {P0 = alpha;}
```
(2.2)

as this will aid understanding by removing irrelevant programming detail. The reason for calling the pixel intensity P0 will become clear later.

Another simple imaging operation is to copy an image from one space to another. This is achieved, without changing the contents of the original space P, by the routine:

```
for all pixels in image do {Q0 = P0;}
```
(2.3)

A more interesting operation is that of inverting the image, as in the process of converting a photographic negative to a positive. This process is represented as follows:

```
for all pixels in image do {Q0 = 255 - P0;}
```
(2.4)

In this case, it is assumed that pixel intensity values lie within the range 0–255, as is commonly true for frame stores that represent each pixel as one byte of information. Note that such intensity values are commonly unsigned and this is assumed generally in what follows.

There are many operations of these types. Some other simple operations are those that shift the image left, right, up, down, or diagonally. They are easy to implement if the new local intensity is made identical to that at a neighboring location in the original image. It is evident how this would be expressed in the double suffix notation used in the original C++ routine. In the new shortened

---

[2]Readers who are unfamiliar with C++ or Java, which is similar at this level of programming, should refer to Stroustrup (1991) and Schildt (1995).

Table 2.1 C++ Notation

Notation	Meaning
++	increment the preceding variable.
[ ]	add array index after a variable.
[ ][ ]	add two array indices after a variable; the last is the faster running index.
(int)	changes the following variable to integer type.
(float)	changes the following variable to floating point.
{ }	encloses a sequence of instructions.
if ( ) { };	basic conditional statement: ( ) encloses the condition; { } encloses the instructions to be executed.
if ( ) { }; else if ( ) { }; ... ; else { };	the most general type of conditional statement.
while ( ) { }	common type of iterated loop.
do { } while ( );	another common type of iterated loop.
do { } until ( );	"until" means the same as "while not." This is often a convenient notation, although it is not strict C++ .
for (; ;) { };	here the conditional statement ( ) has three arguments separated by semicolons: they are respectively the initial condition; the terminating condition; and the incrementation operation.
=	forces equality (literally: "takes the value").
==	tests for equality in a conditional expression.
<=	$\leq$
>=	$\geq$
!=	$\neq$
!	logical not.
&&	logical and.
||	logical or.
//	indicates that the remainder of the line is a comment.
/* ... */	brackets enclosing a comment.
A0 ... A8 B0 ... B8 C0 ... C8	bit image variables in 3×3 window.[a]
P0 ... P8 Q0 ... Q8 R0 ... R8	byte image variables in 3×3 window.[a]
P[0], ...	equivalent to P0, ...

[a]These predefined variables denote special syntax not available in C++, but useful for simplifying the image processing algorithms presented in Chapters 2, 3, 4, 7 and 9.
Note: The purpose of this table is to show what is meant by the various C++ commands and instructions used in this book. It is not intended to be comprehensive. The aim is merely to be helpful to the reader. In general, only notation that differs between C++ and other commonly used languages such as Pascal is included, in order to eliminate possible ambiguity or confusion.

notation, it is necessary to name neighboring pixels in some convenient way, and we here employ the following simple scheme:

$$
\begin{array}{ccc}
P4 & P3 & P2 \\
P5 & P0 & P1 \\
P6 & P7 & P8
\end{array}
$$

with a similar scheme for other image spaces. With this notation, it is easy to express a left shift of an image as follows:

```
for all pixels in image do {Q0 = P1;}
```
(2.5)

Similarly, a shift down to the bottom right is expressed as:

```
for all pixels in image do {Q0 = P4;}
```
(2.6)

It will now be clear why P0 and Q0 were chosen for the basic notation of pixel intensity: the "0" denotes the central pixel in the "neighborhood" or "window," and corresponds to zero shift when copying from one space to another. However, the type of window operation presented above is much more powerful than single-pixel operations, and we shall see many examples of it in what follows. Meanwhile, note that it can give rise to difficulties around the boundaries of the image: we shall return to this point in Section 2.4.

There is a whole range of possible operations associated with modifying images in such a way as to match them to the requirements of a human viewer. For example, adding a constant intensity makes the image brighter:

```
for all pixels in image do {Q0 = P0 + beta;}
```
(2.7)

and the image can be made darker in the same way. A more interesting operation is to stretch the contrast of a dull image:

```
for all pixels in image do {Q0 = P0 * gamma + beta;}
```
(2.8)

where gamma > 1. In practice (as for Fig. 2.6), it is necessary to ensure that intensities do not result that are outside the normal range, e.g., by using an operation of the form:

```
for all pixels in image do {
   QQ = P0 * gamma + beta;
   if (QQ < 0) Q0 = 0;
   else if (QQ > 255) Q0 = 255;
   else Q0 = QQ;
}
```
(2.9)

Most practical situations demand more sophisticated transfer functions—either nonlinear or piecewise linear—but such complexities are ignored here.

FIGURE 2.6

Contrast stretching: effect of increasing the contrast in the image of Fig. 2.1(a) by a factor of two and adjusting the mean intensity level appropriately. The interior of the jug can now be seen more easily. Note, however, that there is no additional information in the new image.

A further simple operation that is often applied to grayscale images is that of thresholding to convert to a binary image. This topic is covered in more detail later, since it is widely used to detect objects in images. However, our purpose here is to look on it as another basic imaging operation. It can be implemented using the routine:[3]

```
for all pixels in image do {
    if (P0 > thresh) A0 = 1; else A0 = 0;                    (2.10)
}
```

If, as very often happens, objects appear as dark objects on a light background, it is easier to visualize the subsequent binary processing operations by inverting the thresholded image using a routine such as:

```
for all pixels in image do {A0 = 1 − A0;}                    (2.11)
```

---

[3]The first few letters of the alphabet (A, B, C, …) are used consistently to denote binary image spaces, and later letters of the alphabet (P, Q, R, …) to denote grayscale images (Table 2.1). In software, these variables are assumed to be predeclared, and in hardware (e.g., frame store) terms they are taken to refer to dedicated memory spaces containing only the necessary 1 or 8 bits per pixel. The intricacies of data transfer between variables of different types are important considerations that are not addressed in detail here. It is sufficient to assume that both A0 = P0 and P0 = A0 correspond to a single-bit transfer, except that in the latter case the top 7 bits are assigned the value 0.

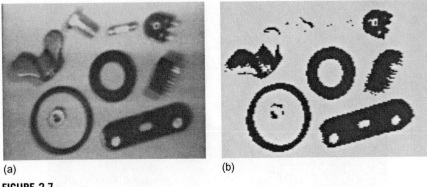

(a)   (b)

FIGURE 2.7

Thresholding of grayscale images: (a) 128×128 pixel grayscale image of a collection
of parts; (b) effect of thresholding the image.

However, it would be more usual to combine the two operations into a single
routine of the form:

```
for all pixels in image do {
    if(P0>thresh) A0 = 0; else A0 = 1;
}
```
(2.12)

To display the resulting image in a form as close as possible to the original,
it can be reinverted and given the full range of intensity values (intensity values
0 and 1 being scarcely visible):

```
for all pixels in image do { R0 = 255*(1 − A0);}
```
(2.13)

Figure 2.7 shows the effect of these two operations.

2.2.2 Basic Operations on Binary Images

Once the image has been thresholded, a wide range of binary imaging operations
become possible. Only a few such operations are covered here, with the aim of
being instructive rather than comprehensive. With this in mind, a routine may be
written for shrinking dark-thresholded objects (Fig. 2.8(a)) that are here represented by a set of 1's in a background of 0's:

```
for all pixels in image do {
    sigma = A1 + A2 + A3 + A4 + A5 + A6 + A7 + A8;
    if (A0 == 0) B0 = 0;
    else if (sigma < 8) B0 = 0;
    else B0 = 1;
}
```
(2.14)

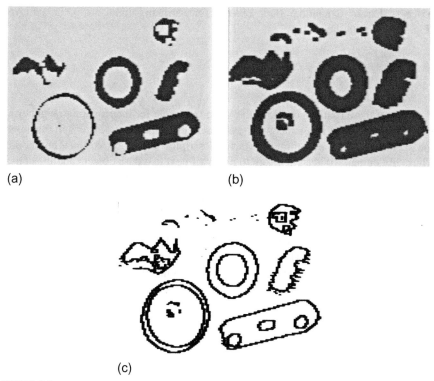

(a)   (b)

(c)

FIGURE 2.8

Simple operations applied to binary images: (a) effect of shrinking the dark-thresholded
objects appearing in Fig. 2.7(b); (b) effect of expanding these dark objects; (c) result
of applying an edge location routine. Note that the shrink, expand, and edge routines
are applied to the *dark* objects: this implies that the intensities are initially inverted
as part of the thresholding operation and then reinverted as part of the display operation
(see text).

In fact, the logic of this routine can be simplified to give the following more
compact version:

```
for all pixels in image do {
    sigma = A1 + A2 + A3 + A4 + A5 + A6 + A7 + A8;
    if (sigma < 8) B0 = 0; else B0 = A0;
}
```

(2.15)

Note that the process of shrinking dark objects also expands light objects,
including the light background. It also expands holes in dark objects. The opposite

process, that of expanding dark objects (or shrinking light ones), is achieved (Fig. 2.8(b)) with the routine:[4]

```
for all pixels in image do {
    sigma = A1 + A2 + A3 + A4 + A5 + A6 + A7 + A8;
    if (sigma > 0) B0 = 1; else B0 = A0;
}
```
(2.16)

Each of these routines employs the same technique for interrogating neighboring pixels in the original image: as will be apparent on numerous occasions in this book, the sigma value is a useful and powerful descriptor for 3×3 pixel neighborhoods. Thus, "if (sigma > 0)" can be taken to mean "if next to a dark object" and the consequence can be read as "then expand it." Similarly, "if (sigma < 8)" can be taken to mean "if next to a light object" or "if next to light background," and the consequence can be read as "then expand the light background into the dark object."

The process of finding the edge of a binary object has several possible interpretations. Clearly, it can be assumed that an edge point has a sigma value in the range 1–7 inclusive. However, it may be defined as being within the object, within the background or in either position. Taking the definition that the edge of an object has to lie within the object (Fig. 2.8(c)), the following edge-finding routine for binary images results:

```
for all pixels in image do {
    sigma = A1 + A2 + A3 + A4 + A5 + A6 + A7 + A8;
    if (sigma == 8) B0 = 0; else B0 = A0;
}
```
(2.17)

This strategy amounts to canceling out object pixels that are not on the edge. For this and a number of other algorithms (including the shrink and expand algorithms already encountered), a thorough analysis of exactly which pixels should be set to 1 and 0 (or which should be retained and which eliminated) involves drawing up tables of the form:

		sigma	
		0–7	8
A0	0	0	0
	1	1	0

This reflects the fact that algorithm specification includes a recognition phase and an action phase, i.e., it is necessary first to locate situations within an image where (for example) edges are to be marked or noise eliminated, and then action must be taken to implement the change.

---

[4]The processes of shrinking and expanding are widely known by the respective terms "erosion" and "dilation." (See also Chapter 7.)

Another function that can usefully be performed on binary images is the removal of "salt and pepper" noise, i.e., noise which appears as a light spot on a dark background or a dark spot on a light background. The first problem to be solved is that of recognizing such noise spots; the second is the simpler one of correcting the intensity value. For the first of these tasks the sigma value is again useful. To remove salt noise (which has binary value 0 in our convention), we arrive at the following routine:

```
for all pixels in image do {
    sigma = A1 + A2 + A3 + A4 + A5 + A6 + A7 + A8;
    if (sigma == 8) B0 = 1; else B0 = A0;
}
```
(2.18)

which can be read as leaving the pixel intensity unchanged unless it is proven to be a salt noise spot. The corresponding routine for removing pepper noise (binary value 1) is:

```
for all pixels in image do {
    sigma = A1 + A2 + A3 + A4 + A5 + A6 + A7 + A8;
    if (sigma == 0) B0 = 0; else B0 = A0;
}
```
(2.19)

Combining these two routines into one operation (Fig. 2.9(a)) gives:

```
for all pixels in image do {
    sigma = A1 + A2 + A3 + A4 + A5 + A6 + A7 + A8;
    if (sigma == 0) B0 = 0;
    else if (sigma == 8) B0 = 1;
    else B0 = A0;
}
```
(2.20)

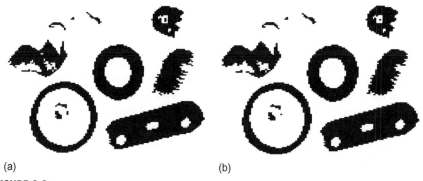

(a)   (b)

FIGURE 2.9

Simple binary noise removal operations: (a) result of applying a "salt and pepper" noise removal operation to the thresholded image in Fig. 2.7(b). (b) Result of applying a less stringent noise removal routine: this is effective in cutting down the jagged spurs that appear on some of the objects.

The routine can be made less stringent in its specification of noise pixels, so that it removes spurs on objects and background: this is achieved (Fig. 2.9(b)) by a variant such as:

```
for all pixels in image do {
    sigma = A1 + A2 + A3 + A4 + A5 + A6 + A7 + A8;
    if (sigma < 2) B0 = 0;
    else if (sigma > 6) B0 = 1;
    else B0 = A0;
}
```
$$(2.21)$$

As before, if there is any doubt about the algorithm, its specification should be set up rigorously—as in the following table:

		sigma		
		0 or 1	2–6	7 or 8
A0	0	0	0	1
	1	0	1	1

There are many other simple operations that can usefully be applied to binary images and some of them are dealt with in Chapter 9.

2.3 CONVOLUTIONS AND POINT SPREAD FUNCTIONS

Convolution is a powerful and widely used technique in image processing and other areas of science. It appears in many applications throughout this book and it is therefore useful to introduce it at an early stage. We start by defining the convolution of two functions $f(x)$ and $g(x)$ as the integral:

$$f(x) * g(x) = \int_{-\infty}^{\infty} f(u)g(x-u)\, du \qquad (2.22)$$

The action of this integral is normally described as the result of applying a point spread function $g(x)$ to all points of a function $f(x)$ and accumulating the contributions at every point. It is significant that if the point spread function (PSF) is very narrow,[5] then the convolution is identical to the original function $f(x)$. This makes it natural to think of the function $f(x)$ as having been spread out under the influence of $g(x)$. This argument may give the impression that convolution necessarily blurs the original function but this is not always so if, for example, the PSF has a distribution of positive and negative values.

When convolution is applied to digital images, the above formulation changes in two ways: (i) a double integral must be used in respect of the two dimensions

---

[5]Formally, it can be a delta function, which is infinite at one point and zero elsewhere while having an integral of unity.

and (ii) integration must be changed into discrete summation. The new form of the convolution is:

$$F(x, y) = f(x, y) * g(x, y) = \sum_i \sum_j f(i, j)g(x - i, y - j) \qquad (2.23)$$

where g is now referred to as a spatial convolution mask. The fact that the mask has to be inverted before it is applied is inconvenient for visualizing the process of convolution—particularly when matching operations are involved, e.g., for corner location (see Chapter 6). In this book, we therefore present only pre-inverted masks of the form:

$$h(x, y) = g(-x, -y) \qquad (2.24)$$

Convolution can then be calculated using the more intuitive formula:

$$F(x, y) = \sum_i \sum_j f(x + i, y + j)h(i, j) \qquad (2.25)$$

which involves multiplying corresponding values in the modified mask and the neighborhood under consideration. Re-expressing this result for a 3×3 neighborhood and writing the mask coefficients in the form:

$$\begin{bmatrix} h4 & h3 & h2 \\ h5 & h0 & h1 \\ h6 & h7 & h8 \end{bmatrix}$$

the algorithm can be obtained in terms of our earlier notation:

```
for all pixels in image do {
    Q0 = P0 * h0 + P1 * h1 + P2 * h2 + P3 * h3 + P4 * h4
       + P5 * h5 + P6 * h6 + P7 * h7 + P8 * h8;
}
```
(2.26)

We are now in a position to apply convolution to a real situation. At this stage we attempt to suppress noise by averaging over nearby pixels. A simple way of achieving this is to use the convolution mask:

$$\frac{1}{9} \begin{bmatrix} 1 & 1 & 1 \\ 1 & 1 & 1 \\ 1 & 1 & 1 \end{bmatrix}$$

where the number in front of the mask weighs all the coefficients in the mask and is inserted to ensure that applying the convolution does not alter the mean intensity in the image. As hinted above, this particular convolution has the effect of blurring the image as well as reducing the noise level (Fig. 2.10). More will be said about this in the next chapter.

FIGURE 2.10

Noise suppression by neighborhood averaging achieved by convolving the original image of Fig. 2.1(a) with a uniform mask within a 3×3 neighborhood. Note that noise is suppressed only at the expense of introducing significant blurring.

The above discussion makes it clear that convolutions are linear operators. In fact, they are the most general spatially invariant linear operators that can be applied to a signal such as an image. Note that linearity is often of interest in that it permits mathematical analysis to be performed that would otherwise be intractable.

2.4 SEQUENTIAL VERSUS PARALLEL OPERATIONS

It will be noticed that most of the operations defined so far have started with an image in one space and finished with an image in a different space. Unfortunately, many of the operations will not work satisfactorily if we do not use separate input and output spaces in this way. This is because they are inherently "parallel processing" routines. This term is used as these are the types of process that would be performed by a parallel computer possessing a number of processing elements equal to the number of pixels in the image, so that all the pixels are processed simultaneously. If a serial computer is to *simulate* the operation of a parallel computer, then it must have separate input and output image spaces and rigorously work in such a way that it uses the original image values to compute the output pixel values. This means that an operation such as the following cannot be an ideal parallel process:

```
for all pixels in image do {
    sigma = A1 + A2 + A3 + A4 + A5 + A6 + A7 + A8;
    if (sigma < 8) A0 = 0; else A0 = A0;
}
```
(2.27)

This is so because, when the operation is half completed, the output pixel intensity will depend not only on some of the unprocessed pixel values but also on some that have already been processed. For example, if the computer makes a normal (forward) TV raster scan through the image, the situation at a general point in the scan will be

$$
\begin{array}{ccc}
\sqrt{} & \sqrt{} & \sqrt{} \\
\sqrt{} & \times & \times \\
\times & \times & \times
\end{array}
$$

where the ticked pixels have already been processed and the others have not. As a result, the above operation will shrink all objects to nothing.

A much simpler illustration of this is obtained by attempting to shift an image to the right using the following routine:

```
for all pixels in image do {P0 = P5;}
```
(2.28)

In fact, all this achieves is to fill up the image with values corresponding to those off its left edge,[6] whatever they are assumed to be. Thus, we have shown that the shifting process is inherently parallel.

It will be seen in Chapter 9 that there are some processes that are inherently sequential—i.e., the processed pixel has to be returned immediately to the *original* image space. Meanwhile, note that not all of the routines described so far need to be restricted rigorously to parallel processing. In particular, all single-pixel routines (essentially those that only refer to the single pixel in a 1×1 neighborhood) can validly be performed as if they were sequential in nature. Such routines include the following intensity adjustment and thresholding operations:

```
for all pixels in image do {P0 = P0 * gamma + beta;}
```
(2.29)

```
for all pixels in image do {if (P0 > thresh) P0 = 1; else P0 = 0;}
```
(2.30)

These remarks are intended to act as a warning. In general, it is safest to design algorithms that are exclusively parallel processes unless there is a definite need to make them sequential. It will be seen later how this need can arise.

---

[6]Note that when the computer is performing a 3×3 (or larger) window operation, it has to assume some value for off-image pixel intensities: usually whatever value is selected will be inaccurate, and so the final processed image will contain a border that is also inaccurate. This will be so whether the off-image pixel addresses are trapped in software or in specially designed circuitry in the frame store.

2.5 CONCLUDING REMARKS

This chapter has introduced a compact notation for representing imaging operations and has demonstrated some basic parallel processing routines. The following chapter extends this work to see how noise suppression can be achieved in grayscale images. This leads on to more advanced image analysis work that is directly relevant to machine vision applications. In particular, Chapter 4 studies in more detail the thresholding of grayscale images, building on the work of Section 2.2.1, while Chapter 9 studies object shape analysis in binary images.

> Pixel–pixel operations can be used to make radical changes in digital images. However, this chapter has shown that window–pixel operations are far more powerful, and capable of performing all manner of size- and shape-changing operations, as well as eliminating noise. But caveat emptor—sequential operations can have some odd effects if adventitiously applied.

2.6 BIBLIOGRAPHICAL AND HISTORICAL NOTES

Since the aim of this chapter is not to cover the most recent material but to provide a succinct overview of basic techniques, it will not be surprising that most of the topics discussed were developed well over 20 years ago and have been used by a large number of workers in many areas. For example, thresholding of grayscale images was first reported at least as long ago as 1960, while shrinking and expanding of binary picture objects date from a similar period. Discussion of the origins of other techniques is curtailed: for further detail the reader is referred to the texts by, e.g., Gonzalez and Woods (2008), Nixon and Aguado (2008), Petrou and Petrou (2010), and Sonka et al. (2007). We also refer to two texts that cover programming aspects of image processing in some depth: Parker (1994), which covers C programming, and Whelan and Molloy (2001), which covers Java programming. More specialized texts will be referred to in the following chapters.

2.7 PROBLEMS

1. Derive an algorithm for finding the edges of binary picture objects by applying a shrink operation and combining the result with the original image. Is the result the same as that obtained using the edge-finding routine (Eq. 2.17)? Prove your statement rigorously by drawing up suitable algorithm tables as in Section 2.2.2.

2. In a certain frame store, each off-image pixel can be taken to have either the value 0 or the intensity of the nearest image pixel. Which of the two will give the more meaningful results for (a) shrinking, (b) expanding, and (c) a blurring convolution?

3. Suppose the noise elimination routines of equations (2.20) and (2.21) were reimplemented as sequential algorithms. Show that the action of the first would be unchanged, whereas the second would produce very odd effects on some binary images.

3

Basic Image Filtering Operations

Image filtering involves the application of window operations that achieve useful effects, such as noise removal or image enhancement. This chapter is concerned particularly with what can be achieved with quite basic approaches, such as application of local mean, median, or mode filters to digital images. The focus is on grayscale images, although some aspects of color processing are also covered.

Look out for:

- what can be achieved by low-pass filtering in the spatial frequency domain.
- how the same process can be carried out by convolution in the spatial domain.
- the problem of impulse noise and what can be achieved with a limiting filter.
- the value of median, mode, and rank order filters.
- how computational load can be reduced.
- the distinction between image enhancement and image restoration.
- the distortions produced by standard filters—mean, Gaussian, median, mode, and rank order filters.

This chapter takes pain to delve into the properties of a variety of standard types of filter, because it is necessary to know both what they can achieve and what their limitations are. In fact, the edge shifts produced by most of these filters are small but predictable, and therefore correctable in principle. The exception is the rank order filter, for which the shifts can be large—but this is the *advantage* of this type of filter, which is at the core of mathematical morphology (see Chapter 7).

3.1 INTRODUCTION

Chapter 2 is concerned with simple imaging operations, including such problems as thresholding grayscale images and suppressing noise in binary images. In this chapter, the discussion is extended to noise suppression and enhancement in grayscale images. Although these types of operation can for the most part be avoided in industrial applications of vision, it is useful to examine them in some depth because of their wide use in a variety of other image processing applications and because they set the scene for much of what follows. In addition, some fundamental issues come to light which are of vital importance.

It has already been seen that noise can arise in real images and it is hence necessary to have sound techniques for suppressing it. Commonly, in electrical engineering applications, noise is removed by means of low-pass or other filters that operate in the frequency domain (Rosie, 1966). Applying these filters to 1-D time-varying analog signals is straightforward, since it is necessary only to place them at suitable stages in the sequence of black boxes through which the signals pass. For digital signals the situation is more complicated, since the frequency transform of the signal must first be computed, then the low-pass filter applied, and finally the signal obtained from the modified transform by converting back to the time domain. Thus, two Fourier transforms have to be computed, although modifying the signal while it is in the frequency domain is a straightforward task (Fig. 3.1). In fact, the amount of processing involved in computing the discrete Fourier transform of a signal represented by N samples is of order N^2 (we shall write this as $O(N^2)$), although the amount of computation can be cut down to $O(N \log_2 N)$ by employing the fast Fourier transform (FFT) (Gonzalez and Woods, 1992). This then becomes a practical approach for the elimination of noise.

When applying these ideas to images we must first note that the signal is a spatial rather than a time-varying quantity and must be filtered in the spatial frequency domain. Mathematically this makes no real difference, but there are nevertheless significant problems. First, there is no satisfactory analog shortcut and the whole process has to be carried out digitally (we here ignore optical processing methods despite their obvious power, speed, and high resolution, because they are by no means trivial to marry with digital computer technology). Second,

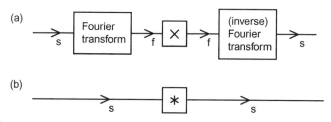

FIGURE 3.1

Low-pass filtering for noise suppression: s, spatial domain; f, spatial frequency domain; ×, multiplication by low-pass characteristic; *, convolution with Fourier transform of low-pass characteristics. (a) Low-pass filtering achieved most simply, by a process of multiplication in the (spatial) frequency domain; (b) low-pass filtering achieved by a process of convolution. Note that (a) may require more computation overall, because of the two Fourier transforms that have to be performed.

for an $N{\times}N$ pixel image, the number of operations required to compute a Fourier transform is $O(N^3)$ and the FFT only reduces this to $O(N^2 \log_2 N)$, so the amount of computation is quite considerable (here it is assumed that the 2-D transforms are implemented by successive passes of 1-D transforms: see Gonzalez and Woods, 1992). Note also that two Fourier transforms are required for the purpose of noise suppression (Fig. 3.1). Nevertheless, in many imaging applications it is worth proceeding in this way, because not only noise can be removed but also TV scan lines and other artifacts can be filtered out. This situation applies particularly in remote sensing and space technology. However, in industrial applications the emphasis is always on real-time processing, so in many cases it is not practicable to remove noise by spatial frequency domain operations. A further problem is that low-pass filtering is suited to removing Gaussian noise, but distorts the image if it is used to remove impulse noise.

Section 3.2 discusses Gaussian smoothing, in both the spatial frequency and the spatial domains. The subsequent three sections introduce median filters, mode filters, and then general rank order filters, and contrast their main properties and uses. In Section 3.6, consideration is given to reducing computational load, with particular reference to the median filter. Section 3.7 introduces the sharp–unsharp masking technique, which provides a rather simple route to image enhancement. Then follow a number of sections that concentrate on the edge shifts produced by the various filters. In the case of the median filter, the discrete theory (Section 3.9) is much more exact than the continuum model (Section 3.8). All edge shifts are quite small, except for rank order filters (Section 3.12): these are treated fairly fully because of their relevance to widely used morphological operators (Chapter 7) where the shifts are turned to advantage. Finally, Section 3.14 gives a brief discussion on the application of filters to color images.

3.2 NOISE SUPPRESSION BY GAUSSIAN SMOOTHING

Low-pass filtering is normally thought of as the elimination of signal components with high spatial frequencies, and it is therefore natural to carry it out in the spatial frequency domain. Nevertheless, it is possible to implement it directly in the spatial domain. That this is possible is due to the well-known fact (Rosie, 1966) that multiplying a signal by a function in the spatial frequency domain is equivalent to convolving it with the Fourier transform of the function in the spatial domain (Fig. 3.1). If the final convolving function in the spatial domain is sufficiently narrow, then the amount of computation involved will not be excessive: in this way a satisfactory implementation of the low-pass filter can be sought. It now remains to find a suitable convolving function.

If the low-pass filter is to have a sharp cut-off, then its transform in image space will be oscillatory: an extreme example of this is the sinc $(\sin x/x)$ function, which is the spatial transform of a low-pass filter of rectangular profile (Rosie,

1966). Oscillatory convolving functions are unsatisfactory since they can introduce halos around objects, hence distorting the image quite grossly. Marr and Hildreth (1980) suggested that the right types of filter to apply to images are those that are well-behaved (nonoscillatory) both in the frequency and in the spatial domain. Gaussian filters are able to fulfill this criterion optimally: they have identical forms in the spatial and spatial frequency domains. In 1-D, these forms are:

$$f(x) = \frac{1}{(2\pi\sigma^2)^{1/2}} \exp\left(-\frac{x^2}{2\sigma^2}\right) \tag{3.1}$$

$$F(\omega) = \exp\left(-\frac{1}{2}\sigma^2\omega^2\right) \tag{3.2}$$

Thus, the type of spatial convolving operator required for the purpose of noise suppression by low-pass filtering is one that approximates to a Gaussian profile. Many such approximations appear in the literature: these vary with the size of the neighborhood chosen and in the precise values of the convolution mask coefficients.

One of the most common is the following mask, first introduced in Chapter 2, which is used more for simplicity of computation than for its fidelity to a Gaussian profile:

$$\frac{1}{9}\begin{bmatrix} 1 & 1 & 1 \\ 1 & 1 & 1 \\ 1 & 1 & 1 \end{bmatrix}$$

Another commonly used mask, which approximates more closely to a Gaussian profile, is the following:

$$\frac{1}{16}\begin{bmatrix} 1 & 2 & 1 \\ 2 & 4 & 2 \\ 1 & 2 & 1 \end{bmatrix}$$

In both cases, the coefficients that precede the mask are used to weight all the mask coefficients: as mentioned in Section 2.3, these weights are chosen so that applying the convolution to an image does not affect the average image intensity. These two convolution masks probably account for over 80% of all discrete approximations to a Gaussian. Note that as they operate within a 3×3 neighborhood, they are reasonably narrow and hence incur a relatively small computational load.

Let us next study the properties of this type of operator, deferring for now consideration of Gaussian operators in larger neighborhoods. First, imagine such an operator is applied to a noisy image whose intensity is inherently uniform. Then clearly noise is suppressed, as it is now averaged over nine pixels. This averaging model is obvious for the first of the two masks above but in fact applies equally to the second mask, once it is accepted that the averaging effect is differently distributed in accordance with the improved approximation to a Gaussian profile.

Although this example shows that noise is suppressed, it is clear that the signal will also be affected. This problem arises only where the signal is initially non-uniform: indeed, if the image intensity is constant, or if the intensity map approximates to a plane, there is again no problem. However, if the signal is uniform over one part of a neighborhood and rises in another part of it, as is bound to occur adjacent to the edge of an object, then the object will make itself felt at the center of the neighborhood in the filtered image (see Fig. 3.2). As a result, the edges of objects become somewhat blurred. Looking at the operator as a "mixing operator" that forms a new picture by mixing together the intensities of pixels fairly close to each other, it is intuitively obvious why blurring occurs.

It is also apparent from a spatial frequency viewpoint why blurring should occur. Basically, we are aiming to give the signal a sharp cut-off in the spatial frequency domain, and as a result it will become slightly blurred in the spatial domain. Clearly, the blurring effect can be reduced by using the narrowest possible approximation to a Gaussian convolution filter, but at the same time the noise suppression properties of the filter are lessened. Assuming that the image was initially digitized at roughly the correct spatial resolution, it will not normally be appropriate to smooth it using convolution masks larger than 3×3 or at most 5×5 pixels (here we ignore methods of analyzing images that use a number of versions of the image with different spatial resolutions: see for example, Babaud et al., 1986).

Overall, low-pass filtering and Gaussian smoothing are not appropriate for the applications considered here because of the blurring effects they introduce. Note also that where interference occurs, it can give rise to impulse or "spike" noise (corresponding to a number of individual pixels having totally the wrong intensities): merely averaging this noise over a larger neighborhood can make the situation worse, since the spikes will be smeared over a sizeable number of pixels and will distort the intensity values of all of them. This consideration is important as it leads naturally to the concepts of limit and median filtering.

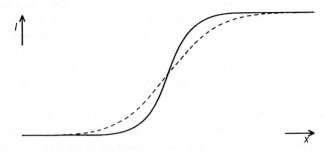

FIGURE 3.2

Blurring of object edges by simple Gaussian convolutions. The simple Gaussian convolution can be regarded as a grayscale neighborhood "mixing" operator, hence explaining why blurring arises.

3.3 MEDIAN FILTERS

The idea explored here is to locate the pixels in the image that have extreme and therefore highly improbable intensities and to ignore their actual intensities, replacing them with more suitable values. This is akin to drawing a graph through a set of plots and ignoring the plots that are evidently a long way from the best fit curve. An obvious way of achieving this is to apply a "limit" filter that prevents any pixel having an intensity outside the intensity range of its neighbors:

```
for all pixels in image do {
  minP = min (P1, P2, P3, P4, P5, P6, P7, P8);
  maxP = max(P1, P2, P3, P4, P5, P6, P7, P8);
  if (P0<minP) Q0 = minP;
  else if (P0>maxP) Q0 = maxP;
  else Q0 = P0;
}
```
$$(3.3)$$

To develop this technique, it is necessary to examine the local intensity distribution within a particular neighborhood. Points at the extremes of the distribution are quite likely to have arisen from impulse noise. So it is sensible not only to eliminate these points, as in the limit filter, but also to try taking the process further, removing equal areas at either end of the distribution and ending with the median. Thus, we arrive at the median filter, which takes all the local-intensity distributions and generates a new image corresponding to the set of median values. As the preceding argument indicates, the median filter is excellent at impulse noise suppression and this is amply confirmed in practice (see Fig. 3.3).

FIGURE 3.3

Effect of applying a 3×3 median filter to the image of Fig. 2.1(a). Note the slight loss of fine detail and the rather "softened" appearance of the whole image.

In view of the blurring caused by Gaussian smoothing operators, it is pertinent to ask whether the median filter also induces blurring. In fact, Fig. 3.3 shows that any blurring is only marginal, although there is some slight loss of fine detail that can give the resulting pictures a "softened" appearance. Theoretical discussion of this point is deferred for now; the lack of blur makes good the main deficiency of the Gaussian smoothing filter and results in the median filter being perhaps the most widely used filter in general image processing applications.

There are many ways of implementing the median filter: Table 3.1 reproduces only an obvious algorithm that essentially implements the above description. The notation of Chapter 2 is used but is augmented in order to permit the nine pixels in a 3×3 neighborhood to be accessed in turn with a running suffix (specifically, P0 to P8 are written as P[m] where m runs from 0 to 8).

The operation of the algorithm is as follows: first, the histogram array is cleared and the image is scanned, generating a new image in Q-space; then, for each neighborhood, the histogram of intensity values is constructed; then the median is found; and, finally, points in the histogram array that have been incremented are cleared. This last feature eliminates the need to clear the whole histogram and hence saves computation. Unlike the general situation in which the median of a distribution is being located, only one (half) scan through the distribution is required, since the total area is known in advance (in this case it is 9).

As is clear from the above discussion, methods of computing the median involve pixel intensity sorting operations. If a bubble sort (Gonnet, 1984) were used for this purpose, then up to $O(n^4)$ operations would be required for an $n \times n$ neighborhood, compared with some 256 operations for the histogram method described above. Thus, sorting methods such as the bubble sort are faster for small neighborhoods where n is 3 or 4 but not for neighborhoods where n is greater than 5, or where pixel intensity values are more restricted.

Much of the discussion of the median filter in the literature is concerned with saving computation (Narendra, 1978; Huang et al., 1979; Danielsson, 1981). In

---

Table 3.1 An Implementation of the Median Filter

```
for (i = 0;  i <= 255;  i++)  hist[i] = 0;
for all pixels in image do {
    for (m = 0;  m <= 8;  m++)  hist[ P[m] ]++;
    i = 0;  sum = 0;
    while (sum < 5) {
        sum = sum + hist[i];
        i = i + 1;
    }
    Q0 = i − 1;
    for (m = 0;  m <= 8;  m++)  hist[ P[m] ] = 0;
}
```

particular, it has been noticed that, on proceeding from one neighborhood to the next, relatively few new pixels are encountered: this means that the new median value can be found by updating the old value rather than starting from scratch (Huang et al., 1979).

3.4 MODE FILTERS

Having considered the mean and the median of the local intensity distribution as candidate intensity values for noise smoothing filters, it also seems relevant to consider the mode of the distribution. Indeed, we might imagine that this is if anything more important than the mean or the median, since the mode represents the most probable value of any distribution.

However, a tedious problem arises as soon as we attempt to apply this idea. The local intensity distribution is calculated from relatively few pixel intensity values (Fig. 3.4). This means that instead of a smooth intensity distribution whose mode is easily located, we are almost certain to have a multimodal distribution whose highest point does not indicate the position of the *underlying* mode. Clearly the distribution needs to be smoothed out considerably before the mode is computed. Another tedious problem is that the width of the distribution varies widely from neighborhood to neighborhood (e.g., from close to zero to close to 256), so that it is difficult to know quite how much to smooth the distribution in any instance. For these reasons, it is likely to be better to choose an indirect measure of the position of the mode rather than to attempt to measure it directly.

In fact, the position of the mode can be estimated with reasonable accuracy once the median has been located (Davies, 1984a, 1988c). To understand the technique, it is necessary to consider how local intensity distributions of various sorts arise in practical situations. At most positions in an image, variations in pixel intensity are generated by steady changes in background illumination, or by steady variations in surface orientation, or else by noise. Thus, a symmetrical unimodal local intensity distribution is to be expected. It is well known that the mean, median, and mode are coincident in such cases. More problematic is what

FIGURE 3.4

The sparse nature of the local intensity histogram for a small neighborhood. This situation clearly causes significant problems for estimation of the mode. It also has definite implications for rigorous estimation of the underlying median, assuming that the observed intensities are only noisy samples of the ideal intensity pattern (see Section 3.8.3).

Source: © *IEEE 1984*

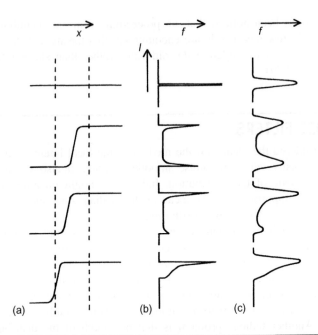

FIGURE 3.5

Local models of image data near the edge of an object: (a) cross-sections of an edge falling in the vicinity of a filter neighborhood; (b) corresponding local intensity distributions when very little image noise is present; (c) situation when the noise level is increased.

Source: © IEEE 1984

happens to the intensity variation near the edge of an object in the image. Here the local intensity distribution is unlikely to be symmetrical and, more important, it may not even be unimodal. In fact, near an edge the distribution is in general inherently *bimodal*, since the neighborhood contains pixels with intensities corresponding to the values they would have on either side of the edge (Fig. 3.5). Considering the image as a whole, this will be the most likely alternative to a symmetrical unimodal distribution, any further possibilities such as trimodal distributions being rare and of varied causes (e.g., odd glints on the edges of metal objects) which are outside the scope of the present discussion.[1]

If the neighborhood straddles an edge and the local intensity distribution is bimodal, the larger peak position should clearly be selected as the most probable intensity value. A good strategy for finding the larger peak is to eliminate the smaller peak. If we knew the position of the mode, we could find where to truncate the smaller peak by first finding which extreme of the distribution was closer to the mode, and then moving an equal distance to the opposite side of the mode

---

[1]Here we are ignoring the effects of noise and just considering the underlying image signal.

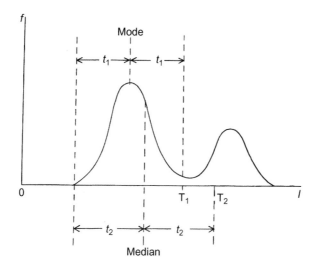

FIGURE 3.6

Rationale for the method of truncation. The obvious position at which to truncate the distribution is T_1. Since the position of the mode is not initially known, it is suboptimal but safe to truncate instead at T_2.

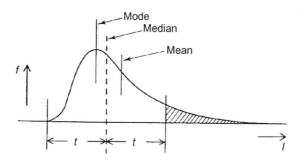

FIGURE 3.7

Relative positions of the mode, median, and mean for a typical unimodal distribution. This ordering is unchanged for a bimodal distribution, as long as it can be approximated by two Gaussian distributions of similar width.

Source: © IEEE 1984

(Fig. 3.6). Since we start off *not* knowing the position of the mode, one option is to use the position of the median as an estimator of the position of the mode, and then to use that position to find where to truncate the distribution. Since it invariably happens that the three means take the order mean, median, and mode (see Fig. 3.7), except when distributions are badly behaved or multimodal, this method

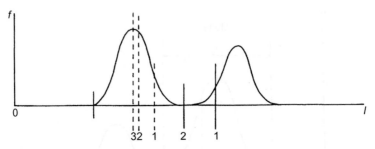

FIGURE 3.8

Iterative truncation of the local intensity distribution. Here the median converges on the mode within three iterations of the truncation procedure. This is possible since at each stage the mode of the new truncated distribution remains the same as that of the previous distribution.

FIGURE 3.9

Effect of a single application of 3×3 truncated median filter to the image of Fig. 2.1(a).

is cautious in the sense that it truncates less of the distribution than the required amount: this makes it a safe method to use. When we now find the median of the truncated distribution, the position is much closer to the mode than the original median was, a good proportion of the second peak being removed (Fig. 3.8). Iteration could be used to find an even closer approximation to the position of the mode. However, the method gives a marked enhancement in the image even when this is not done (Fig. 3.9).

We next examine more closely the properties of the "truncated median filter" (TMF) described above. The median filter is highly successful at removing noise,

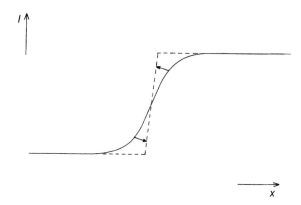

FIGURE 3.10

Image enhancement performed by the mode filter. Here the onset of the edge is pushed laterally by the action of the mode filter within one neighborhood; since the same happens from the other side within an adjacent neighborhood, the actual position of the edge is unchanged in first order. The overall effect is to sharpen the edge.

whereas the TMF not only removes noise but also enhances the image so that edges become sharper. Figure 3.10 makes it clear why this should happen. Basically, at a location even very slightly to one side of an edge, a majority of the pixel intensities contribute to the larger peak and the TMF ignores the pixel intensities contributing to the smaller peak. Thus, the TMF makes an informed binary choice about which side of the edge it is on. At first this seems to mean that it pushes a nearby edge further away. However, it must be remembered that it actually "pushes the edge away" from both sides, and the result is that its sides are made sharper and object outlines are crispened up. Particularly striking is the effect of applying the TMF to an image a number of times, when objects start to become segmented into regions of fairly uniform intensity (Fig. 3.11). The complete algorithm for achieving this is outlined in Table 3.2.

This problem has been dealt with at some length for a number of reasons. First, the mode filter has not hitherto received the attention it deserves. Second, the median filter seems to be used fairly universally, often without very much justification or thought. Third, all these filters show what markedly different characteristics are available merely by analyzing the contents of the local intensity distribution and ignoring totally where in the neighborhood the different intensities appear: it is perhaps remarkable that there is sufficient information in the local intensity distribution for this to be possible. All this shows the danger of applying operators that have been derived in an *ad hoc* manner without first making a specification of what is required and then designing an operator with the required characteristics. In fact, the situation appears to be that if we want a filter that has maximum impulse noise suppression capability, then we should use a median filter; and if we want a filter that enhances images by sharpening edges,

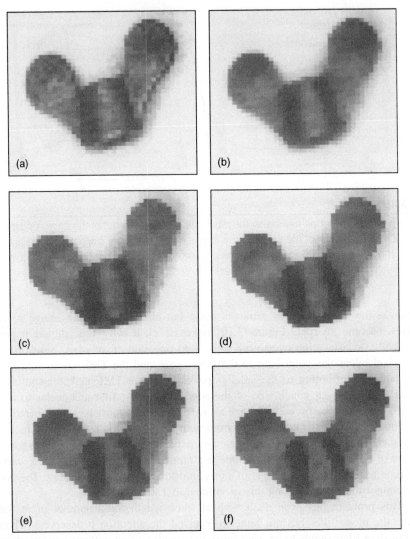

FIGURE 3.11

Results of repeated action of the truncated median filter: (a) the original, moderately noisy picture; (b) effect of a 3×3 median filter; (c)—(f) effect of 1—4 passes of the basic truncated median filter, respectively.

Source: © IEEE 1984

then we should use a mode filter or TMF (note that the TMF should be an improvement on the mode filter in that it is more cautious very close to an edge transition, where noise prevents an exact judgement being made as to which side of the edge a pixel is on: see Davies (1984a, 1988c)).

Table 3.2 Outline of Algorithm for Implementing the Truncated Median Filter

```
do { // as many passes over image as necessary
      for all pixels in image do {
            compute local intensity distribution;
            do { // iterate to improve estimate of mode
                  find minimum, median, and maximum intensity values;
                  decide from which end local intensity distribution should
                    be truncated;
                  deduce where local intensity distribution should be
                    truncated;
                  truncate local intensity distribution;
                  find median of truncated local intensity distribution;
            } until median sufficiently close to mode of local distribution;
            transfer estimate of mode to output image space;
      }
} until sufficient enhancement of image;
```

Comments:

(i) *The outermost and innermost loops can normally be omitted (i.e., they need to be executed once only).*

(ii) *The final estimate of the position of the mode can be performed by simple averaging instead of computing the median: this has been found to save computation with negligible loss of accuracy.*

(iii) *Instead of the minimum and maximum intensity values, the positions of the outermost octiles (for example) may be used to give more stable estimates of the extremes of the local intensity distribution.*

While considering enhancement, attention has been restricted to filters based on the local intensity distribution: there are many filters that enhance images without the aid of the local intensity distribution (Lev et al., 1977; Nagao and Matsuyama, 1979), but they are not within the scope of this chapter. Note that the method of "sharp–unsharp masking" (Section 3.7) performs an enhancement function, although its main purpose is to restore images that have inadvertently become blurred, e.g., by a hazy atmosphere or defocussed camera.

Finally, while this section has concentrated on the grayscale properties of mode filters, Charles and Davies (2003a, 2004) have shown how to devise versions of the TMF that operate on color images. Typical results are shown in Fig. 3.12. In addition, Fig. 3.13 shows that the TMF has the useful property of being able to eliminate very large amounts of impulse noise from images—significantly more than a median filter—in spite of being designated as an image enhancement filter.

(a) (b)

(c) (d)

FIGURE 3.12

Color filtering of brightly colored objects. (a) Original color image of some sweets. (b) Vector median filtered version. (c) Vector mode filtered version. (d) Version to which a mode filter has been applied to each color channel separately. Note that (b) and (c) show no evidence of color bleeding, although it is strongly evident in (d). It is most noticeable as isolated pink pixels, plus a few green pixels, around the yellow sweets. For further details on color bleeding, see Section 3.14 (see also color Plate 5).

Source: © *RPS 2004*

3.5 RANK ORDER FILTERS

The principle employed in rank order filters is to take all the intensity values in a given neighborhood, to place these in order of increasing value, and finally to select the rth of the n values and return this value as the filter local output value. Clearly, n rank order filters can be specified in terms of the value r that is used, but these filters are intrinsically nonlinear, i.e., the output intensity cannot be expressed as a linear sum of the component intensities within the neighborhood. In particular, the median filter (for which $r = (n+1)/2$, and which is only defined

(a)

(b)

(c)

FIGURE 3.13

Color filtering of images containing substantial impulse noise. (a) Version of the Lena image containing 70% random color impulse noise. (b) Effect of applying a vector median filter, and (c) effect of applying a vector mode filter. While the mode filter is designed more for enhancement than for noise suppression, it has been found to perform remarkably well at this task when the noise level is very high (see also color Plate 6).

Source: © RPS 2004

if n is odd)[2] does not normally give the same output image as a mean filter: indeed, it is well known that the mean and median of a distribution are in general only coincident for symmetrical distributions. Note that minimum and maximum filters (corresponding to $r = 1$ and $r = n$, respectively) are also often classed as morphological filters (see Chapter 7).

---

[2]If n is even, it is usual to take the mean of the central two values in the distribution as representing the median.

3.6 REDUCING COMPUTATIONAL LOAD

Significant efforts have been made to speedup the operation of the Gaussian filter since implementations in large neighborhoods require considerable amounts of computation (Wiejak et al., 1985). For example, smoothing an image of 256×256 pixels using a 30×30 Gaussian convolution mask involves 64 million basic operations. For such a basic operation as smoothing, this is unacceptable. However, it is possible to cut down the amount of computation drastically, since a 2-D Gaussian convolution can be factorized into two 1-D Gaussian convolutions, which can be applied in turn:

$$\exp\left(-\frac{r^2}{2\sigma^2}\right) = \exp\left(-\frac{x^2}{2\sigma^2}\right) \exp\left(-\frac{y^2}{2\sigma^2}\right) \tag{3.4}$$

It is important to realize that the decomposition is rigorously provable and is not an approximation: we shall refer to this below in the context of the median filter. Meanwhile, the decompositions for the two 3×3 Gaussian filters we discussed earlier are:

$$\frac{1}{9}\begin{bmatrix} 1 & 1 & 1 \\ 1 & 1 & 1 \\ 1 & 1 & 1 \end{bmatrix} = \frac{1}{3}\begin{bmatrix} 1 \\ 1 \\ 1 \end{bmatrix} \frac{1}{3}\begin{bmatrix} 1 & 1 & 1 \end{bmatrix} \tag{3.5}$$

and

$$\frac{1}{16}\begin{bmatrix} 1 & 2 & 1 \\ 2 & 4 & 2 \\ 1 & 2 & 1 \end{bmatrix} = \frac{1}{4}\begin{bmatrix} 1 \\ 2 \\ 1 \end{bmatrix} \frac{1}{4}\begin{bmatrix} 1 & 2 & 1 \end{bmatrix} \tag{3.6}$$

Overall, this approach replaces a single $n \times n$ operator whose load is $O(n^2)$ with two operators of load $O(n)$, and ignoring scanning and other overheads, the saving factor is $n/2$. Hence, for $n > 2$, there will always be a useful saving.

In fact, it is not possible to decompose the median filter in the same way without making approximations. However, it is quite common to try to perform a similar function by applying two 1-D median filters in turn (Narendra, 1978). Although the effect is similar in its outlier rejection properties to the standard 2-D median filter, the following example confirms that the two are not exactly equivalent:

Original image segment:

```
0 0 0 0 0 0
0 0 1 1 2 0
0 0 2 2 0 0
0 0 0 0 0 0
```

After applying a 3×3 filter:

$$
\begin{matrix}
0 & 0 & 0 & 0 & 0 & 0 \\
0 & 0 & 0 & 1 & 0 & 0 \\
0 & 0 & 0 & 1 & 0 & 0 \\
0 & 0 & 0 & 0 & 0 & 0
\end{matrix}
$$

After applying a 3×1 and then a 1×3 filter:

$$
\begin{matrix}
0 & 0 & 0 & 0 & 0 & 0 \\
0 & 0 & 1 & 1 & 1 & 0 \\
0 & 0 & 2 & 2 & 0 & 0 \\
0 & 0 & 0 & 0 & 0 & 0
\end{matrix}
\qquad
\begin{matrix}
0 & 0 & 0 & 0 & 0 & 0 \\
0 & 0 & 1 & 1 & 0 & 0 \\
0 & 0 & 1 & 1 & 0 & 0 \\
0 & 0 & 0 & 0 & 0 & 0
\end{matrix}
$$

After applying a 3×1 and then a 1×3 filter:

$$
\begin{matrix}
0 & 0 & 0 & 0 & 0 & 0 \\
0 & 0 & 1 & 1 & 0 & 0 \\
0 & 0 & 1 & 1 & 0 & 0 \\
0 & 0 & 0 & 0 & 0 & 0
\end{matrix}
\qquad
\begin{matrix}
0 & 0 & 0 & 0 & 0 & 0 \\
0 & 0 & 1 & 1 & 0 & 0 \\
0 & 0 & 1 & 1 & 0 & 0 \\
0 & 0 & 0 & 0 & 0 & 0
\end{matrix}
$$

By chance, the final results in the last two cases are the same, but this is a relatively rare occurrence.

In general, spots and streaks not more than one pixel wide are eliminated quite effectively by the original or by the separated forms of the filter. Larger filters should effectively eliminate wider spots and streaks, although exact functional equivalence between the original and its separated forms is not to be expected, as has been indicated.

Finally, the problem of inexact decomposition is not an exclusive property of nonlinear filters; many linear filters cannot be decomposed exactly either: this is evident because the number of independent coefficients in an $n \times n$ mask is n^2, which is much greater than the total number in an $n \times 1$ and a $1 \times n$ component mask.

3.7 SHARP–UNSHARP MASKING

When images are blurred either before or as part of the process of acquisition, it is possible frequently to restore them substantially to their ideal state. Properly, this is achieved by making a model of the blurring process and applying an inverse transformation that is intended to cancel the blurring. This is a complex task to carry out rigorously, but in some cases a rather simple method called sharp–unsharp masking is able to produce significant improvement (Gonzalez and Woods, 1992). As indicated in Fig. 3.14, this technique involves first obtaining an even more blurred version of the image (e.g., with the aid of a Gaussian filter) and then subtracting this image from the original. Note that the amount of

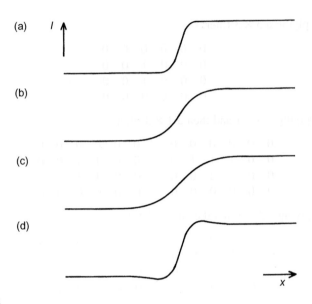

FIGURE 3.14

The principle of sharp—unsharp masking: (a) cross-section of an idealized edge; (b) observed edge; (c) artificially blurred version of (b); (d) result of subtracting a proportion of (c) from (b).

artificial blurring to apply and the proportion of the blurred image to subtract are rather arbitrary quantities that are normally adjusted by eye. Thus, the method is better categorized under the heading "enhancement" than "restoration," as it is not the precise mathematical technique normally understood by the latter term. Of such enhancement techniques, Hall (1979) states: "Much of the art of enhancement is knowing when to stop."

3.8 SHIFTS INTRODUCED BY MEDIAN FILTERS

Despite knowing the main characteristics of the different types of filter there are still some unknown factors. In particular, it is often important (especially when making precision measurements on manufactured components) to ensure that noise is removed in such a way that object locations and sizes are unchanged. However, at this point the following two problems arise.

First, it has been assumed that the intensity profile of an edge is symmetrical. If this is so, then the mean, median, and mode of the local intensity distribution will be coincident and there will clearly be no overall bias for any of them. However, when the edge profile is asymmetrical, it will not be obvious in the absence of a detailed model of the situation what the result will be for any of the

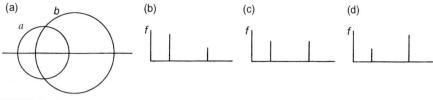

FIGURE 3.15

Variation in local intensity distribution with position of neighborhood: (a) neighborhood of radius *a* overlapping a dark circular object of radius *b*; (b)–(d) intensity distributions *I* when the separations of the centers are respectively less than, equal to, or greater than the distance *d* for which the object bisects the area of the neighborhood.

filters. The situation is even more involved when significant noise is present (Yang and Huang, 1981; Bovik et al., 1987). Since the problem is so data-dependent, it is not profitable to consider it further here.

The second problem concerns the situation for a curved edge. In this case, there is again a variety of possibilities, and filters employing the different means will modify the edge position in ways that depend markedly on its shape. In robot vision applications, the median filter is the one we are most likely to use because its main purpose is to suppress noise without introducing blurring. Hence, the bias produced by this type of filter is worth considering in some detail. This is done in the next subsection.

3.8.1 Continuum Model of Median Shifts

This section takes the case of a continuous image (i.e., a nondiscrete lattice), assuming (i) that the image is binary, (ii) that neighborhoods are exactly circular, and (iii) that images are noise-free. To proceed we notice that binary edges have symmetrical cross-sections, while straight edges extend this symmetry into 2-D: hence applying a median filter in a (symmetrical) circular neighborhood cannot pull a straight edge to one side or the other.

Now consider what happens when the filter is applied to an edge that is not straight. If, for example, the edge is circular, the local intensity distribution will contain two peaks whose relative sizes will vary with the precise position of the neighborhood (Fig. 3.15). At some position the sizes of the two peaks will be identical. Clearly, this happens when the center of the neighborhood is situated at a point where the output of the median filter changes from dark to light (or *vice versa*). Thus, the median filter produces an inward shift toward the center of a circular object (or the center of curvature), whether the object is dark on a light background or light on a dark background.

To calculate the magnitude of this effect, we need to find at what distance *d* from the center of a circular object (of radius *b*) the area of a circular neighborhood (of radius *a*) is bisected by the object boundary.

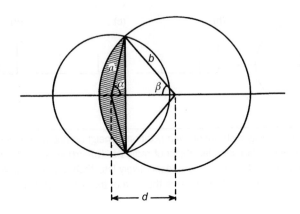

FIGURE 3.16

Geometry for calculating neighborhood and object overlap.

From Fig. 3.16 the area of the sector of angle 2β is βb^2, while the area of the triangle of angle 2β is $b^2\sin\beta\cos\beta$. Hence, the area of the segment shown shaded is:

$$B = b^2(\beta - \sin\beta \cos\beta) \tag{3.7}$$

Making a similar calculation of the area A of a circular segment of radius a and angle 2α, the area of overlap (Fig. 3.16) between the circular neighborhood of radius a and the circular object of radius b may be deduced as:

$$C = A + B \tag{3.8}$$

For a median filter this is equal to $\pi a^2/2$. Hence:

$$F = a^2(\alpha - \sin\alpha \ \cos\alpha) + b^2(\beta - \sin\beta \ \cos\beta) - \frac{\pi a^2}{2} = 0 \tag{3.9}$$

where

$$a^2 = b^2 + d^2 - 2bd \ \cos\beta \tag{3.10}$$

and

$$b^2 = a^2 + d^2 - 2ad \ \cos\alpha \tag{3.11}$$

To solve this set of equations, we take a given value of d, deduce values of α and β, calculate the value of F, and then adjust the value of d until $F = 0$. Since d is the modified value of b obtained after filtering, the shift produced by the filtering process is:

$$D = b - d \tag{3.12}$$

The results of doing this computation numerically have been found by Davies (1989b). As expected, $D \to 0$ as $b \to \infty$ or $a \to 0$. Conversely, the shift becomes very large as a first approaches and then exceeds b. Note, however, that when

$a > \sqrt{2}b$ the object is ignored, being small enough to be regarded as irrelevant noise by the filter: beyond this point it has no effect at all on the final image. The maximum edge shift before the object finally disappears is $(2 - \sqrt{2})b \approx 0.586b$.

It is instructive to approximate the above equations for the case when edge curvature is small, i.e., $a \ll b$. Under these conditions, β is small, $\alpha \approx \pi/2$ and $d \approx b$. Hence, we find:

$$\beta \approx \frac{a}{b} \qquad (3.13)$$

After some manipulation the edge shift D is obtained in the form:

$$D \approx \frac{a^2}{6b} = \frac{\kappa a^2}{6} \qquad (3.14)$$

$\kappa = 1/b$ being the local curvature. In Chapter 6, this equation is found to be useful for estimating the signals from a median-based corner detector.

3.8.2 Generalization to Grayscale Images

To extend these results to grayscale images, first consider the effect of applying a median filter near a smooth step edge in 1-D. Here the median filter gives zero shift, since for equal distances from the center to either end of the neighborhood there are equal numbers of higher and lower intensity values and hence equal areas under the corresponding portions of the intensity histogram. Clearly this is always valid where the intensity increases monotonically from one end of the neighborhood to the other—a property first pointed out by Gallagher and Wise (1981) [for more recent discussions on related "root" (invariance) properties of signals under median filtering, see Fitch et al. (1985) and Heinonen and Neuvo (1987)].

Next, it is clear that for 2-D images, the situation is again unchanged in the vicinity of a straight edge, since the situation remains highly symmetrical. Hence the median filter gives zero shift, as in the binary case.

For curved boundaries, the situation has to be considered carefully for grayscale edges, which, unlike binary edges, have finite slope. When boundaries are roughly circular, contours of constant intensity often appear as in Fig. 3.17. To find how a median filter acts, we merely need to identify the contour of median intensity (in 2-D the median intensity value labels a whole contour) that divides the area of the neighborhood into two equal parts. The geometry of the situation is identical to that already examined in Section 3.8.1: the main difference here is that for every position of the neighborhood, there is a corresponding median contour with its own particular value of shift depending on the curvature. Intriguingly, the formulae already deduced may immediately be applied for calculating the shift for each contour. Figure 3.17 shows an idealized case in which the contours of constant intensity have similar curvature, so that they are all moved

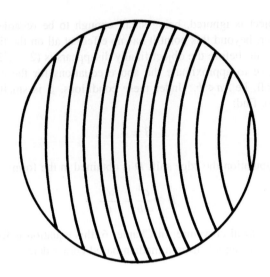

FIGURE 3.17

Contours of constant intensity on the edge of a large circular object, as seen within a small circular neighborhood.

inward by similar amounts. This means that, to a first approximation, the edges of the object retain their cross-sectional profile as it becomes smaller.

For grayscale images, the shifts predicted by this theory (with certain additional corrections: see Davies (1989b)) agree with experimental shifts within approximately 10% for a large range of circle sizes in a discrete lattice (see Fig. 3.18). Paradoxically, the agreement is less perfect for binary images, since circles of certain sizes show stability effects (akin to median root behavior): these effects tend to average out for grayscale images, owing to the presence of many contours of different sizes at different gray levels. Overall, the edge shifts obtained with median filters are now quite well understood. Figures 3.19 and 3.20 give some indication of the magnitudes of these shifts in practical situations. Note that once image detail such as a small hole or screw thread has been eliminated by a filter, it is not possible to apply any edge shift correction formula to recover it, although for larger features such formulae are useful for deducing true edge positions.

3.8.3 Problems with Statistics

Thus far it has been seen that computations of the position of the mode are made more difficult because of the sparse statistics of the local intensity distribution. In fact this also affects the median calculation. Suppose the median value happens to be well spaced near the center of the distribution (Fig. 3.4). Then a small error in

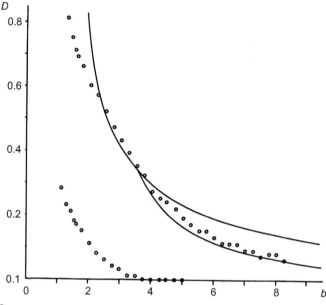

FIGURE 3.18

Edge shifts for 5 × 5 median filter applied to a grayscale image. The upper set of plots represents the experimental results and the upper continuous curve is derived from the theory of Section 3.8.1. The lower continuous curve is derived from a more accurate model (Davies, 1989b). The lower set of plots represents the much reduced shifts obtained with the "detail-preserving" type of filter (see Section 3.16).

this one intensity value is immediately reflected in full when calculating the median: i.e., the poor statistics have biased the median in a particular way. Ideally, what is required is a stable estimator of the median of the underlying distribution. Thus, the distribution should be made smoother before arriving at a specific value for the median. In practice, this procedure adds significant computational load to the filter calculation and is commonly not carried out. As a consequence the median filter tends to result in runs of constant intensity, thereby giving images the "softened" appearance noted earlier. This is apparent on studying the following 1-D example:

```
Original:   0  0  1  0  0  1  1  2  1  2  2  1  2  3  3  4  3  2  2  3
Filtered:   ?  0  0  0  0  1  1  1  2  2  2  2  2  3  3  3  3  2  2  ?
```

Although histogram smoothing is not commonly carried out, some workers have felt it necessary to adjust the relative weights of the various pixels in the neighborhood according to their distance from the central pixel (Akey and Mitchell, 1984). This mimics what happens for a Gaussian filter and is theoretically

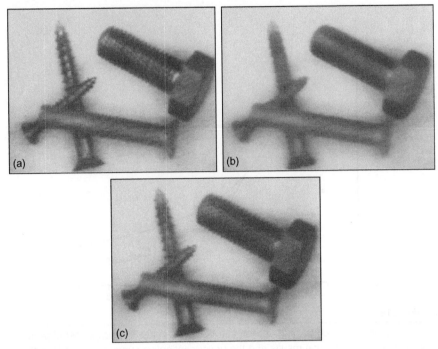

FIGURE 3.19

Edge smoothing property of the median filter: (a) Original image; (b) median filter smoothing of irregularities, in particular those around the boundaries (note how the threads on the screws are virtually eliminated although detail larger in scale than half the filter area is preserved), using a 21-element filter operating within a 5 × 5 neighborhood on a 128 × 128 pixel image of 6-bit gray scale; (c) effect of the detail-preserving filter (see Section 3.16).

necessary, although it is not generally implemented. However, there have recently been further developments on this front (e.g., see Charles and Davies, 2003b).

3.9 DISCRETE MODEL OF MEDIAN SHIFTS

To produce a discrete model of median shifts we need to recognize explicitly the positions of the pixels within the chosen neighborhood. We approximate by assuming that the intensity of any pixel is the mean intensity over the whole pixel and is represented by a sample positioned at the center of the pixel. We start by examining the case of a 3 × 3 neighborhood, and proceed by taking the underlying analog intensity variation to have contours of curvature κ, as shown in

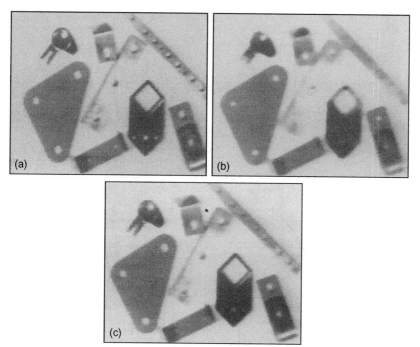

FIGURE 3.20

Circular holes in metal objects before and after filtering: (a) original 128×128 pixel image with 6-bit gray scale; (b) 5×5 median-filtered image: the diminution in size of the hole is clearly visible and such distortions would have to be corrected when taking measurement from real filtered images of this type; (c) result using a detail-preserving filter: some distortions are present although the overall result is much better than in (b).

Fig. 3.17. Following what happened in the continuum case, it will not matter whether the contours of constant intensity are those of a step edge or those of a slowly varying slant edge: it is what happens at the median contour that determines the shift that arises.

The starting point is that zero shift occurs for $\kappa = 0$. Next, if κ is even minutely greater than zero, the center pixel will not necessarily be the median pixel. Consider the case when the circular median intensity contour passes close to the center of the neighborhood at a small angle θ to the positive x-axis (Figs. 3.21 and 3.22(a)). In that case, the filter will produce a definite shift, whose value is:

$$D_\theta \approx \frac{1}{2}\kappa a_0^2 - a_0\theta = \frac{1}{2}\kappa - \theta \qquad (3.15)$$

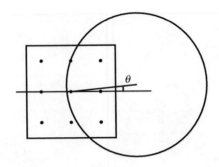

FIGURE 3.21

Geometry for calculation of median shifts on the discrete model.

Source: © IEE 1999

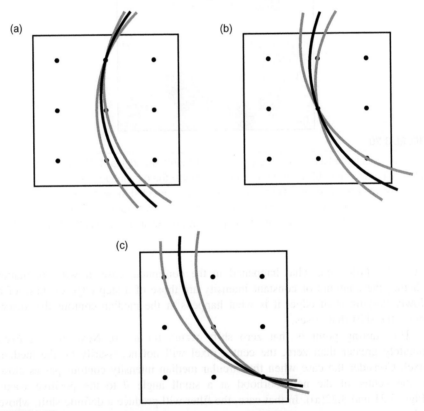

FIGURE 3.22

Geometry for calculation of median shifts at low κ. These three diagrams show the positions of the median pixels and the ranges of orientations of circular intensity contours for which they apply, (a) for low θ, (b) for intermediate θ, and (c) for high θ.

Source: © IEE 1999

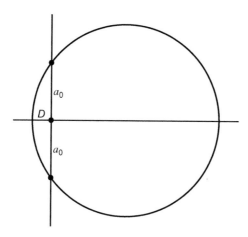

FIGURE 3.23

Geometry for calculation of shift when the median contour passes through the centers of two pixels.

where a_0 is the inter-pixel separation—here taken as unity. The first term arises from the following simple result for the geometry of the circle of radius $b = 1/\kappa$ in Fig. 3.23:

$$a_0^2 = D \times (2b - D) \approx 2Db \qquad (3.16)$$

It is too tedious to recount here the complete analysis for the variation in D_θ for all θ. Suffice it to say that when it has been carried out, it is necessary to average it over all θ for each value of κ. When this is done (Davies, 1999f) the agreement between theory and experiment is essentially exact over a wide range of values of κ, as shown in Fig. 3.24: the reason for the discrepancy of high values of κ is due to the limited intensity gradients that occur at edges in grayscale images.

Overall, the problem of median shifts is now well understood, and is fully explained using the discrete model. The continuum model turns out to be capable of giving accurate results only in the limiting case where a and b ($= 1/\kappa$) are many pixels in size.

3.10 SHIFTS INTRODUCED BY MODE FILTERS

In this section we consider the shifts produced by mode filters in continuous images. As in the cases of median filters, straight edges with symmetrical profiles cannot be shifted by mode filters, because of symmetry. We proceed to the two paradigm cases—step edges and slant edges with circular boundaries. Again, the effects of noise will be ignored as we are considering the intrinsic rather than the noise-induced behavior of the mode filtering operation.

The situation for a curved step edge can again be understood by appealing to Fig. 3.15. The result for the mode also has to be identical to that for the median,

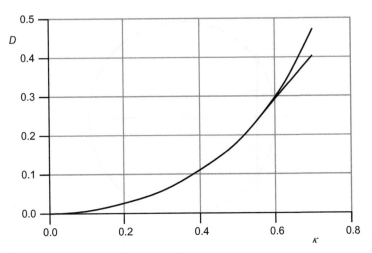

FIGURE 3.24

Comparisons of 3 × 3 median shifts. The lower solid curve shows the nonapproximated results of the discrete model (Davies, 1999f): the upper solid curve shows the results of experiments on grayscale circles.

Source: © IEE 1999

because the local intensity distribution is exactly symmetric and bimodal at the point where the median filter switches from a left-hand to a right-hand decision: at that point the mode must give the same result, since the median and the mode are coincident for a symmetric distribution. Hence, we conclude that the mode also gives a shift of $1/6\ \kappa a^2$ for a curved step edge.

Next, we calculate edge shifts in the case where smoothly varying intensity functions exist—or within the confines of a small neighborhood—*appear* to be smoothly varying. In this case the calculation is especially simple (Davies, 1997a). Using the geometry of Fig. 3.17, we consider the intensity pattern within a circular neighborhood C. Of all the circular intensity contours appearing within C, the one possessing the most frequently occurring intensity, as selected by a mode filter, is the longest. Clearly, this is the one (M) whose ends are at opposite ends of a diameter of C. To estimate the shift in this case, all we need to do is to calculate the position of M, and determine its distance from the center of C. To proceed, we use the well-known formula relating the lengths of parts of intersecting chords of a circle, which gives:[3]

$$a^2 = D(2b - D) \approx 2Db \qquad (3.17)$$

---

[3]This equation is a more general form of equation (3.16), as a can have any value; the proof follows similarly, on replacing a_0 by a in Fig. 3.23.

Table 3.3 Summary of Edge Shifts for Neighborhood Averaging Filters

Edge Type	Filter		
	Mean	**Median**	**Mode**
Step	$\frac{1}{6}\kappa a^2$	$\frac{1}{6}\kappa a^2$	$\frac{1}{6}\kappa a^2$
Intermediate	$\sim\frac{1}{7}\kappa a^2$	$\frac{1}{6}\kappa a^2$	$\frac{1}{2}\kappa a^2$
Linear	$\frac{1}{8}\kappa a^2$	$\frac{1}{6}\kappa a^2$	$\frac{1}{2}\kappa a^2$

© *IEE 1999.*

Hence:

$$D = \frac{1}{2}\kappa a^2 \qquad (3.18)$$

i.e., there is a *right* shift of the contour, toward the local center of curvature, of $1/2\ \kappa a^2$. If we regard this set of contours as forming part of a grayscale edge profile, then the mode filter shifts the edge through $1/2\ \kappa a^2$ toward the center of curvature.

Some comments on the marked difference between the cases of step edges and linear intensity profiles are called for. This is all the more interesting as the median filter produces identical shifts, of $1/6\ \kappa a^2$, for the two profiles (see Table 3.3). In fact, of all the cases listed in Table 3.3, the outstanding one is the large shift for a mode filter operating on a linear intensity profile: what is special in this case is that the result relies on a single extreme contour length rather than an average of lengths amounting to an area measure. Hence, it is not surprising that the mode filter gives an exceptionally large shift in this case.

Finally, we note that edge shifts are not avoided merely by choosing an alternative method of neighborhood averaging, but rather that they are intrinsic to the averaging process, and can be avoided only by specially designed operators (e.g., see Greenhill and Davies, 1994a).

3.11 SHIFTS INTRODUCED BY MEAN AND GAUSSIAN FILTERS

In this section, we consider the shifts produced by mean and Gaussian filters in continuous images. As in the cases of median and mode filters, straight edges with symmetrical profiles cannot be shifted by mean and Gaussian filters, because of symmetry. We again consider the two paradigm cases—step edges and slant edges with circular boundaries. Again, we will ignore the effects of noise as we are considering the intrinsic rather than the noise-induced behavior of the filters.

The situation for a curved step edge can again be understood by appealing to Fig. 3.15, and is identical to that for the median and mode filters, and follows because of the symmetry of the local intensity distribution at the point where the filter switches from a left-hand to a right-hand decision: at that point the mean filter must give the same result, since all three statistics coincide for a symmetric distribution. Hence, it also gives a shift of $1/6\ \kappa a^2$ for a curved step edge.

In the case of a smoothly varying slant edge, the result for the mean filter has to be calculated by integrating over the area of the neighborhood. The results cannot be obtained by intuitive or simple geometric or intuitive arguments, and here we merely quote the shift for the mean filter as being $1/8\ \kappa a^2$.

These considerations complete the entries in Table 3.3. The results for Gaussian filters do not differ substantially from those for mean filters, but have to be obtained by integration, taking account of the Gaussian weighting function (Davies, 1991b). A general point is that all such filters have similar shifting effects because they all incorporate a measure of signal averaging: the shifting effect is not avoided simply by employing a different central-seeking statistic to perform the averaging.

3.12 SHIFTS INTRODUCED BY RANK ORDER FILTERS

This section is particularly concerned with rank order filters (Bovik et al., 1983), which form a whole family of filters that can be applied to digital images—often in combination with other filters of the family—in order to give a variety of effects (Goetcherian, 1980; Hodgson et al., 1985): other notable members of the family are max and min filters. Because rank order filters generalize the concept of the median filter, it is relevant to study the types of distortion they produce on straight and curved intensity contours. It should also be pointed out that these filters are of central importance in the design of filters for morphological image analysis and measurement. In addition, it has been pointed out that they have some advantages when used for this purpose in that they help to suppress noise (Harvey and Marshall, 1995) (although the effect vanishes in the special cases of max and min filters).

Section 3.12.1 examines the reasons underlying the shifts produced by rank order filters and makes calculations of their extent for rectangular neighborhoods. It then generalizes these results to circular neighborhoods and goes on to examine the extent to which the theoretical predictions are borne out in practice by measurements of the shifts produced by 5×5 rank order filters on circular disks of varying sizes. It will be taken as axiomatic that the application of rank order filters produces edge shifts on real images (they are well attested in the case of max, min, and median filters): the main question to be answered here is the exact numerical extent of these shifts and how they may be modeled for general rank order filters.

3.12.1 **Shifts in Rectangular Neighborhoods**

In common with previous work in this area we here concentrate on the ideal noiseless case, in which the filter operates within a small neighborhood, over which the signal is basically a monotonically increasing intensity function in some direction. The most complex intensity variation that will be considered is that in which the intensity contours are curved with curvature κ. In spite of this simplified configuration, valuable statements can still be made about the level of distortion likely to be produced in practice by rank order filters.

Because of the complexity of the calculations that arise in the case of rank order filters, which involve an additional parameter *vis-à-vis* the median filter, it is worth studying their properties first for the simple case of rectangular neighborhoods (Davies, 2000f). Let us presume that a rank order filter is being applied in a situation in which straight intensity contours are aligned parallel to the short sides of a rectangular neighborhood that we initially take to be a $1 \times n$ array of pixels (Fig. 3.25). In this case, we can assume without loss of generality that the successive pixels within the neighborhood will have *increasing* values of intensity. We next take the basic property of the rank order filter (effectively or in fact) to construct an intensity histogram of the local intensity distribution and return the value of the rth of the n intensity values within the neighborhood. This means that the rank order filter selects an intensity that has physical separation B from the lowest intensity pixel of the neighborhood and C from the highest intensity pixel, where:

$$B = r - 1 \tag{3.19}$$

$$C = n - r \tag{3.20}$$

$$A = B + C = n - 1 \tag{3.21}$$

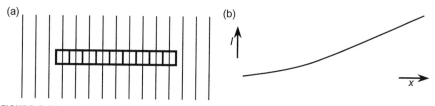

FIGURE 3.25

Basic situation for a rank order filter in a rectangular neighborhood. This figure illustrates the problem of applying a rank order filter within a rectangular neighborhood consisting of a $1 \times n$ array of pixels. The intensity is taken to increase monotonically from left to right, as in (b); the intensity contours in (a) are taken to be parallel to the short sides of the neighborhood.

Source: © *RPS 2000*

Table 3.4 Properties of the Three Paradigm Filters

Filter	r	η	B	C	D
Median	$\frac{1}{2}(n+1)$	0	$\frac{1}{2}A$	$\frac{1}{2}A$	0
Max	n	-1	A	0	$-\frac{1}{2}(n-1)$
Min	1	1	0	A	$\frac{1}{2}(n-1)$

© RPS 2000.

These definitions underline that a rank order filter will in general produce a D-pixel shift, whose value is:

$$D = \frac{1}{2}(n+1) - r \qquad (3.22)$$

Before proceeding further, it will be useful to introduce a parameter η that is more symmetric than r, and has value $+1$ at $r = 1$ and -1 at $r = n$:

$$\eta = (n - 2r + 1)/(n - 1) \qquad (3.23)$$

Using this parameter in preference to r, we can write down new formulae for B, C, D:

$$B = \frac{1}{2}A(1 - \eta) \qquad (3.24)$$

$$C = \frac{1}{2}A(1 + \eta) \qquad (3.25)$$

$$D = \frac{1}{2}\eta(n - 1) \qquad (3.26)$$

The properties of the three paradigm filters are summarized in Table 3.4 in terms of these parameters.

We now proceed to a continuum model, assuming a large number of pixels in any neighborhood (i.e., $n \to \infty$). The main difference will be that we shall specify distance in terms of the half-length a of the neighborhood rather than in terms of numbers of pixels:

$$D = \eta a \qquad (3.27)$$

Next note that this formulation is independent of the width of the neighborhood, so long as the latter is rectangular. We now generalize the situation by taking the neighborhood to be rectangular and of dimensions $2a$ by $2\tilde{a}$ (Fig. 3.26).

The next task is to determine the result of a curvature $\kappa = 1/b$ in the intensity contours. Here we approximate the equation of a circle of radius b, with its diameter on the positive x-axis and passing through the origin, as:

$$x = \frac{y^2}{2b} \tag{3.28}$$

We can integrate the area under an intensity contour (see Fig. 3.26) as follows:

$$K = \int_{-\tilde{a}}^{\tilde{a}} x \, dy = \left(\frac{1}{2b}\right) \int_{-\tilde{a}}^{\tilde{a}} y^2 \, dy = \left(\frac{1}{2b}\right) \left[\frac{y^3}{3}\right]_{-\tilde{a}}^{\tilde{a}}$$

$$= \frac{\tilde{a}^3}{3b} = \frac{1}{3}\kappa\tilde{a}^3 \tag{3.29}$$

We deduce that the shift D is given by:

$$B = 2\tilde{a}(a - D) + \frac{1}{3}\kappa\tilde{a}^3 \tag{3.30}$$

$$C = 2\tilde{a}(a + D) - \frac{1}{3}\kappa\tilde{a}^3 \tag{3.31}$$

$$\eta = \frac{(C - B)}{A} = \frac{4\tilde{a}D - (2/3)\kappa\tilde{a}^3}{A} \tag{3.32}$$

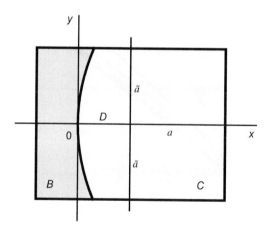

FIGURE 3.26

Geometry of a rectangular neighborhood with curved intensity contours. Here the neighborhood is a general rectangular neighborhood of dimensions $2a \times 2\tilde{a}$. Again, the intensity is taken to increase monotonically from left to right; the intensity contours are taken to be parallel and in this case are curved with identical curvature κ. x and y axes needed for area calculations are also shown. B and C represent the areas of the two shaded regions on either side of the thick intensity contour.

Source: © RPS 2000

where

$$A = 4a\tilde{a} \tag{3.33}$$

Hence:

$$D = \frac{\eta A}{4\tilde{a}} + \frac{1}{6}\kappa\tilde{a}^2 = \eta a + \frac{1}{6}\kappa\tilde{a}^2 \tag{3.34}$$

What is important about this equation is that it shows that the effects of rank order and curvature can be calculated and summed separately, the first term being that obtained above for the case of zero curvature and the second term being exactly that calculated for a median filter when the intensity contour is of length $2\tilde{a}$ (the earlier calculation (Davies, 1989b) related to a circular neighborhood). Thus, in principle we merely need to recompute the first term for any appropriate shape of neighborhood. However, various complications arise, particularly in the case of high curvature contours. These have been dealt with successfully, with the results shown in Figs. 3.27 and 3.28 (Davies, 2000f).

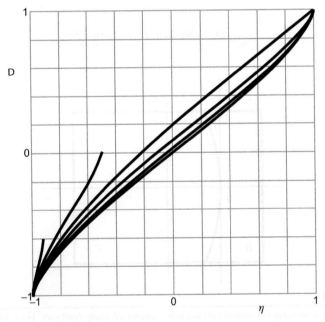

FIGURE 3.27

Graphs of shift D against rank order parameter η for various κ. This diagram summarizes the operation of rank order filters, with graphs, bottom to top, respectively, for $\kappa = 0$, $0.2/a$, $0.5/a$, $1/a$, $2/a$, $5/a$. Note that graphs for which $b < a$ ($\kappa > 1/a$) apply for restricted ranges of η and D (see Section 3.12.1). A multiplier of a must be included in the D-values.

As the above theory is based on a continuum model, it is not perfect, and it does indicate only the main features of the practical situation. In particular, when the curvatures are very high, they may arise from spots that are entirely within the neighborhood, and then there is the possibility that they will be completely eliminated by the rank order filter (note that noise points are entirely eliminated by a median filter, which indeed is the prime practical use of that type of filter). Correspondingly, the assumptions made in the model break down when there is no intersection of the circular neighborhood and the intensity contour of radius $b = 1/\kappa$.

Some of the conclusions of this work are quite important. In particular, the result for a median filter is the special case that arises when $\eta = 0$ and is in agreement

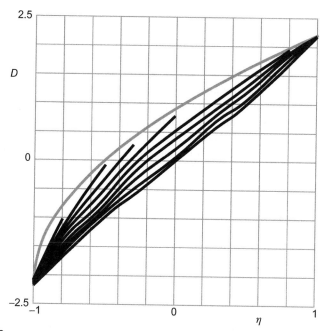

FIGURE 3.28

Shifts obtained for a typical discrete neighborhood. These shifts were obtained for rank order filters operating within a truncated 5×5 neighborhood when applied to eight discrete circular disks with radii ranging from 10.0 down to 1.25 pixels, the mean curvatures being 0.1–0.8 in steps of 0.1; the lowest curve was obtained by averaging the responses from circular disks of radius ± 20.0 pixels, with curvatures ± 0.05, and to the given scale are indistinguishable from the result that would be obtained with zero curvature. The uppermost curve represents the theoretical limiting value. However, because of the directional effects that occur in the discrete case, the upper limit is actually lower than indicated by this curve (see text).

with the calculations of Section 3.8. Next, the max and min filters are also special cases and occur for $\eta = -1$ and 1, respectively. In these limiting cases, the shifts are $D = -a$ and a, respectively, the results being independent of κ: this is as might be expected since the value of $\tilde{a}$ is zero in each case. Between the max and min filters and the median filter, there is a continuous gradation of performance, with very significant but opposite shifts for the max and min filters, and the two basic effects canceling out for median filters—although the cancellation is only exact for straight contours. The full situation is summarized in Fig. 3.27.

3.13 THE ROLE OF FILTERS IN INDUSTRIAL APPLICATIONS OF VISION

It has been shown above how the median filter can successfully remove noise and artifacts such as spots and streaks from images. Unfortunately, many useful features such as fine lines and important points and holes are effectively indistinguishable from spots and streaks. In addition, it has been seen that the median filter "softens" pictures by removing fine detail. It is also found to clip corners of objects—another generally undesirable trait (but see Chapter 6). Finally, although it does not blur edges, it can still shift them slightly. In fact, shifting of curved edges seems to be a general characteristic of noise suppression filters.

Such distortions are quite alarming and mitigate against the indiscriminate use of filters. If applied in situations where accurate measurements are to be made on images, particular care must be taken to test whether the data are being biased in any way. Although it is possible to make suitable corrections to the data, it seems a good general policy to employ noise removal filters only where they are absolutely essential for object visibility. The alternative is to employ edge detection and other operators that automatically suppress noise as an integral part of their function. This is the general approach taken in subsequent chapters: indeed, it is one of the principles underlined in this book that algorithms should be "robust" against noise or other artifacts that might upset measurements. There is quite significant scope for the design of robust algorithms, since images contain so much information that it is normally possible to arrange for erroneous information to be ignored.

3.14 COLOR IN IMAGE FILTERING

In Chapter 2, it was indicated that color often adds to the complexity of image analysis algorithms, and could also add to the associated computational costs. From these points of view color might, except for applications such as assessing the ripeness of fruit, be regarded as an irrelevant luxury. Nevertheless, in the field of image processing and image filtering, where good quality images have to be presented to human operators, it is a vital concern. In fact, in recent years much effort has been devoted to the development of effective color filtering algorithms. Here we shall consider mainly median and related impulse noise filtering procedures.

Perhaps the first point to note is that median filtering is defined in terms of sorting operations and is thus undefined in the color domain, which normally contains three dimensions. However, a simple solution is to apply a standard median filter to each of the color channels, and then to reassemble the color image. Unfortunately, this approach leads to certain problems, the most obvious one being that of color "bleeding" (Fig. 3.12). This occurs when an impulse noise point appears in just one of the channels and is situated near an edge or other image feature. The case of an impulse noise point near an edge is hereby illustrated in simplified form:

$$
\begin{array}{llllllllllll}
\text{Original:} & 0 & 0 & 0 & 0 & 1 & 0 & 1 & 1 & 1 & 1 & 1 & 1 \\
\text{Filtered:} & ? & 0 & 0 & 0 & 0 & 1 & 1 & 1 & 1 & 1 & 1 & ?
\end{array}
$$

We see that a 3-element median filter eliminates the impulse noise point but at the same time moves the edge toward it. The end result for a color image is that the edge will be tinted with the color of the impulse noise point.

Fortunately, there is a standard solution to this problem. First, note that it is possible to express single-channel median filtering as the minimization of a distance metric, and this metric is trivially extendible to three color channels (or indeed any number of channels). The relevant single channel metric is:[4]

$$
\text{median} = \arg\ \min_i \sum_j |d_{ij}| \tag{3.35}
$$

where d_{ij} is the distance between sample points i and j in the single-channel (gray scale) space. In the three-color domain, the metric is readily extended to:

$$
\text{median} = \arg\ \min_i \sum_j |\tilde{d}_{ij}| \tag{3.36}
$$

where $\tilde{d}_{ij}$ is the generalized distance between sample points i and j, and we typically take the L_2 norm to define the distance measure for three colors:

$$
\tilde{d}_{ij} = \left[\sum_{k=1}^{3} (I_{i,k} - I_{j,k})^2 \right]^{1/2} \tag{3.37}
$$

Here $\mathbf{I}_j$, $\mathbf{I}_j$ are RGB vectors and $I_{i,k}$, $I_{j,k}$ ($k = 1, 2, 3$) are their color components.

---

[4]"arg min" is a standard mathematical term that means the argument (here pixel intensity) corresponding to the index (here i) giving rise to the minimum value of the expression in Eq. (3.36) (here $\sum_j |d_{ij}|$).

While the resulting vector median filter (VMF) no longer treats the individual color components separately, it is by no means guaranteed to eliminate color bleeding completely. In fact, like the standard median, it replaces any noisy intensity $\mathbf{I}_n$, (including color) by the intensity $\mathbf{I}_j$ of another pixel that exists in the same window—rather than by an ideal intensity $\mathbf{I}$. Hence, color bleeding is only reduced, but not eliminated. If indeed there is a confluence of colors at any one point in an image, even in the absence of any impulse noise there is the possibility that these sorts of algorithms will become confused and inadvertently introduce small amounts of color bleeding: ultimately, the effect is due to the increased dimensionality of the data, which means that the algorithm has to contend with a greatly increased number of possible outcomes in spite of being an *ad hoc* procedure that does not embody specific understanding of images.

Figure 3.12 demonstrates the nature of color bleeding, albeit in the case of mode filtering: this figure shows vector median and vector mode filters to be remarkably free from color bleeding, but the same does not apply to scalar mode filters—for similar reasons to those indicated above for median filters.

3.15 CONCLUDING REMARKS

Although this chapter has dwelt on the implementation of noise suppression and image enhancement operators based on the local intensity distribution, it has made certain other points. In particular, it has shown the need to make a specification of the required imaging process and only then to work out the algorithm design strategy. Not only does this ensure that the algorithm will perform its function effectively, but also it should make it possible to optimize the algorithm for various practical criteria including speed, storage, and other parameters of interest. In addition, this chapter has demonstrated that any undesirable properties of the particular design strategy chosen (such as the inadvertent shifting of edges) should be sought and dealt with. Next, it has demonstrated a number of fundamental problems to do with imaging in discrete lattices—not least being problems of statistics that arise with small pixel neighborhoods. Finally, the large edge shifts of certain types of rank order filter are particularly important because they are turned to advantage in morphological operators (Chapter 7).

The next chapter moves on to a particularly vital problem in machine vision—that of segmenting images in order to find where objects are situated. This work builds on what has been learnt in the present chapter about edge profiles and how they are "seen" by neighborhood operators.

Median filters have long been used to eliminate impulse noise without blurring edges. However, this chapter has shown that significant shifting of edges can result from use of median filters, and this property extends to mode filters and a fortiori to rank order filters— so much so that the latter form the basis of morphological processing.

3.16 BIBLIOGRAPHICAL AND HISTORICAL NOTES

Much of the work of this chapter has built on a paper by the author (Davies, 1988c), which rests on considerable earlier work on Gaussian, median, and other rank order filters (Hodgson et al., 1985; Duin et al., 1986). Note that the edge shifts that occur for median filters are not limited to this type of filter but apply almost equally to mean filters (Davies, 1991b). In addition, other inaccuracies have been found with median filters and methods have been found to correct them (Davies, 1992e).

The early literature hardly mentions mode filters, presumably because of the difficulty of finding simple mode estimators that are not unduly confused by noise and which still operate rapidly. Indeed, only one early reference has been found (Coleman and Andrews, 1979), although it has been backed up by later work (e.g., Evans and Nixon, 1995; Griffin, 2000). Other work referred to here is that on decomposing Gaussian and median filters (Narendra, 1978; Wiejak et al., 1985), and the many papers on fast implementation of median filters (e.g., Narendra, 1978; Huang et al., 1979; Danielsson, 1981; Davies, 1992a).

Considerable efforts have been devoted to studying the "root" behavior of the median filter, i.e., the result of applying median filtering operations until no further change occurs. In fact, much of this work has been carried out on 1-D signals, including cardiac and speech waveforms, rather than on images (Gallagher and Wise, 1981; Fitch et al., 1985; Heinonen and Neuvo, 1987). Root behavior is of interest as it relates to the underlying structure of signals, although its realization involves considerable amounts of processing. Some of the work on filtering aims to improve on rather than to emulate the median filter. Work of this type includes the detail-preserving filters of Heinonen and others (Nieminen et al., 1987) and relates to the lower set of plots in Fig. 3.18. See also the neural network approach to this topic (e.g., Greenhill and Davies, 1994a). More recent work on nonlinear filtering appears in Marshall et al. (1998): see Marshall (2004) for a new design method for weighted order statistics filters.

The author (Davies, 1987c) has reported methods of optimizing linear smoothing filters in small neighborhoods by minimizing the total error in fitting them to a continuous Gaussian function: a balance has to be struck between subpixel errors within the neighborhood and errors that arise from the proportion of the distribution that lies outside the neighborhood (Fig. 3.29).

With the advent of extremely low cost color frame grabbers on PCs, and the widespread use of digital cameras, digital color images have become ubiquitous, and this has extended to (or even necessitated) much research on color filtering. A useful summary of work in this area up to 1998 appears in Sangwine and Horne (1998). More recent work on vector (color) filtering includes that of Lukac (2003). Charles and Davies (2003b) describe new distance-weighted median filters and their application to color images. They also extend the author's earlier mode filter work to color images (Charles and Davies, 2003a, 2004). Davies's (2000e) theorem shows that restricting a multichannel (color) filter output to the

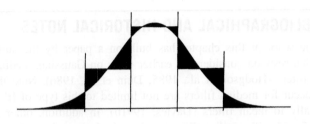

FIGURE 3.29

Approximating a discrete to a continuous Gaussian. This diagram shows how a balance needs to be struck between subpixel errors and those arising from the truncated part of the function.

vector value of one of the input sample points (i.e., from the current window in the image) will increase the inaccuracy present in the final image, for a large proportion of pixels: since this represents the usual vector median strategy that is employed to minimize color bleeding, the effectiveness of the current generation of color filter algorithms needs to be looked at further.

Davies has further analyzed the distortions and edge shifts produced by a range of rank order, mean, and mode filters, and has produced a unified review of the subject (Davies, 2003e). In the case of median filters, it proved possible, and necessary for high accuracy, to produce a discrete model of the situation (Davies, 2003c), rather than extending the continuum model described much earlier (Davies, 1989b).

3.16.1 More Recent Developments

The 2000s have seen a new approach to filtering via "switched" types of filter that judge whether or not any pixel is corrupted by impulse noise: if the latter, they use a method such as the median or vector-median filter to eliminate it; if the former, they adopt a policy of zero change by using the original pixel intensity or color. The zero change policy is useful because it helps maintain image sharpness and fidelity. An early example of this approach was the work of Eng and Ma (2001): see Chen et al. (2009) and Smolka (2010) for recent, more sophisticated versions of this concept (Smolka's version falls in the category of a "peer group switching filter").

Davies (2007b) has studied the properties of the generalized (nonvector) median filter that has the capability for eliminating even more noise than the VMF while not being targeted so specifically at eliminating color bleeding. He demonstrates ways of implementing the filter so that it runs sufficiently rapidly to make it a viable alternative to the VMF.

Celebi (2009) has shown how to reduce the computational needs of directional vector filters based on order statistics without significant loss of accuracy. At another end of the scale, Rabbani and Gazor (2010) have shown how to reduce additive Gaussian noise by using local mixture models; they find that of the wavelet types of local representation, the discrete complex wavelet transform is preferable in terms of both peak noise performance and computational cost.

3.17 PROBLEMS

1. Draw up a table showing the numbers of operations required to implement a median filter in various sizes of the neighborhood. Include in your table (i) results for a straight bubble sort of all n^2 pixels, (ii) results for bubble sorts in separated $1 \times n$ and $n \times 1$ neighborhoods, and (iii) results for the histogram method of Section 3.3. Discuss the results, taking account of possible computational overheads.

2. Show how to perform a median filtering operation on a binary image. Show also that if a set of binary images is formed by thresholding a grayscale image at various levels, and each of these binary images is median filtered, then a grayscale image can be reconstructed that is a median filtered version of the original grayscale image. Consider to what extent the reduced amount of computation in filtering a binary image compensates for the number of separate thresholded images to be filtered.

3. An "extremum" filter is an image-parallel operation that assigns every pixel the intensity value closer to the two extreme values in its local intensity distribution. Show that it should be possible to use such a filter to enhance images. What would be the *disadvantage* of such a filter?

4. Under what conditions is a 1-D signal that has been filtered once by a median filter a root signal? What truth is there in the statement that a straight edge in an image is neither shifted nor blurred by a median filter, whatever its cross-section?

5. **a.** Explain the action of the following median filtering algorithm:

```
for all pixels in image do{
  for (i = 0; i <= 255; i++) hist[i] = 0;
  for (m = 0; m <= 8; m++) hist[P[m]] ++;
  i = 0; sum = 0;
  while (sum < 5) {
    sum = sum + hist[i];
    i = i + 1;
  }
  Q0 = i − 1;
}
```

   b. Show how this algorithm can be speeded up (i) by a more efficient histogram clearing technique and (ii) by calculating the minimum intensity in each 3×3 window. In each case, estimate approximately how much the algorithm will be speeded up.

   c. Explain why a median filter is able to smooth images without introducing blurring.

d. A 1-D cross-section of an image has the following intensity profile:

1 2 1 1 2 3 0 2 2 3 1 1 2 2 9 2 2 8 8 8 7 8 8 7 9 9 9

Apply (i) a 3-element median filter and (ii) a 5-element median filter to this profile. With the aid of these examples, show that median filters tend to produce "runs" of constant values in 1-D profiles. Show also that under some circumstances an edge in the profile can be shifted by a nearby spike: give a rule showing when this is likely to occur for an n-element median filter in one dimension.

6. a. A *mode* filter is defined as one in which the new pixel intensity at any pixel takes the most probable value in the local intensity distribution of a window placed around that pixel in the original image space. Show for a grayscale image that a mode filter will, if anything, sharpen the image, while a *mean* filter will tend to blur the image.

b. A *max* filter is one that takes the maximum value of the local intensity distribution in a window around each pixel. Explain what will be seen when a max filter is applied to an image. Consider whether any similar effects are liable to happen when a mode filter is applied to an image.

c. Explain the purpose of a *median* filter. Why are 2-D median filters sometimes implemented as two 1-D median filters applied in sequence?

d. Contrast the behavior of 5-element 1-D mean, max, and median filters as applied to the following waveform:[5]

0 1 1 2 3 2 2 0 2 3 9 3 2 4 4 6 5 6 7 0 8 8 9 1 1 8 9

e. Work out what would happen if the 1-D median filter were applied many times, starting with this waveform.

7. a. Determine the effect of applying (i) a 3 × 3 median filter and (ii) a 5 × 5 median filter to the portion of an image shown in Fig. 3.30.

```
0 0 0 0 0 0 0 0 0 0
0 0 0 0 0 0 0 2 0 0
0 1 0 0 0 0 0 0 0 0
0 0 1 0 0 0 0 0 0 0
0 0 0 0 9 9 9 9 9 9
0 0 0 0 9 9 8 9 9 9
0 0 1 0 9 8 9 9 7 9
0 0 0 0 7 9 9 8 9 9
0 1 0 8 9 9 9 9 9 9
0 0 0 0 9 9 9 9 9 9
0 0 0 0 8 9 9 9 9 9
```

FIGURE 3.30

Portion of image for tests of median filter.

[5]For the mean filter, give the nearest integer value in each case.

b. Show that it should be possible to develop a corner detector based on the properties of these median filters. What advantages or disadvantages might result from employing this design strategy?

8. a. Distinguish between *mean* and *median filtering*. Explain why a mean filter would be expected to blur an image, while a median filter would not have this effect. Illustrate your answer by showing what happens in the following 1-D case with a window of size 1×3:

1 1 1 1 2 1 1 2 3 4 4 0 4 4 4 5 6 7 6 5 4 3 3

b. Give a complete median filter algorithm based on histograms and operating within a 3×3 window. Explain why it operates relatively slowly.

c. A computer language has the *max(a, b)* operation as standard. Show how it may be used to find the maximum intensity within a 3×3 window. Show also how it may be used to find the median by successively replacing the maximum values by zeros. If the *max(a, b)* operation is about the same speed as the $a + b$ operation, determine whether the median can be found any faster by this method.

d. Discuss whether splitting a 3×3 median operation into 1×3 and 3×1 median operations is likely to be effective at eliminating impulse noise in images. How would the speed of this approach be affected by use of the *max(a, b)* operation?

9. a. Determine the result of applying a 3-element median filter to the following 1-D signals:

    i. 0 0 0 0 0 1 0 1 1 1 1 1 1 1
    ii. 2 1 2 3 2 1 2 2 3 2 4 3 3 4
    iii. 1 1 2 3 3 4 5 8 6 6 7 8 9 9

b. What general lessons can be learnt from the results? In the first case, consider also the corresponding situation for a grayscale edge in a 2-D image.

c. 2-D median filters are sometimes implemented as two 1-D median filters applied in sequence in order to improve the speed of processing. Estimate the gain in speed that could be achieved in this way for (i) a 3×3 median filter, (ii) a 7×7 median filter, and (iii) in the general case.

4

Thresholding Techniques

One of the important practical aims of image processing is the demarcation of objects appearing in digital images. This process is called segmentation, and a good approximation to it can often be achieved by thresholding. Broadly, this involves separating the dark and light regions of the image, and thus identifying dark objects on a light background (or vice versa). This chapter discusses the effectiveness of this idea and the means for achieving it.

Look out for:

- the segmentation, region-growing and thresholding concepts.
- the problem of threshold selection.
- the limitations of global thresholding.
- problems in the form of shadows or glints (highlights).
- the possibility of modeling the image background.
- the idea of adaptive thresholding.
- the rigorous Chow and Kaneko approach.
- what can be achieved with simple local adaptive thresholding algorithms.
- more thoroughgoing variance, entropy-based, and maximum likelihood methods.
- the possibility of modeling images by multilevel thresholding.
- the value of the global valley transformation.
- how thresholds can be found in unimodal distributions.

Thresholding is limited in what it can achieve, and there are severe difficulties in automatically estimating the optimum threshold—as evidenced by the many available techniques that have been devised for the purpose. In fact, segmentation is an ill-posed problem, and it is misleading that the human eye appears to perform thresholding reliably. Nevertheless, there are instances where the task can be simplified, for example, by suitable lighting schemes, so that thresholding becomes effective. Hence, it is a useful technique that needs to be included in the toolbox of available algorithms for use when appropriate. However, edge detection (Chapter 5) provides an alternative highly effective means to key into complex image data.

4.1 INTRODUCTION

One of the first tasks to be undertaken in vision applications is to segment objects from their backgrounds. When objects are large and do not possess very much surface detail, segmentation can be imagined as splitting the image into a number of regions each having a high level of uniformity in some parameter such as brightness, color, texture or even motion. Hence, it should be straightforward to separate objects from one another and from their background, and also to discern the different facets of solid objects such as cubes.

Unfortunately, the concept of segmentation presented above is an idealization that is sometimes reasonably accurate, but more often in the real world, it is an invention of the human mind, generalized inaccurately from certain simple cases. This problem arises because of the ability of the eye to understand real scenes at a glance, and hence to segment and perceive objects within images in the form they are known to have. Introspection is not a good way of devising vision algorithms, and it must not be overlooked that segmentation is actually one of the central and most difficult practical problems of machine vision.

Thus, the common view of segmentation as looking for regions possessing some degree of uniformity is to a large extent invalid. There are many examples of this in the world of 3-D objects: one is a sphere lit from one direction, the brightness in this case changes continuously over the surface so that there is no distinct region of uniformity; another is a cube where the direction of the lighting may lead to several of the facets having equal brightness values so that it is impossible from intensity data alone to segment the image completely as desired.

Nevertheless, there is sufficient correctness in the concept of segmentation by uniformity measures for it to be worth pursuing for practical applications. The reason is that in many (especially industrial) applications, only a very restricted range and number of objects are involved, and in addition it is possible to have almost complete control over the lighting and the general environment. The fact that a particular method may not be completely general need not be problematic, since by employing tools that are appropriate for the task in hand, a cost-effective solution will have been achieved in that case at least. However, in practical situations, there is clearly a tension between simple cost-effective solutions and general-purpose but more computationally expensive solutions; this tension must always be kept in mind in severely practical subjects such as machine vision.

4.2 REGION-GROWING METHODS

The segmentation idea outlined in Section 4.1 leads naturally to the region-growing technique (Zucker, 1976b). Here, pixels of like intensity (or other suitable property) are successively grouped together to form larger and larger regions until the whole image has been segmented. Clearly, there have to be rules about not combining adjacent pixels that differ too much in intensity, while permitting combinations for which intensity changes gradually because of variations in background illumination over the field of view. However, this is not enough to make a viable strategy, and in practice the technique has to include the facility not only to merge regions together but also to split them if they become too large and inhomogeneous (Horowitz and Pavlidis, 1974). Particular problems are noise and sharp edges and lines that form disconnected boundaries, and for which it is difficult to formulate simple criteria to decide whether they form true region boundaries. In remote sensing applications, for example, it is often difficult to separate fields rigorously when hedges are broken and do not give continuous lines: in such applications, segmentation may have to be performed interactively, with a human operator helping the computer. Hall (1979) found that in practice regions tend to grow too far,[1] so that to make the technique work well it is necessary to limit their growth with the aid of edge detection schemes.

Thus, the region-growing approach to segmentation turns out to be quite complex to apply in practice. In addition, region-growing schemes usually operate iteratively, gradually refining hypotheses about which pixels belong to which regions. The technique is complicated because, carried out properly, it involves global as well as local image operations. Thus, each pixel intensity will in principle have to be examined many times, and as a result the process tends to be quite computation intensive. For this reason, it is not considered further here, since we are often more interested in methods involving low computational load that are amenable to real-time implementation.

4.3 THRESHOLDING

If background lighting is arranged so as to be fairly uniform, and we are looking for rather flat objects that can be silhouetted against a contrasting background, segmentation can be achieved simply by thresholding the image at a particular intensity level. This possibility was apparent from Fig. 2.2. In such cases, the complexities of the region-growing approach are bypassed. The process of thresholding has already been covered in Chapter 2, the basic result being that the initial grayscale image is converted into a binary image in which objects appear as black figures on a white background, or as white figures on a black background. Further analysis of the image then devolves into analysis of the shapes and dimensions of the figures: at this stage, object identification should be straightforward. Chapter 9 concentrates on such tasks. Meanwhile, there is one outstanding problem—how to devise an automatic procedure for determining the optimum thresholding level.

---

[1]Clearly, there is a danger that even one small break could join two regions into a single larger one.

4.3.1 Finding a Suitable Threshold

One simple technique for finding a suitable threshold arises in situations such as optical character recognition (OCR) where the proportion of the background that is occupied by objects (i.e., print) is relatively constant in a variety of conditions. A preliminary analysis of relevant picture statistics then permits subsequent thresholds to be set by insisting on a fixed proportion of dark and light in a sequence of images (Doyle, 1962). In practice, a series of experiments is performed in which the thresholded image is examined as the threshold is adjusted, and the best result ascertained by eye: at that stage, the proportions of dark and light in the image are measured. Unfortunately, any changes in noise level following the original measurement will upset such a scheme, since they will affect the relative amounts of dark and light in the image. However, this is frequently a useful technique in industrial applications, especially when particular details within an object are to be examined: typical examples of this are holes in mechanical components such as brackets (note that the mark— space ratio for objects may well vary substantially on a production line, but the proportion of hole area *within* the object outline would not be expected to vary).

The technique that is most frequently employed for determining thresholds involves analyzing the histogram of intensity levels in the digitized image (Fig. 4.1): if a significant minimum is found, it is interpreted as the required threshold value (Weska, 1978). Clearly, the assumption being made here is that the peak on the left of the histogram corresponds to dark objects, and the peak on the right corresponds to light background (here it is assumed that, as in many industrial applications, objects appear dark on a light background).

This method is subject to the following major difficulties:

1. the valley may be so broad that it is difficult to locate a significant minimum.
2. there may be a number of minima because of the type of detail in the image, and selecting the most significant one will be difficult.
3. noise within the valley may inhibit location of the optimum position.
4. there may be no clearly visible valley in the distribution because noise may be excessive or because the background lighting may vary appreciably over the image.
5. either of the major peaks in the histogram (usually due to the background) may be much larger than the other and this will then bias the position of the minimum.
6. the histogram may be inherently multimodal, making it difficult to determine which is the relevant thresholding level.

Perhaps the worst of these problems is the last point: that is, if the histogram is inherently multimodal, and we are trying to employ a single threshold, then we are applying what is essentially an *ad hoc* technique to obtain a meaningful result. In general, such efforts are unlikely to succeed, and this is clearly a case where full image interpretation must be performed before we could be sure that the results are valid. Ideally, thresholding rules have to be formed after many images

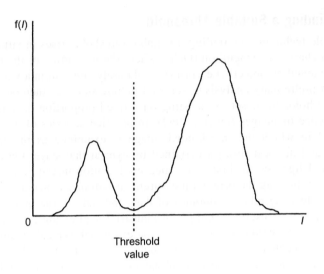

FIGURE 4.1

Idealized histogram of pixel intensity levels in an image. The large peak on the right results from the light background; the smaller peak on the left is due to dark foreground objects. The minimum of the distribution provides a convenient intensity value to use as a threshold.

have been analyzed. In what follows such problems of meaningfulness are eschewed and attention is concentrated on how best to find a genuine single threshold when its position is obscured as indicated by problems 1–5 above (which can be ascribed to image "clutter," noise, and lighting variations).

4.3.2 Tackling the Problem of Bias in Threshold Selection

This section considers problem 5 of Section 4.3.1—that of eliminating the bias in the selection of thresholds that arises when one peak in the histogram is larger than the other. First, note that if the relative heights of the peaks are known, this effectively eliminates the problem, since the "fixed proportion" method of threshold selection outlined above can be used. However, this is not normally possible. A more useful approach is to prevent bias by weighting down the extreme values of the intensity distribution and weighting up the intermediate values in some way. To achieve this, note that the intermediate values are special in that they correspond to object edges. Hence, a good basic strategy is to find positions in the image where there is a significant intensity gradient—corresponding to pixels in the regions of edges—and to analyze the intensity values of these locations while ignoring other points in the image.

One way of dealing with this is to construct "scattergrams" in which pixel properties are plotted on a 2-D map with intensity variation along one axis and

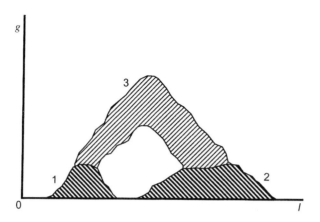

FIGURE 4.2

Scattergram showing the frequency of occurrence of various combinations of pixel intensity I and intensity gradient magnitude g in an idealized image. There are three main populated regions of interest: (1) a low-I, low-g region; (2) a high-I, low-g region; and (3) a medium-I, high-g region. Analysis of the scattergram sometimes provides useful information on how to segment the image.

intensity gradient magnitude variation along the other. As indicated in Fig. 4.2, there are three main populated regions on the map: (1) a low-intensity, low-gradient region corresponding to the dark objects; (2) a high-intensity, low-gradient region corresponding to the background; and (3) a medium-intensity, high-gradient region corresponding to object edges (Panda and Rosenfeld, 1978). By analyzing how these regions merge into each other, it is sometimes possible to obtain better results than can be obtained using simple thresholding. In particular, by examining the situation for moderate values of gradient, bias may be reduced, as indicated above. However, instead of constructing a scattergram, we can try weighting the plots in the intensity histogram in such a way as to minimize threshold bias: this possibility is discussed in the following section.

4.3.2.1 Methods Based on Finding a Valley in the Intensity Distribution

This section considers how to weight the intensity distribution using a parameter other than the intensity gradient, in order to locate accurately the valley in the intensity distribution. A simple strategy is first to locate all pixels that have a significant intensity gradient, and then to find the intensity histogram not only of these pixels but also of nearby pixels. This means that the two main modes in the intensity distribution are still attenuated very markedly and hence the bias in the valley position is significantly reduced. Indeed, the numbers of background and foreground pixels that are now being examined are very similar, so the bias from the relatively large number of background pixels is virtually eliminated (note that if the modes are modeled as two Gaussian distributions of equal widths and they also have equal heights, then the minimum lies exactly halfway between them).

Although obvious, this approach clearly includes the edge pixels themselves, which tend to fill the valley between the two modes. For the best results, the points of highest gradient must actually be removed from the intensity histogram. A well-attested way of achieving this is to weight pixels in the intensity histogram according to their response to a Laplacian filter (Weska et al., 1974). Since such a filter gives an isotropic estimate of the second derivative of the image intensity (i.e., the magnitude of the first derivative of the intensity gradient), it is zero where intensity gradient magnitude is high: hence, it gives such locations zero weight, but it nevertheless weights up those locations on the shoulders of edges. It has been found that this approach is very good at estimating where to place a threshold within a wide valley in the intensity histogram (Weska et al., 1974).

4.3.3 Summary

It has been shown that available techniques are able to provide values at which intensity thresholding can be applied, but they do not themselves solve the problems caused by uneven lighting. They are even less capable of coping with glints, shadows and image clutter. Unfortunately, these artifacts are common in most real situations (Figs. 4.3–4.5) and are only eliminated with difficulty in practice. Indeed, in industrial applications where shiny metal components are involved, glints are the rule rather than the exception, while shadows can seldom be avoided with any sort of object. Even flat objects are liable to have quite strong shadow contours around them because of the particular placement of lights. Lighting problems are studied in detail in Chapter 25. Meanwhile, note that glints and shadows can only be allowed for properly in a two-stage image analysis system, where tentative assignments are made first, and these are firmed up by exact explanation of all pixel intensities. We now return to the problem of making the most of the thresholding technique, by finding how variations in background lighting can be allowed for.

4.4 ADAPTIVE THRESHOLDING

The problem that arises when illumination is not sufficiently uniform may be tackled by permitting the threshold to vary adaptively (or "dynamically") over the whole image. In principle, there are several ways of achieving this. One involves modeling the background within the image. Another is to work out a local threshold value for each pixel by examining the range of intensities in its neighborhood. A third approach is to split the image into subimages and deal

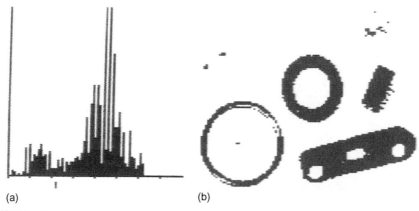

(a)   (b)

FIGURE 4.3

Histogram for the image shown in Fig. 2.7(a). Note that the histogram is not particularly close to the ideal form of Fig. 4.1. Hence, the threshold obtained from (a) (indicated by the short line beneath the scale) does not give ideal results with all the objects in the binarized image (b). Nevertheless, the results are better than for the arbitrarily thresholded image of Fig. 2.7(b).

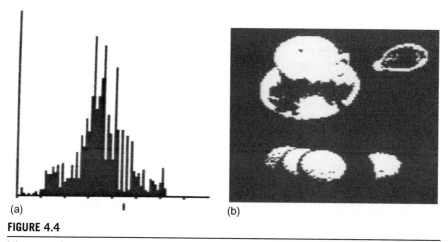

(a)   (b)

FIGURE 4.4

Histogram for the image shown in Fig. 2.1(a). The histogram is not at all close to the idealized form, and the results of thresholding (b) are not a particularly useful aid to interpretation.

with them independently. Although "obvious," the last method will clearly run into problems at the boundaries between subimages, and by the time these problems have been solved, it will look more like one of the other two methods.

The problem can sometimes be solved rather neatly in the following way. On some occasions—such as in automated assembly applications—it is possible to

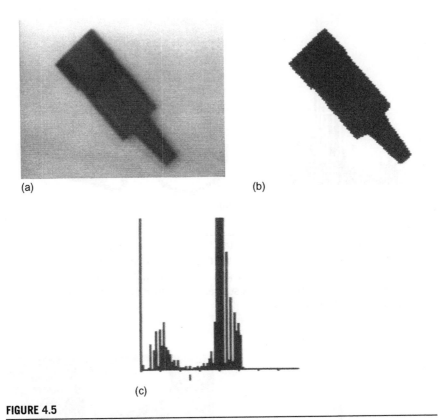

(a)   (b)

(c)

FIGURE 4.5

A picture with more ideal properties. (a) Image of a plug that has been lit fairly uniformly. The histogram (c) approximates to the ideal form, and the result of thresholding (b) is acceptable. However, much of the structure of the plug is lost during binarization.

obtain an image of the background in the absence of any objects. This appears to solve the problem of adaptive thresholding in a rigorous manner, since the tedious task of modeling the background has already been carried out. However, caution is needed because objects bring with them not only shadows (which can in some sense be regarded as part of the objects) but also an additional effect due to the reflections they cast over the background and other objects. This additional effect is nonlinear in the sense that it is necessary to add not only the difference between the object and the background intensity in each case but also an intensity that depends on the products of the reflectances of pairs of objects. These considerations mean that using the no-object background as the equivalent background when several objects are present is ultimately invalid. However, as a first approximation, it is frequently possible to assume an equivalence. If this proves impracticable, there is no option but to model the background from the actual image to be segmented.

On other occasions, the background intensity may be rather slowly varying, in which case it may be possible to model it by the following technique (this is a form of Hough transform—see Chapter 11). First, an equation is selected, which can act as a reasonable approximation to the intensity function, for example, a quadratic variation:

$$I = a + bx + cy + dx^2 + exy + fy^2 \qquad (4.1)$$

Next, a parameter space for the six variables a, b, c, d, e, f is constructed; then each pixel in the image is taken in turn and all sets of values of the parameters that could have given rise to the pixel intensity value are accumulated in parameter space. Finally, a peak is sought in parameter space, which represents an optimal fit to the background model. So far it appears that this has been carried out only for a linear variation, the analysis being simplified initially by considering only the differences in intensities of pairs of points in image space (Nixon, 1985). Note that a sufficient number of pairs of points must be considered so that the peak in parameter space resulting from background pairs is sufficiently well populated.

4.4.1 The Chow and Kaneko Approach

As early as 1972, Chow and Kaneko introduced what is widely recognized as the standard technique for dynamic thresholding: the technique performs a thoroughgoing analysis of the background intensity variation, making few compromises to save computation (Chow and Kaneko, 1972). In this method, the image is divided into a regular array of overlapping subimages and individual intensity histograms are constructed for each one. Those that are unimodal are ignored since they are assumed not to provide any useful information that can help in modeling the background intensity variation. However, the bimodal distributions are well suited to this task: these are individually fitted to pairs of Gaussian distributions of adjustable height and width and the threshold values are located. Thresholds are then found, by interpolation, for the unimodal distributions. Finally, a second stage of interpolation is necessary to find the correct thresholding value at each pixel.

One problem with this approach is that if the individual subimages are made very small in an effort to model the background illumination more exactly, the statistics of the individual distributions become worse, their minima become less well defined and the thresholds deduced from them are no longer statistically significant. This means that it does not pay to make subimages too small and that ultimately only a certain level of accuracy can be achieved in modeling the background in this way. Clearly, the situation is highly data dependent, but it might be expected that little would be gained by reducing the subimage size below 32×32 pixels. Chow and Kaneko employed 256×256 pixel images and divided these into a 7×7 array of 64×64 pixel subimages with 50% overlap.

Overall, this approach involves considerable computation, and in real-time applications it may well not be viable for this reason.

4.4.2 **Local Thresholding Methods**

The other approach mentioned earlier is particularly useful for finding local thresholds. It involves analyzing intensities in the neighborhood of each pixel to determine the optimum local thresholding level. Ideally, the Chow and Kaneko histogramming technique would be repeated at each pixel, but this would significantly increase the computational load of this already computationally intensive technique. Thus, it is necessary to obtain the vital information by an efficient sampling procedure. One simple means for achieving this is to take a suitably computed function of nearby intensity values as the threshold: often the mean of the local intensity distribution is taken because this is a simple statistic and gives good results in some cases. For example, in astronomical images, stars have been thresholded in this way. Niblack (1985) reported a case in which a proportion of the local standard deviation was added to the mean to give a more suitable threshold value, the reason (presumably) being to help suppress noise (clearly, addition is appropriate where bright objects such as stars are to be located, whereas subtraction is more appropriate in the case of dark objects).

Another statistic that is frequently used is the mean of the maximum and minimum values in the local intensity distribution. The justification for this is that whatever the sizes of the two main peaks of the distribution, this statistic often gives a reasonable estimate of the position of the histogram minimum. The theory presented earlier shows that this method will only be accurate if (a) the intensity profiles of object edges are symmetrical, (b) noise acts uniformly everywhere in the image so that the widths of the two peaks of the distribution are similar, and (c) the heights of the two distributions do not differ markedly. Sometimes these assumptions are definitely invalid—e.g., when looking for (dark) cracks in eggs or other products. In such cases, the mean and maximum of the local intensity distribution can be found and a threshold deduced using the statistic

$$T = mean - (maximum - mean) \tag{4.2}$$

where the strategy is to estimate the lowest intensity in the bright background assuming the distribution of noise is symmetrical (Fig. 4.6): use of the mean here is realistic only if the crack is narrow and does not affect the value of the mean significantly. If it does, then the statistic can be adjusted by use of an *ad hoc* parameter:

$$T = mean - k(maximum - mean) \tag{4.3}$$

where k may be as low as 0.5 (Plummer and Dale, 1984).

This method is essentially the same as that of Niblack (1985), but the computational load in estimating the standard deviation is minimized. Each of the last two techniques relies on finding local extrema of intensity. Using these measures helps save computation, but they are clearly somewhat unreliable because of the effects of noise. If this is a serious problem, quartiles or other statistics of the distribution may be used. The alternative of prefiltering the image to remove noise is unlikely to work for crack thresholding, since cracks will almost certainly be removed at the same time as the noise. A better strategy is to form an image of

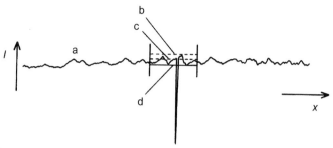

FIGURE 4.6

Method for thresholding the crack in an egg. a, Intensity profile of an egg in the vicinity of a crack: the crack is assumed to appear dark (e.g., under oblique lighting); b, local maximum of intensity on the surface of the egg; c, local mean intensity. Eq. (4.2) gives a useful estimator T of the thresholding level d.

T-values obtained using Eq. (4.2) or (4.3): smoothing this image should then permit the initial image to be thresholded effectively.

Unfortunately, all these methods work well only if the size of the neighborhood selected for estimating the required threshold is large enough to span a significant amount of foreground and background. In many practical cases, this is not possible and the method then adjusts itself erroneously, for example, so that it finds darker spots within dark objects as well as segmenting the dark objects themselves. However, there are certain applications where there is little risk of this occurring. One notable case is that of OCR. Here the widths of character limbs are likely to be known in advance and should not vary substantially. If this is so, then a neighborhood size can be chosen to span or at least sample both character and background, and it is thus possible to threshold the characters highly efficiently using a simple functional test of the type described above. The effectiveness of this procedure (Table 4.1) is demonstrated in Fig. 4.7.

Finally, before leaving this topic, note that hysteresis thresholding is a type of adaptive thresholding—effectively permitting the threshold value to vary locally: this topic is investigated in Section 5.10.

4.5 MORE THOROUGHGOING APPROACHES TO THRESHOLD SELECTION

At this point, we return to global threshold selection and describe some important approaches that have a rigorous mathematical basis. The first of these is variance-based thresholding, the second is entropy-based thresholding, and the third is maximum likelihood thresholding. All three are widely used, the second having achieved an increasingly wide following over the past 20–30 years, and the third is a more broad-based technique that has its roots in statistical pattern recognition—a subject that is covered in Chapter 24.

Table 4.1 A Simple Algorithm for Adaptively Thresholding Print

```
minrange = 255 / 5;
/* minimum likely difference in intensity between print and background:
this parameter can be preset manually or "learnt" by a previous routine */
for all pixels in image do {
    find minimum and maximum of local intensity distribution;
    range = maximum − minimum;
    if (range > minrange)
        T = (minimum + maximum)/2; // print is visible in neighborhood
    else T = maximum − minrange/2; // neighborhood is all white
    if (P0 > T) Q0 = 255; else Q0 = 0; // now binarize print
}
```

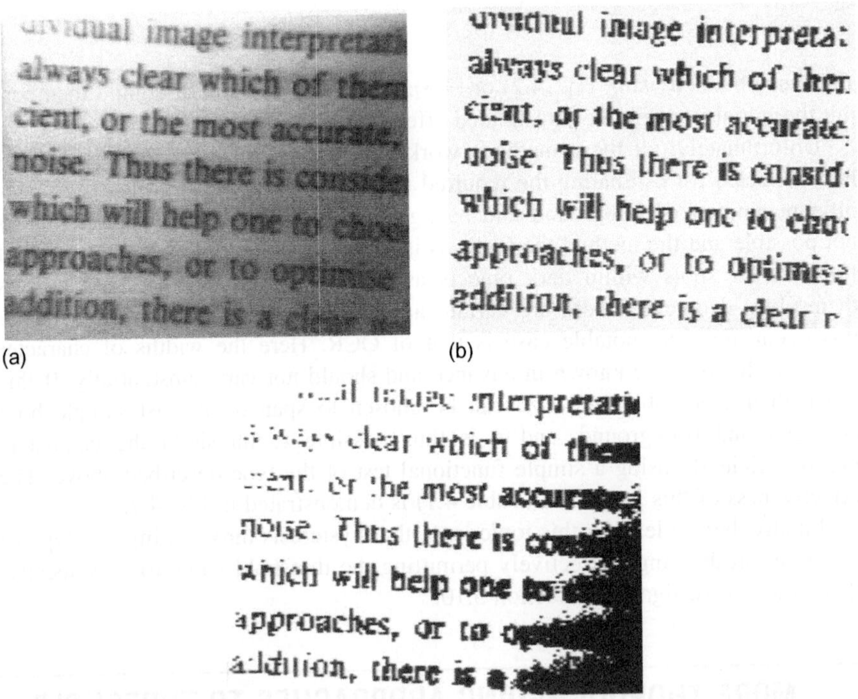

(a)

(b)

(c)

FIGURE 4.7

Effectiveness of local thresholding on printed text. Here, a simple local thresholding procedure (Table 4.1), operating within a 3×3 neighborhood, is used to binarize the image of a piece of printed text (a). Despite the poor illumination, binarization is performed quite effectively (b). Note the complete absence of isolated noise points in (b), while by contrast the dots on all the i's are accurately reproduced. The best that could be achieved by uniform thresholding is shown in (c).

4.5.1 **Variance-Based Thresholding**

The standard approach to thresholding outlined earlier involved finding the neck of the global image intensity histogram. However, this is impracticable when the dark peak of the histogram is minuscule in size, as it will then be hidden among the noise in the histogram and it will not be possible to extract it with the usual algorithms.

A good many investigators have studied this sort of problem (e.g., Otsu, 1979; Kittler et al., 1985; Sahoo et al., 1988; Abutaleb, 1989): among the most well-known approaches are the variance-based methods. In these methods, the image intensity histogram is analyzed to find where it can best be partitioned to optimize criteria based on ratios of the within-class, between-class, and total variance. The simplest approach (Otsu, 1979) is to calculate the between-class variance, as will now be described.

First, we assume that the image has a grayscale resolution of L gray levels. The number of pixels with gray level i is written as n_i, so the total number of pixels in the image is $N = n_1 + n_2 + \cdots + n_L$. Thus, the probability of a pixel having gray level i is:

$$p_i = \frac{n_i}{N} \tag{4.4}$$

where

$$p_i \geq 0 \quad \sum_{i=1}^{L} p_i = 1 \tag{4.5}$$

For ranges of intensities up to and above the threshold value k, we can now calculate the between-class variance σ_B^2 and the total variance σ_T^2:

$$\sigma_B^2 = \pi_0 \left(\mu_0 - \mu_T\right)^2 + \pi_1 \left(\mu_1 - \mu_T\right)^2 \tag{4.6}$$

$$\sigma_T^2 = \sum_{i=1}^{L} \left(i - \mu_T\right)^2 p_i \tag{4.7}$$

where

$$\pi_0 = \sum_{i=1}^{k} p_i \quad \pi_1 = \sum_{i=k+1}^{L} p_i = 1 - \pi_0 \tag{4.8}$$

$$\mu_0 = \sum_{i=1}^{k} i p_i / \pi_0 \quad \mu_1 = \sum_{i=k+1}^{L} i p_i / \pi_1 \quad \mu_T = \sum_{i=1}^{L} i p_i \tag{4.9}$$

Making use of the latter definitions, the formula for the between-class variance can be simplified to:

$$\sigma_B^2 = \pi_0 \pi_1 \left(\mu_1 - \mu_0\right)^2 \tag{4.10}$$

For a single threshold, the criterion to be maximized is the ratio of the between-class variance to the total variance:

$$\eta = \frac{\sigma_B^2}{\sigma_T^2} \tag{4.11}$$

However, the total variance is constant for a given image histogram, so maximizing η simplifies to maximizing the between-class variance.

The method can readily be extended to the dual threshold case $1 \leq k_1 \leq k_2 \leq L$, where the resultant classes, $C_0, C_1,$ and C_2, have respective gray-level ranges of $[1, ..., k_1]$, $[k_1 + 1, ..., k_2]$, and $[k_2 + 1, ..., L]$.

In some situations (e.g., Hannah et al., 1995), this approach is still not sensitive enough to cope with histogram noise, and more sophisticated methods must be used. One such technique is that of entropy-based thresholding, which has become firmly embedded in the subject (Pun, 1980; Kapur et al., 1985; Abutaleb, 1989; Brink, 1992). For further insight into the performance of the between-class variance method (BCVM), see Section 4.7.

4.5.2 Entropy-Based Thresholding

Entropy measures of thresholding are based on the concept of entropy. The entropy statistic is high if a variable is well distributed over the available range, and low if it is well ordered and narrowly distributed: specifically, entropy is a measure of disorder, and is zero for a perfectly ordered system. The concept of entropy thresholding is to threshold at an intensity for which the sum of the entropies of the two intensity probability distributions thereby separated is maximized. The reason for this is to obtain the greatest reduction in entropy—i.e., the greatest increase in order—by applying the threshold: in other words, the most appropriate threshold level is the one that imposes the greatest order on the system, and thus leads to the most meaningful result.

To proceed, the intensity probability distribution is again divided into two classes—those with gray levels up to the threshold value k and those with gray levels above k (Kapur et al., 1985). This leads to two probability distributions A and B:

$$A: \quad \frac{p_1}{P_k}, \frac{p_2}{P_k}, \ldots, \frac{p_k}{P_k} \tag{4.12}$$

$$B: \quad \frac{p_{k+1}}{1 - P_k}, \frac{p_{k+2}}{1 - P_k}, \ldots, \frac{p_L}{1 - P_k} \tag{4.13}$$

where

$$P_k = \sum_{i=1}^{k} p_i \qquad 1 - P_k = \sum_{i=k+1}^{L} p_i \tag{4.14}$$

The entropies for each class are given by:

$$H(A) = - \sum_{i=1}^{k} \frac{p_i}{P_k} \ln \frac{p_i}{P_k} \tag{4.15}$$

$$H(B) = - \sum_{i=k+1}^{L} \frac{p_i}{1 - P_k} \ln \frac{p_i}{1 - P_k} \tag{4.16}$$

and the total entropy is:

$$H(k) = H(A) + H(B) \tag{4.17}$$

Substitution leads to the final formula:

$$H(k) = \ln \left(\sum_{i=1}^{k} p_i \right) + \ln \left(\sum_{i=k+1}^{L} p_i \right) - \frac{\sum_{i=1}^{k} p_i \ln p_i}{\sum_{i=1}^{k} p_i} - \frac{\sum_{i=k+1}^{L} p_i \ln p_i}{\sum_{i=k+1}^{L} p_i} \tag{4.18}$$

and it is this parameter that has to be maximized.

This approach can give very good results—see, e.g., Hannah et al. (1995). Again, it is straightforwardly extended to dual thresholds, but we shall not go into the details here (Kapur et al., 1985). In fact, probabilistic analysis to find mathematically ideal dual thresholds may not be the best approach in practical situations: an alternative technique for determining dual thresholds sequentially has been devised by Hannah et al. (1995), and applied to an X-ray inspection task—as described in Chapter 20.

4.5.3 Maximum Likelihood Thresholding

When dealing with distributions such as intensity histograms, it is important to compare the actual data with the data that might be expected from a previously constructed model based on a training set: this is in agreement with the methods of statistical pattern recognition (see Chapter 24), which takes full account of prior probabilities. For this purpose, one option is to model the training set data using a known distribution function such as a Gaussian. The latter has many advantages, including its accessibility to relatively straightforward mathematical analysis. In addition, it is specifiable in terms of two well-known parameters—the mean and standard deviation—which are easily measured in practical situations. Indeed, for any Gaussian distribution, we have:

$$p_i(x) = \frac{1}{(2\pi\sigma_i^2)^{1/2}} \exp \left[-\frac{(x - \mu_i)^2}{2\sigma_i^2} \right] \tag{4.19}$$

where the suffix i refers to a specific distribution, and of course when thresholding is being carried out, there is a supposition that two such distributions are involved. Applying the respective *a priori* class probabilities P_1, P_2 (Chapter 24), careful analysis (Gonzalez and Woods, 1992) shows that the condition $p_1(x) = p_2(x)$ reduces to the form:

$$x^2 \left(\frac{1}{\sigma_1^2} - \frac{1}{\sigma_2^2} \right) - 2x \left(\frac{\mu_1}{\sigma_1^2} - \frac{\mu_2}{\sigma_2^2} \right) + \left(\frac{\mu_1^2}{\sigma_1^2} - \frac{\mu_2^2}{\sigma_2^2} \right) + 2 \log \left(\frac{P_2 \sigma_1}{P_1 \sigma_2} \right) = 0 \qquad (4.20)$$

Note that, in general, this equation has two solutions,[2] implying the need for two thresholds, although when $\sigma_1 = \sigma_2$ there is a single solution:

$$x = \frac{1}{2} (\mu_1 + \mu_2) + \frac{\sigma^2}{\mu_1 - \mu_2} \ln \left(\frac{P_2}{P_1} \right) \qquad (4.21)$$

In addition, when the prior probabilities for the two classes are equal, the equation reduces to the altogether simpler and more obvious form:

$$x = \frac{1}{2} (\mu_1 + \mu_2) \qquad (4.22)$$

Of all the methods described in this chapter, only the maximum likelihood method makes use of *a priori* probabilities. While this makes it look as if it is the only rigorous method, and indeed that all other methods are automatically erroneous and biased in their estimations, this is not the actual position. The reason lies in the fact that the other methods incorporate actual frequencies of sample data, which embody the *a priori* probabilities (see Section 24.4). Hence, the other methods should give correct results. Nevertheless, it is refreshing to see *a priori* probabilities brought in explicitly, as this gives a greater confidence of getting unbiased results in any doubtful situations.

4.6 THE GLOBAL VALLEY APPROACH TO THRESHOLDING

An important disadvantage of the many approaches to threshold estimation, including particularly entropy thresholding and its variants, is that it is often unclear how they will react to unusual or demanding situations, such as where multiple thresholds have to be found in the same image (Kapur et al., 1985;

---

[2]The reason for the existence of two solutions is that one solution represents a threshold in the area of overlap between the two Gaussians; the other solution is mathematically unavoidable, and lies at either very high or very low intensities. It is this latter solution that disappears when the two Gaussians have equal variance, as the distributions clearly never cross again. In any case, it seems unlikely that the distributions being modeled would in practice approximate so well to Gaussians that the non-central solution could ever be important—i.e. it is essentially a mathematical fiction that needs to be eliminated from consideration.

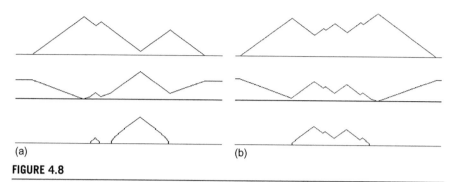

(a)   (b)

FIGURE 4.8

Result of applying global minimization algorithm to 1-D data sets. (a) A basic two-peak structure. (b) A basic multimode structure. Top trace: original 1-D data sets. Middle trace: results from Eq. (4.23). Bottom trace: results from Eq. (4.24).

Source: © IET 2008

Hannah et al., 1995; Tao et al., 2003; Wang and Bai, 2003; Sezgin and Sankur, 2004). Added to this, there is the risk that the more complex approaches will miss important aspects of the original data. The global valley approach (Davies, 2007a) aimed to provide a rigorous means of going back to basics to find global valleys of intensity histograms in such a way as to embody the intrinsic meaning of the data.

The top trace of Fig. 4.8(a) shows the basic situation—where thresholding is effective and the optimum threshold should be simple to locate. However, the intensity histogram often contains such a welter of peaks and valleys that even the human eye, with its huge capability for analysis "at a glance," can be confused—especially when it is necessary to identify global valley positions rather than local minima of lesser significance. The situation is made clearer by the example shown in the top trace of Fig. 4.8(b). Here valley 1 (numbering from the left) is lower than valley 3, but valley 3 is deeper in the sense that it has two high peaks immediately around it; however, valley 1 also lies between the highest two peaks, and in that sense it is the *globally* deepest valley in the distribution.

Clearly, to judge global valley deepness, we need a mathematical criterion so that comparisons between all the valleys can be carried out unambiguously. To proceed, for any potential global valley point (call it point j), we need to look at all points (i) on the left of it to find the highest peak to the left and all points (k) on the right of it to find the highest peak to the right, before we can construct a suitable criterion value for point j. Hence, we need to take the maximum over all points i and the maximum over all points k. Furthermore, we need to do this for all points j, and for each of them we need to consider only points i ($i < j$) and points k ($k > j$), and take account of the corresponding heights h_i, h_j, h_k in the distribution. The maximum must then be taken for a criterion function C_j of general form $\max_{i,k} \{Q(h_i - h_j, h_k - h_j)\}$. An obvious criterion function of this form employs the arithmetic mean. However, to avoid complications from negative heights, we introduce a sign function $s(\cdot)$ such that $s(u) = u$ if $u > 0$ and $s(u) = 0$ if $u \leq 0$. The result is the following function:

$$F_j = \max_{i,k} \left\{ \tfrac{1}{2}\left[s(h_i - h_j) + s(h_k - h_j)\right] \right\} \tag{4.23}$$

When this is applied to the top trace of Fig. 4.8(a), the result is a distribution (middle trace of Fig. 4.8(a)) that has a maximum at the required valley position. In addition, the values of i and k corresponding to this maximum are the first and third peak positions in the original intensity distribution. The sign function $s(\cdot)$ has the effect of preventing negative responses that would complicate the situation unnecessarily.

While the function F used above is straightforward to apply and employs linear expressions that are often attractive in permitting in-depth analysis, it results in pedestals at either end of the output distribution: these could complicate the situation when there are many peaks and valleys. Fortunately, the geometric mean is not subject to this problem, and so it is the one that is adopted in the global valley method (GVM). Thus, we use the following function instead of F_j:

$$K_j = \max_{i,k} \left\{ \left[s(h_i - h_j)s(h_k - h_j)\right]^{1/2} \right\} \tag{4.24}$$

Note that the arithmetic and geometric means are very similar when the two arguments are nearly equal, but deviate a lot when the two arguments are dissimilar: it is the dissimilar case that applies at the ends of the distribution, where it is required to suppress a potential valley that has only one peak near to it, and the geometric mean then offers a sound advantage over the arithmetic mean. These ideas are further made clear in Fig. 4.8(b).

Overall, the rationale for this approach is that we are looking for the most significant valley in an intensity distribution, corresponding to an optimum discriminating point between, for example, dark objects and light background in the original image. While in some cases the situation is obvious (Fig. 4.8(a)), in general it is difficult to sort out a confusing set of peaks and valleys and in particular to identify global valleys. So the concept embodied in Eq. (4.24) is that of aiming to guarantee an optimal global solution by automatic means. Clearly, by analysis of the output distribution, it is also possible to find a whole range of maxima corresponding to global valley positions in the input distribution: to this extent, the method is able to cope with multimode distributions and to find multiple threshold positions.

With all histogramming methods, it is necessary to take due account of local noise in the distribution, as it could lead to inaccurate results. Hence, the K distribution is smoothed before proceeding with further analysis to locate thresholds.

Another important factor is the amount of computation required for this approach. While it at first appears that a computationally intensive scan over all possible sets of sampling points i, j, k is required to obtain the optimal solution, it turns out that with care the computational load can be reduced from $O(N^3)$ to $O(N)$, where N is the number of gray levels in the intensity distribution.

4.7 PRACTICAL RESULTS OBTAINED USING THE GLOBAL VALLEY METHOD

The ideas presented in Section 4.6 are next tested using Fig. 4.9(a). Starting with this image, the following sequence of operations is applied: (a) an intensity histogram is generated (top trace in Fig. 4.9(d)); (b) the function K is applied (middle trace in Fig. 4.9(d)); (c) the output distribution is smoothed (bottom trace in Fig. 4.9(d)); (d) peaks are located (see the short vertical lines at the bottom of Fig. 4.9(d)); (e) the most significant peaks are chosen as threshold levels (here all eight are selected); (f) a new image is generated by applying the mean of the adjacent threshold intensity levels. The result (Fig. 4.9(b)) is a reasonably segmented likeness of the original image, albeit with clear limitations in the cloud regions— simply because accurate renditions of these would require a rather full range of gray levels, and thresholding is not appropriate in such regions. However, what is significant is the ease with which the approach automatically incorporates multi-level thresholding of multimode intensity distributions—a point that has been a difficulty with entropy thresholding, for example (Hannah et al., 1995). Finally, Fig. 4.9(c) gives a comparison with the maximum BCVM of Otsu (1979), which has recently undergone something of a resurgence of popularity and use, partly as a result of the ease with which it can be used for the systematic generation of multi-level thresholds (Liao et al., 2001; Otsu, 1979).

The reconstructability of the method (in the sense that much of the image is reconstructed so well that it is difficult to distinguish from the original) is an indication of success in that it is clear that the information removed was by no means arbitrary, but was actually redundant and unhelpful. This property is also evident in Fig. 4.10, which shows the application of the method to the well-known Lena image.

The basic criterion used for smoothing is that of reducing noise as far as possible without eliminating relevant thresholding points. To achieve this, repeated convolutions of the K distributions are made with a three-element $\frac{1}{4}$[1 2 1] kernel until an appropriate amount of smoothing is obtained. Note that the GVM peaks are by no means static. In particular, as smoothing progresses, they gradually move and then merge, as can be seen in the bottom traces in Fig. 4.10(f–h). Just before merger, there is often a rapid movement to align the merging peaks. To cope with this and to find suitable thresholding levels, a useful heuristic was to move one quarter of the way from the merged position to the next merger position

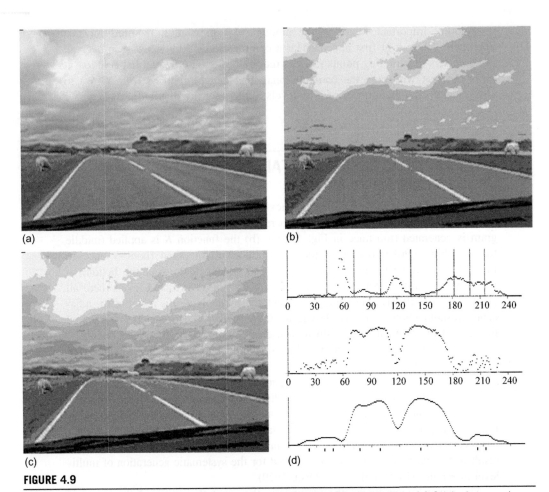

(a)

(b)

(c)

(d)

FIGURE 4.9

Result of applying the global valley algorithm to a multimode intensity distribution. (a) Original grayscale image. (b) Reconstituted image after multiple thresholding using the eight peaks in the output distribution. (d) Top: original intensity histogram for (a). Middle: result of applying the global valley transformation. Bottom: result of smoothing. The eight short vertical lines at the very bottom indicate the peak positions. In (d), the intensity scale is 0–255; the vertical scale is normalized to a maximum height indicated by the height of the vertical axis. *Note:* the three traces are computed 25 times more accurately than the rounded values displayed, so the peak locations *are* determined as accurately as indicated. For comparison, (c) shows the result of applying the between-class variance method to the same image: the eight thresholds are indicated by vertical lines in the top trace of (d).

Source: © *IET 2008*

FIGURE 4.10

Multilevel thresholding of the Lena image. For the original Lena grayscale image, see 'Miscellaneous' at the USC-SIPI Image Database*. (a) Result of applying the between-class variance method (BCVM) to original image. (b)–(d) Results of applying the global valley method to original image, producing, respectively, bi-level, tri-level, and five-level images. (e) Top: intensity histogram of original image: the vertical line indicates the bi-level threshold selected by the BCVM. Bottom: the resulting K distribution. (f)–(h) The upper traces show smoothed versions of the K distribution, with short vertical lines indicating, respectively, one, two, or four threshold positions; the lower traces show threshold positions resulting from progressive smoothing of the K distribution: note that these are scaled and some are truncated as indicated by the horizontal gray line at the top; the horizontal dotted lines show how sets of threshold values are selected automatically (see text).

Source: © *IET 2008*

*http://sipi.usc.edu/database/database.php (website accessed 13 December 2011).

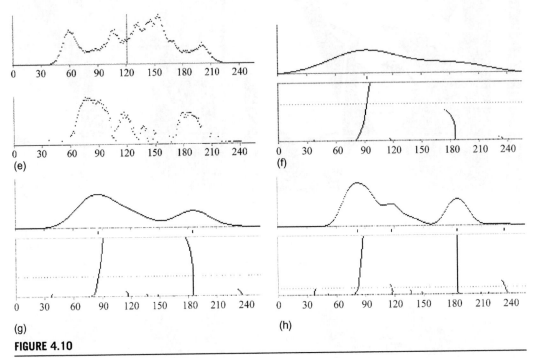

FIGURE 4.10

(Continued)

(see horizontal dotted lines in Fig. 4.10(f–h)). To clarify the process, the basic GVM algorithm is given in Fig. 4.11.

Figure 4.10(f–h) gives three examples of smoothing until 1, 2 or 4 thresholding points are produced (these give bi-level, tri-level, and five-level thresholding).[3] These lead to the images shown in Fig. 4.10(b–d) (note particularly that the light shaded region on Lena's nose is very stable and noise-free).

We concentrate next on a specific advantage of the GVM: that it produces robust judgments of minority intensities at the ends of the intensity range. Effectively, it amplifies such regions of the distribution and provides highly stable image segmentations: see, in particular, the under-vehicle shadows located in Fig. 23.1(d) and the ergot contaminant located in Fig. 21.2(d).[4] That the

---

[3]There is a potential confusion here: as smoothing proceeds, the number of GVM thresholds progressively *decreases*. Hence the ordering of the respective images and traces in subfigures (f)–(h) appears inverted from this point of view. However, it is the logical order for the BCVM for which computation increases approximately exponentially with the number of thresholds.

[4]Note that these represent important vehicle guidance and inspection tasks: (1) use of under-vehicle shadows is a promising technique for locating vehicles on the road ahead (Liu et al., 2007); (2) ergot is poisonous and it is important to locate it amongst wheat or other grains that are to be used for human consumption (Davies, 2003b).

```
scan = 0;
do {
  numberofpeaks = 0;
    for (all intensity values in distribution) {
      if (peak found) {
        peakposition[scan, numberofpeaks] = intensity;
        numberofpeaks ++;
      }
    }
    if (numberofpeaks == requirednumber) {
      if (previousnumberofpeaks > numberofpeaks) lowestscan = scan;
      else highestscan = scan;
    }
    previousnumberofpeaks = numberofpeaks;
    apply incremental smoothing kernel to distribution;
    scan ++;
} while (numberofpeaks > 0);

optimumscan = (lowestscan*3 + highestscan)/4;
for (all peaks up to requirednumber)
   bestpeakposition[peak] = peakposition[optimumscan, peak];
```

FIGURE 4.11

Basic global valley algorithm. This version of the algorithm assumes that the required number of peaks (*required number*) is known in advance, although the optimum amount of smoothing is unknown. Here, the latter is estimated by taking a weighted mean of the lowest and highest numbers of smoothing scans that yield the required number of peaks. The final line of the algorithm gives the required number of peaks in the best positions. While this form of the algorithm obtains positions for a specific required number of peaks, the underlying process also maps out a complete set of stability graphs because it proceeds until the number of peaks is zero. For further details, see Section 4.7.

Source: © IET 2008

GVM is able to make sense of the exceptionally noisy K distribution shown in Fig. 21.2(d) seems rather remarkable.

Comparing the GVM results with those of the BCVM (see Fig. 4.10(a, e)), we see that the bi-level BCVM threshold appears to lie in an *a priori* quite reasonable position in the intensity histogram: however, closer examination shows that the performance of the BCVM approximates to splitting the active area of the histogram into equal parts, corresponding to finding an approximate median. This means that for nearly unimodal histograms, it has much less chance of leading to optimal segmentations. This view of its operation is supported by tests (Fig. 4.12) made on idealized histograms, which show that it is unable to locate the bottom of the valley. It is also noteworthy that, unlike the GVM, the multilevel BCVM sometimes misses thresholds at the ends of the range of intensities (see, e.g., the vertical lines in the top trace of Fig. 4.9(d)).

Overall, it has been found that the GVM produces significantly more stable thresholds than the BCVM, that it is less prone to producing noisy boundaries in the thresholded images, and that its results tend to be more meaningful.

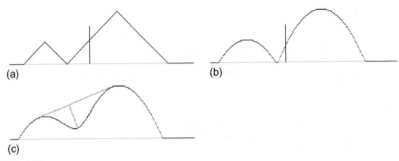

FIGURE 4.12

Results of applying the between-class variance method (BCVM) and concavity analysis in idealized cases. Applying the BCVM to (a) a triangular histogram and (b) a parabolic histogram. The vertical lines indicate the bi-level threshold selected by the BCVM: note that in each case it lies well away from the obvious global minimum of the histogram. (c) Finding thresholds by concavity analysis. The technique forms the convex hull of the distribution, takes each joining line, and uses the foot of the longest normal as an indicator of the threshold position. This approach is often highly effective but tends to give a result closer to the main peak than the optimum minimum location.

Source: © IET 2008

In fact, the BCVM tends to split intensity distributions rather blindly into approximately equal areas: although its mathematical formulation does not explicitly aim at this, it often seems to have essentially this effect.

4.8 HISTOGRAM CONCAVITY ANALYSIS

In this section, we briefly consider previous work on histogram concavity analysis. Rosin (2001) described how a simple geometrical construction (Fig. 4.12(c)) could be used to identify a suitable bi-level threshold. The technique depends on the histogram having a "corner," which is then easily identified, but when the corner is less well defined, bias can creep in and it becomes necessary to model the histogram distributions to obtain systematic corrections to the thresholding point. The approach will work for true unimodal distributions (including those produced by grey-scale edge images) or for "nearly" unimodal distributions where there is a very weak mode in addition to the main mode. For true unimodal distributions, the GVM will not work because one of the component signals in function K is zero: in such cases, it is imperative to use a method such as that described by Rosin—although others have been described over a long period—by Rosenfeld and de la Torre (1983), Tsai (1995), and others. For nearly unimodal distributions, the Rosin approach gives some intrinsic bias, as indicated in Fig. 4.12(c): but it is presumably possible in many applications to perform modeling to overcome this problem. However, the need for modeling does not seem to arise with the GVM—as has already been demonstrated (see particularly Figs. 21.2 and 23.1).

4.9 CONCLUDING REMARKS

Sections 4.3 and 4.4 have revealed a number of factors that are crucial to the process of thresholding. First, the need to avoid bias in threshold selection by arranging roughly equal populations in the dark and light regions of the intensity histogram; second, the need to work with a small subimage (or neighborhood) size so that the intensity histogram has a well-defined valley despite variations in illumination; and third, the need for subimages to be sufficiently large so that statistics are reliable, permitting the valley to be located accurately.

Unfortunately, these conditions are not compatible and compromises are needed in practical situations. In particular, it is generally not possible to find a neighborhood size that can be applied everywhere in an image, on all occasions yielding roughly equal populations of dark and light pixels. Indeed, if the chosen size is small enough to span edges ideally, hence yielding unbiased local thresholds, it will be valueless inside large objects. Attempting to avoid this situation by resorting to alternative methods of threshold calculation does not solve the problem since inherent to such methods is a built-in region size. It is therefore not surprising that a number of workers have opted for variable resolution and hierarchical techniques in an attempt to make thresholding more effective (Wu et al., 1982; Wermser et al., 1984; Kittler et al., 1985).

At this stage, we call into question the complications involved in such thresholding procedures—which become even worse when intensity distributions start to become multimodal. Note that the overall procedure is to find local intensity gradients in order to obtain accurate, unbiased estimates of thresholds so that it then becomes possible to take a horizontal slice through a grayscale image and hence, ultimately, find "vertical" (i.e., spatial) boundaries within the image. Why not use the gradients *directly* to estimate the boundary positions? Such an approach, for example, leads to no problems from large regions where intensity histograms are essentially unimodal, although it would be foolish to pretend that there are no other problems (see Chapters 5 and 10).

On the whole, the author takes the view that many approaches (region-growing, thresholding, edge detection, etc.), *taken to the limits of approximation*, will give equally good results. After all, they are all limited by the same physical effects— image noise, variability of lighting, presence of shadows, etc. However, some methods are easier to coax into working well, or need minimal computation, or have other useful properties such as robustness. Thus, thresholding can be a highly efficient means of aiding the interpretation of certain types of image: but as soon as image complexity rises above a certain critical level, it suddenly becomes more effective and considerably less complicated to rely on edge detection. This is studied in the next chapter. Meanwhile, we must not overlook the possibility of easing the thresholding task by optimizing the lighting system and ensuring that any worktable or conveyor is kept clean and white: this turns out to be a viable approach in a surprisingly large number of industrial applications.

The end result of thresholding is a set of silhouettes representing the shapes of objects: these constitute a "binarized" version of the original image. Many techniques exist for performing binary shape analysis, and some of these are described in Chapter 9. Meanwhile, note that many features of the original scene—e.g., texture, grooves or other surface structure—will not be present in the binarized image. Although the use of multiple thresholds to generate a number of binarized versions of the original image can preserve relevant information present in the original image, this approach tends to be clumsy and impracticable, and sooner or later one may be forced to return to the original grayscale image for the required data.

> Thresholding is among the simplest of image processing operations and is an intrinsically appealing way of performing segmentation. While the approach is clearly limited, it would be a mistake to ignore it and its recent developments, which provide useful tools for the programmer's toolkit.

4.10 BIBLIOGRAPHICAL AND HISTORICAL NOTES

Segmentation by thresholding started many years ago from simple beginnings, and in recent years has been refined into a set of mature procedures. Among the notable early methods is the paradigm but computation-intensive Chow and Kaneko method (1972), which has been outlined in Section 4.4.1. Nakagawa and Rosenfeld (1979) studied the method and developed it for cases of trimodal distributions but without improving computational load.

Fu and Mui (1981) provided a useful general survey on image segmentation: which was updated by Haralick and Shapiro (1985). These papers review many topics that could not be covered in this chapter due to space reasons—which also applies for Sahoo et al.'s (1988) valuable survey of thresholding techniques. Nevertheless, it is worth emphasizing the point made by Fu and Mui (1981) that "All the region extraction techniques process the pictures in an iterative manner and usually involve a great expenditure in computation time and memory."

As hinted in Section 4.4, thresholding (particularly local adaptive thresholding) has had many applications in optical character recognition. Among the earliest were the algorithms described by Bartz (1968) and Ullmann (1974): also two highly effective algorithms have been described by White and Rohrer (1983).

During the 1980s, the entropy approach to automatic thresholding evolved (e.g., Pun, 1981; Kapur et al., 1985; Abutaleb, 1989; Pal and Pal, 1989): this approach (Section 4.5.2) proved highly effective, and its development continued during the 1990s (e.g., Hannah et al., 1995).

In the 2000s, the entropy approach to threshold selection has remained important, in respect both of conventional region location and ascertaining the transition region between objects and background to make the segmentation process more reliable (Yan et al., 2003). In one instance, it was found useful to employ fuzzy entropy and genetic algorithms (Tao et al., 2003). Wang and Bai (2003) have shown how threshold selection may be made more reliable by clustering the intensities of boundary pixels, while ensuring that a continuous rather than a discrete boundary is considered (the problem is that in images that approximate to binary images over restricted regions, the edge points will lie preferentially in the object or the background, not neatly between both). However, in complex outdoor scenes and for many medical images such as brain scans, thresholding alone will not be sufficient, and resort may even have to be made to graph matching (Chapter 14) to produce the best results—reflecting the important fact that segmentation is necessarily a high-level rather than a low-level process (Wang and Siskind, 2003). In rather less demanding cases, deformable model-guided split-and-merge techniques may, on the other hand, still be sufficient (Liu and Sclaroff, 2004).

4.10.1 More Recent Developments

Sezgin and Sankur (2004) give a thorough review and assessment of work on thresholding prior to 2004. More recently, there has been continued interest in thresholding in the case of unimodal (Coudray et al., 2010; Medina-Carnicer et al., 2011) and near-unimodal histograms (Davies, 2007a, 2008b): the latter case is covered fairly fully in Sections 4.6 and 4.7. In the case of Coudray et al. (2010), the aim is to threshold intensity gradient histograms in order to locate edges reliably: the approach taken is to model the contribution from noise as a Rayleigh distribution and then to devise heuristics for analyzing the overall distribution. With the same aim, Medina-Carnicer et al. (2011) show that applying a histogram transformation improves the performance of the Otsu (1979) and Rosin (2001) methods. Li et al. (2011) adopt the novel approach of constraining the gray-level ranges considered by the thresholding algorithm in such a way as to weaken gray-level changes in both foreground and background, thus simplifying the original image and making the intensity histogram more closely bimodal. After that several thresholding methods are found to operate more reliably. Ng (2006) describes a revised version of the Otsu (1979) method that operates well for unimodal distributions, and which is useful for defect detection. This "valley emphasis" method works by applying a weight to the Otsu threshold calculation. Overall, several of the recent developments can be construed as applying transformations or other improvements to older methods to make them more sophisticated and accurate: in fact none is highly complex in any theoretical way. Finally, it may seem somewhat surprising that, after so many decades, thresholding is still something of a "hot" subject: the driving force for this is its extreme simplicity and high level of utility.

4.11 PROBLEMS

1. Using the ideas outlined in Section 4.3.2, model the intensity distribution obtained by finding all the edge pixels in an image and including also all pixels adjacent to these pixels. Show that while this gives a sharper valley than for the original intensity distribution, it is not as sharp as for pixels located by the Laplacian operator.

2. Consider whether it is more accurate to estimate a suitable threshold for a bimodal, dual-Gaussian distribution by (a) finding the position of the minimum, or (b) finding the mean of the two peak positions. What corrections could be made by taking account of the magnitudes of the peaks?

3. Obtain a complete derivation of Eq. (4.20). Show that, in general (as stated in Section 4.5.3), it has two solutions. What is the physical reason for this? How can it have only one solution when $\sigma_1 = \sigma_2$?

4. Prove the statement made in Section 4.6 that the computational load of the histogram analysis for the global value method can be reduced from $O(N^3)$ to $O(N)$. Show also that the number of passes over the histogram required to achieve this is at most 2.

Edge Detection

5

Edge detection provides an intrinsically more rigorous means than thresholding for initiating image segmentation. However, there is a large history of *ad hoc* edge detection algorithms, and this chapter aims to distinguish what is principled from what is *ad hoc* and to provide theoretical and practical knowledge underpinning available techniques.

Look out for:

- the variety of template matching operators that have been used for edge detection—e.g., the Prewitt, Kirsch, and Robinson operators.
- the differential gradient approach to edge detection—exemplified by the Roberts, Sobel, and Frei–Chen operators.
- theory explaining the performance of the template matching operators.
- methods for the optimal design of differential gradient operators and the value of "circular" operators.
- tradeoffs between resolution, noise suppression capability, location accuracy, and orientation accuracy.
- the distinction between edge enhancement and edge detection.
- outlines of more modern operators—the Canny and Laplacian-based operators.
- the use of active contour models (snakes) for modeling object boundaries.
- the "graph cut" approach to object segmentation.

In discussing the process of edge detection, this chapter shows that it is possible to estimate edge orientation with surprising accuracy within a small window—the secret being the considerable information residing in the grayscale values. High orientation accuracy turns out to be of particular value when using the Hough transform to locate extended objects in digital images—as will be seen in several chapters in Part 2 of this book.

5.1 INTRODUCTION

In Chapter 4, segmentation has been tackled by the general approach of finding regions of uniformity in images—on the basis that the areas found in this way would have a fair likelihood of coinciding with the surfaces and facets of objects. The most computationally efficient means of following this approach was that of thresholding but for real images. This turns out to be failure-prone or quite difficult to implement satisfactorily. Indeed, to make it work well seems to require a multiresolution or hierarchical approach, coupled with sensitive measures for obtaining suitable local thresholds. Such measures have to take account of local intensity gradients as well as pixel intensities, and the possibility of proceeding more simply—by taking account of intensity gradients alone—was suggested.

In fact, edge detection has long been an alternative path to image segmentation and is the method pursued in this chapter. Whichever way is inherently the better approach, edge detection has the additional advantage in that it immediately reduces by a large factor (typically around 100) the considerable redundancy of most image data: this is useful because it significantly reduces both the space needed to store the information and the amount of processing subsequently required to analyze it.

Edge detection has gone through an evolution spanning well over 30 years. Two main methods of edge detection have been apparent over this period, the first of these being the template matching (TM) approach and the second being the differential gradient (DG) approach. In either case the aim is to find where the intensity gradient magnitude g is sufficiently large to be taken as a reliable indicator of the edge of an object. Then g can be thresholded in a similar way in which intensity has been thresholded in Chapter 4 (in fact, we shall see that it is possible to look for local maxima of g instead of, or as well as, thresholding it). The TM and DG methods differ mainly in how they proceed to estimate g locally; however, there are also important differences in how they determine local edge orientation, which is an important variable in certain object detection schemes.

Later in the chapter we look at the Canny operator, which was much more rigorously designed than previous edge detectors. Then we consider Laplacian-based operators before moving on to study active contour models or "snakes." Finally, we outline the "graph cut" approach to object segmentation: this makes use of intensity gradient information to zone in on object regions, thereby in a sense embodying both the edge detection and the region growing paradigms and ending up with ideal, provably unique solutions.

Before proceeding to discuss the performance of the various edge detection operators, note that there are a variety of types of edge, including in particular the "sudden step" edge, the "slanted step" edge, the "planar" edge, and various intermediate edge profiles (see Fig. 5.1). This chapter considers edges of the types shown in Fig. 5.1(a)–(d): edges of the types shown in Fig. 5.1(e) and (f) are much rarer; an example being shown in Fig. 11.4(a).

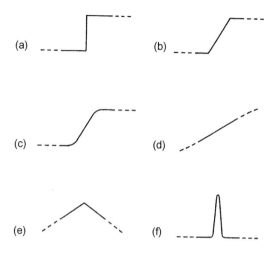

FIGURE 5.1

Edge models: (a) sudden step edge; (b) slanted step edge; (c) smooth step edge; (d) planar edge; (e) roof edge; and (f) line edge. The effective profiles of edge models are nonzero only within the stated neighborhood. The slanted step and the smooth step are approximations to realistic edge profiles: the sudden step and the planar edge are extreme forms that are useful for comparisons (see text).

5.2 BASIC THEORY OF EDGE DETECTION

Both DG and TM operators estimate local intensity gradients with the aid of suitable convolution masks. In the case of the DG type of operator, only two such masks are required—for the x and y directions. In the TM case, it is usual to employ up to 12 convolution masks capable of estimating local components of gradient in different directions (Prewitt, 1970; Kirsch, 1971; Robinson, 1977; Abdou and Pratt, 1979).

In the TM approach, the local edge gradient magnitude (for short, the edge "magnitude") is approximated by taking the maximum of the responses for the component masks:

$$g = \max \left(g_i : i = 1, \ldots, n \right) \qquad (5.1)$$

where n is usually 8 or 12.

In the DG approach, the local edge magnitude may be computed vectorially using the nonlinear transformation:

$$g = \left(g_x^2 + g_y^2 \right)^{1/2} \qquad (5.2)$$

To save computational effort, it is common practice (Abdou and Pratt, 1979) to approximate this formula by one of the simpler forms:

$$g = |g_x| + |g_y| \tag{5.3}$$

or

$$g = \max\left(|g_x|, |g_y|\right) \tag{5.4}$$

which are, on average, equally accurate (Föglein, 1983).

In the TM approach, edge orientation is estimated simply as that of the mask giving rise to the largest value of gradient in Eq. (5.1). In the DG approach, it is estimated vectorially by the more complex equation:

$$\theta = \arctan\frac{g_y}{g_x} \tag{5.5}$$

Clearly, DG equations ((5.2) and (5.5)) require considerably more computation than TM equation (5.1), although they are more accurate. However, in some situations orientation information is not required; in addition, image contrast may vary widely, so there may appear to be little gain from thresholding a more accurate estimate of g. This may explain why so many workers have employed the TM instead of the DG approach. Since both approaches essentially involve estimation of local intensity gradients, it is not surprising that TM masks often turn out to be identical to DG masks (see Tables 5.1 and 5.2).

Table 5.1 Masks of Well-known Differential Edge Operators

(a) Masks for the Roberts 2 × 2 operator

$$R_{x'} = \begin{bmatrix} 0 & 1 \\ -1 & 0 \end{bmatrix} \qquad R_{y'} = \begin{bmatrix} 1 & 0 \\ 0 & -1 \end{bmatrix}$$

(b) Masks for the Sobel 3 × 3 operator

$$S_x = \begin{bmatrix} -1 & 0 & 1 \\ -2 & 0 & 2 \\ -1 & 0 & 1 \end{bmatrix} \qquad S_y = \begin{bmatrix} 1 & 2 & 1 \\ 0 & 0 & 0 \\ -1 & -2 & -1 \end{bmatrix}$$

(c) Masks for the Prewitt 3 × 3 "smoothed gradient" operator

$$P_x = \begin{bmatrix} -1 & 0 & 1 \\ -1 & 0 & 1 \\ -1 & 0 & 1 \end{bmatrix} \qquad P_y = \begin{bmatrix} 1 & 1 & 1 \\ 0 & 0 & 0 \\ -1 & -1 & -1 \end{bmatrix}$$

In this table masks are presented in an intuitive format (viz. coefficients increasing in the positive x and y directions) by rotating the normal convolution format through 180°. This convention is employed throughout this chapter. The Roberts 2 × 2 operator masks (a) can be taken as being referred to axes x' and y' at 45° to the usual x and y axes.

5.3 THE TEMPLATE MATCHING APPROACH

Table 5.2 shows four sets of well-known TM masks for edge detection. These masks were originally (Prewitt, 1970; Kirsch, 1971; Robinson, 1977) introduced on an intuitive basis, starting in two cases from the DG masks shown in Table 5.1. In all cases the eight masks of each set are obtained from a given mask by permuting the mask coefficients cyclically. By symmetry, this is a good strategy for even permutations, but symmetry alone does not justify it for odd permutations: the situation is explored in more detail below.

Note first that four of the "3-level" and four of the "5-level" masks can be generated from the other four of their set by sign inversion. This means that in either case only four convolutions need to be performed at each pixel neighborhood, thereby saving computation. This is an obvious procedure if the basic idea of the TM approach is regarded as one of comparing intensity gradients in the eight directions. The two operators that do not employ this strategy were developed much earlier on some unknown intuitive basis.

Before proceeding, we note the rationale behind the Robinson "5-level" masks. These were intended (Robinson, 1977) to emphasize the weights of diagonal edges in order to compensate for the characteristics of the human eye, which tends to enhance vertical and horizontal lines in images. Normally, image analysis is concerned with computer interpretation of images, and an isotropic set of responses is required. Thus, the "5-level" operator is a special-purpose one that need not be discussed further.

Table 5.2 Masks of Well-known 3×3 Template Matching Edge Operators

	0°	45°
(a) Prewitt masks	$\begin{bmatrix} -1 & 1 & 1 \\ -1 & -2 & 1 \\ -1 & 1 & 1 \end{bmatrix}$	$\begin{bmatrix} 1 & 1 & 1 \\ -1 & -2 & 1 \\ -1 & -1 & 1 \end{bmatrix}$
(b) Kirsch masks	$\begin{bmatrix} -3 & -3 & 5 \\ -3 & 0 & 5 \\ -3 & -3 & 5 \end{bmatrix}$	$\begin{bmatrix} -3 & 5 & 5 \\ -3 & 0 & 5 \\ -3 & -3 & -3 \end{bmatrix}$
(c) Robinson "3-level" masks	$\begin{bmatrix} -1 & 0 & 1 \\ -1 & 0 & 1 \\ -1 & 0 & 1 \end{bmatrix}$	$\begin{bmatrix} 0 & 1 & 1 \\ -1 & 0 & 1 \\ -1 & -1 & 1 \end{bmatrix}$
(d) Robinson "5-level" masks	$\begin{bmatrix} -1 & 0 & 1 \\ -2 & 0 & 2 \\ -1 & 0 & 1 \end{bmatrix}$	$\begin{bmatrix} 0 & 1 & 2 \\ -1 & 0 & 1 \\ -2 & 1 & 0 \end{bmatrix}$

The table illustrates only two of the eight masks in each set; the remaining masks can in each case be generated by symmetry operations. For the 3-level and 5-level operators, four of the eight available masks are inverted versions of the other four (see text).

These considerations show that the four template operators mentioned above have limited theoretical justification. It is therefore worth studying the situation in more depth (see Section 5.4).

5.4 THEORY OF 3 × 3 TEMPLATE OPERATORS

In what follows, it is assumed that eight masks are to be used, with angles differing by 45°. In addition, four of the masks differ from the others only in sign, since this seems unlikely to result in any loss of performance. Symmetry requirements then lead to the following masks for 0° and 45°, respectively.

$$\begin{bmatrix} -A & 0 & A \\ -B & 0 & B \\ -A & 0 & A \end{bmatrix} \quad \begin{bmatrix} 0 & C & D \\ -C & 0 & C \\ -D & -C & 0 \end{bmatrix}$$

It is clearly of great importance to design masks so that they give consistent responses in different directions. To find how this affects the mask coefficients, we employ the strategy of ensuring that intensity gradients follow the rules of vector addition. If the pixel intensity values within a 3 × 3 neighborhood are:

$$\begin{vmatrix} a & b & c \\ d & e & f \\ g & h & i \end{vmatrix}$$

the above masks will give the following estimates of gradient in the 0°, 90°, and 45° directions:

$$g_0 = A(c + i - a - g) + B(f - d) \tag{5.6}$$

$$g_{90} = A(a + c - g - i) + B(b - h) \tag{5.7}$$

$$g_{45} = C(b + f - d - h) + D(c - g) \tag{5.8}$$

If vector addition is to be valid, then:

$$g_{45} = \frac{g_0 + g_{90}}{\sqrt{2}} \tag{5.9}$$

Equating coefficients of $a, b, \ldots, i$ leads to the self-consistent pair of conditions:

$$C = \frac{B}{\sqrt{2}} \tag{5.10}$$

$$D = A\sqrt{2} \tag{5.11}$$

A further requirement is for the 0° and 45° masks to give equal responses at 22.5°. This can be shown to lead to the formula:

$$\frac{B}{A} = \sqrt{2}\,\frac{9t^2 - \left(14 - 4\sqrt{2}\right)t + 1}{t^2 - \left(10 - 4\sqrt{2}\right)t + 1} \tag{5.12}$$

where $t = \tan 22.5°$, so that:

$$\frac{B}{A} = \frac{13\sqrt{2} - 4}{7} = 2.055 \tag{5.13}$$

We can now summarize our findings with regard to the design of TM masks. First, obtaining sets of masks by permuting coefficients "cyclically" in a square neighborhood is *ad hoc* and cannot be relied upon to produce useful results. Next, following the rules of vector addition and the need to obtain consistent responses in different directions, we have shown that ideal TM masks need to closely match the Sobel coefficients; we have also rigorously derived an accurate value for the ratio B/A.

Having obtained some insight into the process of designing TM masks for edge detection, we next move on to study the design of DG masks.

5.5 THE DESIGN OF DIFFERENTIAL GRADIENT OPERATORS

This section studies the design of DG operators. These include the Roberts 2×2 operator and the Sobel and Prewitt 3×3 operators (Roberts, 1965; Prewitt, 1970; for the Sobel operator see Pringle, 1969, Duda and Hart, 1973, p. 271) (see Table 5.1). The Prewitt or "gradient smoothing" type of operator has been extended to larger pixel neighborhoods by Prewitt (1970) and others (Brooks, 1978; Haralick, 1980) (see Table 5.3). In these instances the basic rationale is to model local edges by the best fitting plane over a convenient size of neighborhood. Mathematically, this amounts to obtaining suitably weighted averages to estimate slope in the x and y directions. As pointed out by Haralick (1980), the use of equally weighted averages to measure slope in a given direction is incorrect: the proper weightings to use are given by the masks listed in Table 5.3. Thus, the Roberts and Prewitt operators are apparently optimal, whereas the Sobel operator is not. This point is discussed in more detail below.

A full discussion of the edge detection problem involves consideration of the accuracy with which edge magnitude and orientation can be estimated when the local intensity pattern cannot be assumed to be planar. In fact, there have been a number of analyses of the angular dependencies of edge detection operators for a step edge approximation. In particular, O'Gorman (1978) considered the variation of estimated versus actual angle resulting from a step edge observed within a

Table 5.3 Masks for Estimating Components of Gradient in Square Neighborhoods

	M_x	M_y
(a) 2×2 neighborhood	$\begin{bmatrix} -1 & 1 \\ -1 & 1 \end{bmatrix}$	$\begin{bmatrix} 1 & 1 \\ -1 & -1 \end{bmatrix}$
(b) 3×3 neighborhood	$\begin{bmatrix} -1 & 0 & 1 \\ -1 & 0 & 1 \\ -1 & 0 & 1 \end{bmatrix}$	$\begin{bmatrix} 1 & 1 & 1 \\ 0 & 0 & 0 \\ -1 & -1 & -1 \end{bmatrix}$
(c) 4×4 neighborhood	$\begin{bmatrix} -3 & -1 & 1 & 3 \\ -3 & -1 & 1 & 3 \\ -3 & -1 & 1 & 3 \\ -3 & -1 & 1 & 3 \end{bmatrix}$	$\begin{bmatrix} 3 & 3 & 3 & 3 \\ 1 & 1 & 1 & 1 \\ -1 & -1 & -1 & -1 \\ -3 & -3 & -3 & -3 \end{bmatrix}$
(d) 5×5 neighborhood	$\begin{bmatrix} -2 & -1 & 0 & 1 & 2 \\ -2 & -1 & 0 & 1 & 2 \\ -2 & -1 & 0 & 1 & 2 \\ -2 & -1 & 0 & 1 & 2 \\ -2 & -1 & 0 & 1 & 2 \end{bmatrix}$	$\begin{bmatrix} 2 & 2 & 2 & 2 & 2 \\ 1 & 1 & 1 & 1 & 1 \\ 0 & 0 & 0 & 0 & 0 \\ -1 & -1 & -1 & -1 & -1 \\ -2 & -2 & -2 & -2 & -2 \end{bmatrix}$

The masks provided in this table can be regarded as extended Prewitt masks. The 3×3 masks are Prewitt masks, included in this table for completeness. In all cases weighting factors have been omitted in the interests of simplicity, as they are throughout this chapter.

square neighborhood (see also Brooks, 1978): note that the case considered was that of a continuum rather than a discrete lattice of pixels. This was found to lead to a smooth variation with angular error varying from zero at 0° and 45° to a maximum of 6.63° at 28.37° (where the estimated orientation was 21.74°), the variation for angles outside this range being replicated by symmetry. Abdou and Pratt (1979) obtained similar variations for the Sobel and Prewitt operators in a discrete lattice, the respective maximum angular errors being 1.36° and 7.38° (Davies, 1984b). It seems that the Sobel operator has angular accuracy that is close to optimal because it is close to being a "truly circular" operator. This point is discussed in more detail in Section 5.6.

5.6 THE CONCEPT OF A CIRCULAR OPERATOR

It has been stated above that when step edge orientation is estimated in a square neighborhood, an error of up to 6.63° can result. Such an error does not arise with a planar edge approximation, since fitting of a plane to a planar edge profile within a square window can be carried out exactly. Errors appear only when the edge profile differs from the ideal planar form, within the square neighborhood— with the step edge probably being something of a "worst case."

One way to limit errors in the estimation of edge orientation might be to restrict observation of the edge to a circular neighborhood. In the continuous case this is sufficient to reduce the error to zero for all orientations, since symmetry dictates that there is only one way of fitting a plane to a step edge within a circular neighborhood, assuming that all planes pass through the same central point; the estimated orientation θ is then equal to the actual angle φ. A rigorous calculation along the lines indicated by Brooks (1976), which results in the following formula for a square neighborhood (O'Gorman, 1978):

$$\tan \theta = \frac{2 \tan \varphi}{3 - \tan^2 \varphi} \quad 0° \le \varphi \le 45° \tag{5.14}$$

leads to the following formula:

$$\tan \theta = \tan \varphi, \quad \text{i.e., } \theta = \varphi \tag{5.15}$$

for a circular neighborhood (Davies, 1984b). Similarly, zero angular error results from fitting a plane to an edge of *any* profile within a circular neighborhood, in the continuous approximation. Indeed, for an edge surface of arbitrary shape, the only problem is whether the mathematical best fit plane coincides with one that is subjectively desirable (and, if not, a fixed angular correction will be required). Ignoring such cases, the basic problem is how to approximate a circular neighborhood in a digitized image of small dimensions, containing typically 3×3 or 5×5 pixels.

To proceed systematically, we first recall a fundamental principle stated by Haralick (1980):

> the fact that the slopes in two orthogonal directions determine the slope in any direction is well known in vector calculus. However, it seems not to be so well known in the image processing community.

Essentially, appropriate estimates of slopes in two orthogonal directions permit the slope in any direction to be computed. For this principle to apply, appropriate estimates of the slopes have first to be made: if the components of slope are inappropriate, they will not act as components of true vectors and the resulting estimates of edge orientation will be in error. This appears to be the main source of error with the Prewitt and other operators—it is not so much that the components of slope are in any instance incorrect, but rather that they are inappropriate for the purpose of vector computation since *they do not match one another adequately in the required way* (Davies, 1984b).

Following the arguments for the continuous case discussed earlier, slopes must be rigorously estimated within a circular neighborhood. Then the operator design problem devolves into determining how best it is to simulate a circular neighborhood on a discrete lattice so that errors are minimized. To carry this out, it is necessary to apply a close to circular weighting while computing the masks, so that correlations between the gradient weighting and circular weighting factors are taken properly into account.

5.7 DETAILED IMPLEMENTATION OF CIRCULAR OPERATORS

In practice, the task of computing angular variations and error curves has to be tackled numerically, dividing each pixel in the neighborhood into arrays of suitably small subpixels. Each subpixel is then assigned a gradient weighting (equal to the x or y displacement) and a neighborhood weighting (equal to 1 for inside and 0 for outside a circle of radius r). Clearly, the angular accuracy of "circular" differential gradient edge detection operators must depend on the radius of the circular neighborhood. In particular, poor accuracy would be expected for small values of r and reasonable accuracy for large values of r, as the discrete neighborhood approaches a continuum.

The results of such a study are presented in Fig. 5.2. The variations depicted represent RMS angular errors (Fig. 5.2(a)) and maximum angular errors (Fig. 5.2(b)) in the estimation of edge orientation. The structures on each variation are surprisingly smooth: they are so closely related and systematic that they can only represent statistics of the arrangement of pixels in neighborhoods of various sizes. Details of these statistics are discussed in the next section.

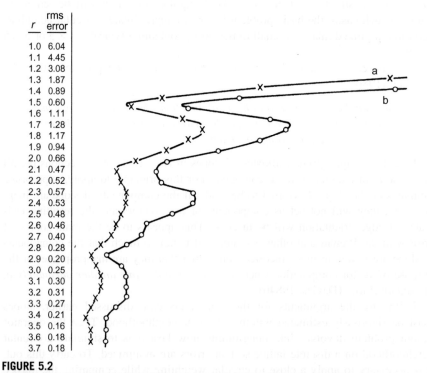

r	rms error
1.0	6.04
1.1	4.45
1.2	3.08
1.3	1.87
1.4	0.89
1.5	0.60
1.6	1.11
1.7	1.28
1.8	1.17
1.9	0.94
2.0	0.66
2.1	0.47
2.2	0.52
2.3	0.57
2.4	0.53
2.5	0.48
2.6	0.46
2.7	0.40
2.8	0.28
2.9	0.20
3.0	0.25
3.1	0.30
3.2	0.31
3.3	0.27
3.4	0.21
3.5	0.16
3.6	0.18
3.7	0.18

FIGURE 5.2

Variations in angular error as a function of radius r: (a) RMS angular error and (b) maximum angular error.

Overall, three features of Fig. 5.2 are noteworthy. First, as expected, there is a general trend to zero angular error as r tends to infinity. Second, there is a very marked periodic variation, with particularly good accuracy resulting where the circular operators best match the tessellation of the digital lattice. The third feature of interest is the fact that errors do not vanish for any finite value of r—clearly, the constraints of the problem do not permit more than the minimization of errors. These curves show that it is possible to generate a family of optimal operators (at the minima of the error curves), the first of which corresponds closely to an operator (the Sobel operator) that is known to be nearly optimal.

The variations shown in Fig. 5.2 can be explained (Davies, 1984b) as pixel centers lying in well-packed or "closed" bands approximating to continua—indicated by the low error points in Fig. 5.2—between which centers would be more loosely packed. Thus, we get the "closed band" operators listed in Table 5.4; their angular variations appear in Table 5.5. It is seen that the Sobel operator, which is already the most accurate of the 3×3 edge gradient operators suggested previously, can be made some 30% more accurate by adjusting its coefficients to make it more circular. In addition, the closed bands idea indicates that the corner pixels of 5×5 or larger operators are best removed altogether: not only does this require less computation, but also it actually improves performance. It also seems likely that this situation would apply for many other operators and would not be specific to edge detection.

Table 5.4 Masks of "Closed Band" Differential Gradient Edge Operators

(a) Band containing shells a–c (effective radius = 1.500)

$$\begin{bmatrix} -0.464 & 0.000 & 0.464 \\ -0.959 & 0.000 & 0.959 \\ -0.464 & 0.000 & 0.464 \end{bmatrix}$$

(b) Band containing shells a–e (effective radius = 2.121)

$$\begin{bmatrix} 0.000 & -0.294 & 0.000 & 0.294 & 0.000 \\ -0.582 & -1.000 & 0.000 & 1.000 & 0.582 \\ -1.085 & -1.000 & 0.000 & 1.000 & 1.085 \\ -0.582 & -1.000 & 0.000 & 1.000 & 0.582 \\ 0.000 & -0.294 & 0.000 & 0.294 & 0.000 \end{bmatrix}$$

(c) Band containing shells a–h (effective radius = 2.915)

$$\begin{bmatrix} 0.000 & 0.000 & -0.191 & 0.000 & 0.191 & 0.000 & 0.000 \\ 0.000 & -1.085 & -1.000 & 0.000 & 1.000 & 1.085 & 0.000 \\ -0.585 & -2.000 & -1.000 & 0.000 & 1.000 & 2.000 & 0.585 \\ -1.083 & -2.000 & -1.000 & 0.000 & 1.000 & 2.000 & 1.083 \\ -0.585 & -2.000 & -1.000 & 0.000 & 1.000 & 2.000 & 0.585 \\ 0.000 & -1.085 & -1.000 & 0.000 & 1.000 & 1.085 & 0.000 \\ 0.000 & 0.000 & -0.191 & 0.000 & 0.191 & 0.000 & 0.000 \end{bmatrix}$$

In all cases only the x-mask is shown: the y-mask may be obtained by a trivial symmetry operation. Mask coefficients are accurate to ~0.003 but would in normal practical applications be rounded to one- or two-figure accuracy.

Table 5.5 Angular Variations for the Best Operators Tested

Actual Angle (°)	Estimated Angle (°)[a]					
	Prew	**Sob**	**a–c**	**circ**	**a–e**	**a–h**
0	0.00	0.00	0.00	0.00	0.00	0.00
5	3.32	4.97	5.05	5.14	5.42	5.22
10	6.67	9.95	10.11	10.30	10.81	10.28
15	10.13	15.00	15.24	15.52	15.83	14.81
20	13.69	19.99	20.29	20.64	20.07	19.73
25	17.72	24.42	24.73	25.10	24.62	25.00
30	22.62	28.86	29.14	29.48	29.89	30.02
35	28.69	33.64	33.86	34.13	35.43	34.86
40	35.94	38.87	39.00	39.15	40.30	39.71
45	45.00	45.00	45.00	45.00	45.00	45.00
RMS error	5.18	0.73	0.60	0.53	0.47	0.19

Prew, *Prewitt*; Sob, *Sobel*; a–c, *theoretical optimum—closed band containing shells a–c*; circ, *actual optimum circular operator (as defined by the first minimum in Fig. 5.2)*; a–e, *theoretical optimum—closed band containing shells a–e*; a–h, *theoretical optimum—closed band containing shells a–h*.
[a]*Values are accurate to within* ~0.02° *in each case.*

Before leaving this topic, note that the optimal 3×3 masks obtained above numerically by consideration of circular operators are very close to those obtained purely analytically in Section 5.4, for TM masks, following the rules of vector addition. In the latter case a value of 2.055 was obtained for the ratio of the two mask coefficients, whereas for circular operators the value $0.959/0.464 = 2.067 \pm 0.015$ is obtained. Clearly this is no accident, and it is very satisfying that a coefficient that was formerly regarded as *ad hoc* (Kittler, 1983) is in fact optimizable and can be obtained in closed form (see Section 5.4).

5.8 THE SYSTEMATIC DESIGN OF DIFFERENTIAL EDGE OPERATORS

The family of "circular" differential gradient edge operators studied in Sections 5.6 and 5.7 incorporates only one design parameter—the radius r. Only a limited number of values of this parameter permit optimum accuracy for estimation of edge orientation to be attained.

It is worth considering what additional properties this one parameter can control and how it should be adjusted during operator design. In fact, it affects signal-to-noise ratio, resolution, measurement accuracy, and computational load. To understand this, note first that signal-to-noise ratio varies linearly with the radius of the circular neighborhood, since signal is proportional to area and

Gaussian noise is proportional to the square root of area. Likewise, the measurement accuracy is determined by the number of pixels over which averaging occurs and hence is proportional to operator radius. Resolution and "scale" also vary with radius, since relevant linear properties of the image are averaged over the active area of the neighborhood. Finally, computational load, and the associated cost of hardware for speeding up the processing, is generally at least in proportion to the number of pixels in the neighborhood, and hence proportional to r^2.

Overall, the fact that four important parameters vary in a fixed way with the radius of the neighborhood means that there are exact tradeoffs between them and that improvements in some are only obtained by losses to others: from an engineering point of view, compromises between them will have to be made according to circumstances.

5.9 PROBLEMS WITH THE ABOVE APPROACH—SOME ALTERNATIVE SCHEMES

Although the above ideas may be interesting, they have their own inherent problems. In particular, they take no account of the displacement E of the edge from the center of the neighborhood, or of the effects of noise in biasing the estimates of edge magnitude and orientation. In fact, it is possible to show that a Sobel operator gives *zero* error in the estimation of step edge orientation under the following condition:

$$|\theta| \leq \arctan\left(\frac{1}{3}\right) \quad \text{and} \quad |E| \leq \frac{(\cos\theta - 3\sin|\theta|)}{2} \tag{5.16}$$

Furthermore, for a 3×3 operator of the form

$$\begin{bmatrix} -1 & 0 & 1 \\ -B & 0 & B \\ -1 & 0 & 1 \end{bmatrix} \qquad \begin{bmatrix} 1 & B & 1 \\ 0 & 0 & 0 \\ -1 & -B & -1 \end{bmatrix}$$

applied to the edge

$$\begin{vmatrix} a & a + h(0.5 - E\sec\theta + \tan\theta) & a+h \\ a & a + h(0.5 - E\sec\theta) & a+h \\ a & a + h(0.5 - E\sec\theta - \tan\theta) & a+h \end{vmatrix}$$

Lyvers and Mitchell (1988) found that the estimated orientation is:

$$\varphi = \arctan\left[\frac{2B\,\tan\theta}{B+2}\right] \tag{5.17}$$

which immediately shows why the Sobel operator should give zero error for a specific range of θ and E. However, this is somewhat misleading, since considerable errors arise outside this region. Not only do they arise when $E = 0$, as assumed in the foregoing sections, but also they vary strongly with E. Indeed, the maximum errors for the Sobel and Prewitt operators rise to 2.90° and 7.43°, respectively in this more general case (the corresponding RMS errors are 1.20° and 4.50°). Hence, a full analysis should be performed to determine how to reduce the maximum and average errors. Lyvers and Mitchell (1988) carried out an empirical analysis and constructed a lookup table with which to correct the orientations estimated by the Sobel operator, the maximum error being reduced to 2.06°.

Another scheme that reduces the error is the moment-based operator of Reeves et al. (1983). This leads to Sobel-like 3 × 3 masks, which are essentially identical to the 3 × 3 masks of Davies (1984b), both having $B = 2.067$ (for $A = 1$). However, the moment method can also be used to estimate the edge position E if additional masks are used to compute second-order moments of intensity. Hence, it is possible to make a very significant improvement in performance by using a 2-D lookup table to estimate orientation: the result is that the maximum error is reduced from 2.83° to 0.135° for 3 × 3 masks and from 0.996° to 0.0042° for 5 × 5 masks.

However, Lyvers and Mitchell (1988) found that much of this additional accuracy is lost in the presence of noise, and RMS standard deviations of edge orientation estimates are already around 0.5° for 3 × 3 operators at 40 dB signal-to-noise ratios. The reasons for this are quite simple. Each pixel intensity has a noise component that induces errors in its weighted mask components; the combined effects of these errors can be estimated assuming that they arise independently, so that their variances add (Davies, 1987c). Thus, noise contributions to the x and y components of gradient can be computed. These provide estimates for the components of noise along and perpendicular to the edge gradient vector (Fig. 5.3): the edge orientation for a Sobel operator turns out to be affected by an amount $\sqrt{12}\sigma/4h$ radians, where σ is the standard deviation on the pixel intensity values and h is the edge contrast. This explains the angular errors given by Lyvers and Mitchell, if Pratt's (2001) definition of signal-to-noise ratio (in dB) is used:

$$S/N = 20 \log_{10} \left(\frac{h}{\sigma} \right) \tag{5.18}$$

A totally different approach to edge detection was developed by Canny (1986). He used functional analysis to derive an optimal function for edge detection, starting with three optimization criteria—good detection, good localization, and only one response per edge under white noise conditions. The analysis is too technical to be discussed in detail here. However, the 1-D function found by Canny is accurately approximated by the derivative of a Gaussian: this is then combined with a Gaussian of identical σ in the perpendicular direction, truncated at 0.001 of its peak value, and split into suitable masks. Underlying this method is the idea of locating edges at local maxima of gradient magnitude for a Gaussian-smoothed image. In addition, the Canny implementation employs a hysteresis operation (Section 5.10) on edge magnitude in order to make edges

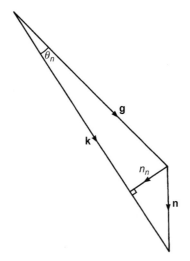

FIGURE 5.3

Calculating angular errors arising from noise: **g**, intensity gradient vector; **n**, noise vector; **k**, resultant of intensity gradient and noise vector; n_n, normal component of noise; θ_n, noise-induced orientation error.

reasonably connected. Finally, a multiple-scale method is employed to analyze the output of the edge detector. It is discussed in more detail below. Lyvers and Mitchell (1988) tested the Canny operator and found it to be significantly less accurate for orientation estimation than for the moment and IDD operators described above. In addition, it needed to be implemented using 180 masks and hence took enormous computation time, although many practical implementations of this operator are much faster than this early paper indicates.[1] One such implementation is described in Section 5.11.

An operator that has been of great historical importance is that of Marr and Hildreth (1980). The motivation for the design of this operator was the modeling of certain psychophysical processes in mammalian vision. The basic rationale is to find the Laplacian of the Gaussian-smoothed ($\nabla^2 G$) image and then to obtain a "raw primal sketch" as a set of zero-crossing lines. The Marr–Hildreth operator does not use any form of threshold since it merely assesses where the $\nabla^2 G$ image passes through zero. This feature is attractive, since working out threshold values is a difficult and unreliable task. However, the Gaussian smoothing procedure can be applied at a variety of scales, and in one sense the scale is a new parameter that substitutes for the threshold. In fact, a major feature of the Marr–Hildreth approach, which has been very influential in later work (Witkin, 1983; Bergholm,

---

[1] In fact, it is nowadays necessary to ask "Which Canny?", as there are a great many implementations of it, and this leads to problems for any realistic comparison between operators.

1986), is the fact that zero crossings can be obtained at several scales, giving the potential for more powerful semantic processing: clearly, this necessitates finding means for combining all the information in a systematic and meaningful way. This may be carried out by a bottom-up or top-down approach, and there has been much discussion in the literature about methods for carrying out these processes. However, it is worth remarking that in many (especially industrial inspection) applications, one is interested in working at a particular resolution, and considerable savings in computation can then be made. It is also noteworthy that the Marr–Hildreth operator is reputed to require neighborhoods of at least 35×35 for proper implementation (Brady, 1982). Nevertheless, other workers have implemented the operator in much smaller neighborhoods, down to 5×5. Wiejak et al. (1985) showed how to implement the operator using linear smoothing operations to save computation. Lyvers and Mitchell (1988) reported orientation accuracies using the Marr–Hildreth operator that are not especially high ($2.47°$ for a 5×5 operator and $0.912°$ for a 7×7 operator, in the absence of noise).

It has been noted above that those edge detection operators that are applied at different scales lead to different edge maps at different scales. In such cases, certain edges that are present at lower scales disappear at larger scales; in addition, edges that are present at both low and high scales appear shifted or merged at higher scales. Bergholm (1986) demonstrated the occurrence of elimination, shifting, and merging, whereas Yuille and Poggio (1986) showed that edges that are present at low resolution should not disappear at some higher resolution. These aspects of edge location are by now well understood.

In what follows, we first consider hysteresis thresholding, a process already mentioned with regard to the Canny operator. In Section 5.11 we give a fuller appraisal of the Canny operator and show detailed results on real images. Then in Section 5.12 we consider the Laplacian type of operator. In Sections 5.13 and 5.14, we show how the well-known active contour (snake) concept can be used to lead to connected object boundaries. In Section 5.15 we outline the level set approach. Finally, in Section 5.16 we discuss the aims and methodology of the graph cut approach to producing connected object boundaries: it should be noticed that this method essentially dispenses with the usual ideas of edge detection and regional analysis and aims to give an integrated, generalized methodology for segmentation.

5.10 HYSTERESIS THRESHOLDING

The concept of hysteresis thresholding is a general one and can be applied in a range of applications, including both image and signal processing. In fact, the Schmitt trigger is a very widely used electronic circuit for converting a varying voltage into a pulsed (binary) waveform. In the latter case there are two thresholds, and the input has to rise above the upper threshold before the output is allowed to switch on, and has to fall below the lower threshold before the output is allowed to switch off. This gives considerable immunity against noise in the input waveform—far more than where the difference between the upper and lower switching thresholds is zero (the case of zero hysteresis), since then a small amount of noise can cause an undue amount of switching between the upper and lower output levels.

When the concept is applied in image processing it is usually with regard to edge detection, in which case there is an exactly analogous 1-D waveform to be negotiated around the boundary of an object—although, as we shall see, some specifically 2-D complications arise. The basic rule is to threshold the edge at a high level, and then to allow extension of the edge down to a lower level threshold, but only adjacent to points that have already been assigned edge status.

Figure 5.4 shows the results of making tests on the edge gradient image in Fig. 7.4(b). Figure 5.4(a) and (b) shows the result of thresholding at the upper and lower hysteresis levels respectively and Fig. 5.4(c) shows the result of hysteresis thresholding using these two levels. For comparison, Fig. 5.4(d) shows the effect of thresholding at a suitably chosen intermediate level. Note that isolated edge points within the object boundaries are ignored by hysteresis thresholding, although noise spurs can occur and are retained. We can envision the process of hysteresis thresholding in an edge image as the location of points that:

1. form a superset of the upper threshold edge image.
2. form a subset of the lower threshold edge image.
3. form that subset of the lower threshold image that is connected to points in the upper threshold image via the usual rules of connectedness (Chapter 9).

Clearly, edge points survive only if they are seeded by points in the upper threshold image.

Although the result in Fig. 5.4(c) is better than in Fig. 5.4(d), in that gaps in the boundaries are eliminated or reduced in length, in a few cases noise spurs are introduced. Nevertheless, the aim of hysteresis thresholding is to obtain a better balance between false positives and false negatives by exploiting connectedness in the object boundaries. Indeed, if managed correctly, the additional parameter will normally lead to a net (average) reduction in boundary pixel classification error. However, there are few simple guidelines for selection of hysteresis thresholds, apart from the following:

1. Use a pair of hysteresis thresholds that provides immunity against the known range of noise levels.
2. Choose the lower threshold to limit the possible extent of noise spurs (in principle, the lowest threshold subset that contains *all* true boundary points).
3. Select the upper threshold to guarantee as far as possible the seeding of important boundary points (in principle, the highest threshold subset that is connected to *all* true boundary points).

Unfortunately, in the limit of high signal variability, rules 2 and 3 appear to suggest eliminating hysteresis altogether! Ultimately, this means that the only rigorous way of treating the problem is to perform a complete statistical analysis of false positives and false negatives for a large number of images in any new application.

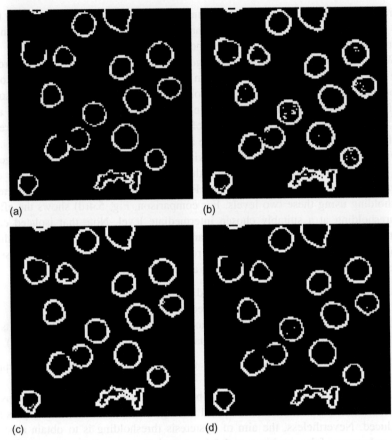

  (a)   (b)

  (c)   (d)

FIGURE 5.4

Effectiveness of hysteresis thresholding. This figure shows tests made on the edge gradient image of Fig. 7.4(b). (a) Effect of thresholding at the upper hysteresis level. (b) Effect of thresholding at the lower hysteresis level. (c) Effect of hysteresis thresholding. (d) Effect of thresholding at an intermediate level.

5.11 THE CANNY OPERATOR

Since its publication in 1986 the Canny operator (Canny, 1986) has become one of the most widely used edge detection operators—and for good reason, as it seeks to get away from a tradition of mask-based operators, many of which can hardly be regarded as "designed", into one that is entirely principled and fully integrated. Intrinsic to the method is that of carefully specifying the spatial bandwidth within which it is expected to work, and also the exclusion of unnecessary thresholds, while permitting thin line structures to emerge and ensuring that they are connected together as far as possible and indeed are meaningful at the particular scale and bandwidth. As a result of these considerations, the method involves a number of stages of processing:

1. Low-pass spatial frequency filtering
2. Application of first-order differential masks
3. Nonmaximum suppression involving subpixel interpolation of pixel intensities
4. Hysteresis thresholding

In principle, low-pass filtering needs to be carried out by Gaussian convolution operators for which the standard deviation (or spatial bandwidth) σ is known and prespecified. Then first-order differential masks need to be applied: for this purpose, the Sobel operator is acceptable. In this context note that the Sobel operator masks can be regarded as convolutions ($\otimes$) of a basic $\begin{bmatrix} -1 & 1 \end{bmatrix}$ type of mask with a $\begin{bmatrix} 1 & 1 \end{bmatrix}$ smoothing mask. Thus, taking the Sobel x-derivative we have:

$$\begin{bmatrix} -1 & 0 & 1 \\ -2 & 0 & 2 \\ -1 & 0 & 1 \end{bmatrix} = \begin{bmatrix} 1 \\ 2 \\ 1 \end{bmatrix} \begin{bmatrix} -1 & 0 & 1 \end{bmatrix} \tag{5.19}$$

where

$$\begin{bmatrix} 1 & 2 & 1 \end{bmatrix} = \begin{bmatrix} 1 & 1 \end{bmatrix} \otimes \begin{bmatrix} 1 & 1 \end{bmatrix} \tag{5.20}$$

and

$$\begin{bmatrix} -1 & 0 & 1 \end{bmatrix} = \begin{bmatrix} -1 & 1 \end{bmatrix} \otimes \begin{bmatrix} 1 & 1 \end{bmatrix} \tag{5.21}$$

These equations make it clear that the Sobel operator itself includes a considerable amount of low-pass filtering, so the amount of additional filtering needed in stage 1 can reasonably be reduced. Another thing to bear in mind is that low-pass filtering can itself be carried out by a smoothing mask of the type shown in Fig. 5.5(b), and it is interesting how close this mask is to the full 2-D Gaussian shown in Fig. 5.5(a). Note also that the bandwidth of the mask in Fig. 5.5(b) is exactly known (it is 0.707), and when combined with that of the Sobel the overall bandwidth becomes almost exactly 1.0.

Next, we turn our attention to stage 3—that of nonmaximum suppression. For this purpose we need to determine the local edge normal direction using Eq. (5.5), and move either way along the normal to determine whether the current location is or is not a local maximum along it. If it is not, we suppress the edge output at the current location, only retaining edge points that are proven local maxima along the edge normal. Since only one point along this direction should be a local maximum, this procedure will necessarily thin the grayscale edges to unit width. Here a slight problem arises in that the edge normal direction will in general not pass through the centers of the adjacent pixels, and the Canny method requires the intensities along the normal to be estimated by interpolation. In a

0.000	0.000	0.004	0.008	0.004	0.000	0.000
0.000	0.016	0.125	0.250	0.125	0.016	0.000
0.004	0.125	1.000	2.000	1.000	0.125	0.004
0.008	0.250	2.000	4.000	2.000	0.250	0.008
0.004	0.125	1.000	2.000	1.000	0.125	0.004
0.000	0.016	0.125	0.250	0.125	0.016	0.000
0.000	0.000	0.004	0.008	0.004	0.000	0.000

1	2	1
2	4	2
1	2	1

(a)   (b)

FIGURE 5.5

Exactness of the well-known 3×3 smoothing kernel. This figure shows the Gaussian-based smoothing kernel (a) that is closest to the well-known 3×3 smoothing kernel (b) over the central (3×3) region. For clarity, neither is normalized by the factor 1/16. The larger Gaussian envelope drops to 0.000 outside the region shown and integrates to 18.128 rather than 16. Hence, the kernel in (b) can be said to approximate a Gaussian within ~13%. Its actual standard deviation is 0.707 compared with 0.849 for the Gaussian.

3×3 neighborhood this is simply achieved, as the edge normal in any octant will have to lie within a given pair of pixels, as shown in Fig. 5.6(a). In a larger neighborhood, interpolation can take place between several pairs of pixels. For example, in a 5×5 neighborhood, it will have to be determined which one of the two pairs is relevant (Fig. 5.6(b)), and an appropriate interpolation formula applied. However, it could be construed that there is no need to use larger neighborhoods, as a 3×3 neighborhood will contain all the relevant information, and given enough presmoothing in stage 1, negligible loss of accuracy will result. Of course, if impulse noise is present, this could lead to serious error, but low-pass filtering is in any case not guaranteed to eliminate impulse noise, so no special loss results from using the smaller neighborhood for nonmaximum suppression. Such considerations need to be examined carefully in the light of the particular image data and the noise it contains. Figure 5.6 shows the two distances l_1 and l_2 that have to be determined. The pixel intensity along the edge normal is given by weighting the corresponding pixel intensities in *inverse* proportion to the distances:

$$I = \frac{l_2 I_1 + l_1 I_2}{l_1 + l_2} = (1 - l_1)I_1 + l_1 I_2 \tag{5.22}$$

where

$$l_1 = \tan \theta \tag{5.23}$$

This brings us to the final stage, that of hysteresis thresholding. By this point as much as possible has been achieved without applying thresholds, and it becomes necessary to take this final step. However, by applying the two hysteresis thresholds, it is intended to limit the damage that can be caused by a single threshold and repair it with another: that is to say, select the upper threshold to ensure capturing edges that are reliable and then select other points that have high likelihood of being viable edge points because they are adjacent to edge points of

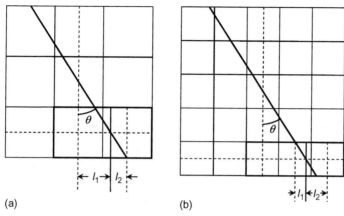

(a)   (b)

FIGURE 5.6

Pixel interpolation in the Canny operator. (a) Interpolation between the two highlighted pixels at the bottom right in a 3×3 neighborhood. (b) Interpolation in a 5×5 neighborhood: note that two possibilities exist for interpolating between pairs of adjacent pixels, the relevant distances being marked for the one on the right.

known reliability. In fact, this is still somewhat *ad hoc*, but in practice it gives quite good results. A simple rule for choice of the lower threshold is that it should be about half the upper threshold. Again, this is only a rule of thumb, and it has to be examined carefully in the light of the particular image data.

Figures 5.7 and 5.8 show results for the Canny operator at various stages. They also show comparisons for various thresholds. In particular, both figures show the effects of (e) hysteresis thresholding, (f) single thresholding at the lower level, and (g) single thresholding at the upper level. The evidence is that hysteresis thresholding is usually more reliable and more coherent than single-level thresholding, in the sense of giving fewer false or misleading results.

5.12 THE LAPLACIAN OPERATOR

An edge detector such as the Sobel is a first derivative operator, whereas the Laplacian is a second derivative operator, and as such it is sensitive only to changes in intensity gradient. In 2-D its standard (mathematical) definition is given by:

$$\nabla^2 = \frac{\partial^2}{\partial x^2} + \frac{\partial^2}{\partial y^2} \tag{5.24}$$

Localized masks for computing Laplacian output can be derived by taking difference of Gaussian (DoG) kernels using two Gaussians of different bandwidths; for details of this procedure, see Section 6.7.3. This gives them an isotropic 2-D

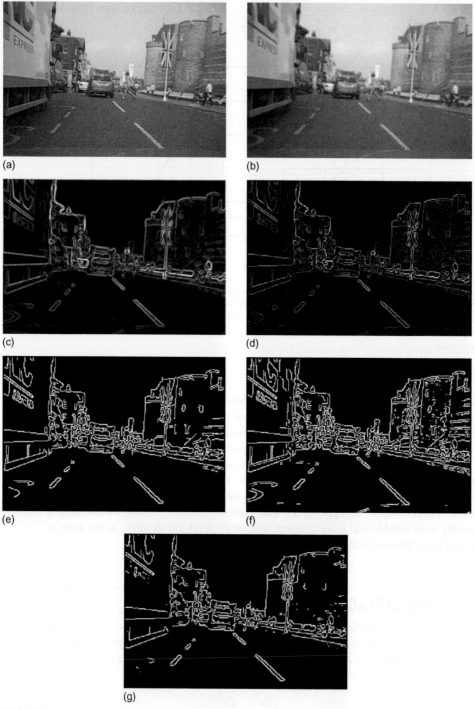

(a)

(b)

(c)

(d)

(e)

(f)

(g)

FIGURE 5.7

Application of the Canny edge detector. (a) Original image. (b) Smoothed image. (c) Result of applying Sobel operator. (d) Result of nonmaximum suppression. (e) Result of hysteresis thresholding. (f) Result of thresholding only at the lower threshold level. (g) Result of thresholding at the upper threshold level. Note that there are fewer false or misleading outputs in (e) than would result from using a single threshold.

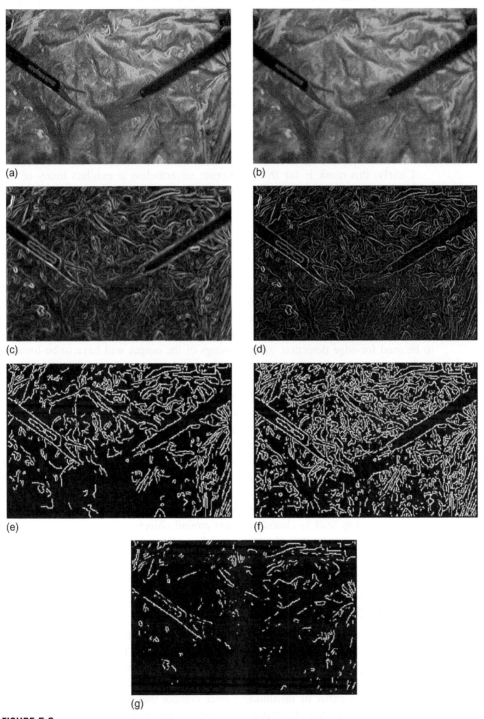

FIGURE 5.8

Another application of the Canny edge detector. (a) Original image. (b) Smoothed image. (c) Result of applying Sobel operator. (d) Result of nonmaximum suppression. (e) Result of hysteresis thresholding. (f) Result of thresholding only at the lower threshold level. (g) Result of thresholding at the upper threshold level. Again there are fewer false or misleading outputs in (e) than would result from using a single threshold.

profile, with a positive center and a negative surround. This shape can be approximated in 3×3 windows by masks such as the following:

$$\begin{bmatrix} -1 & -1 & -1 \\ -1 & 8 & -1 \\ -1 & -1 & -1 \end{bmatrix} \qquad (5.25)$$

Clearly, this mask is far from isotropic: nevertheless it exhibits many of the properties of larger masks, such as DoG kernels, that are much more accurately isotropic.

Here we present only an outline of the properties of this type of operator. These can be seen in Fig. 5.9. First, note that the Laplacian output ranges from positive to negative: hence, in Fig. 5.9(c) it is presented on a medium-gray background, which indicates that on the exact edge of an object the Laplacian output is actually zero, as stated earlier. This is made clearer in Fig. 5.9(d), where the magnitude of the Laplacian output is shown. It is seen that edges are highlighted by strong signals just inside and just outside the edge locations that are located by a Sobel or Canny operator (see Fig. 5.9(b)). Ideally this effect is symmetrical, and if the Laplacian is to be used for edge detection, zero crossings of the output will have to be located. However, in spite of preliminary smoothing of the image (Fig. 5.9(a)), the background in Fig. 5.9(d) has a great deal of noise in the background, and attempting to find zero crossings will therefore lead to a lot of noise being detected in addition to the edge points: in fact, it is well known that differentiation (especially double differentiation, as here) tends to accentuate noise. Nevertheless, this approach has been used highly successfully, usually with DoG operators working in much larger windows. Indeed, with much larger windows there will be a good number of pixels lying very near the zero crossings, and it will be possible to discriminate much more successfully between them and the pixels merely having low Laplacian output. A particular advantage of using Laplacian zero crossings is that theoretically they are bound to lead to closed contours around objects (albeit noise signals will also have their own separate closed contours).

5.13 ACTIVE CONTOURS

Active contour models (also known as "deformable contours" or "snakes") are widely used for systematically refining object contours. The basic concept is to obtain a complete and accurate outline of an object that may be ill-defined in places, whether through lack of contrast, or noise or fuzzy edges. A starting approximation is made, either by instituting a large contour that may be shrunk to size, or a small contour that may be expanded suitably, until its shape matches that of the object. In principle, the initial boundary can be rather arbitrary, whether mostly outside or within the object in question. Then its shape is made to evolve subject to an energy minimization process: on the one hand it is desired to minimize the *external*

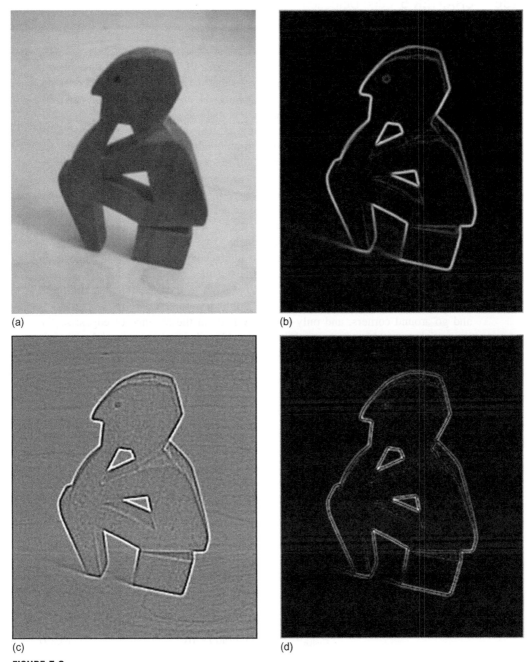

(a)

(b)

(c)

(d)

FIGURE 5.9

Comparison of Sobel and Laplacian outputs. (a) Pre-smoothed version of original image. (b) Result of applying Sobel operator. (c) Result of applying Laplacian operator. Because the Laplacian output can be positive or negative, the output in (c) is displayed relative to a medium (128)-gray-level background. (d) Absolute magnitude Laplacian output. For clarity, (c) and (d) have been presented at increased contrast. Note that the Laplacian output in (d) gives double edges—one just inside and one just outside the edge position indicated by a Sobel or Canny operator. (To find edges using a Laplacian, zero crossings have to be located.) Both the Sobel and the Laplacian used here operate within a 3×3 window.

energy corresponding to imperfections in the degree of fit, and on the other hand it is desired to minimize the *internal* energy, so that the shape of the snake does not become unnecessarily intricate, e.g., by taking on any of the characteristics of image noise. There are also model constraints that are represented in the formulation as contributions to the external energy: typical of such constraints is that of preventing the snake from moving into prohibited regions, such as beyond the image boundary, or, for a moving vehicle, off the region of the road.

The snake's internal energy includes elastic energy, which might be needed to extend or compress it, and bending energy. If no bending energy terms were included, sharp corners and spikes in the snake would be free to occur with no restriction. Similarly, if no elastic energy terms were included, the snake would be permitted to grow or shrink without penalty.

The image data are normally taken to interact with the snake via three main types of image feature—lines, edges, and terminations (the last can be line terminations or corners). Various weights can be given to these features according to the behavior required of the snake. For example, it might be required to hug edges and go around corners, and only to follow lines in the absence of edges: so the line weights would be made much lower than the edge and corner weights.

These considerations lead to the following breakdown of the snake energy:

$$
\begin{aligned}
E_{\text{snake}} &= E_{\text{internal}} + E_{\text{external}} \\
&= E_{\text{internal}} + E_{\text{image}} + E_{\text{constraints}} \\
&= E_{\text{stretch}} + E_{\text{bend}} + E_{\text{line}} + E_{\text{edge}} + E_{\text{term}} + E_{\text{repel}}
\end{aligned}
\tag{5.26}
$$

The energies are written down in terms of small changes in position $\mathbf{x}(s) = (x(s), y(s))$ of each point on the snake, the parameter s being the arc length distance along the snake boundary. Thus, we have:

$$
E_{\text{stretch}} = \int \kappa(s) \|\mathbf{x}_s(s)\|^2 ds
\tag{5.27}
$$

and

$$
E_{\text{bend}} = \int \lambda(s) \|\mathbf{x}_{ss}(s)\|^2 ds
\tag{5.28}
$$

where the suffices s and ss imply first- and second-order differentiation, respectively. Similarly, E_{edge} is calculated in terms of the intensity gradient magnitude $|\text{grad } I|$, leading to:

$$
E_{\text{edge}} = - \int \mu(s) \|\text{grad } I\|^2 ds
\tag{5.29}
$$

where $\mu(s)$ is the edge weighting factor.

The overall snake energy is obtained by summing the energies for all positions on the snake: a set of simultaneous differential equations is then set up to minimize the total energy. This process is not discussed in detail due to the space limitation. Suffice it to say that the equations cannot be solved analytically, and recourse has to be made to iterative numerical solution, during which the shape of the snake evolves from some high-energy initialization state to the final low-energy equilibrium state, defining the contour of interest in the image.

In the general case, there are several possible complications to be tackled:

1. Several snakes may be required to locate an initially unknown number of relevant image contours.
2. Different types of snake will need different initialization conditions.
3. Snakes will sometimes have to split up as they approach contours that turn out to be fragmented.

There are also procedural problems. The intrinsic snake concept is that of well-behaved differentiability. However, lines, edges, and terminations are usually highly localized, so there is no means by which a snake even a few pixels away could be expected to learn about them and hence to move toward them. In these circumstances the snake would "thrash around" and fail to systematically zone in on a contour representing a global minimum of energy. To overcome this problem, smoothing of the image is required, so that edges can communicate with the snake some distance away, and the smoothing must gradually be reduced as the snake nears its target position. Ultimately, the problem is that the algorithm has no high-level appreciation of the overall situation, but merely reacts to a conglomerate of local pieces of information in the image: this makes segmentation using snakes somewhat risky despite the intuitive attractiveness of the concept.

In spite of these potential problems, a valuable feature of the snake concept is that, if set up correctly, the snake can be rendered insensitive to minor discontinuities in a boundary: it is important as this makes the snake capable of negotiating practical situations such as fuzzy or low contrast edges, or places where small artifacts get in the way (this may happen with resistor leads, for example); this capability is possible because the snake energy is set up *globally*—quite unlike the situation for boundary tracking where error propagation can cause wild deviations from the desired path. The reader is referred to the abundant literature on the subject, not only to clarify the basic theory (Kass and Witkin, 1987; Kass et al., 1988), but also to find how it may be made to work well in real situations.

5.14 PRACTICAL RESULTS OBTAINED USING ACTIVE CONTOURS

In this section we briefly explore a simple implementation of the active contour concept. Arguably, the implementation chosen is among the simplest that will work in practical situations while still adhering to the active contour concept. To make it work without undue complication or high levels of computation, a "greedy" algorithm is used, i.e., one that makes local optimizations (energy minimizations) in the expectation that this will result in global optimization. Naturally, it could lead to solutions that do not correspond to absolute minima of the energy function, although this is by no means a problem that is caused solely by using a greedy algorithm, as almost all forms of iterative energy minimization method can fall into this trap.

The first thing to do when devising such an algorithm is to interpret the theory in practical terms. Thus, we rewrite the snake stretch function (Eq. (5.27)) in the discrete form:

$$E_{\text{stretch}} = \sum_{i=1}^{N} \kappa \left\| \mathbf{x}_i - \mathbf{x}_{i+1} \right\|^2 \tag{5.30}$$

where there are N snake points $\mathbf{x}_i$, $i = 1, \ldots, N$: note that this set must be accessed cyclically. In addition, when using a greedy algorithm and updating the position of the ith snake point, the following local form of Eq. (5.30) has to be used:

$$\varepsilon_{\text{stretch},i} = \kappa \left(\left\| \mathbf{x}_i - \mathbf{x}_{i-1} \right\|^2 + \left\| \mathbf{x}_i - \mathbf{x}_{i+1} \right\|^2 \right) \tag{5.31}$$

Unfortunately, although this function causes the snake to be tightened, it can also result in clustering of snake points. To avoid this, the following alternative form can be useful:

$$\varepsilon_{\text{stretch},i} = \kappa \left[\left(d - \left\| \mathbf{x}_i - \mathbf{x}_{i-1} \right\| \right)^2 + \left(d - \left\| \mathbf{x}_i - \mathbf{x}_{i+1} \right\| \right)^2 \right] \tag{5.32}$$

where d is a fixed number representing the smallest likely value of the mean distance between adjacent pairs of snake points for the given type of target object. In the implementation used in Fig. 5.10, d had the noncritical value of 8 pixels; interestingly, this also resulted in faster convergence toward the final form of the snake, as it was encouraged to move further to minimize the magnitudes of the terms in round brackets.

The contour shown in Fig. 5.10 fills the concavity at the top right, but hardly moves into the concavity at the bottom because of a low contrast shadow edge: note that more or less influence by weak edges can readily be obtained by adjusting the grad^2 coefficient μ in Eq. (5.29). Elsewhere the snake ends up with almost exact adherence to the object boundary. The snake shown in the figure employs $p = 40$ points, and $r = 60$ iterations are needed to bring it to its final position. In each iteration, the greedy optimization for each snake point is over an $n \times n$ pixel region with $n = 11$. Overall, the computation time is controlled by and essentially proportional to the quantity prn^2.

The final contour in Fig. 5.10(d) shows the result of using an increased number of initialization points and joining the final locations to give a connected boundary: some of the remaining deficiencies could be reduced by fitting with splines or other means instead of simply joining the dots.

As indicated earlier, this was a simple implementation—so much so that no attempt was made to take corners and bends into account, although in the case shown in Fig. 5.10 no disadvantages or deviations can be seen, except in Fig. 5.10 (d). Clearly, a suitable redesign involving additional energy terms would have to be included to cope with more complex image data. It is interesting that so much

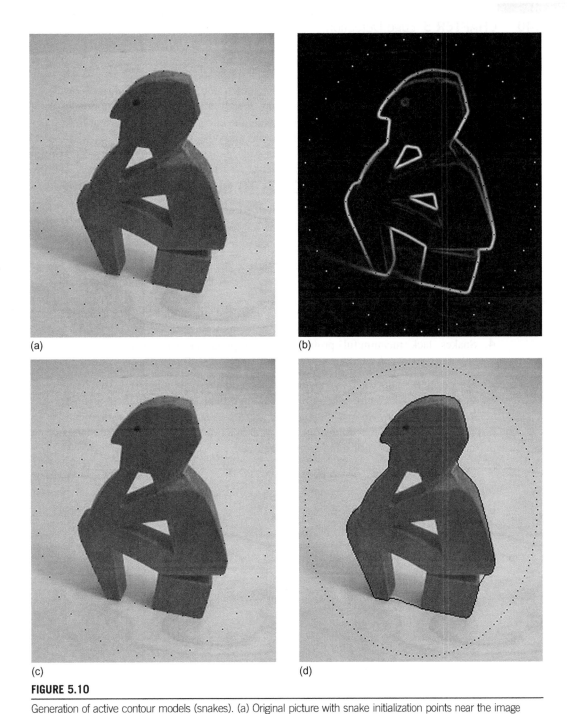

(a)

(b)

(c)

(d)

FIGURE 5.10

Generation of active contour models (snakes). (a) Original picture with snake initialization points near the image boundary; the final snake locations hug the outside of the object but hardly penetrate the large concavity at the bottom: they actually lie approximately along a weak shadow edge. (b) Result of smoothing and application of Sobel operator to (a); the snake algorithm used this image as its input. The snake output is superimposed in black on (b), so that the high degree of co-location with the edge maxima can readily be seen. (c) Intermediate result, after half (30) the total number of iterations (60): this illustrates that after one edge point has been captured, it becomes much easier for other such points to be captured. (d) Result of using an increased number of initialization points and joining the final locations to give a connected boundary: some remanent deficiencies are evident.

can be achieved by just two terms, *viz.* the stretch and edge terms in Eq. (5.26). However, an important factor in getting the greedy algorithm to work optimally one snake point at a time is the need to include the energies for *both* adjacent links (as in Eqs. (5.31) and (5.32)) so as to limit bias and other complications.

5.15 THE LEVEL SET APPROACH TO OBJECT SEGMENTATION

Although the active contour approach described in the previous two sections can be effective in many situations, it nevertheless has several drawbacks (Cremers et al., 2007):

1. There is the possibility of snake self-intersection.
2. Topological changes like splitting or merging of the evolving contour are not allowed.
3. The algorithm is highly dependent on the initialization, and this can result in the snake being biased or getting stuck in local minima.
4. Snakes lack meaningful probabilistic interpretation, so generalizing their action to cover color, texture, or motion is not straightforward.

The level set approach is intended to remedy these deficiencies. The basic approach is to work with whole regions rather than edges, and to evolve an "embedding function" in which contours are represented implicitly rather than directly. In fact, the embedding function is a function $\varphi(\mathbf{x}, t)$ and the contour is defined as the zero level of this function:

$$C(t) = \{\mathbf{x} | \varphi(\mathbf{x}, t) = 0\} \tag{5.33}$$

For a contour that evolves (by gradient descent) along each local normal $\mathbf{n}$ with a speed F, we have:

$$\varphi(C(t), t) = 0 \tag{5.34}$$

which leads to:

$$\frac{\mathrm{d}}{\mathrm{d}t} \varphi(C(t), t) = \nabla\varphi \frac{\partial C}{\partial t} + \frac{\partial \varphi}{\partial t} = F \nabla\varphi \cdot \mathbf{n} + \frac{\partial \varphi}{\partial t} = 0 \tag{5.35}$$

Substituting for $\mathbf{n}$ using:

$$\mathbf{n} = \frac{\nabla\varphi}{|\nabla\varphi|} \tag{5.36}$$

we obtain:

$$\frac{\partial \varphi}{\partial t} = -|\nabla\varphi| F \tag{5.37}$$

Next we need to substitute for F. Following Caselles et al. (1997), we have:

$$\frac{\partial \varphi}{\partial t} = |\nabla \varphi| \, \mathrm{div} \left(g(I) \frac{\nabla \varphi}{|\nabla \varphi|} \right) \tag{5.38}$$

where $g(I)$ is a generalized version of $|\nabla \varphi|$ in the snake potential.

Note that because the contour C is not mentioned explicitly, the updating takes place over all pixels, thereby involving a great many useless calculations: thus, the "narrow band" method was devised to overcome this problem, and involves updating only in a narrow strip around the current contour. However, the need to continually update this strip means that the computational load remains considerable. An alternative approach is the "fast marching" method, which essentially propagates a solution rapidly along an active wavefront, while leaving pixel values frozen behind it. As a result, this method involves maintaining the sign of the speed values F. The Hermes algorithm of Paragios and Deriche (2000) seeks to combine the two approaches. It aims at a *final* solution where all the necessary constraints are fulfilled while maintaining these constraints only loosely at intermediate stages. The overall front propagation algorithm overcomes the four problems mentioned above: in particular, it is able to track nonrigid objects, copes with splitting and merging, and has low computational cost. The paper confirms these claims by showing traffic scenes in which vehicles and pedestrians are successfully tracked.

5.16 THE GRAPH CUT APPROACH TO OBJECT SEGMENTATION

It has been stated several times in both Chapter 4 and this chapter that image segmentation tends to be an unreliable *ad hoc* process if simple uniformity measures are used for locating the extents of regions. For example, region growing techniques are prone to problems such as leaking through weak points in object boundaries, whereas thresholding techniques are sensitive to problems of variable illumination, so again one region will elide into another. Edge-based methods can also become confused at breaks, and edge linking (using hysteresis thresholding or more sophisticated techniques) cannot be relied upon to improve the situation dramatically. The reason for the widespread use of such techniques is their simplicity and speed. However, modern computers are nowadays capable of handling much more powerful segmentation algorithms, and it is key that such algorithms should be robust, effective, and accurate even if they involve increased computation. Hence, there is a trend for segmentation and other vision algorithms to be designed to optimize carefully selected energy functions. We have already seen this with the active contour formalism covered in Section 5.13. The snake algorithm employed in Section 5.14 is "greedy," so it involves many local processes and therefore is not guaranteed to find a single global optimum: however, to a lesser extent this also applies with more rigorously conceived active contour models. Neural network techniques (whether used for segmentation, recognition, or other purposes) normally operate by minimizing energy functions, but again have a tendency to be trapped by local energy minima.

In this context, the graph cut approach to segmentation has become an especially attractive one because its aim is to guarantee exact convergence to the minimum of a global energy function. This is because, in a carefully chosen mathematical milieu, it can be proven that only a single global solution exists. Of course, it is also necessary to relate the particular mathematical milieu to the type of reality arising for practical segmentation tasks.

To achieve this, we first describe an ideal traffic flow problem where every road has an exact, known maximum flow capacity. (It will be most people's experience that any road has a fairly well-defined maximum flow capacity.) Taking a network of roads joining a source s to a sink t, we aim to determine the maximum flow rate between these two terminals. It turns out that there is a theorem (Ford and Fulkerson, 1962) that gives a unique answer to this. We first define a "cut," which is a line passing across the roads between s and t and partitioning the junctions between them into two sets S and T. We then work out the capacity of the cut, this being defined as the sum of the (s to t) capacities of the roads intersected by the cut (Fig. 5.11(a) and (b)). Next, we assess which of all possible cuts has the minimum capacity c: this is called the "minimum cut" C. Interestingly, c is equal to the maximum flow capacity f allowed by the network between s and t. To understand why this is so, suppose first that $f > c$. Then there must be another route between s and t, which is able to carry additional flow. But by definition, any cut has to partition *all* the junctions into two sets S and T, so it has to cross *all possible routes* between s and t. Hence, the premise is violated and the maximum network capacity f cannot be greater than the minimum cut capacity c. Suppose now that $f < c$, then there is no reason why the flow through at least one road cut by C—in particular one that is not saturated—cannot be augmented until the condition no longer holds. (Note that increasing the flow will violate the condition for C first as, by definition, C is the minimum cut.) Hence, by the end of this process, f will be equal to c. Thus, we finally conclude that max flow = min cut. Fig. 5.11(c) and (d) shows how the maximum flow rates can be worked out intuitively by identifying a suitable sequence of paths: the process can be carried out more systematically using the augmenting paths approach outlined below.

Since the max flow—min cut theorem is guaranteed to give a unique global solution, it is worth trying to map it onto the object segmentation task. To achieve this we take all neighbor links (n-links) between pixels (we can envisage these as being nearest neighbor links between pixels, which is highly practical but not essential) and see whether we can identify the minimum cut with the boundary of an object. First, it is clearly necessary to place s inside the object and t outside it, in each case with suitable terminal links (t-links) between the terminal (s or t) and a number of pixels inside and outside the object. It also has to be arranged for the links between pixels to be weighted in such a way as to regulate the flow pattern. Denoting the *i*th pixel intensity by I_i, the flow capacity can be minimized at edges

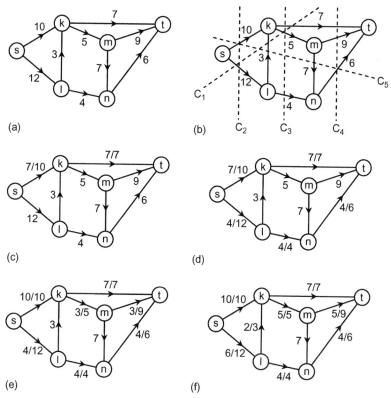

FIGURE 5.11

Achieving maximum flow in a network. (a) A network of roads with a source s, a sink t, and a set of junctions k, l, m, and n. Each road is marked with a number representing its maximum flow capacity. (b) Five cuts C1–C5 are defined, each of which intersects a set of roads and partitions the junctions into two sets S and T. The respective cuts have capacities 24, 22, 16, 22, and 19, making C3 the minimum cut, with capacity $c = 16$. (c) A flow of 7 is initiated via path s–k–t: this is the maximum possible as the capacity of road kt is 7 (the notation "u/v" next to a road means that the actual flow is u and the capacity is v). (d) It is easy to augment the total flow by initiating a maximal flow of 4 in path s–l–n–t. (e) Further thought shows that dual use of road sk is permissible, allowing an additional flow of 3 in path s–k–m–t. (f) A further, longer augmenting path s–l–k–m–t is possible, involving dual use of roads sl, km, and mt. This increases the net flow through the network to 16, as predicted. The actual flow through each of the five cuts shown in (b) is seen to be 16: in case C1 the result is not 18 because we are now considering actual flows rather than capacities (conservation of flow applies to actual flows, but not to capacities). Note that road mn is not useful for optimizing flow.

where $|I_i - I_j|$ is large and maximized where $I_i \approx I_j$. For this purpose a convenient energy function (Boykov and Jolly, 2001) is:

$$E_{ij} = \frac{1}{d_{ij}} \exp\left[\frac{-(I_i - I_j)^2}{2\sigma^2}\right] \tag{5.39}$$

where d_{ij} is the distance between pixels i and j, and is introduced to give greater relevance for pairs of pixels that are closer together. Note that E_{ij} tends to zero for high-intensity gradients, but is low for pairs of pixels with similar intensities. In fact, one of the main requirements is to prevent noise from influencing the locations of object boundaries, so σ^2 can be taken to correspond to the local Gaussian noise energy.

Overall, the idea is to penalize the formation of boundaries in regions of uniformity and to encourage it at locations of high-intensity gradient. Note that the mapping between the max flow–min cut milieu and the optimization of object boundaries is slightly arbitrary, and corresponds to an *analogy*—albeit with a sensible choice of intensity mapping function, a single global solution is still guaranteed to exist (this is mediated more by the algorithm than by the data). In fact, there is another problem as well: that finding a fast algorithm to determine the min cut is nontrivial, so it is possible that approximations might have to be made that are incompatible with identifying an exact global minimum.

The classic Ford and Fulkerson (1962) algorithm worked by starting with a single complete path P from s to t, and determining the n-link (p, q) with the minimum capacity, and then incrementing the flow along path P until link (p, q) was saturated and became the bottleneck for the whole path P. Then "augmenting" paths from s to t were sought in turn, each time incrementing the flow until another bottleneck link became saturated. At each stage the algorithm is most conveniently presented in the form of a *residual network R* consisting of all the unsaturated links in the network (this is essentially a subset of the original network N) and an augmenting path is a path within R. Clearly, one link in the residual network R becomes saturated and is lost to R in each successive iteration: eventually R will become disconnected (no paths from s to t will exist in R) and the total flow from s to t will be a maximum. (Here we ignore the mathematical complication that R will have acquired oppositely orientated (reverse) flows in many of its links: these arise because the residual network is a representation in which the flows in the saturated links are canceled by oppositely directed flows.)

The Ford and Fulkerson (1962) algorithm was long known to be quite slow, and not to be guaranteed to run in polynomial time, except for integer flows. However, two routes for speeding it up came to light later on—first the Dinic (1970) algorithm and second the push-relabel algorithm (Goldberg and Tarjan, 1988). The former uses an augmentation strategy similar to that of the Ford and Fulkerson algorithm, but uses breadth-first search to tackle short paths first. The latter aims to push flows as far as possible from s toward t, taking account of the fact that various links will become saturated. The two approaches give similar flow rates, so here we concentrate on the former.

In fact, the shortest path strategy of the Dinic algorithm is important in permitting it to achieve a worst-case running time complexity of $O(mn^2)$, where n is the number of junctions and m is the number of links in the network. Boykov and Jolly (2001), and others (Boykov and Kolmogorov, 2004; Boykov and Funka-Lea, 2006) developed an even faster algorithm based on the augmented paths approach. Although it has essentially the same worst-case complexity as the Dinic algorithm, it is found to be far faster in practice (at least when applied in typical vision applications). This is because the search trees it uses in each iteration do not have to be developed from scratch but on the contrary capitalize on their previous forms—albeit incorporating novelties such as pruning "orphan" nodes, which turn out to be efficient in practice. This means that graph cut algorithms have now achieved a high degree of efficiency for practical energy minimization tasks. Nevertheless, they have not yet attained absolute supremacy for image segmentation—which may be due partly to the fact that seeds have to be selected to act as terminal nodes, thereby making the approach best suited for interactive use. (Interactive use is not necessarily overly disadvantageous, e.g., when analyzing medical data such as that from MRI brain scans.)

5.17 CONCLUDING REMARKS

The above sections make it clear that the design of edge detection operators has by now been taken to quite an advanced stage, so that edges can be located to subpixel accuracy and orientated to fractions of a degree. In addition, edge maps may be made at several scales and the results correlated to aid image interpretation. Unfortunately, some of the schemes that have been devised to achieve these things (and especially that outlined in the previous section) are fairly complex and tend to consume considerable computation. In many applications this complexity may not be justified because the application requirements are, or can reasonably be made, quite restricted. Furthermore, there is often the need to save computation for real-time implementation. For these reasons, it will often be useful to explore what can be achieved using a single high-resolution detector such as the Sobel operator, which provides a good balance between computational load and orientation accuracy. Indeed, several of the examples in Part 2 of the book have been implemented using this type of operator, which is able to estimate edge orientation to within about 1°. This does not in any way invalidate the latest methods, particularly those involving studies of edges at various scales: such methods come into their own in applications such as general scene analysis, where vision systems are required to cope with largely unconstrained image data.

This chapter has completed another facet of the task of low-level image segmentation. Later chapters move on to consider the shapes of objects that have been found by the thresholding and edge detection schemes discussed in the last two chapters. In particular, Chapter 9 studies shapes by analysis of the regions over which objects extend, whereas Chapter 10 studies shapes by considering their boundary patterns.

Edge detection is perhaps the most widely used means of locating and identifying objects in digital images. Although different edge detection strategies vie with each other for acceptance, this chapter has shown that they obey fundamental laws, such as sensitivity, noise suppression capability, and computation cost all increasing with footprint size.

5.18 BIBLIOGRAPHICAL AND HISTORICAL NOTES

As seen in the first few sections of this chapter, early attempts at edge detection tended to employ numbers of template masks that could locate edges at various orientations. Often these masks were *ad hoc* in nature, and after 1980 this approach finally gave way to the differential gradient approach that had already existed in various forms for a considerable period (see the influential paper by Haralick, 1980).

The Frei–Chen approach is of interest in that it takes a set of nine 3×3 masks forming a complete set within this size of neighborhood—of which one test for brightness, four test for edges, and four test for lines (Frei and Chen, 1977). Although interesting, the Frei–Chen edge masks do not correspond to those devised for optimal edge detection: Lacroix (1988) makes further useful remarks about the approach.

Meanwhile, psychophysical work by Marr (1976), Wilson and Giese (1977), and others provided another line of development for edge detection. This led to the well-known paper by Marr and Hildreth (1980), which was highly influential in the following few years. This spurred others to think of alternative schemes, and the Canny (1986) operator emerged from this vigorous milieu. In fact, the Marr–Hildreth operator was among the first to preprocess images in order to study them at different scales—a technique that has expanded considerably (see, e.g., Yuille and Poggio, 1986) and which will be considered in more depth in Chapter 6. The computational problems of the Marr–Hildreth operator kept others thinking along more traditional lines, and the work by Reeves et al. (1983), Haralick (1984), and Zuniga and Haralick (1987) fell into this category. Lyvers and Mitchell (1988) reviewed many of these papers and made their own suggestions. Another study (Petrou and Kittler, 1988) carried out further work on operator optimization. The work of Sjöberg and Bergholm (1988), which found rules for discerning shadow edges from object edges, is also of interest.

More recently, there was a move to achieving greater robustness and confidence in edge detection by careful elimination of local outliers: in Meer and Georgescu's (2001) method, this was achieved by estimating the gradient vector, suppressing nonmaxima, performing hysteresis thresholding, and integrating with a confidence measure to produce a more general robust result; in fact, each pixel was assigned a confidence value *before* the final two steps of the algorithm. Kim et al. (2004) took this technique a step further and eliminated the need for setting a threshold by using a fuzzy reasoning approach. Similar sentiments were expressed by Yitzhaky and Peli (2003), and they aimed to find an optimal parameter set for edge detectors by ROC and chi-square measures, which actually gave very similar results. Prieto and Allen (2003) designed a similarity metric for edge images, which could be used to test the effectiveness of a variety of edge detectors. They pointed to the fact that metrics need to allow slight latitude in the positions of edges, in order to compare the similarity of edges reliably. They reported a new approach that took into account both displacement of edge positions and edge strengths in determining the similarity between edge images.

Not content with hand-crafted algorithms, Suzuki et al. (2003) devised a back-propagation neural edge enhancer, which undergoes supervised learning on model data to permit it to cope well (in the sense of giving clear, continuous edges) with noisy images: it was found to give results superior to those of conventional algorithms (including Canny, Heuckel, Sobel, and Marr–Hildreth) in similarity tests relative to the desired edges. The disadvantage was a long learning time, although the final execution time was short.

5.18.1 More Recent Developments

Among the most recent developments, Shima et al. (2010) have described the design of more accurate gradient operators on hexagonal lattices. Although the latter are not commonly used, there has long been a special interest in this area because of the greater number of nearest neighbors at equal distances from a given pixel in a hexagonal lattice: this makes certain types of window operation and algorithm more accurate and efficient, and is particularly useful for edge detection and thinning. Ren et al. (2010) have described an improved edge detection algorithm that operates via the fusion of intensity and chromatic difference, thereby making better use of inter-component information in color images.

Cosío et al. (2010) used simplex search in active shape models for improved boundary segmentation: this involves fast numerical optimization to find the most suitable values of nonlinear functions without the need to calculate function derivatives. Their approach typically employs 4 pose parameters and 10 shape parameters for defining a shape such as the prostate. The method significantly increases the range of object poses, and thus results in more accurate boundary segmentation. Chiverton et al. (2008) describe a method that is closely related to the active contour concept: it zones in on objects using parameters relating to foreground similarity and background dissimilarity, and employs a new variational logistic maximum a posteriori (MAP) contextual modeling schema. In this case the (achieved) aim is to permit tracking of moving objects by iterative adaptive matching. Mishra et al. (2011) identify five basic limitations of preexisting active contour methods. Their solution is to decouple the internal and external active contour energies and to update for each of them separately. The method is shown to be faster and to have at least comparable segmentation accuracy to five earlier methods.

Papadakis and Bugeau (2011) underline the power of the graph cut approach (see Section 5.16) by showing that it can be applied to tracking partially occluded objects if predictions to allow for this are included in the formalism. They make the key point that the advantages of the graph cut approach are "its low computational cost and the fact that it converges to the global minimum without getting stuck in local minima."

5.19 PROBLEMS

1. Prove Eqs. (5.12) and (5.13).
2. Check the results quoted in Section 5.9 giving the conditions under which the Sobel operator leads to zero error in the estimation of edge orientation. Proceed to prove Eq. (5.17).

Corner and Interest Point Detection

6

Corner detection is valuable for locating complex objects and for tracking them in 2-D or 3-D. This chapter discusses this detection problem and considers methods that are best suited for the task.

Look out for:

- the ways in which corner features are useful.
- the variety of methods available for corner detection—template matching, the second-order derivative method, the median-based method, the Harris interest point detector.
- where the corner signal is a maximum: how detector bias arises.
- how corner orientation may be estimated.
- why invariant feature detectors are needed: the hierarchy of relevant types of invariance.
- how feature detectors may be made invariant to similarity and affine transformations.
- the need for invariant detectors to embody multiparameter descriptors to help with subsequent matching tasks.
- what criteria can be developed for measuring the performance of conventional and invariant types of feature and feature detector.

Note the variety of methods available for performing interrelated detection tasks. However, different methods have different speeds, accuracies, sensitivities, and degrees of robustness: this chapter aims to bring out all these aspects of the problem.

6.1 INTRODUCTION

This chapter is concerned with the efficient detection of corners. It has been noted in previous chapters that objects are generally located most efficiently from their features. Prominent features include straight lines, circles, arcs, holes, and corners. Corners are particularly important since they may be used to locate and orientate objects and to provide measures of their dimensions; for example, knowledge about orientation will be vital if a robot is to find the best way of picking up an object, while dimensional measurement will be necessary in most inspection applications. Hence efficient, accurate corner detectors are of great relevance in machine vision.

We start this chapter by considering what is perhaps the most obvious detection scheme—that of template matching. Then we move on to other types of detectors, based on the second-order derivatives of the local intensity function; subsequently, we find that median filters can lead to useful corner detectors, with properties similar to those for the second-order derivative based detectors. Next, we consider detectors based on the second moments of the *first* derivatives of the local intensity function. While this will complete the traditional approach to corner detection, it opens the door for consideration of the highly important invariant local feature detectors that have in the past decade or so been developed for matching widely separated views of 3-D scenes, including those containing rapidly moving objects. By the end of the chapter the vital task of considering performance criteria for the various types of corner and feature detector will also be undertaken.

6.2 TEMPLATE MATCHING

Following our experience with template matching methods for edge detection (Chapter 5), it would appear to be straightforward to devise suitable templates for corner detection. These would have the general appearance of corners, and in a 3×3 neighborhood would take forms such as the following:

$$\begin{bmatrix} -4 & 5 & 5 \\ -4 & 5 & 5 \\ -4 & -4 & -4 \end{bmatrix} \quad \begin{bmatrix} 5 & 5 & 5 \\ -4 & 5 & -4 \\ -4 & -4 & -4 \end{bmatrix}$$

The complete set of eight templates would be generated by successive 90° rotations of the first two shown. An alternative set of templates was suggested by Bretschi (1981). As for edge detection templates, the mask coefficients are made to sum to zero, so that corner detection is insensitive to absolute changes in light intensity. Ideally, this set of templates should be able to locate all corners and to estimate their orientation to within 22.5°.

Unfortunately, corners vary very much in a number of their characteristics, including in particular their degree of pointedness,[1] internal angle and the intensity gradient at the boundary. Hence it is quite difficult to design optimal corner detectors. In addition, corners are generally insufficiently pointed for good results to be obtained with the 3×3 template masks shown above. Another problem is that in larger neighborhoods, not only do the masks become larger but also more of them are needed to obtain optimal corner responses, and it rapidly becomes clear that the template matching approach is likely to involve excessive computation for practical corner detection. The alternative is to approach the problem analytically, somehow deducing the ideal response for a corner at any arbitrary orientation, and thereby bypassing the problem of calculating many individual responses to find the one that gives maximum signal. The methods described in the remainder of this chapter embody this alternative philosophy.

6.3 SECOND-ORDER DERIVATIVE SCHEMES

Second-order differential operator approaches have been used widely for corner detection and to mimic the first-order operators used for edge detection. Indeed, the relationship lies deeper than this. By definition, corners in grayscale images occur in regions of rapidly changing intensity levels. By this token they are detected by the same operators that detect edges in images. However, corner pixels are much rarer[2] than edge pixels—by one definition, they arise where two relatively straight-edged fragments intersect. Thus, it is useful to have operators that detect corners *directly*, i.e. *without unnecessarily locating edges*. To achieve this sort of discriminability it is clearly necessary to consider local variations in image intensity up to at least second order. Hence, the local intensity variation is expanded as follows:

$$I(x,y) = I(0,0) + I_x x + I_y y + I_{xx}\frac{x^2}{2} + I_{xy}xy + I_{yy}\frac{y^2}{2} + \cdots \qquad (6.1)$$

where the suffices indicate partial differentiation with respect to x and y and the expansion is performed about the origin X_0 (0,0). The symmetrical matrix of second derivatives is:

$$\mathbf{I}_{(2)} = \begin{bmatrix} I_{xx} & I_{xy} \\ I_{yx} & I_{yy} \end{bmatrix}, \quad \text{where } I_{xy} = I_{yx} \qquad (6.2)$$

---

[1]The term "pointedness" is used as the opposite to "bluntness," the term "sharpness" being reserved for the total angle η through which the boundary turns in the corner region, i.e. π minus the internal angle.

[2]We might imagine a 256×256 image of 64K pixels, of which 1000 ($\sim$2%) lie on edges and a mere 30 ($\sim$0.06%) are situated at corner points.

This gives information on the local curvature at X_0. In fact, a suitable rotation of the coordinate system transforms $\mathbf{I}_{(2)}$ into diagonal form:

$$\tilde{\mathbf{I}}_{(2)} = \begin{bmatrix} I_{\tilde{x}\tilde{x}} & 0 \\ 0 & I_{\tilde{y}\tilde{y}} \end{bmatrix} = \begin{bmatrix} \kappa_1 & 0 \\ 0 & \kappa_2 \end{bmatrix} \tag{6.3}$$

where appropriate derivatives have been reinterpreted as principal curvatures at X_0.

We are particularly interested in rotationally invariant operators and it is significant that the trace and determinant of a matrix such as $\mathbf{I}_{(2)}$ are invariant under rotation. Thus, we obtain the Beaudet (1978) operators:

$$\text{Laplacian} = I_{xx} + I_{yy} = \kappa_1 + \kappa_2 \tag{6.4}$$

and

$$\text{Hessian} = \det(\mathbf{I}_{(2)}) = I_{xx}I_{yy} - I_{xy}^2 = \kappa_1\kappa_2 \tag{6.5}$$

It is well known that the Laplacian operator gives significant responses along lines and edges and hence is not particularly suitable as a corner detector. On the other hand, Beaudet's "DET" operator does not respond to lines and edges but gives significant signals in the vicinity of corners: it should therefore form a useful corner detector. However, DET responds with one sign on one side of a corner and with the opposite sign on the other side of the corner: at the point of real interest—on the point of the corner—it gives a null response. Hence, rather more complicated analysis is required to deduce the presence and exact position of each corner (Dreschler and Nagel, 1981; Nagel, 1983). The problem is clarified in Fig. 6.1. Here the dotted line shows the path of maximum horizontal curvature

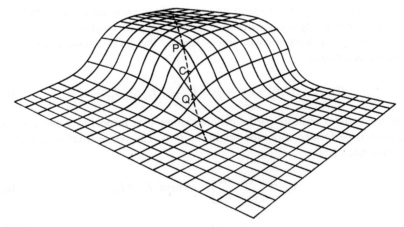

FIGURE 6.1

Sketch of an idealized corner, taken to give a smoothly varying intensity function. The dotted line shows the path of maximum horizontal curvature for various intensity values up the slope. The DET operator gives maximum responses at P and Q, and it is required to find the ideal corner position C where DET gives a null response.

for various intensity values up the slope. The DET operator gives maximum response at positions P and Q on this line, and the parts of the line between P and Q must be explored to find the "ideal" corner point C where DET is zero.

Perhaps to avoid rather complicated procedures of this sort, Kitchen and Rosenfeld (1982) examined a variety of strategies for locating corners, starting from the consideration of local variation in the directions of edges. They found a highly effective operator that estimates the projection of the local rate of change of gradient direction vector along the horizontal edge tangent direction, and showed that it is mathematically identical to calculating the horizontal curvature κ of the intensity function I. To obtain a realistic indication of the strength of a corner they multiplied κ by the magnitude of the local intensity gradient g:

$$C = \kappa g = \kappa (I_x^2 + I_y^2)^{1/2}$$
$$= \frac{I_{xx}I_y^2 - 2I_{xy}I_xI_y + I_{yy}I_x^2}{I_x^2 + I_y^2} \tag{6.6}$$

Finally, they used the heuristic of nonmaximum suppression along the edge normal direction to localize the corner positions further.

In 1983, Nagel was able to show that the Kitchen and Rosenfeld (KR) corner detector using nonmaximum suppression is mathematically virtually identical to the Dreschler and Nagel (DN) corner detector. A year later, Shah and Jain (1984) studied the Zuniga and Haralick (ZH) corner detector (1983) based on a bicubic polynomial model of the intensity function: they showed that this is essentially equivalent to the KR corner detector. However, the ZH corner detector operates rather differently in that it thresholds the intensity gradient and then works with the subset of edge points in the image, only at that stage applying the curvature function as a corner strength criterion. By making edge detection explicit in the operator, the ZH detector eliminates a number of false corners that would otherwise be induced by noise.

The inherent near-equivalence of these three corner detectors need not be overly surprising, since in the end the different methods would be expected to reflect the same underlying physical phenomena (Davies, 1988d). However, it is gratifying that the ultimate result of these rather mathematical formulations is interpretable by something as easy to visualize as horizontal curvature multiplied by intensity gradient.

6.4 A MEDIAN FILTER-BASED CORNER DETECTOR

An entirely different strategy for detecting corners was developed by Paler et al. (1984). It adopts an initially surprising and rather nonmathematical approach based on the properties of the median filter. The technique involves applying a median filter to the input image, and then forming another image that is the difference between the input and the filtered images. This difference image contains a set of signals that are interpreted as local measures of corner strength.

Clearly, it seems risky to apply such a technique since its origin suggests that, far from giving a correct indication of corners, it may instead unearth all the noise in the original image and present this as a set of "corner" signals. Fortunately, analysis shows that these worries may not be too serious. First, in the absence of noise, strong signals are not expected in areas of background; nor are they expected near straight edges, since median filters do not shift or modify such edges significantly (see Chapter 3). However, if a window is moved gradually from a background region until its central pixel is just over a convex object corner, there is no change in the output of the median filter: hence, there is a strong difference signal indicating a corner.

Paler et al. (1984) analyzed the operator in some depth and concluded that the signal strength obtained from it is proportional to (a) the local contrast, and (b) the "sharpness" of the corner. The definition of sharpness they used was that of Wang et al. (1983), meaning the angle η through which the boundary turns. Since it is assumed here that the boundary turns through a significant angle (perhaps the whole angle η) within the filter neighborhood, the difference from the second-order intensity variation approach is a major one. Indeed, it is an implicit assumption in the latter approach that first- and second-order coefficients describe the local intensity characteristics reasonably rigorously, the intensity function being inherently continuous and differentiable. Thus, the second-order methods may give unpredictable results with pointed corners where directions change within the range of a few pixels. Although there is some truth in this, it is worth looking at the similarities between the two approaches to corner detection before considering the differences. We proceed with this in the next subsection.

6.4.1 Analyzing the Operation of the Median Detector

This subsection considers the performance of the median corner detector under conditions where the grayscale intensity varies by only a small amount within the median filter neighborhood region. This permits the performance of the corner detector to be related to low-order derivatives of the intensity variation, so that comparisons can be made with the second-order corner detectors mentioned earlier.

To proceed we assume a continuous analog image and a median filter operating in an idealized circular neighborhood. For simplicity, since we are attempting to relate signal strengths and differential coefficients, noise is ignored. Next, recall (Chapter 3) that for an intensity function that increases monotonically with distance in some arbitrary direction $\tilde{x}$ but which does not vary in the perpendicular direction $\tilde{y}$, the median within the circular window is equal to the value at the center of the neighborhood. This means that the median corner detector gives zero signal if the horizontal curvature is locally zero.

If there is a small horizontal curvature κ, the situation can be modeled by envisaging a set of constant-intensity contours of roughly circular shape and approximately equal curvature, within a circular window of radius a (Fig. 6.2). Consider the contour having the median intensity value. The center of this

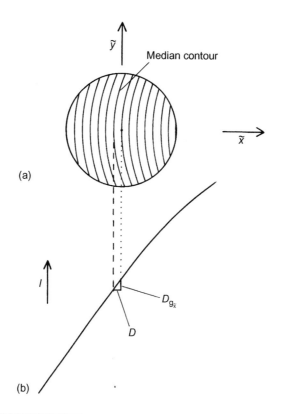

FIGURE 6.2

(a) Contours of constant intensity within a small neighborhood: ideally, these are parallel, circular and of approximately equal curvature (the contour of median intensity does not pass through the center of the neighborhood); (b) cross-section of intensity variation, indicating how the displacement D of the median contour leads to an estimate of corner strength.

Source: © Springer 1988

contour does not pass through the center of the window but is displaced to one side along the negative $\tilde{x}$-axis. Furthermore, the signal obtained from the corner detector depends on this displacement. If the displacement is D, it is easy to see that the corner signal is $Dg_{\tilde{x}}$ because $g_{\tilde{x}}$ allows the intensity change over the distance D to be estimated (Fig. 6.2). The remaining problem is to relate D to the horizontal curvature κ. A formula giving this relation has already been obtained in Chapter 3. The required result is:

$$D = \frac{1}{6}\kappa a^2 \qquad (6.7)$$

so the corner signal is

$$C = Dg_{\tilde{x}} = \frac{1}{6}\kappa g_{\tilde{x}}a^2 \qquad (6.8)$$

Note that C has the dimensions of intensity (contrast), and that the equation may be re-expressed in the form:

$$C = \frac{1}{12}(g_{\tilde{x}}a)\cdot(2a\kappa) \qquad (6.9)$$

so that, as in the formulation of Paler et al. (1984), corner strength is closely related to corner contrast and corner sharpness.

To summarize, the signal from the median-based corner detector is proportional to horizontal curvature and to intensity gradient. Thus, this corner detector gives an identical response to the three second-order intensity variation detectors discussed in Section 6.3, the closest practically being the KR detector. However, this comparison is valid only when second-order variations in intensity give a complete description of the situation. Clearly, the situation might be significantly different where corners are so pointed that they turn through a large proportion of their total angle within the median neighborhood. In addition, the effects of noise might be expected to be rather different in the two cases, as the median filter is particularly good at suppressing impulse noise. Meanwhile, for small horizontal curvatures, there ought to be no difference in the positions at which median and second-order derivative methods locate corners, and accuracy of localization should be identical in the two cases.

6.4.2 Practical Results

Experimental tests with the median approach to corner detection have shown that it is a highly effective procedure (Paler et al., 1984; Davies, 1988d). Corners are detected reliably and signal strength is indeed roughly proportional both to local image contrast and to corner sharpness (see Fig. 6.3). Noise is more apparent for 3×3 implementations and this makes it better to use 5×5 or larger neighborhoods to give good corner discrimination. However, the fact that median operations are slow in large neighborhoods, and that background noise is still evident even in 5×5 neighborhoods, means that the basic median-based approach gives poor performance by comparison with the second-order methods. However, both of these disadvantages are virtually eliminated by using a "skimming" procedure, in which edge points are first located by thresholding the edge gradient, and the edge points are then examined with the median detector to locate the corner points (Davies, 1988d). With this improved method, performance is found to be generally superior to that for the KR method in that corner signals are better localized and accuracy is enhanced. Indeed, the second-order methods appear to give rather fuzzy and blurred signals that contrast with the sharp signals obtained with the improved median approach (Fig. 6.4).

At this stage the reason for the more blurred corner signals obtained using the second-order operators is not clear. Basically, there is no valid rationale for

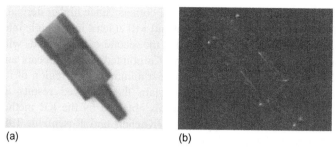

(a)   (b)

FIGURE 6.3

(a) Original off-camera 128×128 6-bit grayscale image; (b) result of applying the median-based corner detector in a 5×5 neighborhood. Note that corner signal strength is roughly proportional both to corner contrast and to corner sharpness.

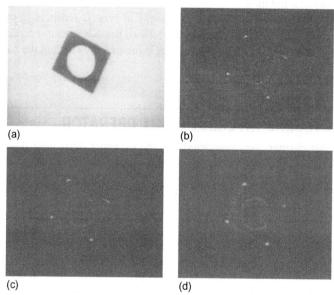

(a)   (b)

(c)   (d)

FIGURE 6.4

Comparison of the median and KR corner detectors: (a) original 128×128 grayscale image; (b) result of applying a median detector; (c) result of including a suitable gradient threshold; (d) result of applying a KR detector. The considerable amount of background noise is saturated out in (a) but is evident from (b). To give a fair comparison between the median and KR detectors, 5×5 neighborhoods are employed in each case, and nonmaximum suppression operations are not applied: the same gradient threshold is used in (c) and (d).

Source: © Springer 1988

applying second-order operators to pointed corners, since higher derivatives of the intensity function will become important and will at least in principle interfere with their operation. However, it is evident that the second-order methods will probably give strong corner signals when the tip of a pointed corner appears anywhere in their neighborhood, so there is likely to be a minimum blur region of radius a for any corner signal. This appears to explain the observed results adequately. However, note that the sharpness of signals obtained by the KR method may be improved by nonmaximum suppression (Kitchen and Rosenfeld, 1982; Nagel, 1983). Furthermore, this technique can also be applied to the output of median-based corner detectors: hence, the fact remains that the median-based method gives inherently better localized signals than the second-order methods.

Overall, the inherent deficiencies of the median-based corner detector can be overcome by incorporating a skimming procedure, and then the method becomes superior to the second-order approaches in giving better localization of corner signals. The underlying reason for the difference in localization properties appears to be that the median-based signal is ultimately sensitive only to the particular few pixels whose intensities fall near the median contour within the window, whereas the second-order operators use typical convolution masks that are in general sensitive to the intensity values of all the pixels within the window. Thus, the KR operator tends to give a strong signal when the tip of a pointed corner is present anywhere in the window.

6.5 THE HARRIS INTEREST POINT OPERATOR

Earlier in this chapter, we considered the second-order derivative type of corner detector that was designed on the basis that corners are ideal, smoothly varying differentiable intensity profiles. We also examined median filter-based detectors: these had a totally different modus operandi and were found to be suitable for processing curved step edges whose profiles were quite likely *not* to be smoothly varying and differentiable. At this point we consider what other strategies are available for corner detection. An important one that has become extremely widely used is the Harris operator. Far from being a second-order derivative type of detector, the Harris operator only takes account of *first*-order derivatives of the intensity function. Thus, there is a question of how it can acquire enough information to detect corners. In this section, we construct a model of its operation in order to throw light on this crucial question.

The Harris operator is defined very simply, in terms of the local components of intensity gradient I_x, I_y in an image. The definition requires a window region to be defined and averages $\langle \cdot \rangle$ are taken over this whole window. We start by computing the following matrix:

$$\Delta = \begin{bmatrix} \langle I_x^2 \rangle & \langle I_x I_y \rangle \\ \langle I_x I_y \rangle & \langle I_y^2 \rangle \end{bmatrix} \qquad (6.10)$$

where the suffixes indicate partial differentiation of the intensity I; we then use the determinant and trace to estimate the corner signal:

$$C = \frac{\det\Delta}{\text{trace}\Delta} \tag{6.11}$$

While this definition involves averages, we shall find it more convenient to work with sums of quadratic products of intensity gradients:

$$\Delta = \begin{bmatrix} \sum I_x^2 & \sum I_x I_y \\ \sum I_x I_y & \sum I_y^2 \end{bmatrix} \tag{6.12}$$

To understand the operation of the detector, first consider its response for a single edge (Fig. 6.5(a)). In fact:

$$\det\Delta = 0 \tag{6.13}$$

because I_x is zero over the whole window region. Note that there is no loss in generality from selecting a horizontal edge, as $\det\Delta$ and $\text{trace}\Delta$ are invariant under rotation of axes.

Next consider the situation in a corner region (Fig. 6.5(b)). Here:

$$\Delta = \begin{bmatrix} l_2 g^2 \sin^2\theta & l_2 g^2 \sin\theta\cos\theta \\ l_2 g^2 \sin\theta\cos\theta & l_2 g^2 \cos^2\theta + l_1 g^2 \end{bmatrix} \tag{6.14}$$

where l_1, l_2 are the lengths of the two edges bounding the corner, and g is the edge contrast, assumed constant over the whole window. We now find:

$$\det\Delta = l_1 l_2 g^4 \sin^2\theta \tag{6.15}$$

and

$$\text{trace}\Delta = (l_1 + l_2)g^2 \tag{6.16}$$

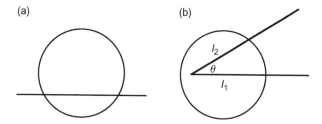

FIGURE 6.5

Case of a straight edge and a general corner. (a) A single straight edge appearing in a circular window. (b) A general corner appearing in a circular window. Circular windows are taken as ideal, in that they will not favor any direction over any other.

Source: © *IET 2005*

so

$$C = \frac{l_1 l_2}{l_1 + l_2} g^2 \sin^2 \theta \qquad (6.17)$$

which may be interpreted as the product of (1) a strength factor λ, which depends on the edge lengths within the window, (2) a contrast factor g^2, and (3) a shape factor $\sin^2 \theta$, which depends on the edge "sharpness" θ. Clearly, C is zero for $\theta = 0$ and $\theta = \pi$, and is a maximum for $\theta = \pi/2$, all these results being intuitively correct and appropriate.

There is a useful theorem about the sets of lengths l_1, l_2 for which the strength factor λ, and thus C, is a maximum. Suppose we set $L = l_1 + l_2 =$ constant. Then $l_1 = L - l_2$, and substituting for l_1 we find:

$$\lambda = \frac{l_1 l_2}{l_1 + l_2} = \left[\frac{L l_2 - l_2^{\,2}}{L} \right] \qquad (6.18)$$

$$\therefore \quad \frac{d\lambda}{dl_2} = \frac{1 - 2l_2}{L} \qquad (6.19)$$

which is zero for $l_2 = L/2$, at which point $l_1 = l_2$. This means that the best way of obtaining maximum corner signal is to place the corner symmetrically within the window, following which the signal can be increased further by moving the corner so that L is maximized (Fig. 6.6).

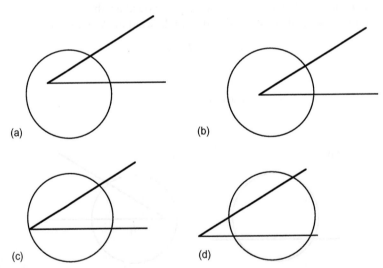

(a)   (b)   (c)   (d)

FIGURE 6.6

Possible geometries for a sharp corner being sampled by a circular window. (a) General case. (b) Symmetrical placement with $l_1 = l_2$ (see notation in Fig. 6.5(b)). (c) Case of maximum signal. (d) Case where the signal is reduced in size as the tip of the corner goes outside the window.

Next, if l_i is small (for either value of i), the corner signal at first increases linearly with l_i, and as noted earlier, the corner detector will ignore a single straight edge on its own.

Finally, the fact that we are exploring the properties of a symmetric matrix, which can be represented using any convenient set of orthogonal axes, means that we can find the eigenvalues and eigenvectors. However, it is illuminating to note that these arise automatically when a symmetrically aligned set of axes is selected along the corner bisectors, as then the off-diagonal elements of the modified Δ matrix acquire two components $(L/2)g^2 \sin(\theta/2) \cos(\theta/2)$ of opposite sign and therefore cancel out. The on-diagonal elements are thus the eigenvalues themselves, and are $(L/2)g^2 \times 2 \cos^2(\theta/2)$, $(L/2)g^2 \times 2 \sin^2(\theta/2)$. Again, if $\theta = 0$ or π, one or the other eigenvalue is zero, so the determinant is zero and the corner signal vanishes; also, the maximum signal occurs for $\theta = \pi/2$.

6.5.1 Corner Signals and Shifts for Various Geometric Configurations

In this section, we seek the conditions for maximum corner signal for corners of different degrees of sharpness. We shall make use of the observation made in the previous section that maximum signal requires that $l_1 = l_2 = L/2$.

First, we take the case when $\theta = 0$: we have already seen that this leads to $C = 0$.

Next, when θ is small, i.e. less than $\pi/2$, we can go on increasing L by moving the corner symmetrically. The optimum is reached exactly as the tip of the corner reaches the far side of the window (Fig. 6.6). We could envisage the corner moving even further, but then the portions of the sides that lie within the window will be moved laterally, so they will become shorter, and the signal will fall (Fig. 6.6(d)).

Now take the case $\theta = \pi/2$. Then we can again proceed as above, and the optimum will still occur when the tip of the corner lies on the far side of the window (Fig. 6.7(a)). However, further increase of θ will result in a different optimum condition (Fig. 6.7(b–d)). In that case the optimum occurs for a reduced shift of the tip of the corner, and occurs when the visible ends of the edges are exactly at opposite ends of a window diameter (Fig. 6.7(d)). Formally, we can see this in the symmetrical case ($l_1 = l_2$) from the following equation:

$$\lambda_{\text{sym}} = \frac{L^2/4}{L} = \frac{L}{4} \tag{6.20}$$

so reduction of L will reduce λ_{sym} and C will fall. This situation continues until $\theta = \pi$, at which point C again falls to zero.

We can now calculate the corner shift produced by the Harris detector. Specifically, the detector places the maximum output signal at the center of the window in the cases where the signal is stated to be "optimum" above. The shift produced has a size equal to radius a of the window for small corner angles, as

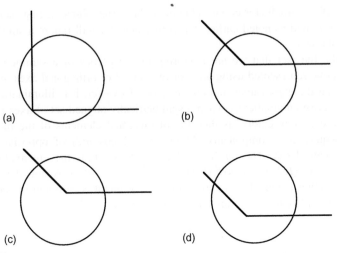

FIGURE 6.7

Possible geometries for right-angle and obtuse corners. (a) Optimum case for a right-angle corner. This is the right-angle case corresponding to that shown in Fig. 6.6(c). (b) General case for an obtuse corner. (c) Symmetrical placement with $l_1 = l_2$. (d) Case of maximum signal. In (d) the edges bounding the corner cross the boundary of the circular window at opposite ends of a diameter.

Source: © IET 2005

then the tip of the corner is symmetrically placed on the boundary of the window. When θ rises above $\pi/2$, simple geometry (Fig. 6.8(a)) shows that the shift is given by:

$$\delta = a \cot \left(\frac{\theta}{2} \right) \tag{6.21}$$

Hence δ starts with value a at $\theta = \pi/2$, and falls to zero as $\theta \to \pi$ (see Fig. 6.8(b)).

6.5.2 Performance with Crossing Points and Junctions

In this section we consider the performance of the Harris operator on other types of feature, which are not normally classed as simple corners. Examples are shown in Figs. 6.9 and 6.10. It turns out that the Harris operator picks these out with much the same efficiency as for corners. We start by considering crossing points.

One of the most important points to note is that many of the same equations apply as for corners, and in particular Eq. (6.17) still applies. However, l_1, l_2 must now be taken as the sum of the edge lengths in each of the two main directions. Here, there is an important point to note that along the two edge directions the signs of the contrast values both reverse at the crossing point. Nevertheless, this does not alter the response, because in Eq. (6.17) the contrast g is squared. So,

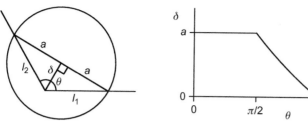

FIGURE 6.8

Geometry for calculating obtuse corner shifts and actual results. (a) Detailed geometry for calculating corner shift for the case shown in Fig. 6.7(d). (b) Graph showing corner shift δ as a function of corner sharpness θ. The left of the graph corresponds to the constant shift of a obtained for sharp corners, while the right shows the varying results for obtuse angles.

Source: © *IET 2005*

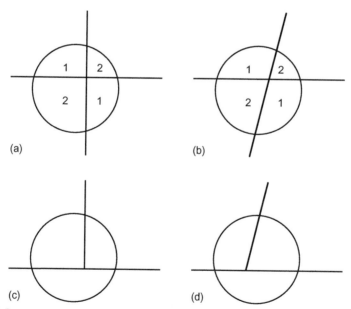

FIGURE 6.9

Other types of interest point. The types of interest point shown in this figure are those that cannot be classed as simple corner points. (a) Crossing point. (b) Oblique crossing point. (c) T-junction. (d) Oblique T-junction. In (a) and (b) the numbers indicate regions of equal, or different, intensity.

Source: © *IET 2005*

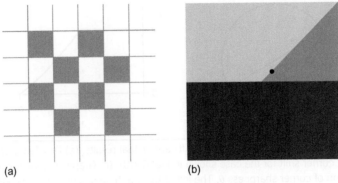

FIGURE 6.10

Effect of the Harris corner detector. (a) A checker-board pattern that gives high responses at each of the edge crossover points. Because of symmetry, the location of the peaks is exactly at the crossover locations. (b) Example of a T-junction. The black dot shows a typical peak location: in this case there is no symmetry to dictate that the peak must occur exactly at the T-junction.

when the window is centered at the crossing point, which is the tip of both constituent corners, the values of l_1, l_2 are doubled.

Another relevant factor is that the corner configuration is now symmetric about the crossing point, so by symmetry this must also be the position of maximum signal. In fact, the global maximum signal must occur when there is a maximum length of both edges within the window, and they must therefore be closely aligned along window diameters. These remarks apply both for $\pi/2$ and for oblique crossovers (Fig. 6.9(a and b) and 6.10(a)).

We now consider another case that arises fairly often—the T-junction interest point. This can be either a $\pi/2$ or an oblique junction. Such cases are more general than corners and the crossing point junctions discussed above, in that they are mediated by three regions with three different intensities (Figs. 6.9(c and d) and 6.10(b)). A complete analysis of the situation for all these cases cannot be undertaken here. Instead, we consider the interesting case of a high contrast edge that is reached but not crossed by a low contrast edge. In this case, the additional intensity breaks the symmetry of the junction, so that not only does the corner peak not lie on the junction point, but also there will be a small lateral movement of the peak. However, if the low contrast edge has much lower contrast than the other two, the lateral shift will be minimal. To calculate the corner signal, we first generalize Eq. (6.17) to take into account the fact that one line will have higher contrast than the other:

$$C' = \frac{l_1 l_2 g_1^2 g_2^2}{l_1 g_1^2 + l_2 g_2^2} \sin^2 \theta \qquad (6.22)$$

where l_1 is taken as the straight edge with high contrast g_1 and l_2 is the straight edge with low contrast g_2. Proceeding as before we find that the optimal signal occurs where $l_1|g_1| = l_2|g_2|$. Interestingly, this can mean that the maximum signal occurs on the low contrast edge, in a highly asymmetric way (Fig. 6.10(b)). Part of the motivation of this study was the observation that the Harris operator peaks shown in the literature (e.g. Shen and Wang, 2002) often seem to be localized at such points, though oddly this does not seem to have been remarked upon before 2005 when the author noted and explained the phenomenon (Davies, 2005). While apparently trivial it is actually important, as measurement bias can mislead and/or be the cause of error in subsequent algorithms. However, here the bias is known, systematic and calculable, and can be allowed for when the operator is used in practice.

Note that the Harris operator[3] is often called an "interest" operator, as it detects not only corners but also other interesting points such as crossovers and T-junctions: and we have seen that there is good reason why this happens. Indeed, it is difficult to imagine that a second-order derivative signal would give sizeable signals in these other cases, as the coherence of the second derivative would be largely absent, or even identically zero in the case of a crossover.

6.5.3 Different Forms of the Harris Operator

In this section we consider the different forms the Harris operator can take. The form in Eq. (6.11) is due to Noble (1988) who actually gave the inverse of this expression and included a small positive constant in the denominator to prevent divide-by-zero situations. However, the original Harris operator had the rather different form:

$$C = \det\Delta - k(\text{trace}\Delta)^2 \tag{6.23}$$

where $k \approx 0.04$. Ignoring the constant, we find that the analysis presented above remains virtually unchanged, particularly concerning the optimal signal and the localization bias. The term involving k was added by Harris and Stevens (1988) in order to limit the number of false positives due to prominent edges. In principle, isolated edges should have no such effect, because as shown earlier, they lead to $\det \Delta = 0$. However, noise or clutter can affect this by introducing short extraneous edges that interact with any existing strong edges to constitute pseudo-corners:[4] so k is adjusted empirically to minimize the number of false positives. Searching the literature shows that in practice workers almost invariably give k a value close to 0.04 or 0.05: in fact, Rocket has investigated this, and has found that: (a) making k equal to 0.04 rather than zero drastically cuts down the number

---

[3]Note also that the Harris operator often used to be called the Plessey operator, after the company at which it was originally developed.

[4]In the absence of explanations in the literature, this seems to be the most reasonable interpretation of the situation.

of false positives due to edges; (b) there appears to be an optimum value for k that is actually much closer to 0.05 than to 0.04, but definitely below 0.06, the k response function being a smoothly varying curve (Rocket, 2003). Nevertheless, we must expect the optimum value of k to vary with the image data.

Interestingly, in tests carried out using the Harris operator, the form given in Eq. (6.11) was used without any attempt to introduce a term in k (though divide-by-zero was taken care of), with the results shown in Fig. 6.11. Excessive numbers of false positives due to edges were not evident, though possibly this was so because of the lack of sensitivity to that effect with this particular type of data.

Finally, it should be pointed out that, when making direct theoretical comparisons between the Harris and other operators (such as the second-order derivative and median-based operators), the square roots of the expressions in Eqs. (6.11) and (6.17) will need to be taken to ensure that the result is directly proportional to edge contrast g.

6.6 CORNER ORIENTATION

This chapter has so far considered the problem of corner detection as relating merely to corner location. However, of the possible point features by which objects might be detected, corners differ from holes in that they are not isotropic, and hence are able to provide orientation information. Such information can be used by procedures that collate the information from various features in order to deduce the presence and positions of objects containing them. In Chapter 14, it will be seen that orientation information is valuable in immediately eliminating a large number of possible interpretations of an image, and hence of quickly narrowing down the search problem and saving computation.

Clearly, when corners are not particularly pointed (Fig. 6.12), or are detected within rather small neighborhoods, the accuracy of orientation will be somewhat restricted.[5] However, orientation errors will seldom be worse than 45°, and will generally be less than 20°. Although these accuracies are far worse than those (around 1°) for edge orientation (see Chapter 5), nonetheless they provide valuable constraints on possible interpretations of an image.

Here we consider only simple means of estimating corner orientation. Basically, once a corner has been located accurately, it is a rather trivial matter to estimate its orientation from that of the intensity gradient at that location. This estimate can be made more accurate by finding the mean intensity gradient over a small region surrounding the estimated corner position, i.e. using the components $\langle I_x \rangle$ and $\langle I_y \rangle$.

---

[5]Clearly, accuracy of corner location will also suffer. However, a way of overcoming this problem will be described in Chapter 13, by making use of the generalized Hough transform.

(a)

(b)

(c)

(d)

(e)

(f)

(g)

(h)

FIGURE 6.11

Application of the Harris interest point detector. (a) Original image. (b) Interest point feature strength. (c) Map of interest points showing only those giving greatest response over a distance of 5 pixels: (d) their placement in the original image. (e and f) Corresponding results for interest points giving greatest response

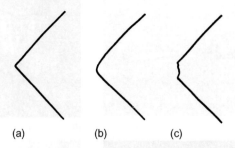

(a)   (b)   (c)

FIGURE 6.12

Types of corner: (a) pointed; (b) rounded; (c) chipped. Corners of type (a) are normal with metal components, those of type (b) are usual with biscuits and other food products, whereas those of type (c) are common with food products but rarer with metal parts.

Source: © IEE 1988

6.7 LOCAL INVARIANT FEATURE DETECTORS AND DESCRIPTORS

The discussion in the foregoing sections covered corner and interest point detectors useful for general-purpose object location, i.e. finding objects from their features. The specifications of the detectors were that they should be sensitive, reliable and accurate, so that there would be little chance of missing any object containing them, and so that object location would be accurate. In the context of the object inference schemes described in Part 2—and particularly in Chapter 14—it does not matter if some features are missing or whether additional noise or clutter features arise, as the inferential schemes are sufficiently robust to be able to find the objects in spite of this. However, the whole context was essentially the 2-D situation where it was good enough to imagine that the objects were nearly flat, or had nearly flat faces, so that 3-D perspective types of distortion could be avoided. Even so, in 3-D, corners appear as corners from almost any viewpoint, so robust inference algorithms should still be able to perform object location. However, when viewing objects from quite different directions in

FIGURE 6.11 (Continued)

over a distance of 7 pixels. (g and h) Later frames in the sequence (also using maximum responses over a distance of 7 pixels), showing a high consistency of feature identification, which is important for tracking purposes. Note that interest points really do indicate locations of interest—corners, people's feet, ends of white road markings, and castle window and battlement features. Also, the greater the significance as measured by the pixel suppression range, the greater relevance the feature tends to have.

3-D, appearance can change dramatically, so it becomes extremely difficult to recognize them, even if all the features are present in the images. Thus, we arrive at the concept of viewing over wide baselines. In the case of binocular vision that takes two views over quite a narrow baseline ($\sim$7 cm for human eyes), the difference between the views is necessary in order to convey depth information, but it is rarely so great that features recognized in one view cannot be re-identified in the second view (though when huge numbers of similar, e.g. textural, features occur, as when viewing a piece of material, this may not apply). On the other hand, when objects are viewed on a wide baseline, as happens after significant motion has occurred, the angular separation between the views may be as large as 50°. If even larger angular separations occur, there will be much less possibility of recognition. While this sort of situation can be tackled by memorizing sequences of views of objects, here we concentrate primarily on what can be discerned from local features that are seen in wide baseline views of up to $\sim$50°.

At this point we have established the need to be able to recognize local features from wide baseline views as far apart as 50° so that objects can be recognized and tracked or found in databases without especial difficulty. Clearly, the corner and interest point detectors described thus far have no special provision for this. To achieve this aim, additional criteria have to be fulfilled. The first is that feature detection must be consistent and repeatable in spite of substantial change of viewpoint. The second is that features must embody descriptions of their localities so that there is high probability that the same physical feature will be positively identified in each of the views. Imagine that each image contains $\sim$1000 corner features. Then there will be $\sim$1 million potential feature matches between two views. While a robust inference scheme could perform the match in the case of flat objects, the situation becomes so much more demanding for general views of 3-D objects that matching might not be possible, either at all, or more likely, within a reasonable time—or without large numbers of ambiguities occurring. So, it is highly important to minimize the feature matching task. Indeed, ideally, if a rich enough descriptor is provided for each feature, feature matching might be reducible to one-to-one between views. At this stage we are extremely far from this possibility, as the corners that we have detected can so far only be characterized on the basis of their enclosed angle and intensity or color (and note that the first of these parameters will generally be substantially changed by the altered viewing angle). In what follows we consider in turn the two requirements of consistent, repeatable feature detection, and feature description.

Broadly speaking, obtaining consistent, repeatable feature detection involves allowing for and normalizing the variations between views. The obvious candidates for normalization are scale, affine distortion and perspective distortion. Of these, the first is straightforward, the second difficult, and the third impractical to implement. This is because of the number of parameters that need to be estimated for each feature. Bearing in mind that local features are necessarily small, the accuracy with which the parameters can be estimated decreases rapidly with increase in their number. Figure 6.13 shows how various transformations affect a

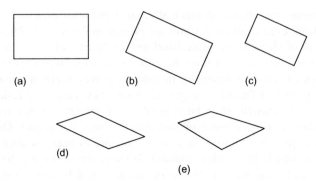

FIGURE 6.13

Effects of various transformations on a convex 2-D shape. (a) Original shape. (b) Effect of Euclidean transform (translation + rotation). (c) Effect of similarity transform (change of scale). (d) Effect of affine transform (stretch + shear). (e) Effect of perspective transform. Note that in (d) parallel lines still remain parallel: this is not in general the case after a projective transform, as indicated in (e). Overall, each of the transforms illustrated is a generalization of the previous one. The respective numbers of degrees of freedom are 3, 4, 6, 8: in the last case each of the four points is independent and has two degrees of freedom, though there are constraints, such as convexity having to be maintained.

2-D shape. The following equations respectively define Euclidean, similarity (scale variation), and affine transformations:

$$\begin{bmatrix} x' \\ y' \end{bmatrix} = \begin{bmatrix} r_{11} & r_{12} \\ r_{21} & r_{22} \end{bmatrix} \begin{bmatrix} x \\ y \end{bmatrix} + \begin{bmatrix} t_1 \\ t_2 \end{bmatrix} \tag{6.24}$$

$$\begin{bmatrix} x' \\ y' \end{bmatrix} = \begin{bmatrix} sr_{11} & sr_{12} \\ sr_{21} & sr_{22} \end{bmatrix} \begin{bmatrix} x \\ y \end{bmatrix} + \begin{bmatrix} t_1 \\ t_2 \end{bmatrix} \tag{6.25}$$

$$\begin{bmatrix} x' \\ y' \end{bmatrix} = \begin{bmatrix} a_{11} & a_{12} \\ a_{21} & a_{22} \end{bmatrix} \begin{bmatrix} x \\ y \end{bmatrix} + \begin{bmatrix} t_1 \\ t_2 \end{bmatrix} \tag{6.26}$$

where rotation takes place through an angle theta, and the rotation matrix is:

$$\begin{bmatrix} r_{11} & r_{12} \\ r_{21} & r_{22} \end{bmatrix} = \begin{bmatrix} \cos\theta & -\sin\theta \\ \sin\theta & \cos\theta \end{bmatrix} \tag{6.27}$$

Euclidean transformations allow translation and rotation operations and have three degrees of freedom (DoF); *additionally*, similarity transformations include scaling operations and have four DoF; *additionally*, affine transformations include stretching and shearing operations, have six DoF, and are the most complex of the transformations that make parallel lines transform into parallel lines; projective transformations are much more complex, have eight DoF, and include operations that (a) make parallel lines nonparallel, and (b) change *ratios* of lengths on straight lines. The steady increase in the number of parameters is what mitigates against estimation of perspective distortions in the feature points: in fact, it also tends to reduce accuracy for the scale parameter when estimating full affine distortion.

6.7.1 Harris Scale and Affine-Invariant Detectors and Descriptors

Before proceeding to consider the above ideas in more detail, note that feature detectors such as the Harris operator already estimate location and orientation, so normalization for translation and rotation is already allowed for. This leaves scale as the next candidate for normalization. Here, the basic concept is to apply a given feature detector at various scales, using larger and larger masks. In the case of the Harris operator, there are two relevant scales: one is the edge detection (differentiation) scale σ_D and the other is the overall feature (integration) scale σ_I. In practice, these need to be linked together (this involves little loss of generality) so that $\sigma_I = \gamma \sigma_D$, where γ has a suitable value in the range $0-1$ (typically ~ 0.5). σ_I then represents the scale of the overall operator. The approach is now to vary σ_I and to find the value that provides the best match of the operator to the local image data: the best match (extremum value) is the one representing the local image structure: it is intended to be independent of image resolution, which is arbitrary. In fact, the resulting "scale-adapted" Harris operator rarely attains true maxima over scales in such a ("scale-space") representation (Mikolajczyk and Schmid, 2004); this is because a corner appears as a corner over a wide range of scales (Tuytelaars and Mikolajczyk, 2008). To achieve an optimal scale for matching, a totally different approach is applied: that is to use the Harris operator to locate a suitable feature point, and then to examine its surroundings to find the ideal scale, using a Laplacian operator. The scale of the latter is then adjusted to determine, in a matched filter (i.e. optimum signal-to-noise ratio) way, when the profile of the Laplacian most accurately matches the local image structure (Fig. 6.14). The required operator is called a Laplacian of Gaussian (LoG). It corresponds to smoothing the image using a Gaussian and then applying the Laplacian $\nabla^2 = \partial^2/\partial x^2 + \partial^2/\partial y^2$ (see Chapter 5), and results in the following combined isotropic convolution operator:[6]

$$LoG = \frac{(r^2 - 2\sigma^2)}{\sigma^4(2\pi\sigma^2)}\exp\left(-\frac{r^2}{2\sigma^2}\right) = \frac{(r^2 - 2\sigma^2)}{\sigma^4}G(\sigma) \qquad (6.28)$$

where

$$G(\sigma) = \frac{1}{2\pi\sigma^2}\exp\left(-\frac{r^2}{2\sigma^2}\right) \qquad (6.29)$$

Having optimized this operator, we know the scale of the corner, and also its location and 2-D orientation. This means that when comparing two such corner

---

[6]Note that convolution ($\otimes$) is associative, so we have $\nabla^2 \otimes (G \otimes I) = (\nabla^2 \otimes G) \otimes I = LoG \otimes I$.

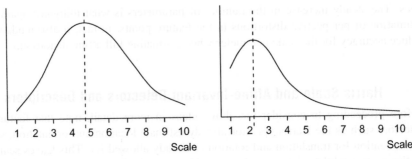

FIGURE 6.14

Scaling graphs for two objects that are being matched. The scaling graph on the left has an extremum at 4.7, and the one on the right has an extremum at 2.2. This shows that the best match occurs when the ratios of their scaling factors are approximately 2.14:1. The vertical scales of the graphs do not come into the optimization calculation.

features we can maintain translation, rotation, and scale invariance. To obtain affine invariance we estimate the affine shape of the corner neighborhood. Examining the Harris matrix Eq. (6.10), we rewrite it in the scale-adapted form:

$$\Delta = \sigma_D^2 G(\sigma_1) \otimes \begin{bmatrix} I_x^2(\sigma_D) & I_x(\sigma_D)I_y(\sigma_D) \\ I_x(\sigma_D)I_y(\sigma_D) & I_y^2(\sigma_D) \end{bmatrix} \tag{6.30}$$

where

$$I_x(\sigma_D) = \frac{\partial}{\partial x} G(\sigma_D) \otimes I \tag{6.31}$$

and similarly for $I_y(\sigma_D)$. These equations take full account of the differentiation and integration scales σ_D, σ_I. Then for each scale of the scale-adapted Harris operator, we repeat the process that was applied while determining the scale using the Laplacian, this time iteratively determining the best-fit ellipse (rather than circle) profile that fits the local intensity pattern. In fact, in spite of starting at separate scales, the resulting elliptic fits are, for well-defined corner structures, highly consistent and a robust average can be selected. The corresponding ellipse will (compared with the original circle) be stretched by different amounts in two perpendicular directions: the degrees of stretch and skew are the output affine parameters.

The final step is to normalize the feature by transforming it so that the elliptic profile becomes isotropic, circular and therefore affine invariant (i.e. the affine deformation is nullified). This corresponds to equalizing the eigenvalues of the optimum scale-adapted second-order matrix, Eq. (6.30).

When comparing two corners we require invariant parameters. To obtain these descriptors, it is necessary to determine Gaussian derivatives of the local neighborhood of the interest points, computed on the transformed isotropic feature profile. Clearly, the Gaussian derivatives have to be adjusted for a standardized isotropic profile size, and they have to be normalized to intensity variations by dividing the higher order derivatives by the first-order derivative (i.e. the average intensity gradient in the neighborhood). In the work of Mikolajczyk and Schmid (2004), descriptors of dimension 12 were obtained by using derivatives up to fourth order. (There are two first-order derivatives, three second order, four third order, and five fourth order: excluding the first-order derivatives, this leaves a total of 12 up to fourth order.) This set of descriptors proved highly effective for identifying corresponding pairs of features in widely different views of up to $\sim 40°$ angular separation with better than 40% repeatability, and in the affine case up to $\sim 70°$ with up to 40% repeatability. In addition, the localization accuracy for Harris–Laplace dropped off more or less linearly with angular separation, becoming excessive above 40°, whereas that for Harris–Affine remained at an acceptable level (~ 1.5 pixel error). The Harris–Laplace was described as having a breakdown point at a viewpoint change of 40°.

6.7.2 Hessian Scale and Affine-Invariant Detectors and Descriptors

Over the same period that scale and affine-invariant detectors and descriptors based on the Harris operator were developed, investigations of similar operators based on the Hessian operator were being undertaken. Here it is useful to recall that the Harris operator is defined in terms of first derivatives of the intensity function I, while the Hessian operator (see Eq. (6.5)) is defined in terms of the second derivatives of I. Thus, we can consider the Harris operator as being edge-based, and the Hessian operator as being blob-based. This matters for two reasons. One is that the two types of operator might, and do, bring in different information about objects and hence to some extent they are complementary. The other is that the Hessian is better matched than the Harris to the Laplacian scale estimator: indeed, the Hessian arises from the determinant and the Laplacian from the trace of the matrix of second-order derivatives (Eq. (6.2)). The better matching of the Hessian to the Laplacian results in improved scale selection accuracy for this operator (Mikolajczyk and Schmid, 2005). The other details of the Hessian–Laplacian and Hessian–Affine operators are similar to those for the corresponding Harris operators and will not be discussed in more detail here. However, it is worth remarking that in all four cases there are typically 200–3000 detected regions per image depending on the content (Mikolajczyk and Schmid, 2005).

6.7.3 The SIFT Operator

Lowe's scale invariant feature transform (widely known as "SIFT") was first introduced in 1999, a much fuller account being given by Lowe (2004). While being restricted to a scale invariant version, it is important for two reasons: (1) for impressing on the vision community the existence, importance, and value of invariant types of detector; and (2) for demonstrating the richness that feature descriptors can bring to feature matching. For estimating scale, the SIFT operator uses the same basic principle as for the Harris and Hessian-based operators outlined above. However, it differs in using the Difference of Gaussians (DoG) instead of the Laplacian of Gaussians (LoG), in order to save computation. This possibility is seen by differentiating G with respect to σ in Eq. (6.29):

$$\frac{\partial G}{\partial \sigma} = \left(\frac{r^2}{\sigma^3} - \frac{2}{\sigma}\right) G(\sigma) = \sigma \, LoG \qquad (6.32)$$

which means that we can approximate LoG as the difference of Gaussians of two scales:

$$LoG \approx \frac{G(\sigma') - G(\sigma)}{\sigma(\sigma' - \sigma)} = \frac{G(k\sigma) - G(\sigma)}{(k-1)\sigma^2} \qquad (6.33)$$

where use of the constant scale factor k permits scale normalization to be carried out easily between scales.

In fact, it is in the design of the descriptors that SIFT is particularly different from the Harris and Hessian-based detectors. Here the operator divides the support region, at each scale, into a 16×16 sample array and estimates the intensity gradient orientations for each of these. They are then grouped into sets of sixteen 4×4 sub-arrays and orientation histograms are generated for each of these, the directions being restricted to one of eight directions. The final output is a 4×4 array of histograms each containing entries for eight directions—amounting to a total output dimensionality of $4 \times 4 \times 8 = 128$.

The overall detector is found (Mikolajczyk, 2002) to be more repeatable than Harris—Affine and to retain a final matching accuracy above 50% out to a 50° angular separation. However, because of the limited stability of Harris—Affine, Lowe (2004) recommends the approach of Pritchard and Heidrich (2003) of including additional SIFT features with 60° viewpoint separation during training. We defer further discussion of the performance of this detector to Section 6.7.6.

6.7.4 The SURF Operator

The development of SIFT stimulated efforts to produce an effective invariant feature detector that was also highly efficient and required a smaller descriptor than the large one employed by SIFT. An important operator in this mold was the speeded-up robust features (SURF) method of Bay et al. (2006, 2008). This was based on the Hessian—Laplace operator. In order to increase speed, several measures were taken: (1) the integral image approach was used to perform rapid computation of the Hessian and was also used during scale-space analysis; (2) the DoG was used in place of the LoG for assessing scale; (3) sums of Haar wavelets were used in place of gradient histograms, resulting in a descriptor dimensionality of 64—half that of SIFT; (4) the sign of the Laplacian was used at the matching stage; (5) various reduced forms of the operator were used to adapt it to different situations, notably an "upright" version capable of recognizing features within $\pm 15°$ of those pertaining to an upright stance, as occurs for outdoor buildings and other objects. By maintaining a rigorous, robust design, the operator was described as outperforming SIFT, and also proved capable of estimating 3-D object orientation within fractions of a degree and certainly more accurately than SIFT, Harris–Laplace, and Hessian–Laplace.

Of some importance for this implementation is the integral image approach (Simard et al., 1999): this was brought prominently to light by Viola and Jones (2001), but maybe not utilized as much as one might expect since then. This is extremely simple, yet radical in the levels of speedup it can bring. It involves computing an integral image I_Σ, which is an image that retains sums of all pixel intensities encountered so far in a single scan over the input image:

$$I_\Sigma(x,y) = \sum_{i=0}^{i \leq x} \sum_{j=0}^{j \leq y} I(i,j) \tag{6.34}$$

This not only permits any pixel intensity in the original image to be recovered:

$$I(i,j) = I_\Sigma(i,j) - I_\Sigma(i-1,j) - I_\Sigma(i,j-1) + I_\Sigma(i-1,j-1) \tag{6.35}$$

but also allows the sum of the pixel intensities in any upright rectangular block, such as those ranging from $x = i$ to $i + a$ and $y = j$ to $j + b$ within block D in Fig. 6.15, to be utilized:

$$\begin{aligned} \sum_D I &= \sum_A I - \sum_{A,B} I - \sum_{A,C} I + \sum_{A,B,C,D} I \\ &= I_\Sigma(i,j) - I_\Sigma(i+a,j) - I_\Sigma(i,j+b) + I_\Sigma(i+a,j+b) \end{aligned} \tag{6.36}$$

The method is exceptionally well adapted to computing Haar filters that typically consist of arrays containing blocks of identical values, for example:

$$\begin{bmatrix} -1 & -1 & 1 & 1 & 1 & 1 & -1 & -1 \\ -1 & -1 & 1 & 1 & 1 & 1 & -1 & -1 \\ -1 & -1 & 1 & 1 & 1 & 1 & -1 & -1 \\ -1 & -1 & 1 & 1 & 1 & 1 & -1 & -1 \end{bmatrix}$$

Note that once the integral image has been computed, it permits summations to be made over any block merely by performing four additions—taking a miniscule time that is independent of the size of the block. A simple generalization to the 3-D box filter (Simard et al., 1999) is also possible, and this is used in computing within scale-space in the SURF implementation.

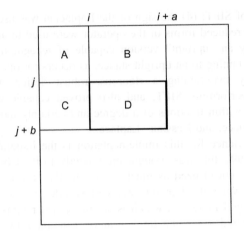

FIGURE 6.15

The integral image concept. Here block D can be considered as made up by taking block A + B + C + D, then subtracting block A + B and block C, in the latter case by subtracting A + C and adding A: see text for an exact mathematical treatment.

6.7.5 Maximally Stable Extremal Regions

There is one class of invariant feature that does not fall into the pattern covered in the preceding sections—the invariant region type of feature. Among the most important examples of this type of feature is the maximally stable extremal region (MSER). The method analyzes regions with increasing ranges of intensity, and aims to determine those that are extremal in a particularly stable way. (Recall that finding extrema is a powerful general method for locating invariant features, as we have already seen in Section 6.7.1.)

The method (Matas et al., 2002) starts by taking pixels of zero intensity and progressively adding pixels with higher intensity levels, at each stage monitoring the regions that form. At each stage largest connected regions or "connected components" will represent extremal regions.[7] As more and more gray levels are added, the connected component regions will grow and some initially separate ones will merge. Maximally stable extremal regions are those connected components that are close to stable (as conveniently measured by their area) over a range of intensities, i.e. each MSER is represented by the position of a local intensity minimum in the rate of change of the area function. Interestingly, relative area change is an affine-invariant property, so finding MSER regions guarantees both scale and affine

---

[7]See Section 9.3 for a full explanation of connected components and their computation. Meanwhile, assume that a "connected component" means a region *containing everything that is connected to any part of that region*. Hence, by definition, a connected component is an extremal region.

invariance. In fact, because of the way intensities are handled in this method, the results are also independent of monotonic transformation of image intensities.

While every MSER can be regarded as a connected component of a threshold image, global thresholding is not carried out, and optimality is judged on the basis of the stability of the connected components that are located. Intrinsically, MSERs have arbitrary shapes, though for matching purposes they can be converted into ellipses of appropriate areas, orientations and moments. Perhaps surprisingly for affine features, they can be computed highly efficiently in times that are nearly linear in the number of pixels; they also have good repeatability, though they are quite sensitive to image blur. This last problem is to be expected, given the dependence on individual gray levels and the precision with which connected components analysis is carried out; in fact, this difficulty has been addressed in a recent extension of the work (Perdoch et al., 2007).

6.7.6 Comparison of the Various Invariant Feature Detectors

While there are many more invariant feature detectors than we have been able to cover here (salient regions, IBR, FAST, SFOP, . . .) and many variants of them (GLOH, PCA-SIFT, . . .), we next concentrate on comparisons between them. In fact, most of the papers describing new detectors make comparisons with older detectors, but often with limited datasets. Here we outline the conclusions of Ehsan et al. (2010), who compared SIFT, SURF, Harris−Laplace, Harris−Affine, Hessian−Laplace, and Hessian−Affine, using the following datasets: Bark, Bikes, Boat, Graffiti, Leuven, Trees, UBC, and Wall (viz. eight sequences of six images).[8] Table 6.1 presents the results in a modified form with SURF placed after Hessian−Laplace, as it is based on the latter.

Apart from Table 6.1, Ehsan et al. (2010) showed the results of using three different criteria for judging repeatability of feature detector performance. The first was the standard repeatability criterion:

$$C_0 = \frac{N_{\text{rep}}}{\min(N_1, N_2)} \qquad (6.37)$$

where N_1 is the total number of points detected in the first image, N_2 is the total number of points detected in the second image, and N_{rep} is the number of repeated points.

They emphasized that it has been remarked (Tuytelaars and Mikolajczyk, 2008) that repeatability "does not guarantee high performance in a given application." They reasoned that this was due in part to comparing features within adjacent pairs of images rather than over whole image sequences: specifically, they

---

[8]See Oxford DataSets at: http://www.robots.ox.ac.uk/~vgg/research/affine/ (website accessed 19 April 2011).

Table 6.1 Comparison of Invariant Feature Detectors

Datasets	SIFT	Harris–Laplace	Hessian–Laplace	SURF	Harris–Affine	Hessian–Affine	Total
Bark							9
Bikes							14
Boat							12
Graffiti							10
Leuven							12
Trees							13
UBC							17
Wall							14
total	16	14	18	20	15	18	101

The totals give some indication of the overall capabilities of the detectors, and of the complexity of the individual datasets. However, the detector totals must be interpreted in the light of the highest level of invariance achievable—viz. scale or affine.

recommended that each image should be compared taking the first frame of the sequence as a reference and using the following criterion:

$$C_1 = \frac{N_{rep}}{N_{ref}} \tag{6.38}$$

Nevertheless, they also proposed a more symmetric measure of repeatability:

$$C_2 = \frac{N_{rep}}{(N_{ref} + N_c)} \tag{6.39}$$

where N_c is the total number of points detected in the current frame. This proved to be a less harsh and more realistic criterion when compared with the trends of the observed ground truth for an image sequence. Further evidence for moving away from the standard repeatability criterion C_0 is that it rewards failure to detect features (because decrease in N_1 or N_2 will, if anything, *raise* the value of C_0). This suggests altering C_0 to use the maximum instead of the minimum. However, using either the maximum or the minimum tends to emphasize extreme results, leading to nonrobust measures. From this point of view the most appropriate measure has to be C_2. In fact, this criterion gave optimal results and low error probability measures when run against ground truth using Pearson's correlation coefficients (Ehsan et al., 2010). Using C_2, an important result was the dominance of the Hessian-based detectors, which is already very evident in Table 6.1 (derived from their Table 2), where the three Hessian-based totals are 18, 18, 20, vis-à-vis 14, 15, 16 for the others (interestingly, the situation is even more polarized in favor of Hessian-based detectors when the datasets giving the best and worst results—Bark and UBC—are ignored). Note that Tuytelaars and Mikolajczyk (2008) did not come out so strongly in favor of the Hessian-based

Table 6.2 Performance Evaluation of Various Feature Detectors

Detector	Invariance	Repeatability	Accuracy	Robustness	Efficiency	Total
Harris	Rotation	▬	▬	▬	▪	11
Hessian	Rotation	▪	▪	▪	▪	7
SIFT	Scale	▪	▪	▪	▪	8
Harris–Laplace	Scale	▬	▬	▪	▪	9
Hessian–Laplace	Scale	▬	▬	▬	▪	10
SURF	Scale	▪	▪	▪	▬	9
Harris–Affine	Affine	▬	▬	▪	▪	10
Hessian–Affine	Affine	▬	▬	▬	▪	11
MSER	Affine	▬	▬	▪	▬	11

The totals give some indication of the overall capabilities of the detectors: however, they must be interpreted in the light of the highest level of invariance achievable (column 2).

detectors, but this was probably because their analysis of datasets was not so extensive; also, Ehsan et al. (2010) were the first to look at image sequences so rigorously using C_2.

The review by Tuytelaars and Mikolajczyk (2008) is of great value in evaluating performance using several disparate criteria, viz. repeatability, localization accuracy, robustness and efficiency. Some of their results are shown in Table 6.2—notably those for all the feature detectors covered in Table 6.1 and those for the single-scale Harris and Hessian, and for the MSER detector (Matas et al., 2002) mentioned earlier. They also make the following valuable observations:

1. Scale invariant operators can normally be dealt with adequately by a robustness capability for viewpoint changes of less than 30°, as affine deformations only rise above those due to variations in object appearance beyond that level.
2. In different applications, different feature properties may be important, and thus success depends largely on appropriate selection of features.
3. Repeatability may not always be the most important feature performance characteristic: not only is it hard to define and measure but robustness to small appearance variations matters more.
4. There is a need for work focussing on complementarity of features, leading either to complementary detectors or to detectors providing complementary features.

In the last respect, note that some ground work has recently been carried out by Ehsan et al. (2011) to measure the coverage of interest point detectors. They identify the recent SFOP scale invariant feature transform (Förstner et al., 2009) as the most outstanding detector in this respect, either used on its own or in conjunction with others. Ehsan et al.'s new criterion for coverage C is based on the harmonic mean so as not to overemphasize nearby features:

$$C = \frac{N(N-1)}{\sum_{i=1}^{N-1} \sum_{j>1}^{N} (1/d_{ij})} \tag{6.40}$$

(In this formula d_{ij} is the Euclidean distance between feature points i and j.)

Note that a detector which has high coverage is not guaranteed to be sound: after all, a detector giving a random selection of feature points might fare well on this count. Hence, a coverage criterion can only come into its own when it is used to select complementary types of feature and feature detector, or a detector that provides a good mix of types of feature, for which the output selection is known to be sound on other counts (repeatability, robustness, and so on).

6.8 CONCLUDING REMARKS

Corner detection provides a useful start to the process of object location, and to this end is often used in conjunction with the abstract pattern matching approaches discussed in Chapter 14. Apart from the obvious template matching procedure, which is of limited applicability, three main approaches have been described. The first was the second-order derivative approach that includes the KR, DN, and ZH methods—all of which embody the same basic schema; the second was the median-based method, which turned out to be equivalent to the second-order derivative methods in situations where corners have smoothly varying intensity functions; and the third was the Harris detector which is based on the matrix of second moments of the *first* derivatives of the intensity function. Perhaps surprisingly, the latter is able to extract much the same information as the other two approaches, though there are differences, in that the Harris detector is better described as an interest point detector than as a corner detector. In fact, the Harris detector has probably been the most widely used corner and interest point detector of all, and for general purpose (non-3-D) operation this still seems to be the case—in spite of the advent of the SUSAN detector (Smith and Brady, 1997), which is known to be faster and more efficient, but somewhat less resistant to noise.

Interestingly, the situation presented above started changing radically from about 1998, when workers started looking for approaches to object location that were not merely robust to noise, distortion, partial occlusion, and extraneous features, but were able to overcome problems of gross distortion due to viewing the same scene from widely different directions. This "wide baseline" problem, which is prominent with 3-D and motion applications, including tracking moving objects, became the driving force for radical new thinking and development. As we have seen, attempts were made to adapt the Harris operator to this scenario, making it invariant to similarity (scale) and affine transformations, though in the end somewhat more success was achieved by returning to the Hessian operator discussed early in the chapter. Alternative approaches included the MSER approach, which is by no means based on the location of any sort of corner or interest point. Indeed, it harks back to the thresholding methods of Chapter 4. But this is all to the good, as the underlying task is that of segmentation coupled with recognition and identification/matching: division of the subject into watertight topics, such as thresholding, edge detection, and corner detection, has limited validity or at least it is too restrictive to offer the best solutions to the real problems of the subject. In this context it is relevant that the newly evolved feature detectors embody multiparameter descriptors, making them far better suited not only to detection *per se* but also to the more exacting task of wide baseline 3-D matching.

Overall, in this chapter we have seen the corner detector approach transmogrify itself to overcome the problems of viewing objects from directions as far apart as 70°—and with a great deal of success. Remarkably, all this was achieved in little more than a decade—evidence that progress in this subject has been accelerating. Importantly, the old computer adage "garbage in—garbage out" is relevant, because feature detection forms a crucial link between the original pictures and their high-level interpretation.

> This chapter has studied how objects may be detected and located from their corners and interest points. It has developed both the classic approach to detector design and the more recent invariant approaches, which result in multiparameter feature descriptors to aid matching between widely separated views of objects.

6.9 BIBLIOGRAPHICAL AND HISTORICAL NOTES

The subject of corner detection has been developing for over three decades. The scene was set for the development of parallel corner detection algorithms by Beaudet's (1978) work on rotationally invariant image operators. This was soon followed by Dreschler and Nagel's (1981) more sophisticated second-order corner detector: the motivation for this research was to map the motion of cars in traffic scenes, corners providing the key to unambiguous interpretation of image sequences. One year later, Kitchen and Rosenfeld (1982) had completed their study of corner detectors based mainly on edge orientation, and had developed the second-order KR method described in Section 6.3. Years 1983 and 1984 saw the development of the second-order ZH detector and the median-based detector (Zuniga and Haralick, 1983; Paler et al., 1984). Subsequently, the author published work on the detection of blunt corners (see Chapter 13) and on analyzing and improving the median-based detector (Davies, 1988a,d, 1992a). Meanwhile, other methods had been developed, such as the Harris algorithm (Harris and Stephens, 1988; see also Noble, 1988). The Smith and Brady (1997) "SUSAN" algorithm marked a further turning point, needing no assumptions on the corner geometry, as it works by making simple comparisons of local gray levels: this is one of the most cited of all corner detection algorithms.

In the year 2000s further corner detectors were developed. Lüdtke et al. (2002) designed a detector based on a mixture model of edge orientation: in addition to being effective in comparison with the Harris and SUSAN operators, particularly at large opening angles, the method provides accurate angles and strengths for the corners. Olague and Hernández (2002) worked on a unit step edge function (USEF) concept, which is able to model complex corners well: this resulted in adaptable detectors that are able to detect corners with subpixel accuracy. Shen and Wang (2002) described a Hough transform-based detector: as this works in a 1-D parameter space it is fast enough for real-time operation; a useful feature of the paper is the comparison with, and between, the Wang and Brady detector, the Harris detector, and the SUSAN detector. The several example images show that it is difficult to be sure exactly what one is looking for in a corner detector (i.e. corner detection is an ill-posed problem), and that even the well-known detectors sometimes inexplicably fail to find corners in obvious places. Golightly and Jones (2003) present a practical problem in outdoor country scenery: they discuss not only the incidence of false positives and false negatives but also the probability of correct association in corner matching, e.g. during motion.

Rocket (2003) gives a performance assessment of three corner detection algorithms—the KR detector, the median-based detector, and the Harris detector: the results are complex, and the three detectors are found to have very different characteristics. The paper is valuable in showing how to optimize the three methods (not least showing that the Harris detector parameter k should be ~ 0.05), and also because it concentrates on careful research rather than "selling" a new detector. Tissainayagam and Suter (2004) gave an assessment of the performance of corner detectors, with vitally important coverage of point feature (motion) tracking applications. Interestingly, it finds that, in image sequence analysis, the Harris detector is more robust to noise than the SUSAN detector, a possible explanation being that it "has a built-in smoothing function as part of its formulation." Finally, Davies (2005) analyzed the localization properties of the Harris operator: see Section 6.5 for the main results of this work.

While the above discussion covers many of the developments on corner detection, it is not the whole story. This is because in many applications it is not specific corner detectors that are needed but "interest point" detectors, which are capable of detecting any characteristic patterns of intensity that can be used as reliable feature points. In fact, the Harris detector is often called an interest point detector—with good reason, as indicated in Section 6.5.2. Moravec (1977) was among the first to refer to interest points and was followed by Schmid et al. (2000) and many others. However, Sebe and Lew (2003) and Sebe et al. (2003) call them salient points—a term more often reserved for points that attract the attention of the *human* visual system. Overall, it is probably safest to use the Haralick and Shapiro (1993) definition: a point being "interesting" if it is both distinctive and invariant—i.e. it stands out and is invariant to geometric distortions such as might result from moderate changes in scale or viewpoint (note that Haralick and Shapiro also list other desirable properties—stability, uniqueness, and interpretability). The invariance aspect is taken up by Kenney et al. (2003) who show how to remove ill-conditioned points from consideration, to make matching more reliable.

The subject of invariant feature detectors and descriptors has taken little over a decade to develop and over that period it has come a long way. It started with papers by Lindeberg (1998) and Lowe (1999) that indicated the way forward and provided basic techniques. It arose largely because of difficulties in wide baseline stereo work, and with tracking object features over many video frames—because features change their appearance over time and correspondences are easily lost. To proceed, it was necessary first to eliminate the relatively simple problem of features changing in size, thereby necessitating scale invariance (it being implicit that translation and rotation invariance have already been dealt with). Later, improvements became necessary to cope with affine invariance. Thus, Lindeberg's pioneering theory (1998) was soon followed by Lowe's work (1999, 2004) on scale invariant feature transforms (SIFT). This was followed by affine-invariant methods developed by Tuytelaars and Van Gool (2000), Mikolajczyk and Schmid (2002, 2004), Mikolajczyk et al. (2005) and others. In parallel with these developments, work was published on maximally stable extremal regions (Matas et al., 2002) and other extremal methods (e.g. Kadir and Brady, 2001; Kadir et al., 2004).

Much of this work capitalized on the interest point work of Harris and Stephens (1988), and was underpinned by careful in-depth experimental investigations and comparisons (Schmid et al., 2000; Mikolajczyk and Schmid, 2005; Mikolajczyk et al., 2005). Next, the tide turned in other directions, in particular the design of feature detectors that aim at real-time operation—as in the case of the speeded-up robust features (SURF) approach (Bay et al., 2006, 2008).

A review article summarizing the main approaches was published by Tuytelaars and Mikolajczyk in 2008. However, that was by no means the end of the story. As outlined in Section 6.7.6, Ehsan et al. (2010) briefly reviewed the status quo on repeatability of the main features and detectors, and reported on experiments to assess it: their work included two new repeatability criteria, which more realistically reflected the underlying requirements for invariant detectors. In addition, they presented new work (Ehsan et al., 2011) on the coverage of invariant feature detectors, reflecting poignant remarks made in Tuytelaars and Mikolajczyk's (2008) review. It is clear that, with the passing of the first decade of the 2000s, an even more exacting phase of development is under way, with more rigorous performance evaluation: it will no longer be sufficient to produce new invariant feature detectors; instead it will be necessary to integrate them much more fully with the target applications, following rigorous design to ensure that all relevant criteria are being met, and that the tradeoffs between the criteria are much more transparent.

6.9.1 **More Recent Developments**

Since the Tuytelaars and Mikolajczyk's (2008) review, further relevant work on feature detectors and descriptors has emerged. Rosten et al. (2010) have presented the FAST family of corner detectors, which is designed on a new heuristic to be especially fast while at the same time to be highly repeatable; they also review methods for comparing feature detectors and call for less concentration on how a feature detector should do its job than on what performance measure it is required to optimize. Cai et al. (2011) work on a "linear discriminant projections" procedure for reducing the dimensionality of local image descriptors, and manage to bring the SIFT tally down from 128 to just 30. However, they warn that this seems to be achievable only by making the projections specific to the type of image data. With a similar motivation, Teixeira and Corte-Real (2009) quantize the SIFT descriptor to form visual words using a predefined vocabulary, though in this case the vocabulary is structured in the form of a tree; it is constructed using a generic dataset related to the type of object tracking being performed. van de Sande et al. (2010) discuss the generation of color object descriptors. They find that the choice of a single color descriptor for all categories of data is suboptimal: but for unknown data, "OpponentSIFT" (using three sets of SIFT features for the three opponent colors) showed the highest degree of invariance with respect to photometric variations. Zhou et al. (2011) proposed a method to perform descriptor combination and classifier fusion. They cast the problem of object classification into a learning setting, which again means that the method is adaptive and not applicable to new data without retraining. Overall, we see that trying to reduce the original 128 SIFT features (or the equivalent) tends to make such methods specific to particular training data.

6.10 **PROBLEMS**

1. By examining suitable binary images of corners, show that the median corner detector gives a maximal response within the corner boundary rather than half-way down the edge outside the corner. Show how the situation is modified for grayscale images. How will this affect the value of the gradient noise-skimming threshold to be used in the improved median detector?

2. Prove Eq. (6.6), starting with the following formula for curvature:

$$\kappa = \frac{d^2y/dx^2}{[1 + (dy/dx)^2]^{3/2}}$$

Hint: First express dy/dx in terms of the components of intensity gradient, remembering that the intensity gradient vector (I_x, I_y) is oriented along the edge normal; then replace the x, y variation by I_x, I_y variation in the formula for κ.

Mathematical Morphology

Historically, the study of shape took place over a long period of time and resulted in a highly variegated set of algorithms and methods. Over the past 30 years the formalism of mathematical morphology was set up, and provided a background theory into which many of the individual advances could be slotted. This chapter takes a journey through this interesting subject, but aims to steer an intuitive path between the many mathematical theorems, concentrating on finding practically useful results.

Look out for:

- how the concepts of expanding and shrinking are transformed into the more general concepts of dilation and erosion.
- how dilation and erosion operations may be combined to form more complex operations whose properties may be predicted mathematically.
- how the concepts of closing and opening are defined, and how they are used to find defects in binary object shapes, via residue (or "top hat") operations.
- how mathematical morphology is generalized to cover grayscale processing.
- how noise affects morphological grouping operations.

This theory in the present chapter is especially valuable because of the way in which it integrates a range of topics. Once the methodology has been learnt, morphology should be of distinct value in taking the earlier ideas forward and optimizing algorithms that use them.

7.1 INTRODUCTION

In Chapter 2, we have discussed the operations of erosion and dilation: in Chapter 9 we will apply them to the filtering of binary images, and will show that with suitable combinations of these operators it is possible to eliminate certain types of object from images, and also to locate other objects. These possibilities are not fortuitous, but on the contrary reflect fundamental properties of shape, which are dealt with in the subject known as mathematical morphology. This subject has grown up over the past two or three decades, and over the past few years knowledge in this area has become consolidated and is now understood in considerable depth. It is the purpose of this chapter to give some insight into this vital area of study. Note that mathematical morphology is especially important because it provides a backbone for the whole study of shape and thus is able to unify techniques as disparate as noise suppression, shape analysis, feature recognition, skeletonization, convex hull formation, and a host of other topics.

Section 7.2 starts the discussion by extending the concepts of expanding and shrinking first encountered in Section 2.2. Section 7.3 then develops the theory of mathematical morphology, arriving at many important results—with emphasis deliberately being placed on understanding of concepts rather than mathematical rigor. Section 7.4 goes on to show how morphology can be generalized to cope with grayscale images. The chapter also includes a discussion (Section 7.5) on the noise behavior of morphological grouping operations and arrives at a formula explaining the shifts introduced by noise.

7.2 DILATION AND EROSION IN BINARY IMAGES

7.2.1 Dilation and Erosion

As we have seen in Chapter 2, dilation expands objects into the background and is able to eliminate "salt" noise within an object. It can also be used to remove cracks in objects that are less than 3 pixels in width.

In contrast, erosion shrinks binary picture objects, and has the effect of removing "pepper" noise. It also removes thin object "hairs" whose widths are less than 3 pixels.

As we shall see in more detail below, erosion is strongly related to dilation, in that a dilation acting on the inverted input image acts as an erosion, and vice versa.

7.2.2 Cancellation Effects

An obvious question is whether erosions cancel out dilations, or vice versa. We can easily answer this question: for if a dilation has been carried out, salt noise and cracks will have been removed, and once they are gone, erosion cannot bring them back; hence, exact cancellation will not occur in general. Thus, for the set S of object pixels in a general image I, we may write:

$$\text{erode (dilate}(S)) \neq S \tag{7.1}$$

equality only occurring for certain specific types of image (these will lack salt noise, cracks, and fine boundary detail). Similarly, pepper noise or hairs that are eliminated by erosion will not in general be restored by dilation:

$$\text{dilate (erode}(S)) \neq S \tag{7.2}$$

Overall, the most general statements that can be made are:

$$\text{erode (dilate}(S)) \supseteq S \qquad (7.3)$$

$$\text{dilate (erode}(S)) \subseteq S \qquad (7.4)$$

We may note, however, that large objects will be made 1 pixel larger all round by dilation, and will be reduced by 1 pixel all round by erosion, so a considerable amount of cancellation will normally take place when the two operations are applied in sequence. This means that sequences of erosions and dilations provide a good basis for filtering noise and unwanted detail from images.

7.2.3 Modified Dilation and Erosion Operators

It sometimes happens that images contain structures that are aligned more or less along the image axes' directions, and in such cases it is useful to be able to process these structures differently. For example, it might be useful to eliminate fine vertical lines, without altering broad horizontal strips. In that case the following "vertical erosion" operator will be useful:

```
for all pixels in image do {
   sigma = A1 + A5;
   if (sigma < 2) B0 = 0; else B0 = A0;
}
```
(7.5)

although it will be necessary to follow it with a compensating dilation operator[1] so that horizontal strips are not shortened:

```
for all pixels in image do {
   sigma = A1 + A5;
   if (sigma > 0) B0 = 1; else B0 = A0;
}
```
(7.6)

This example demonstrates some of the potential for constructing more powerful types of image filter. To realize these possibilities, we next develop a more general mathematical morphology formalism.

7.3 MATHEMATICAL MORPHOLOGY
7.3.1 Generalized Morphological Dilation

The basis of mathematical morphology is the application of set operations to images and their operators. We start by defining a generalized dilation mask as a set of locations within a 3×3 neighborhood. When referred to the center of the neighborhood as origin, each of these locations causes a shift of the image in the direction defined by the vector from the origin to the location. When several

---

[1]Here and elsewhere in this chapter, any operations required to restore the image to the original image space are not considered or included.

shifts are prescribed by a mask, the 1 locations in the various shifted images are combined by a set union operation.

The simplest example of this type is the identity operation *I*, which leaves the image unchanged:

(Note that we leave the 0s out of this mask, as we are now focussing on the set of elements at the various locations, and set elements are either present or absent.)

The next operation to consider is:

which is a left shift, equivalent to the one discussed in Section 2.2. Combining the above two operations into a single mask:

leads to a horizontal thickening of all objects in the image, by combining it with its left-shifted version. An isotropic thickening of all objects is achieved by the operator:

(clearly, this is equivalent to the dilation operator discussed in Sections 2.2 and 7.2), whereas a symmetrical horizontal thickening operation (see Section 7.2.3) is achieved by the mask:

$$
\begin{array}{|c|c|c|}
\hline
 & & \\
\hline
1 & 1 & 1 \\
\hline
 & & \\
\hline
\end{array}
$$

A rule of such operations is that if we want to guarantee that all the original object pixels are included in the output image, then we must include a 1 at the center (origin) of the mask.

Finally, there is no compulsion for all masks to be 3×3. Indeed, all but one of those listed above are effectively smaller than 3×3, and in more complex cases larger masks could be used. To emphasize this point, and to allow for asymmetrical masks in which the full 3×3 neighborhood is not given, we shall shade the origin—as shown in the above cases.

7.3.2 Generalized Morphological Erosion

We now move on to describe erosion in terms of set operations. The definition is somewhat peculiar in that it involves reverse shifts, but the reason for this will become clear as we proceed. Here the masks define directions as before, but in this case we shift the image in the reverse of each of these directions and perform intersection operations to combine the resulting images. For masks with a single element (as for the identity and shift left operators in Section 7.3.1), the intersection operation is improper and the final result is as for the corresponding dilation operator, but with a reverse shift. For more complex cases, the intersection operation results in objects being reduced in size. Thus, the mask:

has the effect of stripping away the left sides of objects (the object is moved right and *and*ed with itself). Similarly, the mask:

results in an isotropic stripping operation, and is hence identical to the erosion operation described in Section 7.2.1.

7.3.3 Duality Between Dilation and Erosion

We shall write the dilation and erosion operations formally as $A \oplus B$ and $A \ominus B$, respectively, where A is an image and B is the mask of the relevant operation:

$$A \oplus B = \cup_{b \in B} A_b \qquad (7.7)$$

$$A \ominus B = \cap_{b \in B} A_{-b} \qquad (7.8)$$

In these equations, A_b indicates a basic shift operation in the direction of element b of B and A_{-b} indicates the reverse shift operation.

We next prove an important theorem relating the dilation and erosion operations:

$$(A \ominus B)^c = A^c \oplus B^r \qquad (7.9)$$

where A^c represents the complement of A, and B^r represents the reflection of B in its origin. We first note that:[2]

$$x \in A^c \Leftrightarrow x \notin A \qquad (7.10)$$

and

$$b \in B^r \Leftrightarrow -b \in B \qquad (7.11)$$

---

[2]The sign " $\Leftrightarrow$ " means "if and only if," i.e., the statements so connected are equivalent.

We now have:[3]

$$x \in (A \ominus B)^c \Leftrightarrow x \notin A \ominus B$$
$$\Leftrightarrow \exists \, b \text{ such that } x \notin A_{-b}$$
$$\Leftrightarrow \exists \, b \text{ such that } x + b \notin A$$
$$\Leftrightarrow \exists \, b \text{ such that } x + b \in A^c$$
$$\Leftrightarrow \exists \, b \text{ such that } x \in (A^c)_{-b} \qquad (7.12)$$
$$\Leftrightarrow x \in \cup_{b \in B} \, (A^c)_{-b}$$
$$\Leftrightarrow x \in \cup_{b \in B^r} \, (A^c)_b$$
$$\Leftrightarrow x \in A^c \oplus B^r$$

This completes the proof. The related theorem:

$$(A \oplus B)^c = A^c \ominus B^r \qquad (7.13)$$

is proved similarly.

The fact that there are two such closely related theorems, following the related union and intersection definitions of dilation and erosion given above, indicates an important duality between the two operations. Indeed, as stated earlier, erosion of the objects in an image corresponds to dilation of the background, and vice versa. However, the two theorems indicate that this relation is not absolutely trivial, on account of the reflections of the masks required in the two cases. It is perhaps curious that in contrast with the case of the de Morgan rule for complementation of an intersection:

$$(P \cap Q)^c = P^c \cup Q^c \qquad (7.14)$$

the effective complementation of the dilating or eroding mask is its reflection rather than its complement *per se*, while that for the operator is the alternate operator.

7.3.4 Properties of Dilation and Erosion Operators

Dilation and erosion operators have some very important and useful properties. First, note that successive dilations are associative:

$$(A \oplus B) \oplus C = A \oplus (B \oplus C) \qquad (7.15)$$

whereas successive erosions are not. In fact, the corresponding relation for erosions is:

$$(A \ominus B) \ominus C = A \ominus (B \oplus C) \qquad (7.16)$$

Clearly, the apparent symmetry between the two operators is more subtle than their simple origins in expanding and shrinking might indicate.

Next, the property:

$$X \oplus Y = Y \oplus X \qquad (7.17)$$

---

[3]This proof is based on that of Haralick et al. (1987). The sign "∃" means "there exists," and in this context should be interpreted as "there is a value of." The symbol " ∈ " means "is a member of the following set"; the symbol "∉" means "is *not* a member of the following set."

means that the order in which dilations of an image are carried out does not matter, and the same applies to the order in which erosions are carried out:

$$(A \oplus B) \oplus C = (A \oplus C) \oplus B \tag{7.18}$$

$$(A \ominus B) \ominus C = (A \ominus C) \ominus B \tag{7.19}$$

In addition to the above relations, which use only the morphological operators $\oplus$ and $\ominus$, there are many more relations that involve set operations. In the examples that follow, great care must be exercised to note which particular distributive operations are actually valid:

$$A \oplus (B \cup C) = (A \oplus B) \cup (A \oplus C) \tag{7.20}$$

$$A \ominus (B \cup C) = (A \ominus B) \cap (A \ominus C) \tag{7.21}$$

$$(A \cap B) \ominus C = (A \ominus C) \cap (B \ominus C) \tag{7.22}$$

In certain other cases, where equality might *a priori* have been expected, the strongest statements that can be made are typified by the following:

$$A \ominus (B \cap C) \supseteq (A \ominus B) \cup (A \ominus C) \tag{7.23}$$

Note that the associative relations are of value in showing how large dilations and erosions might be factorized so that they can be implemented more efficiently as two smaller dilations and erosions applied in sequence. Similarly, the distributive relations show that a large mask may be split into two separate masks, which may then be applied separately and the resulting images *or*ed together to create the same final image. These approaches can be useful for providing efficient implementations, especially in cases where very large masks are involved. For example, we could dilate an image horizontally and vertically by two separate operations, which would then be merged together—as in the following instance:

1	1	1

followed by

	1	
	1	
	1	

=

1	1	1
1	1	1
1	1	1

Next, let us consider the importance of the identity operation I, which corresponds to a mask with a single 1 at the central ($A0$) position:

By way of example, we take Eqs. (7.20) and (7.21) and replace C by I in each of them. If we write the union of B and I as D, so that mask D is bound to contain a central 1 (i.e., $D \supseteq I$), we have:

$$A \oplus D = A \oplus (B \cup I) = (A \oplus B) \cup (A \oplus I) = (A \oplus B) \cup A \tag{7.24}$$

which always contains A:

$$A \oplus D \supseteq A \tag{7.25}$$

Similarly:

$$A \ominus D = A \ominus (B \cup I) = (A \ominus B) \cap (A \ominus I) = (A \ominus B) \cap A \qquad (7.26)$$

which is always contained within A:

$$A \ominus D \subseteq A \qquad (7.27)$$

Operations (such as dilation by a mask containing a central 1) which give outputs that are guaranteed to contain the inputs are termed *extensive*, whereas those (such as erosion by a mask containing a central 1) for which the outputs are guaranteed to be contained by the inputs are termed *antiextensive*. Clearly, extensive operations extend objects and antiextensive operations contract them, or in either case, leave them unchanged.

Another important type of operation is the *increasing* type of operation. An increasing operation is one, such as union, which preserves order in the size of the objects on which it operates. If object F is small enough to be contained within object G, then applying erosions or dilations will not affect the situation, even though the objects change their sizes and shapes considerably. We can write these conditions in the form:

$$\text{if} \quad F \subseteq G \qquad (7.28)$$

$$\text{then} \quad F \oplus B \subseteq G \oplus B \qquad (7.29)$$

$$\text{and} \quad F \ominus B \subseteq G \ominus B \qquad (7.30)$$

Next, we note that erosion can be used for locating the boundaries of objects in binary images:[4]

$$P = A - (A \ominus B) \qquad (7.31)$$

There are many practical applications of dilation and erosion, which follow particularly from using them together, as we shall see below.

Finally, we explore why the morphological definition of erosion involves a reflection. The idea is so that dilation and erosion are able, under the right circumstances, to cancel each other out. Take the left-shift dilation operation and the right-shift erosion operation. These are both achieved via the mask:

but in the erosion operation it is applied in its reflected form, thereby producing the right shift required to erode the left edge of any object. This makes it clear why an operation of the type $(A \oplus B) \ominus B$ has a chance of canceling to give A.

---

[4]Technically, we are here dealing with sets, and the appropriate set operation is the *andnot* function / rather than minus. However, the latter admirably conveys the required meaning without ambiguity.

More specifically, there must be shifts in opposite directions as well as appropriate subtractions produced by *and*ing instead or *or*ing in order for cancellation to be possible. Of course, in many cases the dilation mask will have 180° rotation symmetry, and then the distinction between B^r and B will be purely academic.

7.3.5 Closing and Opening

Dilation and erosion are basic operators from which many others can be derived. Earlier, we were interested in the possibility of an erosion canceling a dilation and vice versa. Hence, it is an obvious step to define two new operators that express the degree of cancellation: the first is called *closing* since it often has the effect of closing gaps between objects and the other is called *opening* because it often has the effect of opening gaps (Fig. 7.1). Closing (•) and opening (∘) are formally defined by the formulae:

$$A \bullet B = (A \oplus B) \ominus B \tag{7.32}$$

$$A \circ B = (A \ominus B) \oplus B \tag{7.33}$$

Closing is able to eliminate salt noise, narrow cracks or channels, and small holes or concavities.[5] Opening is able to eliminate pepper noise, fine hairs, and small protrusions. Thus, these operators are extremely important for practical applications. Furthermore, by subtracting the derived image from the original image, it is possible to locate many sorts of defect, including those cited above as being eliminated by opening and closing: this possibility makes the two operations even more important. For example, we might use the following operation to locate all the fine hairs in an image:

$$Q = A - A \circ B \tag{7.34}$$

This operator and its dual using opening:

$$R = A \bullet B - A \tag{7.35}$$

are extremely important for defect detection tasks. They are often, respectively, called the white and black top-hat operators.[6] (Practical applications of these two operators include location of solder bridges and cracks in printed circuit board tracks.)

Closing and opening have the interesting property that they are idempotent: this means that repeated application of either operation has no further effect.

---

[5]Here we continue to take the convention that dark objects have become 1s in binary images, and light background or other features have become 0s.

[6]It is dubious whether "top hat" is a very appropriate name for this type of operator: *a priori*, the term "residue function" (or simply "residue") would appear to be better, as it conjures up the right functional connotations.

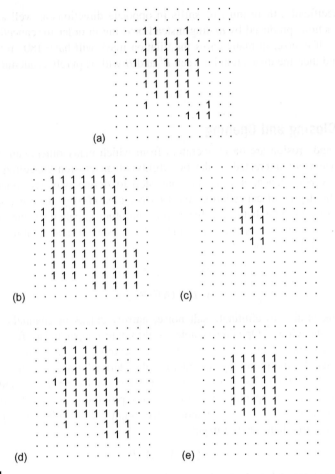

FIGURE 7.1

Results of morphological operations. (a) The original image, (b) the dilated image, (c) the eroded image, (d) the closed image, and (e) the opened image.

Source: © World Scientific 2000

(This property contrasts strongly with what happens when dilation and erosion are applied a number of times.) We can write these results formally as follows:

$$(A \bullet B) \bullet B = A \bullet B \tag{7.36}$$

$$(A \circ B) \circ B = A \circ B \tag{7.37}$$

From a practical point of view these properties are to be expected, since any hole or crack that has been filled in remains filled in, and there is no point in repeating the operation. Similarly, once a hair or protrusion has been removed, it cannot again be removed without first recreating it. Not quite so obvious is the fact that the combined closing and opening operation is idempotent:

$$\{[(A \bullet B) \circ C] \bullet B\} \circ C = (A \bullet B) \circ C \tag{7.38}$$

The same applies to the combined opening and closing operation. A simpler result is the following:

$$(A \oplus B) \circ B = (A \oplus B) \tag{7.39}$$

which shows that there is no point in opening with the same mask that has already been used for dilation: essentially, the first dilation produces some effects that are not reversed by the erosion (in the opening operation), and the second dilation then merely reverses the effects of the erosion. The dual of this result is also valid:

$$(A \ominus B) \bullet B = (A \ominus B) \tag{7.40}$$

There are a number of other properties of closing and opening; among the most important ones are the following set containment properties, which apply when $D \supseteq I$:

$$A \oplus D \supseteq A \bullet D \supseteq A \tag{7.41}$$

$$A \ominus D \subseteq A \circ D \subseteq A \tag{7.42}$$

Thus, closing an image will, if anything, increase the sizes of objects, while opening an image will, if anything, make objects smaller, although there are clear limits on how much change closing and opening operations can induce.

Finally, note that closing and opening are subject to the same duality as for dilation and erosion:

$$(A \bullet B)^c = A^c \circ B^r \tag{7.43}$$

$$(A \circ B)^c = A^c \bullet B^r \tag{7.44}$$

7.3.6 Summary of Basic Morphological Operations

The past few sections have by no means exhausted the properties of the morphological operations, dilate, erode, close, and open. However, these sections have outlined some of their properties and have demonstrated some of the practical results obtained using them. Perhaps the main aim of including the mathematical analysis has been to show that these operations are not *ad hoc* and that their properties are mathematically provable. Furthermore, the analysis has also indicated (a) how sequences of operations can be devised for a number of eventualities and (b) how sequences of operations can be analyzed to save computation (for instance) by taking care not to use idempotent operations repeatedly and by breaking masks down into smaller more efficient ones.

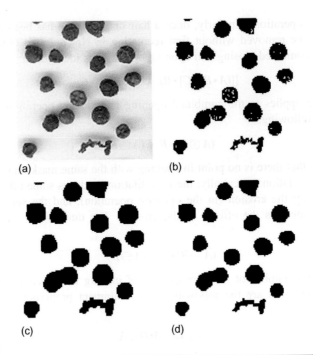

FIGURE 7.2

Use of the closing operation. (a) A peppercorn image, (b) the result of thresholding, (c) the result of applying a 3×3 dilation operation to the object shapes, and (d) the effect of subsequently applying a 3×3 erosion operation. The overall effect of the two operations is a "closing" operation. In this case closing is useful for eliminating the small holes in the objects: this would, e.g., be useful for helping to prevent misleading loops from appearing in skeletons. For this picture, extremely large window operations would be required to group peppercorns into regions.

Source: © World Scientific 2000

Overall, the operations devised here can help to eliminate noise and irrelevant artifacts from images so as to obtain more accurate recognition of shapes; they can also help to identify defects on objects by locating specific features of interest. In addition, they can perform grouping functions such as locating regions of images where small objects such as seeds may reside (Section 7.5). In general, elimination of artifacts is carried out by operations such as closing and opening, while location of such features is carried out by finding how the results of these operations differ from the original image (cf. Eqs. (7.20) and (7.21)); and locating regions where clusters of small objects occur may be achieved by larger scale closing operations. Clearly, care in the choice of scales and mask sizes is of vital importance in the design of complete algorithms for all these tasks. Figures 7.2 and 7.3 illustrate some of these possibilities in the case of a peppercorn image: some of the interest in this image relates to the presence of a twiglet and how it is eliminated from consideration and/or identified.

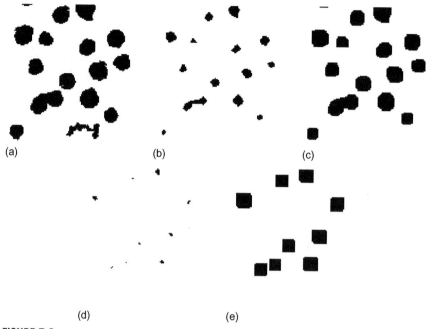

(a)   (b)   (c)

(d)   (e)

FIGURE 7.3

Use of the opening operation. (a) A thresholded peppercorn image. (b) The result of applying a 7×7 erosion operation to the object shapes. (c) The effect of subsequently applying a 7×7 dilation operation. The overall effect of the two operations is an "opening" operation. In this case, opening is useful for eliminating the twiglet. (d) and (e) The same respective operations when applied within an 11×11 window. Here some size filtering of the peppercorns has been achieved and all the peppercorns have been separated— thereby helping with subsequent counting and labeling operations.

Source: © World Scientific 2000

7.4 GRAYSCALE PROCESSING

The generalization of morphology to grayscale images can be achieved in a number of ways. A particularly simple approach is to employ "flat" structuring elements. These perform morphological processing in the same way for each of the gray levels, acting as if the shapes at each level were separate, independent binary images. If dilation is carried out in this way, the result turns out to be identical to the effect of applying a maximum intensity operation of the same shape: i.e., we replace set inclusion by a magnitude comparison; needless to say, this is mathematically identical in action for a normal binary image, but when applied to a grayscale image it neatly generalizes the dilation concept. Similarly, erosion can be carried out by applying a minimum intensity structuring element of the same shape as the original binary structuring element. This discussion assumes that we focus on light objects against dark backgrounds[7]: these will be dilated when the maximum intensity operation is applied, and eroded when the minimum intensity operation is applied; we could of course reverse the convention, depending on what type of objects we are concentrating on at any moment, or in any application. We can summarize the situation as follows:

$$A \oplus B = \max_{b \in B} A_b \qquad (7.45)$$

$$A \ominus B = \min_{b \in B} A_{-b} \qquad (7.46)$$

There are other more complex grayscale analogs of dilation and erosion: these take the form of 3-D structuring elements whose output at any gray level depends not only on the shape of the image at that gray level, but also on the shapes at a number of nearly gray levels. Although such "nonflat" structuring elements are useful, for a good many applications they are not necessary, as flat structuring elements already embody a very considerable amount of generalization relative to the binary case.

Next we consider how edge detection is carried out using grayscale morphology.

7.4.1 Morphological Edge Enhancement

In Section 2.2.2, we have shown how edge detection can be carried out in binary images. We have defined edge detection asymmetrically, in the sense that the edge is the part of the object that is next to the background. This is useful because including the part of the background next to the object would merely have served to make the boundary wider and less precise. However, edge detection in grayscale images does not need to embody such an asymmetry,[8] because it starts by performing edge enhancement and then carrying out a thresholding type of operation—the width being controlled largely by the manner of thresholding and whether nonmaximum suppression or other factors are brought to bear. Here we start by formulating the original binary edge detector in morphological form. Then we make it more symmetric. Finally we generalize it to grayscale operation.

The original binary edge detector may be written in the form (cf. Eq. (7.31)):

$$E = A - (A \ominus B) \qquad (7.47)$$

---

[7]In fact, this is the opposite convention to that employed in Chapter 2, but, as we shall see below, in grayscale processing it is probably more general to focus on intensities rather than on specific objects.

[8]Indeed, any asymmetry would lead to an unnecessary bias and hence inaccuracy in the location of edges.

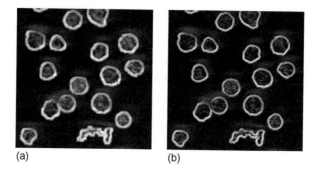

(a)   (b)

FIGURE 7.4

Determination of the morphological gradient of an image. The original image is that of Fig. 7.2(a). (a) The morphological gradient, obtained using 3×3 window operations. (b) The result for a Sobel operator: note that the latter gives less diffuse responses.

Source: © World Scientific 2000

Making it symmetrical merely involves adding the background edge $((A \oplus B) - A)$:

$$G = (A \oplus B) - (A \ominus B) \tag{7.48}$$

To convert to grayscale operation involves employing maximum and minimum operations in place of dilation and erosion. In this case we are concentrating on intensity *per se*, and so these respective assignments of dilation and erosion are used (the alternate arrangement would result in negative edge contrast). Thus, we here use Eqs. (7.45) and (7.46) to *define* dilation and erosion for grayscale processing, so Eq. (7.48) already represents morphological edge enhancement for grayscale images. The argument G is often called the morphological gradient of an image (Fig. 7.4). Note that it is not accompanied by an accurate edge orientation value, although approximate orientations can, of course, be computed by determining which part of the structuring element gives rise to the maximum signal.

7.4.2 Further Remarks on the Generalization to Grayscale Processing

In the previous subsection we found that reinterpreting Eq. (7.48) permitted edge detection to be generalized immediately from binary to gray scale. This is a consequence of the extremely powerful *umbra homomorphism theorem*. This starts with the knowledge that intensity I is a single-valued function of position within the image. This means that it represents a surface in a 3-D (grayscale) space. However, as we have seen, it is useful to take into account the individual gray levels. In particular, we note that the set of pixels of gray level g_i is a subset of the set of pixels of gray level g_{i-1}, where $g_i \geq g_{i-1}$. The important step forward is to interpret the 3-D volume containing all these gray levels under the intensity surface as constituting an umbra—a 3-D shadow region for the relevant part of the surface. In fact, we write the umbra volume of I as $U(I)$, and clearly we also have $I = T(U(I))$, where the operator $T(\cdot)$ recalculates the top surface.

The umbra homomorphism theorem then states that a dilation has to be defined and interpreted as an operation on the umbras:

$$U(I \oplus K) = U(I) \oplus U(K) \qquad (7.49)$$

To find the relevant intensity function we merely need to apply the top-surface operator to the umbra:

$$I \oplus K = T(U(I \oplus K)) = T(U(I) \oplus U(K)) \qquad (7.50)$$

A similar statement applies for erosion.

The next step is to note that the generalization from binary to grayscale dilation using flat structuring elements involved applying a maximum operation in place of a set union operation. The operation can, for the simple case of a 1-D image, be rewritten in the form:

$$(I \oplus K)(x) = \max(I(x - z) + K(z)) \qquad (7.51)$$

where $K(z)$ takes value 0 or 1 only, and $x - z$, z must lie within the domains of I and K. However, another vital step is to notice that this form generalizes to nonbinary $K(z)$, whose values can, e.g., be the integer gray-level values. This makes the dilation operation considerably more powerful, yielding the nonflat structuring element concept—which will clearly also work for 2-D images with full gray scale.

It can be tedious working out the responses of this sort of operation, but it is susceptible to a neat geometric interpretation. If the function $K(z)$ is inverted and turned into a template, this may be run over the image $I(x)$ in such a way as to remain just in contact with it. Thus, the origin of the inverted template will trace out the top surface of the dilated image. The process is depicted in Fig. 7.5 for the case of a triangular structuring element in a 1-D image.

Similar relations apply for erosion, closing, opening, and a variety of set functions. This means that the standard binary morphological relations, Eqs. (7.15)–(7.23), apply for grayscale images as well as for binary images. Furthermore, the dilation–erosion and closing–opening dualities (Eqs. (7.9), (7.13), (7.43), and (7.44)) also apply for grayscale images. These are extremely powerful results, and allow one to apply morphological concepts in an intuitive manner. In that case the practically important factor devolves into choosing the right grayscale structuring element for the application.

Another interesting factor is the possibility of using morphological operations instead of convolutions in the many places where the latter are employed throughout image analysis. We have already seen how edge enhancement and detection can be performed using morphology in place of convolution. In addition, noise suppression by Gaussian smoothing can be replaced by opening and closing operations. However, it must always be borne in mind that convolutions are linear operations, and are thereby grossly restricted, whereas morphological operations

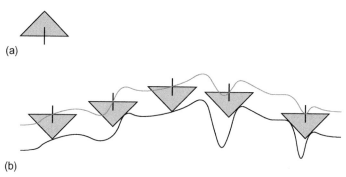

FIGURE 7.5

Dilation of 1-D grayscale image by triangular structuring element. (a) The structuring element, with the vertical line at the bottom indicating the origin of coordinates. (b) The original image (continuous black line), several instances of the inverted structuring element being applied, and the output image (continuous gray line). This geometric construction automatically takes account of the maximum operation in Eq. (7.51). Note that as no part of the structuring element is below the origin in (a), the output intensity is increased at every point in the image.

are highly nonlinear, their very structure embodying multiple "if" statements, so outward appearances of similarity are bound to hide deep differences of operation, effectiveness, and applicability. Consider, e.g., the optimality of the mean filter for suppressing Gaussian noise and the optimality of the median filter (a morphological operator) for eliminating impulse noise.

One example of this is worth including here: when performing edge enhancement prior to edge detection, it is possible to show that if both a differential gradient (e.g., Sobel) operator and a morphological gradient operator are applied to a noise-free image with a steady intensity gradient, the results will be identical, within a constant factor. However, if one impulse noise pixel arises, the maximum or minimum operations of the morphological gradient operator will select this value in calculating the gradient, whereas the differential gradient operator will average its effect over the window, giving significantly less error.

Space prevents grayscale morphological processing from being considered in more detail here (see, e.g., Haralick and Shapiro (1992) and Soille (2003)).

7.5 EFFECT OF NOISE ON MORPHOLOGICAL GROUPING OPERATIONS

Texture analysis is an important area of machine vision, and is relevant not only for segmenting one region of an image from another (as in many remote sensing applications), but also for characterizing regions absolutely—as is necessary when performing surface inspection (e.g., when assessing the paint finish on automobiles). Many methods have been employed for texture analysis. These range from the widely used gray-level co-occurrence matrix approach to Law's texture energy approach and from the use of Markov random fields to fractal modeling (Chapter 8).

In fact, there are approaches that involve even less computation and which are applicable when the textures are particularly simple and the shapes of the basic texture elements are not especially critical. For example, if it is required to locate regions containing small objects, simple morphological operations applied to thresholded versions of the image are often appropriate (Fig. 7.6) (Haralick and Shapiro, 1992; Bangham and Marshall, 1998). Such approaches can be used for locating regions containing seeds, grains, nails, sand, or other materials, either for assessing the overall quantity or spread or for determining whether there are regions that have not yet been covered. The basic operation to be applied is the dilation operation, which combines the individual particles into fully connected regions. This method is suitable not only for connecting individual particles but also for separating regions containing high and low densities of such particles. The expansion characteristic of the dilation operation can be largely canceled by a subsequent erosion operation, using the same morphological kernel. Indeed, if the particles are always convex and well separated, the erosion should exactly cancel the dilation, although in general the combined closing operation is not a null operation, and this is relied upon in the above connecting operation.

FIGURE 7.6

Idealized grouping of small objects into regions such as might be attempted using closing operations.

Source: © IEE 2000

Closing operations have been applied to images of cereal grains containing dark rodent droppings in order to consolidate the droppings (which contain significant speckle—and therefore holes when the images are thresholded) and thus to make them more readily recognizable from their shapes (Davies et al., 1998a). However, the initial result was rather unsatisfactory as dark patches on the grains tend to combine with the dark droppings: this has the effect of distorting the shapes and also makes the objects larger. This problem was partially overcome by a subsequent erosion operation so that the overall procedure is dilate + erode + erode (for further details, see Chapter 21). Originally, adding this final operation seemed to be an *ad hoc* procedure, but on analysis it was found that the size increase actually applies quite generally when segmentation of textures containing different densities of particles is carried out. It is this general effect that we now consider.

7.5.1 Detailed Analysis

Let us take two regions containing small particles with occurrence densities ρ_1 and ρ_2, where $\rho_1 > \rho_2$. In region 1, the mean distance between particles will be d_1 and in region 2, the mean distance will be d_2, where $d_1 < d_2$. If we dilate using a kernel of radius a, where $d_1 < 2a < d_2$, this will tend to connect the particles in region 1 but should leave the particles in region 2 separate. To *ensure* connecting the particles in region 1, we can make $2a$ larger than $\frac{1}{2}(d_1 + d_2)$, but this may risk connecting the particles in region 2 (the risk will be reduced when the subsequent erosion operation is taken into account). Selecting an optimum value of a clearly depends not only on the mean distances d_1 and d_2 but also on their distributions. This is not discussed in detail due to space limitation: here we assume that a suitable selection of a is made, and that it is effective. The problem that is tackled next is whether the size of the final regions matches the *a priori* desired segmentation, i.e., whether any size distortion takes place. We start by taking this to be an essentially 1-D problem, which can be modeled as in Fig. 7.7 (in what follows, the 1-D particle densities are given an x suffix).

Suppose first that $\rho_{2x} = 0$. Then in region 2, the initial dilation will be counteracted exactly (in 1-D) by the subsequent erosion. Next take $\rho_{2x} > 0$: when dilation occurs, a number of particles in region 2 will be enveloped, and the erosion process will not exactly reverse the dilation. If a particle in region 2 is within $2a$ of an outermost particle in region 1, it will merge with region 1, and will remain merged when erosion occurs. The probability P that this will happen is the integral over a distance $2a$ of the probability density for particles in region 2. In addition, when the particles are well separated we can take the probability density as being equal to the mean particle density ρ_{2x}. Hence:

$$P = \int_0^{2a} \rho_{2x} \mathrm{d}x = 2a\rho_{2x} \qquad (7.52)$$

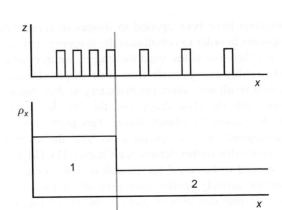

FIGURE 7.7

1-D particle distribution. z indicates the presence of a particle, and ρ_x shows the densities in the two regions.

Source: © IEE 2000

If such an event occurs, then region 1 will be expanded by amounts ranging from a to $3a$, or 0 to $2a$ after erosion, although these figures must be increased by b for particles of width b. Thus, the *mean* increase in size of region 1 after dilation + erosion is $2a\rho_{2x} \times (a + b)$, where we have assumed that the particle density in region 2 remains uniform right up to region 1.

Next we consider what additional erosion operation will be necessary to cancel this increase in size. In fact, we just make the radius $\tilde{a}_{1\text{-D}}$ of the erosion kernel equal to the increase in size:

$$\tilde{a}_{1\text{-D}} = 2a\rho_{2x}(a + b) \tag{7.53}$$

Finally, we must recognize that the required process is 2-D rather than 1-D, and take y to be the lateral axis, normal to the original (1-D) x-axis. For simplicity, we assume that the dilated particles in region 2 are separated laterally, and are not touching or overlapping (Fig. 7.8). As a result, the change of size of region 1 given in Eq. (7.53) will be diluted relative to the 1-D case by the reduced density along the direction (y) of the border between the two regions: i.e., we must multiply the right-hand side of Eq. (7.53) by $b\rho_{2y}$. We now obtain the relevant 2-D equation:

$$\tilde{a}_{2\text{-D}} = 2ab\rho_{2x}\rho_{2y}(a + b) = 2ab\rho_2(a + b) \tag{7.54}$$

where we have finally reverted to the appropriate 2-D *area* particle density ρ_2.

Clearly, for low values of ρ_2 an additional erosion will not be required, whereas for high values of ρ_2 substantial erosion will be necessary, particularly if b is comparable to or larger than a. If $\tilde{a}_{2\text{-D}} < 1$, it will be difficult to provide an accurate correction by applying an erosion operation, and all that can be done is

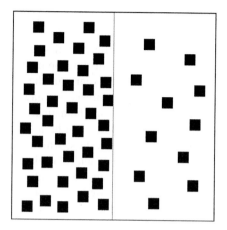

FIGURE 7.8

Model of the incidence of particles in two regions. Region 2 (on the right) has sufficiently low density that the dilated particles will not touch or overlap.

Source: © IEE 2000

to bear in mind that any measurements made from the image will require correction. (Note that if, as often happens, $a \gg 1$, $\tilde{a}_{2\text{-D}}$ could well be at least 1.)

7.5.2 **Discussion**

The work of the author described above (Davies, 2000c) was motivated by analysis of cereal grain images containing rodent droppings, which had to be consolidated by dilation operations to eliminate speckle, followed by erosion operations to restore size.[9] It has been found that if the background contains a low density of small particles that tend, upon dilation, to increase the sizes of the foreground objects, additional erosion operations will in general be required to accurately represent the sizes of the regions. The effect would be similar if impulse noise were present, although theory shows what is observed in practice—that the effect is enhanced if the particles in the background are not negligible in size. The increases in size are proportional to the occurrence density of the particles in the background, and the kernel for the final erosion operation is calculable, the overall process being a necessary measure rather than an *ad hoc* technique.

7.6 **CONCLUDING REMARKS**

Binary images contain all the data needed to analyze the shapes, sizes, positions, and orientations of objects in two dimensions, and thereby to recognize them

---

[9]For further background on this application, see Chapter 21.

and even to inspect them for defects. As we shall see in Chapters 9 and 10, many simple small neighborhood operations exist for processing binary images and moving toward the goals stated earlier. At first sight these may appear a somewhat random set, reflecting historical development rather than systematic analytic tools. However, in the past few years, mathematical morphology has emerged as a unifying theory of shape analysis: we have aimed to give the flavor of the subject in this chapter. In fact, mathematical morphology, as its name suggests, is mathematical in nature, and this can be a source of difficulty,[10] but there are a number of key theorems and results, which are worth remembering: a few of these have been considered here and placed in context. For example, generalized dilation and erosion have acquired a central importance, since further vital concepts and constructs are based on them—closing, opening, template matching, and even connectedness properties (although space has prevented a detailed discussion of its application to the last two topics). For further information on grayscale morphological processing, see Haralick and Shapiro (1992) and Soille (2003).

> Mathematical morphology is one of the standard methodologies that has evolved over the past few decades. This chapter has demonstrated how the mathematical aspects make the subject of shape analysis rigorous and less ad hoc. Its extensions to grayscale image processing are interesting and useful, although at this stage they have not ousted more traditional approaches.

7.7 BIBLIOGRAPHICAL AND HISTORICAL NOTES

The book by Serra (1982) was an important early landmark in the development of morphology. Many subsequent papers helped to lay the mathematical foundations, perhaps the most important and influential being that by Haralick et al. (1987); see also Zhuang and Haralick (1986) for methods for decomposing morphological operators, and Crimmins and Brown (1985) for more practical aspects of shape recognition. The papers by Dougherty and Giardina (1988), Heijmans (1991), and Dougherty and Sinha (1995a, 1995b) were important in the development of methods for grayscale morphological processing, while the work of Huang and Mitchell (1994) on grayscale morphology decomposition and that of Jackway and Deriche (1996) on multiscale morphological operators gave further impetus to the subject.

---

[10]The rigor of mathematics is a cause for celebration, but at the same time it can make the arguments and the results obtained from them less intuitive. On the other hand, the real benefit of mathematics is to leapfrog what is possible by intuition alone and to arrive at results that are new and unexpected.

One problem is that it is by no means obvious how to decide on the sequence of morphological operations that is required in any application. This is an area where genetic algorithms have contributed to the systematic generation of complete systems (see, e.g., Harvey and Marshall, 1994).

Work up to 1998 is reviewed in a useful tutorial paper by Bangham and Marshall (1998): more recently Soille (2003) produced a thoroughgoing volume on the subject. Gil and Kimmel (2002) address the problem of rapid implementation of dilation, erosion, opening, and closing algorithms, and arrive at a new approach based on deterministic calculations: these give low computational complexity for calculating max and min functions, and similar complexities for the four cited filters and other derived filters. The paper goes on to state some open problems, and to suggest how they might be tackled: it is clear that some apparently simple tasks that ought perhaps to have been dealt with before the 2000s are still unsolved—such as how to compute the median in better than $O(\log^2 p)$ time per filtered point (in a $p \times p$ window) or how to *optimally* extend 1-D morphological operations to circular rather than square window (2-D) operations. Note that the new implementation is immediately applicable to determining the morphological gradient of an image.

7.7.1 More Recent Developments

More recently, Bai and Zhou (2010) have designed a top-hat selection transformation for locating and enhancing small dim infrared targets typified by aircraft in the sky. The selection transformation is based on the classical top-hat (residue) operator. A necessary parameter in the analysis is the value of n, the minimum difference in intensity between the target and the background, and methods are given for estimating it. Jiang et al. (2007) also use a residue operator to find thin low-contrast edges. The method uses five basic 5×5 masks to detect edges of the right widths. Very high resistance to noise is demonstrated by the particular combination of techniques applied in this approach. Soille and Vogt (2009) show how binary images may be segmented to identify a range of different types of pattern. These include the following mutually exclusive foreground categories: core, islet, connector (loop and bridge), boundary (perforation and edge), branch, and segmented binary pattern. Lézoray and Charrier (2009) describe a new approach to color image segmentation, by analysis of color projections in 2-D histograms to find the dominant colors: the important factor is that clustering in 2-D histograms can proceed very effectively using standard image processing techniques, including morphological processing. Valero et al. (2010) use directional mathematical morphology to detect roads in remote sensing images. The paper starts by taking roads to be linear connected paths; however, curved road segments and other network details can be dealt with using "path openings" and "path closings" in order to obtain the required structural information.

7.8 PROBLEM

1. A morphological gradient binary edge enhancement operator is defined by the formula:

$$G = (A \oplus B) - (A \ominus B)$$

Using a 1-D model of an edge, or otherwise, shows that this will give wide edges in binary images. If grayscale dilation ($\oplus$) is equated to taking a local *maximum* of the intensity function within a 3×3 window, and grayscale erosion ($\ominus$) is equated to taking a local *minimum* within a 3×3 window, sketch the result of applying the operator G. Show that it is similar in effect to a Sobel edge enhancement operator, if edge orientation effects are ignored by taking the Sobel magnitude:

$$g = (g_x^2 + g_y^2)^{1/2}$$

Texture

8

It is quite easy to understand what a texture is, although somewhat less easy to define it. For many reasons it is useful to be able to classify textures and to distinguish them from one another; it is also useful to be able to determine the boundaries between different textures, as they often signify the boundaries of real objects. This chapter studies the means for achieving these aims.

Look out for:

- basic measures by which textures can be classified—such as regularity, randomness, and directionality.
- problems that arise with "obvious" texture analysis methods, such as autocorrelation.
- the long-standing graylevel co-occurrence matrix method.
- Laws' method and Ade's generalization of it.
- the fact that textures have to be analyzed statistically, because of the random element in their construction.

Texture analysis is a core element in the vision repertoire, just as textures are core components of most images. It therefore seemed most appropriate to include this topic in Part 1 of the book.

8.1 INTRODUCTION

In the foregoing chapters, many aspects of image analysis and recognition have been studied. At the core of these matters has been the concept of segmentation, which involves the splitting of images into regions that have some degree of uniformity, whether in intensity, color, texture, depth, motion, or other relevant attributes. Care has been taken in Chapter 4 to emphasize that such a process will be largely *ad hoc*, since the boundaries produced will not necessarily correspond to those of real objects. Nevertheless, it is important to make the attempt, either as a preliminary to more accurate or iterative demarcation of objects and their facets, or as an end in itself—e.g., to judge the quality of surfaces.

In this chapter, we move on to the study of texture and its measurement. Texture is a difficult property to define: indeed, in 1979, Haralick reported that no satisfactory definition of it had up till then been produced. Perhaps we should not be surprised by this, as the concept has rather separate meanings in the contexts of vision, touch, and taste: furthermore, the ways in which different people understand the terms are highly individual and subjective. Nevertheless, we require a working definition of texture, and in vision the particular aspect we focus on is the variation in intensity of a particular surface or region of an image. Even with this statement we are being indecisive about whether it is the physical object being observed which is being described or the image derived from it. This reflects the fact that it is the roughness of the surface or the structure or composition of the material that originally gives rise to its visual properties. However, in this chapter, we are mainly interested in the interpretation of images, and so we define texture as the characteristic variation in intensity of a region of an image, which should allow us to recognize and describe it and to outline its boundaries (Fig. 8.1).

This definition of texture implies that texture is nonexistent in a surface of uniform intensity and does not say anything about how the intensity might be expected to vary or how we might recognize and describe it. In fact, there are many ways in which intensity might vary, but if the variation does not have sufficient uniformity, the texture may not be characterized sufficiently close to permit recognition or segmentation.

We next consider ways in which intensity might vary. Clearly, it can vary rapidly or slowly, markedly or with low contrast, with a high or low degree of directionality, and with greater or lesser degrees of regularity. This last characteristic is often taken as key: either the textural pattern is regular as for a piece of cloth, or it is random as for a sandy beach or a pile of grass cuttings. However, this ignores the fact that a regular textural pattern is often not wholly regular (again, as for a piece of cloth), or not wholly random (as for a mound of potatoes of similar size). Thus, the degrees of randomness and of regularity will have to be measured and compared when characterizing a texture.

There are more profound things to say about the textures described earlier in this section. Often textures are derived from tiny objects or components that are themselves similar, but that are placed together in ways ranging from purely random to purely regular—be they bricks in a wall, grains of sand, blades of grass, strands of material, stripes on a shirt, wickerwork on a basket, or a host of other items. In texture analysis it is useful to have a name for the similar textural

FIGURE 8.1

The variety of textures obtained from real objects. (a) Bark, (b) wood grain, (c) fir leaves, (d) chick peas, (e) carpet, (f) fabric, (g) stone chips, (h) water. These textures demonstrate the wide variety of familiar textures that are easily recognized from their characteristic intensity patterns.

elements that are replicated over a region of the image: such textural elements are called *texels*. These considerations lead us to characterize textures in the following ways:

1. The texels will have various sizes and degrees of uniformity.
2. The texels will be orientated in various directions.
3. The texels will be spaced at varying distances in different directions.
4. The contrast will have various magnitudes and variations.
5. Various amounts of background may be visible between texels.
6. The variations composing the texture may each have varying degrees of regularity *vis-à-vis* randomness.

It is quite clear from this discussion that a texture is a complicated entity to measure. The reason is primarily that many parameters are likely to be required to characterize it: in addition, when so many parameters are involved, it is difficult to disentangle the available data and measure the individual values or decide the ones that are most relevant for recognition. And of course, the statistical nature of many of the parameters is by no means helpful. However, we have so far only attempted to show how complex the situation can be. In what follows, we attempt to show that quite simple measures can be used to recognize and segment textures in practical situations.

Before proceeding, it is useful to recall that in the analysis of shape, there is a dichotomy between available analysis methods. We could, e.g., use a set of measures such as circularity, aspect ratio, and so on, which would permit a description of the shape, but which would not allow it to be reconstructed; or else we could use descriptors such as skeletons with distance function values, or moments, which would permit full and accurate reconstruction—although the set of descriptors might have been curtailed so that only limited but predictable accuracy was available. In principle, such a reconstruction criterion should be possible with texture. However, in practice there are two levels of reconstruction. In the first, we could reproduce a pattern which, to human eyes, would be indistinguishable from the off-camera texture until one compared the two on a pixel-by-pixel basis. In the second, we could reproduce a textured pattern exactly. The point is that textures are normally partially statistical in nature, so it will not be easy to obtain a pixel-by-pixel match in intensities: nor, in general, will it be worth aiming to do so. Thus, texture analysis generally only aims at obtaining accurate statistical descriptions of textures, from which *apparently* identical textures can be reproduced, if desired.

Very many workers have contributed to, and used, a wide range of approaches for texture analysis over a period of 40 years. The sheer weight of the available material and the statistical nature of it can be daunting for many. Note that Section 8.4 is particularly relevant to practitioners because it describes the Laws' texture energy approach which is intuitive, straightforward to apply in both software and hardware, and highly effective in many application areas. However, Section 8.3 on graylevel co-occurrence matrices (which were important historically) can be omitted on a first reading.

8.2 SOME BASIC APPROACHES TO TEXTURE ANALYSIS

In Section 8.1, texture was defined as the characteristic variation in intensity of a region of an image that should allow us to recognize and describe it and to outline its boundaries. In view of the likely statistical nature of textures, this prompts us to characterize texture by the variance in intensity values taken over the whole region of the texture.[1] However, such an approach will not give a rich enough description of the texture for most purposes and will certainly not provide any possibility of reconstruction: it will also be especially unsuitable in cases where the texels are well defined, or where there is a high degree of periodicity in the texture. On the other hand, for highly periodic textures such as that arise with many textiles, it is natural to consider the use of Fourier analysis. Indeed, in the early days of image analysis, this approach was tested thoroughly, although the results were not always encouraging.

Bajcsy (1973) used a variety of ring and orientated strip filters in the Fourier domain to isolate texture features—an approach that was found to work successfully on natural textures such as grass, sand, and trees. However, there is a general difficulty in using the Fourier power spectrum in that the information is more scattered than might be expected at first. In addition, strong edges and image boundary effects can prevent accurate texture analysis by this method. Perhaps more important is the fact that the Fourier approach is a global one that is difficult to apply successfully to an image that is to be segmented by texture analysis (Weszka et al., 1976).

Autocorrelation is another obvious approach to texture analysis, since it should show up both local intensity variations and also the repeatability of the texture (see Fig. 8.2). An early study was carried out by Kaizer (1955). He examined how many pixels an image has to shift before the autocorrelation function drops to $1/e$ of its initial value and produced a subjective measure of coarseness on this basis. However, Rosenfeld and Troy (1970a, 1970b) later showed that autocorrelation is not a satisfactory measure of coarseness. In addition, autocorrelation is not a very good discriminator of isotropy in natural textures. Hence, workers were quick to take up the co-occurrence matrix approach introduced by Haralick et al. (1973): in fact, this approach not only replaced the use of autocorrelation but during the 1970s also became to a large degree the "standard" approach to texture analysis.

8.3 GRAYLEVEL CO-OCCURRENCE MATRICES

The graylevel co-occurrence matrix approach[2] is based on studies of the statistics of pixel intensity distributions. As hinted above with regard to the variance in

---

[1]We defer for now the problem of finding the region of a texture so that we can compute its characteristics in order to perform a segmentation function. However, some preliminary training of a classifier may clearly be used to overcome this problem for supervised texture segmentation tasks.
[2]This is also frequently called the spatial graylevel dependence matrix (SGLDM) approach.

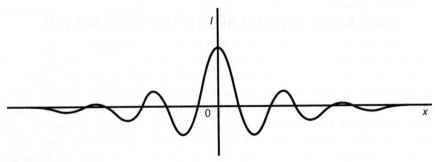

FIGURE 8.2

Use of autocorrelation function for texture analysis. This diagram shows the possible 1-D profile of the autocorrelation function for a piece of material in which the weave is subject to significant spatial variation: note that the periodicity of the autocorrelation function is damped down over quite a short distance.

pixel intensity values, single-pixel statistics do not provide rich enough descriptions of textures for practical applications. Thus, it is natural to consider second-order statistics obtained by considering *pairs* of pixels in certain spatial relations to each other. Hence, co-occurrence matrices are used, which express the relative frequencies (or probabilities) $P(i, j|d,\theta)$ with which two pixels having relative polar coordinates (d, θ) appear with intensities i, j. The co-occurrence matrices provide raw numerical data on the texture, although this data must be condensed to relatively few numbers before it can be used to classify the texture. The early paper by Haralick et al. (1973) gave 14 such measures, and these were used successfully for classification of many types of material (including, e.g., wood, corn, grass, and water). However, Conners and Harlow (1980a) found that only five of these measures were normally used, *viz.* "energy," "entropy," "correlation," "local homogeneity," and "inertia" (note that these names do not provide much indication of the modes of operation of the respective operators).

To obtain a more detailed idea of the operation of the technique, consider the co-occurrence matrix shown in Fig. 8.3. This corresponds to a nearly uniform image containing a single region in which the pixel intensities are subject to an approximately Gaussian noise distribution, the attention being on pairs of pixels at a constant vector distance $\mathbf{d} = (d, \theta)$ from each other. Next, consider the co-occurrence matrix shown in Fig. 8.4, which corresponds to an almost noiseless image with several nearly uniform image regions. In this case, the two pixels in each pair may correspond either to the same image regions or to different ones, although if d is small they will only correspond to adjacent image regions. Thus, we have a set of N on-diagonal patches in the co-occurrence matrix, but only a limited number L of the possible number M of off-diagonal patches linking them, where $M = {}^{N}C_2$ and $L \leq M$ (typically L will be of order N rather than N^2). With textured images, if the texture is not too strong, it may be modeled as noise, and

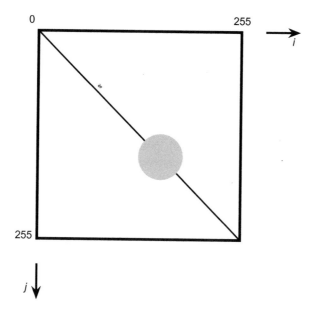

FIGURE 8.3

Co-occurrence matrix for a nearly uniform grayscale image with superimposed Gaussian noise. Here the intensity variation is taken to be almost continuous: normal convention is followed by making the j index increase downward, as for a table of discrete values (cf. Fig. 8.4).

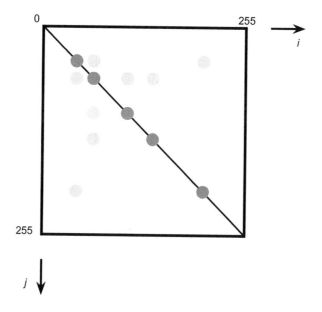

FIGURE 8.4

Co-occurrence matrix for an image with several distinct regions of nearly constant intensity. Again, the leading diagonal of the diagram is from top left to bottom right (cf. Figs. 8.2 and 8.5).

the $N + L$ patches in the image will be larger but still not overlapping. However, in more complex cases, the possibility of segmentation using the co-occurrence matrices will depend on the extent to which **d** can be chosen to prevent the patches from overlapping. Since many textures are directional, careful choice of θ will clearly help with this task, although the optimum value of d will depend on several other characteristics of the texture.

As a further illustration, we consider the small image shown in Fig. 8.5(a). To produce the co-occurrence matrices for a given value of **d**, we merely need to calculate the numbers of cases for which pixels, a distance **d** apart, have intensity values i and j. Here, we content ourselves with the two cases **d** $= (1, 0)$ and **d** $= (1, \pi/2)$. We thus obtain the matrices shown in Fig. 8.5(b) and (c).

This simple example demonstrates that the amount of data in the matrices is liable to be many times more than in the original image—a situation which is exacerbated in more complex cases by the number of values of d and θ that are required to accurately represent the texture. In addition, the number of gray levels will normally be closer to 256 than to 6, and the amount of matrix data varies as the square of this number. Finally, we should notice that the co-occurrence

0	0	0	1
1	1	1	1
2	2	2	3
3	3	4	5

(a)

	0	1	2	3	4	5
0	2	1	0	0	0	0
1	1	3	0	0	0	0
2	0	0	2	1	0	0
3	0	0	1	1	1	0
4	0	0	0	1	0	1
5	0	0	0	0	1	0

(b)

	0	1	2	3	4	5
0	0	3	0	0	0	0
1	3	1	3	1	0	0
2	0	3	0	2	1	0
3	0	1	2	0	0	1
4	0	0	1	0	0	0
5	0	0	0	1	0	0

(c)

FIGURE 8.5

Co-occurrence matrices for a small image. (a) The original image; (b) the resulting co-occurrence matrix for **d** $= (1, 0)$; and (c) the matrix for **d** $= (1, \pi/2)$. Note that even in this simple case, the matrices contain more data than the original image.

matrices merely provide a new representation: they do not themselves solve the recognition problem.

These factors mean that the gray scale has to be compressed into a much smaller set of values and careful choice of specific sample d, θ values must be made: in most cases, it is not at all obvious how such a choice should be made, and it is even more difficult to arrange for it to be made automatically. In addition, various functions of the matrix data must be tested before the texture can be properly characterized and classified.

These problems with the co-occurrence matrix approach have been tackled in many ways: just two are mentioned here. The first is to ignore the distinction between opposite directions in the image, thereby reducing storage by 50%. The second is to work with *differences* between gray levels; this amounts to performing a summation in the co-occurrence matrices along axes parallel to the main diagonal of the matrix. The result is a set of *first*-order *difference* statistics. While these modifications have given some additional impetus to the approach, the 1980s saw a highly significant diversification of methods for the analysis of textures. Of these, Laws' approach (1979, 1980a, 1980b) is important in that it has led to other developments which provide a systematic, adaptive means of tackling texture analysis. This approach is covered in Section 8.4.

8.4 LAWS' TEXTURE ENERGY APPROACH

In 1979 and 1980 Laws presented his novel texture energy approach to texture analysis (Laws, 1979, 1980a, 1980b). This involved the application of simple filters to digital images. The basic filters he used were common Gaussian, edge detector, and Laplacian-type filters, and were designed to highlight points of high "texture energy" in the image. By identifying these high energy points, smoothing the various filtered images, and pooling the information from them, he was able to characterize textures highly efficiently. As remarked earlier, Laws' approach has strongly influenced much subsequent work and it is therefore worth considering it here in some detail.

The Laws' masks are constructed by convolving together just three basic 1×3 masks:

$$L3 = \begin{bmatrix} 1 & 2 & 1 \end{bmatrix} \tag{8.1}$$

$$E3 = \begin{bmatrix} -1 & 0 & 1 \end{bmatrix} \tag{8.2}$$

$$S3 = \begin{bmatrix} -1 & 2 & -1 \end{bmatrix} \tag{8.3}$$

The initial letters of these masks indicate *L*ocal averaging, *E*dge detection, and *S*pot detection. In fact, these basic masks span the entire 1×3 subspace and form

Table 8.1 The Nine 3×3 Laws Masks

$L3^{T}L3$			$L3^{T}E3$			$L3^{T}S3$		
1	2	1	−1	0	1	−1	2	−1
2	4	2	−2	0	2	−2	4	−2
1	2	1	−1	0	1	−1	2	−1
$E3^{T}L3$			$E3^{T}E3$			$E3^{T}S3$		
−1	−2	−1	1	0	−1	1	−2	1
0	0	0	0	0	0	0	0	0
1	2	1	−1	0	1	−1	2	−1
$S3^{T}L3$			$S3^{T}E3$			$S3^{T}S3$		
−1	−2	−1	1	0	−1	1	−2	1
2	4	2	−2	0	2	−2	4	−2
−1	−2	−1	1	0	−1	1	−2	1

a complete set. Similarly, the 1×5 masks obtained by convolving pairs of these 1×3 masks together form a complete set:[3]

$$L5 = \begin{bmatrix} 1 & 4 & 6 & 4 & 1 \end{bmatrix} \tag{8.4}$$

$$E5 = \begin{bmatrix} -1 & -2 & 0 & 2 & 1 \end{bmatrix} \tag{8.5}$$

$$S5 = \begin{bmatrix} -1 & 0 & 2 & 0 & -1 \end{bmatrix} \tag{8.6}$$

$$R5 = \begin{bmatrix} 1 & -4 & 6 & -4 & 1 \end{bmatrix} \tag{8.7}$$

$$W5 = \begin{bmatrix} -1 & 2 & 0 & -2 & 1 \end{bmatrix} \tag{8.8}$$

In Eqs. (8.7) and (8.8), the initial letters indicate *R*ipple detection and *W*ave detection. We can also use matrix multiplication (see also Section 3.6) to combine the 1×3 and a similar set of 3×1 masks to obtain nine 3×3 masks—e.g.:

$$\begin{bmatrix} 1 \\ 2 \\ 1 \end{bmatrix} \begin{bmatrix} -1 & 2 & -1 \end{bmatrix} = \begin{bmatrix} -1 & 2 & -1 \\ -2 & 4 & -2 \\ -1 & 2 & -1 \end{bmatrix} \tag{8.9}$$

The resulting set of masks also forms a complete set (Table 8.1): note that two of these masks are identical to the Sobel operator masks. The corresponding 5×5 masks are entirely similar but are not considered in detail here as all relevant principles are illustrated by the 3×3 masks.

All such sets of masks include one whose components do not average to zero. Thus, it is less useful for texture analysis since it will give results dependent more on image intensity than on texture. The remainder are sensitive to edge points, spots, lines, and combinations of these.

---

[3]In principle nine masks can be formed in this way, but only five of them are distinct.

Having produced images that indicate, e.g., local edginess, the next stage is to deduce the local magnitudes of these quantities. These magnitudes are then smoothed over a fair-sized region rather greater than the basic filter mask size (e.g., Laws used a 15×15 smoothing window after applying his 3×3 masks): the effect of this is to smooth over the gaps between the texture edges and other microfeatures. At this point the image has been transformed into a vector image, each component of which represents energy of a different type. Although Laws (1980b) used both squared magnitudes and absolute magnitudes to estimate texture energy, the former correspond to true energy and give a better response, while the latter are useful in requiring less computation:

$$E(l,m) = \sum_{i=l-p}^{l+p} \sum_{j=m-p}^{m+p} |F(i,j)| \tag{8.10}$$

where $F(i, j)$ is the local magnitude of a typical microfeature, which is smoothed at a general scan position (l, m) in a $(2p + 1) \times (2p + 1)$ window.

A further stage is required to combine the various energies in a number of different ways, providing several outputs that can be fed into a classifier to decide upon the particular type of texture at each pixel location (Fig. 8.6). If necessary, principal components analysis is used at this point to help select a suitable set of intermediate outputs.

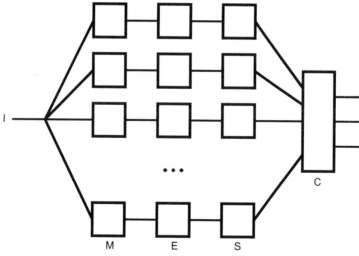

FIGURE 8.6

Basic form for a Laws' texture classifier. Here, I is the incoming image, M represents the microfeature calculation, E the energy calculation, S the smoothing, and C the final classification.

Laws' method resulted in excellent classification accuracy quoted at, e.g., 87% compared with 72% for the co-occurrence matrix method, when applied to a composite texture image of grass, raffia, sand, wool, pigskin, leather, water, and wood (Laws, 1980b). He also found that the histogram equalization normally applied to images to eliminate first-order differences in texture field grayscale distributions gave little improvement in this case.

Research was undertaken by Pietikäinen et al. (1983) to determine whether the precise coefficients used in the Laws' masks are responsible for the performance of his method. They found that so long as the general forms of the masks were retained, performance did not deteriorate, and could in some instances be improved. They were able to confirm that Laws' texture energy measures are more powerful than measures based on pairs of pixels (i.e., co-occurrence matrices).

8.5 ADE'S EIGENFILTER APPROACH

In 1983, Ade investigated the theory underlying the Laws' approach and developed a revised rationale in terms of eigenfilters.[4] He took all possible pairs of pixels within a 3×3 window, and characterized the image intensity data by a 9×9 covariance matrix. He then determined the eigenvectors required to diagonalize this matrix. These correspond to filter masks similar to the Laws' masks, i.e., use of these "eigenfilter" masks produces images that are principal component images for the given texture. Furthermore, each eigenvalue gives the part of the variance of the original image that can be extracted by the corresponding filter. Essentially, the variances give an exhaustive description of a given texture in terms of the texture of the images from which the covariance matrix was originally derived. Clearly, the filters that give rise to low variances can be taken to be relatively unimportant for texture recognition.

It will be useful to illustrate the technique for a 3×3 window. Here, we follow Ade (1983) in numbering the pixels within a 3×3 window in scan order:

1	2	3
4	5	6
7	8	9

This leads to a 9×9 covariance matrix for describing relationships between pixel intensities within a 3×3 window, as stated above. At this point, we recall that we are describing a texture and assuming that its properties are not synchronous with the pixel tessellation, we would expect various coefficients of the covariance matrix $\mathbf{C}$ to be equal: e.g., C_{24} should equal C_{57}; in addition, C_{57} must equal C_{75}.

---

[4]Before reading further, the reader may find it helpful to refer to Section 24.10, where principal components analysis and eigenvalue problems are discussed.

Table 8.2 Spatial Relationships Between Pixels in a 3 × 3 Window

a	b	c	d	e	f	g	h	i	j	k	l	m
9	6	6	4	4	3	3	1	1	2	2	2	2

This table shows the number of occurrences of the spatial relationships between pixels in a 3 × 3 window. Note that a is the diagonal element of the covariance matrix C, and that all others appear twice as many times in C as indicated in the table.

It is worth pursuing this matter, as a reduced number of parameters will lead to increased accuracy in determining the remaining ones. In fact, there are $9\,C_2 = 36$ ways of selecting pairs of pixels, but there are only 12 distinct spatial relationships between pixels if we disregard translations of whole pairs—or 13 if we include the null vector in the set (see Table 8.2). Thus, the covariance matrix (see Section 24.10), whose components include the 13 parameters $a-m$, takes the form:

$$\mathbf{C} = \begin{bmatrix} a & b & f & c & d & k & g & m & h \\ b & a & b & e & c & d & l & g & m \\ f & b & a & j & e & c & i & l & g \\ c & e & j & a & b & f & c & d & k \\ d & c & e & b & a & b & e & c & d \\ k & d & c & f & b & a & j & e & c \\ g & l & i & c & e & j & a & b & f \\ m & g & l & d & c & e & b & a & b \\ h & m & g & k & d & c & f & b & a \end{bmatrix} \qquad (8.11)$$

C is symmetric, and the eigenvalues of a real symmetric covariance matrix are real and positive, and the eigenvectors are mutually orthogonal (see Section 24.10). In addition, the eigenfilters thus produced reflect the proper structure of the texture being studied and are ideally suited to characterizing it. For example, for a texture with a prominent highly directional pattern, there will be one or more high energy eigenvalues with eigenfilters having strong directionality in the corresponding direction.

8.6 APPRAISAL OF THE LAWS AND ADE APPROACHES

At this point, it will be worthwhile to compare the Laws and Ade approaches more carefully. In the Laws approach, standard filters are used, texture energy images are produced, and then principal component analysis may be applied to lead to recognition, whereas in the Ade approach, special filters (the eigenfilters) are applied, incorporating the results of principal component analysis, following which texture energy measures are calculated, and a suitable number of these are applied for recognition.

The Ade approach is superior to the extent that it permits low-value energy components to be eliminated early on, thereby saving computation. For example, in Ade's application, the first five of the nine components contain 99.1% of the total texture energy, so the remainder can be ignored. In addition, it would appear that another two of the components containing respectively 1.9% and 0.7% of the energy could also be ignored, with little loss of recognition accuracy. However, in some applications textures could vary continually, and it may well be inadvisable to fine-tune a method to the particular data pertaining at any one time.[5]

In 1986, Unser developed a more general version of the Ade technique that also covered the methods of Faugeras (1978), Granlund (1980), and Wermser and Liedtke (1982). In this approach, performance is optimized not only for texture classification, but also for discrimination between two textures by simultaneous diagonalization of two covariance matrices. The method was developed further by Unser and Eden (1989, 1990): this work makes a careful analysis of the use of nonlinear detectors. As a result, two levels of nonlinearity are employed, one immediately after the linear filters are designed (by employing a specific Gaussian texture model) to feed the smoothing stage with genuine variance or other suitable measures, and the other after the spatial smoothing stage to counteract the effect of the earlier filter and aiming to provide a feature value that is in the same units as the input signal. In practical terms, this means having the capability for providing an RMS texture signal from each of the linear filter channels.

Overall, the originally intuitive Laws approach emerged during the 1980s as a serious alternative to the co-occurrence matrix approach. It is as well to note that alternative methods that are potentially superior have also been devised—see e.g., the local rank correlation method of Harwood et al. (1985), and the forced-choice method of Vistnes (1989) for finding edges between different textures, which apparently has considerably better accuracy than the Laws approach. Vistnes's (1989) investigation concludes that the Laws approach is limited by (a) the small scale of the masks that can miss larger-scale textural structures and (b) the fact that the texture energy smoothing operation blurs the texture feature values across the edge. The latter finding (or even the worse situation where a third class of texture appears to be located in the region of the border between two textures) has also been noted by Hsiao and Sawchuk (1989, 1990) who applied an improved technique for feature smoothing; they also used probabilistic relaxation for enforcing spatial organization on the resulting data.

---

[5]For example, these remarks apply (1) to textiles, for which the degree of stretch will vary continuously during manufacture, (2) to raw food products, such as beans, whose sizes will vary with the source of supply, and (3) to processed food products, such as cakes, for which the crumbliness will vary with cooking temperature and water vapor content.

8.7 CONCLUDING REMARKS

In this chapter, we have seen the difficulties of analyzing textures: these arise from the potential, and in many cases the frighteningly real complexities of textures—not least from the fact that their properties are often largely statistical in nature. The erstwhile widely used grayscale co-occurrence matrix approach has been seen to have distinct computational shortcomings. First, many co-occurrence matrices are in principle required (with different values of d and θ) in order to adequately describe a given texture; second, the co-occurrence matrices can be very large and, paradoxically, may hold more data than the images they are characterizing—especially if the range of grayscale values is large. In addition, many sets of co-occurrence matrices may be needed to allow for variation of the texture over the image, and if necessary, to initiate segmentation. Hence, co-occurrence matrices need to be significantly compressed, although in most cases it is not at all obvious *a priori* how this should be achieved, and it is even more difficult to arrange for it to be carried out automatically. This probably explains why attention shifted during the 1980s to other approaches, including particularly Laws' technique and its variations (especially that of Ade). Other developments were fractal-based measures, Markov approaches and the Gabor filter technique, although space has prevented a discussion of these methods here: see Section 8.8 for further reading on these topics.

> Textures are recognized and segmented by humans with the same apparent ease as for plain objects. This chapter has shown that texture analysis needs to be sensitive to microstructures and then pulled into macrostructures—with PCA (principal components analysis) being a natural means of finding the optimum structure. The subject has great importance for new applications, such as iris recognition.

8.8 BIBLIOGRAPHICAL AND HISTORICAL NOTES

Early work on texture analysis was carried out by Haralick et al. (1973) and in 1976, Weska and Rosenfeld applied textural analysis to materials' inspection. The area was reviewed by Zucker (1976a) and Haralick (1979), and excellent accounts appear in the books by Ballard and Brown (1982) and Levine (1985).

At the end of the 1970s, the Laws technique (1979, 1980a, 1980b) arrived upon the scene (which had up till then been dominated by the co-occurrence matrix approach) and led to the principal components approach of Ade (1983), which was further developed by Dewaele et al. (1988), Unser and Eden (1989, 1990), and others. The direction taken by Laws was particularly valuable as it showed how texture analysis could be implemented straightforwardly and in a manner consistent with real-time applications such as inspection.

The 1980s also saw other new developments, such as the fractal approach led by Pentland (1984), and a great amount of work on Markov random field models of texture. Here the work of Hansen and Elliott (1982) was very formative, although the names G.R. Cross, H. Derin, D. Geman, S. Geman, and A.K. Jain come up repeatedly in this context. Bajcsy and Liebermann (1976), Witkin (1981), and Kender (1983) pioneered the *shape from texture* concept, which has received considerable attention ever since. Later, much work appeared on the application of neural networks to texture analysis, e.g., Greenhill and Davies (1993) and Patel et al. (1994). A number of reviews and useful comparative studies have been made including Van Gool et al. (1985), du Buf et al. (1990), Ohanian and Dubes (1992), and Reed and du Buf (1993). For further work on texture analysis related to inspection of faults and foreign objects, see Chapter 20.

More recent developments include further work with automated visual inspection in mind (Davies, 2000c; Tsai and Huang, 2003; Ojala et al., 2002; Manthalkar et al., 2003; Pun and Lee, 2003), although several of these papers also cite medical, remote sensing, and other applications. Of these papers, the last three are specifically aimed at rotation invariant texture classification and the last one also aims at scale invariance. In previous years, there has not been quite enough emphasis on rotation invariance, although it was by no means a new topic. Other work (Clerc and Mallat, 2002) was concerned with recovering shape from texture *via* a texture gradient equation, while Ma et al. (2003) were particularly concerned with person identification based on iris textures. Mirmehdi and Petrou (2000) described an in-depth investigation of color texture segmentation. In this context, the importance of "wavelets"[6] as an increasingly used technique of texture analysis with interesting applications (such as, human iris recognition) should be noted (e.g., Daugman, 1993, 2003).

Then, in a particularly exciting advance, Spence et al. (2004) managed to eliminate texture by using photometric stereo to find the underlying surface shape (or "bump map"), following which they were able to perform impressive reconstructions, including texture, from a variety of viewpoints; McGunnigle and Chantler (2003) have shown that this sort of technique is also able to reveal hidden writing on textured surfaces, where only pen pressure marks have been made. Similarly, Pan et al. (2004) have shown how texture can be eliminated from ancient tablets (in particular those made of lead and wood) to reveal clear images of the writing underneath.

8.8.1 More Recent Developments

Over the 2000s, the trend to scale and rotation invariant texture analysis mentioned above has continued, the paper by Janney and Geers (2010) describing an

---

[6]Wavelets are directional filters reminiscent of the Laws edges, bars, waves, and ripples, but have more rigorously defined shapes and envelopes, and are defined in multiresolution sets (Mallat, 1989).

"invariant features of local textures" approach, using a strictly circular 1-D array of sampling, positions around any given position. The method employs Haar wavelets and as a result is computationally efficient. It is applied at multiple scales in order to achieve scale invariance; in addition, intensity normalization is used to make the method illumination as well as scale and rotation invariant.

Two new books have recently been published on this rather specialist subject—by Petrou and Sevilla (2006) and Mirmehdi et al. (2008). The first is a very sound textbook, starting from a low level and progressing through topics not covered in the present volume, such as fractals, Markov random fields, Gibbs distributions, Gabor functions, wavelets, and the Wigner distribution. The second is an edited volume containing chapters by various researchers and providing much new information—as indicated by some of the more novel chapter titles: "TEXEMS: random texture representation and analysis," "3-D texture analysis," "Texture for appearance models," "From dynamic texture to dynamic shape and appearance models," "Divide-and-texture: Hierarchical feature description," "Practical implementation of the trace transform," and "Face analysis using local binary patterns."

Intermediate-Level Vision

2

In Part 2, we study intermediate-level image analysis, which is concerned with obtaining abstract information about images, starting with the images themselves. At this stage, we are less interested in converting one image into another, as in the subject of image processing. In particular, transform methods that have been designed systematically for the purpose will be used.

For the most part, the chapters in Part 2 result in abstract information on the positions and orientations of various image features; they do not aim to provide real-world data, a function that is left to Parts 3 and 4 (thus Part 2 may indicate that a circle exists in one part of an image; it is left to later parts to interpret this as a wheel and to find any defects it may have). However, an exception to this general strategy is the relational descriptor analysis of scenes that appears toward the end of Chapter 14. This approach may be regarded as constituting the foot-hills of "high-level" vision.

Binary Shape Analysis

While binary images contain much less information than their grayscale counterparts, they embody shape and size information that is highly relevant for object recognition. However, this information resides in a digital lattice of pixels, and this results in intricacies appearing in the geometry. This chapter resolves these problems and explores a number of important algorithms for processing shapes.

Look out for:

- the connectedness paradox and how it is resolved.
- object labeling and how labeling conflicts are resolved.
- problems related to measurement in binary images.
- size filtering techniques.
- the convex hull as a means of characterizing shape, and methods for determining it.
- distance functions and how they are obtained using parallel and sequential algorithms.
- the skeleton and how it is found by thinning: the crucial role played by the crossing number, both in determining the skeleton and in analyzing it.
- simple measures for shape recognition, including circularity, and aspect ratio.
- more rigorous measures of shape, including moments and boundary descriptors.

In reality, this chapter almost exclusively covers area-based methods of shape analysis, leaving boundary-based procedures to Chapter 10—though circularity measures and boundary tracking are both covered. However, chapter boundaries cannot be completely exclusive, as any method requires "hooks" that have been laid down in a variety of places, and indeed, it is often valuable to meet a concept before finding out in detail how to put flesh on it.

Returning to the present chapter, it is interesting to note how intricate some of the algorithmic processes are: connectedness, in particular, pervades the whole subject of digital shape analysis and comes with a serious health warning.

9.1 INTRODUCTION

Over the past few decades 2-D shape analysis has provided the main means by which objects are recognized and located in digital images. Fundamentally, 2-D shape has been important because it uniquely characterizes many types of object, from keys to characters, from gaskets to spanners, and from fingerprints to chromosomes, while in addition it can be represented by patterns in simple binary images. Chapter 1 showed how the template matching approach leads to a combinatorial explosion even when fairly basic patterns are to be found, so preliminary analysis of images to find features constitutes a crucial stage in the process of efficient recognition and location. Thus, the capability for binary shape analysis is a very basic requirement for practical visual recognition systems.

In fact, 40 years of progress have provided an enormous range of shape analysis techniques and a correspondingly large range of applications. Clearly, it will be impossible to cover the whole field within the confines of a single chapter—so completeness will not even be attempted (the alternative of a catalog of algorithms and methods, all of which are covered only in brief outline, is eschewed). At one level, the main topics covered are examples with their own intrinsic interest and practical application, and at another level they introduce matters of fundamental principle. Recurring themes are the central importance of connectedness for binary images; the contrasts between local and global operations on images and between different representations of image data; the need to optimize accuracy and computational efficiency; and the compatibility of algorithms and hardware. The chapter starts with a discussion of how connectedness is measured in binary images.

9.2 CONNECTEDNESS IN BINARY IMAGES

This section begins with the assumption that objects have been segmented, by thresholding or other procedures, into sets of 1's in a background of 0's (see Chapters 2–4). At this stage it is important to realize that a second assumption is already being made implicitly—that it is easy to demarcate the boundaries between objects in binary images. However, in an image that is represented digitally in rectangular tessellation, a problem arises with the definition of connectedness. Consider a dumbell-shaped object:[1]

---

[1] All unmarked image points are taken to have the binary value 0.

```
                              1   1
                          1   1   1   1
                              1   1
                          1
                      1
                  1
          1   1
      1   1   1   1
          1   1
```

At its center, this object has a segment of the form

```
          0   1
          1   0
```

which separates two regions of background. At this point diagonally adjacent 1's are regarded as being connected, whereas diagonally adjacent 0's are regarded as disconnected—a set of allocations that seems inconsistent. However, we can hardly accept a situation where a connected diagonal pair of 0's crosses a connected diagonal pair of 1's without causing a break in either case. Similarly, we cannot accept a situation in which a disconnected diagonal pair of 0's crosses a disconnected diagonal pair of 1's without there being a join in either case. Hence, a symmetrical definition of connectedness is not possible and it is conventional to regard diagonal neighbors as connected only if they are foreground, i.e. the foreground is "8-connected" and the background is "4-connected". This convention is followed in the subsequent discussion.

9.3 OBJECT LABELING AND COUNTING

Now we have a consistent definition of connectedness, we can unambiguously demarcate all objects in binary images and should be able to devise algorithms for labeling them uniquely and counting them. Labeling may be achieved by scanning the image sequentially until a 1 is encountered on the first object; a note is then made of the scanning position, and a "propagation" routine is initiated to label the whole of the object with a 1; since the original image space is already in use, a separate image space has to be allocated for labeling. Next, the scan is resumed, ignoring all points already labeled, until another object is found; this is labeled with a 2 in the separate image space. This procedure is continued until the whole image has been scanned and all the objects have been labeled (Fig. 9.1). Implicit in this procedure is the possibility of propagating through a connected object. Suppose at this stage no method is available for limiting the

```
                              1 1 1
                            1 1 1 1              2 2 2
              3 3              1 1 1             2 2 2
              3
                      4 4 4 4 4 4                        5 5
                                      6 6 6              5 5
                                    6 6 6 6 6             5
                                      6 6 6 6           5 5 5
                                    6   6 6 6 6
                      7                 6 6 6 6 6
                        7               6 6 6 6
                      7
              8 8         7                              9 9 9
              8 8       7 7 7 7 7 7                       9 9 9
              8 8                                         9 9 9
```

FIGURE 9.1

A process in which all binary objects are labeled.

field of the propagation routine, so that it has to scan the whole image space. Then the propagation routine takes the form:

```
do {
   for all points in image
      if point is in an object
         and next to a propagating region labelled N
      assign it the label N
} until no further change;
```
(9.1)

the kernel of the do—until loop being expressed more explicitly as:

```
//original image in A-space; labels to be inserted in P-space
for all pixels in image do {
   if((A0 == 1)
      &&((P1 == N)||(P2 == N)||(P3 == N)||(P4 == N)
      ||(P5 == N)||(P6 == N)||(P7 == N)||(P8 == N)))
      P0 = N;
}
```
(9.2)

At this stage a fairly simple type of algorithm for object labeling is obtained, as shown in Table 9.1: the *for forward scan over image do {···}* notation denotes a *sequential* forward raster scan over the image.

Note that the above object counting and labeling routine requires a *minimum* of $2N + 1$ passes over the image space, and in practice the number will be closer to $NW/2$, where W is the average width of the objects: hence, the algorithm is inherently rather inefficient. This prompts us to consider how the number of passes over the image could be reduced to save computation. One possibility would be to scan forward through the image, propagating new labels through objects as they are discovered. While this would work mostly straightforwardly with convex objects, problems would be encountered with objects possessing concavities—e.g. "U" shapes—since different parts of the same object would end with different labels, and also means would have to be devised for coping with "collisions" of labels (e.g. the largest local label could be propagated through the remainder of the object: see Fig. 9.2). Then inconsistencies could be resolved by a reverse scan through the image. However, this procedure will not resolve all problems that can arise, as in the case of more complex (e.g. spiral) objects. In such cases a general parallel propagation, repeatedly applied until no further

Table 9.1 A Simple Algorithm for Object Labeling

```
//start with binary image containing objects in A-space
//clear label space
for all pixels in image do {P0 = 0;}
//start with no objects
N = 0;
/* look for objects using a sequential scan and propagate
    labels through them */
do { //search for an unlabeled object
  found = false;
  for forward scan over image do {
    if ((A0 == 1)&&(P0 == 0)&&not found) {
        N = N + 1;
        P0 = N;
        found = true;
    }
  }
  if (found) //label the object just found
    do {
      finished = true;
      for all pixels in image do {
        if ((A0 == 1)&&(P0 == 0)
          &&((P1 == N)||(P2 == N)||(P3 == N)||(P4 == N)
          ||(P5 == N)||(P6 == N)||(P7 == N)||(P8 == N))) {
            P0 = N;
            finished = false;
        }
      }
    } until finished;
} until not found; //i.e. no (more) objects
//N is the number of objects found and labeled
```

```
                  1 1 1
                  1 1 1       2 2
          3       1 1 1       2 2
        3 3 3     1 1 1   2 2 2 2
        3 3 3 3   1 1 1 2 2 2 2 2
      3 3 3 3 3   1 1 2 2 2 2 2 2
      3 3 3 3 3 3 3 3 3 3 3 3 3 3
      3 3 3 3 3 3 3 3 3 3 3 3 3 3 3
      3 3 3 3 3 3 3 3 3 3 3 3 3 3 3
      3 3 3 3 3 3 3 3 3 3 3 3 3 3 3
        3 3 3 3 3 3 3 3 3 3 3 3 3 3
          3 3 3         3 3 3 3 3 3
4           3           3 3 3 3 3
  4                 3   3 3 3
    4               3
      4 4 4 4 4 4 4
        4 4 4 4 4
```

FIGURE 9.2

Labeling U-shaped objects: a problem that arises in labeling certain types of object if a simple propagation algorithm is used. Some provision has to be made to accept "collisions" of labels although the confusion can be removed by a subsequent stage of processing.

labeling occurs, might be preferable—though as we have seen, such a process is inherently rather computation intensive. However, it is implemented very conveniently on certain types of parallel SIMD processor (see Chapter 26).

Ultimately, the least computationally intensive procedures for propagation involve a different approach: objects and parts of objects are labeled on a *single* sequential pass through the image, at the same time noting which labels coexist on objects. Then the labels are sorted separately, in a stage of abstract information processing, to determine how the initially rather *ad hoc* labels should be interpreted. Finally, the objects are relabeled appropriately in a second pass over the image (in fact, this latter pass is sometimes unnecessary, since the image data are merely labeled in an overcomplex manner and what is needed is simply a key to interpret them). The improved labeling algorithm now takes the form shown in Table 9.2. Clearly this algorithm with its single sequential scan is intrinsically far more efficient than the previous one, although the presence of particular dedicated hardware or a suitable SIMD processor might alter the situation and justify the use of alternative procedures.

Table 9.2 The Improved Algorithm for Object Labeling

```
//clear label space
for all pixels in image do {P0 = 0;}
//start with no objects
N = 0;
//clear the table that is to hold the label coexistence data
for(i = 1; i <= Nmax; i ++)
    for(j = 1; j <= Nmax; j ++)
        coexist[i][j] = false;
//label objects in a single sequential scan
for forward scan over image do {
    if((A0 == 1)
        if((P2 == 0)&&(P3 == 0)&&(P4 == 0)&&(P5 == 0)){
            N = N + 1;
            P0 = N;
        }
        else {
            P0 = max(P2, P3, P4, P5);
            //now note which labels coexist in objects
            coexist[P0][P2] = true;
            coexist[P0][P3] = true;
            coexist[P0][P4] = true;
            coexist[P0][P5] = true;
        }
}
analyze the coexist table and decide ideal labeling scheme;
relabel image if necessary;
```

It will be clear that minor amendments to the above algorithms permit the areas and perimeters of objects to be determined: thus, objects may be labeled by their areas or perimeters instead of by numbers representing their order of appearance in the image. More important, the availability of propagation routines means that objects can be considered in turn in their entirety—if necessary by transferring them individually to separate image spaces or storage areas ready for unencumbered independent analysis. Evidently, if objects appear in individual binary spaces, maximum and minimum spatial coordinates are trivially measurable, centroids can readily be found and more detailed calculations of moments (see below) and other parameters can easily be undertaken.

9.3.1 Solving the Labeling Problem in a More Complex Case

In this section, we add substance to the all too facile statement at the end of Table 9.2—"analyze the coexist table and decide ideal labeling scheme." First, we have to make the task nontrivial by providing a suitable example. Figure 9.3 shows an example image, in which sequential labeling has been carried out in line with the algorithm of Table 9.2. However, one variation has been adopted—of using a minimum rather than a maximum labeling convention, so that the values are in general slightly closer to the eventual ideal labels. (This also serves to demonstrate that there is not just one way of designing a suitable labeling algorithm.) The algorithm itself indicates that the coexist table should now appear as in Table 9.3.

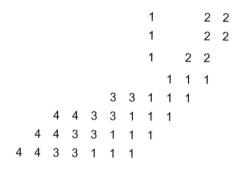

FIGURE 9.3

Solving the labeling problem in a more complex case.

Table 9.3 Coexist Table for the Image of Fig. 9.3

	1	2	3	4	5	6	7	8
1		√	√					
2	√							
3	√			√				
4		√						
5							√	
6								√
7						√		√
8						√	√	

The ticks correspond to clashes of labels.

Table 9.4 Coexist Table with Additional Numerical Information

	1	2	3	4	5	6	7	8
1	1	1	1					
2	1	2						
3	1		3	3				
4			3	4				
5					5		5	
6						6		6
7					5		7	7
8						6	7	8

This coexist table is an enhanced version of Table 9.3. Technically, the numbers along, and below, the leading diagonal are redundant, but nevertheless they speed up the subsequent computation.

However, the whole process of calculating ideal labels can be made more efficient by inserting numbers instead of ticks, and also adding the right numbers along the leading diagonal, as in Table 9.4; for the same reason, the numbers below the leading diagonal, which are technically redundant, are retained here.

The next step is to minimize the entries along the individual rows of the table, as in Table 9.5. Then we minimize along the individual columns (Table 9.6). Then we minimize along rows again (Table 9.7). This process is iterated to completion, which has already happened here after three stages of minimization. We can now read off the final result from the leading diagonal. Note that a further stage of computation is needed to make the resulting labels consecutive integers, starting with unity. However, the procedure needed to achieve this is much more basic and does not need to manipulate a 2-D table of data: this will be left as a simple programming task for the reader.

At this point, some comment on the nature of the process described above will be appropriate. What has happened is that the original image data has effectively been condensed into the minimum space required to express the labels—namely just one entry per original clash. This explains why the table retains the 2-D format

Table 9.5 Coexist Table Redrawn with Minimized Rows

	1	2	3	4	5	6	7	8
1	1	1	1					
2	1	1						
3	1		1	1				
4			3	3				
5					5		5	
6						6		6
7					5		5	5
8						6	6	6

At this stage the table is no longer symmetric.

Table 9.6 Coexist Table Redrawn Again with Minimized Columns

	1	2	3	4	5	6	7	8
1	1	1	1					
2	1	1						
3	1		1	1				
4			1	1				
5					5		5	
6						6		5
7					5		5	5
8						6	5	5

Table 9.7 Coexist Table Redrawn Yet Again with Minimized Rows

	1	2	3	4	5	6	7	8
1	1	1	1					
2	1	1						
3	1		1	1				
4			1	1				
5					5		5	
6						5		5
7					5		5	5
8						5	5	5

At this stage the table is in its final form, and is once again symmetric.

of the original image: lower dimensionality would not permit the image topology to be represented properly. It also explains why minimization has to be carried out, to completion, in two orthogonal directions. On the other hand, the particular implementation, including both above- and below-diagonal elements, is able to minimize computational overheads and finalize the operation in remarkably few iterations.

Finally, it might be felt that too much attention has been devoted to finding connected components of binary images. In fact, this is a highly important topic in practical applications such as industrial inspection, where it is crucial to locate all the objects unambiguously before they can individually be identified and scrutinized. In addition, Fig. 9.3 makes it clear that it is not only U-shaped objects that give problems, but also those that have shape subtleties—as is seen at the left of the upper object in this figure.

9.4 SIZE FILTERING

Before proceeding to study size filtering, we draw attention to the fact that the 8-connected and 4-connected definitions of connectedness lead to the following measures of distance (or "metrics") that apply to pairs of pixels, labeled i and j, in a digital lattice:

$$d_8 = \max(|x_i - x_j|, |y_i - y_j|) \tag{9.3}$$

and

$$d_4 = |x_i - x_j| + |y_i - y_j| \tag{9.4}$$

While the use of the d_4 and d_8 metrics is bound to lead to certain inaccuracies, there is a need to see what can be achieved with the use of local operations in

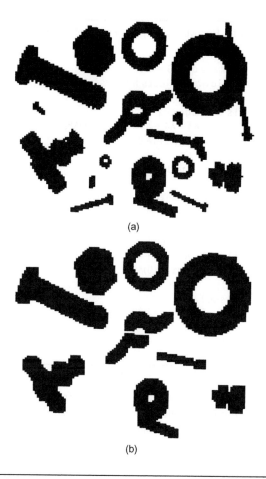

(a)

(b)

FIGURE 9.4

The effect of a simple size filtering procedure. When size filtering is attempted by a set of *N* shrink and *N* expand operations, larger objects are restored approximately to their original size but their shapes frequently become distorted or even fragmented. In this example, (b) shows the effect of applying two shrink and then two expand operations to the image in (a).

binary images. This section studies how simple size filtering operations can be carried out, using merely local (3×3) operations. The basic idea is that small objects may be eliminated by applying a series of shrink operations. In fact, *N* shrink operations will eliminate an object (or those parts of an object) that are $2N$ or fewer pixels across their narrowest dimension. Of course this process shrinks *all* the objects in the image, but in principle a subsequent *N* expand operations will restore the larger objects to their former size.

If complete elimination of small objects is required but perfect retention of larger objects, this will, however, not be achieved by the above procedure, since in many cases the larger objects will be distorted or fragmented by these operations (Fig. 9.4). To recover the larger objects in their *original* form, the proper

Table 9.8 Algorithm for Recovering Original Forms of Shrunken Objects

```
//save original image
for all pixels in image do {C0 = A0;}
//now shrink the original objects N times
  for (i = 1; i < = N; i++){
    for all pixels in image do {
      sigma = A1 + A2 + A3 + A4 + A5 + A6 + A7 + A8;
        if(sigma < 8) B0 = 0; else B0 = A0;
    }
    for all pixels in image do {A0 = B0;}
  }
  //next propagate the shrunken objects using the original image
  do {
      finished = true;
      for all pixels in image do {
        sigma = A1 + A2 + A3 + A4 + A5 + A6 + A7 + A8;
        if ((A0 == 0)&&(sigma > 0)&&(C0 == 1)){
            A0 = 1;
            finished = false;
        }
      }
  } until finished;
```

approach is to use the shrunken versions as "seeds" from which to grow the originals via a propagation process. The algorithm of Table 9.8 is able to achieve this.

Having seen how to remove whole (connected) objects that are everywhere narrower than a specified width, it is possible to devise algorithms for removing any subset of objects that are characterized by a given range of widths: large objects may be filtered out by first removing lesser-sized objects and then performing a logical masking operation with the original image, while intermediate-sized objects may be filtered out by removing a larger subset and then restoring small objects that have previously been stored in a separate image space. Ultimately, all these schemes depend on the availability of the propagation technique, which in turn depends on the internal connectedness of individual objects.

Finally, note that expand operations followed by shrink operations may be useful for joining nearby objects, filling in holes, and so on. Numerous refinements and additions to these simple techniques are possible. A particularly interesting one is to separate the silhouettes of touching objects such as chocolates by a shrinking operation: this then permits them to be counted reliably (Fig. 9.5).

9.5 DISTANCE FUNCTIONS AND THEIR USES

The distance function of an object is a very simple and useful concept in shape analysis. Essentially, each pixel in the object is numbered according to its

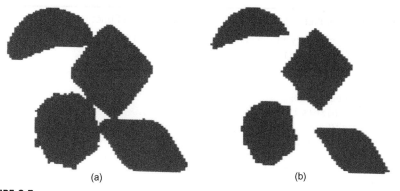

FIGURE 9.5

Separation of touching objects by shrink operations. Here objects (chocolates) in (a) are shrunk (b) in order to separate them so that they may be counted reliably.

```
                        1 1 1
                        1 2 1        1 1
                1       1 2 1        1 1
              1 1 1     1 2 1    1 1 1 1
              1 2 1 1   1 2 1 1 1 2 2 1
          1 1 2 2 1     1 2 2 2 2 2 2 1
          1 2 2 2 1 1   1 2 3 3 3 3 2 1
          1 2 3 2 2 2 2 2 3 4 4 3 2 1 1
          1 2 2 3 3 3 3 3 3 3 3 3 2 2 1
          1 1 2 2 2 2 2 2 2 2 3 3 2 2 1
            1 1 2 1 1 1 1 1 2 2 3 2 1 1
              1 1 1             1 1 2 2 2 1
      1             1           1 1 2 1 1
        1                     1   1 1 1
          1                 1
          1 1 1 1 1 1 1
          1 1 1 1 1
```

FIGURE 9.6

The distance function of a binary shape: the value at every pixel is the distance (in the d_8 metric) from the background.

distance from the background. As usual, background pixels are taken as 0's; then edge pixels are counted as 1's; object pixels next to 1's become 2's; next to those are the 3's; and so on throughout all the object pixels (Fig. 9.6).

The parallel algorithm of Table 9.9 finds the distance function of binary objects by propagation. Note that this algorithm performs a final pass in which nothing happens; this is inevitable if we are to be certain that the process will run to completion.

It is possible to perform the propagation of a distance function with far fewer operations if sequential processing is used. In 1-D, the basic idea would be to build up ramps within objects using a routine like the following:

```
for all pixels in a row of pixels do {Q0 = A0 * 255;}
  for forward scan over the row of pixels do
    if(Q0 > Q5 + 1) Q0 = Q5 + 1;
```

Table 9.9 A parallel Algorithm for Propagating Distance Functions

```
//Start with binary image containing objects in A-space
for all pixels in image do{Q0 = A0 * 255;}
N = 0;
do {
    finished = true;
    for all pixels in image do {
        if((Q0 == 255) //in object and no answer yet
            &&((Q1 == N)||(Q2 == N)||(Q3 == N)||(Q4 == N)
            ||(Q5 == N)||(Q6 == N)||(Q7 == N)||(Q8 == N))) {
                //next to an N
                Q0 = N + 1;
                finished = false; //some action has been taken

        }
    }
    N = N + 1;
} until finished;
```

Next, we need to insist on double-sided ramps within objects, both horizontally and vertically. This is elegantly achieved using two sequential operations, one being a normal forward raster scan and the other being a reverse raster scan:

```
for all pixels in image do {Q0 = A0 * 255;}
for forward scan over image do {
  minplusone = min(Q2,Q3,Q4,Q5) + 1;
  if(Q0 > minplusone) Q0 = minplusone;
}
for reverse scan over image do {
  minplusone = min(Q6,Q7,Q8,Q1) + 1;
  if(Q0 > minplusone) Q0 = minplusone;
}
```
(9.5)

Note the compact notation being used to distinguish between forward and reverse raster scans over the image: *for forward scan over image do {···}* denotes a forward raster scan, while *for reverse scan over image do {···}* denotes a reverse raster scan. A more succinct version of this algorithm is the following:

```
for all pixels in image do {Q0 = A0 * 255;}
for forward scan over image do {
  Q0 = min(Q0 − 1,Q2,Q3,Q4,Q5) + 1;
}
for reverse scan over image do {
  Q0 = min(Q0 − 1,Q6,Q7,Q8,Q1) + 1;
}
```
(9.6)

Before moving on, it will be useful to emphasize the value of sequential processing for propagating distance functions. In fact, when this sequential algorithm is run on a serial computer, it will be $O(N)$ times *faster* than the corresponding parallel algorithm running on a serial computer, but $O(N)$ times *slower* than the

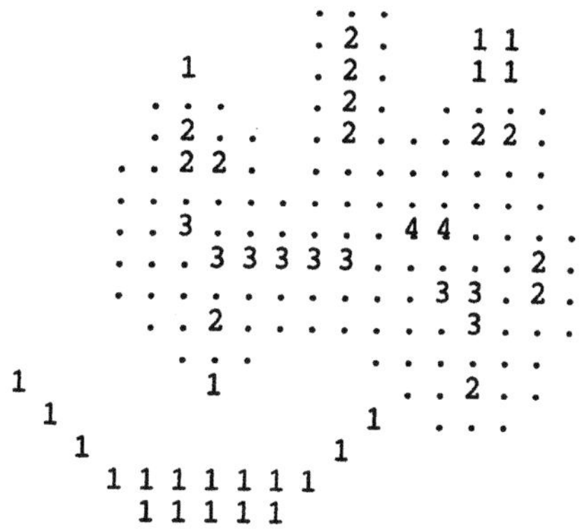

FIGURE 9.7

Local maxima of the distance function of the shape shown in Fig. 9.6, the remainder of the shape being indicated by dots and the background being blank. Notice that the local maxima group themselves into clusters each containing points of equal distance function value, while clusters of different values are clearly separated.

same parallel algorithm running on a parallel computer, for an $N \times N$ image. While this statement is specific to propagation of distance functions, similar statements can be made about a good many other operations. (For further discussion relating to parallel computers such as SIMD machines, see Chapter 26.)

9.5.1 Local Maxima and Data Compression

An interesting application of distance functions is that of data compression. To achieve this, operations are carried out to locate the pixels that are local maxima of the distance function (Fig. 9.7), since storing these pixel values and positions permits the original image to be regenerated by a process of downward propagation, as shown at the end of this section. Note that although finding the local maxima of the distance function provides the basic information for data compression, the actual compression occurs only when the data are stored as a list of points rather than in the original picture format. In order to locate the local maxima, the following parallel routine may be employed:

```
for all pixels in image do {
  maximum = max(Q1, Q2, Q3, Q4, Q5, Q6, Q7, Q8);
  if((Q0 > 0)&&(Q0 >= maximum)) B0 = 1; else B0 = 0;
}
```
(9.7)

Alternatively, the compressed data can be transferred to a single image space:

```
for all pixels in image do {
  maximum = max(Q1, Q2, Q3, Q4, Q5, Q6, Q7, Q8);
  if((Q0 > 0)&&(Q0 >= maximum)) P0 = Q0; else P0 = 0;
}
```
(9.8)

Table 9.10 A Parallel Algorithm for Recovering Objects From Local Maxima of the Distance Functions

```
//assume that input image is in Q-space, and that non-maximum
   values have value 0
do {
  finished = true;
  for all pixels in image do {
    maxminusone = max(Q0 + 1, Q1, Q2, Q3, Q4, Q5, Q6, Q7, Q8) - 1;
    if(Q0 < maxminusone) {
      Q0 = maxminusone;
      finished = false; //some action has been taken
    }
  }
} until finished;
```

Note that the local maxima that are retained for the purpose of data compression are not absolute maxima but are maximal in the sense of not being adjacent to larger values. If this were not so, insufficient numbers of points would be retained for completely regenerating the original object. As a result of this, it is found that the local maxima group themselves into clusters of connected points, each cluster having a common distance value and being separated from points of different distance values (Fig. 9.7). Thus, the set of local maxima of an object is not a connected subset. This fact has an important bearing on skeleton formation (see Section 9.6.3).

Having seen how data compression may be performed by finding local maxima of the distance function, it is relevant to consider a parallel downward propagation algorithm (Table 9.10) for recovering the shapes of objects from an image into which the values of the local maxima have been inserted. Note again that if it can be assumed that at most N passes are needed to propagate through objects of known maximum width, then the algorithm becomes simply:

```
for(i = 1; i <= N; i++)
  for all pixels in image do {
    Q0 = max(Q0 + 1, Q1, Q2, Q3, Q4, Q5, Q6, Q7, Q8) - 1;
  }
```
(9.9)

9.6 SKELETONS AND THINNING

The skeleton is a powerful analog concept that may be employed for the analysis and description of shapes in binary images. A skeleton may be defined as a connected set of medial lines along the limbs of a figure: for example, in the case of thick hand-drawn characters the skeleton may be supposed to be the path actually traveled by the pen. In fact, the basic idea of the skeleton is that of eliminating redundant information while retaining only the topological information concerning the shape and structure of the object that can help with recognition. In the case of hand-drawn characters, the thickness of the limbs is taken to be irrelevant: it may be constant and therefore carry no useful information, or it may vary randomly and again be of no value for recognition (Fig. 1.2).

The definition presented above leads to the idea of finding the loci of the centers of maximal disks inserted within the object boundary. First, suppose the image space to be a continuum. Then the disks are circles and their centers form loci that may be modeled very conveniently when object boundaries are approximated by linear segments. In fact, sections of the loci fall into three categories:

1. They may be angle bisectors, i.e. lines which bisect corner angles and reach right up to the apexes of corners.
2. They may be lines that lie half-way between boundary lines.
3. They may be parabolas that are equidistant from lines and from the nearest points of other lines—namely, corners where two lines join.

Clearly, categories 1 and 2 are special forms of a more general case.

These ideas lead to unique skeletons for objects with linear boundaries, and the concepts are easily generalizable to curved shapes. In fact, this approach tends to give rather more detail than is commonly required, even the most obtuse corner having a skeleton line going into its apex (Fig. 9.8). Hence, a thresholding scheme is often employed such that skeleton lines only reach into corners having a specified minimum degree of sharpness.

We now have to see how the skeleton concept will work in a digital lattice. Here we are presented with an immediate choice: which metric should we employ? If we select the Euclidean metric (i.e. lattice distance is measured as the Euclidean distance between pairs of pixels), there may be a considerable computational load. If we select the d_8 metric we will immediately lose accuracy but the computational requirements should be more modest (we do not here consider the d_4 metric, since we are dealing with the shapes of foreground objects). In what follows we concentrate on the d_8 metric.

At this stage some thought shows that the application of maximal disks in order to locate skeleton lines amounts essentially to finding the positions of local maxima of the distance function. Unfortunately, as seen in the previous section, the set of local maxima does not form a connected graph within a given object: nor is it necessarily composed of thin lines, and indeed it may at places be 2 pixels wide. Thus, problems arise in trying to use this approach to obtain a connected unit-width skeleton that can conveniently be used to represent the shape of the object. We shall return to this approach again in Section 9.6.3. Meanwhile, however, we pursue an alternative idea—that of thinning.

Thinning is perhaps the simplest approach to skeletonization. It may be defined as the process of systematically stripping away the outermost layers of a figure until only a connected unit-width skeleton remains (see, for example, Fig. 9.9). A number of algorithms are available to implement this process, with

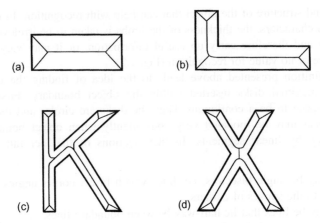

FIGURE 9.8

Four shapes whose boundaries consist entirely of straight line segments. The idealized skeletons go right to the apex of each corner, however obtuse. In certain parts of shapes (b)–(d), the skeleton segments are parts of parabolas rather than straight lines. As a result, the detailed shape of the skeleton (or the approximations produced by most algorithms operating in discrete images) is not exactly what might initially be expected or what would be preferred in certain applications.

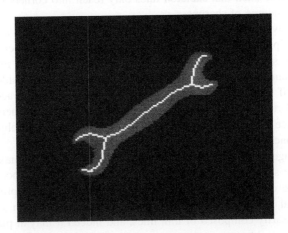

FIGURE 9.9

Typical result of a thinning algorithm operating in a discrete lattice.

varying degrees of accuracy, and we discuss below how a specified level of precision can be achieved and tested for. First, however, it is necessary to discuss the mechanism by which points on the boundary of a figure may validly be removed in thinning algorithms.

```
0 0 0     0 0 0     1 0 0     1 0 0     1 0 1     1 0 0     1 0 1
0 1 0     0 1 1     0 1 1     0 1 0     0 1 0     0 1 0     0 1 0
0 0 0     1 1 1     1 1 1     0 1 0     1 1 1     1 0 1     1 0 1
  0         2         4         4         6         6         8
```

FIGURE 9.10

Some examples of the crossing number values associated with given pixel neighborhood configurations (0, background; 1, foreground).

9.6.1 Crossing Number

The exact mechanism for examining points to determine whether they can be removed in a thinning algorithm must now be considered. This may be decided by reference to the *crossing number* χ (chi) for the 8 pixels around the outside of a particular 3×3 neighborhood. χ is defined as the total number of 0-to-1 and 1-to-0 transitions on going once round the outside of the neighborhood: this number is in fact twice the number of potential connections joining the remainder of the object to the center of the neighborhood (Fig. 9.10). Unfortunately, the formula for χ is made more complex by the 8-connectedness criterion. Basically, we would expect[2]:

$$
\begin{aligned}
\texttt{badchi} = &\texttt{(int)(A1!= A2) +(int)(A2!= A3) +(int)(A3!= A4)} \\
&\texttt{+(int)(A4!= A5) +(int)(A5!= A6) +(int)(A6!= A7)} \\
&\texttt{+(int)(A7!= A8) +(int)(A8!= A1);}
\end{aligned} \quad (9.10)
$$

However, this is incorrect because of the 8-connectedness criterion. For example, in the case

```
0   1   0
0   1   1
1   1   1
```

the formula gives the value 4 for χ instead of 2. The reason is that the isolated 0 in the top right-hand corner does not prevent the adjacent 1's from being joined. It is therefore tempting to use the modified formula:

$$
\texttt{wrongchi} = \texttt{(int)(A1!= A3) +(int)(A3!= A5) +(int)(A5!= A7) +(int)(A7!= A1);} \quad (9.11)
$$

However, this too is wrong, since in the case

```
0   0   1
0   1   0
1   1   1
```

[2]In this section, as in the case of C++, the "(int)" construct is used to convert logical outcomes true and false to integer outcomes 1 and 0.

it gives the answer 2 instead of 4. It is therefore necessary to add four extra terms to deal with isolated 1's in the corners:

$$
\begin{aligned}
\text{chi} = &(\text{int})(A1 != A3) + (\text{int})(A3 != A5) + (\text{int})(A5 != A7) \\
&+ (\text{int})(A7 != A1) \\
&+ 2 * ((\text{int})((A2 > A1)\&\&(A2 > A3)) + (\text{int})((A4 > A3)\&\&(A4 > A5)) \\
&+ (\text{int})((A6 > A5)\&\&(A6 > A7)) + (\text{int})((A8 > A7)\&\&(A8 > A1)));
\end{aligned}
\tag{9.12}
$$

This (now correct) formula for crossing number gives values 0, 2, 4, 6 or 8 in different cases (Fig. 9.10). The rule for removing points during thinning is that points may only be removed if they are at those positions at the boundary of an object where χ is 2. When χ is greater than 2, the point *must* be retained, as it forms a vital connected point between two or more parts of the object; in addition, when it is 0 it must be retained since removing it would create a hole.

Finally, there is one more condition that must be fulfilled before a point can be removed during thinning—that the sum σ (sigma) of the eight pixel values around the outside of the 3×3 neighborhood (see Chapter 2) must not be equal to 1. The reason for this is to preserve line ends, as in the following cases:

$$
\begin{array}{ccc}
0 & 0 & 0 \\
0 & 1 & 0 \\
0 & 1 & 0
\end{array}
\qquad
\begin{array}{ccc}
0 & 0 & 0 \\
0 & 1 & 0 \\
0 & 0 & 1
\end{array}
$$

Clearly, if line ends are eroded as thinning proceeds, the final skeleton will not represent the shape of an object (including the relative dimensions of its limbs) at all accurately (however, it is possible that we might sometimes wish to shrink an object while preserving connectedness, in which case this extra condition need not be implemented). Having covered these basics, we are now in a position to devise complete thinning algorithms.

9.6.2 Parallel and Sequential Implementations of Thinning

Thinning is "essentially sequential" in that it is easiest to ensure that connectedness is maintained by arranging that only one point may be removed at a time. As indicated above, this is achieved by checking before removing a point that it has a crossing number of 2. Now imagine applying the "obvious" sequential algorithm of Table 9.11 to a binary image. Assuming a normal forward raster scan, the result of this process is to produce a highly distorted skeleton, consisting of lines along the right-hand and bottom edges of objects. It may now be seen that the $\chi = 2$ condition is necessary but not sufficient, since it says nothing about the order in which points are removed. To produce a skeleton that is unbiased, giving a set of truly medial lines, it is necessary to remove points as evenly as possible around the object boundary. A scheme that helps with this involves a novel processing sequence: mark edge points on the first pass over an image; on the second pass, strip points sequentially as in the above algorithm, *but only where they have already been marked*; then mark a new set of edge points; then perform another stripping pass; then repeat this marking and stripping sequence until no further

---

Table 9.11 An "Obvious" Sequential Thinning Algorithm

```
do {
   finished = true;
   for forward scan over image do {
      sigma = A1 + A2 + A3 + A4 + A5 + A6 + A7 + A8;
      chi = (int)(A1 != A3) + (int)(A3 != A5) + (int)(A5 != A7)
            + (int)(A7 != A1)
            + 2 * ((int)((A2 > A1)&&(A2 > A3)) + (int)((A4 > A3)&&(A4 > A5))
            + (int)((A6 > A5)&&(A6 > A7)) + (int)((A8 > A7)&&(A8 > A1)));
      if((A0 == 1)&&(chi == 2)&&(sigma != 1)){
         A0 = 0;
         finished = false; //some action has been taken
      }
   }
} until finished;
```

---

change occurs. An early algorithm working on this principle is that of Beun (1973).

While the novel sequential thinning algorithm described above can be used to produce a reasonable skeleton, it would be far better if the stripping action could be performed symmetrically around the object, thereby removing any possible skeletal bias. In this respect a parallel algorithm should have a distinct advantage. However, parallel algorithms result in several points being removed at once: this means that lines 2 pixels wide will disappear (since masks operating in a 3×3 neighborhood cannot "see" enough of the object to judge whether a point may validly be removed or not), and as a result shapes can become disconnected. The general principle for avoiding this problem is to strip points lying on different parts of the boundary in different passes, so that there is no risk of causing breaks. In fact, there is a very large number of ways of achieving this, by applying different masks and conditions to characterize different parts of the boundary. If boundaries were always convex the problem would no doubt be reduced; however, boundaries can be very convoluted and are subject to quantization noise, so the problem is a complex one. With so many potential solutions to the problem, we concentrate here on one that can conveniently be analyzed and which gives acceptable results.

The method discussed is that of removing north, south, east, and west points cyclically until thinning is complete. North points are defined as the following:

$$
\begin{array}{ccc}
\times & 0 & \times \\
\times & 1 & \times \\
\times & 1 & \times
\end{array}
$$

where $\times$ means either a 0 or a 1: south, east, and west points are defined similarly. It is easy to show that all north points for which $\chi = 2$ and $\sigma \neq 1$ may be removed in parallel without any risk of causing a break in the skeleton—and similarly for south, east, and west points. Thus, a possible format for a parallel thinning algorithm in rectangular tessellation is the following:

```
do {
    strip appropriate north points;
    strip appropriate south points;
    strip appropriate east points;
    strip appropriate west points;
} until no further change;
```
(9.13)

where the basic parallel routine for stripping "appropriate" north points is:

```
for all pixels in image do {
    sigma = A1 + A2 + A3 + A4 + A5 + A6 + A7 + A8;
    chi = (int)(A1!=A3) +(int)(A3!=A5) +(int)(A5!=A7)
        +(int)(A7!=A1)
        + 2 *((int)((A2 > A1)&&(A2 > A3)) +(int)((A4 > A3)&&(A4 > A5))
        +(int)((A6 > A5)&&(A6 > A7)) +(int)((A8 > A7)&&(A8 > A1)));
    if((A3 == 0)&&(A0 == 1)&&(A7 == 1) //north point
        &&(chi == 2)&&(sigma != 1))
        B0 = 0;
    else B0 = A0;
}
```
(9.14)

(but extra code needs to be inserted to detect whether any changes have been made in a given pass over the image).

Algorithms of the above type can be highly effective, although their design tends to be rather intuitive and *ad hoc*. In a survey made by the author in 1981 (Davies and Plummer, 1981), a great many such algorithms exhibited problems. Ignoring cases where the algorithm design was insufficiently rigorous to maintain connectedness, four other problems were evident:

1. the problem of skeletal bias;
2. the problem of eliminating skeletal lines along certain limbs;
3. the problem of introducing "noise spurs";
4. the problem of slow speed of operation.

In fact, problems 2 and 3 are opposites in many ways: if an algorithm is designed to suppress noise spurs, it is liable to eliminate skeletal lines in some circumstances; contrariwise, if an algorithm is designed never to eliminate skeletal lines, it is unlikely to be able to suppress noise spurs. This situation arises since the masks and conditions for performing thinning are intuitive and *ad hoc*, and therefore have no basis for discriminating between valid and invalid skeletal lines: ultimately this is because it is difficult to build overt global models of reality into purely local operators. In a similar way, algorithms that proceed with caution, i.e. which do not remove object points in the fear of making an error or causing bias, tend to be slower in operation than they might otherwise be. Again, it is difficult to design algorithms that can make correct global decisions rapidly via intuitively designed local operators. Hence, a totally different approach is needed if solving one of the above problems is not to cause difficulties with the others. Such an alternative approach is discussed in the next section.

9.6.3 Guided Thinning

This section returns to the ideas of Section 9.5.1, where it was found that the local maxima of the distance function do not form an ideal skeleton because they appear in clusters and are not connected. In addition, the clusters are often two pixels wide. On the plus side, the clusters are accurately in the correct positions and should therefore not be subject to skeletal bias. Hence, an ideal skeleton should result if (a) the clusters could be reconnected appropriately and (b) the resulting structure could be reduced to unit width—though, of course, a unit-width skeleton can only be perfectly unbiased where the object is an odd number of pixels wide.

A simple means of reconnecting the clusters is to use them to guide a conventional thinning algorithm (see Section 9.6.2). As a first stage, thinning is allowed to proceed normally but with the proviso that no cluster points may be removed. This gives a connected graph which at certain places is 2 pixels wide. Then a routine is applied to strip the graph down to unit width. At this stage an unbiased skeleton (within 1/2 pixel) should result. The main problem here is the presence of noise spurs. The opportunity now exists to eliminate these systematically by applying suitable global rules. A simple rule is that of eliminating lines on the skeletal graph that terminate in a local maximum of value (say) 1 (or, better, stripping them back to the next local maximum), since such a local maximum corresponds to fairly trivial detail on the boundary of the object. Thus, the level of detail that is ignored can be programmed into the system (Davies and Plummer, 1981). The whole guided thinning process is shown in Fig. 9.11.

9.6.4 A Comment on the Nature of the Skeleton

At the beginning of Section 9.6, the case of character recognition was taken as an example and it was stated that the skeleton may be supposed to be the path traveled by the pen in drawing out the character. However, in one important respect this is not valid. The reason is seen both in the analog reasoning and from the results of thinning algorithms. Take the case of a letter K. The vertical limb on the left of the skeleton theoretically consists of two linear segments joined by two parabolic segments leading into the junction (Fig. 9.8). This limb will only become straight if a higher level model is used to constrain the result.

9.6.5 Skeleton Node Analysis

Skeleton node analysis may be carried out very simply with the aid of the crossing number concept. Points in the middle of a skeletal line have a crossing

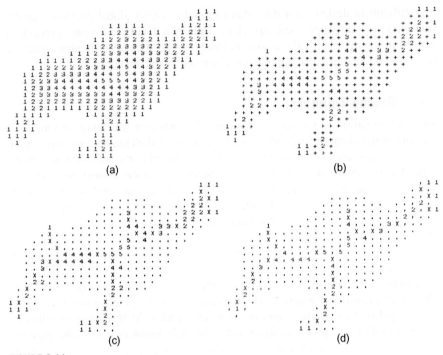

FIGURE 9.11

Results of a guided thinning algorithm: (a) distance function on the original shape; (b) set of local maxima; (c) set of local maxima now connected by a simple thinning algorithm; (d) final thinned skeleton. The effect of removing noise spurs systematically, by cutting limbs terminating in a 1 back to the next local maximum, is easily discernible from the result in (d): the general shape of the object is not perturbed by this process.

number of 4; points at the end of a line have crossing number 2; points at skeletal "T" junctions have a crossing number of 6; and points at skeletal "X" junctions have a crossing number of 8. However, there is a situation to beware of—places which look like a "+" junction:

$$
\begin{matrix}
0 & 1 & 0 \\
1 & 1 & 1 \\
0 & 1 & 0
\end{matrix}
$$

In such places the crossing number is actually 0 (see formula), although the pattern is definitely that of a cross. At first the situation seems to be that there is insufficient resolution in a 3×3 neighborhood to identify a "+" cross, the best option being to look for this particular pattern of 0's and 1's and use a more sophisticated construct than the 3×3 crossing number to check whether or not a cross is present. The problem is that of distinguishing between two situations such as:

```
0  0  0  0  0        0  0  1  0  0
0  0  1  0  0        1  0  1  0  0
0  1  1  1  0  and   1  1  1  1  1
0  0  1  0  0        0  0  1  0  0
0  0  0  0  0        0  0  0  1  0
```

However, further analysis shows that the first of these two cases would be thinned down to a dot (or a short bar), so that if a "+" node appears on the final skeleton (as in the second case) it actually signifies that a cross is present despite the contrary value of χ. Davies and Celano (1993) have shown that the proper measure to use in such cases is the *modified* crossing number $\chi_{skel} = 2\sigma$, this crossing number being different from χ because it is required not to test whether points can be eliminated from the skeleton, but to ascertain the meaning of points that are at that stage *known* to lie on the final skeleton. Note that χ_{skel} can have values as high as 16—it is not restricted to the range 0–8!

Finally, note that sometimes insufficient resolution really is a problem, in that a cross with a shallow crossing angle appears as two "T" junctions:

```
0  0  0  0  0  0  0  1
1  1  1  0  0  1  1  0
0  0  0  1  1  0  0  0
0  1  1  0  0  1  1  1
1  0  0  0  0  0  0  0
```

Clearly, resolution makes it impossible to recognize an asterisk or more complex figure from its crossing number, within a 3×3 neighborhood. Probably, the best solution is to label junctions tentatively, then to consider all the junction labels in the image, and to analyze whether a particular local combination of junctions should be reinterpreted—e.g. two "T" junctions may be deduced to form a cross. This is especially important in view of the distortions that appear on a skeleton in the region of a junction (see Section 9.6.4).

9.6.6 Application of Skeletons for Shape Recognition

Shape analysis may be carried out simply and conveniently by analysis of skeleton shapes and dimensions. Clearly, study of the nodes of a skeleton (points for which there are other than two skeletal neighbors) can lead to the classification of simple shapes but not, for example, discrimination of all block capitals from each other. Many classification schemes exist which can complete the analysis, in terms of limb lengths and positions, and methods for achieving this are touched on in later chapters.

A similar situation exists for the analysis of the shapes of chromosomes, which take the form of a cross or a "V". For small industrial components, more detailed shape analysis is called for; this can still be approached with the skeleton technique, by examination of distance function values along the lines of the skeleton. In general, shape analysis using the skeleton proceeds by examination in turn of nodes, limb lengths and orientations, and distance function values, until the required level of characterization is obtained.

The particular importance of the skeleton as an aid in the analysis of connected shapes is not only that it is invariant under translations and rotations but also that it embodies what is for many purposes a highly convenient representation of the figure that (with the distance function values) essentially carries all the original information. If the original shape of an object can be deduced exactly from a representation, this is generally a good sign since it means that it is not merely an *ad hoc* descriptor of shape but that considerable reliance may be placed on it (compare other methods such as the circularity measure—see Section 9.7).

9.7 OTHER MEASURES FOR SHAPE RECOGNITION

There are many simple tests of shape that can be made to confirm the identity of objects or to check for items such as defects. These include measurements of product area and perimeter, length of maximum dimension, moments relative to the centroid, number and area of holes, area and dimensions of the convex hull (see below) and enclosing rectangle, number of sharp corners, number of intersections with a check circle and angles between intersections (Fig. 9.12), and numbers and types of skeleton nodes.

The list would not be complete without a mention of the widely used shape measure $C = area/perimeter^2$. This quantity is often called "circularity" or "compactness," since it has a maximum value of $1/4\pi$ for a circle, decreases as shapes become more irregular, and approaches zero for long narrow objects: alternatively, its reciprocal is sometimes used, being called "complexity" since it increases in size as shapes become more complex. Note that both measures are dimensionless so that they are independent of size and are therefore sensitive only

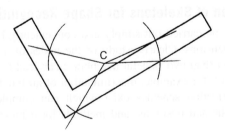

FIGURE 9.12

Rapid product inspection by polar checking.

to the shape of an object. Other dimensionless measures of this type include rectangularity and aspect ratio.

All these measures have the property of characterizing a shape but not of describing it uniquely. Thus, it is easy to see that there are in general many different shapes having the same values of parameters such as circularity. Hence, these rather *ad hoc* measures are on the whole less valuable than approaches such as skeletonization (Section 9.6) or moments (see below) that can be used to represent and reproduce a shape to any required degree of precision. Nevertheless, rigorous checking of even one measured number to high precision often permits a machined part to be identified positively.

The use of moments for shape analysis was mentioned above: these are widely used and should be covered in more detail. In fact, moment approximations provide a rigorous means of describing 2-D shapes, and take the form of series expansions of the type:

$$M_{pq} = \sum_x \sum_y x^p y^q f(x, y) \tag{9.15}$$

for a picture function $f(x,y)$; such a series may be curtailed when the approximation is sufficiently accurate. By referring axes to the centroid of the shape, moments can be constructed that are position-invariant: they can also be normalized so that they are invariant under rotation and change of scale (Hu, 1962; see also Wong and Hall, 1978). The main value of using moment descriptors is that in certain applications the number of parameters may be made small without significant loss of precision—although the number required may not be clear without tests being made on a range of relevant shapes. Moments can prove particularly valuable in describing shapes such as cams and other fairly round objects, although they have also been used in a variety of other applications including aeroplane silhouette recognition (Dudani et al., 1977).

The convex hull was also mentioned above and has also been used as the basis for sophisticated, complete descriptions of shapes. The *convex hull* is defined as the smallest convex shape that contains the original shape (it may be envisaged as the shape contained by an elastic band placed around the original shape). The *convex deficiency* is defined as the shape that has to be added to a given shape to create the convex hull (Fig. 9.13). The convex hull may be used as a simple approximation providing a rapid indication of the extent of an object. A fuller description of the shape of an object may be obtained by means of *concavity trees*: here the convex hull of an object is first obtained with its convex deficiencies, then the convex hulls and deficiencies of the obtained convex deficiencies are found, then the convex hulls and deficiencies of these convex deficiencies— and so on until all the derived shapes are convex, or until an adequate approximation to the original shape is obtained. Thus, a tree is formed that can be used for systematic shape analysis and recognition (Fig. 9.14). We shall not dwell on this approach beyond noting its inherent utility and that at its core is the need for a reliable means of determining the convex hull of a shape.

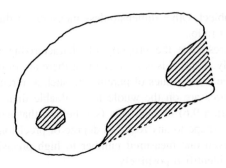

FIGURE 9.13

Convex hull and convex deficiency. The convex hull is the shape enclosed on placing an elastic band around an object. The shaded portion is the convex deficiency that is added to the shape to create the convex hull.

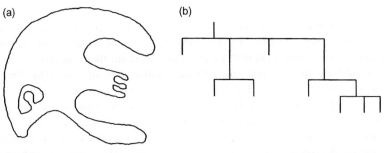

(a)   (b)

FIGURE 9.14

A simple shape and its concavity tree. The shape in (a) has been analyzed by repeated formation of convex hulls and convex deficiencies until all the constituent regions are convex (see text). The tree representing the entire process as shown in (b): at each node, the branch on the left is the convex hull and the branches on the right are convex deficiencies.

A simple strategy for obtaining the convex hull is to repeatedly fill in the center pixel of all neighborhoods that exhibit a concavity, including each of the following:

$$
\begin{array}{ccc}
1 & 1 & 1 \\
1 & 0 & 0 \\
0 & 0 & 0
\end{array}
\qquad
\begin{array}{ccc}
0 & 1 & 1 \\
1 & 0 & 0 \\
0 & 0 & 0
\end{array}
$$

until no further change occurs. In fact, the shapes obtained by the above approach are larger than ideal convex hulls and approximate to octagonal (or degenerate octagonal) shapes. Hence more complex algorithms are required to generate convex hulls, a useful approach involving the use of boundary tracking to search for positions on the boundary that have common tangent lines.

9.8 BOUNDARY TRACKING PROCEDURES

The preceding sections have described methods of analyzing shape on the basis of body representations such as skeletons and moments. However, an important approach has so far been omitted—the use of boundary pattern analysis. This approach has the potential advantage of requiring considerably reduced computation, since the number of pixels to be examined is equal to the number of pixels on the boundary of any object rather than the much larger number of pixels within the boundary. Before proper use can be made of boundary pattern analysis techniques, means must be found for tracking systematically around the boundaries of all the objects in an image: in addition, care must be taken not to ignore any holes that are present or any objects within holes.

In one sense the problem has been analyzed already, in that the object labeling algorithm of Section 9.3 systematically visits and propagates through all objects in the image. All that is required now is some means of tracking round object boundaries once they have been encountered. Quite clearly it will be useful to mark in a separate image space all points that have been tracked: alternatively, an object boundary image may be constructed and the tracking performed in this space, all tracked points being eliminated as they are passed.

In the latter procedure, objects having unit width in certain places may become disconnected. Hence, we ignore this approach and adopt the previous one. There is still a problem when objects have unit-width sections, since these can cause badly designed tracking algorithms to choose a wrong path, going back around the previous section instead of on to the next (Fig. 9.15). To avoid this circumstance it is best to adopt the following strategy:

1. track round each boundary, keeping to the left path consistently;
2. stop the tracking procedure only when passing through the starting point in the original direction (or passing through the first two points in the same order).

Apart from necessary initialization at the start, a suitable tracking procedure is given in Table 9.12.

Having seen how to track around the boundaries of objects in binary images, we are now in a position to embark on boundary pattern analysis. This is done in Chapter 10.

9.9 CONCLUDING REMARKS

This chapter has concentrated on rather traditional methods of performing image analysis—using image processing techniques. This has led naturally to area representations of objects, including for example moment and convex hull-based

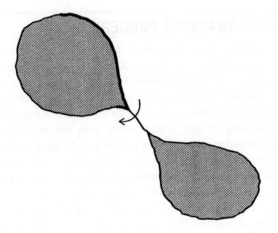

FIGURE 9.15

A problem with an over-simple boundary tracking algorithm: the boundary tracking procedure takes a short-cut across a unit-width boundary segment instead of continuing and keeping to the left path at all times.

Table 9.12 Basic Procedure for Tracking Around a Single Object

```
do {
    //find direction to move next
    start with current tracking direction;
    reverse it;
    do {
      rotate tracking direction clockwise
    } until the next 1 is met on outer pixels of 3×3 neighborhood;
    record this as new current direction;
    move one pixel along this direction;
    increment boundary index;
    store current position in boundary list;
} until (position == original position)&&(direction == original direction)
```

---

schemes, although the skeleton approach appeared as a rather special case in that it converts objects into graphical structures. An alternative schema is to represent shapes by their boundary patterns, after applying suitable tracking algorithms: this latter approach is considered in the following chapter. Meanwhile, connectedness has been an underlying theme in the present chapter, objects being separated from each other by regions of background, thereby permitting objects to be considered and characterized individually. Connectedness has been seen to involve rather more intricate computations than might have been expected, and this necessitates great care in the design of algorithms: this must partly explain why after so many years, new thinning algorithms are still being developed (e.g. Kwok, 1989; Choy et al., 1995) (ultimately, these complexities arise because global properties of images are being computed by purely local means).

Although it will turn out that boundary pattern analysis is in certain ways more attractive than region pattern analysis, this comparison cannot be completely divorced from considerations of the hardware the algorithms have to run on. In this respect, note that many of the algorithms of this chapter can be performed efficiently on SIMD processors, which have one processing element per pixel (see Chapter 26), whereas boundary pattern analysis will be seen to match better the capabilities of more conventional serial computers.

> Shape analysis can be attempted by boundary or region representations. Both are deeply affected by connectedness and related metric issues for a digital lattice of pixels. This chapter has shown that these issues are only solved by carefully incorporating global knowledge alongside local information—e.g. by use of distance transforms.

9.10 BIBLIOGRAPHICAL AND HISTORICAL NOTES

The development of shape analysis techniques has been particularly extensive: hence, only a brief perusal of the history is attempted here. The all-important theory of connectedness and the related concept of adjacency in digital images was developed largely by Rosenfeld (see for example, Rosenfeld, 1970). The connectedness concept led to the idea of a distance function in a digital picture (Rosenfeld and Pfaltz, 1966, 1968), and the related skeleton concept (Pfaltz and Rosenfeld, 1967). However, the basic idea of a skeleton dates from the classic work by Blum (1967)—see also Blum and Nagel (1978). Important work on thinning has been carried out by Arcelli et al. (1975, 1981), Arcelli and di Baja (1985) and parallels work by Davies and Plummer (1981). The latter paper demonstrates possibilities for limb pruning, and a rigorous method for testing the results of *any* thinning algorithm, however generated, and in particular for detecting skeletal bias. More recently, Arcelli and Ramella (1995) have reconsidered the problem of skeletons in grayscale images. There have also been important developments to generalize the distance function concept and to make distance functions uniform and isotropic: see for example Huttenlocher et al. (1993). The design of a modified crossing number χ_{skel} for the analysis of skeletal shape dates from the same period: as pointed out in Section 9.6.5, χ_{skel} is different from χ since it evaluates the *remaining* (i.e. skeletal) points rather than points that *might* be eliminated from the skeleton (Davies and Celano, 1993).

Sklansky has carried out much work on convexity and convex hull determination (see for example, Sklansky, 1970; Sklansky et al., 1976), while Batchelor (1979) developed concavity trees for shape description. Haralick et al. (1987) have generalized the underlying mathematical (morphological) concepts, including the case of grayscale analysis. Use of invariant moments for pattern recognition dates from the two seminal papers by Hu (1961, 1962). Pavlidis has drawn attention to the importance of unambiguous ("asymptotic") shape representation schemes (Pavlidis, 1980)—as distinct from *ad hoc* sets of shape measures.

In the early 2000s, skeletons maintained their interest and utility, becoming if anything more precise by reference to exact analog shapes (Kégl and Krzyżak, 2002), and giving rise to the concept of a shock graph, which characterizes the result of the much earlier grass-fire transformation (Blum, 1967) more rigorously (Giblin and Kimia, 2003). Wavelet transforms have also been used to implement skeletons more accurately in the discrete domain (Tang and You, 2003). In contrast, shape matching has been carried out using self-similarity analysis coupled with tree representation—an approach that has been especially valuable for tracking articulated object shapes, including human profiles and hand motions (Geiger et al., 2003). It is interesting to see graphical analysis of skeletonized hand-written character shapes performed taking account of catastrophe theory (Chakravarty and Kompella, 2003): this is relevant because (a) critical points—where points of inflection exist—can be deformed into pairs of points each corresponding to a curvature maximum plus a minimum; (b) crossing of t's can be actual or non-crossing; and (c) loops can turn into cusps or corners (many other possibilities also exist). The point is that methods are needed for mapping between *variations* of shapes rather than making snap judgements as to classification (this corresponds to the difference between scientific understanding of process and *ad hoc* engineering).

9.10.1 More Recent Developments

More recently, increased attention has been devoted to processing skeletons and using them for object matching and classification. Bai and Latecki (2008) discuss how to prune skeletons meaningfully, by ensuring that endpoints of skeleton branches correspond to visual parts of objects (such as all the legs of a horse). Once this has been achieved, it should be possible to match objects (such as horses) in spite of any articulations or contour deformations that may have taken place. The method is found to permit much more efficient matching and to be more resistant to partial occlusion: this is because meaningfulness is built into the final skeleton, while minor intricacies (which may originally have been due to noise forming tiny holes in the object) will have been eliminated. This approach is potentially useful for tracking, stereo matching, and database matching. Ward and Hamarneh (2010) attend to the order in which skeleton branches should be pruned. They report on several pruning algorithms and quantify their performance in terms of denoizing, classification, and within-class skeleton similarity measures. The work is important because of the well-known fact that the medial axis transform is unstable with respect to minor perturbations on the boundary of a shape: this means that before skeletons can be used reliably, noise spurs need to be pruned so that they correspond to the underlying shapes.

9.11 PROBLEMS

1. Write the full C++ routine required to sort the lists of labels, to be inserted at the end of the algorithm of Table 9.2.

2. Show that, as stated in Section 9.6.2 for a parallel thinning algorithm, all north points may be removed in parallel without any risk of causing a break in the skeleton.

3. Describe methods for locating, labeling and counting objects in binary images. You should consider whether methods based on propagation from a "seed" pixel, or those based on progressively shrinking a skeleton to a point, would provide the more efficient means for achieving the stated aims. Give examples for objects of various shapes.

4. **a.** Give a simple one-pass algorithm for labeling the objects appearing in a binary image, making clear the role played by connectedness. Give examples showing how this basic algorithm goes wrong with real objects: illustrate your answer with clear pixel diagrams, which show the numbers of labels that can appear on objects of different shapes.

   b. Show how a table-orientated approach can be used to eliminate multiple labels in objects. Make clear how the table is set up and what numbers have to be inserted into it. Are the number of iterations needed to analyze the table similar to the number that would be needed in a multi-pass labeling algorithm taking place entirely within the original image? Consider how the *real* gain in using a table to analyze the labels arises.

5. **a.** Using the following notation for a 3 × 3 window:

A4	A3	A2
A5	A0	A1
A6	A7	A8

   work out the effect of the following algorithm on a binary image containing small foreground objects of various shapes:

```
do{
  for all pixels in image do {
    sum = (int)(A1 + A3 == 2) + (int)(A3 + A5 == 2)
        + (int)(A5 + A7 == 2) + (int)(A7 + A1 == 2);
    if(sum > 0) B0 = 1; else B0 = A0;
  }
  for all pixels in image do {A0 = B0;}
} until no further change;
```

   b. Show in detail how to implement the *do ... until no further change* function in this algorithm.

6. **a.** Give a simple algorithm operating in a 3 × 3 window for generating a *rectangular* convex hull around each object in a binary image. Include

in your algorithm any necessary code for implementing the required *do ... until no further change* function.

b. A more sophisticated algorithm for finding accurate convex hulls is to be designed. Explain why this would employ a boundary tracking procedure. State the general strategy of an algorithm for tracking around the boundaries of objects in binary images and write a program for implementing it.

c. Suggest a strategy for designing the complete convex hull algorithm and indicate how rapidly you would expect it to operate, in terms of the size of the image.

7. a. Explain the meaning of the term *distance function*. Give examples of the distance functions of simple shapes, including that shown in Fig. 9.16.

b. Rapid image transmission is to be performed by sending only the coordinates and values of the local maxima of the distance functions. Give a complete algorithm for finding the local maxima of the distance functions in an image, and devise a further algorithm for reconstructing the original binary image.

c. Discuss which of the following sets of data would give more compressed versions of the binary picture object shown in Fig. 9.16:

   i. The list of local maxima coordinates and values.

   ii. A list of the coordinates of the boundary points of the object.

  iii. A list consisting of one point on the boundary and the relative directions (each expressed as a 3-bit code) between each *pair* of boundary points on tracking round the boundary.

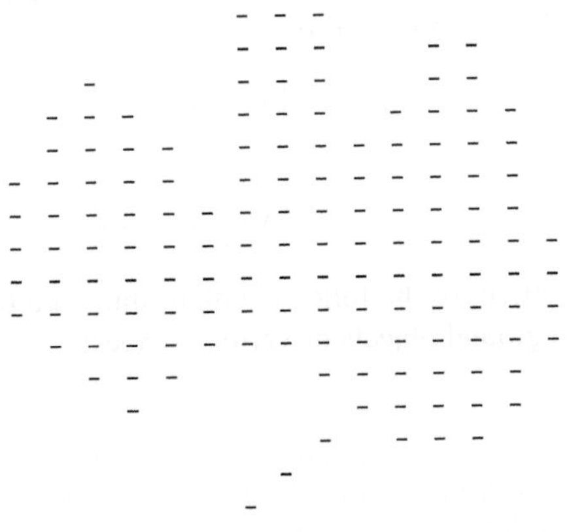

FIGURE 9.16

Binary picture object for shape analysis tests.

8. a. What is the *distance function* of a binary image? Illustrate your answer for the case where a 128×128 image P contains just the object shown in Fig. 9.17. How many passes of (i) a parallel algorithm and (ii) a sequential algorithm would be required to find the distance function of the image?

b. Give a complete parallel *or* sequential algorithm for finding the distance function, and explain how it operates.

c. Image P is to be transmitted rapidly by determining the coordinates of locations where the distance function is locally maximum. Indicate the positions of the local maxima, and explain how the image could be reconstituted at the receiver.

d. Determine the compression factor η if image P is transmitted in this way.

```
1 1 1 1 1 1 1 1 1 1 1 1 1 1 1 1 1
1 1 1 1 1 1 1 1 1 1 1 1 1 1 1 1 1 1
1 1 1 1 1 1 1 1 1 1 1 1 1 1 1 1 1 1 1
1 1 1 1 1 1 1 1 1 1 1 1 1 1 1 1 1 1 1 1
1 1 1 1 1 1 1 1 1 1 1 1 1 1 1 1 1 1 1 1
  1 1 1 1 1 1 1 1 1 1 1 1 1 1 1 1 1 1 1
                      1 1 1 1 1 1 1 1
                      1 1 1 1 1 1 1 1
                      1 1 1 1 1 1 1 1
                      1 1 1 1 1 1 1 1
                      1 1 1 1 1 1 1 1
                      1 1 1 1 1 1 1 1 1 1
```

FIGURE 9.17

Binary picture object for distance function analysis.

Show that η can be increased by eliminating some of the local maxima before transmission, and estimate the resulting increase in η.

9. a. The local maxima of the distance function can be defined in the following two ways:

   i. Pixels whose values are greater than those of all the neighboring pixels.

   ii. Pixels whose values are greater than *or equal to* the values of all the neighboring pixels. Which definition of the local maxima would be more useful for reproducing the original object shapes? Why is this?

b. Give an algorithm that is capable of reproducing the original object shapes from the local maxima of the distance function, and explain how it operates.

c. Explain the *run-length encoding* approach to image compression. Compare the run-length encoding and local maxima methods for compressing binary images. Explain why the one method would be expected to lead to a greater degree of compression with some types of image while the other method would be expected to be better with other types of image.

10. a. Explain how propagation of a distance function may be carried out using a parallel algorithm. Give in full a simpler algorithm that operates using two sequential passes over the image.

b. It has been suggested that a four-pass sequential algorithm will be even faster than the two-pass algorithm, as each pass can use just a 1-D window involving at most three pixels. Write down the code for *one* typical pass of the algorithm.

c. Estimate the approximate speeds of these three algorithms for computing the distance function, in the case of an $N \times N$ pixel image. Assume a conventional serial computer is used to perform the computation.

11. Small dark insects are to be located among cereal grains. The insects approximate to rectangular bars of dimensions 20×7 pixels, and the cereal grains are approximately elliptical with dimensions 40×24 pixels. The proposed algorithm design strategy is: (i) apply an edge detector which will mark all the edge points in the image as 0's in a 1's background, (ii) propagate a distance function in the *background* region, (iii) locate the local maxima of the distance function, (iv) analyze the values of the local maxima, and (v) carry out necessary further processing to identify the nearly parallel sides of the insects. Explain how to design stages (iv) and (v) of the algorithm in order to identify the insects, ignoring the cereal grains. Assume that the image is not so large that the distance function will overflow the byte limit. Determine how robust the method will be if the edge is broken in a few places.

12. Give the general strategy of an algorithm for tracking around the boundaries of objects in binary images. If the tracker has reached a boundary point with crossing number $\chi = 2$ and neighborhood

$$\begin{array}{ccc} 0 & 0 & 1 \\ 0 & 1 & 1 \\ 1 & 1 & 1 \end{array}$$

decide in which direction it should now proceed. Hence, give a complete procedure for determining the direction code of the next position on the boundary for cases where $\chi = 2$. Take the direction codes starting *from* the current pixel (*) as being specified by the following diagram:

$$\begin{array}{ccc} 4 & 3 & 2 \\ 5 & * & 1 \\ 6 & 7 & 8 \end{array}$$

How should the procedure be modified for cases where $\chi \neq 2$?

13. **a.** Explain the principles involved in tracking around the boundaries of objects in binary images to produce reliable outlines. Outline an algorithm which can be used for this purpose, assuming it is to get its information from a normal 3×3 window.

b. A binary image is obtained and the data in it is to be compressed as much as possible. The following range of algorithms is to be tested for this purpose:

i. the boundary image;

ii. the skeleton image;

iii. the image of the local maxima of the distance function;

iv. the image of a suitably chosen subset of the local maxima of the distance function;

v. a set of run-length data, i.e. a series of numbers obtained by counting runs of 0's, then runs of 1's, then runs of 0's, and so on, in a continuous line-by-line scan over the image.

c. In (i) and (ii) the lines may be encoded using chain code, i.e. giving the *coordinates* of the first point met, and the *direction* of each subsequent point using the direction codes 1−8 defined relative to the current position C by:

$$
\begin{array}{ccc}
4 & 3 & 2 \\
5 & C & 1 \\
6 & 7 & 8
\end{array}
$$

d. With the aid of suitably chosen examples, discuss which of the methods of data compression should be most suitable for different types of data. Wherever possible, give numerical estimates to support your arguments.

e. Indicate what you would expect to happen if noise were added to initially noise-free input images.

14. Test the two-mask strategy outlined in Section 9.7 for obtaining the convex hulls of binary picture objects. Confirm that it operates consistently, and give a geometrical construction that predicts the final shapes it produces. What happens if either the first or the second mask is used on its own? Show that the two-mask strategy will operate both as a sequential and as a parallel algorithm. Devise a version of the algorithm that does not permit nearby shapes to be merged.

10 Boundary Pattern Analysis

Recognition of binary objects by boundary pattern analysis should be a straightforward process, but this chapter shows that there are a number of problems to be overcome. In particular, any boundary distortions such as those due to breakage or several objects being in contact may result in total failure of the matching process. This chapter discusses the problems and their solution.

Look out for:

- the centroidal profile approach and its limitations.
- how the method may be speeded up.
- how recognition based on the (s, ψ) boundary plot is significantly more robust.
- how the (s, ψ) plot leads on to the more convenient (s, κ) plot.
- the relation between κ and ψ.
- more rigorous ways of dealing with the occlusion problem.
- discussion of the accuracy of boundary length measures.

Disparate ways are available for representing object boundaries and for measuring and recognizing objects using them. All the methods are subject to the same ultimate difficulties—particularly that of managing occlusion (which necessarily removes relevant data), and that of inaccuracy in the pixel-based description. Sound ways of managing occlusion are indicated in Section 10.6. This work presages later chapters where methods such as the Hough transform are widely employed for robust object recognition.

10.1 INTRODUCTION

Chapter 4 has shown how thresholding may be used to binarize grayscale images and hence to present objects as 2-D shapes. However, that method of segmenting objects is successful only when considerable care is taken with lighting and when the object can be presented conveniently, e.g., as dark patches on a light background. In other cases, adaptive thresholding may help to achieve the same end: as an alternative, edge detection schemes can be applied, which are generally rather more resistant to problems of lighting. Nevertheless, thresholding of edge-enhanced images still gives certain problems: in particular, edges may peter out in some places and may be thick in others (Fig. 10.1). For many purposes, the output of an edge detector is ideally a connected unit-width line around the periphery of an object and steps need to be taken to convert edges to this form—if this has not already been achieved using a Canny or other operator employing nonmaximum suppression (see Chapter 5).

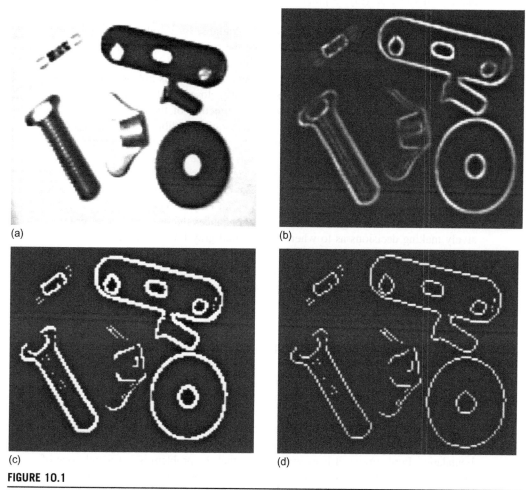

(a)

(b)

(c)

(d)

FIGURE 10.1

Some problems with edges. The edge-enhanced image (b) from an original image (a) is thresholded as in (c). The edges so detected are found to peter out in some places and to be thick in other places. A thinning algorithm is able to reduce the edges to unit thickness (d) but *ad hoc* (i.e., not model-based) linking algorithms are liable to produce erroneous results (not shown).

Thinning algorithms can be used to reduce edges to unit thickness while maintaining connectedness (Fig. 10.1(d)). Many algorithms have been developed for this purpose, and the main problems here are: (a) slight bias and inaccuracy due to uneven stripping of pixels, especially in view of the fact that even the best algorithm can only produce a line that is locally within 1/2 pixel of the ideal position; and (b) introduction of a certain number of noise spurs. The first problem can be minimized by using grayscale edge thinning algorithms, which act directly on the original grayscale edge-enhanced image (e.g., Paler and Kittler, 1983). Noise spurs around object boundaries can be eliminated quite efficiently by removing lines that are shorter than (say) 3 pixels. Overall, the major problem to be dealt with is that of reconnecting boundaries that have become fragmented during thresholding.

A number of rather *ad hoc* schemes are available for relinking broken boundaries: e.g., line ends may be extended along their existing directions—a *very* limited procedure since there are (at least for binary edges) only eight possible directions, and it is quite possible for the extended line ends not to meet. Another approach is to join line ends that are sufficiently close together and point in similar directions, both to each other *and* to the direction of the vector between the two ends. In fact, this approach can be made quite credible in principle, but in practice it can lead to all sorts of problems as it is still *ad hoc* and not model driven. Hence, adjacent lines that arise from genuine surface markings and shadows may be arbitrarily linked together by such algorithms. In many situations, it is therefore best if the process is model driven—e.g., by finding the best fit to some appropriate idealized boundary such as an ellipse. Yet another approach is that of relaxation labeling, which iteratively enhances the original image, progressively making decisions as to where the original gray levels reinforce each other. Thus, edge linking is permitted only where evidence is available in the original image that this is appropriate. A similar but computationally more efficient line of attack is the hysteresis thresholding method described in Chapter 5. Here intensity gradients above a certain upper threshold are taken to give definite indication of edge positions, whereas those above a second, lower threshold are taken to indicate edges only if they are adjacent to positions that have already been accepted as edges (for a more detailed analysis, see Section 5.10).

It may be thought that the Marr–Hildreth and related (Laplacian-based) edge detectors do not run into these problems, because they give edge contours that are necessarily connected. However, the result of using methods that force connectedness is that sometimes (e.g., when edges are diffuse, or of low contrast, so that image noise is an important factor) parts of a contour will lack meaning; indeed, a contour may meander over such regions following noise rather than useful object boundaries. Furthermore, it is as well to note that the problem is not merely one of pulling low-level signals out of noise, but rather that sometimes there is no signal at all present that could be enhanced to a meaningful level. Reasons for this include lighting being such as to give zero contrast (as, e.g., when a cube is lit in such a way that two faces have equal brightness) and occlusion. Lack of spatial resolution can also cause problems by merging together several lines on an object.

In what follows, it is assumed that all of these problems have been overcome by sufficient care with the lighting scheme, appropriate digitization, and other means. It is also assumed that suitable thinning and linking algorithms have been applied so that all objects are outlined by connected unit-width boundary lines. At this stage, it should be possible to locate the objects from the boundary image, and to identify and orientate them accurately.

10.2 BOUNDARY TRACKING PROCEDURES

Before objects can be matched with their boundary patterns, means must be found for tracking systematically around the boundaries of all the objects in an image. Means have already been demonstrated for achieving this in the case of regions such as those that result from intensity thresholding routines (Chapter 9). However, if a connected unit-width boundary is formed by an alternative process such as edge detection, the problem of tracking is much simpler, since it is necessary only to move repeatedly to the next edge pixel. Clearly, it is necessary to ensure that we (a) never reverse direction, (b) know when we have been round the whole boundary once, and (c) record which object boundaries have been encountered. As when tracking around regions, we must ensure that in each case we end by passing through the starting point in the same direction.

10.3 CENTROIDAL PROFILES

The substantial matching problems that occur with 2-D template matching make it attractive to attempt to locate objects in a less demanding search space. In fact, it is possible to achieve this very simply by matching the boundary of each object in a single dimension. Perhaps the most obvious such scheme uses an (r, θ) plot. Here the centroid of the object is first located,[1] and then a polar coordinate system is set up relative to this point and the object boundary is plotted as an (r, θ) graph—often called a "centroidal profile" (Fig. 10.2). Next, the 1-D graph so obtained is matched against the corresponding graph for an idealized object of the same type. Since objects generally have arbitrary orientation, it is necessary to "slide" the idealized graph along that obtained from the image data until the best match is obtained. The match for each possible orientation α_j of the object is commonly tested by measuring the differences in radial distance between the boundary graph B and the template graph T for various values of θ and summing their squares to give a difference measure D_j for the quality of the fit:

$$D_j = \sum_i [r_B(\theta_i) - r_T(\theta_i + \alpha_j)]^2 \qquad (10.1)$$

---

[1]Note that the position of the centroid is deducible directly from the list of boundary pixel coordinates—there is no need to start with a region-based description of the object for this purpose.

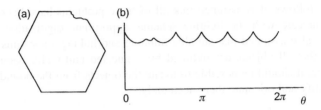

FIGURE 10.2

Centroidal profiles for object recognition and scrutiny: (a) hexagonal nut shape in which one corner has been damaged; (b) centroidal profile, which permits both straightforward identification of the object and detailed scrutiny of its shape.

Alternatively, the absolute magnitudes of the differences are used:

$$D_j = \sum_i \left| r_B(\theta_i) - r_T(\theta_i + \alpha_j) \right| \qquad (10.2)$$

The latter measure has the advantage of being easier to compute and less biased by extreme or erroneous difference values. Note that the basic 2-D matching operation has now been reduced to 1-D, and if we need work in $1°$ steps, the orientation indices i and j will each have to range over 360 values. The result is that the number of operations that are required to test each object drops to around 360^2 (i.e., $\sim 100,000$), so boundary pattern analysis should give a very substantial saving in computation.

The 1-D boundary pattern matching approach described above is able to identify objects and also to find their orientations. In fact, initial location of the centroid of the object also solves one other part of the problem as specified at the end of Section 10.1. At this stage, it may be noted that the matching process also leads to the possibility of inspecting the object's shape as an inherent part of the process (Fig. 10.2). In principle, this combination of capabilities makes the centroidal profile technique quite powerful.

Finally, note that the method is able to cope with objects of identical shapes but different sizes. This is achieved by using the maximum value of r to normalize the profile, giving a variation (ρ, θ) where $\rho = r/r_{max}$.

10.4 PROBLEMS WITH THE CENTROIDAL PROFILE APPROACH

In practice, there are several problems with the procedure outlined in Section 10.3. First, any major defect or occlusion of the object boundary can cause the centroid to be moved away from its true position, and the matching process will be largely spoiled (Fig. 10.3). Thus, instead of concluding that this is an object of type X with a specific part of its boundary damaged, the algorithm will most probably not recognize it at all. Such behavior would be inadequate in many

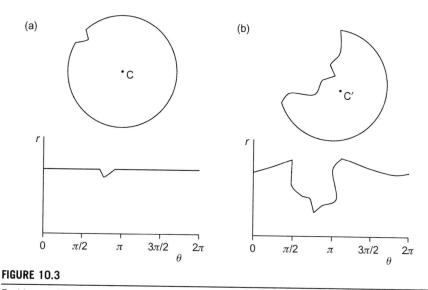

FIGURE 10.3

Problems with the centroidal profile descriptor. (a) A circular object with a minor defect on its boundary; its centroidal profile appears beneath it. (b) The same object, this time with a gross defect: because the centroid is shifted to C', the *whole* of the centroidal profile is grossly distorted.

automated inspection applications, where positive identification and fault-finding are required, and the object would have to be rejected without a satisfactory diagnosis being made.

Second, the (r, θ) plot will be multivalued for a certain class of object (Fig. 10.4). This has the effect of making the matching process partly 2-D and leads to complication and excessive computation.

Third, the very variable spacing of the pixels when plotted in (r, θ) space is a source of complication. It leads to the requirement for considerable smoothing of the 1-D plots, especially in regions where the boundary comes close to the centroid—as for elongated objects such as spanners or screwdrivers (Fig. 10.5); however, in other places accuracy will be greater than necessary and the overall process will be wasteful. The problem arises because quantization should ideally be uniform along the θ-axis so that the two templates can be moved conveniently relative to one another to find the orientation of best match.

Finally, computation times can still be quite significant, so some timesaving procedure is required.

10.4.1 Some Solutions

All four of the above-described problems can be tackled in one way or another, with varying degrees of success. The first problem, that of coping with occlusions

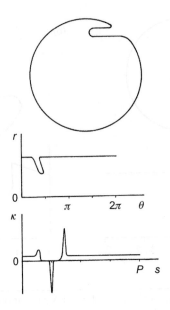

FIGURE 10.4

Boundary pattern analysis via (r, θ) and (s, κ) plots.

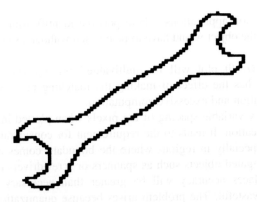

FIGURE 10.5

A problem in obtaining a centroidal profile for elongated objects. This figure highlights the pixels around the boundary of an elongated object—a spanner—showing that it will be difficult to obtain an accurate centroidal profile for the region near the centroid.

and gross defects, is probably the most fundamental and the most resistant to satisfactory solution. For the method to work successfully, a stable reference point must be found within the object. The centroid is naturally a good candidate for this since the averaging inherent in its location tends to eliminate most forms of noise or minor defect: however, major distortions such as those arising from breakages or occlusions are bound to affect it adversely. The centroid of the boundary is no better, and may also be less successful at suppressing noise. Other possible candidates are the positions of prominent features such as corners, holes, centers of arcs, and so on. In general, the smaller such a feature, the more likely it is to be missed altogether in the event of a breakage or occlusion, although the larger such a feature, the more likely it is to be affected by the defect. In fact, circular arcs can be located accurately (at their centers) even if they are partly occluded (see Chapter 12), so these features are very useful for leading to suitable reference points. A set of symmetrically placed holes may sometimes be suitable, since even if one of them is obscured, one of the others is likely to be visible and can act as a reference point.

Clearly, such features can help the method to work adequately but their presence also calls into question the value of the 1-D boundary pattern matching procedure, since they make it likely that superior methods can be used for object recognition (see later chapters in Part 2). For the present, we therefore accept that (a) severe complications arise when part of an object is missing or occluded and (b) it may be possible to provide some degree of resistance to such problems by using a prominent feature as a reference point instead of the centroid. Indeed, the only significant *further* change that is required to cope with occlusions is that difference $(r_B - r_T)$ values greater than (say) 3 pixels should be ignored, and the best match then becomes one for which the greatest number of values of θ gives good agreement between B and T.

The second problem, of multivalued (r, θ) plots, is solved very simply by employing the heuristic of taking the smallest value of r for any given θ and then proceeding with matching as normal (here it is assumed that the boundaries of any holes present within the boundary of the object are dealt with separately, information about any object and its holes being collated at the end of the recognition process). This *ad hoc* procedure should in fact be acceptable when making a preliminary match of the object to its 1-D template, and may be discarded at a later stage when the orientation of the object is known accurately.

The third problem described above arises because of uneven spacing of the pixel boundaries along the θ dimension of the (r, θ) graph. To some extent this problem can be avoided by deciding in advance on the permissible values of θ and querying a list of boundary points to find which has the closest θ to each permissible value. Some local smoothing of the ordered set of boundary points can be undertaken but this is in principle unnecessary, since for a connected boundary, there will always be 1 pixel, which is closest to a line from the centroid at a given value of θ.

The two-stage approach to matching hinted in Section 10.3 can also be used to help with the last of the problems mentioned above—the need to speed up the processing. First, a coarse match is obtained between the object and its 1-D template by taking θ in relatively large steps of (say) 5° and ignoring intermediate angles in both the image data and the template, and then a better match is obtained by making fine adjustments to the orientations, obtaining a match to within 1°. In this way, the coarse match is obtained perhaps 20 times faster than the previous full match, whereas the final fine match takes a relatively short time, since very few distinct orientations have to be tested.

This two-stage process can be optimized by making a few simple calculations. The coarse match is given by increasing the θ incrementation step to $\delta\theta$, so the computational load is proportional to $(360/\delta\theta)^2$, whereas the load for the fine match is proportional to $360\delta\theta$, giving a total load of:

$$\lambda = \left(\frac{360}{\delta\theta}\right)^2 + 360\delta\theta \qquad (10.3)$$

This should be compared with the original load of:

$$\lambda_0 = 360^2 \qquad (10.4)$$

Hence, the load is reduced (and the algorithm speeded up) by the factor:

$$\eta = \frac{\lambda_0}{\lambda} = \frac{1}{(1/\delta\theta)^2 + (\delta\theta/360)} \qquad (10.5)$$

This is a maximum for $d\eta/d\delta\theta = 0$, giving:

$$\delta\theta = \sqrt[3]{2 \times 360} \approx 9° \qquad (10.6)$$

In practice, this value of $\delta\theta$ is rather large and there is a risk that the coarse match will give such a poor fit that the object will not be identified: hence, values of $\delta\theta$ in the range 2–5° are more usual (see, e.g., Berman et al., 1985). Note that the optimum value of η is 26.8 and this reduces only to 18.6 for $\delta\theta = 5°$, although it goes down to 3.9 for $\delta\theta = 2°$.

Another way of approaching the problem is to search the (r, θ) graph for some characteristic feature such as a sharp corner (this step constituting the coarse match), and then to perform a fine match around the object orientation so deduced. Clearly, there are possibilities of error here, in situations where objects have several similar features—as in the case of a rectangle: however, the individual trials are relatively inexpensive and so it is worth invoking this procedure if the object possesses appropriate well-defined features. Note that it is possible to use the position of the maximum value, r_{max}, as an orientating feature, but this is frequently inappropriate because a smooth maximum gives a relatively large angular error.

10.5 THE (s, ψ) PLOT

It can be seen from the above considerations that boundary pattern analysis should usually be practicable except when problems from occlusions and gross defects can be expected. However, these latter problems do give motivation for employing alternative methods if these can be found. In fact, the (s, ψ) graph has proved particularly popular since it is inherently better suited than the (r, θ) graph to situations where defects and occlusions may occur. In addition, it does not suffer from the multiple values encountered by the (r, θ) method.

The (s, ψ) graph does not require prior estimation of the centroid or some other reference point since it is computed directly from the boundary, in the form of a plot of the tangential orientation ψ as a function of boundary distance s. The method is not without its problems and, in particular, distance along the boundary needs to be measured accurately. The commonly used approach is to count horizontal and vertical steps as unit distance and to take diagonal steps as distance $\sqrt{2}$; in fact, this idea must be regarded as a rather *ad hoc* solution, and the situation is discussed further in Section 10.7.

When considering application of the (s, ψ) graph for object recognition, it will immediately be noticed that the graph has a ψ value that increases by 2π for each circuit of the boundary, i.e., $\psi(s)$ is not periodic in s. The result is that the graph becomes essentially 2-D, i.e., the shape has to be matched by moving the ideal object template both along the s-axis and along the ψ-axis directions. Ideally, the template could be moved diagonally along the direction of the graph. However, noise and other deviations of the actual shape relative to the ideal shape mean that in practice the match must be at least partly 2-D, hence adding to the computational load.

One way of tackling this problem is to make a comparison with the shape of a circle of the same boundary length P. Thus, an $(s, \Delta\psi)$ graph is plotted, which reflects the difference $\Delta\psi$ between the ψ expected for the shape and that expected for a circle of circumference P:

$$\Delta\psi = \psi - \left(\frac{2\pi s}{P}\right) \tag{10.7}$$

This expression helps to keep the graph 1-D, since $\Delta\psi$ automatically resets itself to its initial value after one circuit of the boundary (i.e., $\Delta\psi$ is periodic in s).

Next note that the $\Delta\psi(s)$ variation depends on the starting position where $s = 0$ and this is randomly sited on the boundary. It is useful to eliminate this dependence, and this may be achieved by subtracting from $\Delta\psi$ its mean value μ. This gives the new variable:

$$\tilde{\psi} = \psi - \left(\frac{2\pi s}{P}\right) - \mu \tag{10.8}$$

At this stage, the graph is completely 1-D and is also periodic, being similar in these respects to an (r, θ) graph. Matching should now reduce to the straightforward task of sliding the template along the $\tilde{\psi}(s)$ graph until a good fit is achieved.

At this point, there should be no problems so long as (a) the scale of the object is known and (b) occlusions or other disturbances cannot occur. Suppose next that the scale is unknown: then the perimeter P may be used to normalize the value of s. If, however, occlusions *can* occur, then no reliance can be placed on P for normalizing s and hence the method cannot be guaranteed to work. This problem does not arise if the scale of the object is known, since a standard perimeter P_T can be assumed. However, the possibility of occlusion gives further problems, which are discussed in the Section 10.6.

Another way in which the problem of nonperiodic $\psi(s)$ can be solved is by replacing ψ by its derivative $d\psi/ds$. Then the problem of constantly expanding ψ (which results in its increase by 2π after each circuit of the boundary) is eliminated—the addition of 2π to ψ does not affect $d\psi/ds$ locally, since $d(\psi+2\pi)/ds = d\psi/ds$. Note that $d\psi/ds$ is actually the local curvature function $\kappa(s)$ (see Fig. 10.4), so the resulting graph has a simple physical interpretation. Unfortunately, this version of the method has its own problems in that κ approaches infinity at any sharp corner. For industrial components, which frequently have sharp corners, this is a genuine practical difficulty and may be tackled by approximating adjacent gradients and ensuring that κ integrates to the correct value in the region of a corner (Hall, 1979).

Many workers take the (s, κ) graph idea further and expand $\kappa(s)$ as a Fourier series:

$$\kappa(s) = \sum_{n=-\infty}^{\infty} c_n \exp\left(\frac{2\pi i n s}{P}\right) \tag{10.9}$$

This results in the well-known Fourier descriptor method. In this method, shapes are analyzed in terms of a series of Fourier descriptor components, which are truncated to zero after a sufficient number of terms. Unfortunately, the amount of computation involved in this approach is considerable and there is a tendency to approximate curves with relatively few terms. In industrial applications, where computations have to be performed in real time, this can generate problems, so it is often more appropriate to match to the basic (s, κ) graph. In this way, critical measurements between object features can be made with adequate accuracy in real time.

10.6 TACKLING THE PROBLEMS OF OCCLUSION

Whatever means are used for tackling the problem of continuously increasing ψ, problems still arise when occlusions occur. However, the approach is not immediately invalidated by missing sections of boundary as it is for the basic (r, θ) method. As noted in Section 10.5, a major effect of occlusions is that the perimeter of the object is altered, so P can no longer be used to indicate its scale. Hence, the latter has to be known in advance: this is assumed in what follows. Another practical result of occlusions is that certain sections correspond correctly to parts of the object, whereas other sections correspond to parts of occluding objects; alternatively, they may correspond to unpredictable boundary segments where damage has occurred. Note that if the overall boundary is that of two overlapping objects, the observed perimeter P_B will be greater than the ideal perimeter P_T.

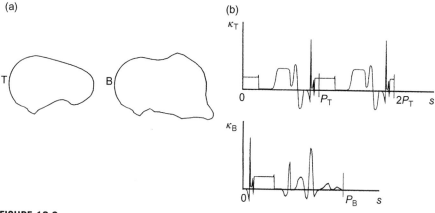

FIGURE 10.6

Matching a template against a distorted boundary. When a boundary B is broken (or partly occluded) but continuous, it is necessary to attempt to match between B and a template T that is doubled to length $2P_T$, to allow for T being severed at any point: (a) the basic problem and (b) matching in (s, κ) space.

Segmenting the boundary between relevant and irrelevant sections is, *a priori*, a difficult task. However, a useful strategy is to start by making positive matches wherever possible and to ignore irrelevant sections—i.e., try to match as usual, ignoring any section of the boundary that is a bad fit. We can imagine achieving a match by sliding the template T along the boundary B. However, a problem arises since T is periodic in s and should not be cut off at the ends of the range $0 \le s \le P_T$. As a result, it is necessary to attempt to match over a length $2P_T$. At first sight, it might be thought that the situation ought to be symmetrical between B and T. However, T is known in advance, whereas B is partly unknown, there being the possibility of one or more breaks in the ideal boundary into which foreign boundary segments have been included. Indeed, the positions of the breaks are unknown, so it is necessary to try matching the whole of T at all positions on B. Taking a length $2P_T$ in testing for a match effectively permits the required break to arise in T at any relevant position: (see Fig. 10.6).

When carrying out the match, we basically use the difference measure:

$$D_{jk} = \sum_i [\psi_B(s_i) - (\psi_T(s_i + s_k) + \alpha_j)]^2 \qquad (10.10)$$

where j and k are the match parameters for orientation and boundary displacement, respectively. Note that the resulting D_{jk} is roughly proportional to the length L of the boundary over which the fit is reasonable. Unfortunately, this means that the measure D_{jk} appears to *improve* as L decreases; hence, when variable occlusions can occur, the best match must be taken as the one for which the greatest length L gives good agreement between B and T (this may be measured

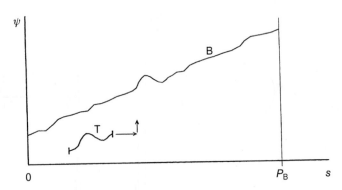

FIGURE 10.7

Matching a short template to part of a boundary. A short template T, corresponding to part of an idealized boundary, is matched against the observed boundary B. Strictly speaking, matching in (s, ψ) space is 2-D, although there is very little uncertainty in the vertical (orientation) dimension.

as the greatest number of values of s in the sum of Eq. (10.10), which gives good agreement between B and T, i.e., the sum over all i such that the difference in square brackets is numerically less than, say, 5°).

If the boundary is occluded in more than one place, then L is at most the largest single length of unoccluded boundary (not the total length of unoccluded boundary), since the separate segments will in general be "out of phase" with the template. This is a disadvantage when trying to obtain an accurate result, since extraneous matches add noise, which degrades the fit that is obtainable—hence adding to the risk that the object will not be identified and reducing accuracy of registration. This suggests that it might be better to use only short sections of the boundary template for matching. Indeed, this strategy can be advantageous since speed is enhanced and registration accuracy can be retained or even improved by careful selection of salient features (note that nonsalient features such as smooth curved segments could have originated from many places on object boundaries and are not very helpful for identifying and accurately locating objects: hence, it is reasonable to ignore them). In this version of the method, we now have $P_T < P_B$, and it is necessary to match over a length P_T rather than $2P_T$, since T is no longer periodic (Fig. 10.7). Once various segments have been located, the boundary can be reassembled as closely as possible into its full form, and at that stage defects, occlusions, and other distortions can be recognized overtly and recorded. Reassembly of the object boundaries can be performed by techniques such as the Hough transform and relational pattern matching techniques (see Chapters 13 and 14). Work of this type has been carried out by Turney et al. (1985), who found that the salient features should be short boundary segments where corners and other distinctive "kinks" occur.

Before leaving this topic, note that $\tilde{\psi}$ can no longer be used when occlusions are present, since although the perimeter can be assumed to be known, the mean value of $\Delta\psi$ (Eq. (10.8)) cannot be deduced. Hence, the matching task reverts to a 2-D search (although, as stated earlier, very little unrestrained search in the ψ direction need be made). However, in the case when small salient features are being sought, it is a reasonable working assumption that no occlusion occurs in any of the individual cases—a feature is either entirely present or entirely absent. Hence, the average slope $\overline{\psi}$ over T can validly be computed (Fig. 10.7) and this again reduces the search to 1-D (Turney et al., 1985).

Overall, missing sections of object boundaries necessitate a fundamental rethink as to how boundary pattern analysis should be carried out. For quite small defects the (r, θ) method is sufficiently robust but in less trivial cases it is vital to use some form of the (s, ψ) approach, while for really gross occlusions it is not particularly useful to try to match for the full boundary: rather it is better to attempt to match small salient features. This sets the scene for the Hough transform and relational pattern matching techniques of later chapters.

10.7 ACCURACY OF BOUNDARY LENGTH MEASURES

Next we examine the accuracy of the idea expressed earlier, that adjacent pixels on an 8-connected curve should be regarded as separated by 1 pixel if the vector joining them is aligned along the major axes and by $\sqrt{2}$ pixels if the vector is in a diagonal orientation. In general, this estimator overestimates the distance along the boundary. The reason for this is quite simple to see by appealing to the following pair of situations:

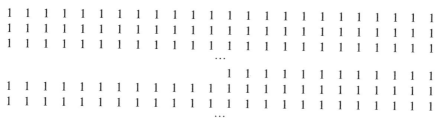

In either case we are considering only the top of the object. In the first example, the boundary length along the top of the object is exactly that given by the rule. However, in the second case the estimated length is increased by amount $\sqrt{2} - 1$ because of the step. Now as the length of the top of the object tends to large values, say p pixels, the actual length approximates to p, whereas the estimated length is $p + \sqrt{2} - 1$; thus, a definite error exists. Indeed, this error initially increases in importance as p decreases, since the actual length of the top of the object (when there is one step) is still:

$$L = (1 + p^2)^{1/2} \approx p \qquad (10.11)$$

so the fractional error is:

$$\xi \approx \frac{\sqrt{2} - 1}{p} \qquad (10.12)$$

which increases as p becomes smaller.

This result can be construed as meaning that the fractional error ξ in estimating boundary length increases initially as the boundary orientation ψ increases from zero. A similar effect occurs as the orientation decreases from 45°. Thus, the ξ variation possesses a maximum between 0° and 45°. This systematic overestimation of boundary length may be eliminated by employing an improved model in which the length per pixel is s_m along the major axis directions and s_d in diagonal directions. A complete calculation (Kulpa, 1977; see also Dorst and Smeulders, 1987) shows that:

$$s_m = 0.948 \tag{10.13}$$

and

$$s_d = 1.343 \tag{10.14}$$

It is perhaps surprising that this solution corresponds to a ratio s_d/s_m that is still equal to $\sqrt{2}$, although the arguments given above make it obvious that s_m should be less than unity.

Unfortunately, an estimator that has just two free parameters can still permit quite large errors in estimating the perimeters of individual objects. To reduce this problem, it is necessary to perform more detailed modeling of the step pattern around the boundary (Koplowitz and Bruckstein, 1989), which seems certain to increase the computational load significantly.

It is important to underline that the basis of this work is to estimate the length of the original continuous boundary rather than that of the digitized boundary: furthermore, it must be noted that the digitization process loses information, so the best that can be done is to obtain the best estimate of the original boundary length. Thus, employing the values 0.948 and 1.343 given above, rather than the values 1 and $\sqrt{2}$, reduces the estimated errors in boundary length measurement from 6.6% to 2.3%—but then only under certain assumptions about correlations between orientations at neighboring boundary pixels (Dorst and Smeulders, 1987).

10.8 CONCLUDING REMARKS

This chapter is concerned with boundary pattern analysis. The boundary patterns were imagined to arise from edge detection operations that have been processed to make them connected and of unit width. However, if intensity thresholding methods were employed for segmenting images, boundary tracking procedures would also permit the boundary pattern analysis methods of this chapter to be used. Conversely, if edge detection operations led to the production of connected boundaries, these could be filled in by suitable algorithms (which are more tricky to devise than might at first be imagined) (Ali and Burge, 1988) and converted to regions to which the binary shape analysis methods of Chapter 9 could be applied. Hence, shapes are representable in region or boundary form: if they initially arise in one representation, they may be converted to the alternate representation. This means that boundary or regional means may be employed for shape analysis, as appropriate.

An important factor here is that a positive advantage is often gained by employing boundary pattern analysis, since computation should be inherently lower (in proportion to the numbers of pixels that are required to describe the shapes in the two representations). Another important determining factor that is discussed in the present chapter is that of occlusion. If occlusions are present, then several of the methods described in Chapter 9 will operate incorrectly—as also happens for the basic centroidal profile method described in Section 10.3. The (s, ψ) method then provides a good starting point. As has been seen, this is best applied to detect small salient boundary features, which can then be reassembled into whole objects by relational pattern matching techniques (see especially Chapter 14).

A variety of boundary representations is available for shape analysis. However, this chapter has shown that intuitive schemes raise fundamental robustness issues: these will only be resolved later on by forgoing deduction in favor of inference. Underlying analog shape estimation in a digital lattice is also an issue.

10.9 BIBLIOGRAPHICAL AND HISTORICAL NOTES

Many of the techniques described in this chapter have been known since the early days of image analysis. Boundary tracking has been known since 1961 when Freeman introduced his chain code. Indeed, Freeman (1974) is responsible for much subsequent work in this area. Freeman (1978) introduced the notion of segmenting boundaries at "critical points" in order to facilitate matching: suitable critical points were corners (discontinuities in curvature), points of inflection, and curvature maxima. This work is clearly strongly related to that of Turney et al. (1985). Early work on Fourier boundary descriptors using the (r, θ) and (s, ψ) approaches was carried out by Rutovitz (1970), Barrow and Popplestone (1971), and Zahn and Roskies (1972); another notable paper in this area is by Persoon and Fu (1977). In an interesting development, Lin and Chellappa (1987) were able to classify partial (i.e., nonclosed) 2-D curves using Fourier descriptors.

At the beginning of the chapter it was noted that there are significant problems in obtaining a thin connected boundary for every object in an image. Since 1988, the concept of active contour models (or "snakes") solved many of these problems. See Chapter 5 for an introduction to snakes and Chapter 22 for their application to vehicle location.

It is worth remarking on the increased attention to accuracy evident over the past 20−30 years: this is seen, e.g., in the length estimators for digitized boundaries discussed in Section 10.7 (see Kulpa, 1977; Dorst and Smeulders, 1987; Beckers and Smeulders, 1989; Koplowitz and Bruckstein, 1989; Davies, 1991). For a later update on the topic, see Coeurjolly and Klette (2004).

In recent times, there has been an emphasis on characterizing and classifying *families* of shapes rather than just individual isolated shapes: (see in particular Cootes et al. (1992), Amit (2002), and Jacinto et al. (2003)). Klassen et al. (2004)

provide a further example of this in their analysis of planar (boundary) shapes using geodesic paths between the various shapes of the family. In their work, they employ the Surrey fish database (Mokhtarian et al., 1996). The same general idea is also manifest in the self-similarity analysis and matching approach of Geiger et al. (2003), which they used for human profile and hand motion analysis. Horng (2003) describes an adaptive smoothing approach for fitting digital planar curves with line segments and circular arcs. The motivation for this approach is to obtain significantly greater accuracy than can be achieved with the widely used polygonal approximation, yet with lower computational load than the spline fitting type of approach. It can also be imagined that any fine accuracy restriction imposed by a line plus circular arc model will have little relevance in a discrete lattice of pixels. da Gama Leitão and Stolfi (2002) have developed a multiscale contour-based method for matching and reassembling 2-D fragmented objects. Although this method is targeted at reassembly of pottery fragments in archeology, the authors imply that it is also likely to be of value in forensic science, in art conservation and in assessing the causes of failure of mechanical parts following fatigue and the like.

Two useful books are available that cover the subjects of shape and shape analysis in rather different ways: one is by Costa and Cesar (2000) and the other is by Mokhtarian and Bober (2003). The former is fairly general in coverage, but emphasizes Fourier methods, wavelets, and multiscale methods. The latter sets up a scale-space (especially curvature scale-space) representation (which is multiscale in nature), and develops the subject quite widely from there.

10.9.1 More Recent Developments

Ghosh and Petkov (2005) have described problems relating to the robust interpretation of incomplete object boundaries. They discuss the problems in relation to the ICR test—*viz.* assessing recognition rate performance as a function of the percentage of the contour retained, where deletions may occur either as segment deletions, or as occlusions, or as random pixel deletions. Experiments showed that occlusion was the most, and random pixel deletion the least, serious problem. Mori et al. (2005) considered problems relating to 3-D shape recognition from multiple 2-D views. They found that "shape contexts" were particularly important for efficient matching in such situations, shape contexts corresponding to representing shapes by a set of n samples on an object and examining the distribution of relative positions. This technique permitted shape matching to take place efficiently in two stages—fast pruning of possibilities followed by detailed matching.

10.10 PROBLEMS

1. Devise a program for finding a thinned (8-connected) boundary of an object in a binary image.

2. a. Describe the centroidal profile approach to shape analysis. Illustrate your answer for a circle, a square, a triangle, and defective versions of these shapes.
   b. Obtain a general formula expressing the shape a straight line presents in the centroidal profile.
   c. Show that there are two means of recognizing objects from their centroidal profiles, one involving analysis of the profile and the other involving comparison with a template.
   d. Show how the latter approach can be speeded up by implementing it in two stages, first at low resolution, and then at full resolution. If the low resolution has $1/n$ of the detail of the full resolution, obtain a formula for the total computational load. Estimate from the formula for what value of n the load will be minimized. Assume that the full angular resolution involves 360 one-degree steps.

3. a. Give a simple algorithm for eliminating salt and pepper noise in binary images, and show how it can be extended to eliminate short spurs on objects.
   b. Show that a similar effect can be achieved by a "shrink" + "expand" type of procedure. Discuss how much such procedures affect the shapes of objects: give examples illustrating your arguments, and try to quantify exactly what sizes and shapes of object would be completely eliminated by such procedures.
   c. Describe the (r, θ) graph method for describing the shapes of objects. Show that applying 1-D median filtering operations to such graphs can be used to smooth the described object shapes. Would you expect this approach to be more or less effective at smoothing object boundaries than methods based on shrinking and expanding?

4. a. Outline the (r, θ) graph method for recognition of objects in two dimensions, and state its main advantages and limitations. Describe the shape of the (r, θ) graph for an equilateral triangle.
   b. Write down a complete algorithm, operating in a 3×3 window, for producing an approximation to the convex hull of a 2-D object. Show that a more accurate approximation to the convex hull can be obtained by joining humps with straight lines in an (r, θ) graph of the object. Give reasons why the result for the latter case will only be an approximation, and suggest how an exact convex hull might be obtained.

5. An alternative approach to shape analysis involves measuring distance around the boundary of any object and estimating increments of distance as 1 unit when progressing to the next pixel in a horizontal or vertical direction and $\sqrt{2}$, units when progressing in a diagonal direction. Taking a square of side 20 pixels, which is aligned parallel to the image axes, and rotating it through small angles show that distance around the boundary of the square is not estimated accurately by the $1:\sqrt{2}$, model. Show that a similar effect occurs when the square is orientated at about $45°$ to the image axes. Suggest ways in which this problem might be tackled.

11

Line Detection

Detection of macroscopic features is an important part of image analysis and visual pattern recognition. Of particular interest is the identification of straight lines in images, as these are ubiquitous—appearing both on manufactured parts and in the built environment. The Hough transform (HT) provides the means for locating these features highly robustly in digital images. This chapter describes the processes and principles needed to achieve this.

Look out for:

- the basic HT technique for locating straight lines in images.
- alternative parametrizations of the HT.
- how lines can be localized along their length.
- how final line fitting can be made more accurate.
- why the HT is robust against noise and background clutter.
- how speed of processing can be improved.
- the RANSAC approach to line fitting.
- the relative efficiencies of RANSAC and the HT.
- how laparoscopic tools may be located.

While this chapter provides interesting methods for detecting line features in images, they may appear somewhat specialized. However, later chapters will show that both the HT and RANSAC (RANdom SAmpling Consensus) have much wider applicability than such arguments might suggest.

11.1 INTRODUCTION

Straight edges are among the most common features of the modern world, arising in perhaps the majority of manufactured objects and components, not least in the very buildings in which we live. Yet it is arguable whether true straight lines ever arise in the natural state: possibly the only example of their appearance in virgin outdoor scenes is the horizon—although even this is clearly seen from space as a circular boundary. The surface of water is essentially planar, although it is important to realize that this is a deduction. The fact remains that straight lines seldom appear in completely natural scenes. Be all this as it may, it is clearly vital both in city pictures and in the factory to have effective means of detecting straight edges. This chapter studies available methods for locating these important features.

Historically, the Hough transform (HT) has been the main means of detecting straight edges, and since the method was originally invented (Hough, 1962) it has been developed and refined for this purpose. Hence this chapter concentrates on this particular technique; it also prepares the ground for applying the HT to the detection of circles, ellipses, corners, etc. in the next few chapters. We start by examining the original Hough scheme, even though it is now seen to be wasteful in computation, since important principles are involved. By the end of the chapter we shall see that the HT is not alone in its capabilities for line detection: RANSAC is also highly capable in this direction. In fact, both approaches have their advantages and limitations, as the discussion in Section 11.6 will show.

11.2 APPLICATION OF THE HOUGH TRANSFORM TO LINE DETECTION

The basic concept involved in locating lines by the HT is point–line duality. A point P can be defined either as a pair of coordinates or in terms of the set of lines passing through it. The concept starts to make sense if we consider a set of collinear points P_i, then list the sets of lines passing through each of them, and finally note that there is just one line that is common to all these sets. Thus, it is possible to find the line containing all the points P_i merely by eliminating those that are not multiple hits. Indeed, it is easy to see that if a number of noise points Q_j are intermingled with the signal points P_i, the method will be able to discriminate the collinear points from among the noise points at the same time as finding the line containing them, merely by searching for multiple hits. Thus, the method is inherently robust against noise, as indeed it is in discriminating against currently unwanted signals such as circles.

In fact, the duality goes further. For just as a point can define (or be defined by) a set of lines, so a line can define (or be defined by) a set of points, as is obvious from the above argument. This makes the above approach to line detection a mathematically elegant one and it is perhaps surprising that the Hough detection scheme was first published as a patent (Hough, 1962) of an electronic apparatus for detecting the tracks of high-energy particles, rather than as a paper in a learned journal.

The form in which the method was originally applied involves parametrizing lines using the slope–intercept equation:

$$y = mx + c \tag{11.1}$$

Every point on a straight edge is then plotted as a line in (m, c) space corresponding to all the (m, c) values consistent with its coordinates, and lines are detected in this space. The embarrassment of unlimited ranges of the (m, c) values (near-vertical lines require near-infinite values of these parameters) is overcome by using two sets of plots, the first corresponding to slopes of less than 1.0 and the second to slopes of 1.0 or more; in the latter case, Eq. (11.1) is replaced by the form:

$$x = \tilde{m}x + \tilde{c} \qquad (11.2)$$

where

$$\tilde{m} = \frac{1}{m} \qquad (11.3)$$

The need for this rather wasteful device was removed by the Duda and Hart (1972) approach, which replaces the slope−intercept formulation with the so-called "normal" (θ, ρ) form for the straight line (see Fig. 11.1):

$$\rho = x \cos \theta + y \sin \theta \qquad (11.4)$$

To apply the method using this form, the set of lines passing through each point P_i is represented as a set of sine curves in (θ, ρ) space: for example, for point $P_1(x_1, y_1)$, the sine curve has equation:

$$\rho = x_1 \cos \theta + y_1 \sin \theta \qquad (11.5)$$

Then multiple hits in (θ, ρ) space indicate, *via* their (θ, ρ) values, the presence of lines in the original image.

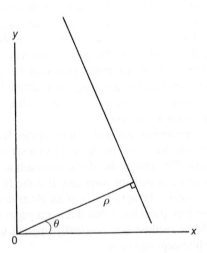

FIGURE 11.1

Normal (θ, ρ) parametrization of a straight line.

Each of the methods described above has the feature that it employs an "abstract" parameter space in which multiple hits are sought. Above we talked about "plotting" points in parameter space, but in fact the means of looking for hits is to seek peaks that have been built by *accumulation* of data from various sources. Although it might be possible to search for hits by logical operations such as use of the AND function, the Hough method gains considerably by *accumulating evidence* for events by a *voting scheme*. It will be seen below that this is the source of the method's high degree of robustness.

Although the methods described above are mathematically elegant and are capable of detecting lines (or sets of collinear points—which may be completely isolated from each other) amid considerable interfering signals and noise, they are subject to considerable computational problems. The reason for this is that every prominent point[1] in the original image gives rise to a great many votes in parameter space, so for a 256×256 image the (m, c) parametrization requires 256 votes to be accumulated, while the (θ, ρ) parametrization requires a similar number— 360 if the θ quantization is to be fine enough to resolve $1°$ changes in line orientation.

A vital key to overcoming this problem was discovered by O'Gorman and Clowes (1976), who noted that points on lines are usually not isolated but instead are joined in fragments that are sufficiently long that their directions can be measured. Supposing that direction is known with high accuracy, it is then sufficient to accumulate just one vote in parameter space for every potential line point (in fact, if the local gradient is known with lesser accuracy then parameter space can be quantized more coarsely—say in $10°$ steps (O'Gorman and Clowes, 1976)—and again a single vote per point can be accumulated). Clearly, this method is much more modest in its computational requirements and it was soon adopted by other workers.

However, the computational load is still quite substantial: not only is a large two-dimensional (2-D) storage area needed but this must be searched carefully for significant peaks—a tedious task if short line segments are being sought. Various means have been tried for cutting down computation further. Dudani and Luk (1978) tackled the problem by trying to separate out the θ and ρ estimations. They accumulated votes first in a 1-D parameter space for θ—i.e., a histogram of θ values (it must not be forgotten that such a histogram is itself a simple form of HT).[2] Having found suitable peaks in the θ histogram, they then built a ρ histogram for all the points that contributed votes to a given θ peak, and repeated this for all θ peaks. Thus, two 1-D spaces replace the original 2-D parameter space, with very significant savings in storage and load. However, two-stage methods of

---

[1]For the present purpose it does not matter in what way these points are prominent. They may in fact be edge points, dark specks, centers of holes, and so on. Later we shall consistently take them to be edge points.

[2]It is now common for any process to be called an HT if it involves accumulating votes in a parameter space, with the intention of searching for significant peaks to find properties of the original data.

this type tend to be less accurate since the first stage is less selective: biased θ values may result from pairs of lines that would be well separated in a 2-D space. In addition, any error in estimating θ values is propagated to the ρ determination stage, making values of ρ even less accurate. For this reason Dudani and Luk added a final least-squares fitting stage to complete the accurate analysis of straight edges present in the image.

From a practical point of view, to proceed with either of the above methods of line detection, it is first necessary to obtain the local components of intensity gradient, and then to deduce the gradient magnitude g and threshold it to locate each edge pixel in the image. θ may be estimated using the arctan function in conjunction with the local edge gradient components g_x, g_y:

$$\theta = \arctan\left(\frac{g_y}{g_x}\right) \tag{11.6}$$

As the arctan function has period π, $\pm\pi$ may have to be added to obtain a principal value in the range $-\pi$ to $+\pi$: this can be decided from the signs of g_x and g_y.[3] Once θ is known, ρ can be found from Eq. (11.4).

Finally, note that straight lines and straight edges are different and need to be detected differently. (Straight *edges* are probably more common and appear as object boundaries, whereas straight *lines* are typified by telephone wires in outdoor scenes.) In fact, we have concentrated above on using the HT to locate straight edges, starting with edge detectors. Straight line segments may be located using Laplacian-type operators and their orientations are defined over a range $0-180°$ rather than $0-360°$: this makes HT design subtly different. For concreteness, in the remainder of this chapter, we concentrate on straight *edge* detection.

11.3 THE FOOT-OF-NORMAL METHOD

An alternative means of saving computation (Davies, 1986b) eliminates the use of trigonometric functions such as arctan by employing a different parametrization scheme. As noted earlier, the methods so far described all employ abstract parameter spaces in which points bear no immediately obvious visual relation to image space. In the alternative scheme, the parameter space is a second image space, which is congruent[4] to image space.

This type of parameter space is obtained in the following way. First, each edge fragment in the image is produced much as required previously so that ρ can be measured, but this time the foot of the normal from the origin is itself taken as a voting

---

[3]Note that in C++, the basic arctan function is *atan*, with a single argument, which should be g_y/g_x used as indicated above. However, the C++ *atan2* function has two arguments, and if g_y and g_x are used respectively for these, the function automatically returns an angle in the range $-\pi$ to π.
[4]That is, parameter space is like image space, *and* each point in parameter space holds information that is immediately relevant to the corresponding point in image space.

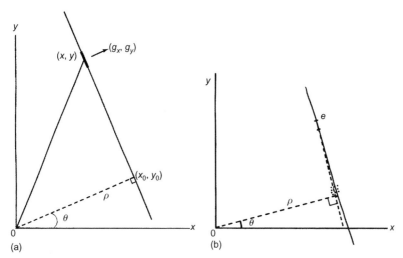

FIGURE 11.2

Image space parametrization of a straight line: (a) parameters involved in the calculation (see text); (b) buildup of foot-of-normal positions in parameter space for a more practical situation, where the line is not exactly straight: e is a typical edge fragment leading to a single vote in the parameter space.

Source: (b) © Unicom 1988

position in parameter space (Fig. 11.1). Clearly, the foot-of-normal position embodies all the information previously carried by the ρ and θ values, and mathematically the methods are essentially equivalent. However, the details differ, as will be seen.

The detection of straight edges starts with the analysis of (a) local pixel coordinates (x, y) and (b) the corresponding local components of intensity gradient (g_x, g_y) for each edge pixel. Taking (x_0, y_0) as the foot of the normal from the origin to the relevant line (produced if necessary—see Fig. 11.2), it is found that

$$\frac{g_y}{g_x} = \frac{y_0}{x_0} \tag{11.7}$$

$$(x - x_0)x_0 + (y - y_0)y_0 = 0 \tag{11.8}$$

These two equations are sufficient to compute the two coordinates (x_0, y_0). Solving for x_0 and y_0 gives

$$x_0 = v g_x \tag{11.9}$$

$$y_0 = v g_y \tag{11.10}$$

where

$$v = \frac{x g_x + y g_y}{g_x^2 + g_y^2} \tag{11.11}$$

Note that these expressions involve only additions, multiplications, and just one division, so voting can be carried out efficiently using this formulation.

11.3.1 Application of the Foot-of-Normal Method

Although the foot-of-normal method is mathematically similar to the (θ, ρ) method, it is unable to determine line orientation directly with quite the same degree of accuracy. This is because the orientation accuracy depends on the fractional accuracy in determining ρ—which in turn depends on the absolute magnitude of ρ. Hence, for small ρ the orientation of a line that is predicted from the position of the peak in parameter space will be relatively inaccurate, even though the *position* of the foot-of-normal is known accurately. However, accurate values of line orientation can always be found by identifying the points that contributed to a given peak in the foot-of-normal parameter space and making them contribute to a θ histogram, from which line orientation may be determined accurately.

Typical results with the above method are shown in Fig. 11.3. Here, it was applied in subimages of size 64×64 within 128×128 images. Clearly, some of the objects in these pictures are grossly overdetermined by their straight edges, so low ρ values are not a major problem. For those peaks where $\rho > 10$, line orientation is estimated within approximately $2°$; as a result, these objects are located within 1 pixel and orientated within $1°$ by this technique, without the necessity for θ histograms. Figure 11.4(a) and (b) contain some line segments that are not detected. This is due partly to their relatively low contrast, higher noise levels, above average fuzziness, or short length. However, it is also due to the thresholds set on the initial edge detector and on the final peak detector: when these were set at lower values, additional lines were detected but other noise peaks also became prominent in parameter space, and each of these needed to be checked in detail to confirm the presence of the corresponding line in the image. This is one aspect of a general problem that arises right through the field of image analysis.

11.4 LONGITUDINAL LINE LOCALIZATION

The preceding sections have provided a variety of means for locating lines in digital images and finding their orientations. However, these methods are insensitive to where along an infinite idealized line an observed segment appears. The reason for this is that the fit includes only two parameters. There is some advantage to be gained in this, in that partial occlusion of a line does not prevent its detection. Indeed, if several segments of a line are visible, they can all contribute to the peak in parameter space, hence improving sensitivity. On the other hand, for full image interpretation, it is useful to have information about the longitudinal placement of line segments.

This is achieved by a further stage of processing. The additional stage involves finding which points contributed to each peak in the main parameter space and carrying out connectivity analysis in each case. Dudani and Luk (1978)

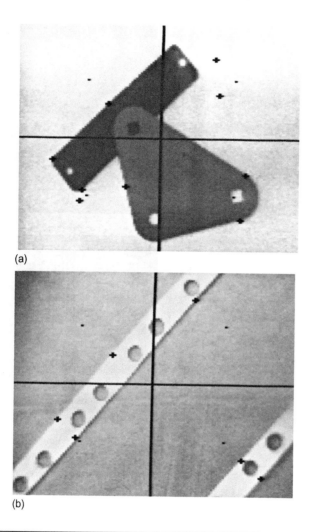

(a)

(b)

FIGURE 11.3

Results of image space parametrization of mechanical parts. The dot at the center of each quadrant is the origin used for computing the image–space transform. The crosses are the positions of peaks in parameter space that mark the individual straight edges (produced if necessary). For further explanation, see text.

Source: (a) © Unicom 1988

called this process "*xy*–grouping." It is not vital that the line segments should be 4- or 8-connected—just that there should be sufficient points on them so that adjacent points are within a threshold distance apart, i.e., groups of points are merged if they are within the prespecified distance (typically, 5 pixels). Finally, segments shorter than a certain minimum length (also typically ∼5 pixels) can be ignored as too insignificant to help with image interpretation.

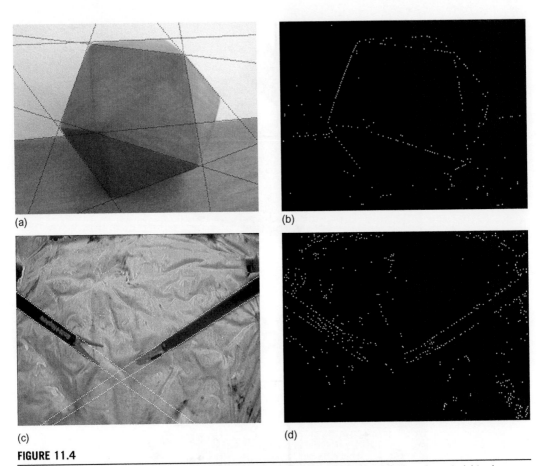

(a) (b)
(c) (d)

FIGURE 11.4

Straight line location using the RANSAC technique. (a) An original grayscale image with various straight edges located using the RANSAC technique; (b) the edge points fed to RANSAC for (a). These were isolated points that were local maxima of the gradient image. (c) The straight edges of a pair of laparoscopic tools—a cutter and a gripper—which have been located by RANSAC. (d) The points fed to RANSAC for (c). In (a) three edges of the icosahedron are missed. This is because they are roof edges with low-contrast and low-intensity gradient. RANSAC missed a fourth edge because of a lower limit placed on the level of support (see text).

11.5 FINAL LINE FITTING

Dudani and Luk (1978) found it useful to include a least-squares fitting stage to complete the analysis of the straight edges present in an image. This is carried out by taking the x, y coordinates of points on the line as input data and minimizing the sum of squares of distances from the data points to the best fit line. Their reason for adding such a stage was to eliminate local edge orientation accuracy as a source of error. This motivation is generally reasonable if the highest possible accuracy is required (e.g., obtaining line orientation to significantly better than 1°). However, many workers have criticized the use of the least-squares technique, since it tends to weight up the contribution of less accurate points—including those points that have nothing to do with the line in question, but which arise from image "clutter." This criticism is probably justified, although the least-squares technique is convenient in yielding to mathematical analysis and in this case in giving explicit equations for θ and ρ in terms of the input data (Dudani and Luk, 1978). Dudani and Luk obtained the endpoints by reference to the best-fit line thus obtained. To understand the limitations of the least-squares technique, see the Appendix A.

11.6 USING RANSAC FOR STRAIGHT LINE DETECTION

RANSAC is an alternative model-based search schema that can often be used instead of the HT. In fact, it is highly effective when used for line detection, which is why the method is introduced here. The strategy can be construed as a voting scheme, but it is used in a different way from that in the HT. The latter operates by building up the evidence for instances of target objects in the form of votes in parameter space, and then making decisions about their existence (or by making hypotheses about their existence that are then finally checked out). RANSAC operates by making a sequence of hypotheses about the target objects, and determines the support for each of them by counting how many data points agree with them. As might be expected, for any potential target object, only the hypotheses with the maximum support are retained at each stage. This results in more compact information storage than for the HT: i.e., for RANSAC a list of hypotheses is held in current memory, whereas for the HT a whole parameter space, which is usually only sparsely populated, is held in memory. Thus, the RANSAC data are abstract lists, whereas the HT data can often be viewed as pictures in parameter space—as in the case of the foot-of-normal line detector. None of this prevents the RANSAC output being displayed (e.g., as straight lines) in image space, nor does it prevent the HT being accumulated using a list representation.

To explain RANSAC in more detail, we take the case of line detection. As for the HT, we start by applying an edge detector and locating all the edge points in the image. As we shall see, RANSAC operates best with a limited number of points, so it is useful to find the edge points that are local maxima of the intensity gradient image. (This does not correspond to the type of nonmaximum suppression used in the Canny operator, which produces thin connected *strings* of edge points but to individual *isolated* points: we shall return to this point below.) Next, to form a straight line hypothesis, all that is necessary is to take any pair of edge points from the list of N that remain after applying the local maximum operation. For each hypothesis we run through the list of N points finding how many points M support the hypothesis. Then we take more hypotheses (more pairs of edge

Table 11.1 Basic RANSAC Algorithm for Finding the Line with Greatest Support

```
Mmax = 0;
for all pairs of edge points do {
    find equation of line defined by the two points i, j;
    M = 0;
    for all N points in list do
        if (point k is within threshold distance d of line) M++;
    if (M > Mmax) {
        Mmax = M;
        imax = i;
        jmax = j;
        //this records the hypothesis giving the maximum support so far
    }
}
/* if Mmax > 0, (x[imax], y[imax]) and (x[jmax], y[jmax]) will be the coordinates of
the points defining the line having greatest support */
```

This algorithm only returns one line: in fact it returns the specific line model that has greatest support, for the line that has greatest support. Lines with less support are in the end ignored.

points) and at each stage retain only the one giving maximum support M_{max}. This process is shown in Table 11.1.

The algorithm in Table 11.1 corresponds to finding the center of the highest peak in parameter space in the case of the HT. To find all the lines in the image, the most obvious strategy is the following: find the first line, then eliminate all the points that gave it support; then find the next line and eliminate all the points that gave it support; and so on until all the points have been eliminated from the list. The process may be written more compactly in the form:

```
repeat {
    find line;
    eliminate support;
}
until no data points remain;
```

Such a strategy carries the problem that if lines cross each other, support for the second line (which will necessarily be the weaker one) could be lesser support than it deserves. However, this should only be a serious disadvantage if the image is severely cluttered with lines. Nevertheless, the process is sequential and as such the results (i.e., the exact line locations) will depend on the order in which lines are eliminated, as the support regions will be minutely altered at each stage. Overall, the interpretation of complex images almost certainly has to proceed sequentially, and there is significant evidence that the human eye—brain system interprets images in this way, following early cues in order to progressively make sense of the data. Interestingly, the HT seems to escape from this by the potential capability for *parallel* identification of peaks. While for simple images this may well be true, for complicated images containing many overlapping edges, there will again be the need for sequential analysis of the type envisaged above (e.g., see the back-projection method of Gerig and Klein, described in Section 13.11). The point is that with the particular list representation employed by RANSAC, we are *immediately* confronted with the problem of how to identify multiple targets, whereas for the reasons given above this does not immediately happen with the HT. Finally, on the plus side, successive elimination of support points necessarily makes it progressively easier and less computation intensive to find subsequent target objects. But the process itself is not cost-free, as the whole RANSAC procedure in Table 11.1 has to be run twice per line in order to identify the support points that have to be eliminated.

Next, we consider the computational load of the RANSAC process. If there are N edge points, the number of potential lines will be $^{N}C_2$, corresponding to a computational load of $O(N^2)$. However, finding the support for each line will involve $O(N)$ operations, so the overall computational load will be $O(N^3)$. In addition, the need to eliminate the support points for each line found will require computation proportional to the number of lines n, amounting to only $O(nN)$, and this will have little effect on the overall computational load.

One point that has not yet been made is that all N edge points will not arise from straight lines: some will arise from lines, some from curves, some from general background clutter, and some from noise. To limit the number of false positives, it will be useful to set a support threshold M_{thr} such that potential lines for which $M > M_{thr}$ are most likely to be true straight lines, while others are most likely to be artifacts, such as parts of curves or noise points. Thus, the RANSAC procedure can be terminated when M_{max} drops below M_{thr}. Of course, it may be required to retain only "significant" lines, e.g., those having length greater than L pixels. In that case, analysis of each line could allow many more points to be eliminated as the RANSAC algorithm proceeds. Another factor is whether hypotheses corresponding to pairs whose points are too close together should be taken into account. In particular, it might be considered that points closer together than 5 pixels would be superfluous as they would be likely to have much reduced chance of pointing along the direction of a line. However, it turns out that RANSAC is fail-safe in this respect, and there is some gain from keeping pairs with quite small separations, as some of the resulting hypotheses can actually be more accurate than any others. Overall, restricting pairs by their separations can be a useful way of reducing computational load, bearing in mind that $O(N^3)$ is rather high. Here, we should recall that the load for the HT is $O(N^2)$ during voting, if pairs of points are used, or $O(N)$, if single edge points and their gradients are used instead.

As we have just seen, RANSAC does not compare well with the HT regarding computational load, so it is better to employ RANSAC when N can be reduced in some way. This is why it is useful to use N local maxima rather than a full list of edge points in the form of strings of edge points generated by nonmaximum suppression or *a fortiori*, those existing before nonmaximum suppression. Indeed, there is much to be gained by repeated random sampling[5] from the full list until sufficient hypotheses have been tested to be confident that all significant lines have been detected. In this way, the computational load is reduced from $O(N^3)$ to $O(N^2)$ or even $O(N)$ (it is difficult to predict the resulting computational complexity: in any case, the achievable computational load will be highly data dependent). Confidence that all significant lines have been detected can be obtained by estimating the risk that a significant line will be missed because no representative pair of points lying on the line was considered. This aspect of the problem will be examined more fully in Section A.6 of the appendix on Robust Statistics.

Before proceeding further, we shall briefly consider another way of reducing computational load. That is, by using the connected strings of edge points resulting from nonmaximum suppression, but eliminating those that are very short and coding the longer ones as isolated points every p pixels, where $p \approx 10$. In this way, there would be far fewer points than for our earlier paradigm, and also those that are employed would have increased coherence and probability of lying on straight edges; hence there would be a concentration on high quality data, while the $O(N^3)$ computation factor would be dramatically reduced. Clearly, in high noise situations this would not work well. It is left to the reader to judge how well it would work for the sets of edges located by the Canny operator in Figs. 5.7 and 5.8.

We are now in a position to consider actual results obtained by applying RANSAC to straight line detection. In the tests described, pairs of points were employed as hypotheses, and all edge points were local maxima of the intensity gradient. The cases shown in Fig. 11.4 correspond to detection of a block of wood in the shape of an icosahedron, and a pair of laparoscopic tools with parallel sides. Note that one line on the right of Fig. 11.4(a) was missed because a lower limit had to be placed on the level of support for each line. This was necessary because below this level of support the number of chance collinearities rose dramatically even for the relatively small number of edge points shown in Fig. 11.4(b), leading to a sharp rise in the number of false-positive lines. Figs. 23.2 and 23.3 show RANSAC being used to locate road lane markings. The same version of RANSAC was used in all the above cases, albeit in the case of Fig. 23.3 a refinement was added to allow improved elimination of points on lines that had already been located (this point will be clarified at the end of this section). Overall, this set of examples shows that RANSAC is a highly important contender for location of straight lines in digital images. Not discussed here is the fact that RANSAC is useful for obtaining robust fits to many other types of shape in 2-D and in 3-D.

It should be mentioned that one characteristic of RANSAC is that it is less influenced by aliasing along straight lines than the HT. This is because HT peaks tend to be fragmented by aliasing, so the best hypotheses can be difficult to obtain

---

[5]This corresponds to the original reason for the term RANSAC, which stands for RANdom SAmpling Consensus, "consensus" indicating that any hypothesis has to form a consensus with the available support data.

without excessive smoothing of the image. The reason why RANSAC wins in this context is that it does not rely on individual hypotheses being accurate: rather it relies on enough hypotheses easily being generatable, and by the same token, discardable.

Finally, we return to the above comment about obtaining improved deletion of points on lines that have already been located. Suppose the cross-section of a line is characterized by a (lateral) Gaussian distribution of edge points. As a true Gaussian extends to infinity in either direction, the support region is not well defined, but for high accuracy it is reasonable to take it as the region within $\pm\sigma$ of the centerline of the line. However, if just these points are eliminated, the remaining points near the line could later on give rise to alternative lines or combine with other points to lead to false positives. It is therefore better (Mastorakis and Davies, 2011) to make the "delete distance" d_d larger than the "fit distance" d_f that is used for support during detection, e.g., $d_d = 2\sigma$ or 3σ, where $d_f = \sigma$. ($d_d \approx 3\sigma$ can be considered to be close to optimal because 99.9% of the samples in a Gaussian distribution lie within $\pm 3\sigma$.) Figure 23.3 shows instances in which the fit distance is 3 pixels and the delete distance has values of 3, 6, 10, and 11 pixels, showing the advantages that can be gained by making d_d significantly greater than d_f. Figure 23.4 shows a flowchart of the algorithm used in this case.

11.7 LOCATION OF LAPAROSCOPIC TOOLS

Section 11.6 showed how RANSAC can provide a highly efficient means for locating straight edges in digital pictures, and gave an example of its use to locate the handles of laparoscopic tools. These are used for various forms of "keyhole" surgery: specifically, one tool (e.g., a cutter) might be inserted through one incision and another (e.g., a gripper) through a second incision. Additional incisions are needed for viewing *via* a laparoscope that employs optical fiber technology, and for inflating the cavity—e.g., the abdominal or chest cavity. In this section, we consider what information can be obtained *via* the laparoscope.

Figure 11.4(c) shows laparoscopic gripper and cutter tools being located in a simulated flesh background. The latter will normally be a wet surface that is largely red and will exhibit many regions that are close to being specularly reflective. The large variations in intensity that occur under these conditions make the scene quite difficult to interpret. While the surgeon who is in control of the instruments can learn a lot about the scene through tactile feedback and thus bolster his understanding of it, other people viewing the scene, e.g., on a TV monitor or computer, are liable to find it highly confusing. The same will apply for any computer attempting to interpret, analyze, or record the progress of an operation. These latter tasks are potentially important for logging operations, for training other doctors, for communicating with specialists elsewhere, or for analyzing the progress of the operation during any subsequent debriefing. It would therefore be useful if the exact locations,

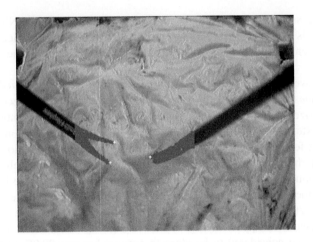

FIGURE 11.5

Tips of laparoscopic tools located from the parts highlighted in gray.

orientations, and other parameters of the tools could be determined at least relative to the frame of reference of the laparoscope. To this end, RANSAC has provided important 2-D data on the location of the tool handles. Assessing the vanishing points of the pairs of lines from the handles also provides 3-D information. Clearly, further information can be obtained from the coordinates of the ends of the tools.

To identify the ends of the tools, the ends of the handles are first located. This is a straightforward task requiring knowledge of the exact ends of the RANSAC support regions for the tool handle edges. The remainder of the tool ends can now be located by initial approximate prediction, adaptive thresholding, and connected components analysis (Fig. 11.5), special attention being paid to accurate location of the tips of the tool ends. In Fig. 11.5, this is achieved within ~1 pixel for the gripper on the left, and slightly less accurately for the closed cutter on the right. If the gripper had been open, accuracy would have been similar to that for the gripper. Note that, because of the complex intensity patterns in the background, it would have been difficult to locate the tool ends without first identifying the tool handles.

Each of the laparoscopic tools referred to above has (X, Y, Z) position coordinates, together with rotations ψ within the image plane (x, y), θ away from the image plane (toward the Z-axis), and φ about the axis of the handle; in addition, each tool end has an opening angle α (Fig. 11.6). It is bound to be difficult to obtain all seven parameters with any great accuracy from a single monocular view. However, in principle, using an exact CAD model of the tool end, this should be possible with ~15−20° accuracy for the angles. The 2-D information about the centerlines and widths of the handles, the convergence of the handle edges, and the exact positions of the tips of the tools, should together permit such a 3-D analysis to be carried out. Here, we have concentrated on the 2-D analysis: details of the relevant 3-D background theory needed to proceed further can be found in Part 3.

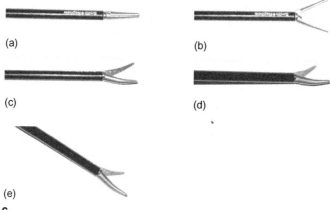

FIGURE 11.6

Orientation parameters for a laparoscopic tool. (a) A gripper tool with closed jaws. (b) Gripper with jaws separated by an angle α. (c) Gripper rotated through an angle φ about a horizontal axis. (d) Gripper tipped through an angle θ away from the image plane. (e) Gripper rotated through an angle ψ about the camera optical axis.

11.8 CONCLUDING REMARKS

This chapter has described a variety of techniques for finding straight lines and straight edges in digital images. Several of these were based on the HT, which is important because it permits systematic extraction of global data from images and has the capability to ignore "local" problems, due for example to occlusions and noise. This is what is required for "intermediate-level" processing, as will be seen repeatedly in later chapters.

The specific techniques covered have involved various parametrizations of a straight line, and means for improving efficiency and accuracy. In particular, speed is improved by using a two-stage line finding procedure—a method that is useful in other applications of the HT, as will be seen in later chapters. Accuracy tends to be cut down by such two-stage processing because of error propagation and because the first stage is liable to be subject to too many interfering signals. However, it is possible to improve the accuracy of approximate solutions by using least-squares refinement procedures.

In fact, by the end of the chapter it became clear that the RANSAC approach also has line fitting capabilities, which are for some purposes superior to those of the HT, although RANSAC tends to be more computation intensive (with N edge points it has a computational load of $O(N^3)$ rather than $O(N^2)$). Suffice it to say that the choice of which approach to use will depend on the exact type of image data, including levels of noise and background clutter.

> The Hough transform is one way of inferring the presence of objects from their feature points, and RANSAC is another. Both methods can be said to use voting schemes to select best-fit lines, although only the HT employs a parameter space representation; and both methods are highly robust as they focus only on positive evidence for the existence of objects.

11.9 BIBLIOGRAPHICAL AND HISTORICAL NOTES

The HT was developed in 1962 (Hough, 1962) with the aim of finding (straight) particle tracks in high-energy nuclear physics and was brought into the mainstream image analysis literature much later by Rosenfeld (1969). Duda and Hart (1972) developed the method further and applied it to the detection of lines and curves in digital pictures. O'Gorman and Clowes (1976) soon developed a Hough-based scheme for finding lines efficiently, by making use of edge orientation information, at much the same time that Kimme et al. (1975) applied the same method (apparently independently) to the efficient location of circles. Many of the ideas for fast effective line finding described in this chapter arose in a paper by Dudani and Luk (1978). The author's foot-of-normal method (Davies, 1986b) was developed much later. During the 1990s, work in this area progressed further—see for example Atiquzzaman and Akhtar's (1994) method for the efficient determination of lines together with their end coordinates and lengths; Lutton et al.'s (1994) application of the transform to the determination of vanishing points; and Kamat-Sadekar and Ganesan's (1998) extensions of the technique to cover the reliable detection of multiple line segments, particularly in relation to road scene analysis. Other applications of the HT are covered in the next two chapters.

   Some mention should be made of the related Radon transform. This is formed by integrating the picture function $I(x, y)$ along infinitely thin straight strips of the image, with normal coordinate parameters (θ, ρ), and recording the results in a (θ, ρ) parameter space. The Radon transform is a generalization of the Hough transform for line detection (Deans, 1981). In fact, for straight lines the Radon transform reduces to the Duda and Hart (1972) form of the HT that, as remarked earlier, involves considerable computation. For this reason the Radon transform is not covered in depth in this book. The transforms of real lines have a characteristic "butterfly" shape (a pencil of segments passing through the corresponding peak) in parameter space. This phenomenon has been investigated by Leavers and Boyce (1987), who have devised special 3×3 convolution filters for sensitively detecting these peaks.

   Contrary to the view of some that the HT is completely worked over and no longer a worthwhile topic for research, there has been strong continuing interest in it. Indeed, the computational difficulties of the method reflect underlying matching problems that are inescapable in computer vision, so development of methods must continue. Thus, Schaffalitsky and Zisserman (2000) carried out an interesting extension of earlier ideas on vanishing lines and points by considering the case of repeated lines, such as those occurring on certain types of fences and brick buildings; Song et al. (2002) developed HT methods for coping with the problems of fuzzy edges and noise in large-sized images; and Guru et al. (2004) demonstrated viable alternatives to the HT, based for example on heuristic search achieved by small eigenvalue analysis.

11.9.1 More Recent Developments

In 2010, advances were still being made in the application of the HT. In particular, Chung et al. (2010) developed an orientation-based elimination strategy that they have shown to be more efficient than previous line-determination methods based on the HT. It operates by dividing edge pixels into sets with small (typically $10°$) ranges of orientation, and for each of these, it carries out the process of line detection. Since this process involves a parameter space of reduced size, both storage and search times are reduced.

The RANSAC procedure was published by Fischler and Bolles in 1981. This must be one of the most cited papers in computer vision, and the method must be one of the most used (more even than the HT, because it only relies on the existence of suitable hypotheses, *however* obtained). The original paper used it for tackling the full perspective n-point fitting problem in 3-D (see Chapter 16). Clarke et al. (1996) used it for locating and tracking straight lines. Borkar et al. (2009) used it for locating lane markings on roads, and Mastorakis and Davies (2011) developed it further for the same purpose. Interestingly, Borkar et al. used a low-resolution HT to feed RANSAC and followed it by least-squares fitting of the inliers. The paper does not report on how much was gained by this three-stage approach, either in accuracy or in reliability. (If enough hypotheses are employed—and there is certainly no lack of these in such an application—both the HT and least-squares fitting might be avoided, but here optimization for speed may make the inclusion of least squares essential.) For further discussion of RANSAC, see Chapter 23 and Appendix A.

11.10 PROBLEMS

1. **a.** In the foot-of-normal HT, straight edges are found by locating the foot of the normal F (x_f, y_f) from an origin O $(0, 0)$ at the center of the image to an extended line containing each edge fragment E (x, y), and placing a vote at F in a separate image space.

   b. By examining the possible positions of lines within a square image and the resulting foot-of-normal positions, determine the exact extent of the parameter space that is theoretically required to form the HT.

   c. Would this form of the HT be expected to be (i) more or less robust and (ii) more or less computation intensive than the (ρ, θ) HT for line location?

2. **a.** Why is it sometimes stated that a HT generates *hypotheses* rather than actual solutions for object location? Is this statement justified?

b. A new type of HT is to be devised for detecting straight lines. It will take every edge fragment in the image and extend it in either direction until it meets the boundary of the image, and then accumulate a vote at each position. Thus *two* peaks should result for every line. Explain why finding these peaks will require *less* computation than for the standard HT, but that deducing the presence of lines will then require *extra* computation. How will these amounts of computation be likely to vary with (i) the size of the image and (ii) the number of lines in the image?

c. Discuss whether this approach would be an improvement on the standard approach for straight line location, and whether it would have any disadvantages.

3. a. Describe the *Hough transform* approach to object location. Explain its advantages relative to the *centroidal* (r, θ) *plot* approach. Illustrate your answer with reference to location of circles of known radius R.

b. Describe how the Hough transform may be used to locate straight edges. Explain what is seen in the parameter space if many curved edges also appear in the original image.

c. Explain what happens if the image contains a square object of arbitrary size and *nothing* else. How would you deduce from the information in parameter space that a square object is present in the image? Give the main features of an algorithm to decide that a square object is present and to locate it.

d. Examine in detail whether an algorithm using the strategy described in (c) would become confused if (i) parts of some sides of the square were occluded; (ii) one or more sides of the square were missing; (iii) several squares appeared in the image; (iv) several of these complications occurred together.

e. How important is it to this type of algorithm to have edge detectors that are capable of accurately determining edge orientation? Describe a type of edge detector that is capable of achieving this.

Circle and Ellipse Detection

12

As in the case of straight lines, circle features occur very widely in digital images of manufactured objects. In fact, they can conveniently be located using the Hough transform approach. This also applies for ellipses, which may appear in their own right or as oblique views of circular objects. However, locating ellipses is more complicated than locating circles because of the greater number of parameters that are involved.

Look out for:

- the basic Hough transform technique for locating circular objects in images.
- how the method can be adapted when circle radius is unknown.
- how accuracy of center location can be improved.
- how speed of processing can be increased.
- the basic diameter bisection method for ellipse detection.
- the chord—tangent method for ellipse detection.
- means of testing a shape to confirm that it is an ellipse.
- how small holes may be detected.
- how the human iris may be located.

This chapter shows how the Hough transform is able to provide a useful means for detecting objects of selected shape in digital images. Again the method is seen to rely on the accumulation of votes in a parameter space and the robustness of the technique to result from concentration on *positive* evidence for the objects.

12.1 INTRODUCTION

Location of round objects is important in many areas of image analysis but it is especially important in industrial applications such as automatic inspection and assembly. In the food industry alone, a very sizable number of products are round—biscuits, cakes, pizzas, pies, jellies, oranges, and so on (Davies, 1984c). In the automotive industry, many circular components are used—washers, wheels, pistons, heads of bolts, and so on, while round holes of various sizes appear in such items as casings and cylinder blocks. In addition, buttons and many other everyday objects are round. Of course, when round objects are viewed obliquely, they appear elliptical; furthermore, certain other objects are *actually* elliptical. This makes it clear that we need algorithms that are capable of finding both circles and ellipses. Finally, objects can frequently be located by their holes, so finding round holes or features is part of the larger problem: this chapter addresses various aspects of this problem.

An important facet of this work is how well object location algorithms cope in the presence of artifacts such as shadows and noise. In particular, the paradigm represented by Table 12.1 has been shown in Chapter 10 to be insufficiently robust to cope in such situations. This chapter shows that the Hough transform (HT) technique is able to overcome many of these problems. Indeed, it is found to be particularly good at dealing with all sorts of difficulties, including quite severe occlusions. It achieves this not by adding robustness but by having robustness built in as an integral part of the technique.

The application of the HT to circle detection is one of the most straightforward uses of the technique. However, there are several enhancements and adaptations that can be applied in order to improve accuracy and speed of operation, and in addition to make the method work efficiently when detecting circles with a range of sizes. These modifications are studied after covering the basic HT technique. Versions of the Hough transform that can perform ellipse detection are then considered. Finally, after a short section on an important application—that of human iris location—the topic of hole detection is considered.

Table 12.1 Procedure for Finding Objects Using (r, θ) Boundary Graphs

1. Locate edges within the image
2. Link broken edges
3. Thin thick edges
4. Track around object outlines
5. Generate a set of (r, θ) plots
6. Match (r, θ) plots to standard templates

This procedure is not sufficiently robust with many types of real data, e.g., in the presence of noise, distortions in product shape, and so on: in fact, it is quite common to find the tracking procedure veering off and tracking around shadows or other artifacts.

12.2 HOUGH-BASED SCHEMES FOR CIRCULAR OBJECT DETECTION

In the original HT method for finding circles (Duda and Hart, 1972), the intensity gradient is first estimated at all locations in the image and then thresholded to give the positions of significant edges. Then the positions of all possible center locations—namely all points a distance R away from every edge pixel—are accumulated in parameter space, R being the anticipated circle radius. Parameter space can be a general storage area but when looking for circles it is convenient to make it congruent to image space: in that case, possible circle centers are accumulated in a new plane of image space. Finally, parameter space is searched for peaks that correspond to the centers of circular objects. Since edges have nonzero width and noise will always interfere with the process of peak location, accurate center location requires the use of suitable averaging procedures (Davies, 1984c; Brown, 1984).

This approach clearly requires a very large number of points to be accumulated in parameter space and so a revised form of the method has now become standard: in this approach, locally available edge orientation information at each edge pixel is used to enable the exact positions of circle centers to be estimated (Kimme et al., 1975). This is achieved by moving a distance R along the edge normal at each edge location. Thus, the number of points accumulated is equal to the number of edge pixels in the image:[1] this represents a significant saving in computational load. For this to be possible, the edge detection operator that is employed must be highly accurate. Fortunately, the Sobel operator is able to estimate edge orientation to $1°$ and is very simple to apply (Chapter 5). Thus, the revised form of the transform is viable in practice.

As was seen in Chapter 5, once the Sobel convolution masks have been applied, the local components of intensity gradient g_x and g_y are available, and the magnitude and orientation of the local intensity gradient vector can be computed using the formulae:

$$g = \left(g_x^2 + g_y^2 \right)^{1/2} \tag{12.1}$$

and

$$\theta = \arctan\left(\frac{g_y}{g_x} \right) \tag{12.2}$$

However, use of the arctan operation is not necessary when estimating center location coordinates (x_c, y_c) since the trigonometric functions can be made to cancel out:

$$x_c = x - R\left(\frac{g_x}{g} \right) \tag{12.3}$$

---

[1]We assume that objects arc known to be *either* lighter *or* darker than the background, so that it is only necessary to move along the edge normal in one direction.

$$y_c = y - R\left(\frac{g_y}{g}\right) \tag{12.4}$$

the values of $\cos\theta$ and $\sin\theta$ being given by:

$$\cos\theta = \frac{g_x}{g} \tag{12.5}$$

$$\sin\theta = \frac{g_y}{g} \tag{12.6}$$

In addition, the usual edge thinning and edge linking operations—which normally require considerable amounts of processing—can be avoided if a little extra smoothing of the cluster of candidate center points is performed (Davies, 1984c) (Table 12.2). Thus, this Hough-based approach can be a very efficient one for locating the centers of circular objects, virtually all the superfluous operations having been eliminated, leaving only edge detection, location of candidate center points, and center point averaging to be carried out. In addition, the method is highly robust, so if part of the boundary of an object is obscured or distorted, the object center is still located accurately. In fact, the results are often quite impressive (see, e.g., Figs. 12.1 and 12.2). The reason for this useful property is clear from Fig. 12.3.

The efficiency of the above technique means that it takes slightly less time to perform the actual HT part of the calculation than to evaluate and threshold the intensity gradient over the whole image. Part of the reason for this is that the edge detector operates within a 3×3 neighborhood and necessitates some 12 pixel accesses, four multiplications, eight additions, two subtractions, and an operation for the evaluation of the square root of sum of squares (Eq. (12.1)). As seen in Chapter 5, the latter type of operation is commonly approximated by taking a sum or maximum of absolute magnitudes of the component intensity gradients in order to estimate the magnitude of the local intensity gradient vector.

Table 12.2 A Hough-Based Procedure for Locating Circular Objects

1. Locate edges within the image
2. Link broken edges
3. Thin thick edges
4. For every edge pixel, find a candidate center point
5. Locate all clusters of candidate centers
6. Average each cluster to find accurate center locations

This procedure is particularly robust. It is largely unaffected by shadows, image noise, shape distortions, and product defects. Note that stages 1–3 of the procedure are identical to stages 1–3 in Table 12.1. However, in the Hough-based method, computation can be saved, and accuracy actually increased, by omitting stages 2 and 3.

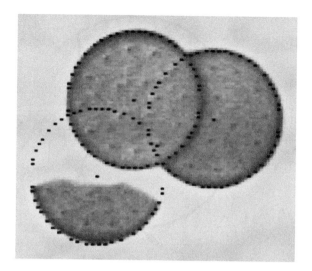

FIGURE 12.1

Location of broken and overlapping biscuits, showing the robustness of the center location technique. Accuracy is indicated by the black dots, which are each within 1/2 pixel of the radial distance from the center.

Source: © IFS Publications Ltd 1984

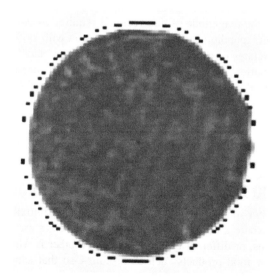

FIGURE 12.2

Location of a biscuit with a distortion, showing a chocolate-coated biscuit with excess chocolate on one edge. Note that the computed center has not been "pulled" sideways by the protuberances. For clarity, the black dots are marked 2 pixels outside the normal radial distance.

Source: © IFS Publications Ltd 1984

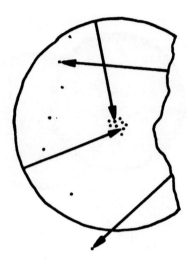

FIGURE 12.3

Robustness of the Hough transform when locating the center of a circular object. The circular part of the boundary gives candidate center points that focus on the true center, whereas the irregular broken boundary gives candidate center points at random positions.

Source: © Unicom 1988

However, this type of shortcut is not advisable in the present context, since an accurate value for the magnitude of this vector is required in order to compute the position of the corresponding candidate center location with sufficient precision.

Overall, the dictates of accuracy imply that candidate center location requires significant computation. However, substantial increases in speed are still possible by software means alone, as will be seen later in the chapter. Meanwhile, we consider the problems that arise when images contain circles of many different radii, or for one reason or another radii are not known in advance.

12.3 THE PROBLEM OF UNKNOWN CIRCLE RADIUS

There are a number of situations where circle radius is initially unknown. One such situation is where a number of circles of various sizes are being sought—as in the case of coins, or different types of washer. Another is where the circle size is variable—as for food products such as biscuits—so that some tolerance must be built into the system. In general, all circular objects have to be found and their radii measured. In such cases, the standard technique is to accumulate candidate center points simultaneously in a number of parameter planes in a suitably augmented parameter space, each plane corresponding to one possible radius value. The centers of the peaks detected in parameter space give not only the location of each circle in two dimensions but also its radius. Although this scheme is entirely viable in principle, there are several problems in practice:

1. Many more points have to be accumulated in parameter space.
2. Parameter space requires much more storage.
3. Significantly greater computational effort is involved in searching parameter space for peaks.

To some extent this is to be expected, since the augmented scheme enables more objects to be detected directly in the original image.

It is shown below that the last two problems may largely be eliminated. This is achieved by using just one parameter plane to store all the information for locating circles of different radii, i.e., accumulating not just one point per edge pixel but a whole line of points along the direction of the edge normal in this one plane. In practice, the line need not be extended indefinitely in either direction but only over the restricted range of radii over which circular objects or holes might be expected.

Even with this restriction, a large number of points are being accumulated in a single parameter plane, and it might be thought initially that this would lead to such a proliferation of points that almost any "blob" shape would lead to a peak in parameter space, which might be interpreted as a circle center. However, this is not so and significant peaks normally result only from genuine circles and some closely related shapes.

To understand the situation, consider how a sizeable peak can arise at a particular position in parameter space. This can happen only when a large number of radial vectors from this position meet the boundary of the object normally. In the absence of discontinuities in the boundary, a contiguous set of boundary points can only be normal to radius vectors if they lie on the arc of a circle (indeed, a circle could be *defined* as a locus of points that are normal to the radius vector and form a thin connected line). If a limited number of discontinuities are permitted in the boundary, it may be deduced that shapes like that of a fan[2] will also be detected using this scheme. Since it is in any case useful to have a scheme that can detect such shapes, the main problem is that there will be some ambiguity in interpretation—i.e., does peak P in parameter space arise from a circle or a fan? In practice, it is quite straightforward to distinguish between such shapes with relatively little additional computation, the really important problem being to cut down the amount of computation needed to key into the initially unstructured image data. Indeed, it is often a good strategy to prescreen the image to eliminate most of it from further detailed consideration and then to analyze the remaining data with tools having much greater discrimination: this two-stage template matching procedure frequently leads to significant savings in computation (VanderBrug and Rosenfeld, 1977; Davies, 1988g).

---

[2] A four-blade fan shape bounded by two concentric circles with eight equally spaced radial lines joining them.

A more significant problem arises because of errors in the measurement of local edge orientation. As stated earlier, edge detection operators such as the Sobel introduce an inherent inaccuracy of about $1°$. Image noise typically adds a further $1°$ error, and for objects of radius 60 pixels, the result is an overall uncertainty of about 2 pixels in the estimation of center location. This makes the scheme slightly less specific in its capability for detecting objects.

Overall, the scheme is likely to accept the following object shapes during its prescreening stage:

1. Circles of various sizes
2. Shapes such as fans, which contain arcs of circles
3. Partly occluded or broken versions of these shapes

Sometimes, objects of type 2 will be known *not* to be present. Alternatively, they may readily be identified, e.g., with the aid of corner detectors. Hence, problems of ambiguity may be nonexistent or easy to eliminate in practice. As the situation is highly application-dependent, we will eschew further discussion of this point here.

Next, note that the information on radial distance has been lost by accumulating all votes in a single parameter plane. Hence, a further stage of analysis is needed to measure object radius. This extra stage of analysis normally involves negligible additional computation, because the search space has been narrowed down so severely by the initial circle location procedure. The radial histogram technique (Chapter 20) can be used to measure the radius: in addition, it can be used to perform a final object inspection function.

12.3.1 Some Practical Results

The method described above works much as expected, the main problems arising with small circular objects (of radii less than about 20 pixels) at low resolution (Davies, 1988b). Essentially, the problems are those of lack of discrimination in the precise shapes of small objects (see Figs. 12.4 and 12.5), as anticipated above. As suggested earlier, this can frequently be turned to advantage in that the method becomes a circular feature detector for small radii (see Fig. 12.5, where a wing nut is located).

As required, objects are detected reliably even when they are partly occluded. However, occlusions can result in the centers of small objects being "pulled" laterally (Fig. 12.4). More generally, it is clear from Figs. 12.4 and 12.5 that high accuracy of center location cannot be expected when a single parameter plane is used to detect objects over a large range of sizes: hence, it is best to cut down the voting range as far as possible.

Overall, there is a tradeoff between speed and accuracy with this approach. However, the results confirm that it is possible to locate objects of various radii within a significantly conflated parameter space, thereby making substantial

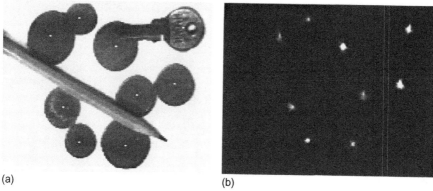

(a)   (b)

FIGURE 12.4

(a) Simultaneous location of coins and a key with the modified Hough scheme: the various radii range from 10 to 17 pixels and (b) transform used to compute the centers indicated in (a). Detection efficiency is unimpaired by partial occlusions, shape distortions, and glints. However, displacements of some centers are apparent; in particular, one coin (top left) has only two arcs showing the fact that one of these is distorted, giving a lower curvature, leads to a displacement of the computed center. The shape distortions are due to rather uneven illumination.

savings in storage and computation—even though the total number of votes that have to be accumulated in parameter space is not itself reduced.

12.4 THE PROBLEM OF ACCURATE CENTER LOCATION

Section 12.3 analyzed the problem of how to cope efficiently with images containing circles of unknown or variable size. This section examines how centers of circles may be located with high (preferably subpixel) accuracy. This problem is relevant since the alternative of increased resolution would entail the processing of many more pixels. Hence, it will be of advantage if high accuracy can be attained with images of low or moderate resolution.

There are a number of causes of inaccuracy in locating the centers of circles using the HT:

1. Natural edge width may add errors to any estimates of the location of the center.
2. Image noise may cause the radial position of the edge to become modified.
3. Image noise may cause variation in the estimates of local edge orientation.
4. The object itself may be distorted, in which case the most accurate and robust estimate of circle center is what is required (i.e., the algorithm should be able to ignore the distortion and take a useful average from the remainder of the boundary).

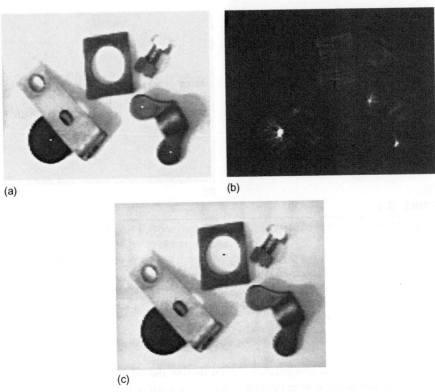

(a)

(b)

(c)

FIGURE 12.5

(a) Accurate simultaneous detection of a lens cap and a wing nut when radii are assumed to range from 4 to 17 pixels; (b) response in parameter space that arises with such a range of radii: note the overlap of the transforms from the lens cap and the bracket; (c) hole detection in the image of (a) when radii are assumed to fall in the range −26 to −9 pixels (negative radii are used since holes are taken to be objects of negative contrast): clearly, in *this* image a smaller range of negative radii could have been employed.

5. The object may appear distorted because of inadequacies in the method of illumination.
6. The edge orientation operator will have a significant inherent inaccuracy of its own, which contributes to inaccuracy in the estimation of center location.

Evidently, it is necessary to minimize the effects of all these potential sources of error. In applying the HT, it is usual to bear in mind that all possible center locations that could have given rise to the currently observed edge pixel should be accumulated in parameter space. Therefore, we should accumulate in parameter space not a single candidate center point corresponding to the given edge pixel, but a point spread function (PSF), which may be approximated by a Gaussian error function: this will generally have different radial and transverse

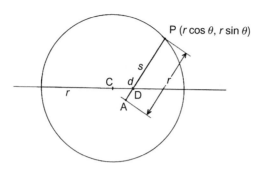

FIGURE 12.6

Arrangement for obtaining improved center approximation.

standard deviations. The radial standard deviation will commonly be 1 pixel—corresponding to the expected width of the edge—whereas the transverse standard deviation will frequently be at least 2 pixels, and for objects of moderate (60 pixels) radius the intrinsic orientation accuracy of the edge detection operator contributes at least 1 pixel to this standard deviation.

When the center need only be estimated to within 1 pixel, these considerations hardly matter. However, when it is desired to locate center coordinates to within 0.1 pixel, each PSF will have to contain 100 points, all of which will have to be accumulated in a parameter space of much increased spatial resolution, or its equivalent,[3] if the required accuracy is to be attained. However, the following section outlines a technique that can cut down the amount of computation from this sort of level, without significant loss of accuracy.

12.4.1 A Solution Requiring Minimal Computation

A potential key to increasing accuracy of center location arises from the observation that most of the inaccuracy in calculating the position of the center is due to transverse rather than radial errors. Hence, it is reasonable to concentrate on eliminating transverse errors.

This immediately leads to the following strategy: find a point D in the region of the center and use it to obtain a better approximation A to the center by moving from the current edge pixel P a distance equal to the expected radius r in the direction of D (Fig. 12.6). Then repeat the process 1 edge pixel at a time until all edge pixels have been taken into account. Although intuition indicates that this strategy will lead to a good final approximation to the center, the edge points need to be taken in random order to prevent the bias of the initial center approximation from

---

[3]For example, some abstract list structure might be employed, which effectively builds up to high resolution only where needed (see also the adaptive HT scheme described in Chapter 13).

permanently affecting the final result.[4] In addition, some averaging of the intermediate results is needed to minimize the effects of boundary noise, and edge points that give extreme predictions for the center approximation (e.g., deviations of more than ∼2 pixels) have to be discounted.

Overall, there are two main stages of calculation in the whole algorithm:

1. Find the position of the center to within ∼1 pixel by the usual Hough technique.
2. Use the iterative procedure to obtain the final result, with an accuracy of 0.1 pixels.

Tests of this algorithm on accurately made circular objects led to accuracies of center location in the region of 0.1 pixels (Davies, 1988e). It proved difficult to increase the accuracy further because of the problem of setting up lighting systems of sufficient uniformity, i.e., the limit was set by practicalities of illumination rather than by the algorithm itself.

Finally, a by-product of the approach is that radius r can be obtained with exceptionally high accuracy—generally within 0.05 pixels (Davies, 1988e).

12.5 OVERCOMING THE SPEED PROBLEM

Section 12.4 studied how the accuracy of the HT circle detection scheme could be improved substantially, with modest additional computational cost. This section examines how circle detection may be carried out with significant improvement in speed. To achieve this, two methods are tried: (1) sampling the image data and (2) using a simpler edge detector. The most appropriate strategy for (1) appears to be to look only at every nth line in the image, whereas that for (2) involves using a small two-element neighborhood while searching for edges (Davies, 1987f). Although this approach will lose the capability for estimating edge orientation, it will still permit horizontal and vertical chords of a circle to be bisected, thereby leading to values for the center coordinates x_c, y_c. It also involves much less computation, the multiplications and square root calculations, and most of the divisions being eliminated or replaced by two-element differencing operations, thereby giving a further gain in speed.

12.5.1 More Detailed Estimates of Speed

To help understand the situation, this section estimates the gain in speed that should result by applying the strategy described above. First, the amount of computation involved in the original Hough-based approach is modeled by:

$$T_0 = N^2 s + S t_0 \qquad (12.7)$$

---

[4]This problem arises because the method is intrinsically sequential rather than parallel, and it is necessary to compensate for this.

where T_0 is the total time taken to run the algorithm on an $N \times N$ pixel image; s is the time taken per pixel to compute the Sobel operator and to threshold the intensity gradient g; S is the number of edge pixels that are detected; and t_0 is the time taken to compute the position of a candidate center point.

Next, the amount of computation in the basic chord bisection approach is modeled as:

$$T = 2(N^2 q + Qt) \qquad (12.8)$$

where T is the total time taken to run the algorithm; q is the time taken per pixel to compute a two-element x or y edge detection operator and to perform thresholding; Q is the number of edge pixels that are detected in one or other scan direction; and t is the time taken to compute the position of a candidate center point coordinate (either x_c or y_c). In Eq. (12.8), the factor 2 results because scanning occurs in both horizontal and vertical directions. If the image data are now sampled by examining only a proportion α of all possible horizontal and vertical scan lines, the overall gain in speed from using the chord bisection scheme should be:

$$G = \frac{N^2 s + St_0}{2\alpha(N^2 q + Qt)} \qquad (12.9)$$

Typical values of relevant parameters for (say) a biscuit of radius 32 pixels in a 128×128 pixel image are listed below:

$N^2 = 16,384$	$t_0/s \approx 1$
$S \approx Q \approx 200$	$s/q \approx 6$
$\alpha \approx 1/3$	$t_0/t \approx 5$

Hence

$$G \approx \frac{s}{2\alpha q} \approx 9 \qquad (12.10)$$

Broadly, this corresponds to a gain ~ 3 from sampling and a further gain ~ 3 from applying a much simplified edge detector. However, if the sampling factor $1/\alpha$ could be increased further, greater gains in speed could be obtained—in principle without limit, but in practice the situation is governed by how robust the algorithm really is.

12.5.2 Robustness

Robustness can be considered relative to two factors. The first is the amount of noise in the image and the second is the amount of signal distortion that can be tolerated. Fortunately, both the original HT and the chord bisection approach lead to peak finding situations, and if there is any distortion of the object shape, then points are thrown into relatively random locations in parameter space and consequently do not have a significant direct impact on the accuracy of peak location.

However, they do have an indirect impact in that the signal-to-noise ratio is reduced, so that accuracy is impaired. In fact, if a fraction β of the original signal is removed, leaving a fraction $\gamma = 1 - \beta$, either due to such distortions or occlusions or else by the deliberate sampling procedures already outlined, then the number of independent measurements of the center location drops to a fraction γ of the optimum. This means that the accuracy of estimation of the center location drops to a fraction $\sqrt{\gamma}$ of the optimum. Since noise affects the optimum accuracy directly, we have shown the result of both major factors governing robustness.

What is important here is that the effect of sampling is substantially the same as that of signal distortion, so that the more distortion that must be tolerated, the higher the value α has to have. This principle applies both to the original Hough approach and to the chord bisection algorithm. However, the latter does its peak finding in a different way—*via* 1-D rather than 2-D averaging processes. As a result, it turns out to be somewhat more robust than the standard HT in its capability for accepting a high degree of sampling.

This gain in capability for accepting sampling carries with it a set of related disadvantages: highly textured objects may not easily be located by the method; high noise levels may confuse it via the production of too many false edges; and (since it operates by specific x and y scanning mechanisms) there must be a sufficiently small amount of occlusion and other gross distortion that a significant number of scans (both horizontally and vertically) pass through the object. Ultimately, this means that the method will not tolerate more than about one-quarter of the circumference of the object being absent.

12.5.3 Practical Results

Tests (Davies, 1987f) with the image in Fig. 12.7 show that gains in speed of more than 25 can be obtained, with values of α down to less than 0.1 (i.e., every 10th horizontal and vertical line scanned). The results for broken circular products (Figs. 12.8 and 12.9) are self-explanatory; they indicate the limits to which the method can be taken. An outline of the complete algorithm is given in Table 12.3 (note the relatively straightforward problem of disambiguating the results if there happen to be several peaks).

Figure 12.10 shows the effect of adjusting the threshold in the two-element edge detector. The result of setting it too low is seen in Fig. 12.10(a). Here the surface texture of the object has triggered the edge detector, and the chord midpoints give rise to a series of false estimates of the center coordinates. Figure 12.10(b) shows the result of setting the threshold at too high a level, so that the number of estimates of center coordinates is reduced and sensitivity suffers. Although the images in Fig. 12.7 were obtained with the threshold adjusted intuitively, a more rigorous approach can be taken by optimizing a suitable criterion function. There are two obvious functions: (1) the number of accurate center predictions n and (2) the speed–sensitivity product. The latter can be written in the form $\sqrt{n}/T$, where T is the execution time. The two methods of optimization make little difference in the

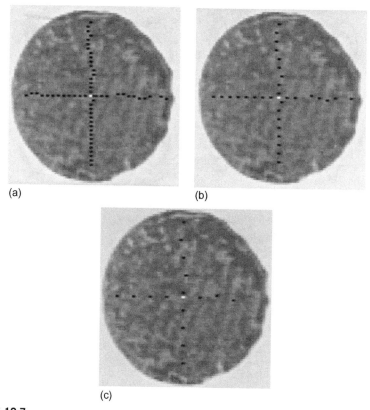

(a)

(b)

(c)

FIGURE 12.7

Successful object location using the chord bisection algorithm for the same initial image, using successive step sizes of 2, 4, and 8 pixels. The black dots show the positions of the horizontal and vertical chord bisectors, and the white dot shows the position found for the center.

example shown in Fig. 12.10. However, had there been a strong regular texture on the object, the situation would have been rather different.

12.5.4 Summary

The center location procedure described above is more than an order of magnitude faster than the standard Hough-based approach and often as much as 25 times faster. This could be quite important in exacting real-time applications. Robustness is so good that the method tolerates at least one-quarter of the circumference of an object being absent, making it adequate for many real applications. In addition, it is entirely clear what types of image data would be likely to confuse the algorithm.

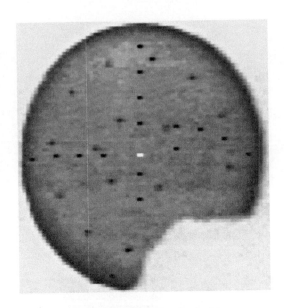

FIGURE 12.8

Successful location of a broken object using the chord bisection algorithm: only about one-quarter of the ideal boundary is missing.

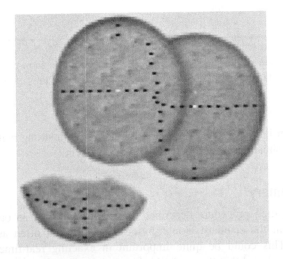

FIGURE 12.9

A test on overlapping and broken biscuits: the overlapping objects are successfully located, albeit with some difficulty, but there is no chance of finding the center of the broken biscuit since over one-half of the ideal boundary is missing.

Table 12.3 Outline of the Fast Center-Finding Algorithm

```
y = 0;
do {
      scan horizontal line y looking for start and end of each object;
      calculate midpoints of horizontal object segments;
      accumulate midpoints in 1-D parameter space (x space);
      // note that the same space, x space, is used for all lines y
      y = y + d;
} until y > ymax;

x = 0;
do {
      scan vertical line x looking for start and end of each object;
      calculate midpoints of vertical object segments;
      accumulate midpoints in 1-D parameter space (y space);
      // note that the same space, y space, is used for all lines x
      x = x + d;
} until x > xmax;

find peaks in x space;
find peaks in y space;
test all possible object centres arising from these peaks;
// the last step is necessary only if ∃ > 1 peak in each space
// d is the horizontal and vertical step-size ( = 1/α)
```

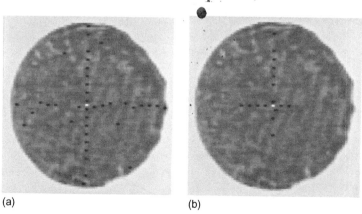

(a)   (b)

FIGURE 12.10

Effect of misadjustment of the gradient threshold: (a) effect of setting the threshold too low, so that surface texture confuses the algorithm and (b) loss of sensitivity on setting the threshold too high.

Not only is robustness built into the algorithm in such a way as to emulate the standard Hough-based approach, but it is possible to interpret the method as being a Hough-based approach, since the center coordinates are each accumulated in their own 1-D "parameter spaces." These considerations give further insight into the robustness of the standard Hough technique.

12.6 ELLIPSE DETECTION

The problem of detecting ellipses may seem only marginally more complex than that of detecting circles—as eccentricity is only a single parameter. However, eccentricity destroys the symmetry of the circle, so the direction of the major axis also has to be defined. As a result, five parameters are required to describe an ellipse, and ellipse detection has to take account of this, either explicitly or implicitly. In spite of this, one method for ellipse detection is especially simple and straightforward to implement, i.e., the diameter bisection method, which is described next.

12.6.1 The Diameter Bisection Method

The diameter bisection method of Tsuji and Matsumoto (1978) is very simple in concept. First, a list is compiled of all the edge points in the image. Then, the list is sorted to find those that are antiparallel, so that they could lie at opposite ends of ellipse diameters; next, the positions of the center points of the connecting lines for all such pairs are taken as voting positions in parameter space (Fig. 12.11). As for circle location, the parameter space that is used for this purpose is congruent to image space. Finally, the positions of significant peaks in parameter space are located to identify possible ellipse centers.

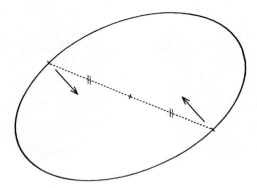

FIGURE 12.11

Principle of the diameter bisection method. A pair of points is located for which the edge orientations are antiparallel. If such a pair of points lies on an ellipse, the midpoint of the line joining the points will be situated at the center of the ellipse.

Naturally, in an image containing many ellipses and other shapes, there will be very many pairs of antiparallel edge points and for most of these the center points of the connecting lines will lead to nonuseful votes in parameter space. Clearly, such clutter leads to wasted computation. However, it is a principle of the HT that votes must be accumulated in parameter space at all points that *could in principle* lead to correct object center location: it is left to the peak finder to find the voting positions that are most likely to correspond to object centers.

Not only does clutter lead to wasted computation, but the method itself also is computationally expensive. This is because it examines all *pairs* of edge points, and there are many more such pairs than there are edge points (m edge points lead to $^mC_2 \approx m^2/2$ pairs of edge points). Indeed, since there are likely to be at least 1000 edge points in a typical image, the computational problems can be formidable.

Interestingly, the basic method is not particularly discriminating about ellipses. It picks out many symmetrical shapes—any indeed that possess 180° rotation symmetry, including rectangles, ellipses, circles, or superellipses (these have equations of the form $x^s/a^s + y^s/b^s = 1$, ellipses being a special case). In addition, the basic scheme sometimes gives rise to a number of false identifications even in an image in which only ellipses are present (Fig. 12.12). However, Tsuji and Matsumoto (1978) also proposed a technique by which true ellipses can be distinguished. The basis of the technique is the property of an ellipse that the lengths of perpendicular semidiameters OP, OQ obey the relation:

$$\frac{1}{OP^2} + \frac{1}{OQ^2} = \frac{1}{R^2} = \text{constant} \tag{12.11}$$

To proceed, the set of edge points that contribute to a given peak in parameter space is used to construct a histogram of R values (the latter being obtained from

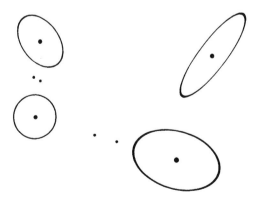

FIGURE 12.12

Result of using the basic diameter bisection method. The larger dots show true ellipse centers found by the method, whereas the smaller dots show positions at which false alarms commonly occur. Such false alarms are eliminated by applying the test described in the text.

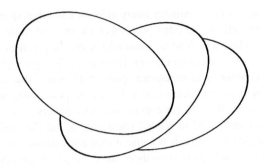

FIGURE 12.13

Limitations of the diameter bisection method: of the three ellipses shown, only the lowest one cannot be located by the diameter bisection method.

Source: © Unicom 1988

Eq. (12.11)). If a significant peak is found in this histogram, then there is clear evidence of an ellipse at the specified location in the image. If two or more such peaks are found, then there is evidence of a corresponding number of concentric ellipses in the image. If, however, no such peaks are found, then a rectangle, superellipse, or other symmetrical shape may be present and each of these would need its own identifying test.

The method obviously relies on there being an appreciable number of pairs of edge points on an ellipse lying at opposite ends of diameters: hence, there are strict limits on the amount of the boundary that must be visible (Fig. 12.13). Finally, it should not go unnoticed that the method wastes the signal available from unmatched edge points. These considerations have led to a search for further methods of ellipse detection.

12.6.2 The Chord–Tangent Method

The chord–tangent method was devised by Yuen et al. (1988) and makes use of another simple geometric property of the ellipse. Again pairs of edge points are taken in turn, and for each point of the pair, tangents to the ellipse are constructed and found to cross at T, the midpoint of the connecting line is found at M, and then the equation of line TM is calculated and all points that lie on the portion MK of this line are accumulated in parameter space (Fig. 12.14) (clearly, T and the center of the ellipse lie on the opposite sides of M). Finally, peak location proceeds as before.

The proof that this method is correct is trivial. Symmetry ensures that the method works for circles, and projective properties then ensure that it also works for ellipses: under projection, straight lines project into straight lines, midpoints into midpoints, tangents into tangents, and circles into ellipses; in addition, it is

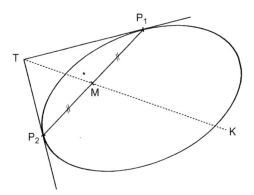

FIGURE 12.14

Principle of the chord–tangent method. The tangents at P_1 and P_2 meet at T and the midpoint of P_1P_2 is M. The center C of the ellipse lies on the line TM produced. Note that M lies between C and T. Hence, the transform for points P_1 and P_2 need only include the portion MK of this line.

always possible to find a viewpoint such that a circle can be projected into a given ellipse.

Unfortunately, this method suffers from significantly increased computation, since so many points have to be accumulated in parameter space. This is obviously the price to be paid for greater applicability. However, computation can be minimized in at least three ways: (1) cutting down the lengths of the lines of votes accumulated in parameter space by taking account of the expected sizes and spacings of ellipses; (2) not pairing edge points initially if they are too close together or too far apart; and (3) eliminating edge points once they have been identified as belonging to a particular ellipse.

12.6.3 Finding the Remaining Ellipse Parameters

Although the methods described above are designed to locate the center coordinates of ellipses, a more formal approach is required to determine other ellipse parameters. Accordingly, we write the equation of an ellipse in the form:

$$Ax^2 + 2Hxy + By^2 + 2Gx + 2Fy + C = 0 \qquad (12.12)$$

an ellipse being distinguished from a hyperbola by the additional condition:

$$AB > H^2 \qquad (12.13)$$

This condition guarantees that A can never be zero and that the ellipse equation may without loss of generality be rewritten with $A = 1$. This leaves five parameters, which can be related to the position of the ellipse, its orientation, and its size and shape (or eccentricity).

Having located the center of the ellipse, we may select a new origin of coordinates at its center (x_c, y_c); the equation then takes the form:

$$x'^2 + 2Hx'y' + By'^2 + C' = 0 \qquad (12.14)$$

where

$$x' = x - x_c; \quad y' = y - y_c \qquad (12.15)$$

It now remains to fit to Eq. (12.14) the edge points that gave evidence for the ellipse center under consideration. The problem will normally be vastly overdetermined. Hence, an obvious approach is the method of least squares. Unfortunately, this technique tends to be very sensitive to outlier points and is therefore liable to be inaccurate. An alternative is to employ some form of Hough transformation. Here we follow Tsukune and Goto (1983) by differentiating Eq. (12.14):

$$x' + \frac{By'}{dx'} + H\left(y' + \frac{x'dy'}{dx'}\right) = 0 \qquad (12.16)$$

Then dy'/dx' can be determined from the local edge orientation at (x', y') and a set of points accumulated in the new (H, B) parameter space. When a peak is eventually located in (H, B) space, the relevant data (a subset of a subset of the original set of edge points) can be used with Eq. (12.14) to obtain a histogram of C' values, from which the final parameter for the ellipse can be obtained.

The following formulae are needed to determine the orientation θ and semi-axes a and b of an ellipse in terms of H, B, and C':

$$\theta = \frac{1}{2}\arctan\left(\frac{2H}{1-B}\right) \qquad (12.17)$$

$$a^2 = \frac{-2C'}{(B+1) - [(B-1)^2 + 4H^2]^{1/2}} \qquad (12.18)$$

$$b^2 = \frac{-2C'}{(B+1) + [(B-1)^2 + 4H^2]^{1/2}} \qquad (12.19)$$

Mathematically, θ is the angle of rotation that diagonalizes the second-order terms in Eq. (12.14); having performed this diagonalization, the ellipse is then essentially in the standard form $\tilde{x}^2/a^2 + \tilde{y}^2/b^2 = 1$, so a and b are determined.

Note that the above method finds the five ellipse parameters in *three* stages: first the positional coordinates are obtained, then the orientation, and finally the size and eccentricity.[5] This three-stage calculation involves less computation but compounds any errors—in addition, edge orientation errors, although low, become a limiting factor. For this reason, Yuen et al. (1988) tackled the problem by speeding up the HT procedure itself rather than by avoiding a direct assault on

---

[5]Strictly, the eccentricity is $e = (1 - b^2/a^2)^{1/2}$, but in most cases we are more interested in the ratio of semiminor to semimajor axes, b/a.

Eq. (12.14): i.e., they aimed at a fast implementation of a thoroughgoing second stage, which finds all the parameters of Eq. (12.14) in one 3-D parameter space.

It is now clear that reasonably optimal means are available for finding the orientation and semi-axes of an ellipse once its position is known: the weak point in the process appears to be that of finding the ellipse initially. Indeed, the two approaches for achieving this that have been described above are particularly computation intensive, mainly because they examine all pairs of edge points; a possible alternative is to apply the generalized Hough transform (GHT), which locates objects by taking edge points singly: this possibility will be considered in Chapter 13.

12.7 HUMAN IRIS LOCATION

Human iris location is an important application of computer vision for three reasons: (1) it provides a useful cue for human face analysis; (2) it can be used for the determination of gaze direction; and (3) it is useful in its own right for biometric purposes, that is to say, for identifying individuals almost uniquely. The latter possibility has already been noted in Chapter 8 where textural methods for iris recognition are outlined and some key references are given. More details of human face location and analysis will be given in Chapter 17. Here we concentrate on iris location using the Hough transform.

In fact, we can tackle the iris location and recognition task reasonably straightforwardly. First, if the head has been located with reasonable accuracy, then it can form a region of interest, inside which the iris can be sought. In a front view with the eyes looking ahead, the iris will be seen as a round often high-contrast object, and can be located straightforwardly with the aid of a HT (Ma et al., 2003). In some cases, this will be less easy because the iris is relatively light and the color may not be distinctive—although a lot will depend on the quality of the illumination. Perhaps more important, in some human subjects, the eyelid and lower contour of the eye may partially overlap the iris (Fig. 12.15(a)), making it more difficult to identify, although, as confirmed below, HTs are capable of coping with a fair degree of occlusion.

Note that the iris will appear elliptical if the eyes are not facing directly ahead; in addition, the shape of the eye is far from spherical and the horizontal diameter is larger than the vertical diameter—again making the iris appear elliptical (Wang and Sung, 2001). In either case, the iris can still be detected using a Hough transform. Furthermore, once this has been done, it should be possible to estimate the direction of gaze with a reasonable degree of accuracy (Gong et al., 2000), thereby taking us further than mere recognition. (The fact that measurement of ellipse eccentricity would lead to an ambiguity in the gaze direction can be offset by measuring the position of the ellipse on the eyeball.) Finally, Toennies et al. (2002) showed that the HT can be used to localize the irises for real-time applications, in spite of quite substantial partial occlusion by the eyelid and lower contour of the eye.

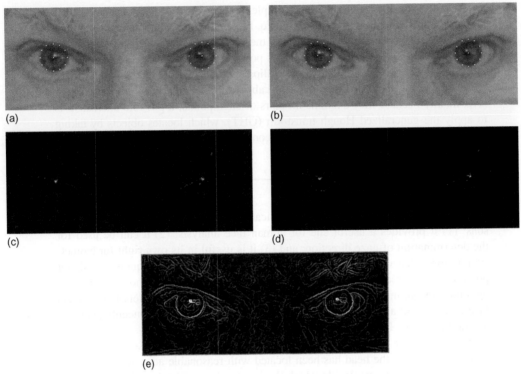

FIGURE 12.15

Iris location using the Hough transform. (a) Original image of the eye region of the face with irises located by a uniformly weighted Hough transform (HT). (b) Irises located using a gradient-weighted HT. (c) and (d) The respective HTs used to locate the irises in (a) and (b). (e) The Canny operator (incorporating smoothing, nonmaximum suppression, and hysteresis) is used to obtain the initial edge image in both cases. The sharper peaks obtained using the gradient weighting permit the irises to be located significantly more robustly and accurately. Note the number of additional edges in (e) that are able to produce substantive numbers of additional votes, which interfere with those from the irises.

A number of the points made above are illustrated by the example in Fig. 12.15. Far from being a trivial application of the HT, there are a surprisingly large number of edges in the eye region: these produce significant numbers of additional votes, which interfere with those from the irises. These arise from the upper and lower eyelids and the folds of skin around them. Hence, the accuracy of the method is not assured: this makes gradient weighting (see Section 13.6) especially valuable. The radii of the irises shown in Fig. 12.15 are about 17.5 pixels, and there is no particular evidence of ellipticity in their shape. However, more accurate location of the iris and measurement of ellipticity in order to estimate orientation (e.g., to determine the angle of gaze) require considerably greater resolution, with iris radii approaching 100 pixels, whereas pictures that analyze iris texture patterns for biometric purposes require even larger radii.

12.8 HOLE DETECTION

At this stage, an obvious question is whether the usual methods for circular object detection can be applied to hole detection. In principle, this is undoubtedly practicable, in many cases the only change arising because holes have negative contrast. This means that if the HT technique is used, votes will have to be accumulated on moving a distance $-R$ along the direction of the local edge normal. There is no difficulty with this for large holes, but as the holes become smaller, complications tend to occur because of poor lighting inside the holes. As a result, large regions of shadow are likely to occur, and the contrast is liable to be weak and variable. We shall not consider the general situation further as it is so dependent on size, illumination, and other factors, such as 3-D shape and the material within which the hole appears.

Nevertheless, for small holes a few pixels across, detection can be important, as they constitute point features, and these can be valuable for precise object location, as happens for biscuits, hinges, brackets, and a host of other manufactured parts (see Chapter 14). In fact, circular holes may even be preferable to corners for object location, as they are, ideally, isotropic and thus less liable to bias when used for measurement.

Very small holes can be detected using template matching. This involves applying a Laplacian type of operator, and in a 3×3 neighborhood this may take the form of one of the following convolution masks:

$$\begin{bmatrix} -1 & -1 & -1 \\ -1 & 8 & -1 \\ -1 & -1 & -1 \end{bmatrix} \qquad \begin{bmatrix} -2 & -3 & -2 \\ -3 & 20 & -3 \\ -2 & -3 & -2 \end{bmatrix}$$

Having applied the convolution, the resulting image is thresholded to indicate regions where holes might be present. Note that the coefficients in each of the above masks sum to zero so that they are insensitive to varying levels of background illumination.

If the holes in question are larger than $1-2$ pixels in diameter, the above masks will prove inadequate. Clearly, larger masks are required if larger holes are to be detected efficiently at their centers. For much larger holes, the masks that are required become impracticably large, and the HT approach becomes a more attractive option.

12.9 CONCLUDING REMARKS

This chapter has described techniques for circle and ellipse detection, starting with the HT approach. Although the HT is found to be effective and highly robust against occlusions, noise, and other artifacts, it incurs considerable storage and computation—especially if it is required to locate circles of unknown radius or if high accuracy is required. Techniques have been described whereby the latter two problems can be tackled much more efficiently; in addition, a method has been described for markedly reducing the computational load involved in circle detection. The general circle location scheme is a type of HT but although the other two methods are related to the HT, they are distinct methods, which draw on the HT for inspiration. As with the HT, these methods achieve robustness as an integral part of their design—i.e., robustness is not included as an afterthought—so they achieve known levels of robustness.

Two HT-based schemes for ellipse detection have also been described—the diameter bisection method and the chord–tangent method. A further approach to ellipse detection, based on the generalized HT, will be covered in Chapter 13. At that point, further lessons will be drawn on the efficacies of the various methods.

As in the case of line detection, a trend running through the design of circle and ellipse detection schemes is the deliberate splitting of algorithms into two or more stages. This is useful for keying into the important and relevant parts of an image prior to finely discriminating one type of object or feature from another, or prior to measuring dimensions or other characteristics accurately. Indeed, the concept can be taken further, in that the efficiencies of all the algorithms discussed in this chapter have been improved by searching first for edge features in the image. The concept of two-stage template matching is therefore deep seated in the methodology of the subject and is developed further in later chapters. Although two-stage template matching is a standard means of increasing efficiency (VanderBrug and Rosenfeld, 1977; Davies, 1988g), it is not obvious that efficiency can always be increased in this way. It appears to be in the nature of the subject that ingenuity is needed to discover means of achieving this.

> The Hough transform is very straightforwardly applied to circle detection, for which it achieves an impressive level of robustness. It is also easily applied to ellipse detection. Practical issues include accuracy, speed, and storage requirements—some of which can be improved by employing parameter spaces of reduced dimension, although this can affect the specificity that can be achieved.

12.10 BIBLIOGRAPHICAL AND HISTORICAL NOTES

The Hough transform was developed in 1962 and first applied to circle detection by Duda and Hart (1972). However, the now standard HT technique, which makes use of edge orientation information to reduce computation, only emerged 3 years later (Kimme et al., 1975). The author's work on circle detection for automated inspection required real-time implementation and also high accuracy. This spurred the development of the three main techniques described in Sections 12.3–12.5 (Davies, 1987f, 1988b, 1988e). In addition, the author has considered the effect of noise on edge orientation computations, showing in particular their effect in reducing the accuracy of center location (Davies, 1987e).

Yuen et al. (1989) reviewed various existing methods for circle detection using the HT. In general, their results confirmed the efficiency of the method described in Section 12.3 when circle radius is unknown, although they found that the two-stage process that was involved can sometimes lead to slight loss of robustness. It appears that this problem can be reduced in some cases by using a modified version of the algorithm of Gerig and Klein (1986). Note that the Gerig and Klein approach is itself a two-stage procedure: it is discussed in detail in Chapter 13. More recently, Pan et al. (1995) have increased the speed of computation of the HT by prior grouping of edge pixels into arcs, for an underground pipe inspection application.

The two-stage template matching technique and related approaches for increasing search efficiency in digital images were known by 1977 (Nagel and Rosenfeld, 1972; Rosenfeld and VanderBrug, 1977; VanderBrug and Rosenfeld, 1977) and have undergone further development since then—especially in relation to particular applications such as those described in this chapter (Davies, 1988g).

Later, Atherton and Kerbyson (1999) (see also Kerbyson and Atherton, 1995) showed how to find circles of arbitrary radius in a single parameter space using the novel procedure of coding radius as a phase parameter and then performing accumulation with the resulting phase-coded annular kernel. Using this approach, they attained higher accuracy with noisy images; Goulermas and Liatsis (1998) showed how the HT could be fine tuned for the detection of fuzzy circular objects such as overlapping bubbles by using genetic algorithms. In effect, the latter are able to sample the solution space with very high efficiency and hand over cleaner data to the following HT.

The ellipse detection sections are based particularly on the work of Tsuji and Matsumoto (1978), Tsukune and Goto (1983), and Yuen et al. (1988); for a fourth method (Davies, 1989a) using the GHT idea of Ballard (1981) in order to save computation, see Chapter 13. The contrasts between these methods are many and intricate, as this chapter has shown. In particular, the idea of saving dimensionality in the implementation of the GHT appears also in a general circle detector (Davies, 1988b). At that point in time, the necessity for a multistage approach to determination of ellipse parameters seemed proven, although somewhat surprisingly the optimum number of such stages was just two.

Later algorithms represented moves to greater degrees of robustness with real data by explicit inclusion of errors and error propagation (Ellis et al., 1992); increased attention was subsequently given to the verification stage of the Hough approach (Ser and Siu, 1995). In addition, work was carried out on the detection of superellipses, which are shapes intermediate in shape between ellipses and rectangles, although the technique used (Rosin and West, 1995) was that of segmentation trees rather than HTs (nonspecific detection of superellipses can of course be achieved by the diameter bisection method (see Section 12.6.1); see also Rosin (2000)).

For cereal grain inspection, with typical flow rates in excess of 300 grains per second, ultrafast algorithms were needed and the resulting algorithms were limiting cases of chord-based versions of the HT (Davies, 1999b, 1999d); a related approach was adopted by Xie and Ji (2002) for their efficient ellipse detection method; Lei and Wong (1999) employed a method that was based on symmetry, and this was found to be able to detect parabolas and hyperbolas as well as ellipses:[6] it was also reported as being more stable than other methods since it does not have to calculate tangents or curvatures; the latter advantage has also been reported by Sewisy and Leberl (2001). The fact that even in the 2000s, basic new ellipse detection schemes are being developed says something about the science of image analysis: even today the toolbox of algorithms is incomplete, and the science of how to choose between items in the toolbox, or how, *systematically*, to develop new items for the toolbox, is immature. Further, although all the parameters for specification of such a toolbox may be known, knowledge about the possible tradeoffs between them is still limited.

12.10.1 More Recent Developments

Much work has recently been carried out on iris detection using the HT. Jang et al. (2008) were particularly concerned with overlap of the iris region by the upper and lower eyelids, and used a parabolic version of the HT to accurately locate their boundaries, taking special care to limit the computational load. Li et al. (2010) used a circular HT to locate the iris and a RANSAC-like technique for locating the upper and lower eyelids, again using a parabolic model for the latter: their approach was designed to cope with very noisy iris images. Chen et al. (2010) used a circular HT to locate the iris and a straight line HT to locate up to two line segment approximations to the boundaries of each of the eyelids. Cauchie et al. (2008) produced a new version of the HT to accurately locate common circle centers from circular or partial circle segments, and demonstrated its value for iris location. Min and Park (2009) used a circular HT for iris detection, a parabolic HT for eyelid detection, and followed this with eyelash detection using thresholding.

Finally, we summarize the work carried out by Guo et al. (2009) to overcome the problems of dense sets of edges in textured regions. To reduce the impact of such edges, a measure of isotropic surround suppression was introduced: the resulting algorithm gave small weights to edges in texture regions and large weights to edges on strong and clear boundaries when accumulating votes in Hough space. The approach gave good results when locating straight lines in scenes containing man-made structures such as buildings.

---

[6]Note that while this is advantageous in some applications, the lack of discrimination could prove to be a disadvantage in other applications.

12.11 PROBLEMS

1. Prove the result of Section 12.4.1 that as D approaches C and d approaches zero, the shape of the locus becomes a circle on DC as diameter.

2. **a.** Describe the use of the Hough transform for circular object detection, assuming that object size is known in advance. Show also how a method for detecting ellipses could be adapted for detecting circles of unknown size.

   b. A new method is suggested for circle location that involves scanning the image both horizontally and vertically. In each case, the midpoints of chords are determined and their x or y coordinates are accumulated in separate 1-D histograms. Show that these can be regarded as simple types of Hough transform, from which the positions of circles can be deduced. Discuss whether any problems would arise with this approach; consider also whether it would lead to any advantages relative to the standard Hough transform for circle detection.

   c. A further method is suggested for circle location. This again involves scanning the image horizontally, but in this case, for every chord that is found, an estimate is immediately made of the *two* points at which the center could lie, and votes are placed at those locations. Work out the geometry of this method, and determine whether this method is faster than the method outlined in (b). Determine whether the method has any disadvantages compared with method described in (b)?

3. Determine which of the methods described in this chapter will detect (a) hyperbolas, (b) curves of the form $Ax^3 + By^3 = 1$, and (c) curves of the form $Ax^4 + Bx + Cy^4 = 1$.

4. Prove Eq. (12.11) for an ellipse. *Hint*: Write the coordinates of P and Q in suitable parametric forms, and then use the fact that OP⊥OQ to eliminate one of the parameters from the left-hand side of the equation.

5. Describe the *diameter bisection* and *chord–tangent* methods for the location of ellipses in images, and compare their properties. Justify the use of the chord–tangent method by proving its validity for circle detection and then extending the proof to cover ellipse detection.

6. Round coins of a variety of sizes are to be located, identified, and sorted in a vending machine. Discuss whether the chord–tangent method should be used for this purpose instead of the usual form of Hough transform circle location scheme operating within a 3D (x, y, r) parameter space.

7. Outline each of the following methods for locating ellipses using the Hough transform: (a) the *diameter bisection* method and (b) the *chord–tangent* method. Explain the principles on which these methods rely. Determine which is more robust and compare the amounts of computation each requires.

8. For the diameter bisection method, searching through lists of edge points with the right orientations can take excessive computation. It is suggested that a

two-stage approach might speed up the process: (a) load the edge points into a table, which may be addressed by orientation and (b) look up the right edge points by feeding appropriate orientations into the table. Estimate how much this would be likely to speed up the diameter bisection method.

9. It is found that the diameter bisection method sometimes becomes confused when several ellipses appear in the same image, and generates false "centers" that are not situated at the centers of any ellipses. It is also found that certain other shapes are detected by the diameter bisection method. Ascertain in each case quite what the method is sensitive to, and consider ways in which these problems might be overcome.

The Hough Transform and Its Nature

13

It has already been seen that the Hough transform can be used to locate straight line, circle, and ellipse features in digital images. It would be useful to know whether the method can be generalized to cover all shapes and whether it is always as robust as it is for the original three examples. This chapter discusses these questions, showing that the method can be generalized and is broadly able to retain its robustness properties.

Look out for:

- the generalized Hough transform technique.
- its relation to spatial matched filtering.
- how sensitivity is optimized by gradient rather than uniform weighting.
- use of the generalized HT for ellipse detection.
- how speed can be improved by the use of a universal lookup table.
- how the computational loads of the various HT techniques can be estimated.
- the value of the Gerig and Klein back-projection technique in cutting down the effects of extraneous clutter.

This chapter describes the generalized Hough transform and generalizes our view of the HT as a generic computer vision technique. It also makes order calculations of computational load for three HT-based methods of ellipse detection.

13.1 INTRODUCTION

In Chapters 11 and 12, it has been seen that the Hough transform (HT) is of great importance for the detection of features such as lines, circles, and ellipses, and for finding relevant image parameters. This makes it worthwhile to see the extent to which the method can be generalized so that it can detect arbitrary shapes. The works of Merlin and Farber (1975) and Ballard (1981) were crucial historically and led to the development of the generalized Hough transform (GHT). The GHT is studied in this chapter, showing first how it is implemented and then examining how it is optimized and adapted to particular types of image data. This requires us to go back to first principles, taking spatial matched filtering as a starting point.

Having developed the relevant theory, it is applied to the important case of ellipse detection, showing in particular how computational load may be minimized. Finally, the computational problems of the GHT and HT are examined more generally.

13.2 THE GENERALIZED HOUGH TRANSFORM

This section shows how the standard Hough technique is generalized so that it can detect arbitrary shapes. In principle, it is trivial to achieve this. First, we need to select a localization point L within a template of the idealized shape. Then, we need to arrange such that, instead of moving from an edge point a *fixed* distance R directly along the local edge normal to arrive at the center, as for circles, we move an appropriate *variable* distance R in a variable direction φ so as to arrive at L: R and φ are now functions of the local edge normal direction θ (Fig. 13.1). Under these circumstances, votes will peak at the preselected object localization point L. The functions $R(\theta)$ and $\varphi(\theta)$ can be stored analytically in the computer algorithm, or for completely arbitrary shapes they may be stored as lookup tables. In either case, the scheme is beautifully simple in principle but two complications arise in practice. The first arises because some shapes have features such as concavities and holes, so several values of R and φ are required for certain values of θ (Fig. 13.2). The second arises because we are going from an isotropic shape (a circle) to an anisotropic shape, which may be in a completely arbitrary orientation.

To cope with the first of these complications, the lookup table (usually called the "R-table") must contain a list of the positions **r**, relative to L, of all points on

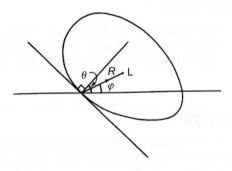

FIGURE 13.1

Computation of the generalized Hough transform.

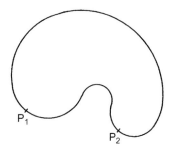

FIGURE 13.2

A shape exhibiting a concavity: certain values of θ correspond to several points on the boundary and hence require several values of R and φ—as for points: P_1 and P_2.

the boundary of the object for each possible value of edge orientation θ (or a similar effect must be achieved analytically). Then, on encountering an edge fragment in the image whose orientation is θ, estimates of the position of L may be obtained by moving a distance (or distances) $\mathbf{R} = -\mathbf{r}$ from the given edge fragment. Clearly, if the R-table has multivalued entries (i.e., several values of $\mathbf{r}$ for certain values of θ), only one of these entries (for given θ) can give a correct estimate of the position of L. However, at least the method is guaranteed to give optimum sensitivity, since all relevant edge fragments contribute to the peak at L in parameter space. This property of optimal sensitivity reflects the fact that the GHT is a form of spatial matched filter: it is analyzed in more detail below.

The second complication arises because any shape other than circle is anisotropic. Since in most applications (including industrial applications such as automated assembly) object orientations are initially unknown, the algorithm has to obtain its own information on object orientation. This means adding an extra dimension in parameter space (Ballard, 1981). Then each edge point contributes a vote in each plane in parameter space at a position given by that expected for an object of given shape and orientation. Finally, the whole of parameter space is searched for peaks, the highest points indicating both the locations of objects and their orientations. Clearly, if object size is also a parameter, the problem becomes far worse but this complication is ignored here (although the method described in Section 12.3 is clearly relevant).

The changes made in proceeding to the GHT leave it just as robust as the HT circle detector described previously. This gives an incentive to improve the GHT so as to limit the computational problems in practical situations. In particular, the size of the parameter space must be cut down drastically both to save storage and to curtail the associated search task. Considerable ingenuity has been devoted to devising alternative schemes and adaptations to achieve this. Important cases are those of ellipse detection and polygon detection, and in each of these, definite advances have been made: ellipse detection is covered in Chapter 12 and for polygon detection, see Davies (1989a). Here we proceed with some more basic studies of the GHT.

13.3 SETTING UP THE GENERALIZED HOUGH TRANSFORM— SOME RELEVANT QUESTIONS

The next few sections explore the theory underpinning the GHT, with the aim of clarifying how to optimize it systematically for specific circumstances. It is relevant to ask what is happening when a GHT is being computed. Although the HT has been shown to be equivalent to template matching (Stockman and Agrawala, 1977) and also to spatial matched filtering (Sklansky, 1978), further clarification is required. In particular, the following three problems (Davies, 1987a) need to be addressed:

1. *The parameter space weighting problem*: In introducing the GHT, Ballard mentioned the possibility of weighing points in parameter space according to the magnitudes of the intensity gradients at the various edge pixels. But when should gradient weighting be used in preference to uniform weighting?

2. *The threshold selection problem*: When using the GHT to locate an object, edge pixels are detected and used to compute candidate positions for the localization point L (see Section 13.2). To achieve this, it is necessary to threshold the edge gradient magnitude. How should the threshold be chosen?

3. *The sensitivity problem*: Optimum sensitivity in detecting objects does not automatically provide optimum sensitivity in locating objects, and vice versa. How should the GHT be optimized for these two criteria?

To understand the situation and solve these problems, it is necessary to go back to first principles. Section 13.4 starts discussion on this.

13.4 SPATIAL MATCHED FILTERING IN IMAGES

To discuss the questions posed in Section 13.3, it is necessary to analyze the process of spatial matched filtering. In principle, this is the ideal method of detecting objects, since it is well known (Rosie, 1966) that a filter that is matched to a given signal detects it with optimum signal-to-noise ratio under white noise[1] conditions (North, 1943; Turin, 1960). (For a more recent discussion of this topic, see Davies (1993).)

Mathematically, using a matched filter is identical to correlation with a signal (or "template") of the same shape as the one to be detected (Rosie, 1966). Here "shape" is a general term meaning the amplitude of the signal as a function of time or spatial location.

---

[1]White noise is noise that has equal power at all frequencies. In image science, white noise is understood to have equal power at all *spatial* frequencies. The significance of this is that noise at different pixels is completely uncorrelated but is subject to the same grayscale probability distribution, i.e., it has potentially the same range of amplitudes at all pixels.

When applying correlation in image analysis, changes in background illumination cause large changes in signal from one image to another and from one part of an image to another. The varying background level prevents straightforward peak detection in convolution space. The matched filter optimizes signal-to-noise ratio only in the presence of white noise. This is likely to be a good approximation in the case of radar signals, whereas this is not generally true in the case of images. For ideal detection, the signal should be passed through a "noise-whitening filter" (Turin, 1960), which in the case of objects in images is usually some form of high-pass filter: this must be applied prior to correlation analysis. However, this is likely to be a computationally expensive operation.

If we are to make correlation work with near optimal sensitivity but without introducing a lot of computation, other techniques must be employed. In the template matching context, the following possibilities suggest themselves:

1. Adjust templates so that they have a mean value of zero to suppress the effects of varying levels of illumination in first order.
2. Break up templates into a number of smaller templates each having a zero mean. Then as the sizes of subtemplates approach zero, the effects of varying levels of illumination will tend to zero in second order.
3. Apply a threshold to the signals arising from each of the subtemplates so as to suppress those that are less than the expected variation in signal level.

If these possibilities fail, only two further strategies appear to be available:

1. Readjust the lighting system—an important option in industrial inspection applications, although it may give little improvement when a number of objects can cast shadows or reflect light over each other.
2. Use a more "intelligent" (e.g., context sensitive) object detection algorithm, although this will almost certainly be computation intensive.

13.5 FROM SPATIAL MATCHED FILTERS TO GENERALIZED HOUGH TRANSFORMS

To proceed, we note that items 1–3 listed in Section 13.3 essentially amount to a specification of the GHT. First, breaking up the templates into small subtemplates each having a zero mean and then thresholding them are analogous, and in many cases identical, to a process of edge detection (see, e.g., the templates used in the Sobel and similar operators). Next, locating objects by peak detection in parameter space clearly corresponds to the process of reconstructing whole template information from the subtemplate (edge location) data. What is important here is that these ideas reveal how the GHT is related to the spatial matched filter. Basically, *the GHT can be described as a spatial matched filter that has been modified, with the effect of including integral noise whitening, by breaking down the main template into small zero-mean templates and thresholding the individual responses before object detection.*

Small templates do not permit edge orientation to be estimated as accurately as large ones. Although the Sobel edge detector is in principle accurate to about 1° (see Chapter 5), there is a deleterious effect if the edge of the object is fuzzy. In such a case, it is not possible to make the subtemplates very small, and an intermediate size should be chosen that gives a suitable compromise between accuracy and sensitivity.

Employing zero-mean templates results in the absolute signal level being reduced to zero and only local relative signal levels being retained. Thus, the GHT is not a true spatial matched filter: in particular, it suppresses the signal from the bulk of the object, retaining only that which is near its boundary. As a result, the GHT is highly sensitive to object position but is not optimized for object detection.

Thresholding of subtemplate responses has much the same effect as employing zero-mean templates, although it may remove a small proportion of the signal giving positional information. This makes the GHT even less like an ideal spatial matched filter and further reduces the sensitivity of object detection. The thresholding process is particularly important in the present context since it provides a means of saving computational effort *without losing significant positional information.* On its own this characteristic of the GHT would correspond to a type of perimeter template around the outside of an object (see Fig. 13.3). This must not be taken as excluding all of the interior of the object, since any high-contrast edges within the object will facilitate location.

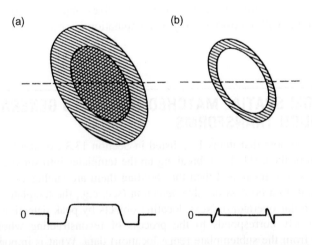

FIGURE 13.3

The idea of a perimeter template: both the original spatial matched filter template (a) and the corresponding "perimeter template" (b) have a zero mean (see text). The lower illustrations show the cross-sections along the dashed lines.

13.6 GRADIENT WEIGHTING VERSUS UNIFORM WEIGHTING

The first problem described in Section 13.3 is that of how best to weight plots in parameter space in relation to the respective edge gradient magnitudes. To find an answer to this problem, it should now only be necessary to go back to the spatial matched filter case to find the ideal solution, and then to determine the corresponding solution for the GHT in the light of the discussion in Section 13.4. First, note that the responses to the subtemplates (or to the perimeter template) are proportional to edge gradient magnitude. With a spatial matched filter, signals are detected optimally by templates of the same shape. Each contribution to the spatial matched filter response is then proportional to the local magnitude of the signal and to that of the template. In view of the correspondence between (a) using a spatial matched filter to locate objects by looking for peaks in convolution space and (b) using a GHT to locate objects by looking for peaks in parameter space, we should use weights proportional to the gradients of the edge points *and* the *a priori* edge gradients.

There are two ways in which the choice of weighting is important. First, the use of uniform weighting implies that all edge pixels whose gradient magnitudes are above threshold will effectively have them reduced to the threshold value, so that the signal will be curtailed: this can mean that the signal-to-noise ratio of high-contrast objects will be reduced significantly. Second, the widths of edges of high-contrast objects will be broadened in a crude way by uniform weighting (see Fig. 13.4) but under gradient weighting this broadening will be controlled, giving a roughly Gaussian edge profile. Thus, the peak in parameter space will be narrower and more rounded, and the object reference point L can be located more easily and with greater accuracy. This effect is visible in Fig. 13.5, which also shows the relatively increased noise level that results from uniform weighting.

Note also that low gradient magnitudes correspond to edges of poorly known location, whereas high values correspond to sharply defined edges. Thus, the accuracy of the information relevant to object location is proportional to the magnitude of the gradient at each of the edge pixels, and appropriate weighting should therefore be used.

13.6.1 Calculation of Sensitivity and Computational Load

The aim of this subsection is to underline the above-described ideas by working out formulae for sensitivity and computational load. It is assumed that p objects of size around $n \times n$ are being sought in an image of size $N \times N$.

Correlation requires $N^2 n^2$ operations to compute the convolutions for all possible positions of the object in the image. Using the perimeter template, the number of basic operations is reduced to $\sim N^2 n$, corresponding to the reduced number of pixels in the template. The GHT requires $\sim N^2$ operations to locate the edge pixels, plus a further $\sim pn$ operations to accumulate the points in parameter space.

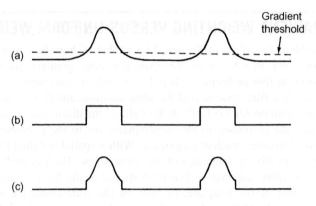

FIGURE 13.4

Effective gradient magnitude as a function of position within a section across an object of moderate contrast, thresholded at a fairly low level: (a) gradient magnitude for original image data and gradient thresholding level; (b) uniform weighting: the effective widths of edges are broadened rather crudely, adding significantly to the difficulty of locating the peak in parameter space; (c) gradient weighting: the position of the peak in parameter space can be estimated in a manner that is basically limited by the shape of the gradient profile for the original image data.

The situation for sensitivity is rather different. With correlation, the results for n^2 pixels are summed, giving a signal proportional to n^2, although the noise (assumed to be independent at every pixel) is proportional to n: this is because of the well-known result that the noise powers of various independent noise components are additive (Rosie, 1966). Overall, this results in the signal-to-noise ratio being proportional to n. The perimeter template possesses only $\sim n$ pixels, and here the overall result is that the signal-to-noise ratio is proportional to $\sqrt{n}$. The situation for the GHT is inherently identical to that for the perimeter template method, so long as plots in parameter space are weighted proportional to edge gradient g multiplied by *a priori* edge gradient G. It is now necessary to compute the constant of proportionality α. Take s as the average signal, equal to the intensity (assumed to be roughly uniform) over the body of the object, and S as the magnitude of a full matched filter template. In the same units, g (and G) is the magnitude of the signal within the perimeter template. Then $\alpha = 1/sS$. This means that the perimeter template method and the GHT method lose sensitivity for two reasons—first they look at less of the available signal and second they look where the signal is low. For a high value of gradient magnitude, which occurs for a step edge (where most of the variation in intensity takes place within the range of 1 pixel), the values of g and G saturate out, so that they are nearly equal to s and S (see Fig. 13.6). Under these conditions, the perimeter template method and the GHT have sensitivities that depend only on the value of n.

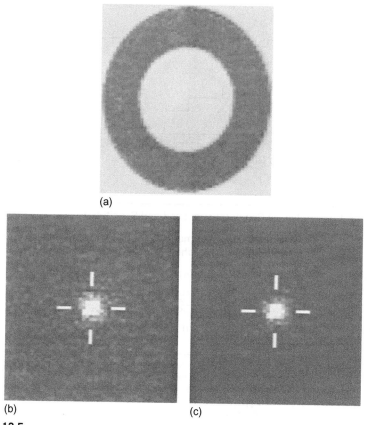

(a)

(b)   (c)

FIGURE 13.5

Results of applying the two types of weighting to a real image: (a) original image;
(b) results in parameter space for uniform weighting; and (c) results for gradient
weighting. The peaks (which arise from the outer edges of the washer) are normalized to
the same levels in the two cases: the increased level of noise in (b) is readily apparent. In
this example, the gradient threshold is set at a low level (around 10% of the maximum
level) so that low-contrast objects can also be detected.

Table 13.1 summarizes the situation discussed above. The oft-quoted statement that the computational load of the GHT is proportional to the number of
perimeter pixels, rather than to the much greater number of pixels within
the body of an object, is only an approximation. In addition, this saving is
not obtained without cost: in particular, the sensitivity (signal-to-noise ratio)
is reduced (at best) as the square root of object area/perimeter (note that
area and perimeter are measured in the same units, so it is valid to find their
ratio).

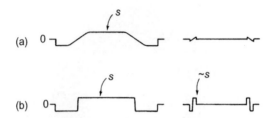

FIGURE 13.6

Effect of edge gradient on perimeter template signal: (a) low edge gradient: signal is proportional to gradient and (b) high edge gradient: signal saturates at value of s.

Table 13.1 Formulae for Computational Load and Sensitivity[a]			
	Template Matching	**Perimeter Template Matching**	**Generalized Hough Transform**
Number of operations	$O(N^2n^2)$	$O(N^2n)$	$O(N^2) + O(pn)$
Sensitivity	$O(n)$	$O\left(\dfrac{\sqrt{n}gG}{sS}\right)$	$O\left(\dfrac{\sqrt{n}gG}{sS}\right)$
Maximum sensitivity[b]	$O(n)$	$O(\sqrt{n})$	$O(\sqrt{n})$

[a]This table gives formulae for computational load and sensitivity when p objects of size $n \times n$ are sought in an image of size $N \times N$. The intensity of the image within the whole object template is taken as s and the value for the ideal template is taken as S: corresponding values for intensity gradient within the perimeter template are g and G.
[b]Maximum sensitivity refers to the case of a step edge, for which $g \approx s$ and $G \approx S$ (see Fig. 13.6).

Finally, the absolute sensitivity for the GHT varies as gG. As contrast changes so that $g \rightarrow g'$, we see that $gG \rightarrow g'G$, i.e., sensitivity changes by a factor of g'/g. Hence, theory predicts that sensitivity is proportional to contrast. Although this result might have been anticipated, we now see that it is valid only under conditions of gradient weighting.

13.7 SUMMARY

The above sections examined the GHT and found a number of factors involved in optimizing it, as follows.

1. Each point in parameter space should be weighted in proportion to the intensity gradient at the edge pixel giving rise to it, *and* in proportion to the *a*

priori gradient, if sensitivity is to be optimized, particularly for objects of moderate-to-high contrast.

2. The ultimate reason for using the GHT is to save computation. The main means by which this is achieved is by ignoring pixels having low magnitudes of intensity gradient. If the threshold of gradient magnitude is set too high, fewer objects are in general detected; if it is set too low, computational savings are diminished. Suitable means are required for setting the threshold but little reduction in computation is possible if the highest sensitivity in a low-contrast image is to be retained.

3. The GHT is inherently optimized for the location of objects in an image but is not optimized for the detection of objects. This means that it may miss low-contrast objects, which are detectable by other methods that take the whole area of an object into account. However, this consideration is often unimportant in applications where signal-to-noise ratio is less of a problem than finding objects quickly in an uncluttered environment.

Overall, it is clear that the GHT is a spatial matched filter only in a particular sense, and as a result it has suboptimal sensitivity. The main advantage of the technique is that it is highly efficient, overall computational load in principle being proportional to the relatively few pixels on the perimeters of objects rather than to the much greater numbers of pixels within them. In addition, by concentrating on the boundaries of objects, the GHT retains its power to locate objects accurately. It is thus important to distinguish clearly between sensitivity in *detecting* objects and sensitivity in *locating* them.

13.8 USE OF THE GHT FOR ELLIPSE DETECTION

It has already been seen that when the GHT is used to detect anisotropic objects, there is an intrinsic need to employ a large number of planes in parameter space. However, it is shown below that by accumulating the votes for all possible orientations in a *single* plane in parameter space, significant savings in computation can sometimes be made. Basically, the idea is largely to reduce the considerable storage requirements of the GHT by using only one instead of 360 planes in parameter space while significantly reducing the computation involved in the final search for peaks. Such a scheme could have concomitant disadvantages such as the production of spurious peaks, and this aspect will have to be examined carefully.

To achieve these aims, it is necessary to analyze the shape of the point spread function (PSF) to be accumulated for each edge pixel. To demonstrate this, we take the case of ellipses of unknown orientation. We start by taking a general edge fragment at a position defined by ellipse parameter ψ and deducing the bearing of the center of the ellipse relative to the local edge normal (Fig. 13.7).

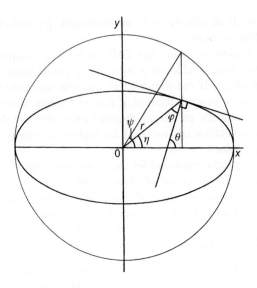

FIGURE 13.7

Geometry of an ellipse and its edge normal.

Working first in an ellipse-based axis system, for an ellipse with semimajor and semiminor axes a and b, respectively, it is clear that:

$$x = a \cos \psi \qquad (13.1)$$

$$y = b \sin \psi \qquad (13.2)$$

Hence

$$\frac{dx}{d\psi} = -a \sin \psi \qquad (13.3)$$

$$\frac{dy}{d\psi} = b \cos \psi \qquad (13.4)$$

giving

$$\frac{dy}{dx} = -\left(\frac{b}{a}\right) \cot \psi \qquad (13.5)$$

Hence, the orientation of the edge normal is given by:

$$\tan \theta = \left(\frac{a}{b}\right) \tan \psi \qquad (13.6)$$

At this point, we wish to deduce the bearing of the center of the ellipse relative to the local edge normal. From Fig. 13.7:

$$\varphi = \theta - \eta \qquad (13.7)$$

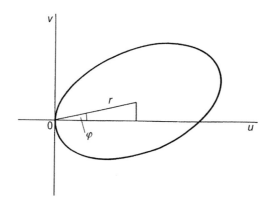

FIGURE 13.8

Geometry for finding the PSF for ellipse detection by forming the locus of the centers of ellipses touching a given edge fragment.

where

$$\tan \eta = \frac{y}{x} = \left(\frac{b}{a}\right) \tan \psi \qquad (13.8)$$

and

$$\tan \varphi = \tan(\theta - \eta)$$
$$= \frac{\tan \theta - \tan \eta}{1 + \tan \theta \tan \eta} \qquad (13.9)$$

Substituting for $\tan \theta$ and $\tan \eta$, and then rearranging, gives:

$$\tan \varphi = \frac{(a^2 - b^2)}{2ab} \sin 2\psi \qquad (13.10)$$

In addition:

$$r^2 = a^2 \cos^2 \psi + b^2 \sin^2 \psi \qquad (13.11)$$

To obtain the PSF for an ellipse of unknown orientation, we now simplify matters by taking the current edge fragment to be at the origin and orientated with its normal along the u-axis (Fig. 13.8). The PSF is then the locus of all possible positions of the center of the ellipse. To find its form, it is merely required to eliminate ψ between Eqs. (13.10) and (13.11). This is facilitated by re-expressing r^2 in double angles (the significance of double angles lies in the 180° rotation symmetry of an ellipse):

$$r^2 = \frac{a^2 + b^2}{2} + \frac{a^2 - b^2}{2} \cos 2\psi \qquad (13.12)$$

After some manipulation, the locus is obtained as:

$$r^4 - r^2(a^2 + b^2) + a^2 b^2 \sec^2 \varphi = 0 \qquad (13.13)$$

which can, in the edge-based coordinate system, also be expressed in the form:

$$v^2 = (a^2 + b^2) - u^2 - a^2 b^2 / u^2 \qquad (13.14)$$

In fact, this is a complex and variable shape, as shown in Fig. 13.9, although for ellipses of low eccentricity, the PSF approximates to an ellipse. This is seen by defining two new parameters:

$$c = \frac{(a + b)}{2} \qquad (13.15)$$

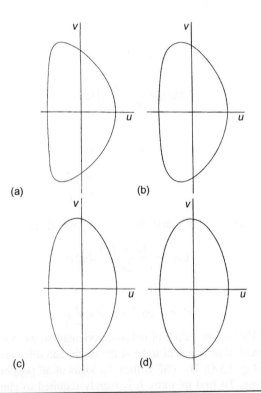

FIGURE 13.9

Typical PSF shapes for detection of ellipses with various eccentricities: (a) ellipse with $a/b = 21.0$ and $c/d = 1.1$, (b) ellipse with $a/b = 5.0$, $c/d = 1.5$, (c) ellipse with $a/b = 2.0$ and $c/d = 3.0$, (d) ellipse with $a/b = 1.4$ and $c/d = 6.0$. Note how the PSF shape approaches a small ellipse of aspect ratio 2.00 as eccentricity tends to zero. The semimajor and semiminor axes of the PSF are $2d$ and d, respectively.

$$d = \frac{(a-b)}{2} \qquad\qquad (13.16)$$

and taking an approximation of small d, the locus is obtained in the form:

$$\frac{(u+c)^2}{d^2} + \frac{v^2}{4d^2} = 1 \qquad\qquad (13.17)$$

As stated above, this approximates to an ellipse, and has semimajor axis $2d$ and semiminor axis d. However, this approximation has restricted validity, applying only when $d \gtrsim 0.1c$, so that $a \gtrsim 1.2b$. However, in practice, where ellipses are small and the PSF is only a few pixels across, it is a reasonable approximation to insist only that $a < 2b$.

Implementation is simplified since a universal lookup table (ULUT) for ellipse detection can be compiled, which is independent of the size and eccentricity of the ellipse to be detected so long as the eccentricity is not excessive. Since ellipses are common features in industrial and other applications—arising both from elliptical objects and from oblique views of circles—this factor should be an important consideration in many applications. Thus, a single ULUT is compiled and stored and it then needs only to be scaled and positioned to produce the PSF in a given instance of ellipse detection.

13.8.1 Practical Details

Having constructed a ULUT for ellipse detection, the detection algorithm has to scale it, position it, and rotate it so that points can be accumulated in parameter space. *A priori*, it would be imagined that a considerable amount of trigonometric computation is involved in this process. However, it is possible to avoid calculating angles directly (e.g., using the arctan function) by always working with sines and cosines; this is rendered possible partly because such edge orientation operators as the Sobel give two components (g_x, g_y) for the intensity gradient vector (this has already been seen to happen in several line and circle detection schemes—see Chapters 11 and 12). Hence, a lot of computation can be saved.

Figure 13.10 shows the result of testing the above scheme on an image of some O-rings lying on a slope of arbitrary direction, whereas Fig. 13.11 shows the result obtained for an elliptical object; the two cases used PSFs containing 50 and 100 votes, respectively. In Fig. 13.10, the O-rings are found accurately and with a fair degree of robustness, i.e., despite overlapping and partial occlusion (up to 40% in one case). In several cases, incidental transforms from points on the inner edges of the O-rings overlap other transforms from points on the outer edges, although only the latter are actually employed usefully here for peak finding. Hence, the scheme is able to overcome problems resulting from additional clutter in parameter space.

Figure 13.10 also shows the arrangement of points in parameter space that results from applying the PSF to every edge point on the boundary of an

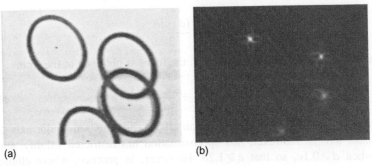

(a) (b)

FIGURE 13.10

Applying the PSF to detection of tilted circles: (a) off-camera 128 × 128 image of a set of circular O-rings on a 45° slope of arbitrary direction; (b) transform in parameter space: note the peculiar shape of the ellipse transform, which is close to a "four-leaf clover" pattern. (a) also indicates the positions of the centers of the O-rings as located from (b): accuracy is limited by the presence of noise, shadows, clutter, and available resolution, to an overall standard deviation of about 0.6 pixels.

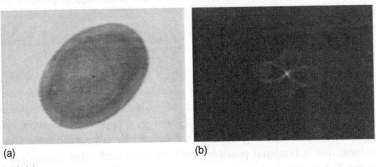

(a) (b)

FIGURE 13.11

Applying the PSF to detection of elliptical objects: (a) off-camera 128 × 128 image of an elliptical bar of soap of arbitrary orientation; (b) transform in parameter space: in this case, the clover-leaf pattern is better resolved. Accuracy of location is limited partly by distortions in the shape of the object but the peak location procedure results in an overall standard deviation of the order of 0.5 pixels.

ellipse: the pattern is somewhat clearer in Fig. 13.11. In either case, it is seen to contain a high degree of structure (curiously, the votes seem to form approximate "four-leaved clover" patterns). For an ideal transform, there would be no structure apart from the main peak, and all points on the PSF *not* falling on the peak at the center of the ellipse would be randomly distributed nearby. Nevertheless, the peak at the center is very well defined and shows that this compressed form of GHT represents a viable solution.

13.9 COMPARING THE VARIOUS METHODS

This section briefly compares the computational loads for the methods of ellipse detection discussed in Section 13.8 and in Chapter 12. To make fair comparisons, we concentrate on ellipse detection *per se* and ignore any additional procedures concerned with (a) finding other ellipse parameters, (b) distinguishing ellipses from other shapes, or (c) separating concentric ellipses. We start by examining the GHT method and the diameter bisection method.

First, suppose that an $N \times N$ pixel image contains p ellipses of identical size given by the parameters a, b, c, and d defined in Section 13.8. By ignoring noise and general background clutter, we shall be favoring the diameter bisection method, as will be seen below. Next, the discussion is simplified by supposing that the computational load resides mainly in the calculation of the positions at which votes should be accumulated in parameter space—the effort involved in locating edge pixels and in locating peaks in parameter space is much smaller.

Under these circumstances, the load for the GHT method may be approximated by the product of the number of edge pixels and the number of points per edge pixel that have to be accumulated in parameter space, the latter being equal to the number of points on the PSF. Hence, the load is proportional to:

$$L_G \approx \frac{p \times 2\pi c \times 2\pi(2d + d)}{2} = 6\pi^2 pcd$$

$$\approx 60pcd \tag{13.18}$$

where the ellipse has been taken to have relatively low eccentricity so that the PSF itself approximates to an ellipse of semiaxes $2d$ and d.

For the diameter bisection method, the actual voting is a minor part of the algorithm—as indeed it is in the GHT method (see the snippet of code listed in Table 13.1). In either case, most of the computational load concerns edge orientation calculations or comparisons. Assuming that these calculations and comparisons involve similar inherent effort, it is fair to assess the load for the diameter bisection method as:

$$L_D \approx {}^{p \times 2\pi c} C_2 \approx \frac{(2\pi pc)^2}{2} \approx 20p^2 c^2 \tag{13.19}$$

Hence

$$\frac{L_D}{L_G} \approx \frac{pc}{3d} \tag{13.20}$$

when a is close to b, as for a circle, $L_G \to 0$ and then the diameter bisection method becomes a poor option. However, in some cases, it is found that a is close to $2b$, so that c is close to $3d$. The ratio of the loads then becomes:

$$\frac{L_D}{L_G} \approx p \tag{13.21}$$

It is possible that p will be as low as 1 in some cases: however, such cases are likely to be rare and are offset by applications where there is significant background image clutter and noise, or where all p ellipses have other edge detail giving irrelevant signals that can be considered as a type of self-induced clutter (see the O-ring example in Fig. 13.10).

It is also possible that some of the pairs of edge points in the diameter bisection method can be excluded before they are considered, e.g., by giving every edge point a range of interaction related to the size of the ellipses. This would tend to reduce the computational load by a factor of the order of (but not as small as) p. However, the computational overhead required for this would not be negligible.

Overall, the GHT method should be significantly faster than the diameter bisection method in most real applications, the diameter bisection method being at a definite disadvantage when image clutter and noise are strong. By comparison, the chord—tangent method always requires more computation than the diameter bisection method, since not only does it examine every pair of edge points but also it generates a *line* of votes in parameter space for each pair.

Against these computational limitations, the different characteristics of the methods must be noted. First, the diameter bisection method is not particularly discriminating, in that it locates many symmetrical shapes, as remarked earlier. The chord—tangent method is selective for ellipses but is not selective about their size or eccentricity. The GHT method is selective about all of these factors. These types of discriminability, or lack of it, can turn out to be advantageous or disadvantageous, depending on the application: hence, we do no more here than draw attention to these different properties. It is also relevant that the diameter bisection method is rather less robust than the other methods. This is so since if one edge point of an antiparallel pair is not detected, then the other point of the pair cannot contribute to detection of the ellipse—a factor that does not apply for the other two methods since they take all edge information into account.

13.10 FAST IMPLEMENTATIONS OF THE HOUGH TRANSFORM

The foregoing sections have shown that the GHT requires considerable computation. The problem arises particularly in respect of the number of planes needed in parameter space to accommodate transforms for different object orientations and sizes. Clearly, significant improvements in speed are needed before the GHT can achieve its potential in practical instances of arbitrary shapes. This section considers some important developments in this area.

The source of the computational problems in the HT is the huge size the parameter space can take. Typically, a single plane parameter space has much the same size as an image plane (this will normally be so in those instances where parameter space is congruent to image space), but when many planes are required to cope with various object orientations and sizes, the number of planes is likely to be multiplied by a factor of around 300 for each extra dimension. Hence, the total storage area will then involve some 10,000 million accumulator cells. Clearly, reducing the resolution might just make it possible to bring this down to 100 million cells, although accuracy will start to suffer. Further, when the HT is stored in such a large space, the search problem involved in locating significant peaks becomes formidable.

Fortunately, data are not stored at all uniformly in parameter space and this provides a key to solving both the storage problem and the subsequent search problem. Indeed, the fact that parameter space is to be searched for the most prominent peaks—in general, the highest and the sharpest ones—means that the process of detection can start during accumulation. Furthermore, initial accumulation can be carried out at relatively low resolution, and then resolution can be increased locally as necessary in order to provide both the required accuracy and the separation of nearby peaks. In this context, it is natural to employ a binary splitting procedure, i.e., to repeatedly double the resolution locally in each of the dimensions of parameter space (e.g., the x dimension, the y dimension, the orientation dimension, and the size dimension, where these exist) until resolution is sufficient: a tree structure may conveniently be used to keep track of the situation.

Such a method (the "fast" HT) was developed by Li and Lavin (1986) and Li et al. (1985). Illingworth and Kittler (1987) found this method to be insufficiently flexible in dealing with real data and produced a revised version (the "adaptive" HT), which permits each dimension in parameter space to change its resolution locally in tune with whatever the data demand, rather than insisting on some previously devised rigid structure. In addition, they employed a 9×9 accumulator array at each resolution rather than the theoretically most efficient 2×2 array, since this was found to permit better judgments to be made on the nature of the local data. This approach seemed to work well with fairly clean images but later, doubts were cast on its effectiveness with complex images (Illingworth and Kittler, 1988). The most serious problem to be overcome here is that at coarse resolutions, extended patterns of votes from several objects can overlap, giving rise to spurious peaks in parameter space. Since all of these peaks have to be checked at all relevant resolutions, the whole process can consume more computation than it saves. Clearly, optimization in multiresolution peak-finding schemes is complex[2] and data-dependent, and so discussion is curtailed here. The reader is

---

[2]Ultimately, the problem of system optimization for analysis of complex images is a difficult one, since in conditions of low signal-to-noise ratio even the eye may find it difficult to interpret an image and may "lock on" to an interpretation that is incorrect. Note that in general image interpretation work, there are many variables to be optimized—sensitivity, efficiency/speed, storage, accuracy, robustness, and so on—and it is seldom valid to consider any of these individually. Often tradeoffs between just two such variables can be examined and optimized but in real situations multivariable tradeoffs should be considered. This is a complex task and it is one of the purposes of this book to show clearly the serious nature of these types of optimization problem, although at the same time it can only guide the reader through a limited number of basic optimization processes.

referred to the original research papers (Li et al., 1985; Li and Lavin, 1986; Illingworth and Kittler, 1987) for implementation details.

An alternative scheme is the hierarchical HT (Princen et al., 1989a); so far it has been applied only to line detection. The scheme can most easily be envisaged by considering the foot-of-normal method of line detection described in Section 11.3. Rather small subimages of just 16×16 pixels are taken first of all, and the foot-of-normal positions determined. Then each foot-of-normal is tagged with its orientation and an identical HT procedure is instigated to generate the foot-of-normal positions for line segments in 32×32 subimages, this procedure being repeated as many times as necessary until the whole image is spanned at once. The paper by Princen et al. discusses the basic procedure in detail and also elaborates necessary schemes for systematically grouping separate line segments into full-length lines in a hierarchical context. An interesting detail is that successful operation of the method requires subimages with 50% overlap to be employed at each level. The overall scheme appears to be as accurate and reliable as the basic HT method but does appear to be faster and to require significantly less storage.

13.11 THE APPROACH OF GERIG AND KLEIN

The Gerig and Klein approach was first demonstrated in the context of circle detection but was only mentioned in passing in Chapter 12. This is because it is an important *approach* that has much wider application than merely to circle detection. The motivation for it has already been noted in Section 13.10—namely, the problem of extended patterns of votes from several objects giving rise to spurious peaks in parameter space.

Ultimately, the reason for the extended pattern of votes is that each edge point in the original image can give rise to a large number of votes in parameter space. The tidy case of detection of circles of known radius is somewhat unusual, as will be seen particularly in Chapters 12 and 17. Hence, in general *most* of the votes in parameter space are in the end unwanted and serve only to confuse. Ideally, we would like a scheme in which each edge point gives rise only to the single vote corresponding to the localization point of the particular boundary on which it is situated. Although this ideal is not initially realizable, it can be engineered by the "back-projection" technique of Gerig and Klein (1986). Here all peaks and other positions in parameter space to which a given edge point contributes are examined, and a new parameter space is built in which only the vote at the strongest of these peaks is retained (there is the greatest probability, but no certainty, that it belongs to the *largest* such peak). This second parameter space thus contains no extraneous clutter and weak peaks are hence found much more easily: this gives objects with highly fragmented or occluded boundaries much more chance of being detected. Overall, the method avoids many of the problems associated with setting arbitrary thresholds on peak height—in principle, no thresholds are required in this approach.

The scheme can be applied to any HT detector that throws multiple votes for each edge point. Thus, it appears to be widely applicable and is capable of improving robustness and reliability at an intrinsic expense of approximately doubling computational effort (however, set against this is the relative ease with which peaks can be located—a factor which is highly data-dependent). Note that the method is another example in which a two-stage process is used for effective recognition.

Other interesting features of the Gerig and Klein method must be omitted here for reasons of space, except to note that, rather oddly, the published scheme ignores edge orientation information as a means of reducing computation.

13.12 **CONCLUDING REMARKS**

The Hough transform was introduced in Chapter 11 as a line detection scheme and then used in Chapter 12 for detecting circles and ellipses. In both chapters, it appeared as a rather cunning method for aiding object detection; although it was seen to offer various advantages, particularly in its robustness in the face of noise and occlusion, there appeared to be no real significance in its rather novel voting scheme. The present chapter has shown that, far from being a trick method, the HT is much more general an approach than originally supposed: indeed, it embodies the properties of the spatial matched filter and is therefore capable of close-to-optimal sensitivity for object detection. However, this does not prevent its implementation from entailing considerable computational load, and significant effort and ingenuity have been devoted to overcoming this problem, both in general and in specific cases. The general case is tackled by the schemes discussed in Sections 13.9 and 13.10. It is important not to underestimate the value of specific solutions, both because such shapes as lines, circles, ellipses, and polygons cover a large proportion of (or approximations to) manufactured objects and because methods for coping with specific cases have a habit (as for the original HT) of becoming more general as workers see possibilities for developing the underlying techniques.

Finally, to further underline the generality of the GHT, it has also been used for optimal location of lines of known length, by emulating a spatial matched filter detector; this result has been applied to the optimal detection of polygons and of corners: for an example of the latter, see Fig. 13.12 (Davies, 1988a, 1989a). For further discussion and critique of the whole HT and GHT approach, see Chapter 27.

Although the Hough transform may appear to have a somewhat arbitrary design, this chapter has shown that it has solid roots in matched filtering, which in turn implies that votes should be gradient weighted for optimal sensitivity. The chapter also contrasts three methods for ellipse detection, showing how computational load may be estimated and minimized.

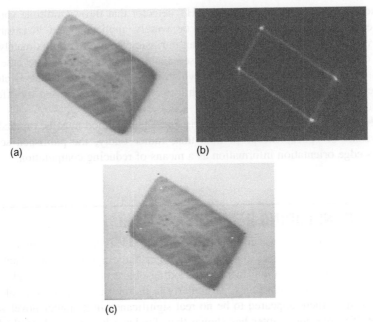

(a) (b)

(c)

FIGURE 13.12

Example of the generalized Hough transform approach to corner detection. (a) Original image of a biscuit (128 × 128 pixels, 64 gray levels). (b) Transform with lateral displacement around 22% of the shorter side. (c) Image with transform peaks located (white dots) and idealized corner positions deduced (black dots). The lateral displacement employed here is close to the optimum for this type of object.

Source: (a,b) © IEE 1988, (c) © Unicom 1988

13.13 BIBLIOGRAPHICAL AND HISTORICAL NOTES

Although the HT was introduced as early as 1962, a number of developments—including especially those of Merlin and Farber (1975) and Kimme et al. (1975)—were required before the GHT was developed (Ballard, 1981). By that time, the HT was already known to be formally equivalent to template matching (Stockman and Agrawala, 1977) and to spatial matched filtering (Sklansky, 1978). However, the questions posed in Section 13.3 were only answered much later (Davies, 1987a), the necessary analysis being reproduced in Sections 13.4–13.7. The author's work (Davies, 1987b, 1987d, 1989b) on line detection by the GHT (not covered here) was aimed particularly at optimizing sensitivity of line detection, although deeper issues of tradeoffs between sensitivity, speed, and accuracy are also involved.

By 1985, the computational load of the HT became the critical factor preventing its more general use—particularly as it could be used for most types of arbitrary shape detection, with well-attested sensitivity and considerable robustness. Following preliminary work by Brown (1984), with the emphasis on hardware implementations of the HT, Li et al. (1985) and Li and Lavin (1986) showed the possibility of much faster peak location by using nonuniformly quantized parameter spaces. This work was developed further by Illingworth and Kittler (1987) and others (see, e.g., Illingworth and Kittler, 1988; Princen et al., 1989a, 1989b; Davies, 1992g). An important development has been the randomized Hough transform (RHT), pioneered by Xu and Oja (1993) among others: it involves casting votes until specific peaks in parameter space become evident, thereby saving unnecessary computation.

Accurate peak location remains an important aspect of the HT approach. Properly, this is the domain of robust statistics, which handles the elimination of outliers (see Appendix A). Davies (1992f) has shown a computationally efficient means of accurately locating HT peaks, and has found why peaks sometimes appear narrower than *a priori* considerations would indicate (Davies, 1992b). Kiryati and Bruckstein (1991) have tackled aliasing effects, which can arise with the HT, and which have the effect of cutting down accuracy.

Over time, the GHT approach has been broadened by geometric hashing, structural indexing, and other approaches (e.g., Lamdan and Wolfson, 1988; Gavrila and Groen, 1992; Califano and Mohan, 1994). At the same time, a probabilistic approach to the subject has been developed (Stephens, 1991), which puts it on a firmer footing. Grimson and Huttenlocher (1990) warn (possibly over-pessimistically) against the blithe use of the GHT for complex object recognition tasks, because of the false peaks that can appear in such cases. For further review of the state of the subject up to 1993, see Leavers (1993).

In various chapters of Part 2, the statement has been made that the HT[3] carries out a search leading to hypotheses that should be checked before a final decision about the presence of an object can be made. However, Princen et al. (1994) show that the performance of the HT can be improved if it is itself regarded as a hypothesis testing framework: this is in line with the concept that the HT is a model-based approach to object location. Other studies have been made about the nature of the HT. In particular, Aguado et al. (2000) consider the intimate relationship between the HT and the principle of duality in shape description: the existence of this relationship underlines the importance of the HT and provides a means for a more general definition of it. Kadyrov and Petrou (2001) have developed the trace transform, which can be regarded as a generalized form of the Radon transform—itself closely related to the Hough transform.

Other workers have used the HT for affine-invariant search: Montiel et al. (2001) made an improvement to reduce the incidence of erroneous evidence in the gathered data, whereas Kimura and Watanabe (2002) made an extension for 2-D shape detection that is less sensitive to the problems of occlusion and broken

---

[3]A similar statement can be made in the case of graph matching methods such as the maximal clique approach to object location (see Chapter 14).

boundaries. Kadyrov and Petrou (2002) have adapted the trace transform to cope with affine parameter estimation.

In a generalization of the work of Atherton and Kerbyson (1999) and of Davies (1987a) on gradient weighting (see Section 13.6), Anil Bharath and his colleagues have examined how to optimize the sensitivity of the HT (private communication, 2004). Their method is particularly valuable in solving the problems of early threshold setting that limit many HT techniques. Similar sentiments come out in a different way in the work of Kesidis and Papamarkos (2000), which maintains the grayscale information throughout the transform process, thereby leading to more exact representations of the original images.

Olson (1999) has shown that localization accuracy can be improved efficiently by transferring local error information into the HT and handling it rigorously. An important finding is that the HT can be subdivided into many sub-problems without decrease in performance. This finding is elaborated in a 3-D model-based vision application where it is shown to lead to reduced false positive rates (Olson, 1998). Wu et al. (2002) extend the 3-D possibilities further by using a 3-D HT to find glasses: first a set of features are located that lie on the same plane, and this is then interpreted as the glasses' rim plane. This approach allows the glasses to be separated from the face, and then they can be located in their entirety.

van Dijck and van der Heijden (2003) develop the geometric hashing method of Lamdan and Wolfson (1988) to perform 3-D correspondence matching using full 3-D hashing. This is found to have advantages in that knowledge of 3-D structure can be used to reduce the number of votes and spurious matches. Tuytelaars et al. (2003) describe how invariant-based matching and HTs can be used to identify regular repetitions in planes appearing within visual (3-D) scenes in spite of perspective skew. The overall system has the ability to reason about consistency and is able to cope with periodicities, mirror symmetries, and reflections about a point.

13.13.1 More Recent Developments

Among the most recent developments are the following. Aragon-Camarasa and Siebert (2010) considered using the GHT for clustering SIFT feature matches. However, it turned out that a *continuous* rather than discretized HT space was needed for this application. This meant that each matched point had to be stored at the full machine precision in a Hough space consisting of a list data structure. Therefore, peak location had to take the form of standard unsupervised clustering algorithms. This was an interesting case where the intended GHT could not follow the standard voting and accumulating procedure. Assheton and Hunter (2011) also deviated sharply from the standard GHT approach when performing pedestrian detection and tracking: they used a shape-based voting algorithm based on Gaussian mixture models. The algorithm was stated to be highly effective for detecting pedestrians based on the silhouette shape. Chung et al. (2010) studied the problem of information retrieval from databases. They produced a region-based solution for object retrieval using the GHT and adaptive image segmentation. A key aspect of the overall scheme was the location of affine-invariant maximally stable extremal regions (MSERs) (see Chapter 6) in the database and query images. Roy et al. (2011) applied the GHT to the detection and verification of seals (stamps) containing lettering and geometric patterns. This is a difficult problem because of the likely presence of noise, interfering text, and signatures as well as incompleteness due to the application of uneven pressure to the stamp. In practice, a seal has to be located using scale and rotation invariant features (particularly text characters); it is then detected as a GHT peak resulting from application of a spatial feature descriptor of neighboring connected component pairs, i.e., in this application, the text characters in the seal are used as basic features for seal detection instead of individual edge or feature points. Memory demands are limited by splitting the R-table into two different lookup tables—the character pair table and the distance table.

13.14 PROBLEMS

1. **a.** Describe the main stages in the application of the HT to locate objects in digital images. What are the particular advantages offered by the HT technique? Give reasons why they arise.

   b. It is said that the HT only leads to *hypotheses* about the presence of objects in images and that they should all be checked independently before making a final decision about the contents of any image. Comment on the accuracy of this statement.

2. Devise a GHT version of the spatial matched filter for detecting lines of known length L. Show that when used to detect an ideal line of length L, it gives a distributed response of length $2L$ that peaks at the center of the line, but when used to detect a partially occluded version of the line, it gives a response that is flat-topped over a range that includes the center of the line.

3. Show how a GHT version of the spatial matched filter can be devised to detect an equilateral triangle, leading to a star-shaped transform that peaks at the center of the triangle. How may this approach be adapted for (a) a general triangle and (b) a regular polygon having N sides?

14

Pattern Matching Techniques

Abstract pattern matching involves stepping back from the image itself and working at a higher level, grouping features in an abstract way to infer the presence of objects. Graph matching has long been a standard approach for achieving this, but in certain circumstances, a suitable adaptation of the generalized Hough transform can actually outperform it. This chapter discusses inference procedures, and goes on to consider relational descriptions of scenes and the various types of search that can be used with image data.

Look out for:

- the match graph approach for identifying objects from their point features.
- how the need for robustness against noise, clutter, and occlusion translates into the requirement for subgraph—subgraph isomorphism.
- the maximal clique paradigm.
- how symmetry can be used to simplify the matching task.
- how the generalized Hough transform can be used for point pattern matching.
- how order calculations can be used to compare the speeds of matching algorithms.
- how relational descriptors may be used for logical analysis of scenes.
- the different types of search algorithm that may be used in scene analysis.

This chapter completes the work of Part 2 by showing how the presence of objects can be inferred from point features as an alternative to edge features. Even with point features, it is found that the Hough transform may sometimes be used with advantage. However, all inference techniques need to be analyzed for computational complexity and suitable optimizations made. The latter lesson carries on with even more force in subsequent work—not least the more complex algorithms used for processing 3-D images in Part 3 of this book.

14.1 INTRODUCTION

In the foregoing chapters, it has been seen how objects having quite simple shapes may be located in digital images via the Hough transform. For more complex shapes, this approach tends to require excessive computation: in general, the way this problem may be overcome is to locate objects by their features. Suitable salient features include small holes, corners, straight, circular, or elliptical segments, and indeed any readily localizable sub-patterns: earlier chapters have shown how such features may be located. However, at some stage, it becomes necessary to find methods for collating the information from the various features, in order to recognize and locate the objects containing them. This task is studied in the present chapter.

It is perhaps easiest to envisage the feature collation problem when the features themselves are unstructured points carrying no directional information—nor indeed any attributes other than their x, y coordinates in the image. Then the object recognition task is often called "point pattern matching."[1] The features that are closest to unstructured points are small holes, such as the "docker" holes in many types of biscuit. Corners can also be considered as points if their other attributes—including sharpness, orientation, etc.—are ignored. In what follows, we start with point features and then see how the attributes of more complex types of feature can be included in recognition schemes.

Overall, it is most efficient to use small high-contrast features for object detection, since the computation involved in searching an image decreases as the template becomes smaller. As is clear from the preceding chapters, the main disadvantage resulting from such an approach to object detection is the loss in sensitivity (in a signal-to-noise sense) due to the greatly impoverished information content of the point feature image. However, the task of identifying objects from a rather small number of point features is far from trivial and frequently involves considerable computation, as will be seen in Sections 14.2–14.4. We start by studying a graph-theoretic approach to point pattern matching, which involves the "maximal clique" concept.

14.2 A GRAPH-THEORETIC APPROACH TO OBJECT LOCATION

This section considers a commonly occurring situation that involves considerable constraints—objects appearing on a horizontal worktable or conveyor at a known distance from the camera. It is also assumed that (a) objects are flat or can appear in only a restricted number of stances in three dimensions, (b) objects are viewed from directly overhead, and (c) perspective distortions are small. In such

---

[1]Note that this term is sometimes used not just for object recognition but also for initial matching of two stereo views of the same scene.

situations, the objects may in principle be identified and located from very few point features. Since such features are taken to have no structure of their own, it will be impossible to locate an object uniquely from a single feature, although positive identification and location would be possible using two features if these were distinguishable and if their distance apart were known. For truly indistinguishable point features, an ambiguity remains for all objects not possessing 180° rotation symmetry. Hence, at least three point features are in general required to locate and identify objects at known range. Clearly, noise and other artifacts such as occlusions modify this conclusion. In fact, when matching a template of the points in an idealized object with the points present in a real image, we may find that:

1. a great many feature points may be present because of multiple instances of the chosen type of object in the image.
2. additional points may be present because of noise or clutter from irrelevant objects and structure in the background.
3. certain points that should be present are missing because of noise or occlusion, or because of defects in the object being sought.

These problems mean that we should in general be attempting to match a subset of the points in the idealized template to various subsets of the points in the image. If the point sets are considered to constitute *graphs* with the point features as *nodes*, the task devolves into the mathematical problem of subgraph–subgraph isomorphism,[2] i.e., finding which subgraphs in the image graph are isomorphic to subgraphs of the idealized template graph. Of course, there may be a large number of matches involving rather few points: these would arise from sets of features that *happen* (see, e.g., item 2 above) to lie at valid distances apart in the original image. The most significant matches will involve a fair number of features and will lead to correct object identification and location. Clearly, a point feature matching scheme will be most successful if it finds the most likely interpretation by searching for solutions with the greatest internal consistency, i.e., with the greatest number of point matches per object.

Unfortunately, the scheme of things presented above is still too simplistic in many applications as it is insufficiently robust against distortions. In particular, optical (e.g., perspective) distortions may arise, or the objects themselves may be distorted, or by resting partly on other objects they may not be quite in the assumed stance: hence, distances between features may not be exactly as expected. These factors mean that some tolerance has to be accepted in the distances between pairs of features and it is common to employ a threshold such that interfeature distances have to agree within this tolerance before matches are accepted as potentially valid. Clearly, distortions lay more strain on the point matching technique and make it all the more necessary to seek solutions with the greatest possible internal consistency. Thus, as many features as possible should

---

[2] Of the same basic shape and structure.

be taken into account in locating and identifying objects. The maximal clique approach is intended to achieve this.

As a start, as many features as possible are identified in the original image and these are numbered in some convenient order such as the order of appearance in a normal TV raster scan. The numbers then have to be matched against the letters corresponding to the features on the idealized object. A systematic way of achieving this is by constructing a *match graph* (or *association graph*) in which the nodes represent feature assignments and arcs joining nodes represent pairwise compatibilities between assignments. To find the best match, it is then necessary to find regions of the match graph where the cross-linkages are maximized. To achieve this, *cliques* are sought within the match graph. A clique is a *complete subgraph*—i.e., one for which all pairs of nodes are connected by arcs. However, the previous arguments indicate that if one clique is completely included within another clique, it is likely that the larger clique represents a better match—and indeed *maximal cliques* can be taken as leading to the most reliable matches between the observed image and the object model.

Figure 14.1(a) illustrates the situation for a general triangle: for simplicity, the figure takes the observed image to contain only one triangle and assumes that lengths match exactly and that no occlusions occur. The match graph in this example is shown in Fig. 14.1(b): there are nine possible feature assignments, six

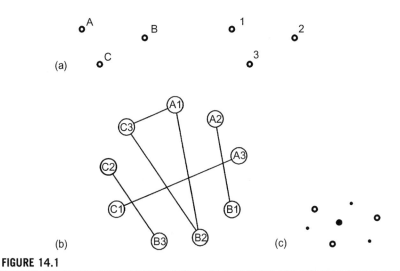

FIGURE 14.1

A simple matching problem—a general triangle: (a) basic labeling of model (*left*) and image (*right*); (b) match graph; (c) placement of votes in parameter space. In (b) the maximal cliques are (1) A1, B2, C3, (2) A2, B1, (3) B3, C2, and (4) C1, A3. In (c) the following notation is used: (o) positions of observed features; (•) positions of votes; (●) position of main voting peak.

Source: © AVC 1988 and Elsevier 1991

valid compatibilities, and four maximal cliques, only the largest corresponding to an exact match.

Figure 14.2(a) shows the situation for the less trivial case of a quadrilateral, the match graph being shown in Fig. 14.2(b). In this case, there are 16 possible feature assignments, 12 valid compatibilities, and 7 maximal cliques. If occlusion of a feature occurs, this will (taken on its own) reduce the number of possible feature assignments and also the number of valid compatibilities: in addition, the number of maximal cliques and the size of the largest maximal clique will be reduced. On the other hand, noise or clutter can add erroneous features. If the latter are at arbitrary distances from existing features, then the number of possible feature assignments will be increased but there will not be any more compatibilities in the match graph, so the latter will have only trivial additional complexity. However, if the extra features appear at *allowed* distances from existing features, this will introduce extra compatibilities into the match graph and make it more tedious to analyze. In the case shown in Fig. 14.3, both types of complication—an occlusion and an additional feature—arise: there are now eight pairwise assignments and six maximal cliques, rather fewer overall than in the original case of Fig. 14.2. However, the important factor is that the largest maximal clique still indicates the most likely interpretation of the image and that the technique is inherently highly robust.

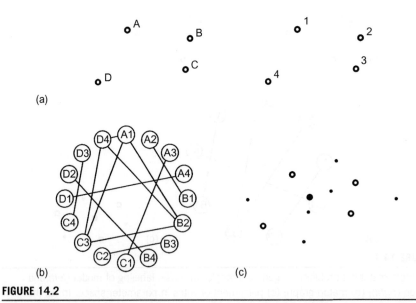

(a)

(b)   (c)

FIGURE 14.2

Another matching problem—a general quadrilateral: (a) basic labeling of model (*left*) and image (*right*), (b) match graph, and (c) placement of votes in parameter space (notation as in Figure 14.1).

Source: © AVC 1988 and Elsevier 1991

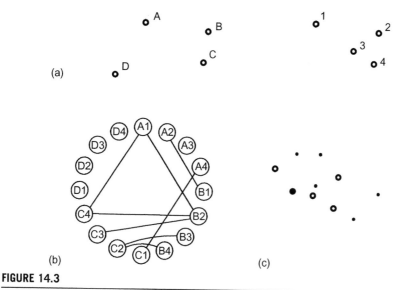

FIGURE 14.3

Matching when one feature is occluded and another is added: (a) basic labeling of model (*left*) and image (*right*), (b) match graph, and (c) placement of votes in parameter space (notation as in Fig. 14.1).

When using methods such as the maximal clique approach that involve repetitive operations, it is useful to look for means of saving computation. In fact, when the objects being sought possess some symmetry, economies can be made. Consider the case of a parallelogram (Fig. 14.4). Here the match graph has 20 valid compatibilities and there are 10 maximal cliques. Of these, the largest two have equal numbers of nodes and *both* identify the parallelogram within a symmetry operation. This means that the maximal clique approach is doing more computation than absolutely necessary: this can be avoided by producing a new "symmetry-reduced" match graph after relabeling the model template in accordance with the symmetry operations (see Fig. 14.5). This gives a much smaller match graph with half the number of pairwise compatibilities and half the number of maximal cliques. In particular, there is only one nontrivial maximal clique: note, however, that its size is not reduced by the application of symmetry.

14.2.1 A Practical Example—Locating Cream Biscuits

Figure 14.6(a) shows one of a pair of cream biscuits, which are to be located from their "docker" holes—this strategy being advantageous since it has the potential for highly accurate product location prior to detailed inspection (in this case the purpose is to locate the biscuits accurately from the holes, and then to

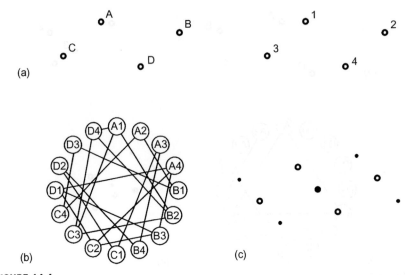

FIGURE 14.4

Matching a figure possessing some symmetry: (a) basic labeling of model (*left*) and image (*right*), (b) match graph, and (c) placement of votes in parameter space (notation as in Fig. 14.1).

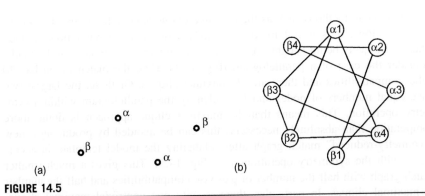

FIGURE 14.5

Using a symmetry-reduced match graph: (a) relabeled model template and (b) symmetry-reduced match graph.

check the alignment of the biscuit wafers and detect any excess cream around the sides of the product). The holes found by a simple template matching routine are indicated in Fig. 14.6(b): the template used is rather small and, as a result, the routine is fairly fast but fails to locate all holes; in addition, it can give false

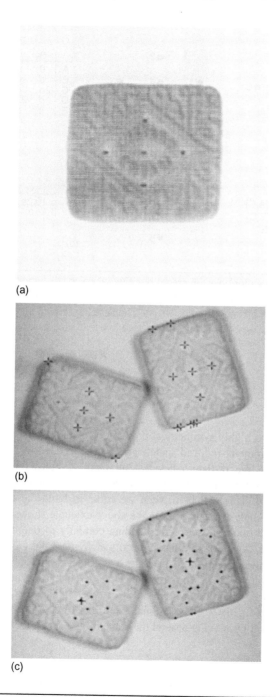

(a)

(b)

(c)

FIGURE 14.6

(a) A typical cream sandwich biscuit; (b) a pair of cream sandwich biscuits with crosses indicating the result of applying a simple hole detection routine; and (c) the two biscuits reliably located by the GHT from the hole data in (b): the isolated small crosses indicate the positions of single votes.

Source: © AVC 1988 and Elsevier 1991

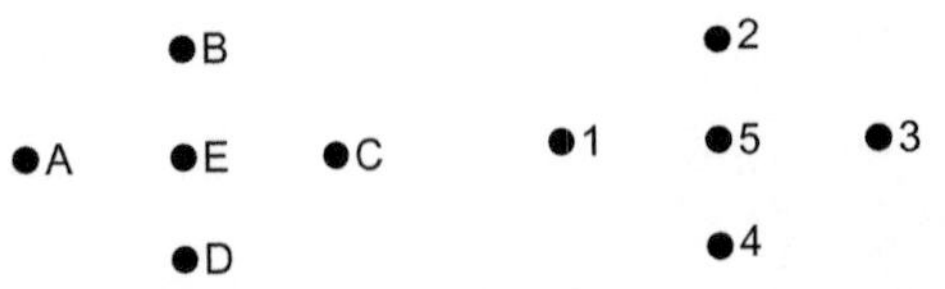

(a)

(b)

	A	B	C	D	E
A	0	d	2a	d	a
B	d	0	d	2b	b
C	2a	d	0	d	a
D	d	2b	d	0	b
E	a	b	a	b	0

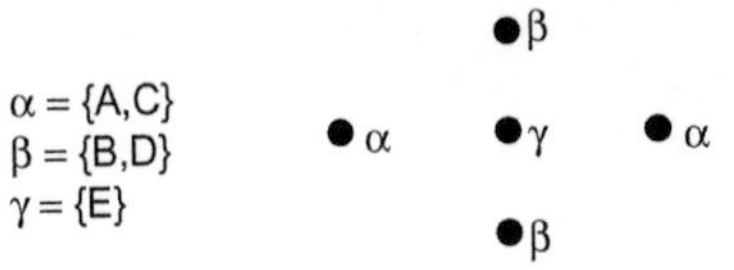

(c)

α = {A,C}
β = {B,D}
γ = {E}

(d)

	α	β	γ
α	0,2a	d	a
β	d	0,2b	b
γ	a	b	0

FIGURE 14.7

Inter-feature distances for holes on cream biscuits: (a) basic labeling of model (*left*) and image (*right*); (b) allowed distance values; (c) revised labeling of model using symmetric set notation; and (d) allowed distance values. The cases of zero interfeature distance in the final table can be ignored as they do not lead to useful matches.

alarms. Hence, an "intelligent" algorithm must be used to analyze the hole location data.

Clearly, this is a highly symmetrical type of object, and so it should be beneficial to employ the symmetry-reduced match graph described above. To proceed, it is helpful to tabulate the distances between all pairs of holes in the object model (Fig. 14.7(b)). Then this table can be regrouped to take account of symmetry operations (Fig. 14.7(d)). This will help when we come to build the match graph for a particular image. Analysis of the data in the above example shows that there are two nontrivial maximal cliques, each corresponding correctly to one of the two biscuits in the image. Note, however, that the reduced match graph does not give a *complete* interpretation of the image: it locates the two objects but it does not confirm uniquely which hole is which. In particular, for a given starting hole of type α, it is not known which is which of the two holes of type β. This can be ascertained by applying simple geometry to the coordinates in order to determine (say) which hole of type β is reached by moving around the center hole γ in a clockwise sense.

14.3 POSSIBILITIES FOR SAVING COMPUTATION

In these examples, the checking of which subgraphs are maximal cliques is a simple problem. However, in real matching tasks it can quickly become unmanageable (the reader is encouraged to draw the match graph for an image containing two objects of seven points!).

Table 14.1 shows what is perhaps the most obvious type of algorithm for finding maximal cliques. It operates by examining in turn all cliques of a given number of nodes and finding what cliques can be constructed from them by adding additional nodes (bearing in mind that any additional nodes must be compatible with all existing nodes in the clique). This permits all cliques in the match graph to be identified. However, an additional step is needed to eliminate (or relabel) all cliques that are included as subgraphs of a new larger clique before it is known which cliques are maximal.

In view of the evident importance of finding maximal cliques, many algorithms have been devised for the purpose. It is probable that the best of these is now close to the fastest possible speed of operation. Unfortunately, the optimum execution time is known to be bounded not by a polynomial in M (for a match graph containing maximal cliques of up to M nodes) but by a much faster varying function. Specifically, the task of finding maximal cliques is akin to the well-known traveling salesman problem and is known to be "NP-complete," implying that it runs in exponential time (see Section 14.4.1). Thus, whatever the run-time is for values of M up to about 6, it will typically be 100 times slower for values of M up to about 10, and 100 times slower again for M greater than $\sim$14. In practical situations, there are several ways of tackling this problem:

1. Use the symmetry-reduced match graph wherever possible.
2. Choose the fastest available maximal clique algorithm.

Table 14.1 A simple Maximal Clique Algorithm

```
set clique size to 2;
// this is the size already included by the match graph
while (newcliques = true) { // new cliques still being found
    increment clique size;
    set newcliques = false;
    for all cliques of previous size {
        set all cliques of previous size to status maxclique;
        for all possible extra nodes
            if extra node is joined to all existing nodes in clique {
                store as a clique of current size;
                set newcliques = true;
            }
    }
    // the larger cliques have now been found
    for all cliques of current size
        for all cliques of previous size
            if all nodes of smaller clique are included in current clique
                set smaller clique to status not maxclique;
    // the subcliques have now been relabelled
}
```

3. Write critical loops of the maximal clique algorithm in machine code.
4. Build special hardware or multiprocessor systems to implement the algorithm.
5. Use the LFF method (see below: this means searching for cliques of small M and then working with an alternative method).
6. Use an alternative sequential strategy, which may however not be guaranteed to find all the objects in the image.
7. Use the GHT approach (see Section 14.4).

Of these methods, the first should be used wherever applicable. Methods 2−4 amount to improving the implementation and are subject to diminishing returns: note that the execution time varies so rapidly with M that even the best software implementations are unlikely to give a practical increase in M of more than 2 (i.e., $M \rightarrow M+2$). Likewise, dedicated hardware implementations may only give increases in M of the order of 4−6. Method 5 is a "shortcut" approach, which proves highly effective in practice. The idea is to search for specific subsets of the features of an object, and then to hypothesize that the object exists and go back to the original image to check that it is actually present. Bolles and Cain (1982) devised this method when looking for hinges in quite complex images. In principle, the method has the disadvantage that the particular subset of an object that is chosen as a cue may be missing because of occlusion or some other artifact. Hence, it may be necessary to look for several such cues on each object. This is an example of further deviation from the matched filter paradigm, which reduces detection sensitivity yet again. The method is called the local-feature-focus (LFF) method because objects are sought by cues or local foci.

The maximal clique approach is a type of exhaustive search procedure and is effectively a parallel algorithm. This has the effect of making it highly robust but is also the source of its slow speed. An alternative is to perform some sort of sequential search for objects, stopping when sufficient confidence is attained in the full or partial interpretation of the image. For example, the search process may be terminated when a match has been obtained for a certain minimum number of features on a given number of objects. Such an approach may be useful in some applications and will generally be considerably faster than the full maximal clique procedure when M is greater than about 6. An analysis of several tree-search algorithms for subgraph isomorphism was carried out by Ullmann (1976): the paper tests algorithms using artificially generated data and it is not clear how they relate to real images. The success or otherwise of all nonexhaustive search algorithms must, however, depend critically on the particular types of image data being analyzed: hence, it is difficult to give further general guidance on this matter (but see Section 14.7 for additional comments on search procedures).

The final method listed above is based on the GHT. In many ways, this provides an ideal solution to the problem since it presents an exhaustive search technique that is essentially equivalent to the maximal clique approach while not falling into the NP-complete category. This may seem contradictory, since any approach to a well-defined mathematical problem should be subject to the mathematical constraints known to be involved in its solution. However, although the abstract maximal clique problem is known to be NP-complete, the *subset* of maximal clique problems that arises from 2-D image-based data may well be solved with less computation by other means, and in particular by a 2-D technique. This special circumstance does appear to be valid but unfortunately it offers no possibility of solving general NP-complete problems by reference to the specific solutions found using the GHT approach! The GHT approach is described in Section 14.4.

14.4 USING THE GENERALIZED HOUGH TRANSFORM FOR FEATURE COLLATION

This section describes how the GHT can be used as an alternative to the maximal clique approach, to collate information from point features in order to find objects. Initially, we consider situations where objects have no symmetries—as for the cases of Figs. 14.1–14.3.

To apply the GHT, we first list all features and then accumulate votes in parameter space at every possible position of a localization point L consistent with each *pair* of features (Fig. 14.8). This strategy is particularly suitable in the present context, as it corresponds to the pairwise assignments used in the maximal clique method. To proceed, it is necessary merely to use the interfeature distance as a lookup parameter in the GHT R-table. For indistinguishable point features, this means that there must be two entries for the position of L for each value of the interfeature distance. Note that we have assumed that no symmetries exist and that all pairs of

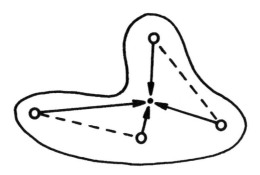

FIGURE 14.8

Method for locating L from pairs of feature positions: each pair of feature points gives two possible voting positions in parameter space, when objects have no symmetries. When symmetries are present, certain pairs of features may give rise to up to four voting positions: this is confirmed on careful examination of Fig. 14.6(c).

Source: © *Unicom 1988*

features have different interfeature distances. If this were not so, then more than two vectors would have to be stored in the R-table per interfeature distance value.

To illustrate the GHT approach outlined above, we first apply it to the triangle example shown in Fig. 14.1. Figure 14.1(c) shows the positions at which votes are accumulated in parameter space. There are four peaks with heights of 3, 1, 1, and 1, it being clear that, in the absence of complicating occlusions and defects, the object is locatable at the peak of maximum size. Next, the method is applied to the general quadrilateral example shown in Fig. 14.2: this leads to seven peaks in parameter space, whose sizes are 6, 1, 1, 1, 1, 1, and 1 (Fig. 14.2(c)).

Close examination of Figs. 14.1–14.3 indicates that every peak in parameter space corresponds to a maximal clique in the match graph. Indeed, there is a one-to-one relation between the two. In the uncomplicated situation being examined here, this is bound to be so for any general arrangement of features within an object, since every pairwise compatibility between features corresponds to two potential object locations, one correct and one that can be correct only from the point of view of that pair of features. Hence, the correct locations all add to give a large maximal clique and a large peak in parameter space, whereas the incorrect ones give maximal cliques each containing two wrong assignments and each corresponding to a false peak of size 1 in parameter space. This situation still applies even when occlusions occur or additional features are present (see Fig. 14.3). The situation is slightly more complicated when symmetries are present, the two methods each deviating in a different way: space does not permit the matter to be explored in depth here but the solution for the case of Fig. 14.4(a) is presented in Fig. 14.4(c). Overall, it seems simplest to assume that there is still a one-to-one relationship between the solutions from the two approaches.

Finally, consider again the example discussed in Section 14.2.1 (Fig. 14.6(a)), this time obtaining a solution by the GHT. Figure 14.6(c) shows the positions of candidate object centers as found by the GHT. The small isolated crosses indicate the positions of single votes, and those very close to the two large crosses lead to voting peaks of weights 10 and 6 at these respective positions. Hence, object location is both accurate and robust, as required.

14.4.1 Computational Load

This subsection compares the computational requirements of the maximal clique and GHT approaches to object location. For simplicity, imagine an image that contains just one wholly visible example of the object being sought. Also, suppose that the object possesses n features and that we are trying to recognize it by seeking all possible pairwise compatibilities, whatever their distance apart (as for all examples discussed in Section 14.2).

For an object possessing n features, the match graph contains n^2 nodes (i.e., possible assignments), and there are $^{n^2}C_2 = n^2(n^2 - 1)/2$ possible pairwise compatibilities to be checked in building the graph. The amount of computation at this stage of the analysis is $O(n^4)$. To this must be added the cost of finding the maximal cliques. Since the problem is NP-complete, the load rises at a rate that is faster than that of polynomial, and probably exponential in n^2 (Gibbons, 1985).

Now consider the cost of getting the GHT to find objects *via* pairwise compatibilities. As has been seen, the total height of all the peaks in parameter space is in general equal to the number of pairwise compatibilities in the match graph. Hence, the computational load is of the same order, $O(n^4)$. Next comes the problem of locating all the peaks in parameter space. In this case, parameter space is congruent to image space. Hence, for an $N \times N$ image only N^2 points have to be visited in parameter space and the computational load is $O(N^2)$. Note, however, that an alternative strategy is available in which a running record is kept of the relatively small numbers of voting positions in parameter space. The computational load for this strategy will be $O(n^4)$: although of a higher *order*, this often represents less computation in practice.

The reader may have noticed that the basic GHT scheme as outlined so far is able to locate objects by their features but does not determine their orientations. However, orientations can be computed by running the algorithm a second time and finding all the assignments that contribute to each peak. Alternatively, the second pass can aim to find a different localization point within each object. In either case, the overall task should be completed in little over twice the time, i.e., still in $O(n^4 + N^2)$ time.

Although the GHT at first appears to solve the maximal clique problem in polynomial time, what it actually achieves is to solve a real-space template matching problem in polynomial time: it does not solve an *abstract* graph-theoretic problem in polynomial time. The overall lesson is that the graph theory representation is not well matched to real space, not that real space can be used to solve abstract NP-complete problems in polynomial time.

14.5 GENERALIZING THE MAXIMAL CLIQUE AND OTHER APPROACHES

This section considers how the graph matching concept can be generalized to cover alternative types of features and also various attributes of features. The earlier discussion was restricted to point features and in particular to small holes, which were supposed to be isotropic. Corners were also taken as point features by ignoring attributes other than position coordinates. Both holes and corners seem to be ideal, in that they give maximum localization and hence maximum accuracy for object location. Straight lines and straight edges at first appear to be rather less well suited to the task. However, more careful thought shows that this is not so. One possibility is to use straight lines to deduce the positions of corners, which can then be used as point features, although this approach is not as powerful as might be hoped because of the abundance of irrelevant line crossings that are thrown up (in this context, note that the maximal clique method is inherently capable of sorting the true corners from the false alarms). A more elegant solution is simply to determine the angles between pairs of lines, referring to a lookup table for each type of object to determine whether each pair of lines should be marked as compatible in the match graph. Once the match graph has been built, an optimal match can be found as before (although ambiguities of scale will arise, which will have to be resolved by further processing). These possibilities significantly generalize the ideas of the foregoing sections.

Other types of feature generally have more than two specifying parameters, one of which may be contrast and the other size. This applies for most holes and circular objects, although for the smallest (i.e., barely resolvable) holes it is sometimes most practicable to take the central dip in intensity as the measured parameter. For straight lines, the relevant size parameter is the length (we here count the line ends, if visible, as points that already have been taken into account). Corners may have a number of attributes, including contrast, color, sharpness, and orientation—none of which is likely to be known to high accuracy. Finally, more complex shapes such as ellipses have orientation, size, and eccentricity, and again contrast or color may be a usable attribute (generally, contrast is an unreliable measure because of possible variations in the background).

In fact, so much information is available that we need to consider how best to use it for locating objects. For convenience, this is discussed in relation to the maximal clique method. In fact, the answer is very simple. When compatibilities are being considered and the arcs are being drawn in the match graph, *any* available information may be taken into account in deciding whether a pair of features in the image matches a pair of features in the object model. In Section 14.2, the discussion was simplified by taking interfeature distances as the only relevant measurements. However, it is quite acceptable to describe the features in the object model more fully and to insist that they all match within prespecified tolerances. For example, holes and corners may be permitted to lead to a match only if the former are of the correct size, the latter are of the correct sharpness and orientation, and the distances between these features are also appropriate. All relevant information has to be held in suitable lookup tables. In general, the gains easily outweigh the losses, since a considerable number of potential interpretations will be eliminated—hence making the match graph significantly simpler and reducing, in many cases by a large factor, the amount of computation that is required to find the maximal cliques. Note, however, that there is a limit to this, since in the absence of occlusions and erroneous tolerances on the additional attributes, the number of nodes in the match graph cannot be less than the square of the number of features in an object, and the number of nodes in a maximal clique will be unchanged.

Thus, extra feature attributes are of very great value in cutting down computation: they are also useful in making interpretation less ambiguous. This latter property is "obvious" but not always realizable. In particular, extra attributes help in this way only if (a) some of the features on an object are missing, through occlusion or for other reasons such as breakage or (b) if the distance tolerances are so large as to make it unclear which features in the image match with those in the model.

Suppose next that the distance attributes become very imprecise, either because of shape distortions or because of unforeseen rotations in 3-D. It is worth enquiring how far we can proceed under these circumstances. In fact, in the limit of low distance accuracy, we may only be able to employ an "adjacent to" descriptor. This parallels the situation in general scene analysis, where use is made of a number of relational descriptors such as "on top of," "to the left of," and so on. Such possibilities are considered in Section 14.6.

14.6 RELATIONAL DESCRIPTORS

The previous section showed how additional attributes could be incorporated into the maximal clique formalism so that the effects of diminishing accuracy of distance measurement could be accommodated. This section considers what happens when the accuracy of distance measurement drops to zero and we are left just with relational attributes such as "adjacent to," "near," "inside," "underneath," "on top of," and so on. We start by taking adjacency as the basic relational attribute. To illustrate the approach, imagine a simple outdoor scene where various rules apply to segmented regions. These rules will be of the type "sky may be adjacent to forest," "forest may be adjacent to field," and so on. Note that "adjacent to" is not transitive, i.e., if P is adjacent to Q, and Q is adjacent to R, this does not imply that P is adjacent to R (in fact, it is quite likely that P will not be adjacent to R, as Q may well separate the two regions completely!).

The rules for a particular type of scene may be summarized as in Table 14.2. Now consider the scene shown in Fig. 14.9. Applying the rules for adjacency from Table 14.2 will be seen to permit four different solutions, in which regions 1−3 may be interpreted as:

1. sky, forest, field (the correct interpretation).
2. sky, forest, sky.
3. field, forest, sky.
4. field, forest, field.

Evidently, there are too few constraints. Possible constraints are the following: *sky is above field and forest, sky is blue, field is green*, and so on. Of these, the first is a binary relation like adjacency, whereas the other two are unary constraints. It is easy to see that two such constraints are required to resolve completely the ambiguity in this particular example.

Paradoxically, adding further regions can make the situation inherently less ambiguous. This is because other regions are less likely to be adjacent to all the original regions, and in addition may act in such a way as to label them uniquely. This is seen in Fig. 14.10, which is interpreted by reference to Table 14.3, although the keys to unique interpretation are the notions that *any small white objects must be sheep*, and *any small dark objects must be birds*. On analyzing

Table 14.2 Adjacency Table for Simple Scene

	Sky	Forest	Field
Sky	—	✓	—
Forest	✓	—	✓
Field	—	✓	—

This table is relevant to the scene in Figure 14.9. Ticks indicate that regions must be adjacent: dashes indicate that regions must not be adjacent (i.e., adjacency is not optional).

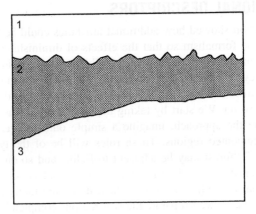

FIGURE 14.9

Regions in a simple scene: 1, sky; 2, forest; 3, field.

the available data, the whole scene can now be interpreted unambiguously. As expected, the maximal clique technique can be applied successfully, although to achieve this, assignments must be taken to be compatible not only when image regions are adjacent and their interpretations are marked as adjacent, but also when image regions are nonadjacent and their interpretations are marked as nonadjacent (the reason for this is easily seen by drawing an analogy between adjacency and distance, adjacent and nonadjacent corresponding respectively to zero, and a significant distance apart). Note that in this scene analysis type of situation, it is *very* tedious to apply the maximal clique method manually because there tend to be a fair number of trivial solutions (i.e., maximal cliques with just a few nodes).

Perhaps oddly, some of the trivial solutions that the maximal clique technique gives rise to are provably *incorrect*. For example, in the above problem, one solution appears to include regions 1 and 2 being forest and sky, respectively, whereas it is clear from the presence of birds that region 1 *cannot* be forest and *must* be

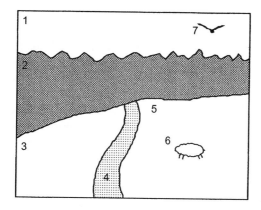

FIGURE 14.10

Regions in a more complex scene: 1, sky; 2, forest; 3, field; 4, road; 5, field; 6, sheep; and 7, bird.

Table 14.3 Adjacency Table for more Complex Scene

	Sky	Forest	Field	Road	Sheep	Bird
Sky	—	√	—	—	—	√
Forest	√	—	√	√	—	—
Field	—	√	—	√	√	—
Road	—	√	√	—	—	—
Sheep	—	—	√	—	—	—
Bird	√	—	—	—	—	—

This table is relevant to the scene in Figure 14.10. Ticks again indicate adjacency and dashes indicate nonadjacency.

sky. Here logic appears to be a stronger arbiter of correctness than an evidence building scheme such as the maximal clique technique. However, on the plus side, the maximal clique technique will take contradictory evidence and produce the best possible response. For example, if noise appears and introduces an invalid "bird" in the forest, it will simply ignore it, whereas a purely logic-based scheme will be unable to find a satisfactory solution: clearly, much depends on the accuracy and universality of the assertions on which these schemes are based.

Logical problems such as the ones outlined above are commonly tackled with the aid of declarative languages such as PROLOG, in which the rules are written explicitly in a standard form not dissimilar to IF statements in English. PROLOG is not discussed in detail here due to the space limitation. However, note that

PROLOG is basically designed to obtain a single "logical" solution where one exists. It can also be instructed to search for all available solutions or for any given number of solutions. When instructed to search for all solutions, it will finally arrive at the same main set of solutions as the maximal clique approach (although, as indicated above, noise can make the situation more complicated). Thus, despite their apparent differences, these approaches are quite similar: it is the implementation of the underlying search problem that is different (PROLOG uses a "depth-first" search strategy (see Section 14.7)). Next, we move on to another scheme—that of relaxation labeling—which has a quite different formulation.

Relaxation labeling is an iterative technique in which evidence is gradually built up for the proper labeling of the solution space—in this case, of the various regions in the scene. Two main possibilities exist. In the first, called *discrete relaxation*, evidence for pairs of labels is examined and a label discarding rule is instigated that eliminates at each iteration those pairs of labels that are currently inconsistent; this technique is applied until no further change in the labeling occurs. The second possibility is that of *probabilistic relaxation*; here the labels are set up numerically to correspond to the probabilities of a given interpretation of each region, i.e., a table is compiled of regions against possible interpretations, each entry being a number representing the probability of that particular interpretation. After providing a suitable set of starting probabilities (possibly all being weighted equally), these are updated iteratively, and in the ideal case, they converge on the value 0 or 1 to give a unique interpretation of the scene. Unfortunately, convergence is by no means guaranteed and can depend on the starting probabilities and the updating rule. A prearranged constraint function defines the underlying process, and this could in principle be either better or worse at matching reality than (for example) the logic programming techniques that are used in PROLOG. Relaxation labeling is a complex optimization process and is not discussed further here. Suffice it to say that the technique often runs into problems of excessive computation. The reader is referred to the seminal paper by Rosenfeld et al. (1976) and other papers mentioned in Section 14.9 for further details.

14.7 SEARCH

The above sections have shown how the maximal clique approach may be used to locate objects in an image, or alternatively to label scenes according to predefined rules about what arrangements of regions are expected in scenes. In either case, the basic process being performed is that of search for solutions that are compatible with the observed data. This search takes place in assignment space, i.e., a space in which all combinations of assignments of observed features with possible interpretations exist. The problem is that of finding one or more valid sets of observed assignments.

It generally happens that the search space is very large, so that an exhaustive search for all solutions would involve enormous computational effort and would take considerable time. Unfortunately, one of the most obvious and appealing methods of obtaining solutions, the maximal clique approach, is NP-complete and can require impracticably large amounts of time to find all the solutions. It is therefore useful to clarify the nature of the maximal clique approach: to achieve this, we first describe the two main categories of search—breadth-first search and depth-first search.

Breadth-first search is a form of search that systematically works down a tree of possibilities, never taking shortcuts to nearby solutions. Depth-first search, in contrast, involves taking as direct a path as possible to individual solutions, stopping the process when a solution is found and backtracking up the tree whenever a wrong decision is found to have been made. It is normal to curtail the depth-first search when sufficient solutions have been found and this means that much of the tree of possibilities will not have been explored. Although breadth-first search can be curtailed similarly when enough solutions have been found, the maximal clique approach as described earlier is in fact a form of breadth-first search that is exhaustive and runs to completion.

In addition to being an exhaustive breadth-first search, the maximal clique approach may be described as being "blind" and "flat"—i.e., it involves neither heuristic nor hierarchical means of guiding the search. In fact, faster search methods involve guiding the search in various ways. First, heuristics are used to specify at various stages in which direction to proceed (which node of the tree to expand), or which paths to ignore (which nodes to prune). Second, the search can be made more "hierarchical," so that it searches first for outline features of a solution, returning later (perhaps in several stages) to fill in the details. Details of these techniques are omitted here. However, an interesting approach was used by Rummel and Beutel (1984): they searched images for industrial components using features such as corners and holes, alternating at various stages between breadth-first and depth-first search by using a heuristic based on a dynamically adjusted parameter: this being computed on the basis of how far the search is still away from its goal and the quality of the fit so far. Rummel and Beutel noted the existence of a tradeoff between speed and accuracy as a "guide factor," based on the number of features required for recognition, is adjusted—the problem being that trying to increase speed introduces some risk of not finding the optimum solution.

14.8 CONCLUDING REMARKS

This chapter has discussed the problem of recognizing objects by their features, and has also considered the related task of scene analysis. The maximal clique approach is seen to be capable of finding solutions to both of these tasks, although ultimately these are search problems, so a much greater range of methods is applicable to each. In particular, blind, flat exhaustive breadth-first search (i.e., the maximal clique method) involves considerable computation and is often best replaced by guided depth-first search, with suitable heuristics being devised to guide the search. In addition, languages such as PROLOG can implement depth-first search, and rather different procedural techniques such as relaxation labeling are available and should be considered (although these are subject to their own complexities and computational problems).

The task of recognizing objects by their features tends to involve considerable computation and the GHT can in some cases provide a satisfactory solution to these problems. When this happens, it is because the graph theory representation is not well matched to the relevant real-space template matching task in the way that the GHT is. Here, recall what was noted in Chapter 13—that the GHT is particularly well suited to object detection in real space as it is one type of spatial matched filter. Indeed, the maximal clique approach can be regarded as a rather inefficient substitute for the GHT form of the spatial matched filter. Furthermore, the LFF method takes a shortcut to save computation and this makes it less like a spatial matched filter, thereby adversely affecting its ability to detect objects among noise and clutter. Note that for scene analysis, where relational descriptors rather than precise dimensional (binary) attributes are involved, the GHT does not provide any very obvious possibilities—probably because we are here dealing with much more abstract data that are linked to real space only at a very high level (but see the work of Kasif et al., 1983).

Finally, note that, on an absolute scale, the graph matching approach takes very little note of detailed image structure, using at most only pairwise feature attributes. This is adequate for 2-D image interpretation but inadequate for situations such as 3-D image analysis where there are more degrees of freedom to contend with (normally three degrees of freedom for position and three for orientation, for each object in the scene). Hence, more specialized and complex approaches need to be taken in such cases: these are examined in Part 3.

> Searching for objects *via* their features is far more efficient than template matching. This chapter has shown that this raises the need to *infer* the presence of objects—a process that can still be computation intensive. Graph matching and generalized Hough transform approaches are robust, although each can lead to ambiguities, so tests of potential solutions need to be made.

14.9 BIBLIOGRAPHICAL AND HISTORICAL NOTES

Graph matching and clique finding algorithms started to appear in the literature around 1970: for an early solution to the graph isomorphism problem, see Corneil and Gottlieb (1970). The subgraph isomorphism problem was tackled soon after by Barrow et al. (1972): see also Ullmann (1976). The double subgraph isomorphism (or subgraph—subgraph isomorphism) problem was commonly tackled by seeking maximal cliques in the match graph, and algorithms for achieving this have been described by Bron and Kerbosch (1973), Osteen and Tou (1973), and Ambler et al. (1975) (note that in 1989, Kehtarnavaz and Mohan reported preferring the algorithm of Osteen and Tou on the grounds of speed). Improved speed has also been achieved using the minimal match graph concept (Davies, 1991a).

Bolles (1979) applied the maximal clique technique to real-world problems (notably the location of engine covers) and showed how operation could be made more robust by taking additional features into account. By 1982, Bolles and Cain had formulated the local-feature-focus method, which (a) searches for restricted sets of features on an object, (b) takes symmetry into account to save computation, and (c) reconsiders the original image data in order to confirm a valid match: the paper gives various criteria for ensuring satisfactory solutions with this type of method.

Not satisfied with the speed of operation of maximal clique methods, other workers have tended to use depth-first search techniques. Rummel and Beutel (1984) developed the idea of alternating between depth-first and breadth-first search as dictated by the data—a powerful approach, although the heuristics that they used for this may well lack generality. Meanwhile, Kasif et al. (1983) showed how a modified GHT (the "relational HT") could be used for graph matching, although their paper gives few practical details. A somewhat different application of the GHT to perform 2-D matching was described in Section 14.4, and has been extended to optimize accuracy (Davies, 1992c). Geometric hashing has been developed to perform similar tasks on objects with complex polygonal shapes (Tsai, 1996).

Relaxation labeling in scenes dates from the seminal paper by Rosenfeld et al. (1976); for later work on relaxation labeling and its use for matching, see Kitchen and Rosenfeld (1979), Hummel and Zucker (1983), and Henderson (1984); for rule-based methods in image understanding, see for example, Hwang et al. (1986); and for preparatory discussion and careful contrasting of these approaches, see Ballard and Brown (1982).

Over the past decades, inexact matching algorithms have acquired increasing predominance over exact matching methods, because of the ubiquitous presence of noise, distortions, and missing or added feature points, together with inaccuracies and thus mismatches of feature attributes. One class of work on inexact (or "error-tolerant") matching considers how structural representations should be compared (Shapiro and Haralick, 1985); this early work on similarity measures shows how the concept of "string edit distance" can be applied to graphical structures (Sanfeliu and Fu, 1983); the formal concept of edit distance was later extended by Bunke and Shearer (1998) and Bunke (1999), who considered and rationalized the cost functions for methods such as graph isomorphisms, subgraph isomorphisms, and maximum common subgraph isomorphisms: choice of cost functions was shown to be of crucial importance to success in each particular dataset, although detailed analysis demonstrated important subtleties in the situation (Bunke, 1999).

Yet another class of work is that on optimization. This has included work on simulated annealing (Herault et al., 1990), genetic search (Cross et al., 1997), and neural processing (Pelillo, 1999). The work of Umeyama (1988) develops the least squares approach using a matrix eigendecomposition method to recover the permutation matrix relating the two graphs being matched. One of the most recent developments has been the use of spectral graph theory[3] to recover the permutation structure. In fact, the Umeyama (1988) approach only matches graphs of the same size. Other related methods have emerged (e.g., Horaud and Sossa, 1995), but they have all suffered from an inability to cope with graphs of different sizes. However, Luo and Hancock (2001) have demonstrated how this particular problem can be overcome—by showing how the graph matching task can be posed as maximum likelihood estimation using the EM algorithm formalism. Hence, singular value decomposition is used efficiently to solve correspondence problems. Ultimately, the method is important because it helps to move graph matching away from a discrete process in which a combinatorial search problem exists toward a continuous optimization problem, which moves systematically towards the optimum solution. It ought to be added that the method works under considerable levels of structural corruption—such as when 50% of the initial entries in the data-graph adjacency matrix are in error (Luo and Hancock, 2001). In a later development, Robles-Kelly and Hancock (2002) managed to achieve the same end, and to achieve even better performance within the spectral graph formalism itself.

Meanwhile, other developments included a fast, phased approach to inexact graph matching (Hlaoui and Wang, 2002), a reproducible kernel Hilbert space (RKHS) interpolator-based graph matching algorithm capable of efficiently matching huge graphs of more than 500 vertices (e.g., those extracted from aerial scenes) on a PC (van Wyk et al., 2002). For a more detailed appraisal of inexact matching algorithms, see Lladós et al. (2001): note that the latter appears in a special section of IEEE Trans. PAMI on *Graph Algorithms and Computer Vision* (Dickinson et al., 2001).

14.9.1 More Recent Developments

Silletti et al. (2011) have devised a variant approach to spectral graph matching in which new similarity measures are applied. The approach permits application to a variety of types of image and yields results that are said to show significant improvements over certain preexisting methods. Gope and Kehtarnavaz (2007) have demonstrated a new method for affine matching between planar point sets. The method makes use of the convex hulls of the point sets and performs matching between them: this is a useful approach because (a) convexity is affine invariant and (b) use of the convex hull is intrinsically robust. Property (b) follows from the fact that convex hulls are only locally altered by point perturbations including insertions and deletions. The method makes use of an enhanced

---

[3]This subject involves analysis of the structural properties of graphs using the eigenvalues and eigenvectors of the adjacency matrix.

modified Haussdorff distance and achieves better results in the presence of noise and occlusion than a number of standard methods. Aguilar et al. (2009) have developed a new "graph transformation matching" algorithm to match points between pairs of images. It validates each match through the spatial configuration of the points by constructing a k-nearest-neighbor graph for each image; vertices that introduce structural dissimilarity between the graphs are iteratively eliminated, thereby yielding a consensus graph representing a correct set of point matches between the images.

14.10 PROBLEMS

1. Find the match graph for a set of features arranged in the form of an isosceles triangle. Find how much simplification occurs by taking account of symmetry and using the symmetry-reduced match graph. Extend your results to the case of a kite (two isosceles triangles arranged symmetrically base to base).

2. Two lino-cutter blades (trapeziums) are to be located from their corners. Consider images in which two corners of one blade are occluded by the other blade. Sketch the possible configurations, counting the number of corners in each case. If corners are treated like point features with no other attributes, show that the match graph will lead to an ambiguous solution. Show further that the ambiguity can in general be eliminated if proper account is taken of corner orientation. Specify how accurately corner orientation would need to be determined for this to be possible.

3. In Problem 2, would the situation be any better if the GHT were used?

4. **a.** Metal flanges are to be located from their holes using a graph matching (maximal clique) technique. Each bar has four identical holes at distances from the narrow end of the bar of 1, 2, 3, and 5 cm, as shown in Fig. 14.11. Draw match graphs for the four different cases in which one of the four holes of a given flange is obscured: determine in each case whether the method is able to locate the metal flange without any error, and whether any ambiguity arises.

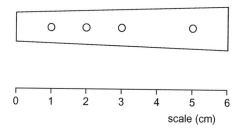

FIGURE 14.11

Metal flanges for location using the GHT.

b. Do your results tally with the results for human perception? How would any error or ambiguity be resolved in practical situations?

5. a. Describe the maximal clique approach to object location. Explain why the largest maximal clique will normally represent the most likely solution to any object location task.

b. If symmetrical objects with four feature points are to be located, show that suitable labeling of the object template will permit the task to be simplified. Does the type of symmetry matter? What happens in the case of a rectangle? What happens in the case of a parallelogram? (In the latter case, see points A, B, C, and D in Fig. 14.12.)

c. A nearly symmetrical object with *five* feature points (see Fig. 14.12) is to be located. This is to be achieved by looking initially for the feature points A, B, C, and D and ignoring the fifth point E. Discuss how the fifth point may be brought into play to finally determine the orientation of the object, using the maximal clique approach. What disadvantage might there be in adopting this two-stage approach?

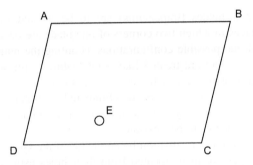

FIGURE 14.12

Object with five feature points.

6. a. What is template matching? Explain why objects are normally located by their features rather than using whole object templates. What are the features that are commonly used for this purpose?

b. Describe templates that can be used for corner and hole detection.

c. An improved type of lino-cutter blade (Fig. 14.13) is to be placed into packs of six by a robot. Show how the robot vision system could locate the blades *either* from their corners *or* from their holes by applying the maximal clique method (i.e., show that *both* schemes would work).

d. After a time it appears that the robot is occasionally confused when the blades overlap. It is then decided to locate the blades from their holes *and* their corners. Show why this helps to eliminate any confusion. Show also how finally distinguishing the corners from the holes can help in extreme cases of overlap.

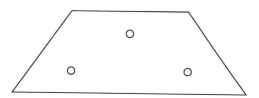

FIGURE 14.13

Symmetrical lino-cutter blade.

7. a. A certain type of lino-cutter blade has four corners and two fixing holes (Fig. 14.14). Blades of this type are to be located using the maximal clique technique. Assume the objects lie on a worktable and that they are viewed orthogonally at a known distance.

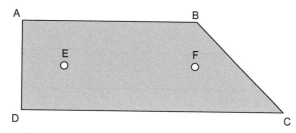

FIGURE 14.14

Non-symmetrical lino-cutter blade.

b. Draw match graphs for the following situations:
  i. The objects are to be located by their holes and their corners, regarding these as *indistinguishable* point features.
  ii. The objects are to be located solely by their corners (i.e., matching corners in the image with corners on an idealized object).
  iii. The objects are to be located solely by their holes.
  iv. The objects are to be located by their holes *and* their corners, but these are to be regarded as *distinguishable* features.
c. Discuss your results with particular reference to:
  i. the robustness that can be achieved.
  ii. the speed of computation.
d. In the latter case, distinguish the time taken to build the basic match graph from the time taken to find all the maximal cliques in it. State any assumptions you make about the time taken to find a maximal clique of m nodes in a match graph of n nodes.
8. a. Decorative biscuits are to be inspected after first locating them from their holes. Show how the maximal clique graph matching technique can be applied to identify and locate the biscuits shown in Fig. 14.15(a), which are of the same size and shape.

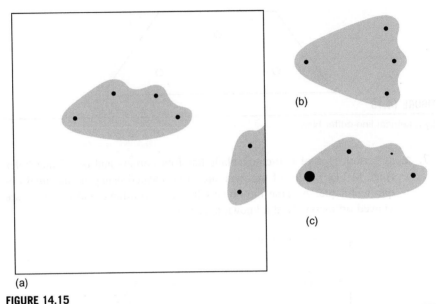

(a)

(b)

(c)

FIGURE 14.15

Decorative biscuits for inspection.

b. Show how the analysis will be affected for biscuits that have an axis of symmetry, as shown in Fig. 14.15(b). Show also how the technique may be modified to simplify the computation for such a case.

c. A more detailed model of the first type of biscuit shows it has holes of *three* sizes, as shown in Fig. 14.15(c). Analyze the situation, and show that a much simplified match graph can be produced from the image data, leading to successful object location.

d. A further matching strategy is devised to make use of the hole size information: matches are *only* shown in the match graph if they arise between pairs of holes of *different* sizes. Determine how successful this strategy is, and discuss whether it is likely to be generally useful, e.g., for objects with increased numbers of features.

e. Work out an optimal object identification strategy, which will be capable of dealing with cases where holes and/or corners are to be used as point features, the holes might have different sizes, the corners might have different angles and orientations, the object surfaces might have different colors or textures, and objects might have larger numbers of features. Make clear what the term "optimal" should be taken to mean in such cases.

9. a. Figure 14.16 shows a 2-D view of a widget with four corners. Explain how the maximal clique technique can be used to locate widgets even if they are partly obscured by various types of object including other widgets.

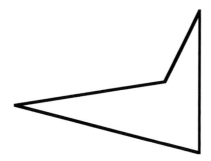

FIGURE 14.16

Diagram of a widget.

b. Explain why the basic algorithm will not distinguish between widgets that are normally presented from those that are upside down. Consider how the basic method could be extended to ensure that a robot only picks up those that are the right way up.

c. The camera used to view the widgets is accidentally jarred and then reset at a different, unknown height above the worktable. State clearly why the usual maximal clique technique will now be unable to identify the widgets. Discuss how the overall program could be modified to make sense of the data and make correct interpretations in which all the widgets are identified. Assume *first* that widgets are the only objects appearing in the scene, and *second* that a variety of other objects may appear.

d. The camera is jarred again and this time is set at a small, unknown angle to the vertical. To be sure of detecting such situations and of correcting for them, a flat calibration object of known shape is to be stuck on the worktable. Decide on a suitable shape and explain how it should be used to make the necessary corrections.

10. Show that flat convex shapes remain convex under affine transformations.

3-D Vision and Motion

3

Part 3 covers the developments needed for an understanding of real scenes, which necessarily contain 3-D objects—a number of which may be in motion. 3-D vision is considerably more complex than 2-D vision, not least because the number of degrees of freedom of an object will typically have increased from three to six, with an accompanying combinatorial increase in the number of scene configurations to be considered.

This part of the book starts (Chapter 15) by airing the problems, before considering the complexities of full perspective projection (Chapter 16). Next, it is useful to see what short cuts can be achieved by taking invariants into account (Chapter 17). Chapter 18 not only deals with camera calibration but also shows how recent research has attempted to avoid the need for explicit calibration by making careful computations that interrelate multiple scenes: here the emphasis is on taking opportunities that permit some of the complexities to be by-passed. Finally, Chapter 19 examines the problems of motion in the context of 3-D vision.

The Three-Dimensional World

15

3-D VISION

Humans are able to employ 3-D vision with consummate ease, and according to conventional wisdom, binocular vision is the key to this success. The truth is more complex than this, and this chapter demonstrates why.

Look out for:

- what can be achieved using binocular vision.
- how the shading of surfaces can be used in place of binocular vision to achieve similar ends.
- how these basic methods provide dimensional information for 3-D scenes but do not immediately lead to object recognition.
- how the process of 3-D object recognition can be tackled by studies of 3-D geometry.

Note that this is an introductory chapter on 3-D vision, designed to give the flavor of the subject and to show its origins in human vision. It will be followed by the other four chapters (Chapters 16–19) that comprise Part 3 of this volume.

At a more detailed level, notice the importance of the epipolar line approach in solving the correspondence problem. The concept is deservedly taken considerably further in Chapter 18, in conjunction with the required mathematical formulation.

15.1 INTRODUCTION

In the foregoing chapters, it has generally been assumed that objects are essentially flat and are viewed from above in such a way that there are only three degrees of freedom—namely, the two associated with position and a further one concerned with orientation. While this approach was adequate for carrying out many useful visual tasks, it is inadequate for interpreting outdoor or factory scenes or even for helping with quite simple robot assembly and inspection tasks. Indeed, over the past few decades a considerable amount of quite sophisticated theory has been developed and backed up by experiment, to find how scenes composed of real 3-D objects can be understood in detail.

In general, this means attempting to interpret scenes in which objects may appear in totally arbitrary positions and orientations—corresponding to six degrees of freedom. Interpreting such scenes, and deducing the translation and orientation parameters of arbitrary sets of objects, takes a substantial amount of computation—partly because of the inherent ambiguity in inferring 3-D information from 2-D images.

A variety of approaches are now available for proceeding with 3-D vision. A single chapter will be unable to describe all of them but the intention here is to provide an overview, outlining the basic principles and classifying the methods according to generality, applicability, and so on. While computer vision need not necessarily mimic the capabilities of the human eye–brain system, much research on 3-D vision has been aimed at biological modeling. This type of research shows that the human visual system makes use of a number of different methods simultaneously, taking appropriate cues from the input data and forming hypotheses about the content of a scene, progressively enhancing these hypotheses until a useful working model of what is present is produced. Thus, individual methods are not expected to work in isolation: rather, they need to provide the model generator with whatever data become available. Clearly, biological machinery of various types will lie idle for much of the time until triggered by specific input stimuli. Computer vision systems are currently less sophisticated than this and tend to be built on specific processing models, so that they can be applied efficiently to more restricted types of image data. In this chapter, we adopt the pragmatic view that particular methods need to be (or have been) developed for specific types of situation, and that they should be used only when appropriate—although some care is taken to elucidate what the appropriate types of applications are.

15.2 3-D VISION—THE VARIETY OF METHODS

One of the most obvious characteristics of the human visual system is that it employs two eyes, and it is well known to the layman that binocular (or "stereo") vision permits depth to be discerned within a scene. However, the loss of vision that results when one eye is shut is relatively insignificant and is by no means a disqualification from driving a car or even an aeroplane. On the contrary, depth can readily be deduced in monocular vision from a plethora of cues that are buried in an image. Naturally, to achieve this, the eye–brain system is able to call on a huge amount of pre-stored data about the physical world and about the types of object in it, be they man-made or natural entities. For example, the size of any car being viewed is strongly constrained; likewise, most objects have highly restricted sizes, both absolutely and in their depths relative to their frontal dimensions. Nevertheless, in a single view of a scene, it is normally impossible to deduce absolute sizes—all the objects and their depths can be scaled up or down by arbitrary factors and this cannot be discerned from a monocular view.

While it is clear that the eye−brain system makes use of a huge database relating to the physical world, there is much that can be learned with negligible prior knowledge, even from a single monocular view. The main key to this is the "shape from shading" concept. For 3-D shape to be deducible from shading information (i.e., from the grayscale intensities in an image), something has to be known about how the scene is lit—the simplest situation being when the scene is illuminated by a single point light source at a known position. Note that indoors a single overhead tungsten light is still the most usual illuminator, while outdoors the sun performs a similar function. In either case, an obvious result is that a single source will illuminate one part of an object and not another—which then remains in shadow—and parts that are orientated in various ways relative to the source and the observer appear with different brightness values, so that orientation can in principle be deduced. In fact, as will be seen below, deduction of orientation and position is not at all trivial and may even be ambiguous. Nevertheless, successful methods have been developed for carrying out this task. One problem that often arises is that the position of the light source is unknown but this information can generally be extracted (at least by the eye) from the scene being examined, so a bootstrapping procedure is then able to unlock the image data gradually and proceed to an interpretation.

While these methods enable the eye to interpret real scenes, it is difficult to say quite to what degree of precision they are carried out. With computer vision, the required precision levels are liable to be higher, although the machine will be aided by knowing exactly where the source of illumination is. However, with computer vision, we can go further and arrange artificial lighting schemes that would not appear in nature, so the computer can acquire an advantage over the human visual system. In particular, a set of light sources can be applied in sequence to the scene—an approach known as photometric stereo—which can in certain cases help the computer to interpret the scene more rigorously and efficiently. In other cases, structured light may be applied. This means projecting onto the scene a pattern of spots or stripes, or even a grid of lines, and measuring their positions in the resulting image. By this means the depth information can be obtained much as for pairs of stereo images.

Finally, a number of methods have been developed for analyzing images on the basis of readily identifiable sets of features. These methods are the 3-D analogs of the graph matching and GHT approaches of Chapter 14. However, they are significantly more complex because they generally involve six degrees of freedom in place of the three assumed throughout Chapter 14. It should also be noted that such methods make strong assumptions about the particular objects to be located within the scene. In general situations, it is unlikely that such assumptions could be made, and so initial analysis of any images must be made on the basis that the entire scene must be mapped out in 3-D, then 3-D models built up, and finally deductions must be made by noticing what relation one part of the scene bears to another part. Note that if a scene is composed from an entirely new set of objects, all that can be done is to *describe* what is present and say perhaps what the set most closely *resembles*: recognition *per se* cannot be performed. Note that scene analysis is—at least from a single monocular image—an inherently ambiguous process: every scene can have a number of possible interpretations and there is evidence that the eye looks for the simplest and most probable explanation rather than an absolute interpretation. Indeed, it is underlined by the many illusions to which the eye–brain system is subject, that decisions must repeatedly be made concerning the most likely interpretation of a scene and that there is some risk that its internal model builder will lock on to an interpretation or part-interpretation that is suboptimal (see the paintings of Escher).

This section has indicated that methods of 3-D vision can be categorized according to whether they start by mapping out the shapes of objects in 3-D space and then attempt to interpret the resulting shapes, or whether they try to identify objects directly from their features. In either case, a knowledge base is ultimately called for. It has also been seen that methods of mapping objects in real space include monocular and binocular methods, although structured lighting can help to offset the deficiencies of employing a single "eye." Laser scanning and ranging techniques must also be included in methods of 3-D mapping, although space precludes detailed discussion of these techniques in this book.

15.3 PROJECTION SCHEMES FOR THREE-DIMENSIONAL VISION

It is common in engineering drawings to provide three views of an object to be manufactured—the plan, the side view, and the elevation. Traditionally these views are simple orthographic (nondistorting) projections of the object—i.e., they are made by taking sets of parallel lines from points on the object to the flat plane on which it is being projected.

However, when objects are viewed by eye or from a camera, rays converge to the lens and so images formed in this way are subject not only to change of scale but also to perspective distortions (Fig. 15.1). This type of projection is called perspective projection, although it includes orthographic projection as the special case of viewing from a distant point. Unfortunately, perspective projections have the disadvantage that they tend to make objects appear more complex than they really are by destroying simple relationships between their features. Thus, parallel edges no longer appear parallel and midpoints no longer appear as such (although

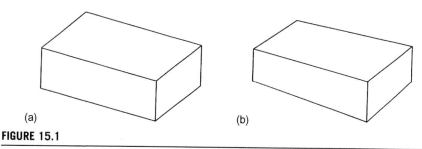

(a)   (b)

FIGURE 15.1

(a) Image of a rectangular box taken using orthographic projection; (b) the same box taken using perspective projection. In (b), note that parallel lines no longer appear parallel, although paradoxically the box appears more realistic.

many useful geometric properties still hold—e.g., a tangent line remains a tangent line and the order of points on a straight line remains unchanged).

In outdoor scenes, it is very common to see lines that are known to be parallel apparently converging toward a vanishing point on the horizon line (Fig. 15.2). In fact, the horizon line is the projection onto the image plane of the line at infinity on the ground plane G: it is the set of all possible vanishing points for parallel lines on G. In general, the vanishing points of a plane P are the projections onto the image plane corresponding to points at infinity in different directions on P. Thus, any plane Q within the field of view may have vanishing points in the image plane, and these will lie on a vanishing line, which is the analog of the horizon line for Q.

Figure 15.3(a) shows how an image is projected into the image plane by a convex (eye or camera) lens at the origin. It is inconvenient to have to consider inverted images and it is a commonly used convention in image analysis to set the center of the lens at the origin (0, 0, 0) and to imagine the image plane to be the plane $Z = f$, f being the focal length of the lens; with this simplified geometry (Fig. 15.3(b)), images in the image plane appear noninverted. Taking a general point in the scene as (X, Y, Z), which appears in the image as (x_1, y_1), perspective projection now gives:

$$(x_1, y_1) = \left(\frac{fX}{Z}, \frac{fY}{Z}\right) \tag{15.1}$$

15.3.1 Binocular Images

Figure 15.4 shows the situation when two lenses are used to obtain a stereo pair of images. In general, the two optical systems do not have parallel optical axes but exhibit a "vergence" (which may be variable, as it is for human eyes), so that they intersect at some point within the scene. Then a general point (X, Y, Z) in the scene has two different pairs of coordinates, (x_1, y_1) and (x_2, y_2), in its two images, which differ both because of the vergence between the optical axes and

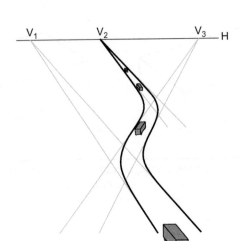

FIGURE 15.2

Vanishing points and the horizon line. This figure shows how parallel lines on the ground plane appear, under perspective projection, to meet at vanishing points V_i on the horizon line H. (Note that V_i and H lie in the *image* plane.) If two parallel lines do not lie on the ground plane, their vanishing point will lie on a different vanishing line. Hence, it should be possible to determine whether any roads are on an incline by computing all the vanishing points for the scene.

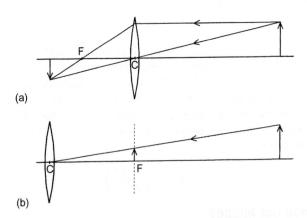

FIGURE 15.3

(a) Projection of an image into the image plane by a convex lens; note that a single image plane only brings objects at a single distance into focus but that for far-off objects the image plane may be taken to be the focal plane, a distance F from the lens; (b) a commonly used convention that imagines the projected image to appear noninverted at a focal plane F in front of the lens. The center of the lens is said to be the center of projection for image formation.

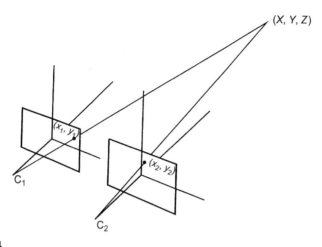

FIGURE 15.4

Stereo imaging using two lenses. The axes of the optical systems are parallel, i.e., there is no "vergence" between the optical axes.

because the baseline b between the lenses causes relative displacement or "disparity" of the points in the two images.

For simplicity, we now take the vergence to be zero, i.e., the optic axes are parallel. Then, with suitable choice of Z-axis on the perpendicular bisector of the baseline b, we obtain two equations:

$$x_1 = \frac{(X + b/2)f}{Z} \tag{15.2}$$

$$x_2 = \frac{(X - b/2)f}{Z} \tag{15.3}$$

so that the disparity is

$$D = x_1 - x_2 = \frac{bf}{Z} \tag{15.4}$$

Rewriting this equation in the form:

$$Z = \frac{bf}{x_1 - x_2} \tag{15.5}$$

now permits the depth Z to be calculated. In fact, computation of Z only requires the disparity for a stereo pair of image points to be found and parameters of the optical systems to be known. However, confirming that both points in a stereo pair actually correspond to the same point in the original scene is in general not at all trivial, and much of the computation in stereo vision is devoted to this task. In addition, to obtain good accuracy in the determination of depth, a large baseline b is required. Unfortunately as b is increased, the correspondence between the images decreases, so it becomes more difficult to find matching points.

15.3.2 The Correspondence Problem

There are two important approaches to finding pairs of points that match in the two images of a stereo pair. One is that of "light striping" (one form of structured lighting), which encodes the two images so that it is easy to see pairs of corresponding points. If a single vertical stripe is used, for every value of y there is in principle only one light stripe point in each image and so the matching problem is solved. We return to this problem in a later section.

The second important approach is to employ epipolar lines. To understand this approach, imagine that we have located a distinctive point in the first image and that we are marking all possible points in the object field which could have given rise to it. This will mark out a line of points at various depths in the scene and, when viewed in the second image plane, a locus of points can be constructed in that plane. This locus is the *epipolar line* corresponding to the original image point in the alternate image (Fig. 15.5). If we now search along the epipolar line for a similarly distinctive point in the second image, the chance of finding the correct match is significantly enhanced. This method has the advantage not only of cutting down the amount of computation required to find corresponding points, but also of reducing significantly the chance of false alarms. Note that the concept of an epipolar line applies to both images—a point in one image gives an epipolar line in the other image. Note also that in the simple geometry of Fig. 15.4, all epipolar lines are parallel to the x-axis, although this is not so in general (in fact, the general situation is that all epipolar lines in one image plane pass through the point that is the image of the projection point of the alternate image plane).

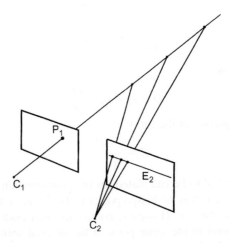

FIGURE 15.5

Geometry of epipolar lines. A point P_1 in one image plane may have arisen from any one of a line of points in the scene, and may appear in the alternate image plane at any point on the so-called epipolar line E_2.

The correspondence problem is rendered considerably more difficult by the fact that there will be points in the scene that give rise to points in one image but not in the other. Such points are either occluded in the one image, or else are so distorted as not to give a recognizable match in the two images (e.g., the different background might mask a corner point in one image while permitting it to stand out in the other). Any attempt to match such points can then only lead to false alarms. Thus, it is necessary to search for consistent sets of solutions in the form of continuous object surfaces in the scene. For this reason, iterative "relaxation" schemes are widely used to implement stereo matching.

Broadly speaking, correspondences are sought by two methods: one is the matching of near-vertical edge points in the two images (near-horizontal edge points do not give the required precision); the other is the matching of local intensity patterns using correlation techniques. Correlation is an expensive operation and in this case is relatively unreliable—principally because intensity patterns frequently appear significantly foreshortened[1] in one or other image and hence are difficult to match reliably. In such cases, the most practical solution is to reduce the baseline; as noted earlier, this has the effect of reducing the accuracy of depth measurement. Further details of these techniques are to be found in Shirai (1987).

Before leaving this topic, we consider in slightly more detail how the problems of visibility mentioned above arise. Figure 15.6 shows a situation in which an object is being observed by two cameras giving stereo images. Clearly, much of the object will not be visible in either image because of self-occlusion, while some feature points will only be visible in one or the other image. Now, consider the order in which the points appear in the two images (Fig. 15.7). The points that are visible appear in the same order as in the scene, and the points that are just going out of sight are those for which the order between the scene and the image is just about to change. Points that provide information about the front surface of the object can thus only bear a simple geometrical relation to each other: in particular, for points not to obscure, or be obscured by, a given point P, they must not lie within a double angular sector defined by P and the centers of projection C_1, C_2 of the two cameras. This region is shown shaded in Fig. 15.7. A surface passing through P for which full depth information can be retrieved must lie entirely within the nonshaded region. (Of course, a new double sector must be considered for each point on the surface being viewed.) Note that the possibility of objects containing holes, or having transparent sections, must not be forgotten (such cases can be detected from differences in the ordering of feature points in the two views—see Fig. 15.7); neither must it be ignored that the foregoing figures represent a single horizontal cross-section of an object that can have totally different shapes and depths in different cross-sections.

---

[1]That is, distorted by the effects of perspective.

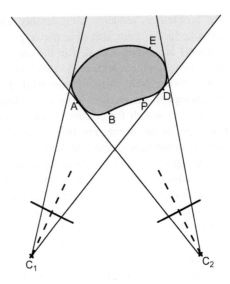

FIGURE 15.6

Visibility of feature points in two stereo views. Here an object is viewed from two directions. Only feature points that appear in both views are of value for depth estimation. This eliminates all points in the shaded region, such as E, from consideration.

15.4 SHAPE FROM SHADING

It was mentioned in Section 15.2 that it is possible to analyze the pattern of intensities in a single (monocular) image and to deduce the shapes of objects from the shading information. The principle underlying this technique is that of modeling the reflectance of objects in the scene as a function of the angles of incidence i and emergence e of light from their surfaces. In fact, a third angle is also involved, and it is called the "phase" g (Fig. 15.8).

A general model of the situation gives the radiance I (light intensity in the image) in terms of the irradiance E (energy per unit area falling on the surface of the object) and the reflectance R:

$$I(x_1, y_1) = E(x, y, z)R(\mathbf{n}, \mathbf{s}, \mathbf{v}) \tag{15.6}$$

It is well known that a number of matt surfaces approximate reasonably well to an ideal Lambertian surface whose reflectance function depends only on the angle of incidence i—i.e., the angles of emergence and phase are immaterial:

$$I = \left(\frac{1}{\pi}\right) E \cos i \tag{15.7}$$

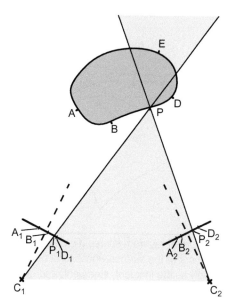

FIGURE 15.7

Ordering of feature points on an object. In the two views of the object shown here, the feature points all appear in the same order A, B, P, D as on the surface of the object. Points for which this would not be valid, such as E, are behind the object and are obscured from view. Relative to a given visible feature P, there is a double sector (shaded) in which feature points must not appear if they are not to obscure the feature under consideration. An exception to these rules might be if the object had a semi-transparent window through which an additional feature T were visible: in that case interpretation would be facilitated by noting that the orderings of the features seen in the two views were different—e.g., A_1, T_1, B_1, P_1, D_1 and A_2, B_2, T_2, P_2, D_2.

For the present purpose, E is regarded as a constant and is combined with other constants for the camera and the optical system (including, e.g., the f-number). In this way, a normalized reflectance is obtained, which in this case is:

$$R = R_0 \cos i = R_0 \mathbf{s} \cdot \mathbf{n}$$

$$= \frac{R_0(1 + pp_s + qq_s)}{(1 + p^2 + q^2)^{1/2}(1 + p_s^2 + q_s^2)^{1/2}} \tag{15.8}$$

where we have used the standard convention of writing orientations in 3-D in terms of p and q values. These are not direction cosines but correspond to the coordinates of the point $(p, q, 1)$ at which a particular direction vector from the origin meets the plane $z = 1$: hence they need suitable normalization, as in Eq. (15.8).

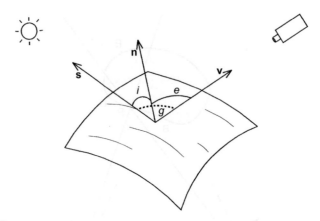

FIGURE 15.8

Geometry of reflection. An incident ray from source direction **s** is reflected along the viewer direction **v** by an element of the surface whose local normal direction is **n**; i, e, and g are defined respectively as the incident, emergent, and phase angles.

Equation (15.8) gives a reflectance map in gradient (p, q) space. We now temporarily set the absolute reflectance value R_0 equal to unity. The reflectance map can be drawn as a set of contours of equal brightness, starting with a point having $R = 1$ at $s = n$, and going down to zero for **n** perpendicular to **s**. When $s = v$, so that the light source is along the viewing direction (here taken to be the direction $p = q = 0$), zero brightness occurs only for infinite distances on the reflectance map $((p^2 + q^2)^{1/2}$ approaching infinity) (Fig. 15.9(a)). In a more general case, when $s \neq v$, zero brightness occurs along a straight line in gradient space (Fig. 15.9(b)). To find the exact shapes of the contours, we can set R at a constant value a, which results in:

$$a(1 + p^2 + q^2)^{1/2}(1 + p_s^2 + q_s^2)^{1/2} = 1 + pp_s + qq_s \qquad (15.9)$$

Squaring this equation clearly gives a quadratic in p and q, which could be simplified by a suitable change of axes. Thus, the contours must be curves of conic section, namely, circles, ellipses, parabolas, hyperbolas, lines, or points (the case of a point arises only when $a = 1$, when we get $p = p_s$, $q = q_s$; and that of a line only if $a = 0$, when we get the equation $1 + pp_s + qq_s = 0$: both of these solutions were implied above).

Unfortunately, object reflectances are not all Lambertian, and an obvious exception is for surfaces that approximate to pure specular reflection. In that case, $e = i$ and $g = i + e$ (**s**, **n**, **v** are coplanar); the only nonzero reflectance position in gradient space is the point representing the bisector of the angle between the

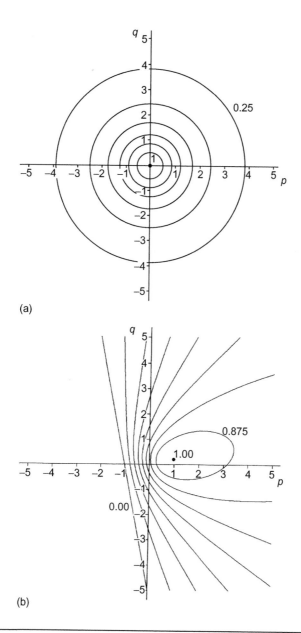

(a)

(b)

FIGURE 15.9

Reflectance maps for Lambertian surfaces: (a) contours of constant intensity plotted in gradient (p, q) space for the case where the source direction **s** (marked by a black dot) is along the viewing direction **v** $(0, 0)$ (the contours are taken in steps of 0.125 between the values shown); (b) the contours that arise where the source direction (p_s, q_s) is at a point (marked by a black dot) in the positive quadrant of (p, q) space: note that there is a well-defined region, bounded by the straight line $1 + pp_s + qq_s = 0$, for which the intensity is zero (the contours are again taken in steps of 0.125).

source direction s (p, q) and the viewing direction v $(0, 0)$—i.e., n is along $s + v$—and very approximately:

$$p \approx \frac{p_s}{2} \tag{15.10}$$

$$q \approx \frac{q_s}{2} \tag{15.11}$$

For less perfect specularity, a peak is obtained around this position. A good approximation to the reflectance of many real surfaces is obtained by modeling them as basically Lambertian but with a strong additional reflectance near the specular reflectance position. Using the Phong (1975) model for the latter component gives:

$$R = R_0 \cos i + R_1 \cos^m \theta \tag{15.12}$$

θ being the angle between the actual emergence direction and the ideal specular reflectance direction.

The resulting contours now have two centers around which to peak: the first is the ideal specular reflection direction ($p \approx p_s/2$, $q \approx q_s/2$) and the second is that of the source direction ($p = p_s$, $q = q_s$). When objects are at all shiny—such as metal, plastic, liquid, or even wood surfaces—the specular peak is quite sharp and rather intense: casual observation may not even indicate the presence of another peak, since Lambertian reflection is so diffuse (Fig. 15.10). In other cases, the specular peak can broaden and become more diffuse: hence it may merge with the Lambertian peak and effectively disappear.

Some remarks should be made about the Phong model employed above. First, it is adapted to different materials by adjusting the values of R_0, R_1, and m. Phong remarks that R_1 typically lies between 10% and 80%, while m is in the range 1–10. However, Rogers (1985) indicates that m may be as high as 50. Note that there is no physical significance in these numbers—the model is simply a phenomenological one. This being so, care should be taken to prevent the $\cos^m \theta$ term from contributing to reflectance estimates when $|\theta| > 90°$. The Phong model is reasonably accurate but has been improved by Cook and Torrance (1982). This is important in computer graphics applications, but the improvement is difficult to apply in computer vision, because of lack of data concerning the reflectances of real objects and because of variability in the current state (cleanliness, degree of polish, etc.) of a given surface. However, the method of photometric stereo gives some possibility of overcoming these problems.

15.5 PHOTOMETRIC STEREO

Photometric stereo is a form of structured lighting that increases the information available from surface reflectance variations. Basically, instead of taking a single monocular image of a scene illuminated from a single source, several images are

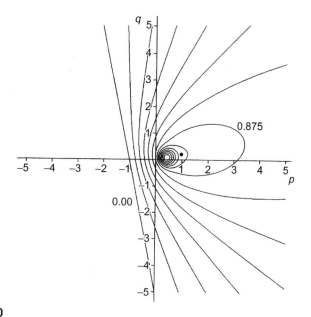

FIGURE 15.10

Reflectance map for a non-Lambertian surface: a modified form of Fig. 15.9(b) for the case where the surface has a marked specular component ($R_0 = 1.0$, $R_1 = 0.8$). Note that the specular peak can have very high intensity (much greater than the maximum value of unity for the Lambertian component). In this case, the specular component is modeled with a $\cos^8\theta$ variation (the contours are again taken in steps of 0.125).

taken, from the same vantage point, with the scene illuminated in turn by separate light sources. These light sources are ideally point sources some distance away in various directions, so that there is in each case a well-defined light source direction from which to measure surface orientation.

The basic idea of photometric stereo is that of cutting down the number of possible positions in gradient space for a given point on the surface of an object. It has already been seen that, for known absolute reflectance R_0, a constant brightness in one image permits the surface orientation to be limited to a curve of conic cross-section in gradient space. This would also be true for a second such image, the curve being a new one if the illuminating source is different. In general, two such conic curves meet at two points, so there is now only a single ambiguity in the gradient of the surface at any given point in the image. To resolve this ambiguity a third source of illumination can be employed (this must not be in the plane containing the first two and the surface point being examined), and the third image gives another curve in gradient space that should pass through the appropriate crossing point of the first two curves (Fig. 15.11). If a third source of

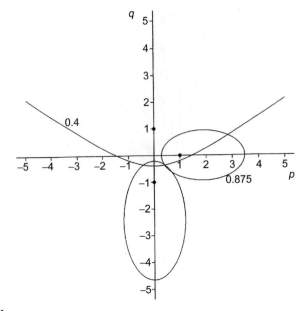

FIGURE 15.11

Obtaining a unique surface orientation by photometric stereo. Three contours of constant intensity arise for different light sources of equal strength: all three contours pass through a single point in (p, q) space and result in a unique solution for the local gradient.

illumination cannot be used, it is sometimes possible to arrange that the inclination of each of the sources is so high that $(p^2 + q^2)^{1/2}$ on the surface is always lower than $(p_s^2 + q_s^2)^{1/2}$ for each of the sources, so that only one interpretation of the data is possible. This method is prone to difficulty, however, since it means that parts of the surface could be in shadow, thereby preventing the gradient for these parts of the surface from being measured. Another possibility is to assume that the surface is reasonably smooth, so that p and q vary continuously over it. This itself ensures that ambiguities are resolved over most of the surface.

However, there are other advantages to be gained from using more than two sources of illumination. One is that information on the absolute surface reflectance can be obtained. Another is that the assumption of a Lambertian surface can be tested. Thus, three sources of illumination ensure that the remaining ambiguity is resolved *and* permit absolute reflectivity to be measured: this is obvious, since if the three contours in gradient space do not pass through the same point, then the absolute reflectivity cannot be unity, so corresponding contours should be sought that do pass through the same point. In practice, the calculation is normally carried out by defining a set of nine matrix components of irradiance, s_{ij} being the jth component of light source vector s_i. Then, in matrix notation:

$$\mathbf{E} = R_0 \mathbf{Sn}$$

(15.13)

where

$$\mathbf{E} = (E_1, E_2, E_3)^{\mathrm{T}} \tag{15.14}$$

and

$$S = \begin{bmatrix} s_{11} & s_{12} & s_{13} \\ s_{21} & s_{22} & s_{23} \\ s_{31} & s_{32} & s_{33} \end{bmatrix} \tag{15.15}$$

Provided that the three vectors $\mathbf{s}_1$, $\mathbf{s}_2$, $\mathbf{s}_3$ are not coplanar, so that S is not a singular matrix, R_0 and $\mathbf{n}$ can now be determined from the formulas:

$$R_0 = \left| S^{-1}\mathbf{E} \right| \tag{15.16}$$

$$\mathbf{n} = \frac{S^{-1}\mathbf{E}}{R_0} \tag{15.17}$$

An interesting special case arises if the three source directions are mutually perpendicular; taking them to be aligned along the respective major axes directions, S is now the unit matrix, so that:

$$R_0 = (E_1^2 + E_2^2 + E_3^2)^{1/2} \tag{15.18}$$

and

$$\mathbf{n} = \frac{(E_1, E_2, E_3)^{\mathrm{T}}}{R_0} \tag{15.19}$$

If four or more images are obtained using further illumination sources, more information can be obtained: e.g., the coefficient of specular reflectance, R_1. In practice, this coefficient varies somewhat randomly with the cleanliness of the surface and it may not be relevant to determine it accurately. More probably, it will be sufficient to check whether significant specularity is present, so that the corresponding region of the surface can be ignored for absolute reflectance calculations. Nonetheless, finding the specularity peak can itself give important surface orientation information, as will be clear from Section 15.4. Note that, although the information from several illumination sources should ideally be collated using least-squares analysis, this method requires significant computation. Hence, it seems better to use the images resulting from further illumination sources as confirmatory—or, instead, to select the three that exhibit the least evidence of specularity as giving the most reliable information on local surface orientation.

15.6 THE ASSUMPTION OF SURFACE SMOOTHNESS

It was hinted above that the assumption of a reasonably smooth surface permits ambiguities to be removed in situations where there are two illuminating sources.

In fact, this method can be used to help analyze the brightness map even for situations where a single source is employed: indeed, the fact that the eye can perform this feat of interpretation indicates that it should be possible to find computer methods for achieving it. Much research has been carried out on this topic and a set of methods is available, although the calculations are complex, iterative, computation intensive procedures. For this reason, they are not studied in depth here: the reader is referred to the volume by Horn (1986) for detailed information on this topic. However, one or two remarks are in order.

First, consider the representation to be employed for this type of analysis. In fact, normal gradient (p, q) space is not very appropriate for the purpose. In particular, it is necessary to average gradient (i.e., the **n**-values) locally within the image; however, (p, q) space is not "linear", in that a simple average of (p, q) values within a window would give biased results. It turns out that a conformal representation of gradient (i.e., one which preserves small shapes) is closer to the ideal, in that the distances between points in such a representation provide better approximations to the relative orientations of surface normals: averaging in such a representation gives reasonably accurate results. The required representation is obtained by a stereographic projection, which maps the unit (Gaussian) sphere onto a plane ($z = 1$) through its north pole but this time using as a projection point not its center but its south pole. This projection has the additional advantage that it projects all possible orientations of a surface onto the plane, not merely those from the northern hemisphere. Hence, backlit objects can be represented conveniently in the same map as used for frontlit objects.

Second, the relaxation methods used to estimate surface orientation have to be provided with accurate boundary conditions: in principle, the more correct the orientations that are presented initially to such procedures, the more quickly and accurately the iterations proceed. There are normally two sets of boundary conditions that can be applied in such programs. One is the set of positions in the image where the surface normal is perpendicular to the viewing direction. The other is the set of positions in the image where the surface normal is perpendicular to the direction of illumination: this set of positions corresponds to the set of shadow edges (Fig. 15.12). Careful analysis of the image must be undertaken to find each set of positions, but once they have been located they provide valuable cues for unlocking the information content of the monocular image, and mapping out surfaces in detail.

Finally, all shapes from shading techniques provide information that initially takes the form of surface orientation maps. Dimensions are not obtainable directly but these can be computed by integration across the image from known starting points. In practice this tends to mean that absolute dimensions are unknown and that dimensional maps are obtainable only if the size of an object is given or if its depth within the scene is known.

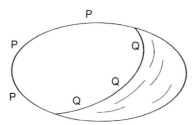

FIGURE 15.12

The two types of boundary condition that can be used in shape-from-shading computations of surface orientation: (i) positions P where the surface normal is perpendicular to the viewing direction; (ii) positions Q where the surface normal is perpendicular to the direction of illumination (i.e., shadow boundaries).

15.7 SHAPE FROM TEXTURE

Texture can be very helpful to the human eye in permitting depth to be perceived. Although textured patterns can be very complex, even the simplest textural elements can carry depth information. Ohta et al. (1981) showed how circular patches on a flat surface viewed more and more obliquely in the distance become first elliptical and then progressively flatter and flatter. At infinite distance, on the horizon line (here defined as the line at infinity in the given plane), they would clearly become very short line segments. To disentangle such textured images sufficiently to deduce depths within the scene, it is first necessary to find the horizon line reliably. This is achieved by taking all pairs of texture elements and deducing from their areas where the horizon line would have to be. To proceed, we make use of the rule:

$$\frac{d_1^3}{d_2^3} = \frac{A_1}{A_2} \tag{15.20}$$

which applies since circles at various depths would give a square law, although the progressive eccentricity also reduces the area linearly in proportion to the depth. This information is accumulated in a separate image space and a line is then fitted to these data: false alarms are eliminated automatically by this Hough-based procedure.

At this stage the original data—the ellipse areas—provide direct information on depth, although some averaging is required to obtain accurate results. Although this type of method has been demonstrated in certain instances, it is in practice highly restricted unless very considerable amounts of computation are performed. Hence it is doubtful whether it can be of general practical use in machine vision applications.

15.8 USE OF STRUCTURED LIGHTING

Structured lighting has already been considered briefly in Section 15.2 as an alternative to stereo for mapping out depth in scenes. Basically, a pattern of light stripes, or other arrangement of light spots or grids, is projected onto the object field. Then these patterns are enhanced in a (generally) single monocular image and analyzed to extract the depth information. To obtain the maximum information the light pattern must be close-knit and the received images must be of very high resolution. When shapes are at all complex, the lines can in places appear so close together that they are unresolvable. It then becomes necessary to separate the elements in the projected pattern, trading resolution and accuracy for reliability of interpretation. Even so, if parts of the objects are along the line of sight, the lines can merge and even cross back and fore, so unambiguous interpretation is never assured. In fact, this is part of a larger problem, in which parts of the object will be obscured from the projected pattern by occluding bodies or by self-occlusion. The method has this feature in common with the shape from shading technique and with stereo vision, which relies on *both* cameras being able to view various parts of the objects simultaneously. Hence, the structured light approach is subject to similar restrictions to those found for other methods of 3-D vision and is not a panacea. Nevertheless, it is a useful technique that is generally simple to set up so as to acquire specific 3-D information that can enable a computer to start the process of cueing into complex images.

Light spots provide perhaps the most obvious form of structured light. However, they are restricted because for each spot, an analysis has to be performed to determine which spot is being viewed: connected lines, in contrast, carry a large amount of coding information with them so that ambiguities are less likely to arise. Grids of lines carry even more coding information but do not necessarily give any more depth information. Indeed, if a pattern of light stripes can be projected, e.g., from the left of the camera so that they are parallel to the y-axis in the observed image, then there is no point in projecting another set of lines parallel to the x-axis, since these merely replicate information that is already available from the rows of pixels in the image—all the depth information is carried by the vertical lines and their horizontal displacements in the image. This analysis assumes that the camera and projected beams are carefully aligned and that no perspective or other distortions are present. In fact, most practical structured

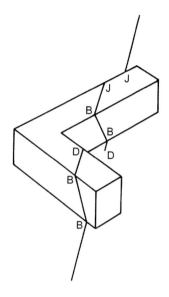

FIGURE 15.13

Three of the structures that are observed when a light stripe is incident on even quite simple shapes: bends (B), jumps (J), and discontinuities (D).

lighting systems in current use employ light stripe patterns rather than spot patterns or full grid patterns.

This section ends with an analysis of the situations that can arise when a single stripe is incident on objects as simple as rectangular blocks. Figure 15.13 shows three types of structure in observed stripes: (a) the effect of a sharp angle being encountered; (b) the effect of "jump edges" at which light stripes jump horizontally and vertically at the same time; and (c) the effect of discontinuous edges at which light stripes jump horizontally but not vertically. The reasons for these circumstances will be obvious from Fig. 15.13. Basically, the problem to be tackled with jump and discontinuous edges is to find whether a given stripe end marks an occluding edge or an occluded edge. The importance of this distinction is that occluding edges mark actual edges of the object being observed, whereas occluded edges may be merely edges of shadow regions and are then not *directly* significant.[2] A simple rule is that, if stripes are projected from the left, the left-hand component of a discontinuous edge will be the occluding edge and the right-hand component will be the occluded edge. Angle edges are located by

---

[2]More precisely, they involve interactions of light with two objects rather than with one, and are therefore more complex to interpret.

applying a Laplacian type of operator that detects the change in orientation of the light stripe.

The ideas outlined above correspond to possible 1-D operators that interpret light stripe information to locate nonvertical edges of objects. The method provides no direct information concerning vertical edges. To obtain such information it is necessary to analyze the information from sets of light stripes. For this purpose 2-D edge operators are required, which collect sufficient data from at least two or three adjacent light stripes. Further details are beyond the scope of this chapter.

Overall, light stripes provide a very useful means of recognizing planes forming the faces of polyhedra and other types of manufactured object. The characteristic sets of parallel lines can be found and demarcated relatively easily, and the fact that the lines usually give rather strong signals means that line tracking techniques can be applied and that algorithms can operate quite rapidly. However, whole-scene interpretation, including inferring the presence and relative positions of different objects, remains a more complex task, as will be seen below.

15.9 THREE-DIMENSIONAL OBJECT RECOGNITION SCHEMES

The methods described so far in this chapter employ various means for finding depth at all places in a scene, and are hence able to map out 3-D surfaces in a fair amount of detail. However, they do not give any clue as to what these surfaces represent. In some situations it may be clear that certain planar surfaces are parts of the background, e.g., the floor and the walls of a room, but in general individual objects will not be inherently identifiable. Indeed, objects tend to merge with each other and with the background, so specific methods are needed to segment the 3-D space map[3] and finally recognize the objects, giving detailed information on their positions and orientations.

Before proceeding to study this problem, notice that further general processing can be carried out to analyze the 3-D shapes. Agin and Binford (1976) and others have developed techniques for likening 3-D shapes to "generalized cylinders," these being like normal (right circular) cylinders but with additional degrees of freedom so that the axes can bend and the cross-sections can vary, both in size and in detailed shape: even an animal like a sheep can be likened to a distorted cylinder. On the whole, this approach is elegant but may not be well adapted to describe many industrial objects, and it is therefore not pursued further here. A simpler approach may be to model the 3-D surfaces as planar, quadratic, cubic, and quartic, and then to try to understand these model surfaces in terms of what

---

[3]This may be defined as an imagined 3-D map showing, without interpretation, the surfaces of all objects in the scene and incorporating all the information from depth or range images. Note that it will generally include only the front surfaces of objects seen from the vantage point of the camera.

is known about existing objects. This approach was adopted by Hall et al. (1982) and was found to be viable, at least for certain quite simple objects such as cups. Shirai (1987) has taken the approach even further so that a whole range of objects can be found and identified in quite complex indoor scenes.

We next consider what we are trying to achieve regarding recognition. First, can recognition be carried out *directly* on the mapped out 3-D surfaces, just as it could for the 2-D images of earlier chapters? Second, if we can bypass the 3-D modeling process, and still recognize objects, might it not be possible to save even more computation and omit the stage of mapping out 3-D surfaces, instead identifying 3-D objects directly in 2-D images? It might even be possible to locate 3-D objects from a single 2-D image.

Consider the first of these problems. When we studied 2-D recognition, many instances were found where the HT approach was of great help. It turned out to give trouble in more complex cases, particularly when attempts were made to find objects where there were more than two or at most three degrees of freedom. Here, however, we have situations where objects normally have six degrees of freedom—three degrees of freedom for translation and another three for rotation. This doubling of the number of free parameters on going from 2-D to 3-D makes the situation far worse, since the search space is proportional in size not to the number of degrees of freedom, but to its exponent: e.g., if each degree of freedom in translation or rotation can have 256 values, the number of possible locations in parameter space changes from 256^3 in 2-D to 256^6 in 3-D. This will be seen to have a very profound effect on object location schemes and tends to make the HT technique difficult to implement. In Section 15.10, we study an interesting approach to the 3-D recognition problem, which uses a subtle combination of 2-D and 3-D techniques.

15.10 HORAUD'S JUNCTION ORIENTATION TECHNIQUE[4]

Horaud's (1987) technique is special in that it uses as its starting point 2-D images of 3-D scenes and "backprojects" them into the scene, with the aim of making interpretations in 3-D rather than 2-D frames of reference. This has the initial effect of increasing mathematical complexity, although in the end useful, more accurate results emerge.

Initially the boundaries of planar surfaces on objects are backprojected. Each boundary line is thus transformed into an "interpretation plane" defined by the center of the camera projection system and the boundary line in the image plane: clearly, the interpretation plane must contain the line that originally projected into the boundary line in the image. Similarly, angles between boundary lines in the image are backprojected into two interpretation planes, which must contain the original two object lines. Finally, junctions between three boundary lines are

---

[4]This and related techniques are sometimes referred to as "shape from angle."

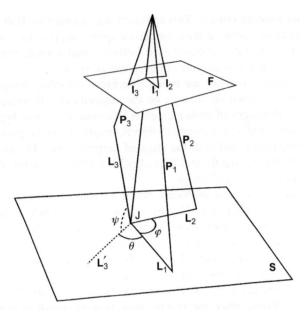

FIGURE 15.14

Geometry for backprojection from junctions: a junction of three lines in an image may be backprojected into three planes, from which the orientation in space of the original corner J may be deduced.

backprojected into three interpretation planes which must contain a corner in the space map (Fig. 15.14). The paper focusses on the backprojection of junctions and shows how measurements of the junction angles in the image relate to those of the original corner; it also shows how the space orientation of the corner can be computed. In fact, it is interesting that the orientation of an object in 3-D can in general be deduced from the appearance of just one of its corners in a single image. This is a powerful result and in principle permits objects to be recognized and located from extremely sparse data.

To understand the method, the mathematics first needs to be set up with some care. Assume that lines L_1, L_2, L_3 meet at a junction in an object, and appear as lines l_1, l_2, l_3 in the image (Fig. 15.14). Take respective interpretation planes containing the three lines and label them by unit vectors P_1, P_2, P_3 along their normals, so that:

$$P_1 \cdot L_1 = 0 \tag{15.21}$$

$$P_2 \cdot L_2 = 0 \tag{15.22}$$

$$P_3 \cdot L_3 = 0 \tag{15.23}$$

In addition, take the space plane containing L_1 and L_2, and label it by a unit vector S along its normal, so that:

$$S_1 \cdot L_1 = 0 \qquad (15.24)$$

$$S_2 \cdot L_2 = 0 \qquad (15.25)$$

Since L_1 is perpendicular to S and P_1, and L_2 is perpendicular to S and P_2, it is found that:

$$L_1 = S \times P_1 \qquad (15.26)$$

$$L_2 = S \times P_2 \qquad (15.27)$$

Note that S is not in general perpendicular to P_1 and P_2, so L_1 and L_2 are not in general unit vectors. Defining φ as the angle between L_1 and L_2, we now have:

$$L_1 \cdot L_2 = L_1 L_2 \cos \varphi \qquad (15.28)$$

which can be re-expressed in the form:

$$(S \times P_1) \cdot (S \times P_2) = |S \times P_1||S \times P_2| \cos \varphi \qquad (15.29)$$

Next we need to consider the junction between L_1, L_2, L_3. To proceed, it is necessary to specify the relative orientations in space of the three lines. θ is the angle between L_1 and the projection L_3' of L_3 on plane S, while ψ is the angle between L_3' and L_3 (Fig. 15.14). Thus, the structure of the junction J is described completely by the three angles φ, θ, ψ. L_3 can now be found in terms of other quantities:

$$L_3 = S \sin \psi + L_1 \cos \theta \cos \psi + (S \times L_1)\sin \theta \cos \psi \qquad (15.30)$$

Applying Eq. (15.23), we find:

$$S \cdot P_3 \sin \psi + L_1 \cdot P_3 \cos \theta \cos \psi + (S \times L_1) \cdot P_3 \sin \theta \cos \psi = 0 \qquad (15.31)$$

Substituting for L_1 from Eq. (15.26), and simplifying, we finally obtain:

$$\begin{aligned}
(S \cdot P_3)|S \times P_1|\sin \psi + S \cdot (P_1 \times P_3)\cos \theta \cos \psi \\
+ (S \cdot P_1)(S \cdot P_3)\sin \theta \cos \psi = (P_1 \cdot P_3)\sin \theta \cos \psi
\end{aligned} \qquad (15.32)$$

Equations (15.31) and (15.34) now exclude the unknown vectors L_1, L_2, L_3 but they retain S, P_1, P_2, P_3 and the three angles φ, θ, ψ. P_1, P_2, P_3 are known from the image geometry, and the angles φ, θ, ψ are presumed to be known from the object geometry; in addition, only two components (α, β) of the unit vector S are independent, so the two equations should be sufficient to determine the orientation of the space plane S. Unfortunately, the two equations are highly nonlinear and it is necessary to solve them numerically. Horaud (1987) achieved this by re-expressing the formulas in the forms:

$$\cos \varphi = f(\alpha, \beta) \qquad (15.33)$$

$$\sin \theta \cos \psi = g_1(\alpha, \beta)\sin \psi + g_2(\alpha, \beta)\cos \theta \cos \psi$$
$$+ g_3(\alpha, \beta)\sin \theta \cos \psi \qquad (15.34)$$

For each image junction, $\mathbf{P}_1$, $\mathbf{P}_2$, $\mathbf{P}_3$ are known and it is possible to evaluate f, g_1, g_2, g_3. Then, assuming a particular interpretation of the junction, values are assigned to φ, θ, ψ and curves giving the relation between α and β are plotted for each equation. Possible orientations for the space plane $\mathbf{S}$ are then given by positions in (α, β) space where the curves cross. Horaud showed that, in general, 0, 1, or 2 solutions are possible. The case of no solutions corresponds to trying to make an impossible match between a corner and an image junction when totally the wrong angles φ, θ, ψ are assumed; one solution is the normal situation; and two solutions arise in the interesting special case when orthographic or near-orthographic projection permits perceptual reversals—i.e., a convex corner is interpreted as a concave corner or *vice versa*. In fact, under orthographic projection the image data from a single corner are insufficient, taken on their own, to give a unique interpretation. In this situation, even the human visual system makes mistakes—as in the case of the well-known Necker cube illusion (see Chapter 16). However, when such cases arise in practical situations, it may be better to take the convex rather than the concave corner interpretation as a working assumption, as it has slightly greater likelihood of being correct.

Horaud has shown that such ambiguities are frequently resolved if the space plane orientation is estimated simultaneously for all the junctions bordering the object face in question, by plotting the α and β values for all such junctions on the same α, β graph. For example, with a cube face on which there are three such junctions, nine curves are coincident at the correct solution, and there are nine points where only two curves cross, indicating false solutions. On the other hand, if the same cube is viewed under conditions approximating very closely to orthographic projection, two solutions with nine coincident curves appear and the situation remains unresolved, as before.

Overall, this technique is important in showing that although lines and angles individually lead to virtually unlimited numbers of possible interpretations of 3-D scenes, junctions lead individually to at most two solutions and any remaining ambiguity can normally be eliminated if junctions on the same face are considered together. As has been seen, the exception to this rule occurs when projection is accurately orthographic, although this is a situation that can often be avoided in practice.

So far we have considered only how a given hypothesis about the scene may be tested: nothing has been said about how assignments of the angles φ, θ, ψ are made to the observed junctions. Horaud's paper discussed this aspect of the work in some depth. In general, the approach is to use a depth-first search technique in which a match is "grown" from the initial most promising junction assignment. In fact, considerable preprocessing of sample data is carried out to find how to rank image features for their utility during depth-first search interpretation. The idea is to order possible alternatives such as linear or circular arcs, convex or concave junctions, and short or long lines. In this way, the tree search becomes more planned and efficient at run time. Generally, the more frequently occurring types of feature should be weighted down in favor of the rarer types of feature, for greater search efficiency. In addition, remember that hypothesis generation is relatively expensive in that it demands a stage of back-projection, as described above. Ideally, this stage need be employed only once for each object (in the case that only a single corner is, initially, considered). Subsequent stages of processing then involve hypothesis verification in which other features of the object are predicted and their presences are sought in the image: if found they are used to refine the existing match; if the match at any stage becomes worse, then the algorithm backtracks and eliminates one or more features and proceeds with other ones. This process is unavoidable, since more than one image feature may be present near a predicted feature.

One of the factors that has been found to make the method converge quickly is the use of grouped rather than individual features, since this tends to decrease the combinatorial explosion in the size of the search. In the present context, this means that attempts should be made to match first all junctions or angles bordering a given object face, and further that a face should be selected that has the greatest number of matchable features around it.

In summary, this approach is successful since it backprojects from the image and then uses geometrical constraints and heuristic assumptions for matching in 3-D space. It is suitable for matching objects that possess planar faces and straight line boundaries, hence giving angle and junction features. However, extending the backprojection technique to situations where object faces are curved and have curved boundaries could be significantly more difficult.

15.11 AN IMPORTANT PARADIGM—LOCATION OF INDUSTRIAL PARTS

In this section, we consider the location of a common class of industrial part: this constitutes an important example that has to be solved in one way or another. Here we go along with the Bolles and Horaud (1986) approach as it leads to sensible solutions and embodies a number of useful didactic lessons. The method starts with a depth map of the scene (obtained in this case using structured lighting).

Figure 15.15 shows in simplified form the type of industrial part being sought in the images. In typical scenes several of these parts may appear jumbled on a worktable, with perhaps three or four being piled on top of each other in some places. In such cases, it is vital that the matching scheme be highly robust if most of the parts are to be found, since even when a part is unoccluded, it appears against a highly cluttered and confusing background. However, the parts themselves have reasonably simple shapes and possess certain salient features. In the

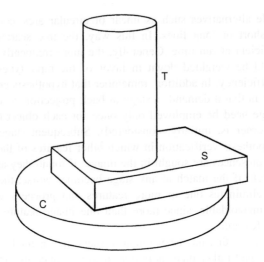

FIGURE 15.15

The essential features of the industrial components located by the 3DPO system of Bolles and Horaud (1986). S, C and T indicate respectively straight and circular dihedral edges and straight tangential edges, all of which are searched for by the system.

particular problem cited, each has a cylindrical base with a concentric cylindrical head, and also a planar shelf is attached symmetrically to the base. To locate such objects, it is natural to attempt to search for circular and straight dihedral edges. In addition, because of the type of data being used, it is useful to search for straight tangential edges, which appear where the sides of curved cylinders are viewed obliquely.

In general, circular dihedral edges appear elliptical, and parameters for five of the six degrees of freedom of the part can be determined by analyzing these edges. The parameter that cannot be determined in this way corresponds to rotation about the axis of symmetry of the cylinder.

Straight dihedral edges also permit five free parameters to be determined, since location of one plane eliminates three degrees of freedom and location of an adjacent plane eliminates a further two degrees of freedom. The parameter that remains undetermined is that of linear motion along the direction of the edge. However, there is also a further ambiguity in that the part may appear either way around on the dihedral edge.

Straight tangential edges determine only four free parameters, since the part is free to rotate about the axis of the cylinder and can also move along the tangential edge. Note that these edges are the most difficult to locate accurately, since range data are subject to greater levels of noise as surfaces curve away from the sensor.

All three of these types of edge are planar. They also provide useful additional information that can help to identify where they are on a part. For example, straight and curved dihedral edges both provide information on the size of the included angle, and the curved edges also give radius values. In fact, curved dihedral edges provide significantly more parametric information about a part than either of the other two types of edge, and therefore they are of most use to form initial hypotheses about the pose (position and orientation) of a part. Having found such an edge, it is necessary to try out various hypotheses about which edge it is, e.g., by searching for other circular dihedral edges at specific relative positions: this is a vital hypothesis verification step. Next, the problem of how to determine the remaining free parameter is solved by searching for the linear straight dihedral edge features from the planar shelf on the part.

At this stage hypothesis generation is complete and the part is essentially found, but hypothesis verification is required (a) to confirm that the part is genuine and not an accidental grouping of independent features in the image, (b) to refine the pose estimate, and (c) to determine the "configuration" of the part, i.e., to what extent it is buried under other parts (making it difficult for a robot to pick it up). When the most accurate pose has been obtained, the overall degree of fit can be considered and the hypothesis rejected if some relevant criterion is not met.

In common with other researchers (Faugeras and Hebert, 1983; Grimson and Lozano-Perez, 1984), Bolles and Horaud took a depth-first tree search as the basic matching strategy. Their scheme uses a minimum number of features to key into the data, first generating hypotheses and then taking care to ensure verification (note that Bolles and Cain (1982) had earlier used this technique in a 2-D part location problem). This contrasts with much work (especially that based on the HT) that makes hypotheses but does not check them. (Note that forming the initial hypotheses is the difficult and computation intensive part of the work. Researchers will therefore write about this aspect of their work and perhaps not state the minor amount of computation that went into confirming that objects had indeed been located. Note also that in much 2-D work, images can be significantly simpler and the size of the peak in parameter space can be so large as to make it virtually certain that an object has been located—thus rendering verification unnecessary.)

15.12 CONCLUDING REMARKS

To the layman, 3-D vision is an obvious and automatic result of the fact that the human visual system is binocular, and presumes both that binocular vision is the only way to arrive at depth maps and that once they have been obtained the subsequent recognition process is trivial. However, what this chapter has actually demonstrated is that neither of these commonly held views is valid. First, there are a good many ways of arriving at depth maps, and some of them are available using monocular vision. Second, the complexity of the mathematical calculations involved in locating objects and the amount of abstract reasoning involved in obtaining robust solutions—plus the need to ensure that the latter are not ambiguous—are taxing even in simple cases, including those where the objects have well-defined salient features.

Despite the diversity of methods covered in this chapter, there are certain important themes: the use of "trigger" features, the value of combining features into groups that are analyzed together, the need for working hypotheses to be generated at an early stage, the use of depth-first heuristic search (combined where appropriate with more rigorous breadth-first evaluation of the possible interpretations), and the detailed verification of hypotheses. All these can be taken as parts of current methodology; *details*, however, vary with the dataset. More specifically, if a new type of industrial part is to be considered, some study must be made of its most salient features: then this causes not only the feature detection scheme to vary but also the heuristics of the search employed—and also the mathematics of the hypothesis mechanism. The reader is referred to the following chapter for further discussion of object recognition under perspective projection.

While the previous two sections have concentrated on object recognition and have perhaps tended to eschew the value of range measurements and depth maps, it is possible that this might give a misleading impression of the situation. In fact, there are many situations where recognition is largely irrelevant but where it is mandatory to map out 3-D surfaces in great detail. Turbine blades, automobile body parts, or even food products such as fruit may need to be measured accurately in 3-D. In such cases, it is known in advance what object is in what position, but some inspection or measurement function has to be carried out and a diagnosis made. In such instances, the methods of structured lighting, stereopsis, or photometric stereo come into their own and are highly effective methods. Ultimately also, one might expect that a robot vision system will have to use all the tricks of the human visual system if it is to be as adaptable and useful when operating in an unconstrained environment rather than at a particular worktable.

This has been a preliminary chapter on 3-D vision, setting the scene for Parts 3 and 4. In particular, Chapter 16 will be devoted to a careful analysis of the distinction between weak and full perspective projection and how this affects the object recognition process; Chapter 17 will aim to show something of the elegance and value of invariants in providing short cuts around some of the complexities of full perspective projection; Chapter 18 will consider camera calibration and will also consider how recent research on interrelating multiple views of a scene has allowed some of the tedium of camera calibration to be by-passed; and Chapter 19 will introduce the topic of motion analysis in 3-D scenes.

> Conventional wisdom indicates that binocular vision is the key to understanding the 3-D world. This chapter has shown that the correspondence problem makes the practice of binocular vision tedious, while the solutions it provides are only depth maps and require further intricate analysis before the 3-D world can fully be understood.

15.13 BIBLIOGRAPHICAL AND HISTORICAL NOTES

As noted earlier in the chapter, the most obvious approach to 3-D perception is to employ a binocular camera system. Burr and Chien (1977) and Arnold (1978) showed how a correspondence could be set up between the two input images by use of edges and edge segments. Forming a correspondence can involve considerable computation. Barnea and Silverman (1972) showed how this problem could be alleviated by passing quickly over unfavorable matches. Likewise, Moravec (1980) devised a coarse-to-fine matching procedure that arrives systematically at an accurate correspondence between images. Marr and Poggio (1979) formulated two constraints—those of uniqueness and continuity—that have to be satisfied in choosing global correspondences: these constraints are important in leading to the simplest available surface interpretation. Ito and Ishii (1986) found that there is something to be gained from three-view stereo in offsetting ambiguity and the effects of occlusions.

The structured lighting approach to 3-D vision was introduced independently by Shirai (1972) and Agin and Binford (1973, 1976) in the form of a single plane of light, while Will and Pennington (1971) developed the grid coding technique. Nitzan et al. (1977) employed an alternative LIDAR (light detecting and ranging) scheme for mapping objects in 3-D; here short light pulses were timed as they traveled to the object surface and back.

Meanwhile, other workers were attempting monocular approaches to 3-D vision. Some basic ideas underlying shape from shading date from as long ago as 1929, with Fesenkov's investigations of the lunar surface: (see also van Digellen, 1951). However, the first shape-from-shading problem to be solved both theoretically and in an operating algorithm appears to have been that of Rindfleisch (1966), also relating to lunar landscapes. Thereafter, Horn systematically tackled the problem both theoretically and with computer investigations, starting with a notable review (1975) and resulting in prominent papers (e.g., Horn, 1977; Ikeuchi and Horn, 1981; Horn and Brooks, 1986), an important book (Horn, 1986) and an edited work (Horn and Brooks, 1989). Interesting papers by other workers in this area include Blake et al. (1985), Bruckstein (1988), and Ferrie and Levine (1989). Woodham (1978, 1980, 1981) must be credited with the photometric stereo idea. Finally, the vital contributions made by workers on computer graphics in this area must not be forgotten—see e.g., Phong (1975) and Cook and Torrance (1982).

The concept of shape-from-texture arose from the work of Gibson (1950) and was developed by Bajcsy and Liebermann (1976), Stevens (1980), and notably by Kender (1980), who carefully explored the underlying theoretical constraints.

The paper by Barrow and Tenenbaum (1981) provides a very readable review of much of this earlier work. The year 1980 marked a turning point, when the emphasis in 3-D vision shifted from mapping out surfaces to interpreting images as sets of 3-D objects. Possibly, this segmentation task could not be tackled earlier because basic tools such as the HT were not sufficiently well developed. The work of Koenderink and van Doorn (1979) and Chakravarty and Freeman (1982) was probably also crucial in providing a framework for interpretation schemes to be developed by using potential 3-D views of objects. The work of Ballard and Sabbah (1983) provided an early breakthrough in segmentation of real objects in 3-D and this was followed by vital further work by Faugeras and Hebert (1983), Silberberg et al. (1984), Bolles and Horaud (1986), Horaud (1987), Pollard et al. (1987), and many others.

Other interesting work includes that of Horaud et al. (1989) on solving the perspective four-point problem (finding the position and orientation of the camera relative to known points): for further references on this topic, see Section 16.6.

Although already a well worked-through topic, research on finding vanishing points proceeded further in the 1990s (e.g., Lutton et al., 1994; Straforini et al., 1993; Shufelt, 1999). Similarly, stereo correlation matching techniques were still under development, to maintain robustness in real-time applications (Lane et al., 1994).

Since 2000, work on stereo vision has continued unabated as a main-line topic (e.g., Lee et al., 2002; Brown et al., 2003), but Horn's approach to shape from shading has been largely superseded. One new technique is the Green's function approach to shape from shading (Torreão, 2001, 2003), while local shape from shading has been used to improve the photometric stereo technique (Sakarya and Erkmen, 2003). Photometric stereo has itself been developed considerably further in a new four-source technique capable of coping with highlights and shadows (Barsky and Petrou, 2003). Another development is the application of shape from shading to radar data—a translation that required significant new theory (Frankot and Chellappa, 1990; Bors et al., 2003). Finally, a thoroughgoing new approach to the whole study of 3-D vision and its dependence on the light field has been initiated (Baker et al., 2003). This paper starts by comparing what can be learned from (a) stereo vision and (b) a shape from silhouette approach (observing object silhouettes from all directions in the given light field). An important conclusion is that the shapes of Lambertian objects can be uniquely determined with n-camera stereo, unless there are regions of constant intensity present. Indeed, constant intensity is found *always* to lead to ambiguity.[5] This paper is important not only in giving a fresh view of the problems of 3-D vision in general, and shape from shading in particular, but also in demonstrating certain open questions.

15.13.1 More Recent Developments

While the complexity of the image acquisition needed for photometric stereo should perhaps have made it relevant only during the early stages of the subject, the opposite now seems to be the case. First, Hernandez et al. (2011) indicate

---

[5]Essentially, this is because there may be a concavity whose light properties outside the concavity hull will be indistinguishable from those of the hull itself (Laurentini, 1994).

why this could be so: if a set of lights of different colors is arranged, there is no need to switch them, as the different color channels can be handled independently. However, this means that normally only three lights can be used, so the Barsky and Petrou (2003) four-light technique cannot be employed, and this makes it difficult to confirm the interpretations obtained using the (usual) minimum number of three lights—an especially important factor when shadows occur. Nevertheless, Hernandez et al. (2011) are able to use regularization methods that cope with as few as two light sources. Wu and Tang (2010) employ the opposite approach of using a dense image set and exploit the resulting data redundancy to determine how well the observations fit a Lambertian model. An expectation maximization approach is used to interpret the data in two stages, concentrating first on surface normals and then on surface properties including orientation discontinuities. The approach is robust and produces good reconstruction results. Goldman et al. (2010) note that most objects are composed of only a small number of fundamental materials: they therefore constrain pixel representations to at most two such materials, and thereby recover not only the shape but also material bidirectional reflectance distribution functions (BRDFs) and weight maps. McGunnigle and Dong (2011) propose a photometric stereo method in which a conventional four-light scheme is augmented with coaxial illumination. Their investigations show that coaxial illumination makes photometric stereo more robust to shadow and specularity.

Chen et al. (2011) devise a fast stereo matching algorithm that uses a global graph-cuts framework, but which is as efficient as some local approaches. By concentrating on region boundaries and cleverly limiting the number of disparity candidates, the number of vertices in the constructed graph is significantly reduced. As a result, promising disparities can readily be selected and partial occlusions can be handled efficiently, thereby improving stereo matching speed.

15.14 PROBLEMS

1. Prove that all epipolar lines in one image plane pass through the point that is the image of the projection point of the alternate image plane.
2. What is the physical significance of the straight line contour in gradient space (see Fig. 15.9(b))?
3. Sketch a curve of the function $\cos^m \theta$. Estimate what the value of m would have to be for 90% of the R_1 component to be reflected within 10° of the direction for pure specular reflection.
4. An alien has three eyes. Does this permit it to perceive or estimate depth more accurately than a human? What would be the best placement for a third eye?
5. A cube is viewed in orthographic projection. Show that although the cube is opaque, it is easy to compute the theoretical position of its centroid in

the image. Show also that the orientation of the cube can be deduced by considering the apparent areas of its faces. If the contrast between the faces becomes so low that only a hexagonal outline is seen, show that ambiguities will arise in our knowledge of the orientation of the cube. Are ambiguities specific to cubes, or do they arise with other shapes? Why?

6. **a.** A feature at (X, Y, Z) appears at locations (x_1, y_1) and (x_2, y_2) in the two images of a binocular imaging system. The image planes of both cameras lie in the same plane, f is the focal length of both camera lenses, and b is the separation of the optical axes of the lenses. Label Fig. 15.16 appropriately; by considering pairs of similar triangles, show that:

$$\frac{Z}{f} = \frac{X + b/2}{x_1} = \frac{X - b/2}{x_2}$$

   b. Hence derive a formula that can be used to determine depth Z from the observed disparity.

7. Give a full proof that the error with which the fractional depth Z in a scene can be computed is (a) proportional to pixel size, (b) proportional to Z, and (c) inversely proportional to the baseline b between the stereo cameras. What other parameter appears in the final formula? Determine under what pair of conditions two very tiny cameras fabricated by nanotechnological methods could still perform viable depth measurement.

8. **a.** Draw a diagram that shows that the ordering of visible points is normally the same in both images seen by a binocular vision system.

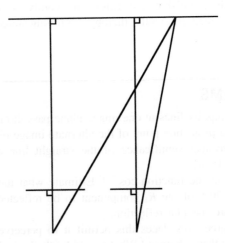

FIGURE 15.16

Geometry of a binocular imaging system.

b. An object has a semi-transparent front surface through which an interior feature F is just visible. Show that the ordering of the features in the two views of the object may be sufficient to prove that F is inside, or perhaps behind, the object.

9. a. State the conditions under which matt surfaces may properly be described as "Lambertian." Show that the normal at a point on a Lambertian surface must lie on a cone of directions whose axis points to the point source of illumination. Show that a minimum of three independent light sources will be needed to identify the exact orientation of a matt surface. Why might four light sources help to determine surface orientation for a surface of unknown or non-ideal properties?

b. Compare the effectiveness of binocular vision and photometric stereo if it is desired to obtain a depth map for each object in a scene. In each case, consider the properties of the object surface and the distance from the observer.

10. a. Compare the properties of matt surfaces with those that exhibit "normal" specular reflection. Matt surfaces are sometimes described as "Lambertian." Describe how the brightness of the surface varies according to the Lambertian model.

b. Show that for a given surface brightness, the orientation of any point on a Lambertian surface must lie on a certain cone of orientations.

c. Three images of a surface are obtained on illuminating it in sequence by three independent point light sources. Show with the aid of a diagram how this can lead to unambiguous estimates of surface orientation. Would surface orientation of any points on the surface *not* be estimated by this method? Are there any constraints on the allowable positions of the three light sources? Would it help if *four* independent point light sources were used instead of three?

d. Discuss whether the surface map that is obtained by shape from shading is identical to that obtained by stereo (binocular) vision. Are the two approaches best applied in the same or different applications? To what extent is the application of structured light able to give better or more accurate information than these basic approaches?

e. Consider what further processing is required before 3-D objects can be recognized by any of these approaches.

16

Tackling the Perspective
n-point Problem

It is possible to recognize 3-D objects from very few point features, even when they are seen in a single view. In fact, the pose of the 3-D object can also be ascertained from a single view. However, ambiguities of interpretation do arise and this chapter discusses the disambiguation problem.

Look out for:

- the distinction between weak and full perspective projection.
- how the "perspective inversion" type of ambiguity arises under weak perspective projection.
- how more serious ambiguities arise under full perspective projection.
- how full perspective projection has the capability to provide more interpretative information than weak perspective projection.
- how coplanarity can impose quite strong constraints on 3-D data, which can be sometimes helpful and at other times an impediment.
- how symmetry can help with 3-D image interpretation.

Note that while this chapter considers only one aspect of 3-D vision, it raises very important issues that are relevant right through the subject of 3-D object recognition.

16.1 INTRODUCTION

This chapter follows on from the previous introductory chapter, and tackles a problem of central importance in the analysis of images from 3-D scenes. It has been kept separate and fairly short so as to focus carefully on relevant factors in the analysis. First, we look at the phenomenon of perspective inversion, which has already been alluded to several times in Chapter 15. Then we refine our ideas on perspective, and proceed to consider the determination of object pose from salient features that are located in the images. It will be useful to consider how many salient features are required for unambiguous determination of pose.

16.2 THE PHENOMENON OF PERSPECTIVE INVERSION

In this section, we study first the phenomenon of *perspective inversion*. This is actually a rather well-known effect that appears in the following "Necker cube" illusion. Consider a wire cube made from 12 pieces of wire welded together at the corners. Looking at it from approximately the direction of one corner, it is difficult to tell which way round the cube is, i.e., which of the opposite corners of the cube is the nearer (Fig. 16.1). Indeed, on looking at the cube for a time, one gradually comes to feel one knows which way round it is, but then it suddenly appears to reverse itself; then that perception remains for some time, until it too reverses itself.[1] This illusion reflects the fact that the brain is making various hypotheses about the scene, and even making decisions based on incomplete evidence about the situation (Gregory, 1971, 1972).

The wire cube illusion could perhaps be regarded as somewhat artificial. But consider instead an aeroplane (Fig. 16.2(a)) that is seen in the distance (Fig. 16.2(b)) against a bright sky. The silhouetting of the object means that its surface details are not visible. In that case interpretation requires that a hypothesis be made about the scene, and it is possible to make the wrong one. Clearly (Fig. 16.2(c)), the aeroplane could be at an angle α (as for P), although it could equally well be at an angle $-\alpha$ (as for Q). The two hypotheses about the orientation of the object are related by the fact that the one can be obtained from the other by reflection in a plane R normal to the viewing direction D.

Strictly, there is only an ambiguity in this case if the object is viewed under orthographic or scaled orthographic projection.[2] However, in the distance, perspective projection approximates to scaled orthographic projection, and it is often difficult to detect the difference.[3] If the aeroplane in Fig. 16.2 were quite near, it would be obvious that one part of the silhouette was nearer, as the perspective would distort it in a particular way. In general, perspective projection will break down symmetries, so searching for symmetries that are known to be present in the object should reveal which way around it is: however, if the object is in the distance, as in Fig. 16.2(b), it will be virtually impossible to see the breakdown.

---

[1] In psychology, this shifting of attention is known as *perceptual reversal*, which is unfortunately rather similar to the term *perspective inversion*, but is actually a much more general effect that leads to a host of other types of optical illusion—see Gregory (1971) and the many illustrations produced by M.C. Escher.

[2] *Scaled orthographic projection* is orthographic projection with the final image scaled in size by a constant factor.

[3] In this case, the object is said to be viewed under *weak perspective projection*. For weak perspective, the depth ΔZ within the object has to be much less than its depth Z in the scene. On the other hand, the perspective scaling factor can be different for each object and will depend on its depth in the scene: so the perspective can validly be locally weak and globally normal.

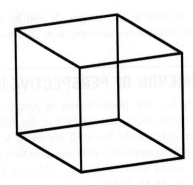

FIGURE 16.1

The phenomenon of perspective inversion. This figure shows a wire cube viewed approximately from the direction of one corner. The phenomenon of perspective inversion makes it difficult to see which of the opposite corners of the cube is the nearer: in fact there are two stable interpretations of the cube, either of which may be perceived at any moment.

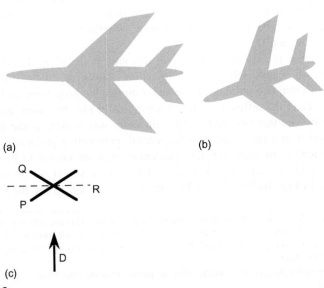

FIGURE 16.2

Perspective inversion for an aeroplane. Here an aeroplane (a) is silhouetted against the sky and appears as in (b). (c) shows the two planes P and Q in which the aeroplane could lie, relative to the direction D of viewing: R is the reflection plane relating the planes P and Q.

Unfortunately, a short-term study of the motion of the aeroplane will not help with interpretation in the case shown in Fig. 16.2(b). Eventually, however, the aeroplane will appear to become smaller or larger, and this will give the additional information needed to resolve the issue.

16.3 AMBIGUITY OF POSE UNDER WEAK PERSPECTIVE PROJECTION

It is instructive to examine to what extent the pose of an object can be deduced under weak perspective projection. We can reduce the above problem to a simplest case in which three points have to be located and identified. Any set of three points is coplanar, and the common plane corresponds to that of the silhouette shown in Fig. 16.2(a) (we assume here that the three points are not collinear, so that they do in fact define a plane). The problem then is to match the corresponding points on the idealized object (Fig. 16.2(a)) with those on the observed object (Fig. 16.2(b)). It is not yet completely obvious that this is possible, or that the solution is unique, even apart from the reflection operation noted earlier. It could be that more than three points will be required—especially if the scale is unknown—or it could be that there are several solutions, even if we ignore the reflection ambiguity. It will be important to see whether it is possible to distinguish the three points in the observed image.

To understand the degree of difficulty, let us briefly consider full perspective projection. In this case, any set of three noncollinear points can be mapped into any other three. This means that it may not be possible to deduce much about the original object just from this information: we will certainly not be able to deduce which point maps to which other point. However, we shall see that the situation is rather less ambiguous when viewing the object under weak perspective projection.

Perhaps the simplest approach (due to Huang et al. as recently as 1995) is to imagine a circle drawn through the original set of points P_1, P_2, P_3 (Fig. 16.3(a)). We then find the centroid C of the set of points and draw additional lines through the points, all passing through C and meeting the circle in another three points Q_1, Q_2, Q_3 (Fig. 16.3(a)). Now in common with orthographic projection, scaled orthographic projection maintains ratios of distances on the same straight line, and weak perspective projection approximates to this. Thus, the distance ratio $P_iC:CQ_i$ remains unchanged after projection. Thus, when we project the whole figure, as in Fig. 16.3(b), we find that the circle has become an ellipse, although all lines remain lines, and all linear distance ratios remain unchanged. The significance of this is as follows. When the points P_1', P_2', P_3' are observed in the image, the centroid C' can be computed, as can the positions of Q_1', Q_2', Q_3'. Thus, we have six points from which to deduce the position and parameters of the ellipse (in fact, five are sufficient). Once the ellipse is known, the orientation of its major axis gives the axis of rotation of the object; while the ratio of the lengths of the minor to

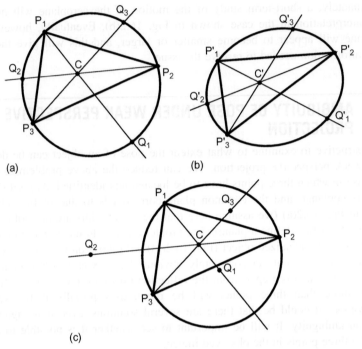

FIGURE 16.3

Determination of pose for three points viewed under weak perspective projection.
(a) shows three feature points P_1, P_2, P_3 that lie on a known type of object. The circle
passing through P_1, P_2, P_3 is drawn, and lines through the points and their centroid C
meet the circle in Q_1, Q_2, Q_3. The ratios $P_iC:CQ_i$ are then deduced. (b) shows the three
points observed under weak perspective as P_1', P_2', P_3', together with their centroid C'
and the three points Q_1', Q_2', Q_3' located using the original distance ratios. An ellipse
drawn through the six points P_1', P_2', P_3', Q_1', Q_2', Q_3' can now be used to determine the
orientation of the plane in which P_1, P_2, P_3 must lie, and also (from the major axis of the
ellipse) the distance of viewing. (c) shows how an erroneous interpretation of the three
points does not permit a circle to be drawn passing through P_1, P_2, P_3, Q_1, Q_2, Q_3 and
hence no ellipse can be found that passes through the observed and the derived points
P_1', P_2', P_3', Q_1', Q_2', Q_3'.

major axes immediately gives the value of cos α. (Notice how the ambiguity in the
sign of α comes up naturally in this calculation.) Finally, the length of the major
axis of the ellipse permits the depth of the object in the scene to be deduced.

   We have now shown that observing three projected points permits a unique
ellipse to be computed passing through them, and when this is back-projected into
a circle, the axis of rotation of the object and the angle of rotation can be deduced,
but not the sign of the angle of rotation. There are two important comments to be

made about the above calculation. The first is that the three distance ratios must be stored in memory, before interpretation of the observed scene can begin. The second is that the order of the three points apparently has to be known before interpretation can be undertaken: otherwise, we will have to perform six computations in which all possible assignments of the distance ratios are tried; furthermore, it might appear from the earlier introductory remarks that several solutions are possible. While there are some instances in which feature points might be distinguishable, there are many cases when they are not (especially in 3-D situations where corner features might vary considerably when viewed from different positions). Thus, the potential ambiguity is important. However, if we can try out each of the six cases, little difficulty will generally arise. For, immediately we deduce the positions Q_1', Q_2', Q_3', we will find that it is not possible in general to fit the six resulting points to an ellipse. The reason is easily seen on returning to the original circle. In that case, if the wrong distance ratios are assigned, the Q_i will clearly not lie on the circle, since the only values of the distance ratios for which the Q_i do lie on the circle are the correctly assigned ones (Fig. 16.3(c)). This means that although computation is wasted testing the incorrect assignments, there appears to be no risk of their leading to ambiguous solutions. Nevertheless, there is one contingency under which things could go wrong. Suppose the original set of points P_1, P_2, P_3 forms an almost perfect equilateral triangle. Then the distance ratios will be very similar, and, taking numerical inaccuracies into account, it may not be clear which ellipse provides the best and most likely fit. This mitigates against taking sets of feature points that form approximately isosceles or equilateral triangles. However, in practice more than three coplanar points will generally be used to optimize the fit, making fortuitous solutions rather unlikely.

Overall, it is fortunate that weak perspective projection requires such weak conditions for the identification of unique (to within a reflection) solutions, especially as full perspective projection demands four points before a unique solution can be found (see below). However, under weak perspective projection additional points lead to greater accuracy but no reduction in the reflection ambiguity: this is because the information content from weak perspective projection is impoverished in the lack of depth cues that could (at least in principle) resolve the ambiguity. To understand this lack of additional information from more than three points under weak perspective projection, note that each additional feature point in the same plane is predetermined once three points have been identified (here we are assuming that the model object with the correct distance ratios can be referred to).

These considerations indicate that we have two potential routes to unique location of objects from limited numbers of feature points. The first is to resort to use of noncoplanar points viewed still under weak perspective projection. The second is to use full perspective projection to view coplanar or noncoplanar sets of feature points. We shall see below that whichever of these options we take, a unique solution demands that a minimum of four feature points be located on any object.

16.4 OBTAINING UNIQUE SOLUTIONS TO THE POSE PROBLEM

The overall situation is summarized in Table 16.1. Looking first at the case of weak perspective projection, the number of solutions only becomes finite for three or more point features. Once three points have been employed, in the coplanar case there is no further reduction in the number of solutions, since (as noted earlier) the positions of any additional points can be deduced from the existing ones. However, this does not apply when the additional points are noncoplanar since they are able to provide just the right information to eliminate any ambiguity (see Fig. 16.4). (Although this might appear to contradict what was said earlier about perspective inversion, note that we are assuming here that the body is rigid and that all its features are at *known* fixed points on it in three dimensions; hence this particular ambiguity no longer applies, except for objects with special symmetries that we shall ignore here—see Fig. 16.4(d).)

Considering next the case of full perspective projection, the number of solutions again becomes finite only for three or more point features. The lack of information provided by three point features means that four solutions are in principle possible (see the example in Fig. 16.5 and the detailed explanation in Section 16.4.1), but the number of solutions drops to one as soon as four coplanar points are employed (the correct solution can be found by making cross checks between subsets of three points, and eliminating inconsistent solutions); when the

Table 16.1 Ambiguities When Estimating Pose from Point Features

Arrangement of the Points	n	WPP	FPP
Coplanar	≤ 2	∞	∞
	3	2	4
	4	2	1
	5	2	1
	≥ 6	2	1
Noncoplanar	≤ 2	∞	∞
	3	2	4
	4	1	2
	5	1	2
	≥ 6	1	1

This table summarizes the numbers of solutions that will be obtained when estimating the pose of a rigid object from point features located in a single image. It is assumed that n point features are detected and identified correctly and in the correct order. The columns WPP and FPP signify weak perspective projection and full perspective projection, respectively. The upper half of the table applies when all n points are coplanar; the lower half of the table applies when the n points are noncoplanar. Note that when n ≤ 3, the results strictly apply only in the coplanar case. However, the top two lines in the lower half of the table are retained for easy comparison.

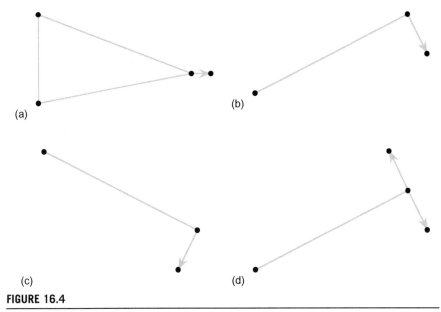

FIGURE 16.4

Determination of pose for four points viewed under weak perspective projection. (a) shows an object containing four noncoplanar points, as seen under weak perspective projection. (b) shows a side view of the object. If the first three points (connected by nonarrowed gray lines) were viewed alone, perspective inversion would give rise to a second interpretation (c). However, the fourth point gives additional information about the pose that permits only one overall interpretation. This would not be the case for an object containing an additional symmetry as in (d), since its reflection would be identical to the original view (not shown).

points are noncoplanar, it is only when six or more points are employed that there is sufficient information to unambiguously determine the pose: there is necessarily no ambiguity with six or more points, as all eleven camera calibration parameters can be deduced from the twelve linear equations that then arise (see Chapter 18). Correspondingly, it is deduced that five noncoplanar points will in general be *insufficient* for all eleven parameters to be deduced, so there will still be some ambiguity in this case.

Next, it should be questioned why the coplanar case is at first ($n = 3$) better[4] under weak perspective projection and then ($n > 3$) better under full perspective projection, while the noncoplanar case is always better, or as good, under weak perspective projection. The reason must be that intrinsically full perspective projection provides more detailed information, but is frustrated by lack of data when there are relatively few points: however, the exact stage at which the additional

---

[4]In this context "better" means less ambiguous, and leading to fewer solutions.

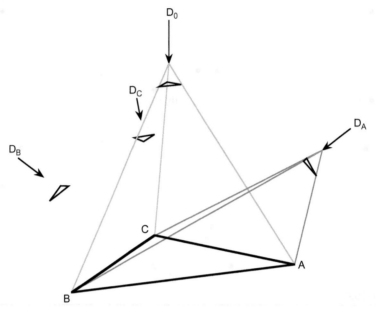

FIGURE 16.5

Ambiguity for three points viewed under full perspective projection. Under full perspective projection, the camera sees three points A, B, C as three directions in space, and this can lead to four-fold ambiguity in interpreting a known object. The figure shows the four possible viewing directions and centers of projection of the camera (indicated by the directions and tips of the bold arrows): in each case the image at each camera is indicated by a small triangle. D_A, D_B, D_C correspond approximately to views from the general directions of A, B, C, respectively.

information becomes available is different in the coplanar and noncoplanar cases. In this respect, it is important to note that when coplanar points are being observed under weak perspective projection, there is never enough information to eliminate the ambiguity.

It should be emphasized that the above discussion assumes that the correspondences between object and image features are all known, i.e., *n* point features are detected and identified correctly and in the correct order. If this is not so, the number of possible solutions could increase substantially, considering the number of possible permutations of quite small numbers of points. This makes it attractive to use the minimum number of features for ascertaining the most probable match (Horaud et al., 1989). Other workers have used heuristics to help reduce the number of possibilities. For example, Tan (1995) used a simple compactness measure (see Section 9.7) to determine which is the most likely geometric solution: extreme obliqueness is perhaps unlikely, and the most likely solution is taken to be the one with highest compactness value. This idea follows on from the extremum principle of Brady and Yuille (1984), which states that the most probable solutions are those nearest to extrema of relevant (e.g., rotation) parameters.[5] In this context, note that coplanar points viewed under weak or full perspective projection always appear in the same cyclic order: this is not trivial to check given the possible distortions of an object, although if a convex polygon can be drawn through the points, the cyclic order around its boundary will not change on projection.[6] However, for noncoplanar points, the pattern of the perceived points can re-order itself almost randomly: this means that a considerably greater number of permutations of the points have to be considered for noncoplanar points than for coplanar points.

Finally, note that the above discussion has concentrated on the existence and uniqueness of solutions to the pose problem. The stability of the solutions has not so far been discussed. However, the concept of stability gives a totally different dimension to the data presented in Table 16.1. In particular, noncoplanar points tend to give more stable solutions to the pose problem. For example, if the plane containing a set of coplanar points is viewed almost head-on ($\alpha \approx 0$), there will be very little information on the exact orientation of the plane, because the changes in lateral displacement of the points will vary as $\cos \alpha$ (see Section 16.2) and there will be no linear term in the Taylor expansion of the orientation dependence.

16.4.1 Solution of the Three-Point Problem

Figure 16.5 showed how four solutions can arise when three point features are viewed under full perspective projection. Here, we briefly explore this situation by considering the relevant equations. Figure 16.5 shows that the camera sees the points as three image points representing three directions in space. This means that we can compute the angles α, β, γ between these three directions. If the distances between the three points A, B, C on the object are the known values D_{AB}, D_{BC}, D_{CA}, we can now apply the cosine rule in an attempt to determine the distances R_A, R_B, R_C of the feature points from the center of projection:

$$D_{BC}^2 = R_B^2 + R_C^2 - 2R_B R_C \cos \alpha \tag{16.1}$$

$$D_{CA}^2 = R_C^2 + R_A^2 - 2R_C R_A \cos \beta \tag{16.2}$$

$$D_{AB}^2 = R_A^2 + R_B^2 - 2R_A R_B \cos \gamma \tag{16.3}$$

Eliminating any two of the variables R_A, R_B, R_C yields an eighth degree equation in the other variable, indicating that eight solutions to the system of equations could be available (Fischler and Bolles, 1981). However, the above cosine rule equations contain only constants and second degree terms: hence, for every positive solution there is another solution that differs only by a sign change in all the

---

[5]Perhaps the simplest way of understanding this principle is obtained by considering a pendulum, whose extreme positions are also its most probable. However, in this case the extremum occurs when the angle α (see Fig. 16.1) is close to zero.
[6]The reason for this is that planar convexity is an invariant of projection.

variables. These solutions correspond to inversion through the center of projection and are hence unrealizable. Thus, there are at most four realizable solutions to the system of equations. In fact, we can quickly demonstrate that there may sometimes be fewer than four solutions: since in some cases, for one or more of the "flipped" positions shown in Fig. 16.5, one of the features could be on the negative side of the center of projection, and hence would be unrealizable.

Before leaving this topic, note that the homogeneity of Eqs. (16.1)–(16.3) implies that observation of the angles α, β, γ permits the orientation of the object to be estimated independently of any knowledge of its scale: in fact, estimation of scale depends directly on estimation of range, and *vice versa*. Thus, knowledge of just one range parameter (e.g., R_A) will permit the scale of the object to be deduced. Alternatively, knowledge of its area will permit the remaining parameters to be deduced. This concept provides a slight generalization of the main results of Sections 16.2 and 16.3, which generally start with the assumption that all the dimensions of the object are known.

16.4.2 Using Symmetric Trapezia for Estimating Pose

One more example will be of interest here. That is the case of four points arranged at the corners of a symmetric trapezium (Tan, 1995). When viewed under weak perspective projection, the mid-points of the parallel sides are easily measured, but under full perspective projection midpoints do not remain midpoints, so the axis of symmetry cannot be obtained in this way. However, producing the skewed sides to meet at S′ and forming the intersection I′ of the diagonals permit the axis of symmetry to be located as the line I′S′ (Fig. 16.6). Thus, we now have not four points but six to describe the perspective view of the trapezium. What is more important is that the axis of symmetry has been located and this is known to be perpendicular to the parallel sides of the trapezium. This is a great help in making the mathematics more tractable and in obtaining solutions quickly so that, e.g., object motion can be tracked in real time. Again, this is a case where object orientation can be deduced straightforwardly, even when the situation is one of strong perspective, and even when the size of the object is unknown. This result is a generalization from that of Haralick (1989) who noted that a single view of a rectangle of unknown size is sufficient to determine its pose. In either case, the range of the object can be found if its area is known, or its size can be deduced if a single range value can be found from other data (see also Section 16.4.1).

16.5 CONCLUDING REMARKS

This chapter has aimed to cover certain aspects of 3-D vision that were not studied in depth in the previous chapter. In particular, it was worth investigating the topic of perspective inversion in some detail and exploring how it was affected by the method of projection. Orthographic projection, scaled orthographic

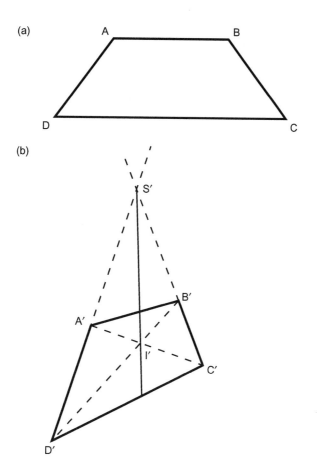

FIGURE 16.6

Trapezium viewed under full perspective projection. (a) shows a symmetrical trapezium, and (b) shows how it appears when viewed under full perspective projection. In spite of the fact that midpoints do not project into midpoints under perspective projection, the two points S′ and I′ on the symmetry axis can be located unambiguously as the intersection of two nonparallel sides and two diagonals, respectively. This gives six points (from which the two midpoints on the symmetry axis can be deduced if necessary), which is sufficient to compute the pose of the object, albeit with a single ambiguity of interpretation (see text).

projection, weak perspective projection, and full perspective projection were considered, and the numbers of object points that would lead to correct or ambiguous interpretations were analyzed. It was found that scaled orthographic projection and its approximation, weak perspective projection, led to straightforward interpretation when four or more noncoplanar points were considered, although the perspective inversion ambiguity remained when all the points were coplanar. This latter ambiguity was resolved with four or more points viewed under full perspective projection. However, in the noncoplanar case, some ambiguity remained until six points were being viewed. The key to understanding the situation is the fact that full perspective projection makes the situation more complex, although at the same time it provides more information, ultimately, to resolve the ambiguity.

Additional problems were found to arise when the points being viewed are indistinguishable, and then a good many solutions may have to be tried before a minimally ambiguous result is obtained. With coplanar points fewer possibilities arise and this leads to less computational complexity: the key to success here is the natural ordering that can arise for points in a plane—e.g., when they form a convex set that can be ordered uniquely around a bounding polygon. In this context, the role that can be played by the extremum principle of Brady and Yuille (1984) in reducing the number of solutions is significant (for further insight on the topic, see Horaud and Brady, 1988).

It is of great relevance to devise methods for rapid interpretation in real-time applications. To achieve this, it is important to work with a minimal set of points and to obtain analytic solutions that move directly to solutions without computationally expensive iterative procedures: for example, Horaud et al. (1989) found an analytic solution for the perspective four-point problem that works both in the general noncoplanar case and in the planar case. Other low computation methods are still being developed, as with pose determination for symmetrical trapezia (Tan, 1995). It should also be noted that understanding is still advancing, as demonstrated by Huang et al.'s (1995) neat geometrical solution to the pose determination problem for three points viewed under weak perspective projection.

This chapter has covered a specific 3-D recognition problem. Chapter 17 covers another—that of invariants, which provides speedy and convenient means of by-passing the difficulties associated with full perspective projection. Chapter 18 aims to finalize the study of 3-D vision by showing how camera calibration can be achieved or, to some extent, circumvented.

> Perspective makes interpretation of images of 3-D scenes intrinsically difficult. However, this chapter has demonstrated that "weak perspective" views of distant objects are much simplified, so objects are commonly located using fewer features: for planar objects a pose ambiguity remains, although it is eliminated under full perspective.

16.6 BIBLIOGRAPHICAL AND HISTORICAL NOTES

The development of solutions to the so-called *perspective n-point* (PnP) *problem* (finding the pose of objects from *n* features under various forms of perspective) has been proceeding for more than two decades, and is by no means complete.

Fischler and Bolles summarized the situation as they saw it in 1981, and several new algorithms were described by them. However, they did not discuss pose determination under weak perspective, and perhaps surprisingly, considering its reduced complexity, this has subsequently been the subject of much research (e.g., Alter, 1994; Huang et al., 1995). Horaud et al. (1989) discussed the problem of finding rapid solutions to the PnP problem by reducing n as far as possible: they also obtained an analytic solution for the case $n = 4$ which should help considerably with real-time implementations. Their solution is related to Horaud's earlier (1987) corner interpretation scheme—described in Section 15.10, while Haralick et al. (1984) provided useful basic theory for matching wire frame objects.

In a later paper, Liu and Wong (1999) described an algorithm for pose estimation from four corresponding points under FPP when the points are not coplanar. Strictly, according to Table 16.1, this will lead to an ambiguity. However, Liu and Wong made the point "that the possibility for the occurrence of multiple solutions in the perspective 4-point problem is much smaller than that in the perspective 3-point problem," so that "using a 4-point model is much more reliable than using a 3-point model;" and they actually only claim "good results." Also, much of the emphasis of the paper is on errors and reliability. Hence, it seems that it is the *scope* for making errors in the sense of misinterpreting the situation that is significantly reduced. Added to this, Liu and Wong's (1999) work involves tracking a known object within a somewhat restricted region of space: this must again cut down the scope for error considerably. Hence it is not clear that their work violates the relevant (FPP; noncoplanar; $n = 4$)[7] entry in Table 16.1, rather than merely making it *unlikely* that a real ambiguity will arise.

Between them, Faugeras (1993), Hartley and Zisserman (2000), Faugeras and Luong (2001), and Forsyth and Ponce (2003) provide good coverage of the whole area of 3-D vision; for an interesting viewpoint on the subject, with particular emphasis on pose refinement, see Sullivan (1992). For further references on specific aspects of 3-D vision, see Sections 15.13, 17.13, and 18.16. (Section 19.10 gives references on motion, but also covers aspects of 3-D vision.)

16.6.1 **More Recent Developments**

Xu et al. (2008) present a new method for tackling the PnP problem. The linear method for the case of four coplanar points is extended to find coarse solutions for the general P3P problem. Once all the accurate solutions for the P3P problem have been found, the algorithm is applied to the general PnP problem. Solution stability issues and possible ambiguities are investigated and experiments are performed to verify the effectiveness of the proposed method. Interestingly, the case of four coplanar points has to be divided into two mutually exclusive cases, in which one point lies, or does not lie, within the triangle presented by the other three points. Lepetit et al. (2008) propose an O(n)

---

[7]See Fischler and Bolles (1981) for the original evidence for this particular ambiguity.

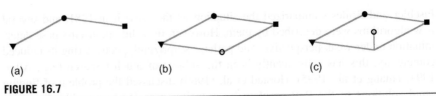

FIGURE 16.7

In this diagram the gray edges are construction lines, *not* parts of the objects. (a) and (b) are completely planar objects.

noniterative solution to the P*n*P problem that is far faster and more accurate than earlier methods and also more stable. The central idea is to express the 3-D points as a weighted sum of four virtual control points and to solve this case in terms of the coordinates. (The computational load of previous methods can be as high as $O(n^5)$ or even $O(n^8)$.)

16.7 PROBLEMS

1. Draw up a complete table of pose ambiguities that arise for *weak* perspective projection, for various numbers of object points identified in the image. Your answer should cover both coplanar points and noncoplanar points, and should make clear in each case how much ambiguity would remain in the limit of an infinite number of object points being seen. Give justification for your results.

2. Distinguish between *full perspective projection* and *weak perspective projection*. Explain how each of these projections presents oblique views of the following real objects: (i) straight lines, (ii) several concurrent lines (i.e., lines meeting in a single point), (iii) parallel lines, (iv) midpoints of lines, (v) tangents to curves, (vi) circles whose centers are marked with a dot. Give justification for your results.

3. Explain each of the following: (i) Why weak perspective projection leads to an ambiguity in viewing an object such as that in Fig. 16.7(a). (ii) Why the ambiguity doesn't disappear for the case of Fig. 16.7(b). (iii) Why the ambiguity *does* disappear in the case of Fig. 16.7(c), if the true nature of the object is known. (iv) Why the ambiguity doesn't occur in the case of Fig. 16.7(b) viewed under full perspective projection. In the last case, illustrate your answer by means of a sketch.

Invariants and Perspective 17

Invariants are important for achieving recognition in both 2-D and 3-D. The basic idea is to identify any parameters that do not vary between different instances of the same object. Unfortunately, perspective projection makes the issue far harder in the general 3-D case. This chapter explores the problem and demonstrates a number of useful techniques. At the same time it explores the problems of perspective projection, with some interesting outcomes.

Look out for:

- how a ratio of distances between features along the same straight line can act as a convenient invariant under weak perspective projection.
- how a ratio of ratios (or "cross-ratio") can act as a convenient invariant under full perspective projection.
- how the cross-ratio type of invariant can rather cunningly be generalized to cover many wider possibilities.
- how the cross-ratio type of invariant seems largely unable to provide invariance outside any given plane.
- vanishing point detection and its relevance to image interpretation.
- use of invariance for initiating face recognition.
- how to optimize views of 2-D pictures to limit perspective distortions.
- problems involved in "stitching" photographs.

While this chapter considers only one aspect of 3-D vision, it is extremely useful both in helping to cue into complex images (see particularly the egomotion example in Fig. 17.4 and the facial analysis example in Section 17.10) and in taking shortcuts around the tedious analysis of 3-D geometry (see, e.g., Sections 17.8 and 17.9).

17.1 INTRODUCTION

Pattern recognition is a complex task and, as stated in Chapter 1, involves the twin processes of discrimination and generalization. Indeed, the latter process is in many ways more important than the first—especially in the initial stages of recognition—since there is so much redundant information in a typical image. Thus, we need to find ways of helping to eliminate invalid matches. This is where the study of invariants comes into its own.

An invariant is a property of an object or class of objects that does not change with changes of viewpoint or object pose and which can therefore be used to help distinguish it from other objects. The procedure is to search for objects with a specific invariant, so that those which do not possess the invariant can immediately be discarded from consideration. An invariant property can be regarded as a necessary condition for an object to be in the chosen class, although in principle only detailed subsequent analysis will confirm the presence of that object in the chosen class. In addition, if an object is found to possess the correct invariant, it will then be profitable to pursue the analysis further and find its pose, size, or other relevant data. Ideally an invariant would uniquely identify an object as being of a particular type or class. Thus, an invariant should not merely be a property that leads to further hypotheses being made about the object, but one that fully characterizes it. However, the difference is a subtle one, more a matter of degree and purpose than an absolute criterion. We shall see later in this chapter the extent of the difference by appealing to a number of specific cases.

Let us first consider an object being viewed from directly overhead at a known distance by a camera whose optical axis is normal to the plane on which the object is lying. We shall assume that the object is flat. Take two-point features on the object, such as corners or small holes. If we measure the image distance between these features, this acts as an invariant, in that:

1. it has a value independent of the translation and orientation parameters of the object;
2. it will be unchanged for different objects of the same type; and
3. it will in general be different from the distance parameters of other objects that might be on the object plane.

Thus, measurement of distance provides a certain lookup or indexing quality that ideally identifies the object uniquely, although further analysis will be required to fully locate it and ascertain its orientation. Hence, distance has all the requirements of an invariant, although it could also be argued that it is only a feature that *helps* to classify objects. Clearly, we are here ignoring an important factor—the effect of imprecision in measurement—due to spatial quantization (or inadequate spatial resolution), noise, lens distortions, and so on; in addition, the effects of partial occlusion or breakage are also being ignored. Most definitely, there is a limit to what can be achieved with a single invariant measure, although in what follows we attempt to reveal what is possible and demonstrate the advantages of employing an invariant-orientated approach.

The above ideas relating to distance as an invariant measure showed it to be useful in suppressing the effects of translations and rotations of objects in 2-D. Hence, it is of little direct value when considering translations and rotations in 3-D. Furthermore, it is not even able to cope with scale variations of objects in 2-D. Moving the camera closer to the object plane and refocussing totally change the situation, and all values of the distance invariant residing in the object indexing table must be changed and the old values ignored. However, a little thought will show that this last problem could be overcome. All we need to do is to take *ratios* of distances. This requires a minimum of three point features to be identified in the image and the inter-feature distances measured. If we call two of these distances d_1 and d_2, then the ratio d_1/d_2 will act as a scale-independent invariant, i.e., we will be able to identify objects using a single indexing operation whatever their 2-D translation, orientation, or apparent size or scale. An alternative to this idea is to measure the angle between pairs of distance vectors, $\cos^{-1}(\mathbf{d}_1.\mathbf{d}_2/|d_1||d_2|)$, which will again be scale invariant.

Of course, this consideration has already been invoked in our earlier work on shape analysis. If objects are subject only to 2-D translations and rotations but not to changes of scale, they can be characterized by their perimeters or areas as well as their normal linear dimensions; furthermore, parameters such as compactness and aspect ratio, which employ dimensionless ratios of image measurements, have been acknowledged in Chapter 9 to overcome the size/scale problem.

Nevertheless, the main motivation for using invariants is to obtain mathematical measures of configurations of object features that are carefully designed to be independent of the viewpoint or coordinate system used, and indeed to not require specific setup or calibration of the image acquisition system. However, it must be emphasized that camera distortions are assumed to be absent or to have been compensated for by suitable post-camera transformations (see Chapter 18).

This chapter proceeds to develop the above ideas and later applies them to vanishing point detection (Sections 17.7–17.9), face recognition (Section 17.10), obtaining optimal views of 2-D pictures, and the "stitching" of digital photographs (Section 17.11). Interestingly, with the small intellectual outlay of the initial sections, these applications emerge with very little additional effort demonstrating the value of the basic theory presented here.

17.2 CROSS-RATIOS: THE "RATIO OF RATIOS" CONCEPT

It would be most useful if we could extend the above ideas to permit indexing for general transformations in 3-D. Indeed, an obvious question is whether finding ratios of ratios of distances will provide suitable invariants and lead to such a generalization. The answer is that ratios of ratios do provide useful further invariants, although going further than this leads to considerable complication, and there are restrictions on what can be achieved with limited computation. In addition, noise ultimately becomes a limiting factor, since so many parameters become involved in the computation of complex invariants that the method ultimately loses steam (it becomes just one of many ways of raising hypotheses and therefore has to compete with other approaches in a manner appropriate to the particular problem application being studied).

We now consider the ratio of ratios approach. Initially, we only examine a set of four collinear points on an object. Figure 17.1 shows such a set of four points (P_1, P_2, P_3, P_4) and a transformation of them (Q_1, Q_2, Q_3, Q_4) such as that produced by an imaging system with optical center C (c, d). Choice of a suitable pair of oblique axes permits the coordinates of the points in the separate sets to be expressed respectively as:

$$(x_1, 0), (x_2, 0), (x_3, 0), (x_4, 0)$$
$$(0, y_1), (0, y_2), (0, y_3), (0, y_4)$$

Taking points P_i, Q_i, we can write the ratio $CQ_i:PQ_i$ both as $c/(-x_i)$ and as $(d - y_i)/y_i$. Hence:

$$\frac{c}{x_i} + \frac{d}{y_i} = 1 \tag{17.1}$$

which must be valid for all i. Subtraction of the ith and jth relations now gives:

$$\frac{c(x_j - x_i)}{x_i x_j} = \frac{-d(y_j - y_i)}{y_i y_j} \tag{17.2}$$

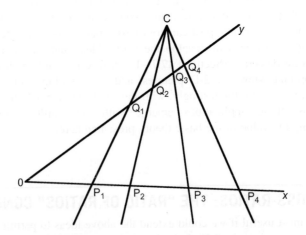

FIGURE 17.1

Perspective transformation of four collinear points. This figure shows four collinear points (P_1, P_2, P_3, P_4) and a transformation of them (Q_1, Q_2, Q_3, Q_4) similar to that produced by an imaging system with optical center C. Such a transformation is called a *perspective transformation*.

Forming a ratio between two such relations will now eliminate the unknowns c and d. For example, we will have:

$$\frac{x_3(x_2 - x_1)}{x_2(x_3 - x_1)} = \frac{y_3(y_2 - y_1)}{y_2(y_3 - y_1)} \tag{17.3}$$

However, the result still contains factors such as x_3/x_2 that depend on absolute position. Hence, it is necessary to form a suitable ratio of such results, which cancels out the effects of absolute positions:

$$\frac{(x_2 - x_4)/(x_3 - x_4)}{(x_2 - x_1)/(x_3 - x_1)} = \frac{(y_2 - y_4)/(y_3 - y_4)}{(y_2 - y_1)/(y_3 - y_1)} \tag{17.4}$$

Thus, our original intuition that a ratio of ratios type of invariant might exist that would cancel out the effects of a perspective transformation is correct. In particular, four collinear points viewed from any perspective viewpoint yield the same value of the cross-ratio, defined as above. The value of the cross-ratio of the four points is written:

$$C(P_1, P_2, P_3, P_4) = \frac{(x_3 - x_1)(x_2 - x_4)}{(x_2 - x_1)(x_3 - x_4)} \tag{17.5}$$

For clarity, we shall write this particular cross-ratio as κ in what follows. Note that there are $4! = 24$ possible ways in which 4 collinear points can be ordered on a straight line, and hence there could be 24 cross-ratios. However, they are not all distinct, and in fact there are only six different values. To verify this we start by interchanging pairs of points:

$$C(P_2, P_1, P_3, P_4) = \frac{(x_3 - x_2)(x_1 - x_4)}{(x_1 - x_2)(x_3 - x_4)} = 1 - \kappa \tag{17.6}$$

$$C(P_1, P_3, P_2, P_4) = \frac{(x_2 - x_1)(x_3 - x_4)}{(x_3 - x_1)(x_2 - x_4)} = \frac{1}{\kappa} \tag{17.7}$$

$$C(P_1, P_2, P_4, P_3) = \frac{(x_4 - x_1)(x_2 - x_3)}{(x_2 - x_1)(x_4 - x_3)} = 1 - \kappa \tag{17.8}$$

$$C(P_4, P_2, P_3, P_1) = \frac{(x_3 - x_4)(x_2 - x_1)}{(x_2 - x_4)(x_3 - x_1)} = \frac{1}{\kappa} \tag{17.9}$$

$$C(P_3, P_2, P_1, P_4) = \frac{(x_1 - x_3)(x_2 - x_4)}{(x_2 - x_3)(x_1 - x_4)} = \frac{\kappa}{\kappa - 1} \tag{17.10}$$

$$C(P_1, P_4, P_3, P_2) = \frac{(x_3 - x_1)(x_4 - x_2)}{(x_4 - x_1)(x_3 - x_2)} = \frac{\kappa}{\kappa - 1} \tag{17.11}$$

These cases provide the main possibilities, but of course interchanging more points will yield a limited number of further values—in particular:

$$C(P_3, P_1, P_2, P_4) = 1 - C(P_1, P_3, P_2, P_4) = 1 - \frac{1}{\kappa} = \frac{\kappa - 1}{\kappa} \tag{17.12}$$

$$C(P_2, P_3, P_1, P_4) = \frac{1}{C(P_2, P_1, P_3, P_4)} = \frac{1}{1 - \kappa} \qquad (17.13)$$

This covers all six cases, and a little thought (based on trying further interchanges of points) will show there can be no others (we can only repeat κ, $1 - \kappa$, $\kappa/(\kappa - 1)$, and their inverses). Of particular interest is the fact that numbering the points in reverse (which would correspond to viewing the line from the other side) leaves the cross-ratio unchanged. Nevertheless, it is inconvenient that the same invariant has six different manifestations, as this implies that six different index values have to be looked up before the class of an object can be ascertained. On the other hand, if points are labeled in order along the line rather than randomly, it should generally be possible to circumvent this situation.

So far we have been able to produce only one projective invariant, and this corresponds to the rather simple case of four collinear points. The usefulness of this measure is augmented considerably when it is noted that four collinear points, taken in conjunction with another point, define a pencil[1] of concurrent coplanar lines passing through the latter point. Clearly, we can assign a unique cross-ratio to this pencil of lines, equal to the cross-ratio of the collinear points on any line passing through them. We can clarify the situation by considering the angles between the various lines (Fig. 17.2). Applying the sine rule four times to determine the four distances in the cross-ratio $C(P_1, P_2, P_3, P_4)$ gives:

$$\frac{x_3 - x_1}{\sin \alpha_{13}} = \frac{OP_1}{\sin \beta_3} \qquad (17.14)$$

$$\frac{x_2 - x_4}{\sin \alpha_{24}} = \frac{OP_4}{\sin \beta_2} \qquad (17.15)$$

$$\frac{x_2 - x_1}{\sin \alpha_{12}} = \frac{OP_1}{\sin \beta_2} \qquad (17.16)$$

$$\frac{x_3 - x_4}{\sin \alpha_{34}} = \frac{OP_4}{\sin \beta_3} \qquad (17.17)$$

Substituting in the cross-ratio formula (Eq. (17.5)) and canceling the factors OP_1, OP_4, $\sin \beta_2$, and $\sin \beta_3$ now give:

$$C(P_1, P_2, P_3, P_4) = \frac{\sin \alpha_{13} \sin \alpha_{24}}{\sin \alpha_{12} \sin \alpha_{34}} \qquad (17.18)$$

Thus, the cross-ratio depends only on the angles of the pencil of lines. It is interesting that appropriate juxtaposition of the sines of the angles gives the final

---

[1] It is a common nomenclature of projective geometry to call a set of concurrent lines a *pencil* (e.g., Semple and Kneebone, 1952).

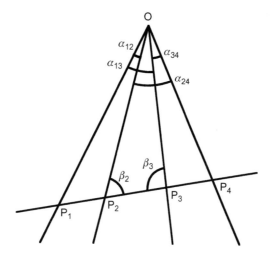

FIGURE 17.2

Geometry for calculation of the cross-ratio of a pencil of lines. The figure shows the geometry required to calculate the cross-ratio of a pencil of lines in terms of the angles between them.

formula for invariance under perspective projection: using the angles themselves would not give the desired degree of mathematical invariance. Indeed, we can immediately see one reason for this: inversion of the direction of any line must leave the situation unchanged, so the formula must be tolerant to adding π to each of the two angles linking the line; this could not be achieved if the angles appeared without suitable trigonometric functions.

We can extend this concept to four concurrent planes since the concurrent lines can be projected into four concurrent planes once a separate axis for the concurrency has been defined. As there are infinitely many such axes, there are infinitely many ways in which sets of planes can be chosen. Thus, the original simple result on collinear points can be extended to a much more general case.

Finally, note that we started by trying to generalize the case of four collinear points, but what we achieved was first to find a dual situation in which points become lines also described by a cross-ratio, and then to find an extension in which planes are described by a cross-ratio. We now return to the case of four collinear points and see how we can extend it in other ways.

17.3 INVARIANTS FOR NONCOLLINEAR POINTS

First, imagine that not all the points are collinear: specifically, let us assume that one point is not in the line of the other three. If this is the case, there is not

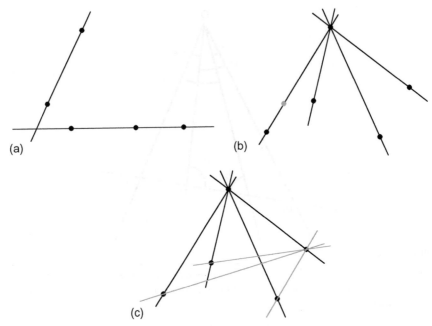

FIGURE 17.3

Calculation of invariants for a set of noncollinear points. Part (a) shows how the addition of a fifth point to a set of four points, one of which is not collinear with the rest, permits the cross-ratio to be calculated. Part (b) shows how the calculation can be extended to any set of noncollinear points; also shown is an additional (gray) point that a single cross-ratio fails to distinguish from other points on the same line. Part (c) shows how any failure to identify a point uniquely can be overcome by calculating the cross-ratio of a second pencil generated from the five original points.

enough information to calculate a cross-ratio. However, if a further coplanar point is available, we can draw an imaginary line between the noncollinear points to intersect their common line in a unique point, which will then permit a cross-ratio to be computed (Fig. 17.3(a)). Nevertheless, this is some way from a general solution to the characterization of a set of noncollinear points. We might inquire how many point features in general position[2] on a plane will be required to calculate an invariant. In fact, the answer is five, since the fact that we can form a cross-ratio from the angles between four lines immediately means that forming a pencil of four lines from five points defines a cross-ratio invariant (Fig. 17.3(b)).

---

[2]Points on a plane that are chosen at random, and that are not collinear or in any special pattern such as a regular polygon, are described as being *in general position*.

While the value of this cross-ratio provides a necessary condition for a match between two sets of five general coplanar points, it could be a fortuitous match because the condition depends only on the relative directions between the various points and the reference point, i.e., any of the nonreference points is only defined to the extent that it lies on a given line. Clearly, two cross-ratios formed by taking two reference points will define the directions of all the remaining points uniquely (Fig. 17.3(c)).

We can now summarize the general result, which stipulates that for five general coplanar points, no three being collinear, two different cross-ratios are required to characterize the shape. These cross-ratios correspond to taking in turn two separate points and producing pencils of lines passing through them and (in each case) the remaining four points (Fig. 17.3(c)). While it might appear that at least five cross-ratios result from this sort of procedure, there are only two functionally independent cross-ratios—essentially because the position of any point is defined once its direction relative to two other points is known.

Next, we consider the problem of finding the ground plane in practical situations—especially that of egomotion including vehicle guidance (Fig. 17.4). Here a set of four collinear points can be observed from one frame to the next. If they are on a single plane, then the cross-ratio will remain constant, but if one is elevated above the ground plane (as, e.g., a bridge or another vehicle), then the cross-ratio will vary over time. Taking a larger number of points, it should clearly be possible to deduce which are on the ground plane and which are not, by using a process of elimination; however, the amount of noise and clutter will determine the computational complexity of the task. Note that this is possible without any calibration of the camera, this being perhaps the main value of concentrating attention on projective invariants. Note also that there is a potential problem regarding irrelevant planes, such as the vertical faces of buildings. The cross-ratio test is so resistant to viewpoint and pose that it merely ascertains whether the points being tested are coplanar. It is only by using a sufficiently large number of independent sets of points that one plane can be discriminated from another (for simplicity we ignore here any subsequent stages of pose analysis that might be carried out).

17.3.1 Further Remarks About the Five-Point Configuration

The above description outlines the principles for solving the five-point invariance problem, but does not show clearly the conditions under which it is guaranteed to operate properly. In fact, these are straightforward to demonstrate. First, the cross-ratio can be expressed in terms of the sines of the angles α_{13}, α_{24}, α_{12}, α_{34}. Next, these can be re-expressed in terms of areas of relevant triangles, using equations typified by the following to express area:

$$\Delta_{513} = \frac{1}{2} a_{51} a_{53} \sin \alpha_{13} \qquad (17.19)$$

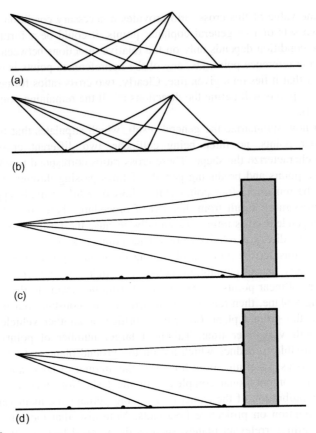

(a)

(b)

(c)

(d)

FIGURE 17.4

Use of cross-ratio for egomotion guidance. Part (a) shows how the cross-ratio for a set of four collinear points can be tracked to confirm that the points are collinear. This suggests that they lie on the ground plane. Part (b) shows a case where the cross-ratio will not be constant. Part (c) shows a case where the cross-ratio is constant, although they actually lie on a plane that is not the ground plane. Part (d) shows a case where all four points lie on planes, yet the cross-ratio will not be constant.

Finally, the area can be re-expressed in terms of the point coordinates in the following way:

$$\Delta_{513} = \frac{1}{2} \begin{vmatrix} p_{5x} & p_{1x} & p_{3x} \\ p_{5y} & p_{1y} & p_{3y} \\ p_{5z} & p_{1z} & p_{3z} \end{vmatrix} = \frac{1}{2} |\mathbf{P}_5\ \mathbf{P}_1\ \mathbf{P}_3| \tag{17.20}$$

Using this notation, a suitable final pair of cross-ratio invariants for the configuration of five points may be written:

$$C_a = \frac{\Delta_{513}\Delta_{524}}{\Delta_{512}\Delta_{534}} \tag{17.21}$$

$$C_b = \frac{\Delta_{124}\Delta_{135}}{\Delta_{123}\Delta_{145}} \tag{17.22}$$

While three more such equations may be written down, these will not be independent of the other two and will not carry any further useful information.

Note that a determinant will go to zero or infinity if the three points it relates to are collinear, corresponding to the situation when the area of the triangle is zero. Clearly, when this happens, any cross-ratio containing this determinant will no longer be able to pass on any useful information. On the other hand, there is actually no further information to pass on, as this now constitutes a special case that is describable by a single cross-ratio: we have reverted to the situation shown in Fig. 17.3(a).

Finally, Fig. 17.3 misses out one further interesting case: the situation of two points and two lines (Fig. 17.5). Constructing a line joining the two points and producing it until it meets the two lines, we then have four points on a single line; thus, the configuration is characterized by a single cross-ratio. Note also that the two lines can be extended until they intersect, and further lines can be constructed from the intersection to meet the two points. This gives a pencil of lines characterized by a single cross-ratio (Fig. 17.5(c)); the latter must have the same value as that computed for the four collinear points.

17.4 INVARIANTS FOR POINTS ON CONICS

These discussions clearly help to build up an understanding of how geometric invariants can be designed to cope with sets of points, lines, and planes in 3-D. Significantly more difficult is the case of curved lines and surfaces, although much headway has now been made with regard to the understanding of conics and certain other surfaces (see Mundy and Zisserman, 1992a). It will not be possible to examine all such cases in depth here. However, it will be useful to consider conic sections and particularly ellipses in more detail.

First, we consider Chasles' theorem, which dates from the 19th century. (The history of projective geometry is quite rich and was initially carried out totally independently of the requirements of machine vision.) Suppose we have four fixed coplanar points F_1, F_2, F_3, F_4 on a conic section curve and one variable point P in the same plane (Fig. 17.6). Then the four lines joining P to the fixed points form a

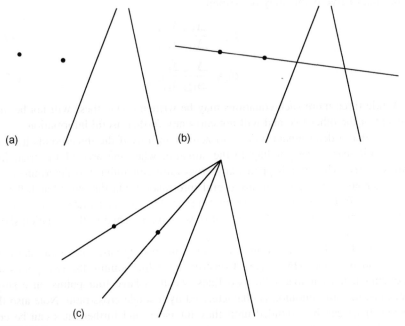

FIGURE 17.5

Cross-ratio for two lines and two points. (a) Basic configuration. (b) How the line joining the two points introduces four collinear points to which a cross-ratio may be applied. (c) How joining the two points to the junction of the two lines creates a pencil of four lines to which a cross-ratio may be applied.

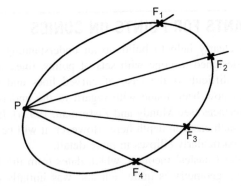

FIGURE 17.6

Definition of a conic using a cross-ratio. Here P is constrained to move so that the cross-ratio of the pencil from P to F_1, F_2, F_3, F_4 remains constant. By Chasles' theorem, P traces out a conic curve.

pencil whose cross-ratio will in general vary with the position of P. Chasles' theorem states that if P now moves so as to keep the cross-ratio constant, then P will trace out a conic section. This clearly provides a means of checking whether a set of points lies on a planar curve such as an ellipse. Note the close analogy with the problem of ground plane detection already mentioned. Again the amount of computation could become excessive if there were a lot of noise or clutter in the image. When the image contains N boundary features that need to be checked out, the problem complexity is intrinsically $O(N^5)$, since there are $O(N^4)$ ways of selecting the first four points, and for each such selection, $N-4$ points must be examined to determine whether they lie on the same conic. However, choice of suitable heuristics would be expected to limit the computation. Note the problem of ensuring that the first four points are tested in the same order around the ellipse, which is liable to be tedious (a) for point features and (b) for disconnected boundary features.

While Chasles' theorem gives an excellent opportunity to use invariants to locate conics in images, it is not at all discriminatory in this. The theorem applies to a general conic, hence it does not immediately permit circles, ellipses, parabolas, or hyperbolas to be distinguished, a fact that would sometimes be a distinct disadvantage. This is an example of a more general problem in pattern recognition system design—of deciding exactly how and in what sequence one object should be differentiated from another. Because of the space constraint, this point is not considered further here.

Finally, we state without proof that conic section curves can all be transformed under perspective projection to other types of conic section, and thus into ellipses; subsequently they can be transformed into circles. Thus, any conic section curve can be transformed projectively into a circle, while the inverse transformation can transform it back again (Mundy and Zisserman, 1992b). This means that simple properties of the circle can frequently be generalized to ellipses or other conic sections. In this context, points to bear in mind are that, after perspective projection, lines intersecting curves do so in the same number of points, and thus tangents transform into tangents and chords into chords, three-point contact (in the case of non-conic curves) remains three-point contact, and so on. Returning to Chasles' theorem, a simple proof in the case of circles will automatically generalize to more complex conic section curves.

In response to this assertion, we can in fact derive Chasles' theorem almost trivially for a circle. Appealing to Fig. 17.7, we see that the angles φ_1, φ_2, φ_3 are equal to the respective angles γ_1, γ_2, γ_3 (angles in the same segment of a circle). Thus, the pencils PF_1, PF_2, PF_3, PF_4 and QF_1, QF_2, QF_3, QF_4 have equal angles, their *relative* directions being superposable. This means that they will have the same cross-ratio, defined by Eq. (17.18). Hence, the cross-ratio of the pencil will remain constant as P traces out the circle. As stated above, the property will automatically generalize to any other conic.

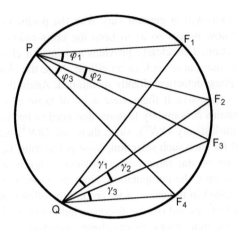

FIGURE 17.7

Proof of Chasles' theorem. This diagram shows that the four points F_1, F_2, F_3, F_4 subtend the same angles at P as they do at the fixed point Q. Thus, the cross-ratio is the same for all points on the circle. This means that Chasles' theorem is valid for a circle.

17.5 DIFFERENTIAL AND SEMI-DIFFERENTIAL INVARIANTS

There have been many attempts to characterize continuous curves by invariants. The obvious way forward is to represent points on a curve in terms of local curve derivatives. If a sufficient number of these can be obtained, invariants can be formed and computed. However, the noise (including digitization noise) that always exists on curves limits the accuracy of higher derivatives, and as a result it is difficult to form useful invariants in this way. In general, the second derivative of the curve function is the highest that can normally be used, and this corresponds to curvature, which is only an invariant for Euclidean transformations (translation and rotation without change of scale).

As a result of this problem, semi-differential invariants are often used instead of differential invariants. They involve considering only a few "distinguished" points on curves, and using these to generate invariants. The most common distinguished points to be used in this way are (Fig. 17.8) the following:

1. Points of inflection
2. Sharp corners on curves
3. Cusps on curves (corners where the bounding tangents are coincident)
4. Bitangent points (points of contact of a line that touches the curve twice)
5. Other points whose locations can be derived from existing distinguished points by geometric constructions

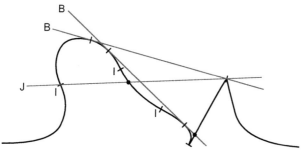

FIGURE 17.8

Means for finding distinguished points on a curve. The two bitangents contact the curve in a total of four bitangent points. Three points of inflection I provide another three distinguished points. A cusp and a corner provide a further two distinguished points (the latter also being a bitangent point). The line marked J contributes a further distinguished point on the curve, as does one of the bitangents: these are marked as large dots rather than as short lines.

Tangent points are unlikely to be suitable, and hence are not included in this list, as a smooth curve will have tangents along its entire length. This is because they are characterized merely by two-point contact between a limiting chord and the curve. However, a point of inflection represents a three-point contact, and this means that it will be reasonably well localized, and its tangent will have a well-defined direction. On the other hand, bitangent points will be even more accurately represented, as the tangent direction will be accurately defined by two well-separated points on the curve (Fig. 17.8): nevertheless bitangent points will still incur some longitudinal error.

Bitangents can be of several sorts: in particular, they can contact the same shape on the same side; they can also cross the body and contact it on both sides. This latter case is more complex and is therefore sometimes discounted in machine vision applications. Nevertheless, it provides a means of finding further invariant reference points on an object. Note that this clearly happens directly, in that the bitangent points are already distinguished points. It also happens indirectly, as the bitangent may cross other reference lines, thereby defining further distinguished points. Figure 17.9 shows several cases of direct and indirect distinguished points, the most accurate of which arise from bitangents, while slightly less accurate ones arise from points of inflection.

Once enough distinguished points and reference lines between them have been found, cross-ratio invariants may be obtained (a) from the incidence of distinguished points lying along suitable reference lines and (b) from pencils of lines drawn from distinguished points to line crossings or to other distinguished points.

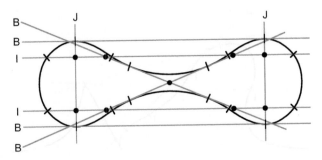

FIGURE 17.9

Means for finding direct and indirect distinguished points for an object. The four lines marked B are bitangents, which contribute twelve bitangent points: two of the bitangents contact the object on opposite sides of its boundary. The two lines marked I arise from points of inflection. The two lines marked J are joins of bitangent points. The nine large dots are indirect distinguished points, which do not lie on the object boundary. Clearly, a good many more indirect distinguished points could be generated, although not all would have accurately defined locations.

A remark is needed to confirm that points of inflection can act as suitable distinguished points that are invariant under perspective transformations. Starting from the premise that perspective transformations preserve straight lines and points arising from crossings between curves and lines, we note that a chord that crosses a curve three times will also cross it three times under perspective projection. This will still apply even when the three crossing points merge into three-point contact.[3] Hence, points of inflection are suitable distinguished points and are perspective-invariant.

This treatment has only dealt with planar curves, and has not covered spatial nonplanar curves. The latter is a significantly more difficult area, as concepts such as bitangents and points of inflection have to be assigned new meaning in this more general domain. It is a subject we shall not be able to broach here.

17.6 SYMMETRIC CROSS-RATIO FUNCTIONS

When applying a cross-ratio to a set of points on a line, it frequently happens that the order of the points on the line is known. For example, this will almost certainly be the case if feature detection of an image is carried out in a forward raster scan. Hence, the only confusion in the ordering will be the direction in which the

---

[3]Three-point contact is distinguishable from two-point contact in that the tangent crosses the curve at the point of contact.

line has been traversed. However, the cross-ratio is independent of the end from which the line is scanned, since $C(P_1, P_2, P_3, P_4) = C(P_4, P_3, P_2, P_1)$. Nevertheless, there are situations in which the ordering of the cross-ratio features will not be known with certainty. This may occur for the situations shown in Figs. 17.3 and 17.5, where the features themselves do not all lie on a single line, where the features are angles, or where the points lie on a conic whose equation is as yet unknown. In such circumstances, it will be useful to have an invariant that takes in all possible orders of the features.

To derive such an invariant, note first that if there is a confusion in the ordering of the points such that the value could be either κ or $(1 - \kappa)$, then we could apply the function $f(\kappa) = \kappa(1 - \kappa)$, which has the property $f(\kappa) = f(1 - \kappa)$, and this will solve the problem. Alternatively, if there is confusion between κ and $1/\kappa$, then we could apply the function $g(\kappa) = \kappa + 1/\kappa$, which has the property $g(\kappa) = g(1/\kappa)$, and again this will solve the problem.

However, if there is potential confusion between the values κ, $(1 - \kappa)$, and $1/\kappa$, the situation becomes more complicated. It is difficult to write down any obvious function that satisfies the double condition $h(\kappa) = h(1 - \kappa) = h(1/\kappa)$, although we may have a soundly based intuition that it will involve symmetric functions such as $f(\kappa)$ and $g(\kappa)$. In fact, the simplest answer seems to be:

$$j(\kappa) = \frac{(1 - \kappa + \kappa^2)^3}{\kappa^2(1 - \kappa)^2} \qquad (17.23)$$

which obeys the symmetry idea as it can be re-expressed in the two forms:

$$j(\kappa) = \frac{[(1 - \kappa(1 - \kappa)]^3}{[\kappa(1 - \kappa)]^2} = \frac{(\kappa + 1/\kappa - 1)^3}{\kappa + 1/\kappa - 2} \qquad (17.24)$$

Fortunately, we need to go no further in our quest to obey the six conditions required to recognize all six cross-ratio values κ, $(1 - \kappa)$, $1/\kappa$, $1/(1 - \kappa)$, $(\kappa - 1)/\kappa$, $\kappa/(\kappa - 1)$. The reason is that they are all deducible from each other by further applications of the initial negation and inversion rules. (The ultimate reason for this is that the operations to transform the function from one to another of the six forms form a group of order six, which is generated from the negation and inversion transforms.)

While this is a powerful result, it does not come without loss. The reason is that there is now a sixfold ambiguity inherent in the solution, so that once we have shown that the set of points satisfies the symmetric cross-ratio function, we still have to make tests to determine which of the six possibilities is the correct one. This is reflected by the complexity of the j-function, which contains a sixth degree polynomial and for every value of j there are six possible values of κ (Fig. 17.10).

The situation can be described by saying that the function $j(\kappa)$ is not "complete," in the sense that this function alone is insufficient to recognize the set of features unambiguously. To underline this, observe that the original cross-ratio is complete: once the value of κ is known, we can uniquely determine the position

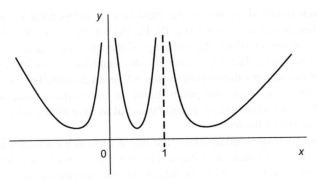

FIGURE 17.10

Symmetric cross-ratio function. This is the function defined by Eq. (17.23).

of one of the points from the other three points. This is obvious from the graph of κ as a function of x (where $x = x_{34}$ gives the position of the fourth point),[4] which is a hyperbola:

$$\kappa = \frac{x_{31}x_{24}}{x_{21}x_{34}} = \frac{x_{31}(x_{23}+x)}{x_{21}x} = \frac{x_{31}x_{23}}{x_{21}}\left(\frac{1}{x} + \frac{1}{x_{23}}\right) \qquad (17.25)$$

17.7 VANISHING POINT DETECTION

In this section, we consider how vanishing points (VPs) can be detected. It is usual to carry this out in two stages: first, we locate all the straight lines in the image; next, we find which of the lines pass through common points. The latter being interpreted as VPs. Finding the lines using the Hough transform should be straightforward, although texture edges will sometimes prevent lines from being located accurately and consistently. Basically, locating the VPs requires a second Hough transform in which whole lines are accumulated in parameter space, leading to well-defined peaks (the VPs) where multiple lines overlap. In practice, the lines of votes will have to be extended to cover all possible vanishing point locations. This procedure is adequate when the VPs appear within the original image space, but it often happens that they will be outside the original image (Fig. 17.11) and may even be situated at infinity. This means that a normal image-like parameter space cannot be used successfully, even if it is extended beyond the original image space. Another problem is that for distant VPs, the peaks in parameter space will be

---

[4]In projective geometry, it is well known that there are three degrees of freedom on a line. The positions of three points on a line are not predictable from other views of the three points, without further information on the viewpoint.

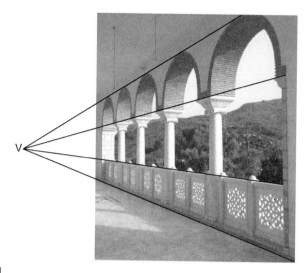

FIGURE 17.11

Position of the vanishing point. In this figure, parallel lines on the arches appear to converge to a vanishing point V outside the image. In general, vanishing points can lie at any distance and may even be situated at infinity.

spread out over a considerable distance, so detection sensitivity will be poor and accuracy of location will be low.

Fortunately, Magee and Aggarwal (1984) found an improved representation for locating VPs. They constructed a unit sphere G, called a Gaussian sphere, around the center of projection of the camera, and used G instead of an extended image plane as a parameter space. In this representation (Fig. 17.12), VPs appear at finite distances even in cases where they would otherwise appear to be at infinity. For this method to work, there has to be a one-to-one correspondence between points in the two representations, and this is clearly valid (note that the back half of the Gaussian sphere is not used). However, the Gaussian sphere representation is not without problems: in particular, many irrelevant votes will be cast from lines that are not parallel in real 3-D space (generally only a small subset of the lines in the image will pass through VPs). To solve this problem, *pairs* of lines are considered in turn, and their crossing points are only accumulated as votes if the lines of each pair are judged likely to originate from parallel lines in 3-D space (e.g., they should have compatible gradients in the image). This procedure drastically limits both the number of votes recorded in parameter space and the number of irrelevant peaks. Nevertheless, the overall cost is still substantial, being proportional to the number of pairs of lines. Thus, if there are N lines, the number of pairs is $^{N}C_{2} = \frac{1}{2} N(N-1)$, so the result is O($N^2$).

The above procedure is important because it provides a highly reliable means for performing the search for VPs and largely discriminating against isolated lines

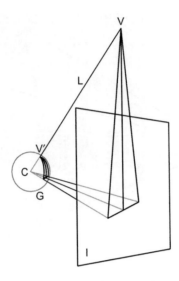

FIGURE 17.12

Detection of vanishing points using the Gaussian sphere. Parallel lines in space lead to converging lines in the image I. While the vanishing point V is here well above the image; it is easily located by projecting the lines onto the Gaussian sphere G. As discussed in the text, G is commonly used as a parameter space for accumulating vanishing point votes. C is the center of projection of the camera lens.

and image clutter. Note that for a moving robot or other system, the correspondences between the VPs seen in successive images will lead to considerably greater certainty in the interpretation of each image.

17.8 MORE ON VANISHING POINTS

One advantage of the cross-ratio is that it can turn up in many situations and on each occasion provide yet another neat result. A further example is when a road or pavement has flagstones whose boundaries are well demarcated and easily measurable. They can then be used to estimate the position of the VP on the ground plane. Imagine viewing the flagstones obliquely from above, with the camera or the eyes aligned horizontally (e.g., as for Fig. 23.12(a)). Then we have the geometry of Fig. 17.13 where the points O, H_1, H_2 lie on the ground plane, whereas O, V_1, V_2, V_3 are in the image plane.[5]

---

[5]Note that slightly oblique measurement of the flagstones, along a line that is not parallel to the sides of the flagstones, still permits the same cross-ratio value to be obtained, as the same angular factor applies to all distances along the line.

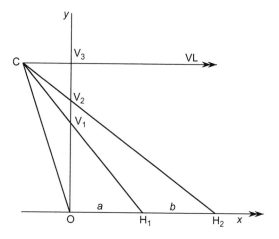

FIGURE 17.13

Geometry for finding the vanishing line from a known pair of spacings. C is the center of projection. VL is the vanishing line direction, which is parallel to the ground plane OH_1H_2. Although the camera plane $OV_1V_2V_3$ is drawn perpendicular to the ground plane, this is not necessary for successful operation of the algorithm (see text).

If we regard C as a center of projection, the cross-ratio formed from the points O, V_1, V_2, V_3 must have the same value as that formed from the points O, H_1, H_2, and infinity in the horizontal direction. Supposing that OH_1 and H_1H_2 have known lengths a and b, equating the cross-ratio values gives:[6]

$$\frac{y_1(y_3 - y_2)}{y_2(y_3 - y_1)} = \frac{x_1}{x_2} = \frac{a}{a+b} \tag{17.26}$$

This allows us to estimate y_3:

$$(a+b)(y_1y_3 - y_1y_2) = ay_2y_3 - ay_2y_1 \tag{17.27}$$

$$\therefore \quad y_3(ay_1 + by_1 - ay_2) = ay_1y_2 + by_1y_2 - ay_1y_2 \tag{17.28}$$

$$\therefore \quad y_3 = \frac{by_1y_2}{ay_1 + by_1 - ay_2} \tag{17.29}$$

If $a = b$ (as is likely to be the case for flagstones):

$$y_3 = \frac{y_1y_2}{2y_1 - y_2} \tag{17.30}$$

---

[6]Note that, in the case of Fig. 17.13, the y values are measured from O rather than from V_3.

Note that this proof does not actually assume that points V_1, V_2, V_3 are vertically above the origin, or that line OH_1H_2 is horizontal, just that these points lie along two coplanar straight lines and that C is in the same plane. Also, note that it is only the ratio of a to b, not their absolute values, that is relevant in this calculation.

Having found y_3, we have calculated the direction of the VP, whether or not the ground plane on which it lies is actually horizontal, and whether or not the camera axis is horizontal.

Finally, note that Eq. (17.30) can be rewritten in the simpler form:

$$\frac{1}{y_3} = \frac{2}{y_2} - \frac{1}{y_1} \qquad (17.31)$$

The inverse factors give some intuition into the processes involved—not least considering the inverse relation $Z = Hf/y$ between distance along the ground plane and image distance from the vanishing line; and similarly, the inverse relation between depth and disparity in Eq. (15.4).

17.9 APPARENT CENTERS OF CIRCLES AND ELLIPSES

It is well known that circles and ellipses project into ellipses (or occasionally into circles). This statement is widely applicable and is valid for orthographic projection, scaled orthographic projection, weak perspective projection, and full perspective projection.

Another factor that can easily be overlooked is what happens to the center of the circle or ellipse under these transformations. It turns out that the ellipse (or circle) center does not project into the ellipse (or circle) center under full perspective projection: there will in general be a small offset (Fig. 17.14).[7]

If the position of the vanishing line of the plane can be identified in the image, the calculation of the offset for a circle is quite simple using the theory in Section 17.8, which applies as the center of a circle bisects its diameter (Fig. 17.15). First, let ε be the shift in the center, d the distance of the center of the ellipse from the vanishing line, and b the length of the semi-minor axis. Next, identify $b + \varepsilon$ with y_1, $2b$ with y_2, and $b + d$ with y_3. Finally, substitute for y_1, y_2, and y_3 in Eq. (17.30). We then obtain the result:

$$\varepsilon = \frac{b^2}{d} \qquad (17.32)$$

---

[7]The fact that this happens may perhaps suggest that ellipses will be slightly distorted under projection. In fact, there is *no* such distortion, and the source of the shift in the center is merely that full perspective projection does not preserve length ratios—and in particular midpoints do not remain midpoints.

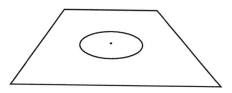

FIGURE 17.14

Projected position of a circle center under full perspective projection. Note that the projected center is not at the center of the ellipse in the image plane.

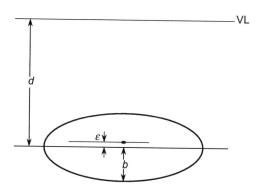

FIGURE 17.15

Geometry for calculating the offset of the circle center. The projected center of the circle is shown as the elongated dot, and the center of the ellipse in the image plane is a distance d below the vanishing line VL.

Note that, unlike the situation in Section 17.8, we are here assuming that y_3 is known and we are using its value to calculate y_1 and hence ε.

If the vanishing line is not known, but the orientation of the plane on which the circle lies is known, and also the orientation of the image plane, then the vanishing line can be deduced, and the calculation can again proceed as above. However, this approach assumes that the camera has been calibrated (see Chapter 18).

The problem of center determination when an ellipse is projected into an ellipse is a little harder to solve: not only is the longitudinal position of the center unknown, but so is the lateral position. Nevertheless, the same basic projective ideas apply. Specifically, let us consider a pair of parallel tangents to the ellipse, which in the image become a pair of lines λ_1, λ_2 meeting on the vanishing line (Fig. 17.16). As the chord joining the contact points of the tangents passes through the center of the original ellipse, and as this property is projectively invariant, the projected center must lie on the chord joining the contact points of

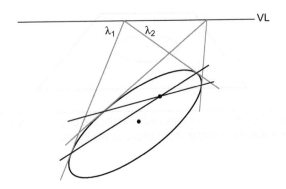

FIGURE 17.16

Construction for calculating the offsets for a projected ellipse. The two lines λ_1, λ_2 from a point on the vanishing line VL touch the ellipse, and the joins of the points of contact for all such line pairs pass through the projected center. (The figure shows just two chords of contact.)

the line pair λ_1, λ_2. As the same applies for all pairs of parallel tangents to the original ellipse, we can straightforwardly locate the projected center of the ellipse in the image plane.[8] For an alternative, numerical analysis of the situation, see Zhang and Wei (2003).

Both the circle result and the ellipse result are important in cases of inspection of mechanical parts, where accurate results of center positions have to be made irrespective of any perspective distortions that may arise. Indeed, circles can also be used for camera calibration purposes, and again high accuracy is required (Heikkilä, 2000).

17.10 THE ROUTE TO FACE RECOGNITION

In a book of this type, it is important to consider the problem of face recognition, which has long been a target for many practitioners of machine vision. In fact, some would say that by now this is a solved problem, but criminologists repeatedly confirm that there is still some distance to go before reliable identification of people can be attained in practical circumstances. Not least, there are problems of

---

[8]Students who are familiar with projective geometry will be able to relate this to the "pole–polar" construction for a conic: in this case the polar line is the vanishing line and its pole is the projected center. In general the pole is not at the center of an ellipse, and will not be so unless the polar is at infinite distance. Indeed, from this point of view, Eq. (17.31) can be understood in terms of "harmonic ranges," y_2 being the "harmonic mean" of y_1 and y_3.

hairstyles, hats, glasses, beards, degree of facial stubble, wildly variable facial expressions, and of course variations in lighting and shadow—and this does not even touch on the problem of deliberate disguise. Added to this, one must never forget that the face is not flat, but part of a solid, albeit malleable object—the head—which can appear in a variety of orientations and positions in space.

These remarks make it clear that full analysis and solution of the face recognition problem would require a whole book on the topic, and several such books have appeared over time (e.g., Gong et al., 2000). However, the topic of facial recognition is itself dated: not only are workers interested in face recognition, but also there is pressure to measure facial expression for reasons as diverse as determining whether a person is telling the truth and finding how to mimic real people as accurately as possible in films (over the next few years, it is possible to anticipate that a good proportion of films will contain no human actors as this has the potential for making them quicker and cheaper to produce). Clearly, medical diagnosis or facial reconstruction can also benefit from facial measurement algorithms, while person verification may well be at least as important as face recognition, so long as it can be done quickly and with minimum error. In this latter respect, recent efforts have moved in the direction of identifying people highly accurately from their iris patterns (e.g., Daugman, 1993, 2003), and even more accurately from their retinal blood vessels[9] (using the methods of retinal angiography). While retinal methods would be rather expensive to implement, e.g. on all-weather ATM machines, the iris method need not be, and much progress has been made in this direction.

These considerations show that a narrow view of face recognition would be rather inappropriate. For this reason we concentrate here on one or two important aspects. Among these are the task of analyzing the face for key features that can then at least provide a proper framework for further work on facial recognition, facial expression, facial verification, and so on, and the task of locating the iris, which will be used both for detailed verification and as an important starting point for facial analysis. In fact, location of the iris can be dealt with reasonably straightforwardly using the Hough transform approach. This has already been covered in Section 12.7.

Other facial features, such as the corners of the eye and mouth, can be found by tracking, snake algorithms, or simply by corner detection. Similarly, the upper and lower contours of the ear and of the nose can be ascertained, thus yielding fixed points that can be used for a multitude of purposes ranging from person identification to recognition of facial expressions. At this point, it is useful to reconsider the fact that the face is part of the head, and that this is a 3-D object.

---

[9]Here some of the main deciding factors are commercial rather than academic, although an important message is that the technical difficulty is only viable where there is a need for the highest security, but in that case "the false acceptance rate for a correctly installed retina scan system falls below 0.0001 percent" (http://ru.computers.toshiba-europe.com; website accessed May 19, 2004).

17.10.1 The Face as Part of a 3-D Object

To start an analysis of the head and face, note that we can define a plane Π containing the outer corners of the eyes and the outer corners of the mouth, if we assume that some odd facial expression has not been adopted. To a very good approximation, it can also be assumed that the inner corners of the eyes will be in the same plane (Fig. 17.17(a–c)). The next step is to estimate the position of the vanishing point V for the three horizontal lines λ_1, λ_2, λ_3 joining these three pairs of features (Fig. 17.17(d)). Once this has been done, it is possible to use the relevant cross-ratio invariants to determine the points that in 3-D lie midway between the two features of each pair. This will give the symmetry line λ_s of the face. It will also be possible to determine the horizontal orientation θ of the facial plane Π, i.e., the angle through which it has been rotated, about a vertical axis, from a full frontal view. The geometry for these calculations is shown in Fig. 17.17(e). Finally, it will be possible to convert the inter-feature distances along λ_1, λ_2, λ_3 to the corresponding full frontal values, taking proper account of perspective, but not yet taking account of the vertical orientation φ of Π, which is still unknown.[10]

In fact, there is insufficient information to estimate the vertical orientation φ, without making further assumptions. Ultimately, this is because the face has no horizontal axis of symmetry. If we can assume that φ is zero (i.e., the head is held neither up nor down, and the camera is on the same level), then we can gain some information on the relative vertical distances of the face, the raw measurements for these being obtained from the intercepts of λ_1, λ_2, λ_3 with the symmetry line λ_s. Alternatively, we can assume average values for the inter-feature distances and deduce the vertical orientation of the face. A further alternative is to make other estimates based on the chin, nose, ears, or hairline, but as these are not guaranteed to be in the facial plane Π, the whole assessment of facial pose may not then be accurate and invariant to perspective effects.

Overall, we are moving toward measurements either of facial pose or of facial inter-feature measurements, with the possibility of obtaining some information on both, even when perspective distortions have to be allowed for (Kamel et al., 1994; Wang et al., 2003). Of course, the analysis will be significantly simpler in the absence of perspective distortions, when the face is viewed from a distance or when a full frontal view is guaranteed. Indeed, the bulk of the work on facial recognition and pose estimation to date has been done in the context of weak perspective, making the analysis altogether simpler. Even then, the possibility of wide varieties of facial expression brings in a great deal of complexity. It should also be noted that the face is not merely a rubber mask (or deformable template) that can be distorted "tidily". The capability for opening and closing the mouth and eyes creates additional nonlinear effects that are not modeled merely by variable stretching of rubber masks.

---

[10]Note that the theory underlying these procedures is closely related to that of Section 17.8. See also Fig. 17.13.

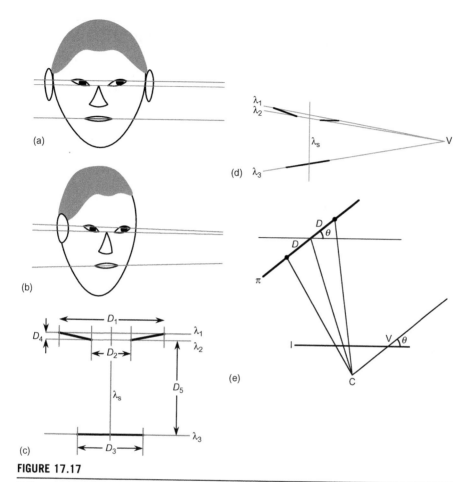

FIGURE 17.17

3-D analysis of facial parameters. (a) Front view of face. (b) Oblique view of face, showing perspective lines for corners of the eye and mouth. (c) Labeling of eye and mouth features and definition of five inter-feature distance parameters. (d) Position of vanishing point under an oblique view. (e) Positions of one pair of features and their midpoint on the facial plane Π. Note how the midpoint is no longer the midpoint in the image plane I when viewed under perspective projection. Note also how the vanishing point V gives the horizontal orientation of Π.

17.11 PERSPECTIVE EFFECTS IN ART AND PHOTOGRAPHY[*]

An artist is painting a picture somewhere on the countryside. Every now and again he looks up from his easel and surveys the scene, then he turns back to his picture and adds a few more touches. He has chosen his location carefully, and has set his easel at the right angle for best effect. We will suppose that he is not aiming to be impressionist, but wishes to present the scene as he sees it. Although his painting is in 2-D, he is able to present all the information needed for others to perceive the scene in 3-D. However, there is a problem. The picture needs to be viewed from precisely the right angle and distance that must correspond to the artist's original viewpoint. Of course, the artist had to rotate his head and mentally rotate the scene between the moments that he viewed it and painted it (he had to do this because the canvas he was painting on was opaque: other artists such as Canaletto have used camera obscura methods to overcome this difficulty). However, we can overcome the problem by temporarily assuming that the canvas is transparent, which significantly simplifies the geometry (Fig. 17.18).

Interestingly, from his viewpoint, the artist could have painted a whole range of pictures of the scene, with his easel set at different angles (Fig. 17.18). All these pictures would be very closely related to each other and in fact would be related by homographies. But each of them would have exactly one proper viewing position and orientation, and when each was viewed from its proper viewing position, exactly the same 3-D regenerative effect would be perceived by the viewer. Thus,

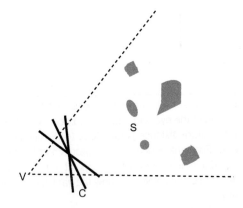

FIGURE 17.18

Effective viewpoint of artist painting a picture. The artist views the scene S from viewpoint V, and paints what he sees on the canvas at C. The picture painted at C could be one of many depending on the orientation of the canvas.

[*]Parts of this Section were inspired by a TV lecture dated 28 December 1978 by Christopher Zeeman entitled "Mathematics into pictures".

the fact that a homography exists between the various views does not change the constraint that each version of a picture is best viewed from a single location.

However, there is a circumstance when this is no longer true. That is, when the scene contains a flat (2-D) surface F that is then to be presented in 2-D. Immediately, we have a homography between the original scene and the canvas, and we also have homographies with all the possible rotated versions of the canvas. But what of the viewing positions? To understand the situation properly, we need to think of the possible viewing points relative to the frame of reference of the canvas C, which we must now regard as fixed, e.g., on a gallery wall W. We can see that as the original canvas is rotated, so the ideal viewing point relative to its location on the gallery wall will rotate, albeit remaining at the distance corresponding to the distance of the artist's eye from the canvas. In fact, as one walks (along a circle) around the painting in the gallery, all the possible pictures that the artist might have painted from his original position unfold before us (Fig. 17.19). They all embody valid perspective distortions and thus all would appear entirely natural. Note that a circular path is appropriate because it corresponds to the (constant) overall angle of the artist's view (angles in the same segment of a circle are equal).

But what of the case not of the flat wall of a house but of a face? In fact, substantial parts of the face approximate to a flat 2-D surface, e.g., the forehead, eyes, cheeks, mouth, and chin. Considering them alone, a considerable range of viewing points would be acceptable. Then there is the human propensity for focussing on the eyes and largely ignoring the rest of the face. If this is done, acceptable views will be obtained by viewing the painting from many directions. Indeed when focussing on the eyes and ignoring the rest, it seems entirely

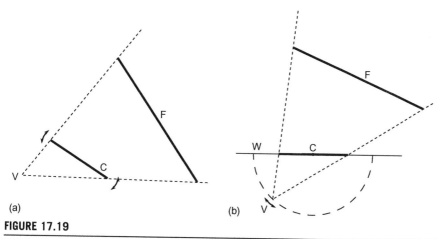

(a)   (b)

FIGURE 17.19

Viewing a painted picture. Process of (a) painting and (b) viewing a picture. As the orientation of the canvas changes in (a), so the proper viewing point V in (b) moves along the path of a circle. For a flat object F, the circular path sweeps out all possible pictures the artist might have painted, and all will be related by homographies (see text).

understandable why people would report after visiting a stately home and seeing a painting of the 17th Earl, that his eyes "followed them around the room."

A further factor is involved in this analysis—the orientation of the face when the picture was painted. If the face was originally at an angle α, the eyes would appear a factor $\cos \alpha$ closer together than in the head-on painting. However, if the canvas were rotated through an angle β, the eyes would be enlarged by a factor $\sec \beta$. Hence, there would be an overall magnification $\cos\alpha/\cos\beta$. Cancellation would occur when $\beta = \alpha$, which corresponds to the canvas being parallel to the face. However, cancellation would also occur when $\beta = -\alpha$, and this corresponds to the canvas and the face being rotated through equal and opposite angles α relative to the final viewing direction (Fig. 17.20). Next, suppose that a certain amount of enlargement or diminution of the apparent distance between the eyes is acceptable (note especially that if one doesn't know the 17th Earl in the painting, some enlargement would be acceptable). The result is that the range of acceptable orientations of the final viewing direction will be increased, with the β distortion tending to cancel the α distortion, the largest distortions for given α occurring at high or low $|\beta|$ (Fig. 17.21). Equalizing these extreme distortions by making $|\alpha| > 0$ would give the maximum permissible range of acceptable orientations (e.g., $\alpha = 20°$, $|\beta| = 0 - 40°$ with $|\beta| = \alpha$ in the middle of the range).

In photography there is also a correct viewing position, but when examining family photographs, there are no exceptional situations where people would look only at the eyes: people would want to look at facial expressions, hairstyles, and so on (they would also be quite sensitive to whether everyone's eyes were open). Unfortunately, photographs of groups often appear distorted around the outside, a factor that could sometimes be partly due to pincushion or barrel distortion (these are lens aberrations—see Chapter 18). However, this effect could also be due to an incorrect viewing position: the camera doesn't lie, but only shows the true

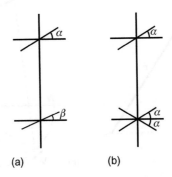

(a)　　　(b)

FIGURE 17.20

Effect of rotations relative to the viewing direction. In (a), α represents the original orientation of the face, and β represents the orientation of the canvas. Part (b) shows the situation both when $\beta = \alpha$ and when $\beta = -\alpha$. In both instances, the two orientation effects cancel and the eyes appear to have their original separation.

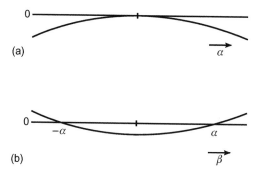

FIGURE 17.21

Effect of varying the viewing orientations. Part (a) shows how the separation of the eyes becomes reduced as $|\alpha|$ increases from zero. Part (b) shows how the change in separation of the eyes for a fixed value of α is first canceled out by increasing $|\beta|$ from zero and then increased so that the change in separation becomes positive. The average magnitude of the change in separation can clearly be lower in (b) than in (a).

geometry according to the perspective it had when the button was pressed. It is a fact that photographs are often viewed at arms length—a distance far greater than the correct viewing distance (Fig. 17.22). If photographs were meant to be viewed at that distance, the photograph should be taken from much further away, to be sure that the camera doesn't lie inadvertently. Of course, there is a complication that people will normally look at the camera, so taking the photograph from a different distance will affect the material content of the scene itself.

Until quite recently, photographs were best taken close up, because of the limited resolution of the film. Nowadays, digital cameras have such phenomenal resolution that there is some advantage from taking pictures further away, or even from a considerable distance, with the aid of a zoom lens. (The latter confers the added advantage that real-life shots can be taken without the subject being embarrassed or even aware they are being taken.) But there is a quite different advantage to be gained from taking photographs much further away: although the correct viewing distance could then be rather larger than ideal from the perspective point of view, all people at all locations in the photograph can be viewed individually without perspective distortions creeping in. Note again that it is common practice for photographs to be handed around, and for each person in them to be scrutinized *individually*—so the overall global composition might well be less important than the individual people who are portrayed (this is all the more true when one knows the people in the photograph, which is much more likely than with a painting of the 17th Earl). Optimization not only of each locality in a photograph but also of the global view is plainly impossible, but taking the photograph from a distance gives a very good compromise (Fig. 17.22(b)). However, taking it from infinity would lead to zero foreshortening of the faces and thereby make them appear flatter. Here a lot depends on whether the lighting provides other cues that can give a good impression of depth.

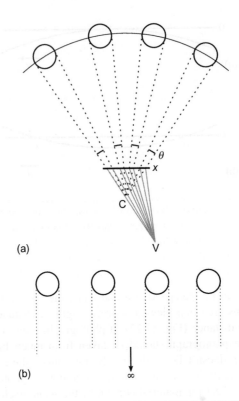

(a)

(b)

FIGURE 17.22

Process of taking and viewing a photograph. (a) The geometry for taking a photograph of a group of people all facing the camera at C. Also shown is the effective viewing location V of the photograph. (Clearly, the photograph would be enlarged before viewing, in which case the part of the illustration just above V would be scaled, but not otherwise changed.) (b) A potentially more ideal way of taking the photograph, from a large distance. By following (b), people examining the photograph would be viewing from an ideal viewing point; in addition, all people shown in the picture could be examined individually without distortion.

Another possibility that digital photography offers is that of automatically stitching together several frames to create a wide scene or panorama. Here best results are obtained if the camera is put into stitching mode, so that it can make exposures constant over a sequence of frames or at least record what they were. Then it could be expected that the edges of various frames would match up without any sudden changes between them. To achieve a proper match, the frames obviously have to overlap, and then special software can be used to find the best set of lines for trimming and stitching. This involves moving the trim lines in such a way—generally in plain background regions—that the breaks will be imperceptible. Of course, some smoothing along the trimmed edges will often help, as long as this process doesn't encroach on regions of totally different intensity. Unfortunately, stitching cannot easily cope with situations containing moving objects, and in this context the author had an interesting early experience of the final country scene containing too many sheep.

The exact algorithms to be used for image stitching are quite intricate. This is because a time-consuming search is required to identify the best trimming lines. The criteria to be used are important, but these obviously involve minimization of intensity and color change along the borders—and there is also advantage in minimizing the intensity and color *gradients* along the borders. Then there are rules for smoothing along the borders, after stitching has been carried out. Here the simplest rule is not to do any smoothing at all, but this rule can be relaxed if the changes in intensity and intensity gradient have successfully been minimized.

There is one major problem with stitching—that scenes containing straight edges that go at angles across the whole scene are almost impossible to deal with (Fig. 17.23). This is because each (flat) frame will have been taken from a different direction in order to obtain a wider overall view. Thus, a straight line, such as a path, will appear straight in each frame, but the orientation will normally have to change at the join (Fig. 17.23): that this is not mere theory is demonstrated clearly by the examples in Fig. 17.24. The only way to overcome the problem is to present the scene on a sphere or cylinder in order to prevent kinks from appearing at the joins. But then the original straight line will become a curved line, especially when the final picture is presented as a flat scene. Here again the specter of the single viewpoint of each original picture is upon us. The best way of handling it seems to be to present it as a picture that is apparently taken from infinity (as for the earlier example of photographing a group of people), so that any straight line will appear straight, at least at every local position, even if a ruler placed along it will show that it is not straight globally. In fact, this is what the special rotating line-scan cameras were able to achieve in the 1960s and earlier, when they were used to take school photographs—and the time taken to gather enough light was often sufficient for at least one small boy to run from one end to the other and be photographed twice!

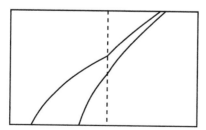

FIGURE 17.23

Effect of stitching two pictures depicting parts of a path. Apparently correct stitching will actually lead to a kink at the join.

FIGURE 17.24

Practical instances of stitching. Part (a) shows that the surmise of Fig. 17.23 is correct. Part (b) shows the result of using an (effectively) overenthusiastic stitching package that manages to avoid a kink, but ends up with perspective nonsense. If the lower end of the road boundary had been visible, the software might have avoided the latter problem but would then have introduced a kink.

17.12 CONCLUDING REMARKS

This chapter has aimed to give some insight into the important subject of invariants and its application in image recognition. The subject takes off when ratios of ratios of distances are considered, and this idea leads in a natural way to the cross-ratio invariant. While its immediate manifestation lies in its application to recognition of the spacings of points on a line, it generalizes immediately to angular spacings for pencils of lines and also to angular separations of concurrent planes. A further extension of the idea is the development of invariants that can describe sets of noncollinear points, and two cross-ratios suffice to characterize a set of five noncollinear points on a plane. The cross-ratio can also be applied to conics. Indeed, Chasles' theorem describes a conic as the locus of points that maintains a pencil of constant cross-ratio with a given set of four points. However, this theorem does not permit one type of conic curve to be distinguished from another.

Many other theorems and types of invariant exist, but space prevents more than a mention being made of them. As an extension to the line and conic examples given in this chapter, invariants have been produced that cover a conic and two coplanar nontangent lines, a conic and two coplanar points, and two coplanar conics. Of particular value is the group approach to the design of invariants (Mundy and Zisserman, 1992a). However, certain mathematically viable invariants, such as those that describe local shape parameters on curves, prove to be too unstable to use in their full generality because of image noise. Nevertheless, semi-differential invariants have been shown (Section 17.5) to be capable of fulfilling essentially the same function.

Next, there is the warning of Åström (1995) that perspective transformations can produce such incredible changes in shape that a duck silhouette can be projected arbitrarily closely into something that looks like a rabbit or a circle, hence upsetting invariant-based recognition.[11] While such reports seem absent from the previous literature, Åström's work indicates that care must be taken to regard recognition via invariants as hypothesis formation that is capable of leading to false alarms.

Overall, the value of invariants lies in making computationally efficient checks of whether points or other features might belong to specific objects. In addition, they achieve this without the necessity[12] for camera calibration or knowing the viewpoint of the camera (although there is an implicit assumption that the camera is Euclidean). While invariants have been known within the vision community for well over 20 years, it is only during the last about 15 years that they have been systematically developed and applied for machine vision. Such is their power that they will undoubtedly assume a stronger and more central role in the future. Nowhere is this power better indicated than by the application to vanishing point detection and face recognition described in Sections 17.7–17.10. Note also the perspective projection problems that led not only to the need for invariants but also to the need for further insight into the problems of viewing and stitching 2-D pictures (Section 17.11).

The problems of perspective projection are ubiquitous in 3-D vision, appearing even in simple situations such as viewing 2-D pictures and stitching digital photographs. However, vital interpretive information is provided by projective invariants that slice right through such complexities and are able to help, e.g., with vanishing point detection and face recognition.

---

[11]It could of course be argued that all recognition methods will be subject to the effects of perspective transformations. However, invariant-based recognition will not flinch from invoking highly extreme transformations that appear to grossly distort the objects in question, whereas more conventional methods are likely to be designed to cope with a reasonable range of expected shape distortions.

[12]Here we assume that the aim is location of specific objects in the image. If the objects are then to be located in the world coordinates, some form of camera calibration will again be needed. However, there are many applications, such as inspection, surveillance, and identification (e.g., of faces or signatures) where location of objects in the *image* can be entirely adequate.

17.13 BIBLIOGRAPHICAL AND HISTORICAL NOTES

The mathematical subject of invariance is very old (cf. the work of Chasles, 1855), but it has only relatively recently been developed systematically for machine vision. Notable in this context is the work of Rothwell, Zisserman, and their coworkers, as reported by Forsyth et al. (1991), Mundy and Zisserman (1992a,b), Rothwell et al. (1992a,b), and Zisserman et al. (1990). In particular, the paper by Forsyth et al. (1991) shows the range of available invariant techniques and discusses the problems of stability which arise in certain cases. The appendix (Mundy and Zisserman, 1992b) on projective geometry for machine vision, which appears in Mundy and Zisserman (1992a), is especially valuable, and provides the background needed for understanding the other papers in the volume. On the whole, the latter volume has a theoretical flavor that demonstrates what ought to be possible using invariants, although comparisons between invariants and other approaches to recognition are perhaps lacking. Thus, it is only by examining whether workers choose to use invariants in real applications that the full story will emerge. In this respect, the paper by Kamel et al. (1994) on face recognition is of great interest, as it shows how invariants helped to achieve more than had been achieved earlier after many attempts using other approaches—specifically in correcting for perspective distortions during face recognition.

Other more recent work appears in a special issue of *Image and Vision Computing* (Mohr and Wu, 1998). In particular, the paper by Van Gool et al. (1998) shows how shadows can be allowed for in aerial images, and the paper by Boufama et al. (1998) shows how invariants can help with object positioning. Startchik et al. (1998) provides a useful demonstration of the semi-differential invariant methods covered in Section 17.5. Maybank (1996) deals with the problem of accuracy with invariants, making the point that this can be serious even for cross-ratios (which contain only four parameters). Another early work, by a totally different set of workers, is Barrett et al. (1991) and contains a number of useful derivations, together with a practical example of aircraft recognition, complete with accuracy assessments.

Rothwell's (1995) book covers the early work on invariance in a thoughtful manner, and the later 3-D books by Hartley and Zisserman (2000) and Faugeras and Luong (2001) integrate the ideas into their structure, but are not always easy to understand by students starting out in the subject. Semple and Kneebone (1952) is a standard work on projective geometry, which is still widely used in its later reprints.

Vanishing point determination has often been considered both in relation to egomotion for mobile robots (Lebègue and Aggarwal, 1993; Shuster et al., 1993) and in general with regard to the vision methodology (Magee and Aggarwal, 1984; Shufelt, 1999; Almansa et al., 2003), which is prone to suffer from inaccuracy when real off-camera data are used in any context. The seminal paper that gave rise to the crucial Gaussian sphere technique was that by Barnard (1983). In an interesting twist, Clark and Mirmehdi (2002, 2003) use VPs to recover text that has been distorted by perspective. The approach permits them to recover paragraph formats; in addition to line spacings, various forms of text justification can be recognized and managed.

17.13.1 More Recent Developments

More recently, Shioyama and Uddin (2004) have used cross-ratio invariants for the reliable location of pedestrian crossings by analyzing multiple crossing points of transverse straight lines with the alternating patterns on the road. Kelly et al. (2005) have used homographies between stereo views to locate shadows and low-lying objects: to achieve this they used the direct linear transformation (see Hartley and Zisserman, 2003) to identify homographies for sets of four or more points. Once homographies are found, they are used to eliminate the corresponding objects from consideration, thereby avoiding costly computation of 3-D depth values from the stereo views for those objects. Rajashekhar et al. (2007) use cross-ratio values to identify man-made structures in images to aid image retrieval. Hough transforms are used to find line structures in images, then feature points on the lines are found and sets of cross-ratio values are computed and presented in the form of histograms (in each case, all six possible cross-ratio values are included in the histograms). It is found that values in the range 0−5 are most suitable for identifying man-made structures, in that the histograms are suitably densely concentrated. Structures such as buildings are well recognized from the histograms, as long as they are quantized with upward of 200 bins. Li and Tan (2010) use a similar approach, but their cross-ratio values occur in continuous streams as outlines of characters or symbols are tracked. The resulting "cross-ratio spectra" allow characters to be recognized even with severe perspective distortions.

In the area of face recognition, An et al. (2010) describe a new illumination normalization model that is able to cope with varied lighting conditions. It works by decomposing the face into a high-frequency part and a low-frequency part: the main innovation is to divide the original intensity pattern by a smoothed version of the low-frequency part (although several other equalization and normalization processes are carried out as well). Hansen and Ji (2010) survey models for eye detection and gaze estimation and summarize the developments that are still needed in this area. Fang et al. (2010) describe a new method of multiscale image stitching. The paper discusses the problems of obtaining global and local alignment. A number of strategies are needed to overcome the various problems, and an iterative processing pipeline is required to integrate the different strategies.

17.14 PROBLEMS

1. Show that the six operations required to transform the cross-ratio κ into the six different values for four points on a line form a group G of order six (see Sections 17.2 and 17.6). Show that G is a noncyclic group, and has two

subgroups of order 2 and 3, respectively. *Hint:* Show that all possible combined operations fall within the same set of six, and also that this set contains the identity operation and the inverses of all the elements of the set.

2. Show that a conic and two points can be used to define an invariant cross-ratio.

3. Show that two conics can be used to define an invariant cross-ratio (a) if they intersect in four points, (b) if they intersect in two points, (c) and if they do not intersect at all, so long as they have common tangents.[13]

4. **a.** Perform a geometric calculation based on the sine rule, which shows that the angles α, β, γ are related to the distances a, b, c in Fig. 17.25 by the equation:

$$\frac{a}{\sin \alpha} \times \frac{c}{\sin \gamma} = \frac{a+b}{\sin(\alpha + \beta)} \times \frac{b+c}{\sin(\beta + \gamma)}$$

b. Show that this leads to a relation between the cross-ratios for various distances on the line and for the sines of various angles. Hence, show that this also leads to the constancy of the cross-ratios on any two lines crossing the pencil of four lines passing through O.

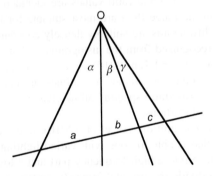

FIGURE 17.25

Geometry for cross-ratio calculation.

5. **a.** Explain the value of using *invariants* in relation to pattern recognition systems. Illustrate your answer by considering the value of thinning algorithms in optical character recognition.

b. The *cross-ratio* of four points (P_1, P_2, P_3, P_4) on a line is defined as the ratio:

$$C(P_1, P_2, P_3, P_4) = \frac{(x_3 - x_1)(x_2 - x_4)}{(x_2 - x_1)(x_3 - x_4)}$$

---

[13]The case of nonintersecting conics with no common tangents requires complex algebra. See, e.g., Rothwell (1995). The possibility of ambiguity and incompleteness is also discussed in Rothwell (1995).

Explain why this is a useful type of invariant for objects viewed under full perspective projection. Show that labeling the points in reverse order will not change the value of the cross-ratio.

c. Give arguments why the cross-ratio concept should also be valid for weak perspective projection. Work out a simpler invariant that is valid for straight lines viewed under weak perspective projection.

d. A flat lino-cutter blade has two parallel sides of different lengths. It is viewed under weak perspective projection. Discuss whether it can be identified from any orientation in 3-D by measuring the lengths of its sides.

6. a. Flagstones are viewed on a pavement, providing a large number of coplanar feature points. Show that a correspondence can be made between five coplanar feature points in two images—however, the camera has been moved between the shots—by checking the values of two cross-ratios.

b. It is required to compose a panorama of a scene by taking a number of photographs and "stitching" them together after making appropriate image transformations. To achieve this, it is necessary to make correspondences between the images. Show that the two cross-ratios type of planar invariant can be used for this purpose, even if the chosen scene features do not lie on a common plane. Determine under what conditions this is possible.

7. Redraw Fig. 17.16 using vanishing points aligned along the observed ellipse axes. Show that the problem of finding the transformed center location now reduces to two 1-D cases and that Eq. (17.32) can be used to obtain the transformed center coordinates.

8. A robot is walking along a path paved with rectangular flagstones. It is able to rotate its camera head so that one set of flagstone lines appears parallel while the other set converges toward a vanishing point. Show that the robot can calculate the position of the vanishing point in two ways: (1) by measuring the varying widths of individual flagstones or (2) by measuring the lengths of adjacent flagstones and proceeding according to Eq. (17.30). In the first case, obtain a formula that could be used to determine the position of the vanishing point. Which is the more general approach? Which would be applicable if the flagstones appeared in a flower garden in random locations and orientations?

18

Image Transformations and Camera Calibration

When setting up a measurement system, it is natural to calibrate it carefully before use. This task has been left to last because (a) it is mathematically more demanding, (b) there are instances where it can be bypassed, (c) it is not always possible to perform the calibration entirely in advance, but rather it has to be updated to a sufficient extent as measurements proceed. This chapter outlines some of the problems of calibration and some of the results of recent research that allow the process to be at least partially bypassed.

Look out for:

- the homogeneous coordinates technique for representing general 3-D positions and transformations.
- "extrinsic" (external world) and "intrinsic" (camera) parameters.
- methods for achieving absolute camera calibration.
- the need for correction of camera lens distortions.
- the idea of a generalized epipolar geometry.
- the "essential" and "fundamental" matrix formulations, relating the observed positions of any point in two camera frames of reference.
- the central position of the eight-point algorithm.
- the possibility of image "rectification."
- the possibility of 3-D reconstruction.

This is one of the key chapters constituting Part 3 of this book. These chapters should be taken together as they involve not merely different topics but also different *aspects* of the subject, and in addition the aim has been to cover them in as gentle an order as possible considering the mathematical complexities involved in extracting 3-D and motion information from 2-D images.

18.1 INTRODUCTION

When images are obtained from 3-D scenes, the exact position and orientation of the camera sensing device is often unknown and there is a need for it to be related to some global frame of reference. This is especially important if accurate measurements of objects are to be made from their images, e.g., in inspection applications. On the other hand, it may in certain cases be possible to dispense with such detailed information—as in the case of a security system for detecting intruders, or a system for counting cars on a motorway. There are also more complicated cases, such as those in which cameras can be rotated or moved on a robot arm or the objects being examined can move freely in space. In such cases, camera calibration becomes a central issue. Before we can consider camera calibration, we need to understand in some detail the transformations that can occur between the original world points and the formation of the final image. We attend to these image transformations in the following section, and then move on to details of camera parameters and camera calibration in the subsequent two sections. Then, in Section 18.5 we consider how any radial distortions of the image introduced by the camera lens can be corrected.

Section 18.6 signals a break with previous work and introduces "multiple view" vision. This topic has become important in recent years, as it uses new theory to bypass the need for formal camera calibration, and makes it possible to update the vision system parameters during actual use. The basis for this work is generalized epipolar geometry: this takes the epipolar line ideas of Section 15.3.2 considerably further. At the core of this new work are the "essential" and "fundamental" matrix formulations, which relate the observed positions of any point in two camera frames of reference. Short sections on image "rectification" (obtaining a new image as it would be seen from an idealized camera position) and 3-D reconstruction follow.

18.2 IMAGE TRANSFORMATIONS

First, we consider the rotations and translations of object points relative to a global frame. After a rotation through an angle θ about the Z-axis (Fig. 18.1), the coordinates of a general point (X, Y) change to:

$$X' = X \cos \theta - Y \sin \theta \tag{18.1}$$

$$Y' = X \sin \theta + Y \cos \theta \tag{18.2}$$

This result is neatly expressed by the matrix equation:

$$\begin{bmatrix} X' \\ Y' \end{bmatrix} = \begin{bmatrix} \cos \theta & -\sin \theta \\ \sin \theta & \cos \theta \end{bmatrix} \begin{bmatrix} X \\ Y \end{bmatrix} \tag{18.3}$$

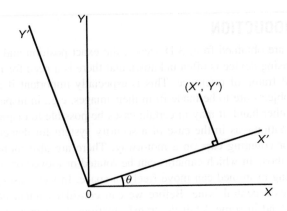

FIGURE 18.1

Effect of a rotation θ about the origin.

Clearly, similar rotations are possible about the X and Y axes. To satisfactorily express rotations in 3-D, we require a more general notation using 3×3 matrices, the matrix for a rotation θ about the Z-axis being:

$$R_Z(\theta) = \begin{bmatrix} \cos\theta & -\sin\theta & 0 \\ \sin\theta & \cos\theta & 0 \\ 0 & 0 & 1 \end{bmatrix} \tag{18.4}$$

Those for rotations ψ about the X-axis and φ about the Y-axis are:

$$R_X(\psi) = \begin{bmatrix} 1 & 0 & 0 \\ 0 & \cos\psi & -\sin\psi \\ 0 & \sin\psi & \cos\psi \end{bmatrix} \tag{18.5}$$

$$R_Y(\varphi) = \begin{bmatrix} \cos\varphi & 0 & \sin\varphi \\ 0 & 1 & 0 \\ -\sin\varphi & 0 & \cos\varphi \end{bmatrix} \tag{18.6}$$

We can make up arbitrary rotations in 3-D by applying sequences of such rotations. Similarly, we can express arbitrary rotations as sequences of rotations about the coordinate axes. Thus $R = R_X(\psi)R_Y(\varphi)R_Z(\theta)$ is a composite rotation in which $R_Z(\theta)$ is applied first, then $R_Y(\varphi)$, and finally $R_X(\psi)$. Rather than multiplying out these matrices, we write down here the general result expressing an arbitrary rotation R:

$$\begin{bmatrix} X' \\ Y' \\ Z' \end{bmatrix} = \begin{bmatrix} R_{11} & R_{12} & R_{13} \\ R_{21} & R_{22} & R_{23} \\ R_{31} & R_{32} & R_{33} \end{bmatrix} \begin{bmatrix} X \\ Y \\ Z \end{bmatrix} \tag{18.7}$$

Note that the rotation matrix R is not completely general: it is orthogonal and thus has the property that $R^{-1} = R^T$.

In contrast with rotation, translation through a distance (T_1, T_2, T_3) is given by:

$$X' = X + T_1 \tag{18.8}$$

$$Y' = Y + T_2 \tag{18.9}$$

$$Z' = Z + T_3 \tag{18.10}$$

which is not expressible in terms of a multiplicative 3×3 matrix. However, just as general rotations can be expressed as rotations about various coordinate axes, so general translations and rotations can be expressed as sequences of basic rotations and translations relative to individual coordinate axes. Thus, it would be most useful to have a notation that unified the mathematical treatment so that a generalized displacement could be expressed as a product of matrices. This is indeed possible if the so-called "homogeneous coordinates" are used. To achieve this, the matrices must be augmented to 4×4. A general rotation can then be expressed in the form:

$$\begin{bmatrix} X' \\ Y' \\ Z' \\ 1 \end{bmatrix} = \begin{bmatrix} R_{11} & R_{12} & R_{13} & 0 \\ R_{21} & R_{22} & R_{23} & 0 \\ R_{31} & R_{32} & R_{33} & 0 \\ 0 & 0 & 0 & 1 \end{bmatrix} \begin{bmatrix} X \\ Y \\ Z \\ 1 \end{bmatrix} \tag{18.11}$$

while the general translation matrix becomes:

$$\begin{bmatrix} X' \\ Y' \\ Z' \\ 1 \end{bmatrix} = \begin{bmatrix} 1 & 0 & 0 & T_1 \\ 0 & 1 & 0 & T_2 \\ 0 & 0 & 1 & T_3 \\ 0 & 0 & 0 & 1 \end{bmatrix} \begin{bmatrix} X \\ Y \\ Z \\ 1 \end{bmatrix} \tag{18.12}$$

The generalized displacement (i.e., translation plus rotation) transformation clearly takes the form:

$$\begin{bmatrix} X' \\ Y' \\ Z' \\ 1 \end{bmatrix} = \begin{bmatrix} R_{11} & R_{12} & R_{13} & T_1 \\ R_{21} & R_{22} & R_{23} & T_2 \\ R_{31} & R_{32} & R_{33} & T_3 \\ 0 & 0 & 0 & 1 \end{bmatrix} \begin{bmatrix} X \\ Y \\ Z \\ 1 \end{bmatrix} \tag{18.13}$$

We now have a convenient notation for expressing generalized transformations including operations other than the translations and rotations that account for the normal motions of rigid bodies. First, we take a scaling in size of an object, the simplest case being given by the matrix:

$$\begin{bmatrix} S & 0 & 0 & 0 \\ 0 & S & 0 & 0 \\ 0 & 0 & S & 0 \\ 0 & 0 & 0 & 1 \end{bmatrix}$$

The more general case:

$$\begin{bmatrix} S_1 & 0 & 0 & 0 \\ 0 & S_2 & 0 & 0 \\ 0 & 0 & S_3 & 0 \\ 0 & 0 & 0 & 1 \end{bmatrix}$$

introduces a shear in which an object line λ will be transformed into a line that is not in general parallel to λ. Skewing is another interesting transformation, being given by linear translations varying from the simple case:

$$\begin{bmatrix} 1 & B & 0 & 0 \\ 0 & 1 & 0 & 0 \\ 0 & 0 & 1 & 0 \\ 0 & 0 & 0 & 1 \end{bmatrix}$$

to the general case:

$$\begin{bmatrix} 1 & B & C & 0 \\ D & 1 & F & 0 \\ G & H & 1 & 0 \\ 0 & 0 & 0 & 1 \end{bmatrix}$$

Rotations can be regarded as combinations of scaling and skewing, and are sometimes implemented as such (Weiman, 1976).

The other simple but interesting case is that of reflection, which is typified by:

$$\begin{bmatrix} 0 & 1 & 0 & 0 \\ 1 & 0 & 0 & 0 \\ 0 & 0 & 1 & 0 \\ 0 & 0 & 0 & 1 \end{bmatrix}$$

This generalizes to other cases of improper rotation where the determinant of the top left 3×3 matrix is -1.

In all the cases discussed above, it will be observed that the bottom row of the generalized displacement matrix is redundant. In fact, we can put this row to good use in certain other types of transformation. Of particular interest in this context is the case of perspective projection. Following Section 15.3, Eq. (15.1), the equations for projection of object points into image points are:

$$x = \frac{fX}{Z} \tag{18.14}$$

$$y = \frac{fY}{Z} \tag{18.15}$$

$$z = f \tag{18.16}$$

We next make full use of the bottom row of the transformation matrix by defining the homogeneous coordinates as $(X_h, Y_h, Z_h, h) = (hX, hY, hZ, h)$, where h is a nonzero constant that we can take to be unity. To proceed, we examine the homogeneous transformation:

$$
\begin{bmatrix} 1 & 0 & 0 & 0 \\ 0 & 1 & 0 & 0 \\ 0 & 0 & 1 & 0 \\ 0 & 0 & 1/f & 0 \end{bmatrix}
\begin{bmatrix} X \\ Y \\ Z \\ 1 \end{bmatrix}
=
\begin{bmatrix} X \\ Y \\ Z \\ Z/f \end{bmatrix}
\tag{18.17}
$$

We see that dividing by the fourth coordinate gives the required values of the transformed Cartesian coordinates $(fX/Z, fY/Z, f)$.

Let us now review this result. First, we have found a 4×4 matrix transformation that operates on 4-D homogeneous coordinates. These do not correspond directly to real coordinates, but real 3-D coordinates can be calculated from them by dividing the first three by the fourth homogeneous coordinate. Thus, there is an arbitrariness in the homogeneous coordinates in that they can all be multiplied by the same constant factor without producing any change in the final interpretation. Likewise, when deriving homogeneous coordinates from real 3-D coordinates, we can employ any convenient constant multiplicative factor h, although we will normally take h to be unity.

The advantage to be gained from use of homogeneous coordinates is the convenience of having a single multiplicative matrix for any transformation, in spite of the fact that perspective transformations are intrinsically nonlinear: thus, a quite complex nonlinear transformation can be reduced to a more straightforward linear transformation. This eases computer calculation of object coordinate transformations and other computations such as those for camera calibration (see below). We may also note that almost every transformation can be inverted by inverting the corresponding homogeneous transformation matrix. The exception is the perspective transformation, for which the fixed value of z leads merely to Z being unknown, and X, Y only being known relative to the value of Z (hence the need for binocular vision or other means of discerning depth in a scene).

18.3 CAMERA CALIBRATION

The above discussion has shown how homogeneous coordinate systems are used to help provide a convenient linear 4×4 matrix representation for 3-D transformations including rigid body translations and rotations, and nonrigid operations including scaling, skewing, and perspective projection. In this last case, it was implicitly assumed that the camera and world coordinate systems are identical, since the image coordinates were expressed in the same frame of reference. However, in general the objects viewed by the camera will have positions that may be known in world coordinates, but that will not *a priori* be known in camera coordinates, since the camera will in general be mounted in a somewhat arbitrary position and will point in a somewhat arbitrary direction. Indeed, it may well be on adjustable gimbals, and may also be motor driven, with no precise calibration system. If the camera is on a robot arm, there are likely to be position sensors that could inform the control system of the camera position and orientation in world coordinates, although the amount of slack may well make the information too imprecise for practical purposes (e.g., to guide the robot toward objects).

These factors mean that the camera system will have to be calibrated very carefully before the images can be used for practical applications, such as robot pick-and-place. A useful approach is to assume a general transformation between the world coordinates and the image seen by the camera under perspective projection, and to locate in the image various calibration points that have been placed in known positions in the scene. If enough such points are available, it should be possible to compute the transformation parameters, and then all image points can be interpreted accurately until recalibration becomes necessary.

The general transformation G takes the form:

$$
\begin{bmatrix} X_H \\ Y_H \\ Z_H \\ H \end{bmatrix} = \begin{bmatrix} G_{11} & G_{12} & G_{13} & G_{14} \\ G_{21} & G_{22} & G_{23} & G_{24} \\ G_{31} & G_{32} & G_{33} & G_{34} \\ G_{41} & G_{42} & G_{43} & G_{44} \end{bmatrix} \begin{bmatrix} X \\ Y \\ Z \\ 1 \end{bmatrix}
\tag{18.18}
$$

where the final Cartesian coordinates appearing in the image are $(x, y, z) = (x, y, f)$, and these are calculated from the first three homogeneous coordinates by dividing by the fourth:

$$
x = \frac{X_H}{H} = \frac{G_{11}X + G_{12}Y + G_{13}Z + G_{14}}{G_{41}X + G_{42}Y + G_{43}Z + G_{44}}
\tag{18.19}
$$

$$
y = \frac{Y_H}{H} = \frac{G_{21}X + G_{22}Y + G_{23}Z + G_{24}}{G_{41}X + G_{42}Y + G_{43}Z + G_{44}}
\tag{18.20}
$$

$$
z = \frac{Z_H}{H} = \frac{G_{31}X + G_{32}Y + G_{33}Z + G_{34}}{G_{41}X + G_{42}Y + G_{43}Z + G_{44}}
\tag{18.21}
$$

However, as we know z, there is no point in determining parameters G_{31}, G_{32}, G_{33}, G_{34}. Accordingly, we proceed to develop the means for finding the other parameters. In fact, because only the ratios of the homogeneous coordinates are meaningful, only the ratios of the G_{ij} values need be computed, and it is usual to take G_{44} as unity: this leaves only 11 parameters to be determined. Multiplying out the first two equations and re-arranging gives:

$$
G_{11}X + G_{12}Y + G_{13}Z + G_{14} - x(G_{41}X + G_{42}Y + G_{43}Z) = x
\tag{18.22}
$$

$$
G_{21}X + G_{22}Y + G_{23}Z + G_{24} - y(G_{41}X + G_{42}Y + G_{43}Z) = y
\tag{18.23}
$$

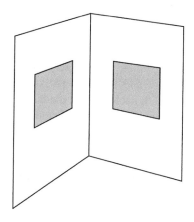

FIGURE 18.2

A convenient near-minimum case for camera calibration. Here two sets of four coplanar points, each set of four being at the corners of a square, provide more than the absolute minimum number of points required for camera calibration.

Noting that a single world point (X, Y, Z), which is known to correspond to image point (x, y), gives us *two* equations of the above form; it requires a minimum of six such points to provide values for all 11 G_{ij} parameters: Figure 18.2 shows a convenient near-minimum case. An important factor is that the world points used for the calculation should lead to independent equations: thus, it is important that they should not be coplanar. More precisely, there must be at least six points, no four of which are coplanar. However, further points are useful in that they lead to over-determination of the parameters and increase the accuracy with which the latter can be computed. There is no reason why the additional points should not be coplanar with existing points: indeed, a common arrangement is to set up a cube so that three of its faces are visible, each face having a pattern of squares with 30–40 easily discerned corner features (as for a Rubic cube).

Least-squares analysis can be used to perform the computation of the 11 parameters, e.g., *via* the pseudo-inverse method. First, the $2n$ equations have to be expressed in matrix form:

$$A\mathbf{g} = \xi \tag{18.24}$$

where A is a $2n \times 11$ matrix of coefficients, which multiplies the G-matrix, now in the form:

$$\mathbf{g} = (G_{11}\ G_{12}\ G_{13}\ G_{14}\ G_{21}\ G_{22}\ G_{23}\ G_{24}\ G_{41}\ G_{42}\ G_{43})^{\mathrm{T}} \tag{18.25}$$

and ξ is a $2n$-element column vector of image coordinates. The pseudo-inverse solution is:

$$\mathbf{g} = A^{\dagger}\xi \tag{18.26}$$

where:

$$A^\dagger = (A^T A)^{-1} A^T \qquad (18.27)$$

The solution is more complex than might have been expected, since a normal matrix inverse is only defined, and can only be computed, for a square matrix. Note that solutions are only obtainable by this method if the matrix $A^T A$ is invertible. For further details of this method, see Golub and van Loan (1983).

18.4 INTRINSIC AND EXTRINSIC PARAMETERS

At this point it is useful to look in more detail at the general transformation leading to camera calibration. When we are calibrating the camera, we are actually trying to bring the camera and world coordinate systems into coincidence. The first step is to move the origin of the world coordinates to the origin of the camera coordinate system. The second step is to rotate the world coordinate system until its axes are coincident with those of the camera coordinate system. The third step is to move the image plane laterally until there is complete agreement between the two coordinate systems (this step is required since it is not known initially which point in the world coordinate system corresponds to the principal point[1] in the image).

There is an important point to be borne in mind during this process. If the camera coordinates are given by **C**, then the translation **T** required in the first step will be $-\mathbf{C}$. Similarly, the rotations that are required will be the inverses of those that correspond to the actual camera orientations. The reason for these reversals is that (for example) rotating an object (here the camera) forward gives the same effect as rotating the axes backward. Thus, all operations have to be carried out with the reverse arguments to those indicated above in Section 18.1. The complete transformation for camera calibration is hence:

$$G = PLRT = \begin{bmatrix} 1 & 0 & 0 & 0 \\ 0 & 1 & 0 & 0 \\ 0 & 0 & 1 & 0 \\ 0 & 0 & 1/f & 0 \end{bmatrix} \begin{bmatrix} 1 & 0 & 0 & t_1 \\ 0 & 1 & 0 & t_2 \\ 0 & 0 & 1 & t_3 \\ 0 & 0 & 0 & 1 \end{bmatrix} \begin{bmatrix} R_{11} & R_{12} & R_{13} & 0 \\ R_{21} & R_{22} & R_{23} & 0 \\ R_{31} & R_{32} & R_{33} & 0 \\ 0 & 0 & 0 & 1 \end{bmatrix} \begin{bmatrix} 1 & 0 & 0 & T_1 \\ 0 & 1 & 0 & T_2 \\ 0 & 0 & 1 & T_3 \\ 0 & 0 & 0 & 1 \end{bmatrix}$$

$$(18.28)$$

where matrix P takes account of the perspective transformation required to form the image. In fact, it is usual to group together the transformations P and L and call them internal camera transformations that include the *intrinsic camera parameters*, while R and T are taken together as external camera transformations corresponding to *extrinsic camera parameters*:

---

[1]The *principal point* is the image point lying on the principal axis of the camera. It is the point in the image that is closest to the center of projection. Correspondingly, the *principal axis* (or *optical axis*) of the camera is the line through the center of projection normal to the image plane.

$$G = G_{\text{internal}} G_{\text{external}} \tag{18.29}$$

where:

$$G_{\text{internal}} = PL = \begin{bmatrix} 1 & 0 & 0 & t_1 \\ 0 & 1 & 0 & t_2 \\ 0 & 0 & 1 & t_3 \\ 0 & 0 & 1/f & t_3/f \end{bmatrix} \rightarrow \begin{bmatrix} 1 & 0 & t_1 \\ 0 & 1 & t_2 \\ 0 & 0 & 1/f \end{bmatrix} \tag{18.30}$$

$$G_{\text{external}} = RT = \begin{bmatrix} \mathbf{R}_1 & \mathbf{R}_1.\mathbf{T} \\ \mathbf{R}_2 & \mathbf{R}_2.\mathbf{T} \\ \mathbf{R}_3 & \mathbf{R}_3.\mathbf{T} \\ 0 & 1 \end{bmatrix} \tag{18.31}$$

In the matrix for G_{internal}, we have assumed that the initial translation matrix T moves the camera's center of projection to the correct position, so that the value of t_3 can be made equal to zero: in that case, the effect of L will indeed be lateral as indicated above. At that point, we can express the (2-D) result in terms of a 3×3 homogeneous coordinate matrix. In the matrix for G_{external}, we have expressed the result succinctly in terms of the rows $\mathbf{R}_1$, $\mathbf{R}_2$, $\mathbf{R}_3$ of R, and have taken dot-products with $\mathbf{T} = (T_1, T_2, T_3)^{\text{T}}$: the (3-D) result is a 4×4 homogeneous coordinate matrix.

Although the above treatment gives a good indication of the underlying meaning of G, it is not general because we have not so far included scaling and skew parameters in the internal matrix. In fact the generalized form of G_{internal} is:

$$G_{\text{internal}} = \begin{bmatrix} s_1 & b_1 & t_1 \\ b_2 & s_2 & t_2 \\ 0 & 0 & 1/f \end{bmatrix} \tag{18.32}$$

Potentially, G_{internal} should include the following:

1. A transform for correcting scaling errors.
2. A transform for correcting translation errors.[2]
3. A transform for correcting sensor skewing errors (due to nonorthogonality of the sensor axes).
4. A transform for correcting sensor shearing errors (due to unequal scaling along the sensor axes).
5. A transform for correcting for unknown sensor orientation within the image plane.

Clearly, translation errors (item 2) are corrected by adjusting t_1 and t_2. All the other adjustments are concerned with the values of the 2×2 submatrix:

$$\begin{bmatrix} s_1 & b_1 \\ b_2 & s_2 \end{bmatrix}$$

---

[2]For this purpose, the origin of the image should be on the principal axis of the camera. Misalignment of the sensor may prevent this point from being at the center of the image.

However, note that application of this matrix performs rotation within the image plane immediately after rotation has been performed in the world coordinates by $G_{external}$, and it is virtually impossible to separate the two rotations. This explains why we now have a total of 6 external and 6 internal parameters totaling 12 rather than the expected 11 parameters (we return to the factor $1/f$ below). As a result, it is better to exclude item 5 in the above list of internal transforms and to subsume it into the external parameters.[3] Since the rotational component in $G_{internal}$ has been excluded, b_1 and b_2 must now be equal, and the internal parameters will be: s_1, s_2, b, t_1, t_2. Note that the factor $1/f$ provides a scaling that cannot be separated from the other scaling factors during camera calibration, without specific (i.e., separate) measurement of f. Thus, we have a total of six parameters from $G_{external}$ and five parameters from $G_{internal}$: which totals 11 and equals the number cited in the previous section.

We next consider the special case where the sensor is known to be Euclidean to a high degree of accuracy. This will mean that $b = b_1 = b_2 = 0$, and $s_1 = s_2$, bringing the number of internal parameters down to three. In addition, if care has been taken over sensor alignment, and there are no other offsets to be allowed for, it may be known that $t_1 = t_2 = 0$. This will bring the total number of internal parameters down to just one, namely $s = s_1 = s_2$, or sf, if we take proper account of the focal length. In this case, there will be a total of seven calibration parameters for the whole camera system, and this may permit it to be set up unambiguously by viewing a known object having four clearly marked features instead of the six that would normally be required (see Section 18.3).

18.5 CORRECTING FOR RADIAL DISTORTIONS

Photographs generally appear so distortion-free that there is a tendency to imagine that camera lenses are virtually perfect. However, it sometimes happens that a photograph will show odd curvatures of straight lines, particularly those appearing around the periphery of the picture. The results commonly take the form of "pincushion" or "barrel" distortion: these terms arise because pincushions have a tendency to be over-extended at the corners, while barrels usually bulge in the middle. In images of paving stones or brick walls, the amount of distortion is usually not more than a few pixels in a total of the order of 512, i.e., typically less than 2%, and this explains why in the absence of particular straight line markers such distortions can be missed (Fig. 18.3). However, it is important both for recognition and for inter-image matching purposes that any distortions should be

---

[3]While doing so may not be ideal, there is no way of separating the two rotational components by purely optical means; only measurements on the internal dimensions of the camera system could determine the internal component, but separation is not likely to be a cogent or even meaningful matter. On the other hand, the internal component is likely to be stable, whereas the external component may be prone to variation if the camera is not mounted securely.

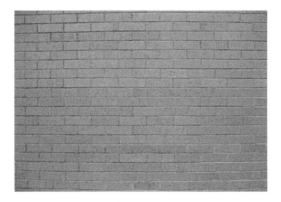

FIGURE 18.3

Photograph of a brick wall showing radial (barrel) distortion.

eliminated. Indeed, image interpretation is nowadays targeted at, and frequently achieves, subpixel accuracy. In addition, disparities between stereo images are in the first order of small quantities, and single pixel errors would lead to significant errors in depth measurement. Hence, it is more the rule than the exception that 3-D image analysis will need to make corrections for barrel or pincushion distortion.

For reasons of symmetry, the distortions that arise in images tend to involve radial expansions or contractions relative to the optical axis—corresponding respectively to pincushion or barrel distortion. As with many types of error, series solutions can be useful. Thus, it is worthwhile to model the distortions as:

$$\mathbf{r}' = \mathbf{r}f(r) = \mathbf{r}(a_0 + a_2 r^2 + a_4 r^4 + a_6 r^6 + \cdots) \qquad (18.33)$$

the odd orders in the brackets canceling to zero, again for reasons of symmetry. It is usual to set a_0 to unity, as this coefficient can be taken up by the scale parameters in the camera calibration matrix.

To fully define the effect, we write the x and y distortions as:

$$x' - x_c = (x - x_c)(1 + a_2 r^2 + a_4 r^4 + a_6 r^6 + \cdots) \qquad (18.34)$$

$$y' - y_c = (y - y_c)(1 + a_2 r^2 + a_4 r^4 + a_6 r^6 + \cdots) \qquad (18.35)$$

Here x and y are measured relative to the position of the optical axis of the lens (x_c, y_c), so $\mathbf{r} = (x - x_c, y - y_c)$, $\mathbf{r}' = (x' - x_c, y' - y_c)$.

As remarked above, the errors to be expected are in the range 2% or less. This means that it is normally sufficiently accurate to take just the first correction term in the expansion and disregard the rest. At the very least, this will introduce such a large improvement in the accuracy that it will be difficult to detect any discrepancies, especially if the image dimensions are 512×512 pixels or less.[4] In addition, computation errors in matrix inversion and convergence of 3-D algorithms will add to the digitization errors, tending further to hide higher powers of radial distortion. Thus, in most cases the latter can be modeled using a single parameter equation:

$$\mathbf{r}' = \mathbf{r}f(r) = \mathbf{r}(1 + a_2 r^2) \tag{18.36}$$

Note that the above theory only models the distortion: clearly, it has to be corrected by the corresponding inverse transformation.

It is instructive to consider the apparent shape of a straight line that appears, e.g., along the top of an image (Fig. 18.3). Take the image dimensions to range over $-x_1 \leq x \leq x_1$, $-y_1 \leq y \leq y_1$, and the optical axis of the camera to be at the center of the image. Then the straight line will have the approximate equation:

$$y' = y_1[1 + a_2(x^2 + y_1^2)] = y_1 + a_2 y_1^3 + a_2 y_1 x^2 \tag{18.37}$$

which represents a parabola. The vertical error at the center of the parabola is $a_2 y_1^3$ and the additional vertical error at the ends is $a_2 y_1 x_1^2$. If the image is square $(x_1 = y_1)$, these two errors are equal (the erroneous impression is given by the parabola shape that the error at $x = 0$ is zero).

Finally, note that digital scanners are very different from single lens cameras, in that their lenses travel along the object space during acquisition. Thus, longitudinal errors are unlikely to arise to anything like the same extent, although lateral errors could in principle be problematic.

18.6 MULTIPLE VIEW VISION

Over the 1990s a considerable advance in 3-D vision was made by examining what could be learnt from uncalibrated cameras using multiple views. At first sight, considering the efforts made in earlier sections of this chapter to understand exactly how cameras should be calibrated, this may seem nonsensical. Nevertheless, there are considerable potential advantages in examining multiple views—not least, many thousands of videotapes are available from uncalibrated cameras, including those used for surveillance and those produced in the film industry. In such cases, as much must be made of the available material as possible, whether or not any regrets over "what might have been" are entertained. However, the need is deeper than this. Many situations exist in which the camera parameters might vary because of thermal variations, or because the zoom or focus setting has been adjusted: and it is impracticable to keep recalibrating a

---

[4]This remark will not apply to many web cameras, which are sold at extremely low prices on the mainly amateur market. While the camera chip and electronics are often very good value, the accompanying low-cost lens may well require extensive correction to ensure that distortion-free measurements are possible.

camera using accurately made test objects. Finally, if multiple (e.g., stereo) cameras are used, each will have to be calibrated separately and the results compared to minimize the combined error: far better to examine the system as a whole, and to calibrate it on the real scenes that are being viewed.

In fact, we have already met some aspects of these aspirations, in the form of invariants that are obtained in sequence by a single camera. For example, if a series of four collinear points are viewed and their cross-ratio is checked, it will be found to be constant as the camera moves forward, changes orientation, or views the points increasingly obliquely—so long as they all remain within the field of view. For this purpose, all that is required to perform the recognition and maintain awareness of the object (the four points) is an uncalibrated but distortion-free camera. By distortion-free we here mean not the ability to correct perspective distortion—which is, after all, the function of the cross-ratio invariant—but the lack of radial distortion, or at least the capability in the following software for eliminating it (see Section 18.5).

To understand how image interpretation can be carried out more generally, using multiple views—whether from the same camera moved to a variety of places, or multiple cameras with overlapping views of the world—we shall need to go back to basics and start afresh with a more general attack on concepts such as binocular vision and epipolar constraints. In particular, two important matrices will be called into play—the "essential" matrix and the "fundamental" matrix. We start with the essential matrix and then generalize the idea to the fundamental matrix. But first we need to look at the geometry of two cameras with general views of the world.

18.7 GENERALIZED EPIPOLAR GEOMETRY

In Section 15.3, we considered the stereo correspondence problem, and had already simplified the task by choosing two cameras whose image planes were not only parallel but in the same plane. This made the geometry of depth perception especially simple, but suppressed possibilities allowed for in the human visual system (HVS), of having a nonzero vergence angle between the two images. Indeed, the HVS is special in adjusting vergence so that the current focus of attention in the field of view has almost zero disparity between the two images, and it seems likely that the HVS estimates depth not merely by measuring disparity but rather by measuring the vergence together with remanent small variations in disparity.

Here we generalize the situation to cover the possibility of disparity coupled with substantial vergence. Figure 18.4 shows the revised geometry. Note first that observation of a real point P in the scene leads to points P_1 and P_2 in the two images; that P_1 could correspond to any point on the epipolar line E_2 in image 2; and similarly, that point P_2 could correspond to any point on the epipolar line E_1

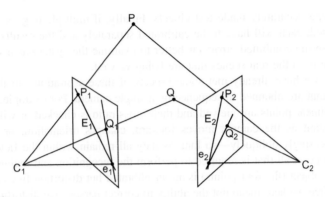

FIGURE 18.4

Generalized imaging of a scene from two viewpoints. In this case there is substantial vergence. All epipolar lines in the left image pass through epipole e_1: of these, only E_1 is shown. Similar comments apply for the right image.

in image 1. Indeed, the so-called "epipolar plane" of P is the plane containing P and the projection points C_1 and C_2 of the two cameras: the epipolar lines (see Section 15.3) are thus the straight lines in which this plane cuts the two image planes. Furthermore, the line joining C_1 and C_2 cuts the image planes in the so-called epipoles e_1 and e_2: these can be regarded as the images of the alternate camera projection points. Note that all epipolar planes pass through points C_1, C_2 and e_1, e_2: this means that all epipolar lines in the two images pass through the respective epipoles. However, if the vergence angle were zero (as in Fig. 18.5), the epipoles would be at infinity in either direction, and all epipolar lines in either image would be parallel, and indeed parallel to the vector **C** from C_1 to C_2.

18.8 THE ESSENTIAL MATRIX

In this section, we start with the vectors $\mathbf{P}_1$, $\mathbf{P}_2$, from C_1, C_2 to P, and also the vector **C** from C_1 to C_2. Vector subtraction gives:

$$\mathbf{P}_2 = \mathbf{P}_1 - \mathbf{C} \tag{18.38}$$

We also know that $\mathbf{P}_1$, $\mathbf{P}_2$, and **C** are coplanar, the condition of coplanarity being:[5]

$$\mathbf{P}_2 \cdot \mathbf{C} \times \mathbf{P}_1 = 0 \tag{18.39}$$

---

[5]This can be thought of as bringing to zero the volume of the parallelepiped with sides $\mathbf{P}_1$, $\mathbf{P}_2$, and **C**.

FIGURE 18.5

Error in locating a feature in space using binocular imaging. The dark shaded regions represent the regions of space that could arise for small errors in the image planes. The crossover region, shaded black, confirms that longitudinal errors will be much larger than lateral errors. A full analysis would involve applying Gaussian or other error functions (see text).

To progress, we need to relate the vectors $\mathbf{P}_1$ and $\mathbf{P}_2$ when these are expressed relative to their own frames of reference. If we take these vectors as having been defined in the C_1 frame of reference, we now re-express $\mathbf{P}_2$ in its own (C_2) frame of reference, by applying a translation $\mathbf{C}$ and a rotation of coordinates expressed as the orthogonal matrix R. This leads to:

$$\mathbf{P}_2' = R\mathbf{P}_2 = R(\mathbf{P}_1 - \mathbf{C}) \tag{18.40}$$

so that:

$$\mathbf{P}_2 = R^{-1}\mathbf{P}_2' = R^{\mathrm{T}}\mathbf{P}_2' \tag{18.41}$$

Substituting in the coplanarity condition gives:

$$(R^{\mathrm{T}}\mathbf{P}_2') \cdot \mathbf{C} \times \mathbf{P}_1 = 0 \tag{18.42}$$

At this point, it is useful to replace the vector product notation by using a skew-symmetric matrix $C_\times$ to denote $\mathbf{C}\times$, where:

$$C_\times = \begin{bmatrix} 0 & -C_z & C_y \\ C_z & 0 & -C_x \\ -C_y & C_x & 0 \end{bmatrix} \tag{18.43}$$

At the same time, we observe the correct matrix formulation of all the vectors by transposing appropriately. We now find that:

$$(R^T \cdot \mathbf{P}_2')^T C_\times \mathbf{P}_1 = 0 \tag{18.44}$$

$$\therefore \quad \mathbf{P}_2'^T R C_\times \mathbf{P}_1 = 0 \tag{18.45}$$

Finally, we obtain the "essential matrix" formulation:

$$\mathbf{P}_2'^T E \mathbf{P}_1 = 0 \tag{18.46}$$

where the essential matrix has been found to be:

$$E = R C_\times \tag{18.47}$$

Equation (18.46) is actually the desired result: it expresses the relation between the observed positions of the same point in the two camera frames of reference. Furthermore, it immediately leads to formulae for the epipolar lines. To see this, first note that in the C_1 camera frame:

$$\mathbf{p}_1 = \left(\frac{f_1}{Z_1}\right) \mathbf{P}_1 \tag{18.48}$$

while in the C_2 camera frame (and expressed in terms of that frame of reference):

$$\mathbf{p}_2' = \left(\frac{f_2}{Z_2}\right) \mathbf{P}_2' \tag{18.49}$$

Eliminating $\mathbf{P}_1$ and $\mathbf{P}_2'$, and dropping the prime (as within the respective image planes the numbers 1 and 2 are sufficient to specify the coordinates unambiguously), we find:

$$\mathbf{p}_2^T E \mathbf{p}_1 = 0 \tag{18.50}$$

as Z_1, Z_2 and f_1, f_2 can be canceled from this matrix equation.

Now note that writing $\mathbf{p}_2^T E = \mathbf{l}_1^T$ and $\mathbf{l}_2 = E\mathbf{p}_1$ leads to the following relations:

$$\mathbf{p}_1^T \mathbf{l}_1 = 0 \tag{18.51}$$

$$\mathbf{p}_2^T \mathbf{l}_2 = 0 \tag{18.52}$$

This means that $\mathbf{l}_2 = E\mathbf{p}_1$ and $\mathbf{l}_1 = E^T\mathbf{p}_2$ are the epipolar lines corresponding to $\mathbf{p}_1$ and $\mathbf{p}_2$, respectively.[6]

Finally, we can find the epipoles from the above formulation. In fact, the epipole lies on every epipolar line within the same image. Thus, $\mathbf{e}_2$ satisfies (can be substituted for $\mathbf{p}_2$ in) equation (18.52), and hence:

$$\mathbf{e}_2^T \mathbf{l}_2 = 0$$

$$\therefore \quad \mathbf{e}_2^T E\mathbf{p}_1 = 0 \quad \text{for all } \mathbf{p}_1.$$

This means that $\mathbf{e}_2^T E = 0$, i.e., $E^T\mathbf{e}_2 = 0$. Similarly, $E\mathbf{e}_1 = 0$.

18.9 THE FUNDAMENTAL MATRIX

Note that in the last part of the essential matrix calculation, we implicitly assumed that the cameras are correctly calibrated. Specifically, $\mathbf{p}_1$ and $\mathbf{p}_2$ are corrected (calibrated) image coordinates. However, there is a need to work with uncalibrated images, using the raw pixel measurements[7]—for all the reasons given in Section 18.6. Applying the camera intrinsic matrices G_1, G_2 to the calibrated image coordinates (Section 18.4), we get the raw image coordinates:

$$\mathbf{q}_1 = G_1\mathbf{p}_1 \tag{18.53}$$

$$\mathbf{q}_2 = G_2\mathbf{p}_2 \tag{18.54}$$

In fact, we here need to go in the reverse direction, so we use the inverse equations:

$$\mathbf{p}_1 = G_1^{-1}\mathbf{q}_1 \tag{18.55}$$

$$\mathbf{p}_2 = G_2^{-1}\mathbf{q}_2 \tag{18.56}$$

Substituting for $\mathbf{p}_1$ and $\mathbf{p}_2$ in equation (18.50), we find the desired equation linking the raw pixel coordinates:

$$\mathbf{q}_2^T (G_2^{-1})^T E G_1^{-1} \mathbf{q}_1 = 0 \tag{18.57}$$

---

[6]Consider a line $\mathbf{l}$ and a point $\mathbf{p}$. $\mathbf{p}^T\mathbf{l} = 0$ means that $\mathbf{p}$ lies on the line $\mathbf{l}$, or dually, $\mathbf{l}$ passes through the point $\mathbf{p}$.

[7]However, any radial distortions need to be eliminated, so as to *idealize* the camera, but not to calibrate it in the sense of Sections 18.3 and 18.4.

which can be expressed as:

$$\mathbf{q}_2^{\mathrm{T}} F \mathbf{q}_1 = 0 \tag{18.58}$$

where

$$F = (G_2^{-1})^{\mathrm{T}} E G_1^{-1} \tag{18.59}$$

F is defined as the "fundamental matrix." Because it contains all the information that would be needed to calibrate the cameras, it contains more free parameters than the essential matrix. However, in other respects the two matrices are intended to convey the same basic information, as is confirmed by the resemblance between the two formulations—Eqs. (18.46) and (18.58).

Finally, just as in the case of the essential matrix, the epipoles are given by $F\mathbf{f}_1 = 0$ and $F^{\mathrm{T}}\mathbf{f}_2 = 0$, although this time in raw image coordinates $\mathbf{f}_1$ and $\mathbf{f}_2$.

18.10 PROPERTIES OF THE ESSENTIAL AND FUNDAMENTAL MATRICES

Next we consider the composition of the essential and fundamental matrices. In particular, note that $C_\times$ is a factor of E and also, indirectly, of F. In fact, they are homogeneous in $C_\times$, so the scale of $\mathbf{C}$ will make no difference to the two matrix formulations (Eqs. (18.46) and (18.58)), only the *direction* of $\mathbf{C}$ being important: indeed, the scales of both E and F are immaterial, and as a result only the relative values of their coefficients are of importance. This means that there are at most only eight independent coefficients in E and F. In fact, in the case of F there are only *seven*, as $C_\times$ is skew-symmetric, and this ensures that it has rank 2 rather than rank 3—a property that is passed on to F. The same argument applies for E, but the lower complexity of E (by virtue of its not containing the image calibration information) means that it has only *five* free parameters. In the latter case, it is easy to see what they are: they arise from the original three translation ($\mathbf{C}$) and three rotation (R) parameters, less the one parameter corresponding to scale.

In this context, note that if $\mathbf{C}$ arises from a translation of a single camera, the same essential matrix will result whatever the scale of $\mathbf{C}$: only the direction of $\mathbf{C}$ actually matters, and the same epipolar lines will result from continued motion in the same direction. In fact, in this case we can interpret the epipoles as foci of expansion or contraction. This underlines the power of this formulation: specifically, it treats motion and displacement as a single entity.

Finally, we should try to understand why there are seven free parameters in the fundamental matrix. The solution is relatively simple. Each epipole requires two parameters to specify it. In addition, three parameters are needed to map any three epipolar lines from one image to the other. But why do just three epipolar lines have to be mapped? This is because the family of epipolar lines is a pencil whose orientations are related by cross-ratios, so once three epipolar lines have been specified the mapping of any other can be deduced. (Knowing the properties of the cross-ratio, it is seen that fewer than three epipolar lines would be insufficient and that more than three would yield no additional information.) This fact is sometimes stated in the following form: a homography (a projective transformation) between two 1-D projective spaces has three degrees of freedom.

18.11 ESTIMATING THE FUNDAMENTAL MATRIX

In the previous section, we showed that the fundamental matrix has seven free parameters. This means that it ought to be possible to estimate it by identifying the same seven features in the two images.[8] However, while this is mathematically possible in principle and a suitable nonlinear algorithm has been devised by Faugeras et al. (1992) to implement it, it has been shown that the computation can be numerically unstable. Essentially, noise acts as an additional variable boosting the effective number of degrees of freedom in the problem to eight. However, a linear algorithm called the *eight-point algorithm* has been devised to overcome the problem. Curiously, this algorithm had been proposed many years earlier by Longuet-Higgins (1981) to estimate the *essential* matrix, but it came into its own when Hartley (1995) showed how to control the errors by first normalizing the values. In addition, by using more than eight points, increased accuracy can be attained, but then a suitable algorithm must be found that can cope with the now overdetermined parameters. Principal components analysis can be used for this, an appropriate procedure being singular value decomposition (SVD).

Apart from noise, gross mismatches in forming trial point correspondences between images can be a source of practical problems. If so, the normal least-squares types of solution can profitably be replaced by the least median of squares robust estimation method (Appendix A).

18.12 AN UPDATE ON THE EIGHT-POINT ALGORITHM

Section 18.11 outlined the value of the eight-point algorithm for estimating the fundamental matrix. Over a period of about 8 years (1995–2003) this essentially became the standard solution to the problem. However, a key contribution by Torr and Fitzgibbon (2003, 2004) has shown that the eight-point algorithm might after all not be the best possible method, since the solutions it obtains depend on the particular coordinate system used for the computation. This is because the

---

[8]Using the minimum number of points in this way carries the health warning that they must be in general position: special configurations of points can lead to numerical instabilities in the computations, total failure to converge, or unnecessary ambiguities in the results. In general, coplanar points are to be avoided.

normalization normally used, namely $\sum_i f_i^2 = 1$,[9] is not invariant to shifts in the coordinate system. In fact, it is by no means obvious how to find an invariant normalization. Nevertheless, Torr and Fitzgibbon's logical analysis of the situation in which they were forced to disregard the affine transform case appropriate for weak perspective, led to the following normalization of F:

$$f_1^2 + f_2^2 + f_4^2 + f_5^2 = K \tag{18.60}$$

where K is a constant and:

$$F = \begin{bmatrix} f_1 & f_2 & f_3 \\ f_4 & f_5 & f_6 \\ f_7 & f_8 & f_9 \end{bmatrix} \tag{18.61}$$

Finally, to determine F, Eq. (18.60) can be applied as a Lagrangian multiplier constraint and this leads to an eigenvector solution for F. Overall, the 8×8 eigenvalue problem solved by the eight-point algorithm is replaced by a 5×5 eigenvalue problem. Furthermore, this approach not only yields the required invariance properties, thus ensuring a more accurate solution, but also it gives a much faster computation that loses significantly fewer tracks in image sequence analysis.

18.13 IMAGE RECTIFICATION

In Section 18.7, we took some pains to generalize the epipolar approach and subsequently arrived at general solutions, corresponding to arbitrary overlapping views of scenes. However, there are distinct advantages in special views obtained from cameras with parallel axes—as in the case of Fig. 18.5 where the vergence is zero. Specifically, it is easier to find correspondences between scenes that are closely related in this way. Unfortunately, such well-prepared pairs of images are not in keeping with the aims promoted in Section 18.6, of insisting on closely aligned and calibrated cameras, and this certainly doesn't apply to frames taken by a single moving camera unless its motion is severely constrained by special means. In fact, the solution is straightforward: take images with uncalibrated cameras, estimate the fundamental matrix, and then apply suitable linear transformations to compute the images for any desired idealized camera positions. The latter technique is called image rectification and ensures, e.g., that the epipolar lines are all parallel to the baseline $\mathbf{C}$ between the centers of projection. This then results in correspondences being found by searching along points with the same ordinate in the alternate image: for a point with coordinates (x_1, y_1) in the first image, search for a matching point (x_2, y_1) in the second image.

---

[9]Early on, Tsai and Huang (1984) suggested the normalization $f_9 = 1$, but this leads to biased solutions, and for example excludes solutions with $f_9 = 0$.

When rectifying an image, it will in general be rotated in 3-D,[10] and the obvious way of achieving this is to transfer each individual pixel to its new location in the rectified image. However, rotations are nonlinear processes and will in some cases have the effect of mapping several pixels into a single pixel; furthermore, a number of pixels may well not have intensity values assigned to them. While the first of these problems could be tackled by some sort of intensity averaging process and the latter problem could be tackled by applying a median or other type of filter to the transformed image, such techniques are insufficiently thoroughgoing to provide accurate, reliable solutions. The *proper* way of overcoming these intrinsic difficulties is to backproject the pixel locations from the transformed image space to the source image, use interpolation to compute the ideal pixel intensities, and then transfer these intensities to the transformed image space.

Bilinear interpolation is used most often in the transformation process. This works by performing interpolation in the x-direction and then in the y-direction. Thus, if the location to be interpolated to is $(x + a, y + b)$ where x and y are integer pixel locations, and $0 \leq a, b \leq 1$, then the interpolated intensities in the x-direction are:

$$I(x + a, y) = (1 - a)I(x, y) + aI(x + 1, y) \tag{18.62}$$

$$I(x + a, y + 1) = (1 - a)I(x, y + 1) + aI(x + 1, y + 1) \tag{18.63}$$

and the final result after interpolating in the y-direction is:

$$\begin{aligned} I(x + a, y + b) = {}&(1 - a)(1 - b)I(x, y) + a(1 - b)I(x + 1, y) \\ &+ (1 - a)bI(x, y + 1) + abI(x + 1, y + 1) \end{aligned} \tag{18.64}$$

The symmetry of the result shows that it makes no difference which axis is chosen for the first pair of interpolations, and this limits the arbitrariness of the method. Note that the method does not assume a locally planar intensity variation in 2-D: this is clear as the value of the $I(x + 1, y + 1)$ intensity is taken into account as well as the other three intensity values. Nevertheless, bilinear interpolation is not a totally ideal solution, as it takes no account of the sampling theorem, and for this reason the bi-cubic interpolation method (which involves more computation) is sometimes used instead. In addition, all such methods introduce slight local blurring of the image as they involve averaging of local intensity values. Overall, transformation processes such as this are bound to result in slight degradation of the image data.

18.14 **3-D RECONSTRUCTION**

In Section 18.10, the fact that F is determined only up to an unknown scale factor (or equivalently that the actual scales of its coefficients as obtained are arbitrary)

---

[10]Of course, it may also be translated and scaled, in which case the effect described here may be even more significant.

was strongly emphasized. This reflected the deliberate avoidance of camera calibration in this work. In practice, this means that if the results of computations of F are to be related back to the real world, the scaling factor must be reinstated. In principle, this can be achieved by viewing a single yardstick: it is unnecessary to view an object such as a Rubik cube, as knowledge of F carries with it a lot of information on relative dimensions in the real world. This factor is important when reconstructing a real scene with a real depth map.

There are a number of methods for image reconstruction, of which perhaps the most obvious is triangulation. This starts by taking two camera positions containing normalized images and projecting rays for a given point P back into the real world until they meet. In fact, attempting to do this meets with an immediate problem: the inaccuracies in the available parameters, coupled with the pixellation of the images, ensure that in most cases rays will not actually meet, as they are skew lines. The best that can be done with skew lines is to determine the position of closest approach. Once this has been found, the bisector of the line of closest approach (which is perpendicular to each of the rays) is, in *this* model, the most accurate estimate of the position of P in space.

Unfortunately, the above model is not guaranteed to give the most accurate prediction of the position of P. This is because perspective projection is a highly nonlinear process: in particular, slight misjudgement of the orientation of the point from either of the images can cause a substantial depth error, coupled with a significant lateral error: so much is indeed obvious from Fig. 18.5. This being so, it has to be asked where the error might still be linear, so that, at that position at least, error calculation can be based on Gaussian distributions.[11] In fact, the errors can be taken to be approximately Gaussian in the images themselves. This means that the point in space that has to be chosen as representing the most accurate interpretation of the data is that which results in the minimum error (in a least mean square sense) when reprojected onto the image planes. Typically, the error obtained using this approach is a factor of two smaller than that for the triangulation method described above (Hartley and Zisserman, 2000).

Finally, it is useful to mention a further type of error that can arise with two cameras. This applies when they both view an object with a smoothly varying boundary. For example, if both cameras are viewing the right-hand edge of a vase of circular cross-section, each will see a different point on the boundary and a discrepancy will arise in the estimated boundary position (Fig. 18.6). It is left as an exercise (Section 18.17) to determine the exact magnitude of such errors. In fact, the error is proportional both to a, where a is the local radius of curvature of the observed boundary, and to Z^{-2}, where Z is the depth in the scene. This means that the error (and the percentage error) tends to zero at large distances, and also that the error falls properly to zero for sharp corners.

---

[11]Here we ignore the possibility of gross errors arising from mismatches between images, which is the subject of further discussion in Section 18.11 and elsewhere.

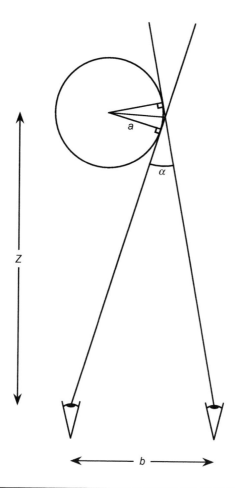

FIGURE 18.6

Lateral estimation error arising with a smoothly varying boundary. The error arises in estimating the boundary position when information from two views is fused in the standard way. *a* is the radius of a vase being observed, α is the disparity in direction of its right-hand boundary, *Z* is its depth in the scene, and *b* is the stereo baseline.

18.15 CONCLUDING REMARKS

This chapter has discussed the transformations required for camera calibration and has outlined how calibration can be achieved. The camera parameters have been classified as "internal" and "external," thereby simplifying the conceptual problem and throwing light on the origins of errors in the system. It has been shown that a minimum of six points is required to perform calibration in the general case where eleven transformation parameters are involved; however, the number of points required might be reduced somewhat in special cases, e.g., where the sensor is known to be Euclidean. Nevertheless, it is normally more important to increase the number of points used for calibration than to attempt to reduce it, since substantial gains in accuracy can be obtained *via* the resulting averaging process.

In an apparent break with the previous work, Section 18.5 introduced multiple view vision. This important topic was seen to rest on generalized epipolar geometry, and led to the essential and fundamental matrix formulations, which relate the observed positions of any point in two camera frames of reference. The importance of the eight-point algorithm for estimating either of these matrices—and particularly the fundamental matrix, which is relevant when the cameras are uncalibrated—was stressed. In addition, the need for accuracy in estimating the fundamental matrix is still a research issue.

> The obvious way of tackling vision problems is to set up a camera and calibrate it, and only then to use it in anger. This chapter has shown how, to a large extent, calibration can be avoided or carried out adaptively "on the fly"—by performing multiple view vision and analyzing the various key matrices that arise from the generalized epipolar problems.

18.16 BIBLIOGRAPHICAL AND HISTORICAL NOTES

One of the first to use the various transformations described in this chapter was Roberts (1965). Important early references for camera calibration are the *Manual of Photogrammetry* (Slama, 1980), Tsai and Huang (1984), and Tsai (1986). Tsai's paper is especially useful in that he provides an extended, highly effective treatment that copes with nonlinear lens distortions. More recent papers on this topic include Haralick (1989), Crowley et al. (1993), Cumani and Guiducci (1995), and Robert (1996): see also Zhang (1995). Note that parametrized plane curves can be used instead of points for the purpose of camera calibration (Haralick and Chu, 1984).

Clearly, camera calibration is an old topic that is revisited every time 3-D vision has to be used for measurement, and otherwise when rigorous analysis of 3-D scenes is called for. The calibration scenario started to undergo a metamorphosis in the early 1990s, when it was realised that much could be learnt without overt calibration, but rather by *comparing* images taken from moving sequences or from multiple views (Faugeras, 1992; Faugeras et al., 1992; Hartley, 1992; Maybank and Faugeras, 1992). In fact, while it was appreciated that much could be learnt without overt calibration, it was not at that stage known how much *might* be learnt, and there ensued a rapid sequence of developments as the frontiers were progressively pushed back (e.g., Hartley, 1995; Hartley, 1997; Luong and Faugeras, 1997). By the late 1990s the fast evolution phase was over, and definitive, albeit quite complex, texts appeared covering these developments

(Hartley and Zisserman, 2000; Faugeras and Luong, 2001; Gruen and Huang, 2001). Nevertheless, many refinements of the standard methods were still emerging (Faugeras et al., 2000; Heikkilä, 2000; Sturm, 2000; Roth and Whitehead, 2002). It is in this light that the innovative insights of Torr and Fitzgibbon (2003, 2004) and Chojnacki et al. (2003) expressing similar but not identical sentiments relating to the eight-point algorithm should be considered.

In retrospect, it is amusing that the early, incisive paper by Longuet-Higgins (1981) presaged many of these developments: while his eight-point algorithm applied specifically to the essential matrix, it was only very much later (Faugeras, 1992; Hartley, 1992) that it was applied to the fundamental matrix, and even later, in a crucial step, that its accuracy was greatly improved by prenormalizing the image data (Hartley, 1997). As already noted, the eight-point algorithm continued to be a focus for new research.

18.16.1 More Recent Developments

Most recently, Gallo et al. (2011) have studied how planes may be fitted to surfaces that are obtained from range data (i.e., sets of data points whose real-world (X, Y, Z) coordinates are approximately known). While RANSAC should provide useful solutions, it sometimes fails when finding pairs of planar patches, and a single plane is fitted to both, with the result that it contains more inliers than the correct models. To cope with this, they devised an alternate form of RANSAC, CC-RANSAC, which only considers the largest connected components of inliers for a given plane hypothesis. The method requires an inlier threshold to be set and this has to be adjusted for the particular application in question. One relevant application is automatic car parking where a single level near a curb has to be identified.

While the eight-point algorithm has become standard for solving the fundamental matrix, the latter only contains seven independent parameters so only identifying the same seven features in two images should be enough to solve it. Bartoli and Sturm (2004) have found that this is realizable if nonlinear estimation is used. The method converges faster than other approaches, although it is somewhat more likely to fall into local minima than methods based on redundant parameters. Fathy et al. (2011) study error criteria for fundamental matrix estimation. They show that the symmetric epipolar distance criterion is biased, and find that of a number of available criteria, the recently developed Kanatani distance criterion (Kanatani et al., 2008) appears to be the most accurate. Ansar and Daniilidis (2003) have devised a novel set of algorithms for linear pose estimation from n points or n lines. The methods will find solutions for cases of $n \geq 4$, for points in general position. While two similar existing noniterative methods exist in the case of estimation from n points (to which the new method is shown to be superior), there is no directly competing case for n lines.

18.17 PROBLEMS

1. For a two-camera stereo system, obtain a formula for the depth error that arises for a given error in disparity. Hence show that the percentage error in depth is numerically equal to the percentage error in disparity. What does this result mean in practical terms? How does the pixellation of the image affect the result?

2. A cylindrical vase with a circular cross-section of local radius a is viewed by two cameras (Fig. 18.6). Obtain a formula giving the error δ in the estimated position of the boundary of the vase. Simplify the calculation by assuming that the boundary is on the perpendicular bisector of the line joining the centers of projection of the two cameras, and hence find α (Fig. 18.6) in terms of b and Z. Determine δ in terms of α and then substitute for α from the previous formula. Hence justify the statements made at the end of Section 18.14.

3. Discuss the potential advantages of trinocular vision in the light of the theory of Section 18.8. What would be the best placement for a third camera? Where should the third camera *not* be placed? Would any gain be achieved by incorporating even more views of a scene?

Motion

19

Motion is another aspect of 3-D vision that humans are able to interpret with ease. This chapter studies the basic theoretical concepts. It is left to Chapters 22 and 23 to apply them to real problems where motion is crucial, including the monitoring of traffic flow and the tracking of people.

Look out for:

- the basic concepts of optical flow, and its limitations.
- the idea of a focus of expansion, and how it leads to the possibility of "structure from motion."
- how motion stereo is achieved.
- the important status of the Kalman filter in motion applications.
- the ways in which invariant features may be used for wide baseline matching.

Note that this introductory chapter on 3-D motion leads to important methods for performing vital surveillance tasks—as will be seen in Chapters 22 and 23.

19.1 INTRODUCTION

This chapter is concerned with the analysis of motion in digital images. For space reasons, it will not be possible to cover the whole subject comprehensively. Instead, the aim is to give the flavor of the subject, airing some of the principles that have proved important over the past two or three decades. Over much of the time, optical flow has been topical. It is appropriate to study it in fair detail because of its importance for surveillance and other applications. Later in the chapter, the use of the Kalman filter for tracking moving objects is discussed, and the use of invariant features such as the scale-invariant feature transform (SIFT) for wide baseline matching, also relevant to motion tracking, is covered.

19.2 OPTICAL FLOW

When scenes contain moving objects, analysis is necessarily more complex than for scenes where everything is stationary, since temporal variations in intensity have to be taken into account. However, intuition suggests that it should be possible—even straightforward—to segment moving objects by virtue of their motion. Image differencing over successive pairs of frames should permit this to be achieved. More careful consideration shows that things are not quite so simple, as illustrated in Fig. 19.1. The reason is that regions of constant intensity give no sign of motion, and edges parallel to the direction of motion give the appearance of not moving. Only edges with a component normal to the direction of motion carry information about the motion. In addition, there is some ambiguity in the direction of the velocity vector. This arises partly because there is too little information available within a small aperture to permit the full velocity vector to be computed (Fig. 19.2). This is hence called the *aperture problem*.

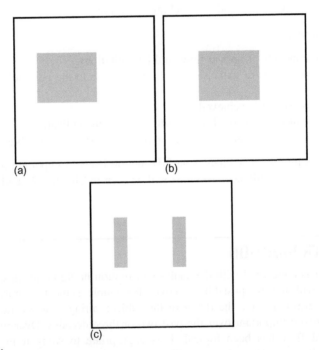

(a)   (b)

(c)

FIGURE 19.1

Effect of image differencing. The figure shows an object that has moved between frames (a) and (b). (c) The result of performing an image differencing operation. Note that the edges parallel to the direction of motion do not show up in the difference image. Also, regions of constant intensity give no sign of motion.

These elementary ideas can be taken further, and they lead to the notion of optical flow, wherein a local operator which is applied at all pixels in the image will lead to a motion vector field that varies smoothly over the whole image. The attraction lies in the use of a local operator, with its limited computational burden. Ideally, it would have an overhead comparable to an edge detector in a normal intensity image—although clearly it will have to be applied locally to pairs of images in an image sequence.

We start by considering the intensity function $I(x, y, t)$ and expanding it in a Taylor series:

$$I(x + dx, y + dy, t + dt) = I(x, y, t) + I_x dx + I_y dy + I_t dt + \cdots \quad (19.1)$$

where second- and higher order terms have been ignored. In this equation, I_x, I_y, and I_t denote respective partial derivatives with respect to x, y, and t.

We next set the local condition that the image has shifted by amount (dx, dy) in time dt so that it is functionally identical at $(x + dx, y + dy, t + dt)$ and (x, y, t):

$$I(x + dx, y + dy, t + dt) = I(x, y, t) \quad (19.2)$$

Hence, we can deduce:

$$I_t = -(I_x \dot{x} + I_y \dot{y}) \quad (19.3)$$

Writing the local velocity $\mathbf{v}$ in the form:

$$\mathbf{v} = (v_x, v_y) = (\dot{x}, \dot{y}) \quad (19.4)$$

we find:

$$I_t = -(I_x v_x + I_y v_y) = -\nabla I \cdot \mathbf{v} \quad (19.5)$$

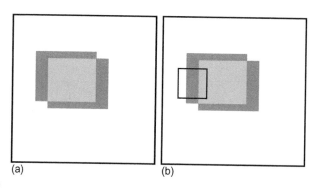

(a)   (b)

FIGURE 19.2

The aperture problem. The figure illustrates the aperture problem. (a) (Dark gray) regions of motion of an object whose central uniform region (light gray) gives no sign of motion. (b) Depiction of how little is visible in a small aperture (black border), thereby leading to ambiguity in the deduced direction of motion of the object.

I_t can be measured by subtracting pairs of images in the input sequence, while ∇I can be estimated by Sobel or other gradient operators. Hence, it should be possible to deduce the velocity field $\mathbf{v}(x, y)$ using the above equation. Unfortunately, this equation is a scalar equation and will not suffice for determining the two local components of the velocity field as we require. There is a further problem with this equation—that the velocity value will depend on the values of both I_t and ∇I, and these quantities are only estimated approximately by the respective differencing operators. In both cases, significant noise will arise, and this will be exacerbated by taking the ratio in order to calculate $\mathbf{v}$.

Let us now return to the problem of computing the full velocity field $\mathbf{v}(x, y)$. All we know about $\mathbf{v}$ is that its components lie on the following line in (v_x, v_y)-space (Fig. 19.3):

$$I_x v_x + I_y v_y + I_t = 0 \tag{19.6}$$

This line is normal to the direction (I_x, I_y), and has a distance from the (velocity) origin that is equal to:

$$|\mathbf{v}| = \frac{-I_t}{(I_x^2 + I_y^2)^{1/2}} \tag{19.7}$$

Clearly, we need to deduce the component of $\mathbf{v}$ along the line given by Eq. (19.6). However, there is no purely local means of achieving this with first derivatives of the intensity function. The accepted solution (Horn and Schunck,

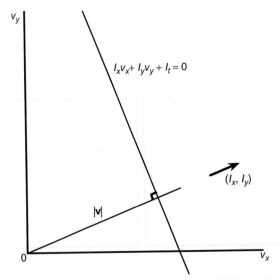

FIGURE 19.3

Computation of the velocity field. The graph shows the line in velocity space on which the velocity vector $\mathbf{v}$ must lie. The line is normal to the direction (I_x, I_y) and its distance from the origin is known to be $|\mathbf{v}|$ (see text).

1981) is to use relaxation labeling to arrive iteratively at a self-consistent solution that minimizes the global error. In principle, this approach will also minimize the noise problem indicated earlier.

In fact, there are still problems with the method. Essentially, these arise as there are liable to be vast expanses of the image where the intensity gradient is low. In that case, only very inaccurate information is available about the velocity component parallel to ∇I, and the whole problem becomes ill-conditioned. On the other hand, in a highly textured image, this situation should not arise (assuming the texture has a large enough grain size to give good differential signals).

Finally, we return to the idea mentioned at the beginning of this section—that edges parallel to the direction of motion would not give useful motion information. Such edges will have edge normals normal to the direction of motion, so ∇I will be normal to $\mathbf{v}$. Thus, from Eq. (19.5), I_t will be zero. In addition, regions of constant intensity will have $\nabla I = 0$, so again I_t will be zero. It is interesting and highly useful that such a simple equation (19.5) embodies all the cases that were suggested earlier on the basis of intuition.

In what follows, we assume that the optical flow (velocity field) image has been computed satisfactorily, that is, without the disadvantages of inaccuracy or ill-conditioning. It must now be interpreted in terms of moving objects and in some cases a moving camera. In fact, we shall ignore motion of the camera by remaining within its frame of reference.

19.3 INTERPRETATION OF OPTICAL FLOW FIELDS

We start by considering a case where no motion is visible. In that case, the velocity field image contains only vectors of zero length (Fig. 19.4(a)). Next, we take a case where one object is moving toward the right, with a simple effect on the velocity field image (Fig. 19.4(b)). Next, we consider the case where the camera is moving forward; in this case, all the stationary objects in the field of view appear to be diverging from a point, which is called the *focus of expansion* (FOE)—see Fig. 19.4(c); this image also shows an object that is moving rapidly past the camera and has its own separate FOE. Figure 19.4(d) shows the case of an object moving directly toward the camera. In this case, its FOE lies within its outline. Similarly, objects that are receding appear to move away from the *focus of contraction*. Next, there are objects that are stationary but rotating about the line of sight. For these, the vector field appears as in Fig. 19.4(e). There is a final case that is also quite simple: an object that is stationary but rotating about an axis normal to the line of sight; if the axis is horizontal, then the features on the object will appear to be moving up or down, while paradoxically the object itself remains stationary (Fig. 19.4(f))—although its outline could oscillate as it rotates.

So far, we have only dealt with cases in which pure translational or pure rotational motion is occurring. If a rotating meteor is rushing past, or a spinning cricket ball is approaching, then both types of motion will occur together. In that

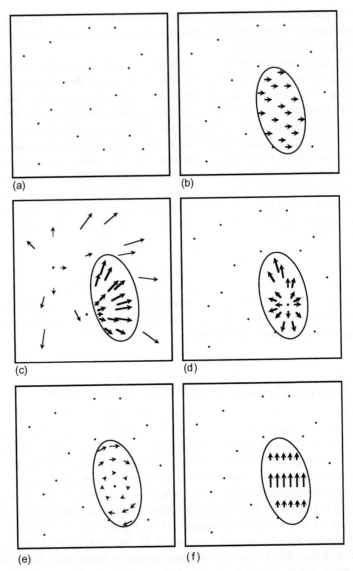

FIGURE 19.4

Interpretation of velocity flow fields. (a) A case where the object features all have zero velocity. (b) A case where an object is moving to the right. (c) A case where the camera is moving into the scene, and the stationary object features appear to be diverging from a focus of expansion (FOE), while a single large object is moving past the camera and away from a separate FOE. In (d), an object is moving directly toward the camera that is stationary: the object's FOE lies within its outline. In (e), an object is rotating about the line of sight to the camera, and in (f), the object is rotating about an axis perpendicular to the line of sight. In all cases, the length of the arrow indicates the magnitude of the velocity vector.

case, unraveling the motion will be far more complex. We shall not solve this problem here, but refer the reader to more specialized texts (e.g., Maybank, 1992). However, the complexity is due to the way depth (Z) creeps into the calculations. First, note that pure rotational motion with rotation about the line of sight does not depend on Z. All we have to measure is the angular velocity, and this can be done quite simply.

19.4 USING FOCUS OF EXPANSION TO AVOID COLLISION

We now take a simple case in which an FOE is located in an image and show how it is possible to deduce the distance of closest approach of the camera to a fixed object of known coordinates. This type of information is valuable for guiding robot arms or robot vehicles and helping to avoid collisions.

In the notation of Chapter 15, we have the following formulas for the location of an image point (x, y, z) resulting from a world point (X, Y, Z):

$$x = \frac{fX}{Z} \tag{19.8}$$

$$y = \frac{fY}{Z} \tag{19.9}$$

$$z = f \tag{19.10}$$

Assuming the camera has a motion vector $(-\dot{X}, -\dot{Y}, -\dot{Z}) = (-u, -v, -w)$, fixed world points will have velocity (u, v, w) relative to the camera. Now a point (X_0, Y_0, Z_0) will after a time t appear to move to $(X, Y, Z) = (X_0 + ut, Y_0 + vt, Z_0 + wt)$ with image coordinates:

$$(x, y) = \left(\frac{f(X_0 + ut)}{Z_0 + wt}, \frac{f(Y_0 + vt)}{Z_0 + wt} \right) \tag{19.11}$$

and as $t \to \infty$ this approaches the focus of expansion F $(fu/w, fv/w)$. This point is in the image, but the true interpretation is that the actual motion of the center of projection of the imaging system is toward the point:

$$\mathbf{p} = \left(\frac{fu}{w}, \frac{fv}{w}, f \right) \tag{19.12}$$

(This is of course consistent with the motion vector (u, v, w) assumed initially.) The distance moved during time t can now be modeled as:

$$\mathbf{X_c} = (X_c, Y_c, Z_c) = \alpha t \mathbf{p} = f\alpha t \left(\frac{u}{w}, \frac{v}{w}, 1 \right) \tag{19.13}$$

where α is a normalization constant. To calculate the distance of closest approach of the camera to the world point $\mathbf{X} = (X, Y, Z)$, we merely specify that the vector

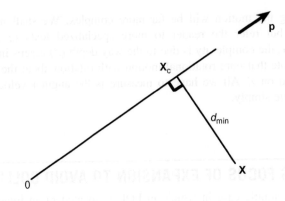

FIGURE 19.5

Calculation of distance of closest approach. Here, the camera is moving from 0 to $\mathbf{X_c}$ in the direction $\mathbf{p}$, not in a direct line to the object at $\mathbf{X}$. d_{min} is the distance of closest approach.

$\mathbf{X_c} - \mathbf{X}$ be perpendicular to $\mathbf{p}$ (Fig. 19.5) so that:

$$(\mathbf{X_c} - \mathbf{X}) \cdot \mathbf{p} = 0 \tag{19.14}$$

$$\text{i.e., } (\alpha t\mathbf{p} - \mathbf{X}) \cdot \mathbf{p} = 0 \tag{19.15}$$

$$\therefore \quad \alpha t\mathbf{p} \cdot \mathbf{p} = \mathbf{X} \cdot \mathbf{p} \tag{19.16}$$

$$\therefore \quad t = \frac{\mathbf{X} \cdot \mathbf{p}}{\alpha(\mathbf{p} \cdot \mathbf{p})} \tag{19.17}$$

Substituting in the equation for $\mathbf{X_c}$ now gives:

$$\mathbf{X_c} = \frac{\mathbf{p}(\mathbf{X} \cdot \mathbf{p})}{\mathbf{p} \cdot \mathbf{p}} \tag{19.18}$$

Hence, the minimum distance of approach is given by:

$$d_{min}^2 = \left[\frac{\mathbf{p}(\mathbf{X} \cdot \mathbf{p})}{(\mathbf{p} \cdot \mathbf{p})} - \mathbf{X} \right]^2 = \frac{(\mathbf{X} \cdot \mathbf{p})^2}{(\mathbf{p} \cdot \mathbf{p})} - \frac{2(\mathbf{X} \cdot \mathbf{p})^2}{(\mathbf{p} \cdot \mathbf{p})} + (\mathbf{X} \cdot \mathbf{X})$$

$$= (\mathbf{X} \cdot \mathbf{X}) - \frac{(\mathbf{X} \cdot \mathbf{p})^2}{(\mathbf{p} \cdot \mathbf{p})} \tag{19.19}$$

which is naturally zero when $\mathbf{p}$ is aligned along $\mathbf{X}$. Clearly, avoidance of collisions requires an estimate of the size of the machine (e.g., robot or vehicle) attached to the camera and the size to be associated with the world point feature $\mathbf{X}$. Finally, note that while $\mathbf{p}$ is obtained from the image data, $\mathbf{X}$ can only be deduced from the image data if the depth Z can be estimated from other information. In fact, this information should be available from time-to-adjacency analysis (see below) if the speed of the camera through space (and specifically w) is known.

19.5 TIME-TO-ADJACENCY ANALYSIS

In this section, we consider the extent to which the depths of objects can be deduced from optical flow. First, note that features on the same object share the same FOE, and this can help us to identify them. But how can we get information on the depths of the various features on the object from optical flow? The basic approach is to start with the coordinates of a general image point (x, y), deduce its flow velocity, and then find an equation linking this with the depth Z.

Taking the general image point (x, y) given in Eq. (19.11), we find:

$$\dot{x} = \frac{f[(Z_0 + wt)u - (X_0 + ut)w]}{(Z_0 + wt)^2}$$
$$= f\frac{(Zu - Xw)}{Z^2} \tag{19.20}$$

and

$$\dot{y} = f\frac{(Zv - Yw)}{Z^2} \tag{19.21}$$

Hence:

$$\frac{\dot{x}}{\dot{y}} = \frac{(Zu - Xw)}{(Zv - Yw)} = \frac{(u/w - X/Z)}{(v/w - Y/Z)}$$
$$= \frac{(x - x_F)}{(y - y_F)} \tag{19.22}$$

This result was to be expected, as the motion of the image point has to be directly away from the focus of expansion (x_F, y_F). Without loss of generality, we now take a set of axes such that the image point considered is moving along the x-axis. Then we have:

$$\dot{y} = 0 \tag{19.23}$$

$$y_F = y = \frac{fY}{Z} \tag{19.24}$$

Defining the distance from the focus of expansion as Δr (see Fig. 19.6), we find:

$$\Delta r = \Delta x = x - x_F = \frac{fX}{Z} - \frac{fu}{w} = \frac{f(Xw - Zu)}{Zw} \tag{19.25}$$

$$\therefore \frac{\Delta r}{\dot{r}} = \frac{\Delta x}{\dot{x}} = -\frac{Z}{w} \tag{19.26}$$

Defining the *time to adjacency* T_a as the time it will take for the origin of the camera coordinate system to arrive at the object point, Eq. (19.26) means that T_a is the same (Z/w) when the object is observed in real-world coordinates as when it is observed in image coordinates $(-\Delta r/\dot{r})$. Hence, it is possible to relate the

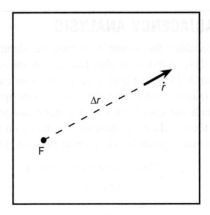

FIGURE 19.6

Calculation of time to adjacency. Here, an object feature is moving directly away from the focus of expansion F with speed $\dot{r}$. At the time of observation, the distance of the feature from F is Δr. These measurements permit the time to adjacency and hence also the relative depth of the feature to be calculated.

optical flow vectors for object points at different depths in the scene. This is important, as the assumption of identical values of w now allows us to determine the relative depths of object points merely from their apparent motion parameters:

$$\frac{Z_1}{Z_2} = \frac{\Delta r_1 / \Delta r_2}{\dot{r}_1 / \dot{r}_2}$$

(19.27)

This is thus the first step in the determination of structure from motion. In this context, note how the implicit assumption that the objects under observation are rigid is included—namely, that all points on the same object are characterized by identical values of w. The assumption of rigidity underlies much of the work on interpretation of motion in images.

19.6 BASIC DIFFICULTIES WITH THE OPTICAL FLOW MODEL

When the optical flow ideas presented above are tried on real images, certain problems arise that are not apparent from the above model. First, not all edge points that should appear in the motion image are actually present. This is due to the contrast between the moving object and the background vanishing locally and limiting visibility. The situation is exactly as for edges that are located by edge detection operators in nonmoving images. The contrast simply drops to a low value in certain localities and the edge peters out. This signals that the edge model, and now the velocity flow model, is limited and such local procedures are *ad hoc* and too impoverished to permit proper segmentation unaided.

Here, we take the view that simple models can be useful, but they become inadequate on certain occasions and robust methods are required to overcome the problems that then arise. Some of the problems were noticed by Horn as early as 1986 (Horn, 1986). First, a smooth sphere may be rotating but the motion will not show up in an optical flow (difference) image. We can if we wish regard this as a simple optical illusion, as the rotation of the sphere may well be invisible to the eye too. Second, a motionless sphere may *appear* to rotate as the light rotates around it. The object is simply subject to the laws of Lambertian optics, and again we may if we wish regard this effect as an optical illusion. (The illusion is relative to the baseline provided by the *normally correct* optical flow model.)

We next return to the optical flow model and see where it could be wrong or misleading. The answer is at once apparent: we stated in writing Eq. (19.2) that we were assuming the *image* is being shifted. Yet it is not images that shift but the objects imaged within them. Thus, we ought to be considering the images of objects moving against a fixed background (or a variable background if the camera is moving). This will then permit us to see how sections of the motion edge can go from high to low contrast and back again in a rather fickle way, which we must nevertheless allow for in our algorithms. With this in mind, it should be permissible to go on using optical flow and difference imaging, even though these concepts have distinctly limited theoretical validity. (For a more thoroughgoing analysis of the underlying theory, see Faugeras, 1993.)

19.7 STEREO FROM MOTION

An interesting aspect of camera motion is that over time the camera sees a succession of images that span a baseline in a similar way to binocular (stereo) images. Thus, it should be possible to obtain depth information by taking two such images and tracking object features between them. The technique is in principle more straightforward than normal stereo imaging in that feature tracking is possible, so the correspondence problem should be nonexistent. However, there is a difficulty in that the object field is viewed from almost the same direction in the succession of images so that the full benefit of the available baseline is not obtained (Fig. 19.7). We can analyze the effect as follows.

First, in the case of camera motion, the equations for lateral displacement in the image depend not only on X but also on Y, although we can make a simplification in the theory by working with R, the radial distance of an object point from the optical axis of the camera, where:

$$R = (X^2 + Y^2)^{1/2} \tag{19.28}$$

We now obtain the radial distances in the two images as:

$$r_1 = \frac{Rf}{Z_1} \tag{19.29}$$

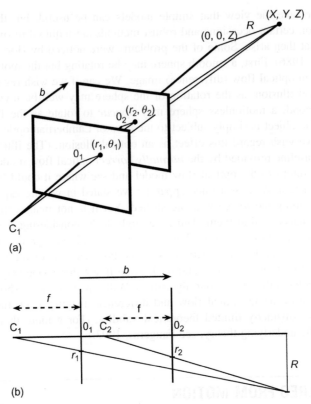

(a)

(b)

FIGURE 19.7

Calculation of stereo from camera motion. (a) Depiction of how stereo imaging can result from camera motion, the vector **b** representing the baseline. (b) The simplified planar geometry required to calculate the disparity. It is assumed that the motion is directly along the optical axis of the camera.

$$r_2 = \frac{Rf}{Z_2} \qquad (19.30)$$

so the disparity is:

$$D = r_2 - r_1 = Rf\left(\frac{1}{Z_2} - \frac{1}{Z_1}\right) \qquad (19.31)$$

Writing the baseline as:

$$b = Z_1 - Z_2 \qquad (19.32)$$

and assuming $b \ll Z_1, Z_2$, and then dropping the suffices, gives:

$$D = \frac{Rbf}{Z^2} \qquad (19.33)$$

While this would appear to mitigate against finding Z without knowing R, we can overcome this problem by observing that:

$$\frac{R}{Z} = \frac{r}{f} \tag{19.34}$$

where r is approximately the mean value $\frac{1}{2}(r_1 + r_2)$. Substituting for R now gives:

$$D = \frac{br}{Z} \tag{19.35}$$

Hence, we can deduce the depth of the object point as:

$$Z = \frac{br}{D} = \frac{br}{(r_2 - r_1)} \tag{19.36}$$

This equation should be compared with Eq. (15.5) representing the normal stereo situation. The important point to note is that for motion stereo, the disparity depends on the radial distance r of the image point from the optical axis of the camera, whereas for normal stereo the disparity is independent of r; as a result, motion stereo gives no depth information for points on the optical axis, and the accuracy of depth information depends on the magnitude of r.

19.8 THE KALMAN FILTER

When tracking moving objects, it is desirable to be able to predict where they will be in future frames, as this will make maximum use of preexisting information and permit the least amount of search in the subsequent frames. It will also serve to offset the problems of temporary occlusion, such as when one vehicle passes behind another, or one person passes behind another, or even when one limb of a person passes behind another. (There are also many military needs for tracking prediction, and others on the sports field.) The obvious equations to employ for this purpose involve sequentially updating the position and the velocity of points on the object being tracked:

$$x_i = x_{i-1} + v_{i-1} \tag{19.37}$$

$$v_i = x_i - x_{i-1} \tag{19.38}$$

assuming, for convenience, a unit time interval between each pair of samples.

In fact, this approach is too crude to yield the best results. First, it is necessary to make three quantities explicit: (1) the raw measurements (e.g., x), (2) the best estimates of the values of the corresponding variables *before* observation (denoted by $^-$), and (3) the best estimates of these same model parameters *following* observation (denoted by $^+$). In addition, it is necessary to include explicit noise terms so that rigorous optimization procedures can be derived for making the best estimates.

In the particular case outlined above, the velocity—and possible variations on it which we shall ignore here for simplicity—constitutes a best estimate model parameter. We include position measurement noise by the parameter u and velocity (model) estimation noise by the parameter w. The above equations now become:

$$x_i^- = x_{i-1}^+ + v_{i-1} + u_{i-1} \tag{19.39}$$

$$v_i^- = v_{i-1}^+ + w_{i-1} \tag{19.40}$$

In the case that the velocity is constant and the noise is Gaussian, we can spot the optimum solutions to this problem:

$$x_i^- = x_{i-1}^+ \tag{19.41}$$

$$\sigma_i^- = \sigma_{i-1}^+ \tag{19.42}$$

these being called the *prediction* equations, and

$$x_i^+ = \frac{x_i/\sigma_i^2 + (x_i^-)/(\sigma_i^-)^2}{1/\sigma_i^2 + 1/(\sigma_i^-)^2} \tag{19.43}$$

$$\sigma_i^+ = \left[\frac{1}{1/\sigma_i^2 + 1/(\sigma_i^-)^2} \right]^{1/2} \tag{19.44}$$

these being called the *correction* equations.[1] (In these equations, $\sigma^\pm$ are the standard deviations for the respective model estimates $x^\pm$, and σ is the standard deviation for the raw measurement x.)

What these equations show is how repeated measurements improve the estimate of the position parameter and the error upon it at each iteration. Note the particularly important feature—that the noise is being modeled as well as the position itself. This permits all positions earlier than $i - 1$ to be forgotten. The fact that there were many such positions, whose values can all be averaged to improve the accuracy of the latest estimate, is of course rolled up into the values of x_i^- and σ_i^-, and eventually into the values for x_i^+ and σ_i^+.

The next problem is how to generalize this result, both to multiple variables and to possibly varying velocity and acceleration. This is the function of the widely used Kalman filter. It achieves this by continuing with a linear approximation and by employing a state vector comprising position, velocity, and acceleration (or other relevant parameters), all in one state vector **s**. This constitutes the dynamic model. The raw measurements x have to be considered separately.

In the general case, the state vector is not updated simply by writing:

$$s_i^- = s_{i-1}^+ \tag{19.45}$$

---

[1]The latter are nothing more than the well-known equations for weighted averages (Cowan, 1998).

but requires a fuller exposition because of the interdependence of position, velocity, and acceleration; hence, we have:[2]

$$s_i^- = K_i s_{i-1}^+ \qquad (19.46)$$

Similarly, the standard deviations σ_i, $\sigma_i^\pm$ in Eqs. (19.42)–(19.44) (or rather, the corresponding variances) have to be replaced by the covariance matrices Σ_i, $\Sigma_i^\pm$, and the equations become significantly more complicated. We will not go into the calculations fully here as they are nontrivial and need several pages to iterate. Suffice it to say that the aim is to produce an optimum linear filter by a least-squares calculation (see, e.g., Maybeck, 1979).

Overall, the Kalman filter is the optimal estimator for a linear system for which the noise is zero mean, white, and Gaussian, although it will often provide good estimates even if the noise is not Gaussian.

Finally, it will be noted that the Kalman filter itself works by averaging processes, which will give erroneous results if any outliers are present. This will certainly occur in most motion applications. Thus, there is a need to test each prediction to determine if it is too far away from reality. If this is the case, it is not unlikely that the object in question has become partially or fully occluded. A simple option is to assume that the object continues in the same motion (albeit with a larger uncertainty as time goes on), and to wait for it to emerge from behind another object. At the very least, it is prudent to keep a number of such possibilities alive for some time, but the extent of this will naturally vary from situation to situation and from application to application.

19.9 WIDE BASELINE MATCHING

The need for wide baseline matching was noted in Chapter 6, where considerable discussion was included on detection of suitable invariant features (see Section 6.7 and its various subsections). The topic has been left until the present chapter because it is relevant for both 3-D vision and motion analysis, and the latter topic has only been covered in this chapter. The wide baseline scenario arises from situations where the same object is viewed from widely different directions, with the result that its appearance may change dramatically so that it may become extremely difficult to recognize. While narrow baseline stereo is the norm for depth estimation using two cameras, wide baselines are common in surveillance applications—e.g., where a pedestrian precinct is viewed by several independent cameras that are widely separated, as described in Chapter 22. They are also the norm when objects are being sought in image databases. However, one of the most likely situations when they occur is with objects that are in motion. While this may appear to be immaterial in surveillance or

---

[2]Some authors write K_{i-1} in this equation, but it is only a matter of definition whether the label matches the previous or the new state.

with driver assistance systems, because every pair of frames will give instances of narrow baseline stereo, it can easily happen that objects will be *temporarily occluded* and come back into view with different orientations or backgrounds; in addition, the *attention* of the software (like that of a human operator) may only be on part of the scene for part of the time. Hence, wide baseline viewing is bound to be a common consequence of motion. Overall, then, wide baseline matching techniques will be needed in a variety of instances of 3-D viewing and motion tracking.

Chapter 6 showed how features could be designed to cover a variety of wide baseline views as far apart as 50°. In these circumstances, an important factor in designing suitable feature detectors is to make them invariant to scale and affine distortions. However, that alone is not enough. The feature detectors must also provide descriptors of each feature that are sufficiently rich in information that matching between views is made as unambiguous as possible. In that way, wide baseline matching has a chance of being highly reliable. Lowe (2004) has found that reliable recognition of objects is possible with as few as three features. Indeed, it is highly important to aim to achieve this when images typically contain several thousand features that come from many different objects as well as background clutter. In this way, the number of false positives is reduced to minimal levels, and there is a high chance of detecting all objects of the chosen type in the input image. That this can be possible is underlined by the richness of Lowe's SIFT features whose descriptors contain 128 parameters. (As discussed in Chapter 6, features devised by other workers may contain fewer parameters, but in the end risk not working in all possible scenarios.)

Granted that wide baseline matching is desirable, and that SIFT and other features have rich descriptor sets, how should the matching actually be achieved? Ideally, all that is necessary is to compare the feature descriptors from each pair of images, and find which ones match well, and which therefore lead to co-recognition of objects in the two views. Clearly, the first requirement is a similarity test for pairs of features. Lowe (2004) achieved this using a nearest neighbor (Euclidean) distance measure in his 128-dimensional descriptor space. He then used a Hough transform to identify clusters of features giving the same interpretations of poses for objects appearing in the two images. Because of the relatively small number of inliers that may occur in this type of situation, he found that the Hough transform approach performed significantly better than RANSAC (Section 11.6). Mikolajczyk and Schmid (2004) used a Mahalanobis distance measure for selecting the most similar descriptors to obtain a set of initial matches; they then used cross-correlation to reject low-score matches; finally, they performed a robust estimation of the transformation between the two images using RANSAC. Tuytelaars and Van Gool (2004) developed this further, using semilocal constraints involving geometric consistency and photometric constraints to refine the selection of matches before (again) relying on RANSAC to perform the final robust estimation of poses. In contrast to the approaches outlined above, Bay et al. (2008) fed the descriptor information to a naive Bayes' classifier working on a "bag-of-words" representation (Dance et al., 2004) in order to perform object recognition. Bay et al. (2008) make no mention of determination of object pose in this application, which was targeted more at recognizing objects in an image database—although it could equally well have been targeted at repeated recognition of cars on the road for which pose would not be especially relevant.

Overall, it is clear that the new regime of utilizing invariant feature detectors with rich descriptors of local image content forms a powerful approach to wide baseline object matching and takes much of the heat out of the subsequent algorithms.

19.10 CONCLUDING REMARKS

In Sections 19.2 and 19.3, we described the formation of optical flow fields and showed how a moving object or a moving camera leads to a focus of expansion. In the case of moving objects, the focus of expansion can be used to decide whether a collision will occur. In addition, analysis of the motion taking account of the position of the focus of expansion led to the possibility of determining structure from motion. Specifically, this can be achieved via time-to-adjacency analysis, which yields the relative depth in terms of the motion parameters measurable directly from the image. We then demonstrated some basic difficulties with the optical flow model, which arise since the motion edge can have a wide range of contrast values, making it difficult to measure motion accurately. In practice, this means that larger time intervals may have to be employed to increase the motion signal. Otherwise, feature-based matching related to that of Chapter 14 can be used. Corners are the features that are most widely used for this purpose because they are ubiquitous and highly localized in 3-D. Space prevents details of this approach from being described here. Details may be found in Barnard and Thompson (1980), Scott (1988), Shah and Jain (1984), and Ullman (1979). However, the value of the Kalman filter for alleviating the difficulties of temporary occlusion has been considered, and the use of invariant features for wide baseline matching (which includes motion tracking applications) has been covered.

Further work on motion as it arises in real applications will be dealt with in Chapters 22 and 23, which address the problems of surveillance and in-vehicle vision systems.

> The obvious way to understand motion is by image differencing and determination of optical flow. This chapter has shown that the "aperture problem" is a difficulty that is avoidable by using corner tracking. Further difficulties are caused by temporary occlusions, thus necessitating techniques such as occlusion reasoning and Kalman filtering.

19.11 BIBLIOGRAPHICAL AND HISTORICAL NOTES

Optical flow has been investigated by many workers over a good many years (see, e.g., Horn and Schunck, 1981; Heikkonen, 1995). A definitive account of the mathematics relating to FOE appeared in 1980 (Longuet-Higgins and Prazdny, 1980). In fact, foci of expansion can be obtained either from the optical flow field or directly (Jain, 1983). The results of Section 19.5 on time-to-adjacency analysis stem originally from the work of Longuet-Higgins and Prazdny (1980), which provides some deep insights into the whole problem of optical flow and the possibilities of using its shear components. Note that numerical solution of the velocity field problem is not trivial; typically, least-squares analysis is required to overcome the effects of measurement inaccuracies and noise, and to finally obtain the required position measurements and motion parameters (Maybank, 1986). Overall, resolving ambiguities of interpretation is one of the main problems and challenges of image sequence analysis (see Longuet-Higgins (1984) for an interesting analysis of ambiguity in the case of a moving plane).

Unfortunately, the substantial and important literature on motion, image sequence analysis, and optical flow, which impinges heavily on 3-D vision, could not be discussed in detail here for reasons of space. For seminal work on these topics, see, for example, Huang (1983), Jain (1983), Nagel (1983, 1986), and Hildreth (1984).

For early work on the use of Kalman filters for tracking, see Marslin et al. (1991). For the huge amount of more recent work on tracking and surveillance of moving objects, including the tracking of people and vehicles, see Chapters 22 and 23 (in fact, Chapter 23 is especially concerned with monitoring moving objects from within vehicles). For recent references on tracking, particle filters, and detection of moving objects, see the bibliographies in Chapters 22 and 23.

For further references on invariant features for wide baseline matching, see Chapter 6.

19.12 PROBLEM

1. Explain why, in Eq. (19.44), the variances are combined in the particular way (in most applications of statistics, variances are combined by addition).

Toward Real-Time Pattern Recognition Systems

4

Part 4 aims to cover the family of needs associated with the production of practical visual pattern recognition systems. Not least among these needs are the requirements of error minimization, real-time operation, and system integration. Hence, this part of the book must cover a disparate variety of topics ranging from analysis of 2-D and 3-D scenes, and the methods of statistical pattern recognition, to applications such as automated visual inspection and vehicle guidance, and consideration of the lighting and hardware systems required to acquire the images and perform the necessary processing in real time. In fact, the chosen topics are so variegated, and interact with each other at such a variety of different levels, that there can be no ideal order of presentation. Nevertheless, the intelligent reader should not have too much difficulty in finding relevant information.

Automated Visual Inspection

20

Humans are good at searching for the unusual and locating faults in manufactured products, but rapidly tire when large number of items have to be scrutinized. The aim of automated visual inspection is to achieve 100% untiring inspection and control of quality. This chapter describes the processes and principles needed to achieve this end, the main limitation being the impossibility of covering the full variety of products in a single chapter.

Look out for:

- the variety of products to be inspected.
- the main categories of inspection.
- how deviations relative to a standard template can be measured.
- the methodology for scrutinizing circular objects.
- the problems of inspecting products exhibiting high levels of variability.
- the principles of X-ray inspection.
- the importance of color in inspection.

Note that this chapter aims to give a broad view of inspection, and is counterbalanced by the following chapter that covers a particular application area in more depth: this approach is appropriate as case studies provide one of the best ways for extending the work to the wide variety of products.

20.1 INTRODUCTION

Thirty years ago, it was already apparent that machine vision would have an important role in the design of automated manufacturing systems. Indeed, it was

525

clear that it would be applied on two main fronts. First, it would be important for helping to control quality during manufacture. Second, it would be vital for providing precise information to assembly robots on the placement of the components and the products being constructed. Note also that it is important for an assembly cell to be flexible, and in this respect, vision is key to adapting from one set of components to another.

In the early days of automated manufacturing, there was a feeling that the vision tasks needed for inspection and assembly would be quite different. Although there was an element of truth in this, it was soon realized that the majority of vision algorithms, such as edge detection and object location, are common to both. Indeed, machine vision is highly generic, and its methods are readily adapted from one application to another.

There are three main aims for automated visual inspection:

1. To check components and products for quality, with a view to rejecting those that are dimensionally inaccurate or otherwise defective.
2. To assess the general quality of production in order to provide feedback to earlier stages of the plant and thereby to correct erroneous trends. For example, in the case of food products, if coatings of jam or chocolate are found to be spreading too rapidly, feedback will need to be provided to reduce the temperature at an earlier stage in the production line.
3. To gather logistics on the operation of the plant, including such parameters as temperatures, variations in product dimensions, and reject rates, in order to help with advance planning.

A further aspect that is important when setting up an inspection system is what can be learned via modalities such as X-rays and thermal imaging, for which the acquired images often resemble visible light images. Similarly, it is relevant to ask what additional information color can provide that is useful or even vital for inspection. Sections 20.10 and 20.11 aim to give answers to some of these questions.

In automated assembly—the other application of vision mentioned above—vision is valuable for monitoring both the positions and rotation parameters of the robot arm and wrist and those of the various components it is working on. Interestingly, an assembly robot should be able to examine the components it is about to assemble, so as to prevent it from attempting impossible tasks such as fitting a screw into a nonexistent hole.

The above discussion broadly confirms that inspection and assembly require basically the same vision algorithms, although there is a potential difference in that a line-scan camera will be more suitable for inspecting components on a conveyer, while an area (whole picture) camera will be better suited for monitoring assembly operations within a workcell. However, once acquired, the images will be much the same in the two cases, so algorithms that are suitable for the one type of task will broadly be suited for the other. Thus, little will be lost by concentrating on automated visual inspection in the remainder of this chapter.

20.2 THE PROCESS OF INSPECTION

Inspection is the process of comparing individual manufactured items against some preestablished standard with a view to maintenance of quality. Before proceeding to study inspection tasks in detail, it is useful to note that the process of inspection commonly takes place in three definable stages:

1. Image acquisition
2. Object location
3. Object scrutiny and measurement

We defer detailed discussion of image acquisition until Chapter 25 and comment here on the relevance of separating the processes of location and scrutiny. This is important because (either on a worktable or on a conveyor) large number of pixels usually have to be examined before a particular product is found, whereas once it has been located, its image frequently contains relatively few pixels and so rather little computational effort need be expended in scrutinizing and measuring it. For example, on a biscuit line, products may be separated by several times the product diameter in two dimensions, so some 100,000 pixels may need to be examined to locate products occupying, say, 5000 pixels. This means that product location is likely to be a much more computation-intensive problem than product scrutiny. Although this is generally true, sampling techniques may permit object location to be performed with much increased efficiency (Chapter 12). Under these circumstances, it is possible that location may be faster than scrutiny, since the latter process, although straightforward, tends to permit no shortcuts and requires all pixels to be examined.

20.3 THE TYPES OF OBJECT TO BE INSPECTED

It is evident from the huge variety of products that are made in factories that there is a correspondingly large variety of inspection tasks to be carried out. However, products fall rather neatly into two main categories. The first is typified by precision metal parts: these have demanding specifications because they are required to fit exactly together, and will usually have standard hole or thread sizes. The second is typified by food products, which vary in appearance between nominally identical samples; for instance, no two apples will look identical. Textiles also fall into the second class, as samples will vary in 3-D shape, closeness of weave, and degree of stretch in any direction. Interestingly, while soldered joints can vary in size and shape, today's electronic components fit more closely into the first category. Broadly, the difference between the two categories is that, for the first, exact dimensional measurement is the prime concern, whereas for the second, appearance to the consumer is more important. We shall look at these two categories more closely in the following sections.

20.3.1 Food Products

At one end of the scale, this category covers raw vegetables, grain, fruit, meat, and fish, and at the other, bread, cakes, chocolate biscuits, pizzas, frozen food packs, and complete set meals. Over time, the food industry has developed toward high added value and quality packaging. Although it is both logical and desirable to inspect food products at every stage of manufacture, starting with the raw materials, in practice it would be too expensive to entertain this aim. This is because the cost of each inspection station, including camera, computer hardware + software, and mechanical rejection hardware, will be sizeable. This usually means that only a single inspection station can be afforded. The main question is then whether it should be placed at the beginning, middle, or end of the line. If at the beginning, expensive processing of low-quality material will be minimized; if at the end, inspection of final appearance in the form that the product will reach the consumer will be monitored and its overall acceptability checked. In fact, there is some gain from inspecting just prior to final packaging, as undersize and, *a fortiori*, oversize products can jam packaging machines—a major problem on food lines. Interestingly, inspecting at the end of the line does not prevent pizzas or other food products from being scrutinized fairly thoroughly, as parts of most of the additives will normally be detectable. Finally, note that chocolate is expensive, so inspection needs to check that minimizing the amount of chocolate cover on biscuits or cakes does not result in incomplete coverage, with consequent consumer dissatisfaction.

It is well known that the human eye can detect many features of objects at a glance. However, vision tasks that seem simple to the eye can take a substantial effort to program on a computer. For example, chocolates often have a jagged "footing" around the base, making it difficult to determine their overall shape or orientation (this may be important if a robot is to place each chocolate in its proper place in a box). As a result, algorithms employing silhouette analysis may be less successful than those examining the full grayscale profile of the object, or in certain cases its 3-D shape.

Returning to packaged meals, these present both an inspection and an assembly problem. A robot or other mechanism will have to place individual items on a plastic tray, and it is clearly preferable that every item should be checked to ensure, for example, that each salad contains an olive or that each cake has a blob of cream.

20.3.2 Precision Components

Many other parts of industry have also progressed to the automatic manufacture and assembly of complex products. It is clearly necessary for items such as washers and O-rings to be tested for size and roundness, and for mains plugs to be examined for the appropriate pins, fuses, and screws. Engines and brake assemblies also have to be checked for numerous possible faults. Perhaps the worst

Table 20.1 Features to be Checked on Precision Components

Dimensions within specified tolerances
Correct positioning, orientation, and alignment
Correct shape, especially roundness, of objects and holes
Whether corners are misshapen, blunted, or chipped
Presence of holes, slots, screws, rivets, etc.
Presence of a thread in screws
Presence of burr and swarf
Pits, scratches, cracks, wear, and other surface marks
Quality of surface finish and texture
Continuity of seams, folds, laps, and other joins

problems arise when items such as flanges, slots, holes, or threads are missing so that further components cannot be fitted properly. In addition, although it might seem certain that a thread, if present, would necessarily have the correct pitch, the author has seen at least one application where this assumption was not justified.

Table 20.1 summarizes some of the common features that need to be checked when dealing with individual precision components. Note that measurement of the extent of any defect, together with knowledge of its inherent seriousness, should permit components to be graded according to quality, thereby saving money for the manufacturer (rejecting all defective items is often too crude an option).

20.3.3 Differing Requirements for Size Measurement

Size measurement is important both in the food industry and in the automotive and small-parts industry. However, the problems in the two cases are often rather different. For example, the diameter of a biscuit can vary within quite wide limits (~5%) without giving rise to undue problems, but when it gets outside this range, there is a serious risk of jamming the packing machine, and the situation must be monitored carefully. In contrast, for mechanical parts, the required precision can vary from 1% for objects such as O-rings to 0.01% for piston heads. This variation clearly makes it difficult to design a truly general-purpose inspection system. However, the manufacturing process often permits little variation in size from one item to the next. Hence, it may be adequate to have a system that is capable of measuring to an accuracy of rather better than 1%, so long as it is capable of checking all the characteristics mentioned in Table 20.1.

For cases where high precision is vital, it is important that accuracy of measurement is proportional to the resolution of the input image. Currently, images of up to 512×512 pixels are common, so accuracy of measurement is basically of the order of 0.2%. Fortunately, grayscale images provide a means of obtaining significantly greater accuracy than indicated by the above arguments, since the exact transition from dark to light at the boundary of an object can be estimated more closely. In addition, averaging techniques (e.g., along the side of a rectangular block of metal) permit accuracies to be increased even further—by a factor $\sqrt{N}$ if N pixel measurements are made. These factors permit measurements to be made to subpixel resolution, sometimes even down to 0.1 pixels, the limit often being set by variations in illumination rather than by the vision algorithms themselves.

20.3.4 Three-Dimensional Objects

Next, note that all real objects are 3-D, although the cost of setting up an inspection station frequently demands that they are examined from one viewpoint in a single 2-D image. This is clearly highly restrictive, and in many cases overrestrictive. Nevertheless, it is generally possible to do an enormous amount of useful checking and measurement from one such image. The clue that this is possible lies in the prodigious capability of the human eye—e.g., to detect at a glance from the play of light on a surface whether or not it is flat. Furthermore, in many cases, products are essentially flat and the information that we are trying to find out about them is simply expressible via their shape or via the presence of some other feature that is detectable in a 2-D image. In cases where 3-D information is required, methods exist for obtaining it from one or more images, for example, via binocular vision or structured lighting, as has already been seen in Chapter 15. More is said about this in Sections 20.4, 20.7 and 20.14.1.

20.3.5 Other Products and Materials for Inspection

This section briefly mentions a few types of product and material that are not fully covered in the foregoing discussion. First, electronic components are increasingly having to be inspected during manufacture, and of these, printed circuit boards (PCBs) and integrated circuits are subject to their own special problems that are currently receiving considerable attention. Second, steel strip and wood inspection are also of great importance. Third, bottle and glass inspection has its own particular intricacies because of the nature of the material, glints being a relevant factor—as also in the case of inspection of cellophane-covered foodpacks. In this chapter, space permits only a short discussion of some of these topics (see Sections 20.7 and 20.8).

20.4 SUMMARY: THE MAIN CATEGORIES OF INSPECTION

Sections 20.1–20.3 have given a general review of the problems of inspection but have not shown how they might be solved. This section takes the analysis a stage further. First, note that the items in Table 20.1 may be classified as *geometrical* (measurement of size and shape—in 2-D or 3-D as necessary), *structural* (whether there are any missing items or foreign objects), and *superficial* (whether the surface has adequate quality). It is evident from Table 20.1 that these three categories are not completely distinct but they are useful for the following discussion.

We start by noting that the methods of object location are also inherently capable of providing geometrical measurements. Distances between relevant edges, holes, and corners can be measured; shapes of boundaries can be checked both absolutely and via their salient features; and alignments can usually be checked quite simply, for example, by finding how closely various straight line segments fit to the sides of a suitably placed rectangle. In addition, shapes of 3-D surfaces can be mapped out by binocular vision, photometric stereo, structured lighting, or other means (see Chapter 15), and subsequently checked for accuracy.

Structural tests can also be made once objects have been located, assuming a database of the features they are supposed to possess is available. In the latter case, it is necessary merely to check whether the features are present in predicted positions. As for foreign objects, these can be looked for via unconstrained search as objects in their own right. Alternatively, they may be found as differences between objects and their idealized forms, as predicted from templates or other data in the database. In either case, the problem is very data dependent and an optimal solution needs to be found for each situation. For example, scratches may be searched for directly as straight line segments.

Tests of surface quality are perhaps more complex. In Chapter 25, methods of lighting are described, which illuminate flat surfaces uniformly, so that variations in brightness may be attributed to surface blemishes. For curved surfaces, it might be hoped that the illumination on the surface would be predictable, and then differences would again indicate surface blemishes. However, in complex cases, there is probably no alternative but to resort to the use of switched lights coupled with rigorous photometric stereo techniques (see Chapter 15). Finally, the problem of checking quality of surface finish is akin to that of ensuring an attractive physical appearance, and this can be highly subjective; this means that inspection algorithms need to be trained to make the right judgments (note that "judgments" are decisions or classifications and so the methods described in Chapter 24 are appropriate).

Overall, accurate object location is a prerequisite to reliable object scrutiny and measurement, for all three main categories of inspection. If a CAD system is available, then providing location information permits an image to be generated that should closely match the observed image, and template matching (or correlation) techniques should in many cases permit the remaining inspection functions to be fulfilled. However, this will not always work without trouble—as in the case where object surfaces have a random or textured component. This means that preliminary analysis of the texture may have to be carried out before relevant templates can be applied—or at least checks made of the maximum and minimum pixel intensities within the product area. More generally, in order to solve this and other problems, some latitude in the degree of fit should be permitted.

It is interesting that the same general technique—that of template matching—arises in the measurement and scrutiny phase as in the object location phase.

However (as remarked earlier), this need not consume as much computational effort as in object location. This is because template matching is difficult when there are many degrees of freedom inherent in the situation, since comparisons with an enormous number of templates may be required. However, when the template is in a standard position relative to the product and when it has been orientated correctly, template matching is much more likely to constitute a practical solution to the inspection task, although the problem is very data dependent.

Despite these considerations, there is a need to find computationally efficient means of performing the necessary checks of parts. The first possibility is to use suitable algorithms to model the image intensity and then to employ the model to check relevant surfaces for flaws and blemishes. Another useful approach is to convert 2-D to 1-D intensity profiles. This approach leads to the radial histogram technique; the latter can conveniently be applied for inspecting the very many objects possessing circular symmetry, as will be seen below. However, we first consider a simple but useful means of checking shapes.

20.5 SHAPE DEVIATIONS RELATIVE TO A STANDARD TEMPLATE

For food and certain other products, an important factor in 2-D shape measurement is the deviation relative to a standard template. Maximum deviations are important because of the need (already referred to) to fit the product into a standard pack. Another useful measure is the area of overflow or underfill relative to the template (Fig. 20.1). For simple shapes that are bounded by circular arcs or straight lines (a category that includes many types of biscuit or bracket), it is straightforward to test whether a particular pixel on or near the boundary is inside the template or outside it. For straight line segments, this is achieved in the following way. Taking the pixel as (x_1, y_1) and the line as having equation:

$$lx + my + n = 0 \tag{20.1}$$

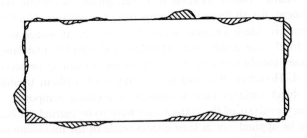

FIGURE 20.1

Measurement of product area relative to a template: in this example, two measurements are taken, indicating, respectively, the areas of overflow and underfill relative to a prespecified rectangular template.

the coordinates of the pixel are substituted in the expression:

$$f(x, y) = lx + my + n \qquad (20.2)$$

The sign will be positive if the pixel lies on one side of the line, negative on the other, and zero if it is on the idealized boundary. Furthermore, the distance on either side of the line is given by the formula:

$$d = \frac{lx_1 + my_1 + n}{(l^2 + m^2)^{1/2}} \qquad (20.3)$$

The same observation about the signs applies to any conic curve if appropriate equations are used, for example, for an ellipse:

$$f(x, y) = \frac{(x - x_c)^2}{a^2} + \frac{(y - y_c)^2}{b^2} - 1 \qquad (20.4)$$

where $f(x, y)$ changes sign on the ellipse boundary. For a circle, the situation is particularly simple, since the distance to the circle center need only be calculated and compared with the idealized radius value. For more complex shapes, deviations need to be measured using centroidal profiles or the other methods described in Chapter 10. However, the method outlined above is useful as it is simple, quick, and reasonably robust, and does not need to employ sequential boundary tracking algorithms. A raster scan over the region of the product is sufficient for the purpose.

20.6 INSPECTION OF CIRCULAR PRODUCTS

Circular objects and holes are so common that it is important to have well-designed techniques for inspecting them. We have already seen (Chapter 12) that they can be located relatively straightforwardly using the Hough transform. For surface scrutiny, it is useful to make use of their rotational symmetry to obtain a 1-D measure of intensity as a function of distance r from the center. However, the resulting "radial histograms" are complicated by the varying number of pixels at different distances r from the center and have to be normalized to eliminate this effect (Davies, 1984c, 1985). Typical results are shown in Figs. 20.2 and 20.3. Note that radial histograms can be used to make accurate measurements of all relevant radii within the product—e.g., both radii of a washer. Because of the averaging of all intensity values for each value of r, accuracy of measurement can be as high as 0.3% for values of r as low as 40 pixels.

As indicated above, the varying numbers of pixels at different radial distances r complicate the problem and prevent the histograms from accurately representing the radial intensity distribution. The obvious way of tackling this problem is to make the independent variable r^2 rather than r. However, it is also necessary to normalize the distribution so that regions of uniform intensity give rise to a uniform radial intensity distribution—a fact made clear by the statistics shown in

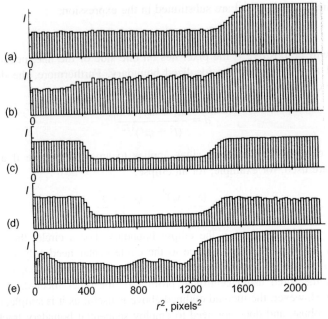

FIGURE 20.2

Practical applications of the radial histogram approach. In all cases, an r^2 base variable is used and histogram columns are individually normalized. These histograms were generated from the original images of Fig. 20.3.

Source: © *IEE 1985*

Fig. 20.4. This means dividing the value for each column of the histogram by the number of pixels contributing to it.

Figure 20.2 shows practical applications of the above theory to various situations, depicted in Fig. 20.3: (see also Fig. 20.5, which relates to the biscuits shown in Fig. 10.1). In particular, note that the radial histogram approach is able to give vital information on various types of defect: the presence or absence of holes in a product such as a washer or a button; whether circular objects are in contact or overlapping; broken objects; "show-through" of biscuit where there are gaps in a chocolate or other coating; and so on. In addition, it is straightforward to derive dimensional measurements from radial histograms. In particular, radii of discs or washers can be obtained to significantly better than 1 pixel accuracy because of the averaging effect of the histogram approach. However, the method is limited here by the accuracy with which the center of the circular region is first located. This underlines the value of the high-accuracy center-finding technique described in Chapter 12. To a certain extent, accuracy is limited by the degree of roundness of the product feature being examined: radius can be measured only to the extent that it is meaningful. In this context, it is emphasized that the

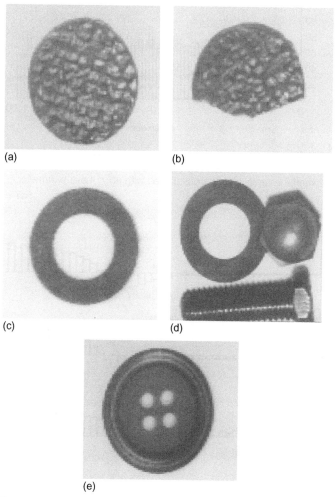

(a)   (b)

(c)   (d)

(e)

FIGURE 20.3

Original images used in generating the radial histograms of Fig. 20.2.

Source: © IEE 1985

combination of techniques described above is not only accurate but also computationally efficient.

Radial histograms are particularly well suited to the scrutiny of symmetrical products that do not exhibit a texture, or for which texture is not prominent and may validly be averaged out. In addition, the radial histogram approach ignores correlations between pixels in the dimension being averaged (i.e., angle); where such correlations are significant, it is not possible to use the approach. An obvious example is the inspection of components such as buttons, where angular

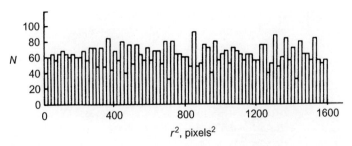

FIGURE 20.4

Pixel statistics for an r^2 histogram base parameter: the pixel statistics are not exactly uniform even when the radial histogram is plotted with an r^2 base parameter.

Source: © IEE 1985

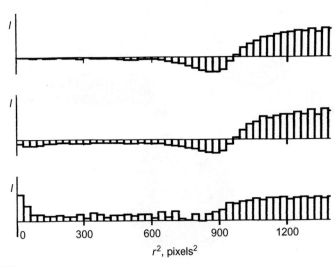

FIGURE 20.5

Radial intensity histograms for the three biscuits of Fig. 12.1: the order is from top to bottom in both figures. Intensity is here measured relative to that at the center of an ideal product.

Source: © IFS Publications Ltd 1984

displacement of one of four button holes will not be detectable unless the hole encroaches on the space of a neighboring hole. Similarly, the method is not able to check the detailed shape of each of the small holes. Clearly, the averaging involved in finding the radial histogram mitigates against such detailed inspection, which is then best carried out by separate direct scrutiny of each of the holes.

It might be imagined that the radial histogram technique is applicable only for symmetrical objects. However, it is also possible to use radial histograms as signatures of intensity patterns in the region of specific salient features. Small holes are suitable salient features but corners are less suitable unless the background is saturated out at a constant value; otherwise, too much variation arises from the background and the technique does not prove viable.

Finally, the radial histogram approach has the useful characteristic of being trainable, since the relevant 1-D templates may be accumulated, averaged over a number of ideal products, and stored in memory, ready for comparisons with less ideal products. This characteristic is valuable, not only for convenience in setting up but also because it permits inspection to be adaptable to cover a range of products.

20.7 INSPECTION OF PRINTED CIRCUITS

Over the past two or three decades, machine vision has been used increasingly in the electronics industry, notably for inspecting PCBs. First, PCBs may be inspected before components are inserted; second, they may be inspected to check that the correct components have been inserted the correct way round; and third, all the soldered joints may be scrutinized. The faults that have to be checked for include touching tracks, whisker bridges, broken tracks (including hairline cracks), and mismatch between pad positions and holes drilled for component insertion. Controlled illumination is required to eliminate glints from the bare metal and to ensure adequate contrast between the metal and its substrate. With adequate control over the lighting, most of the checks (e.g., apart from reading any print on the substrate) may be carried out on a binarized image, and the problem devolves into the checking of shape. This may be tackled by gross template matching—using a logical exclusive-or operation—but this approach requires large data storage and precise registration has to be achieved.

Difficulties with registration errors can largely be avoided if shape analysis is performed by connectedness measurement (using thinning) and morphological processing. For example, if a track disappears or becomes broken after too few erosion operations, then it is too narrow; a similar procedure will check whether tracks are too close together. Likewise, hairline cracks may be detected by dilations followed by tests to check for changes in connectedness.

Alignment of solder pads with component holes is customarily checked by employing a combination of back and front lighting. Powerful back lighting (i.e., from behind the PCB) gives bright spots at the hole positions, whereas front lighting gives sufficient contrast to show the pad positions; it is then necessary to confirm that each hole is, within a suitable tolerance, at the center of its pad. Counting the bright spots from the holes, plus suitable measurements around hole positions (e.g., via radial histogram signatures), permits this process to be performed satisfactorily.

Overall, the main problem with PCB inspection is the resolution required. Typically, images have to be digitized to at least 4000×4000 pixels—as when a

20×20 cm board is being checked to an accuracy of $50\,\mu$m. In addition, suitable inspection systems will typically check each board fully in less than 1 min; they also have to be trainable, to allow for upgrades in the design of the circuit or improvements in the layout; and they should cost no more than approximately £40,000. However, considerable success has been attained with these aims.

To date, the bulk of the work in PCB inspection has concerned the checking of tracks. Nevertheless, useful work has also been carried out on the checking of soldered joints. Here, each joint has to be modeled in 3-D by structured light or other techniques. In one such case, light stripes were used (Nakagawa, 1982), and in another surface reflectance was measured with a fixed lighting scheme (Besl et al., 1985). Note that surface brightness says something about the quality of the soldered joint. This type of problem is probably completely solvable (at least up to the subjective level of a human inspector) but detailed scrutiny of each joint at a resolution of, say, 64×64 pixels may well be required to guarantee that the process is successful, and this implies an enormous amount of computation to cope with the several hundreds (or in some cases thousands) of joints on most PCBs. Hence, Besl et al. needed special hardware to handle the information in the time available.

Similar work is under way on the inspection of integrated circuit masks and die bonds, but space does not permit discussion of this rapidly developing area. For a useful review, see Newman and Jain (1995).

20.8 STEEL STRIP AND WOOD INSPECTION

The problem of inspecting steel strip is one that is very exacting for human operators. First, it is virtually impossible for the human eye to focus on surface faults when the strip is moving past the observer at rates more than 20 m/s; second, several years of experience are required for this sort of work; and third, the conditions in a steel mill are far from congenial, with considerable heat and noise constantly being present. Hence, much work has been done to automate the inspection process (Browne and Norton-Wayne, 1986). At its simplest, this requires straightforward optics and intensity thresholding, although special laser scanning devices have also been developed to facilitate the process (Barker and Brook, 1978).

The problem of wood inspection is more complex, since this natural material is very variable in its characteristics. For example, the grain varies markedly from sample to sample. As a result of this variation, the task of wood inspection is still in its infancy and many problems remain. However, the purpose of wood inspection is reasonably clear: first, to look for cracks, knots, holes, bark inclusions, embedded pine needles, miscoloration, and so on; and ultimately to make full use of this material by identifying regions where strength or appearance is substandard. In addition, the timber may have to be classified as appropriate for different categories of use—furniture, building, outdoor, etc. Overall, wood inspection is something of an art—i.e., it is a highly subjective process—although valiant attempts have been made to solve the problems (e.g., Sobey, 1989). It is notable that in at least one country (Australia), there is a national standard for the inspection of wooden planks.

20.9 INSPECTION OF PRODUCTS WITH HIGH LEVELS OF VARIABILITY

In Sections 20.4–20.8, we have concentrated on certain aspects of inspection—particularly dimensional checking of components ranging from precision parts to food products, and the checking of complex assemblies to confirm the presence of holes, nuts, springs, and so on. These could be regarded as the geometrical aspects of inspection. For the more imprecisely made products such as foodstuffs and textiles, there are greater difficulties as the template against which any product has to be compared is not fixed. Broadly, there are two ways of tackling this problem: one is the use of a range of templates, each of which is acceptable to some degree; the other is the specification of a variety of descriptive parameters. In either case, there will be a number of numerical measurements whose values have to be within prescribed tolerances. Overall, variable products demand greater amounts of checking and computation, and inspection is significantly more demanding. Nowhere is this clearer than for food and textiles, for which the relevant parameters are largely textural. However, "fuzzy" inspection situations can also occur for certain products that might initially be considered as precision components, for example, for electric lamps the contour of the element and the solder pads on the base have significant variability. Thus, this whole area of inspection involves checking that a range of parameters do not fall outside certain prespecified limits on some relevant distribution that *may* be reasonably approximated by a Gaussian parameter.

We have seen above that the inspection task is made significantly more complicated by natural variability in the product, although in the end it seemed best to regard inspection as a process of making measurements that have to be checked statistically. Defects could be detected relative to the templates, either as gross mismatches or else as numerical deviations. And missing parts could likewise be detected since they do not appear at the appropriate positions relative to the templates. Foreign objects would also appear to fit into this pattern, being essentially defects under another name. However, this view is rather too simple for several reasons:

1. Foreign objects are frequently unknown in size, shape, material, or nature.
2. They may appear in the product in a variety of unpredictable positions and orientations.

3. They may have to be detected in a background of texture that is so variable in intensity that they will not stand out.

Overall, it is the unpredictability of foreign objects that can make them difficult to see, especially in textured backgrounds (Fig. 20.6). If one knew their nature in advance, then a special detector could perhaps be designed to locate them. But in many practical situations, the only means of detecting them is to look for the unusual. In fact, the human eye is well tuned to search for the unusual. On the other hand, there are few obvious techniques that can be used to seek it out automatically in digital images. Simple thresholding would work in a variety of

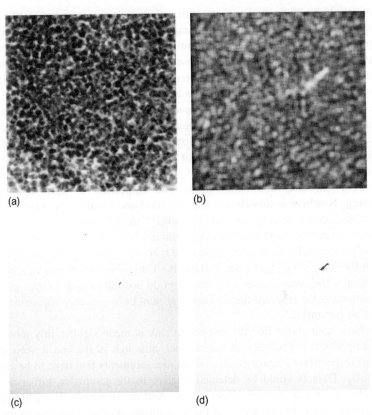

(a)   (b)

(c)   (d)

FIGURE 20.6

Foreign object detection in a packet of frozen vegetables (in this case sweetcorn). (a) The original X-ray image, (b) an image in which texture analysis procedures have been applied to enhance any foreign objects, and (c) and (d) the respective thresholded images. Note the false alarms that are starting to arise in (c), and the increased confidence of detection of the foreign object in (d). For further details, see Patel et al. (1995).

Source: © MCB Univ. Press 1995

practical cases, especially where plain surfaces have to be inspected for scratches, holes, swarf, or dirt. However, looking for extraneous vegetable matter (such as leaves, twigs, or pods) among a sea of peas on a conveyor may be less easy, as the contrast levels may be similar, and the textural cues may not be able to distinguish the shapes sufficiently accurately. Of course, in the latter example, it could be imagined that every pea could be identified by its intrinsic circularity. However, the incidence of occlusion, and the very computation-intensive nature of this approach to inspection, inhibits such an approach. In any case, a method that would detect pods among peas might not detect round stones or small pieces of wood—especially in a grayscale image.

Ultimately, the problem is difficult because the paradigm means of designing sensitive detectors—the matched filter—cannot be used, simply due to the high degree of variability in what has to be detected. With so many degrees of freedom—shape dimensions, size, intensity, texture, and so on—foreign objects can be difficult to detect successfully in complex images. Naturally, a lot depends on the nature of the substrate, and while a plain background might render the task trivial, a textured product substrate may render it impossible, or at least practically impossible in a real-time factory inspection milieu. In general, the solution devolves into not trying to detect the foreign object directly by means of carefully designed matched filters, but in trying to model the intensity pattern of the substrate sufficiently accurately, so that any deviation due to the presence of a foreign object is detected and rendered visible. As hinted above, the approach is to search for the unusual. To achieve this, the basic technique is to identify the 3σ or other appropriate points on all available measures, and initiate rejection when they are exceeded.

There is a fundamental objection to this procedure. If some limit (e.g., 3σ) is assumed, this cannot easily be optimized, since the proper method for achieving this is to find both the distributions—for the background substrate and for the foreign objects—and to obtain a minimum error decision boundary between them. However, in this case, we do not have the distribution corresponding to the foreign objects, so we have to fall back on "reasonable" acceptance limits based on the substrate distribution.

It might appear that this argument is flawed in that the proportion of foreign objects coming along the conveyor is well known. Although this might occasionally be so, the levels of detectability of the foreign objects in the received images will be unknown and will certainly be less than the actual occurrence levels. Hence, arriving at an optimal decision level will be difficult.

However, a far worse problem often exists in practice. The occurrence rates for foreign objects might be almost totally unknown because of (a) their intrinsic variability and (b) their rarity. We might well ask "How often will an elastic band fall onto a conveyor of peas?", but this is a question that is virtually impossible to answer. Maybe it is possible to answer somewhat more accurately the more general question of how often a foreign object of some sort will fall onto the conveyer, but even then the response may well be that somewhere between 1 in

100,000 and 1 in 10 million of bags of peas contain a foreign object. With low levels of risk, the probabilities are extremely difficult to estimate, and indeed there is little available data or other basis on which to calculate them. Consumer complaints can indicate the possible levels of risk, but these arise as individual items, and in any case many customers will not make any fuss and by no means all instances come to light.

With food products, the penalties for not detecting individual foreign objects are not usually especially great. Glass in baby food may be more apocryphal than real, and may be unlikely to cause more than alarm. Similarly, small stones among the vegetables are more of a nuisance than a harm, although cracked teeth could perhaps result in compensation in the £1000 bracket. Far more serious are problems with electric lamps, where a wire emerging from the solder pads is potentially lethal and is a substantive worry for the manufacturer. Litigation for deaths arising from this source could run to a million pounds or so (corresponding to an individual's potential lifetime earnings).

This discussion reveals very clearly that it is the cost rate[1] rather than the error rate that is the important parameter when there is even a remote risk to life and limb. Indeed, it concerned Rodd and his coworkers so much in relation to the inspection of electric lamps than they decided to develop special techniques for ensuring that their algorithms were tested sufficiently (Thomas and Rodd, 1994; Thomas et al., 1995). Computer graphics techniques were used to produce a large number of images with automatically generated variations on the basic defects, and it was checked that the inspection algorithms would always locate them.

20.10 X-RAY INSPECTION

In inspection applications, there is a tension between inspecting products early on, before significant value has been added to a potentially defective product, and at the end of the line, so that the quality of the final products is guaranteed. This consideration applies particularly with food products, where additives such as chocolate can be expensive and constitute substantial waste if the basic product is broken or misshapen. In addition, inspection at the end of the line is especially valuable as oversized products (which may arise if two normal products become stuck together) can jam packing machines. Ideally, it would be beneficial if two inspection stations could be placed on the line in appropriate positions, but if only one can be afforded (the usual situation), it will generally have to be placed at the end of the line.

With many products, it is useful to be absolutely sure about the final quality as the customer will receive it. Thus, there is especial value in inspecting the packaged products. Since the packets will usually be opaque, it will be necessary

---

[1]In fact, the *perceived* cost rate may be even more important, and this can change markedly with reports appearing in the daily press.

to inspect them under X-radiation. This results in substantial expense, since the complete system will include not only the X-ray source and sensing system but also various safety features including heavy shielding. As a result, commercial X-ray food inspection systems rarely cost less than £40,000, and £100,000 is a more typical figure. Such figures do not take account of maintenance costs, and it is also important that the X-ray source and the sensors deteriorate with time so that sensitivity falls, and special calibration procedures have to be invoked.

Fortunately, the X-ray sensors can nowadays take the form of linear photodiode packages constructed using integrated circuit technology.[2] These are placed end to end to span the width of a food conveyor that may be 30–40 cm wide. They act as a line-scan camera that grabs images as the product moves along the conveyor. The main adjustments to be made in such a system are the voltage across the X-ray tube and the electric current passing through it. In food inspection applications, the voltage will be in the range 30–100 kV, and the current will be in the range 3–10 mA. A *basic* commercial system will include thresholding and pixel counting, permitting the detection of small pieces of metal or other hard contaminants, but not soft contaminants. In general, the latter can only be detected if more sophisticated algorithms are used that examine the contrast levels over various regions of the image and arrive at a consensus that foreign objects are present—typically with the help of texture analysis procedures.

The sensitivity of an X-ray detection system depends on a number of factors. Basically, it is highly dependent on the number of photons arriving at the sensors, and this number is proportional to the current passing through the X-ray tube. There are stringent rules on the intensities of X-radiation to which food products may be exposed, but in general, these limits are not approached because of good sensitivity at moderate current levels.

Sensitivity also depends critically on the voltages that are applied to X-ray tubes. In fact, the higher the voltage, the higher the electron energies, the higher the energies of the resulting photons, and the greater the penetrating power of the X-ray beam. However, greater penetrating power is not necessarily an advantage, as the beam will tend to pass through the food without attenuation, and therefore without detecting any foreign objects. While a poorly set up system may well be able to detect quite small pieces of metal without much trouble, detection of small stones and other hard contaminants will be less easy, and detection of soft contaminants will be virtually impossible. Thus, it is necessary to optimize the contrast in the input images.

Unfortunately, X-ray sources provide a wide range of wavelengths, all of which are scattered or absorbed to varying degrees by the intervening substances. In a thick sample, scattering can cause X-radiation to arrive at the detector after passing through material not in a direct line between the X-ray tube and the sensors. This makes a complete analysis of sensitivity rather complicated. In what

---

[2]The X-ray photons are first converted to visible light by passage through a layer of scintillating material.

follows, we will ignore this effect and assume that the bulk of the radiation reaching the sensors follows the direct path from the X-ray source. We will also assume that the radiation is gradually absorbed by the intervening substances, in proportion to its current strength. Thus, we obtain the standard exponential formula for the decay of radiation through the material, which we shall temporarily take to be homogeneous and of thickness z:

$$I = I_0 \exp\left(-\int \mu \, dz\right) = I_0 \exp\left(-\mu \int dz\right) = I_0 e^{-\mu z} \qquad (20.5)$$

where μ depends on the type of material and the penetrating power of the X-radiation. For monochromatic radiation of energy E, we have:

$$\mu = \left(\frac{\rho N}{A}\right)\left[\frac{k_P Z^a}{E^b} + \frac{k_C Z}{E}\right] \qquad (20.6)$$

where ρ is the density of the material, A is its atomic weight, N is Avogadro's number, a and b are numbers depending on the type of material, and k_P and k_C are decay constants resulting from photoelectric and Compton scattering, respectively (see, for example, Eisberg, 1961). It will not be appropriate to examine all the implications of this formula. Instead, we proceed with a rather simplified model that nevertheless shows how to optimize sensitivity:

$$\mu = \frac{\alpha}{E} \qquad (20.7)$$

By substituting into Eq. (20.5), we find:

$$I = I_0 \exp\left(\frac{-\alpha z}{E}\right) \qquad (20.8)$$

If a minute variation in thickness, or a small foreign object, is to be detected sensitively, we need to consider the change in intensity resulting from a change in z or in αz (ultimately, it is the integral of $\mu \, dz$ that is important—see Eq. (20.5)). It will be convenient to relabel the latter quantity as a generalized distance X, and the inverse energy factor as f:

$$I = I_0 \exp(-Xf) \qquad (20.9)$$

$$\therefore \quad \frac{dI}{dX} = -I_0 f \exp(-Xf) \qquad (20.10)$$

so that:

$$\Delta I = -\Delta X \, I_0 f \exp(-Xf) \qquad (20.11)$$

The contrast due to the variation in generalized distance can now be expressed as:

$$\frac{\Delta I}{I} = \frac{-\Delta X \, I_0 f \exp(-Xf)}{I_0 \exp(-Xf)} = -\Delta X f = \frac{-\Delta X}{E} \qquad (20.12)$$

This calculation shows that contrast should improve as the energy of the X-ray photons decreases. However, this result appears wrong, as reducing the photon energy will reduce the penetrating power, and in the end, no radiation will pass through the sample. First, the sensors will not be sufficiently sensitive to detect the radiation. Specifically, noise (including quantization noise) will become the dominating factor. Second, we have ignored the fact that the X-radiation is not monochromatic. We shall content ourselves here with modeling the situation to take account of the latter factor. Assume that the beam has two energies, one fairly low (as above), and one rather high and penetrating. This high-energy component will add a substantially constant value to the overall beam intensity, and will result in a modified expression for the contrast:

$$\frac{\Delta I}{I} = \frac{-\Delta X \, I_0 f \, \exp(-Xf)}{[I_1 + I_0 \exp(-Xf)]} = \frac{-\Delta X \, I_0 f}{[I_1 \exp(Xf) + I_0]} \tag{20.13}$$

To optimize sensitivity, we differentiate with respect to f:

$$\frac{\mathrm{d}(\Delta I / I)}{\mathrm{d}f} = \frac{-\Delta X \, I_0 \{[I_1 \exp(Xf) + I_0] - Xf I_1 \exp(Xf)\}}{[I_1 \exp(Xf) + I_0]^2} \tag{20.14}$$

This is zero when:

$$Xf I_1 = I_1 + I_0 \exp(-Xf) \tag{20.15}$$

$$\text{i.e., } E = \frac{X}{[1 + (I_0/I_1)\exp(-X/E)]} \tag{20.16}$$

When $I_1 \ll I_0$, we have the previous result that optimum sensitivity occurs for low E. However, when $I_1 \gg I_0$, we have the result that optimum sensitivity occurs when $E = X$. In general, this formula gives an optimum X-ray energy that is above zero, in accordance with intuition. In passing, we note that graphical or iterative solutions of Eq. (20.16) are easily obtained.

Finally, we consider the exponential form of the signal given by Eqs. (20.5), (20.8), and (20.9). These are nonlinear in X (i.e., αz), and meaningful image analysis algorithms would tend to require signals that are linear in the relevant physical quantity, namely, X. Thus, it is appropriate to take the logarithm of the signal from the input sensor before proceeding with texture analysis or other procedures:

$$I' = \log I = \log[I_0 \exp(-Xf)] = A - Xf \tag{20.17}$$

where

$$A = \log I_0 \tag{20.18}$$

In this way, doubling the width of the sample doubles the change in intensity, and subsequent (e.g., texture analysis) algorithms can once more be designed on an intuitive basis. (In fact, there is a more fundamental reason for performing this transformation—that it performs an element of noise whitening, which should ultimately help to optimize sensitivity.)

20.10.1 The Dual-Energy Approach to X-Ray Inspection

When inspecting solid objects by means of X-rays, there is often the problem that defects will be masked by variations in the thickness of the objects under scrutiny. This applies both with bags of frozen vegetables and with slabs of meat (and *a fortiori* with human bodies undergoing radiology), to take two relevant examples. What is needed is a means of canceling out the effect of varying thickness. Fortunately, over the past 10 years or so, a dual-energy approach called dual-emission X-ray absorptiometry (DEXA) has been developed for achieving this. The method involves moving the object on a conveyor past two rows of solid-state sensors placed within a millimeter or so of each other so that they generate line-scan images of the same section of the object (due allowance has to be made for the time difference between corresponding sets of signals). Each sensor is made sensitive to different X-ray energies by using different phosphors to convert the X-rays into visible light. To understand how the method works, we use the notation of Eq. (20.5) and define:

$$\eta(t) = \frac{\log\left(I_1(0)/I_1(t)\exp(-\mu_1 z)\right)}{\log\left(I_2(0)/I_2(t)\exp(-\mu_2 z)\right)} \tag{20.19}$$

where $I_1(t)$ and $I_2(t)$ are the time developments of the two sets of received X-ray signals in the absence of any objects. Assuming these are normalized to time $t = 0$ by periodic checking, they cancel from the above equation. Then the log functions cancel the exponential variations, leaving the simple result:

$$\eta(t) = \frac{\mu_1}{\mu_2} = \text{constant} \tag{20.20}$$

Note that this result is completely independent of the sample thickness z, although its "constant" nature depends on the nature of the materials involved, and their homogeneity.

The method is so successful that it is routinely used for scanning humans to measure bone mineral density. A recent application of its use for inspecting meat samples is given by Kröger et al. (2006).

20.11 THE IMPORTANCE OF COLOR IN INSPECTION

In many applications of machine vision, it is not necessary to consider color because almost all that is required can be achieved using grayscale images. For example, many processes devolve into shape analysis and subsequently into statistical pattern recognition. This situation is exemplified by fingerprint analysis and by handwriting and optical character recognition. However, there is one area where color has a big part to play: this is in the picking, inspection, and sorting of fruit. For example, color is very important in the determination of apple quality. Not only is it a prime indicator of ripeness, but also it contributes greatly to physical attractiveness, and thus encourages purchase and consumption.

Although color cameras digitize color into the usual RGB (red, green, blue) channels, humans perceive color differently. As a result, it is better to convert the RGB representation to the HSI (hue, saturation, intensity) domain before assessing the colors of apples and other products.[3] Space prevents a detailed study of the question of color. The reader is referred to more specialized texts for detailed information (e.g., Gonzalez and Woods, 1992; Sangwine and Horne, 1998). However, some brief comments will be useful. Intensity I refers to the total light intensity and is defined by:

$$I = \frac{(R+G+B)}{3} \tag{20.21}$$

Hue H is a measure of the underlying color, and saturation S is a measure of the degree to which it is *not* diluted by white light (S is zero for white light). S is given by the simple formula:

$$S = 1 - \frac{\min(R,G,B)}{I} = 1 - \frac{3\min(R,G,B)}{R+G+B} \tag{20.22}$$

which makes it unity along the sides of the color triangle, and zero for white light ($R = G = B = I$). Note how the equation for S favors none of the R, G, B components. It does not express color but a measure of the proportion of color and differentiation from white.

Hue is defined as an angle of rotation about the central white point $\mathbf{W}$ in the color triangle. It is the angle between the pure red direction (defined by the vector $\mathbf{R}-\mathbf{W}$) and the direction of the color $\mathbf{C}$ in question (defined by the vector $\mathbf{C}-\mathbf{W}$). The derivation of a formula for H is fairly complex and will not be attempted here. Suffice it to say that it may be determined by calculating $\cos H$, which depends on the dot product $(\mathbf{C}-\mathbf{W}).(\mathbf{R}-\mathbf{W})$. The final result is:

$$H = \cos^{-1}\left(\frac{\frac{1}{2}[(R-G)+(R-B)]}{\left[(R-G)^2 + (R-B)(G-B)\right]^{1/2}} \right) \tag{20.23}$$

or 2π minus this value if $B > G$ (Gonzalez and Woods, 1992).

When checking the color of apples, the hue is the important parameter. A rigorous check on the color can be achieved by constructing the hue distribution and comparing it with that for a suitable training set. The most straightforward way to carry out the comparison is to compute the mean and standard deviation of the two distributions to be compared and to perform discriminant analysis assuming

---

[3]Usually, a more important reason for use of HSI is to employ the hue parameter that is independent of the intensity parameter, as the latter is bound to be particularly sensitive to lighting variations.

Gaussian distribution functions. Standard theory (Section 4.5.3, Eqs. (4.19)–(4.22)) then leads to an optimum hue decision threshold.

In the work of Heinemann et al. (1995), discriminant analysis of color based on this approach gave complete agreement between human inspectors and the computer following training on 80 samples and testing on another 66 samples. However, a warning was given about maintaining lighting intensity levels identical to those used for training. In any such pattern recognition system, it is crucial that the training set be representative in every way of the eventual test set.

Finally, note that full color discrimination would require an optimal decision surface to be ascertained in the overall 3-D color space. In general, such decision surfaces are hyperellipses and have to be determined using the Mahalanobis distance measure (see, for example, Webb, 2002). However, in the special case of Gaussian distributions with equal covariance matrices, or more simply with equal isotropic variances, the decision surfaces become hyperplanes.

20.12 BRINGING INSPECTION TO THE FACTORY

The relationship between the producer of vision systems and the industrial user is more complex than might appear at first sight. The user has a need for an inspection system and states his need in a particular way. However, subsequent tests in the factory may show that the initial statement was inaccurate or imprecise, for example, the line manager's requirements[4] may not exactly match those envisaged by the factory management board. Part of the problem lies in the relative importance given to the three disparate functions of inspection mentioned earlier. Another lies in the change of perspective once it is seen exactly what defects the vision system is able to detect. It may be found immediately that one or more of the major defects that a product is subject to may be eliminated by modifications to the manufacturing process; in that case, the need for vision is greatly reduced, and indeed the very process of trying out a vision system may end in its value being undermined and its not being taken up after a trial period. Clearly, this does not detract from the inherent capability of vision systems to perform 100% untiring inspection and to help maintain strict control of quality. However, it must not be forgotten that vision systems are not cheap and that they can in some cases be justified only if they replace a number of human operators. Frequently, a payback period of 2–3 years is specified for installing a vision system.

Textural measurements on products are an attractive proposition for applications in the food and textile industries. Often, textural analysis is written into the prior justification for, and initial specification of, an inspection system. However,

---

[4]In many factories, line managers have the brief of maintaining production at a high level on an hour-by-hour basis, while at the same time keeping track of quality. The tension between these two aims, and particularly the underlying economic constraints, means that on occasion quality is bound to suffer.

what a vision researcher understands by texture and what a line inspector in either of these industries means by it tend to be different. Indeed, what is required of textural measurements varies markedly with the application. The vision researcher may have in mind higher order[5] statistical measures of texture, such as would be useful with a rough irregular surface of no definite periodicity[6]—as in the case of sand or pebbles on a beach, or grass or leaves on a bush. However, the textile manufacturer would be very sensitive to the periodicity of his fabric, and to the presence of faults or overly large gaps in the weave. In such cases, a major problem is likely to be that of minimizing computation so that considerable expanses of fabric, or large numbers of products, can be checked economically at production rates. Similarly, the food manufacturer might be interested in the number and spatial distribution of pieces of pepper on a pizza, while for fish coatings (e.g., batter or breadcrumbs) uniformity will be important and "texture measurement" may end by being interpreted as determining the number of holes per unit area of the coating. Thus, it may sometimes be beneficial to characterize a texture by rather simple counting or uniformity checks instead of higher order statistics. More generally, it is important to keep the inspection system flexible by training on samples so that maximum utility of the production line can be maintained.

With this backcloth to factory requirements, it is clearly vital for the vision researcher to be sensitive to actual rather than idealized needs or the problem as initially specified. There is no substitute for detailed consultation with the line manager and close observation in the factory before setting up a trial system. Then the results from trials need to be considered very carefully to confirm that the system is producing the information that is really required.

20.13 CONCLUDING REMARKS

This chapter has been concerned with the application of computer vision to industrial tasks, and notably to automated visual inspection. The number of relevant applications is exceptionally high, and for that reason, it has been necessary to concentrate on principles and methods rather than on individual cases. The repeated mention of hardware implementation has been a necessary one, since the economics of real-time implementation is often the factor that ultimately decides whether a given system will be installed on a production line. However, speeds of processing are also heavily dependent on the specific algorithms employed, and

---

[5]The zero-order statistic is the mean intensity level; first-order statistics such as variance and skewness are derived from the histogram of intensities; second- and higher order statistics take the form of gray-level co-occurrence matrices, showing the number of times particular gray values appear at two or more pixels in various relative positions. For more discussion on textures and texture analysis, see Chapter 8.

[6]More rigorously, the fabric is intended to have a long-range periodic order that does not occur with sand or grass: in fact, there is a close analogy here with the long- and short-range periodic order for atoms in a crystal and in a liquid, respectively.

these in turn depend on the nature of the image acquisition system—including both the lighting and the camera setup (indeed, the decision of whether to inspect products on a moving conveyor or to bring them to a standstill for more careful scrutiny is perhaps the most fundamental one for implementation). Hence, image acquisition and real-time electronic hardware systems are the main topics of two later chapters (Chapters 25 and 26).

More fundamentally, the reader will have noticed that a major purpose of inspection systems is to make instant decisions on the adequacy of products. Related to this purpose are the often fluid criteria for making such decisions and the need to train the system so that the decisions that are made are at least as good as those that would have arisen with human inspectors. In addition, training schemes are valuable in making inspection systems more general and adaptive, particularly with regard to change of product. Hence, the pattern recognition techniques discussed in Chapter 24 are highly relevant to the process of inspection.

On a different tack, note that much of automated visual inspection falls under the heading of computer-aided manufacture (CAM), of which computer-aided design (CAD) is another part. Nowadays, many manufactured parts can in principle be designed on a computer, visualized on a computer screen, made by computer-controlled (e.g., milling) machines, and inspected by computer—all without human intervention in handling the parts themselves. There is much sense in this computer-integrated manufacture (CIM) concept, since the original design data set is stored in the computer, and therefore it might as well be used (a) to aid the image analysis process that is needed for inspection and (b) as the actual template by which to judge the quality of the goods. After all, why key in a separate set of templates to act as inspection criteria when the specification already exists on a computer? However, some augmentation of the original design information to include valid tolerances is necessary before the dataset is sufficient for implementing a complete CIM system. Also, the purely dimensional input to a numerically controlled milling machine is not generally sufficient—as the frequent references to surface quality in the present chapter indicate.

> Automated industrial inspection is a well-worn application area for vision that severely exercises the reliability, robustness, accuracy, and speed of vision software and hardware. This chapter has discussed the practicalities of this topic, showing how color and other modalities such as X-rays impinge on the basic vision techniques.

20.14 BIBLIOGRAPHICAL AND HISTORICAL NOTES

It is very difficult to provide a bibliography of the enormous number of papers on applications of vision in industry or even in the more restricted area of automated visual inspection. In any case, it can be argued that a book such as this ought to concentrate on principles and to a lesser extent on detailed applications and "mere" history. However, the review article by Newman and Jain (1995) gives a representative overview covering an earlier period.

The overall history of industrial applications of vision has been one of relatively slow beginnings as the potential for visual control became clear, followed only in recent years by explosive growth as methods and techniques evolved and as cost-effective implementations became possible as a result of cheaper computational equipment. In this respect, 1980 marked a turning point, with the instigation of important conferences and symposia, notably that on "Computer Vision and Sensor-Based Robots" held at General Motors Research Laboratories during 1978 (see Dodd and Rossol, 1979), and the Robot Vision and Sensory Controls (ROVISEC) series of conferences organized annually by IFS (Conferences) Ltd, UK, from 1981. In addition, useful compendia of papers were published (e.g., Pugh, 1983), and books outlining relevant principles and practical details (e.g., Batchelor et al., 1985).

Noble (1995) presented an interesting and highly relevant view of the use of machine vision in manufacturing. Davies (1995) developed the same topic by presenting several case studies together with a discussion of some major problems that remained to be tackled in this area.

The 1990s saw considerable interest in X-ray inspection techniques, particularly in the food industry (Boerner and Strecker, 1988; Wagner, 1988; Chan et al., 1990; Penman et al., 1992; Graves et al., 1994; Noble et al., 1994). In the case of X-ray inspection of food, the interest was almost solely in the detection of foreign objects, which could in some cases be injurious to the consumer. Indeed, this was a prime motivation for much work in the author's laboratory (Patel et al., 1994, 1995).

Another topic of growing interest was the automatic visual control of materials such as lace during manufacture, together with high-speed scalloping of lace using lasers (King and Tao, 1995; Yazdi and King, 1998).

A cursory examination of inspection publications reveals growth in emphasis on surface defect inspection, including color assessment, and X-ray inspection of bulk materials and baggage, e.g., at airports. Two journal special issues (Davies and Ip, 1998; Nesi and Trucco, 1999) cover defect inspection, while Tsai and Huang (2003) and Fish et al. (2003) further emphasize the point regarding surface defects.

Work on color inspection includes both food (Heinemann et al., 1995) and pharmaceutical products (Derganc et al., 2003). Work on X-rays includes the location of foreign bodies in food (Patel et al., 1996; Batchelor et al., 2004), the internal inspection of solder joints (Roh et al., 2003), and the examination of baggage (Wang and Evans, 2003). Finally, a recent volume on inspection of natural products (Graves and Batchelor, 2003) has articles on inspection of ceramics, wood, textiles, food, live fish and pigs, and sheep pelts, and embodies work on color and X-ray modalities. In a sense, such work is neither adventurous nor glamorous. Indeed, it involves significant effort to develop the technology and software sufficiently to make it useful for industry—which means this is an exacting type of task, not tied merely to the production of academic ideas.

20.14.1 **More Recent Developments**

More recently, Chao and Tsai (2008) describe an inspection system for detecting surface defects in glass substrates for liquid crystal displays. It is based on an anisotropic diffusion-based smoothing technique. This type of technique is designed to smooth the image in directions parallel to edges, and *not* normal to edges, that is, it is designed to perform edge-preserving smoothing. Thus, it tends to enhance features such as cracks or indentations or breaks in a substance, and to eliminate noise or the effects of minor lighting or lightness variations. By the time this has been achieved, it is much more likely that defects will be locatable by thresholding. In this case, a special new anisotropic diffusion algorithm was designed to improve the situation further. Tsai et al. (2010) had a similar problem—that of looking for micro-cracks in textured solar wafers. Although they used anisotropic diffusion to initiate the process, and followed this with binary thresholding, they completed the task of removing noise and identifying the micro-cracks by use of morphological operations. Mak et al. (2009) had the problem of locating defects in textured materials. Here, texture forms a complex background pattern and this imposes difficulties. In this case, these were solved by applying a special type of neural network—a Gabor wavelet network. Following this, a sequence of morphological operators was applied to locate the defects. In a typical case, this involved 1×7 linear opening, 1×7 linear closing, 3×3 median, 3×3 closing, and finally thresholding. Sun et al. (2010) describe a more general type of system for inspecting electric contacts for a variety of types of surface defect including cracks, breaks, and scratches, using multiple 3-D views. There was some concentration on dimensional measurement, although morphological analysis was also required, including the top-hat transform for locating cracks; edge breaks were located by deviations from circularity. For further details, the reader is referred to the original paper.

Alexandropoulos et al. (2008) worked on template-guided inspection, primarily of under-vehicle views of the exhaust and other vehicle components and structures. Here, the inspection fell into a template registration phase followed by a template differencing phase. In fact, the latter had to contend with noise, illumination variations, shadows, and so on. Problems with change detection were addressed by using a block-based segmentation technique, which contrasted noise and structural variations. Variations were judged relative to statistical significance for the particular block sizes. In general, in block-based change detection, it proved necessary to take into consideration the anticipated scene complexity and adjust the operation parameters accordingly. For example, too small a block size was undesirable as it accentuated edge effects.

Work on X-ray inspection has continued, in particular, using the by now important and well-established dual-energy (DEXA) approach outlined in Section 20.10.1. For a recent application of its use for inspecting meat samples, see Kröger et al. (2006).

Inspection of Cereal Grains

21

Inspection of cereal grains is an exceptionally mundane and repetitive task, and to some extent contrasts with many of the examples given in Chapter 20 in that the emphasis is on detecting contaminants rather than finding manufacturing faults. This chapter presents three case studies that cover the topic and at the same time air and solve relevant theoretical questions.

Look out for:

- the limitations of the immediately obvious technique—global thresholding.
- the value of morphological filtering.
- the relatively unusual need for median filtering as a morphological operation.
- the use of bar (linear feature) detectors for locating insects.
- how the vectorial bar detector operator is optimized.
- how sampling can be used to speed up object location.
- how the outputs of oblique template masks can optimally be combined.

This chapter repeatedly alludes to the problems of achieving real-time operation and largely solves these problems for moderately demanding situations (flow rates up to ~300 items per second): for more demanding situations, dedicated hardware accelerators are needed, as discussed in Chapter 26. At a more detailed level, the balance between false positives (false alarms) and false negatives needs to be optimized, as discussed in Section 24.7.

21.1 INTRODUCTION

Cereal grains are among the most important of the foods we grow. A large proportion of cereal grains is milled and marketed as flour: this is then used to produce bread, cakes, biscuits and many other commodities. Cereal grains can also be processed in a number of other ways (such as crushing), and thus they form the basis of many breakfast cereals. In addition, there are some cereal grains, or cereal kernels, which are eaten with a minimum of processing: rice is obviously in this category—although whole wheat and oat grains are also consumed as decorative additives to "granary" loaves.

Wheat, rice, and other cereal grains are shipped and stored in vast quantities measured in millions of tons, and a large international trade exists to market these commodities. Transport also has to be arranged between farmers, millers, crushers, and the major bakers and marketers. Typically, transit by road or rail is in relatively small loads of up to 20 tons, and grain depots, warehouses, and ports are not unlikely to receive lorries containing such consignments at intervals ranging from 20 min to as little as 3 min. All the necessary transportation and storage result in grains being subject to degradation of various sorts: damage, molds, sprouting, insect infestation, and so on. In addition, the fact that grains are grown on the land and threshed implies the possibility of contamination by rodent droppings, stones, soil, chaff, and foreign grains. Finally, the quality of grain from various sources will not be uniformly high, and varietal purity is an important concern.

These factors mean that ideally, the grain that arrives at any depot should be inspected for a good many possible causes of degradation. This chapter is concerned with grain inspection. In the space of one chapter, we shall not aim to cover all possible methods and modes of inspection. Indeed, this would be impossible as the subject is moving ahead quite fast: not only are inspection methods evolving rapidly, but the standards against which inspection is carried out are also evolving fairly quickly. To some extent, the improvement of automatic inspection methods and the means of implementing them efficiently in hardware are helping to drive the process onward. We shall explore the situation with the aid of three main case studies, and then we shall look at the overall situation.

The first case study involves the examination of grains to locate rodent droppings and molds, such as ergot. The second case study considers how grains may be scrutinized for insects, such as the saw-toothed grain beetle. The third case study is concerned with inspection of the grains themselves. However, this case study is more general, and is less involved with the scrutiny of individual grains than with how efficiently they can be located: this is an important factor when lorry loads of grain are arriving at depots every few minutes—leading to the need to sample of the order of 300 grains per second. As remarked in Chapter 20, object location can involve considerably more computation than object scrutiny and measurement.

21.2 CASE STUDY: LOCATION OF DARK CONTAMINANTS IN CEREALS

As noted above, there is a demand for grain quality to be monitored before processing to produce flour, breakfast cereals, and a myriad of other derived products. Early work in this area was applied mainly to the determination of grain quality

(Ridgway and Chambers, 1996), with concentration on the determination of varietal purity (Zayas and Steele, 1990; Keefe, 1992) and the location of damaged grains (Liao et al., 1994). In fact, there is also the need to detect insect infestations and other important contaminants in grain—while not being confused by up to $\sim 2\%$ "permitted admixture" such as chaff and dust. The inspection work described in this section (Davies et al., 1998a) pays particular attention to the detection of noninsect contaminants. Relevant contaminants in this category include rodent (especially rat and mouse) droppings, molds such as ergot, and foreign seeds. (Note that ergot is poisonous to humans, so locating any instances of it is of especial importance.) In this case study the substrate grain is wheat, and foreign seeds such as rape would be problematic if present in too great a concentration.

Many of the potential contaminants for wheat grains (and for grains of similar general appearance such as barley or oats) are quite dark in color. This means that thresholding is the most obvious approach for locating them. However, there are a number of problems, in that false alarms arise from shadows between grains, dark patches on the grains, rapeseeds, chaff and other admixture components: this means that further recognition procedures have to be invoked to distinguish between the various possibilities. As a result, the thresholding approach is not eventually as attractive as might *a priori* have been thought (Fig. 21.1(a) and (b)). This problem is exacerbated by the extreme speeds of processing required in real applications. For example, a successful device for monitoring lorry loads of grain might well have to analyze a 3-kg sample of grain in 3 min (the typical time between arrival of lorries at a grain terminal), and this would correspond to some 60,000 grains having to be monitored for contaminants in that time. This places a distinct premium on rapid, accurate image analysis.

This case study concentrates on monitoring grain for rodent droppings. As indicated above, these types of contaminant are generally darker than the grain background, but cannot simply be detected by thresholding since there are significant shadows between the grains, which themselves often have dark patches. In addition, the contaminants are speckled because of their inhomogeneous constitution and because of lighting and shadow effects. In spite of these problems, the contaminants are identifiable by human operators because they are relatively large and their shape follows a distinct pattern (e.g., an aspect ratio of three or four to one). Thus, it is the combination of size, shape, relative darkness, and speckle that characterizes the contaminants and differentiates them from the grain substrate.

Designing efficient and rapidly operating algorithms to identify these contaminants is something of a challenge, but an obvious way of tackling it is *via* mathematical morphology—as we shall see in Section 21.2.1.

21.2.1 Application of Morphological and Nonlinear Filters to Locate Rodent Droppings

As indicated above, the obvious approach to the location of rodent droppings is to process thresholded images by erosion and dilation. In this way, shadows between

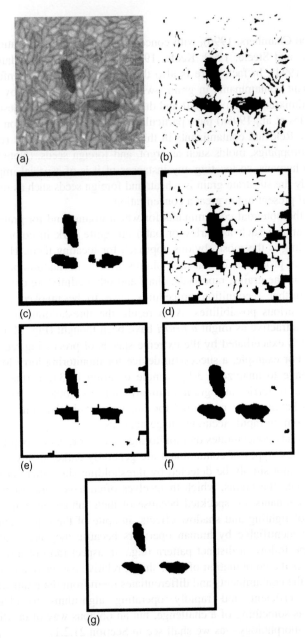

(a) (b)

(c) (d)

(e) (f)

(g)

FIGURE 21.1

Effects of various operations and filters on a grain image. (a) Grain image containing several contaminants (rodent droppings). (b) Thresholded version of (a). (c) Result of erosion and dilation on (b). (d) Result of dilation and erosion on (b). (e) Result of erosion on (d). (f) Result of applying 11×11 median filter to (b). (g) Result of erosion on (f). In all cases, "erosion" means three applications of the basic 3×3 erosion operator, and similarly for "dilation."

Source: © *IEE 1998*

grains, and discoloration of grains would be eliminated by the erosions, and the shapes and sizes of the contaminants restored by the subsequent dilations. The effect of this procedure is shown in Fig. 21.1(c). Note that the method has been successful in eliminating the shadows between the grains, but has been decidedly weak in coping with light regions on the contaminants. Remembering that while considerable uniformity might be expected between grains, the same cannot be said about rodent droppings, whose size, shape, and color vary quite markedly. Hence, the erosion–dilation schema has limited value, although it would probably be successful in most instances. Accordingly other methods of analysis were sought (Davies et al., 1998a).

The second approach is to overcome the problem of fragmentation of the contaminants which arose with the previous approach: this was attempted by first consolidating the contaminants by applying dilation before erosion. The effect of this approach is shown in Fig. 21.1(d). Notice that the result is to consolidate the shadows between grains even more than the shapes of the contaminants. Even when an additional few erosions are applied (Fig. 21.1(e)), the consolidated shadows do not disappear, and are of comparable sizes to the contaminants. Hence, the approach is not viable, and creates more problems than it solves. One possibility is to use the results of the earlier erosion–dilation schema as "seeds" to validate a subset of the dilation–erosion schema. However, this would be far more computation intensive and the results would clearly not be especially impressive (see Fig. 21.1(c) and (e)). Instead, a totally different approach was adopted.

The new approach was to apply a large median filter to the thresholded image, as shown in Fig. 21.1(f). This gives good segmentation of the contaminants, retaining their intrinsic shape to a reasonable degree, and suppresses the shadows between grains quite well. In fact, the shadows immediately around the contaminants enhance the sizes of the latter in the median filtered image, while some shadows further away are consolidated and retained by the median filtering. It was found useful to perform a final erosion operation (Fig. 21.1(g)): this eliminates the extraneous shadows and brings the contaminants back to something like their proper size and shape: although the lengths are slightly curtailed, this is not a severe disadvantage. Overall, the median filtering–erosion schema gave easily the greatest fidelity to the original contaminants, while being particularly successful at eliminating other artifacts (Davies et al., 1998a). In this case, it seems that the median filter is acting as an analytical device that carefully meditates and obtains the final result in a single rigorous stage—thus avoiding the error-propagation inherent in a two-stage process.

Finally, although median filtering is intrinsically more computation intensive than erosion and dilation operations, many methods have been devised for accelerating median operations (see Chapter 3), and indeed the speed aspect was found to be easily soluble within the specification set out above.

21.2.2 Problems with Closing

The above case study was found to require the use of a median filter coupled with a final erosion operation to eliminate artifacts in the background. An earlier test similarly involved a closing operation (a dilation followed by an erosion), followed by a final erosion. In other applications, grains or other small particles are often grouped by applying a closing operation to locate the regions where the particles are situated (Fig. 7.6). It is interesting to speculate whether, in the latter type of approach, closing should occasionally be followed by erosion, and also whether the final erosions used in the tests were no more than *ad hoc* procedures or whether they were vital to achieve the defined goal.

The situation was analyzed by Davies (2000c) and is summarized in Sections 7.5.1 and 7.5.2. The analysis starts by considering two regions containing small particles with occurrence densities ρ_1, ρ_2, where $\rho_1 > \rho_2$ (Figs. 7.7 and 7.8). It finds that a final erosion is indeed required to eliminate a shift in the region boundary, the estimated shift δ being:[1]

$$\delta = 2ab\rho_2(a + b) \tag{21.1}$$

where a is the radius of the dilation kernel and b is the width of the particles in Region 2.

It is important to notice that if $b = 0$, no shift will occur, but for particles of measurable size this is not so. Clearly, if b is comparable to a or if a is much greater than 1 pixel, a substantial final erosion may be required. On the other hand, if b is small, it is possible that the two-dimensional shift will be less than 1 pixel. In that case, it will not be correctable by a subsequent erosion, but due allowance for the shift can be made during subsequent analysis. While in this work, the background artifacts were induced mainly by shadows around and between grains, in other cases impulse noise or light background texture could give similar effects, and care must be taken to select a morphological process that limits any overall shifts in region boundaries. In addition, notice that the whole process can be modeled and the extent of any necessary final erosion estimated accurately. More particularly, the final erosion is a necessary measure, and is not merely an *ad hoc* procedure (Davies, 2000c).

21.2.3 Ergot Detection Using the Global Valley Method

While the early parts of Section 21.2 emphasized the difficulty of locating dark contaminants in cereals, and concluded that, in general, morphological methods would be needed for this purpose, it turns out that ergot can be isolated by thresholding. However, there is an intrinsic difficulty in finding a suitable threshold value, and appealing to the intensity histogram shows that it usually approximates closely to a unimodal distribution with no real indication of where the best

---

[1] A full derivation of this result appears in Section 7.5.1.

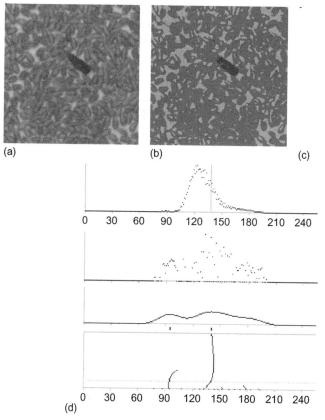

(a)   (b)   (c)

(d)

FIGURE 21.2

Location of ergot among wheat grains. (a) Original image. (b) Doubly thresholded image. (c) Result of only applying the lower threshold. (d) Top to bottom: Intensity histogram of (a). Result of applying the global valley transformation. Result of smoothing the global valley transform. The two thresholds used in (b) being located automatically at the dotted line. For further details, see text in Chapter 4.

Source: © *IET 2008*

thresholding point would be. However, the global valley transformation described in Chapter 4 is highly effective in this situation and leads to a global valley transform which, when smoothed, provides a virtually perfect thresholding value (Fig. 21.2). In fact, the method yields two threshold values, one which is useful for locating ergot and the other which is suited to separating wheat grains from a light conveyor background: clearly, there is no problem ignoring the latter threshold if it is not needed.

The reason why ergot can be detected in this way is that it is generally somewhat darker than rodent droppings. Ergot also differs from rat droppings in that the pieces are rather smaller, although in this respect they are not dissimilar to mouse droppings. However, its most important characteristic is that it is poisonous to humans (rodent droppings are unlikely to be poisonous, even though they are clearly highly undesirable). Hence, a further method that can be employed for detecting this type of contaminant is of some value.

21.3 CASE STUDY: LOCATION OF INSECTS

The case study described in this section pays particular attention to the need to detect insects. Note that insects present an especially serious threat, because they can multiply alarmingly in a short span of time, so greater certainty of detection is vital in this case. This means that a highly discriminating method is required for locating adult insects (Davies et al., 1998b).

Not surprisingly, thresholding initially seemed to be the approach offering the most promise for locating insects, which appear dark relative to the light brown color of the grain. However, early tests on these lines showed that a good many false alarms would result from chaff and other permitted admixture, from less serious contaminants such as rapeseeds, and even from shadows between, and discolorations on, the grains themselves (Fig. 21.3). These considerations led to use of the linear feature detection approach for detecting small adult insects. This proved possible because these insects appear as dark objects with a linear (essentially bar-shaped) structure; hence attempts to detect them by applying bar-shaped masks led ultimately to a linear feature detector that had good sensitivity for a reasonable range of insect sizes. Before proceeding further, we consider the problems of designing linear feature detector masks.

21.3.1 The Vectorial Strategy for Linear Feature Detection

In earlier chapters we found that location of features in 3×3 windows typically required eight template masks to be applied in order to cope with arbitrary orientation. In particular, to detect corners, eight masks were required, while only one is needed to locate small holes because of the high degree of symmetry present in the latter case (Chapter 6). To detect edges by this means, four masks were required because $180°$ rotations correspond to a sign change that effectively eliminates the need for half of the masks: however, edge detection is a special case, as edges are vector quantities with magnitude and direction, and hence they are fully definable using just two component masks. In contrast, a typical line segment detection mask would have the form:

$$\begin{bmatrix} -1 & 2 & -1 \\ -1 & 2 & -1 \\ -1 & 2 & -1 \end{bmatrix}$$

This indicates that four masks are sufficient, the total number being cut from eight to four by the particular rotation symmetry of a line segment. Curiously,

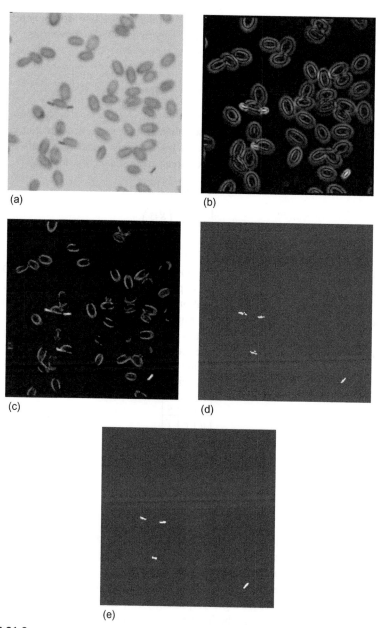

(a)

(b)

(c)

(d)

(e)

FIGURE 21.3

Insects located by line segment detector. (a) Original image. (b) Result of applying vectorial operator. (c) Result of masking (b) using an intensity threshold at a standard high level. (d) Result of thresholding (c). (e) Result of applying TM operation within (d).

while only edges qualify as strict vectors, requiring just two component masks, the same can be made to apply for line segments. To achieve this, a rather unusual set of masks has to be employed:[2]

$$L_0 = A \begin{bmatrix} 0 & -1 & 0 \\ 1 & 0 & 1 \\ 0 & -1 & 0 \end{bmatrix} \quad L_{45} = B \begin{bmatrix} -1 & 0 & 1 \\ 0 & 0 & 0 \\ 1 & 0 & -1 \end{bmatrix} \tag{21.2}$$

The two masks are given different weights, so that some account can be taken of the fact that the nonzero coefficients are respectively 1 and $\sqrt{2}$ pixels from the center pixel in the window. These masks give responses g_0, g_{45} leading to:

$$g = (g_0^2 + g_{45}^2)^{1/2} \tag{21.3}$$

$$\theta = \frac{1}{2} \arctan \left(\frac{g_{45}}{g_0} \right) \tag{21.4}$$

where the additional factor of one half in the orientation arises as the two masks correspond to basic orientations of 0° and 45° rather than 0° and 90° (Davies, 1997c). Nevertheless, these masks can cope with the full complement of orientations as line segments have 180° rotational symmetry and the final orientation should only be determined within the range 0–180°. In Section 21.5, we comment further on the obvious similarities and differences between Eqs. (21.3), (21.4) and those pertaining to edge detection.

Next, we have to select appropriate values of the coefficients A and B. Applying the above masks to a window with the intensity pattern:

a	b	c
d	e	f
g	h	i

leads to the following responses at 0° and 45°:

$$g_0 = A(d + f - b - h) \tag{21.5}$$

$$g_{45} = B(c + g - a - i) \tag{21.6}$$

Using this model, theory and simulations were carried out (Davies, 1997c) for the case of a line segment of width w passing through the center of the 3×3 window, pixel responses being taken in proportion to the area of the line falling within each pixel. These showed that high orientation accuracy occurred for $w = 1.4$, when $B/A = 0.86$, giving a surprisingly low maximum error of just 0.4°. However, in this application, high orientation accuracy was not required, so

[2]The proof of the concepts described here involves describing the signal in a circular path around the current position and modeling it as a sinusoidal variation, which is subsequently interrogated by two quadrature sinusoidal basis functions set at 0° and 45° in real space (but by definition at 90° in the quadrature space). The mapping between orientation θ in the quadrature space and orientation φ in real space leads to the additional factor of one half in Eq. (21.4).

Eq. (21.3) was used on its own to obtain an accurate estimate of line contrast using the two masks. In this way, high sensitivity could be attained in the detection of image features corresponding to insect contaminants.

21.3.2 Designing Linear Feature Detection Masks for Larger Windows

When larger windows are used, it is possible to design the masks more ideally to match the image features that have to be detected, because of the greater resolution then permitted. However, there are many more degrees of freedom in the design, and there is some uncertainty as to how to proceed. The basic principle (Davies et al., 1998b) is to use masks whose profile matches the intensity profile of a linear feature around a ring of radius R centered on the feature, and at the same time follows a particular mathematical model—namely an approximately sinusoidal amplitude variation. For a given linear feature of width w, the sinusoidal model will achieve the best match for a thin ring of radius R_0 for which the two arc lengths within the feature are each one quarter of the circumference $2\pi R_0$ of the ring. Simple geometry (Fig. 21.4) shows that this occurs when:

$$2R_0 \sin\left(\frac{\pi}{4}\right) = w \tag{21.7}$$

$$\therefore \quad R_0 = \frac{w}{\sqrt{2}} \tag{21.8}$$

The width ΔR of the ring should in principle be infinitesimal, but in practice, considering noise and other variations in the dataset, ΔR can validly be up to about 40% of R_0. The other relevant factor is the intensity profile of the ring, and how accurately this has to map to the intensity profile of the linear features to be

FIGURE 21.4

Geometry for application of a thin ring mask. Here two quarters of the ring mask lie within and two outside a rectangular bar feature.

Source: © *World Scientific 2000*

```
 .  . -1 -2 -1  .  .        .  . -1  .  1  .  .
 .  . -1 -2 -1  .  .        . -2 -2  .  2  2  .
 1  1  .  .  .  1  1       -1 -2  .  .  .  2  1
 2  2  .  •  .  2  2        .  .  .  •  .  .  .
 1  1  .  .  .  1  1        1  2  .  .  . -2 -1
 .  . -1 -2 -1  .  .        .  2  2  . -2 -2  .
 .  . -1 -2 -1  .  .        .  .  1  . -1  .  .
```

FIGURE 21.5

Typical (7 × 7) linear feature detection masks.

Source: © EURASIP 1998

located in the image. In many applications, such linear features will not have sharp edges, but will be slightly fuzzy and the sides will have significant and varying intensity gradient. Thus, the actual intensity profile is quite likely to correspond reasonably closely to a true sinusoidal variation. Masks designed on this basis proved close to optimal when experimental tests were made (Davies et al., 1998b). Figure 21.5 shows masks that resulted from this type of design process for one specific value of R (2.5 pixels).

21.3.3 Application to Cereal Inspection

The main class of insect that was targeted in this study (Davies et al., 1998b) was *Oryzaephilus surinamensis* (saw-toothed grain beetle): insects in this class approximate to rectangular bars of about 10×3 pixels, and masks of size 7×7 (Fig. 21.5) proved to be appropriate for identifying the pixels on the centrelines of these insects. In addition, the insects could appear in any orientation, and thus the vectorial approach to template matching was used, employing two masks as outlined above.

Preliminary decisions on the presence of the insects were made by thresholding the output of the vectorial operator. However, the vectorial operator responses for insect-like bars have a low-intensity surround, which is joined to the central response at each end, and between the low-intensity surround and the central response the signal drops close to zero except near the ends (Fig. 21.3(b)). This response pattern is readily understandable because symmetry demands that the signal be zero when the centers of the masks are coincident with the edge of the bar. To avoid problems from the parts of the response pattern outside the object, intensity thresholding (applied separately to the original image) was used to set the output to zero where it indicates no insect could be present (Fig. 21.3(c)).

21.3.4 Experimental Results

In laboratory tests, the vectorial operator was applied to 60 grain images containing a total of 150 insects. The output of the basic enhancement operator

```
2  2  2  2  2  2  2        2  2  3  2  1  0  0        4  2  2  0  0 -1 -2
0  0  0  0  0  0  0        2  2  0  0  0 -1 -1        2  2  0  0 -1 -2 -1
-1 -1 -1 -1 -1 -1 -1       0  0  0 -2 -2 -2 -1        2  0  0 -1 -3 -1  0
-2 -2 -2 -2 -2 -2 -2       0 -2 -2 -2 -2 -2  0        0  0 -1 -2 -1  0  0
-1 -1 -1 -1 -1 -1 -1      -1 -2 -2 -2  0  0  0        0 -1 -3 -1  0  0  2
0  0  0  0  0  0  0       -1 -1  0  0  0  2  2       -1 -2 -1  0  0  2  2
2  2  2  2  2  2  2        0  0  1  2  3  2  2       -2 -1  0  0  2  2  4
```

FIGURE 21.6

7 × 7 template matching masks. There are 8 masks in all, orientated at multiples of 22.5° relative to the x-axis: only three are shown as the others are straightforwardly generated by 90° rotation and reflection (symmetry) operations.

(described in Section 21.3.1) was thresholded to perform the actual detection and then passed to an area discrimination procedure: objects having fewer than 6 pixels were eliminated, as these normally correspond to dark shadows or discolorations on the grains.

The detection threshold was adjusted to minimize the total number of errors. Minor failures were typically due to two insects in contact being interpreted as one, insects being masked by lying along the edges of grains, and the dark edge of a grain appearing similar to an insect.

While adjustment of the detection threshold was found to eliminate either the false positive or two of the false negatives (but not the one in which two insects were in contact), the ultimate reason why the total number of errors could not be reduced further lay in the limited sensitivity of the vectorial operator masks, which contain relatively few coefficients. This was verified by carrying out a test with a set of conventional template matching (TM) masks (Fig. 21.6), which led to a single false negative (due to the two insects being in contact) and no false positives—and a much increased computational load (Davies et al., 2003a). It was accepted that the case of two touching insects would remain a problem that could only be eliminated by introducing a greater level of image understanding—an aspect beyond the scope of this work (although this particular problem could easily be eliminated by noting the excessive length and/or area of this one object).

Next, it was desired to increase the interpretation accuracy of the vectorial operator in order to limit computation, and tests were made of a two-stage system, aimed at combining the speed of the vectorial operator with the accuracy of TM. Accordingly, the vectorial operator was used to create regions of interest for the TM operator. To obtain optimum performance, it was found necessary to decrease the vectorial operator threshold level to reduce the number of false negatives, leaving a slightly increased number of false positives that would subsequently be eliminated by the TM operator. The final result was an error rate exactly matching that of the TM operator, but a considerably reduced overall execution time composed of the adjusted vectorial operator execution time plus the TM execution time for the set of regions of interest. Further speed improvements, and no loss in accuracy, resulted from employing a dilation operation within the vectorial operator to recombine fragmented insects before applying the final TM operator: in this case, the vectorial operator was used without reducing its threshold value. At this stage, the speedup factor on the original TM operator was around seven (Davies et al., 2003a).

Finally, intensity thresholding was used as a preliminary skimming operation on the vectorial operator—where it produced a speedup by a factor of about 20: this brought the overall speedup factor on the original TM algorithm into the range 50–100, being close to 100 where insects are rare. Perhaps more important, the combined algorithm was in line with the target to inspect 3 kg of grain for a variety of contaminants within 3 min without additional fast hardware.

21.4 CASE STUDY: HIGH-SPEED GRAIN LOCATION

It has already been mentioned several times in this book that object location often requires considerable computation, as it involves unconstrained search over the image data: as a result the whole process of automated inspection can be slowed down significantly, and this can be of crucial importance in the design of real-time inspection systems. Indeed, if the scrutiny of particular types of object requires quite simple dimensional measurements to be made, the object location routine can be the bottleneck. This case study is concerned with the high-speed location of objects in 2D images, a topic on which relatively little systematic work has been carried out—at least on the software side—although many studies have been made on the design of hardware accelerators for image processing. As hardware accelerators represent the more expensive option, this case study concentrates on software solutions to the problem, and then specializes it to the case of cereal grains in images.

21.4.1 Extending an Earlier Sampling Approach

Following on from the earlier sampling approach of Section 12.5, in which circular objects were rapidly located by bisecting a limited number of horizontal and vertical chords, a new approach was tried. This method aimed at even greater speed by taking a minimum number of sampling points in the image rather than by scanning along whole lines (Davies, 1998b).

Suppose that we are looking for an object such as that shown in Fig. 21.7(a), whose shape is defined relative to a reference point R as the set of pixels $A = \{\mathbf{r}_i : i = 1 \text{ to } n\}$, n being the number of pixels within the object. If the position of R is $\mathbf{x}_R$, pixel i will appear at $\mathbf{x}_i = \mathbf{x}_R + \mathbf{r}_i$. This means that when a sampling point $\mathbf{x}_s$ gives a positive indication of an object, the location of its reference point R will be $\mathbf{x}_R = \mathbf{x}_s - \mathbf{r}_i$. Thus, the reference point of the object is known to lie at one of the set of points $U_R = \cup_i(\mathbf{x}_s - \mathbf{r}_i)$, so knowledge of its location is naturally incomplete. Indeed, the map of possible reference point locations has the same

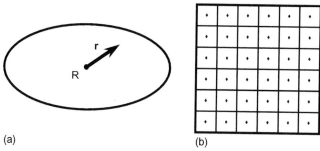

(a)   (b)

FIGURE 21.7

Object shape and method of sampling. (a) Object shape, showing reference point R and vector **r** pointing to a general location $\mathbf{x_R} + \mathbf{r}$. (b) Image and sampling points, with associated tiling squares.

Source: © EURASIP 1998

shape as the original object, but rotated through 180°—because of the minus sign in front of $\mathbf{r}_i$. Furthermore, the fact that reference point positions are only determined within n pixels means that many sampling points will be needed, the minimum number required to cover the whole image clearly being N/n, if there are N pixels in the image. This means that the optimum speedup factor will be $N/(N/n) = n$, as the number of pixels visited in the image is N/n rather than N (Davies, 1997d).

Unfortunately, it will not be possible to find a set of sampling point locations such that the "tiling" produced by the resulting maps of possible reference point positions covers the whole image without overlap. Thus, there will normally be some overlap (and thus loss of efficiency in locating objects) or some gaps (and thus loss of effectiveness in locating objects). Clearly, the set of tiling squares shown in Fig. 21.7(b) will only be fully effective if square objects are to be located.

However, a more serious problem arises because objects may appear in any orientation. This prevents an ideal tiling from being found. Thus, the best that can be achieved is to search the image for a maximal *rotationally invariant* subset of the shape, which must be a circle, as indicated in Fig. 21.8(a). Furthermore, as no perfect tiling for circles exists, the tiling that must be chosen is either a set of hexagons or, more practically, a set of squares. This means that the speedup factor for object location will be significantly less than n, although it will still be substantial.

21.4.2 Application to Grain Inspection

A prime application for this technique is that of fast location of grains on a conveyor in order to scrutinize them for damage, varietal purity, sprouting, molds, etc. Under these circumstances it is best to examine each grain in isolation: specifically, touching or overlapping grains would be more difficult to cope with.

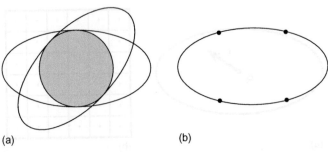

(a)　　　　　　　　　　　　(b)

FIGURE 21.8

Geometry for location of ellipses by sampling. (a) Ellipse in two orientations and maximal rotationally invariant subset (shaded). (b) Horizontal ellipse and geometry showing size relative to largest permitted spacing of sampling points.

Source: © EURASIP 1998

Thus, the grains would need to be spread out with at most 25 grains being visible in any 256×256 image. With so much free image space there would be an intensive search problem, with far more pixels having to be considered than would otherwise be the case. Hence a very fast object location algorithm would be of especial value.

Wheat grains are well approximated by ellipses in which the ratio of semi-major (a) to semi-minor (b) axes is almost exactly 2. The deviation is normally less than 20%, although there may also be quite large apparent differences between the intensity patterns for different grains. Hence this model was used as an algorithm optimization target. First, the (nonideal) $L \times L$ square tiles would appear to have to fit inside the circular maximal rotationally invariant subset of the ellipse, so that $\sqrt{2}L = 2b$, i.e., $L = \sqrt{2}b$. This value should be compared with the larger value $L = (4/\sqrt{5})b$ that could be used if the grains were constrained to lie parallel to the image x-axis—see Fig. 21.8(b) (here we are ignoring the dimensions $2\sqrt{2}b \times \sqrt{2}b$ for optimal *rectangular* sampling tiles).

Another consequence of the difference in shape of the objects being detected (here ellipses) and the tile shape (square) is that the objects may be detected at several sample locations, thereby wasting computation (see Section 21.4.1). A further consequence of this is that we cannot merely count the samples if we wish to count the objects: instead we must relate the multiple object counts together and find the centers of the objects. This also applies if the main goal is to locate the objects for inspection and scrutiny. In the present case, the objects are convex, so we only have to look along the line joining any pair of sampling points to determine whether there is a break and thus whether they correspond to more than one object. We shall return later to the problem of systematic location of object centers.

For ellipses, it is relevant to know how many sample points could give positive indications for any one object. Now the maximum distance between one sampling point and another on an ellipse is $2a$, and for the given eccentricity this is equal to $4b$, which in turn is equal to $2\sqrt{2}L$. Thus, an ellipse of this eccentricity

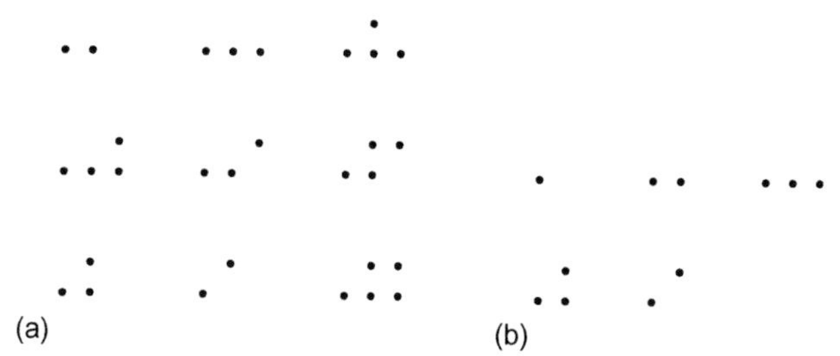

(a)   (b)

FIGURE 21.9

Possible arrangements of positive sampling points for ellipse, (a) with $L = \sqrt{2}b$ and (b) with $L = (4/\sqrt{5})b$.

Source: © EURASIP 1998

could overlap three sample points along the x-axis direction if it were aligned along this direction; alternatively, it could overlap just two sample points along the $45°$ direction if it were aligned along this direction, although it could in that case also overlap just one laterally placed sample point. In an intermediate direction (e.g., at an angle of around arctan 0.5 to the image x-axis), the ellipse could overlap up to five points. Similarly, it is easy to see that the minimum number of positive sample points per ellipse is two. The possible arrangements of positive sample points are presented in Fig. 21.9(a).

Fortunately, the above approach to sampling is over-rigorous. Specifically, we have insisted upon the sampling tile being contained within the ideal (circular) maximal rotationally invariant subset of the shape. However, what is required is that the sampling tile must be of such a size that all possible orientations of the shape are allowed for. In the present example, the limiting case that must be allowed for occurs when the ellipse is orientated parallel to the x-axis, and it must be arranged that it can just pass through four sampling points at the corners of a square so that on any infinitesimal displacement, at least one sampling point is contained within it. For this to be possible, it can be shown that $L = (4/\sqrt{5})b$, the same situation as already depicted in Fig. 21.8(b). This leads to the possible arrangements of positive sampling points shown in Fig. 21.9(b)—a distinct reduction in the average number of positive sampling points, which leads to useful savings in computation (the average number of positive sampling points per ellipse is reduced from ~ 3 to ~ 2).

Object location normally takes considerable computation because it involves an unconstrained search over the whole image space, and in addition there is normally (as in the ellipse location task) the problem that the orientation is unknown. This contrasts with the other crucial aspect of inspection, that of object scrutiny and measurement, in that relatively few pixels have to be examined in detail, requiring relatively little computation. Clearly, the sampling approach outlined above largely eliminates the search aspect of object location, since it quickly eliminates any large tracts of blank background. Nevertheless, there is still the problem of refining the object location phase. One way of approaching this problem is to expand the

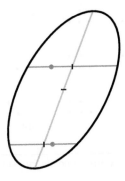

FIGURE 21.10

Illustration of triple bisection algorithm. The round spots are the sampling points, and the short bars are the midpoints of the three chords, the short horizontal bar being at the center of the ellipse.

Source: © EURASIP 1998

positive samples into fuller regions of interest and then to perform a restricted search over these regions. For this purpose we could use the same search tools that we might use over the whole image if sampling were not being performed. However, the preliminary sampling technique is so fast that this approach would not take full advantage of its speed. Instead we could use the following procedure.

For each positive sample, draw a horizontal chord to the boundary of the object, and find the local boundary tangents. Then use the chord–tangent technique (join of tangent intersection to midpoint of chord: see Chapter 12) to determine one line on which the center of an ellipse must lie. Repeat this for the all positive samples, and obtain all possible lines on which ellipse centers must lie. Finally, deduce the possible ellipse center locations, and check each of them in detail in case some correspond to false alarms arising from objects that are close together rather than from genuine self-consistent ellipses. Note that in cases where there is a single positive sampling point, another positive sampling point has to be found (say $L/2$ away from the first).

In fact, a significantly faster approach called the *triple bisection* algorithm was developed (Davies, 1998). Draw horizontal (or vertical) chords through adjacent vertically (or horizontally) separated pairs of positive samples, bisect them, join, and extend the bisector lines, and finally find the midpoints of these bisectors (Fig. 21.10). (In cases where there is a single positive sampling point, another positive sampling point has to be found, say $L/2$ away from the first.) The triple bisection algorithm has the additional advantage of not requiring estimates of tangent directions to be made at the ends of chords, which can prove inaccurate when objects are somewhat fuzzy, as in many grain images. The result of applying this technique to an image containing mostly well separated grains is shown in Figure 21.11: this illustrates that the whole procedure for locating grains by modeling them as ellipses and searching for them by sampling and chord bisection

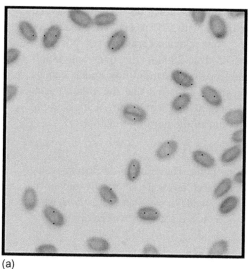 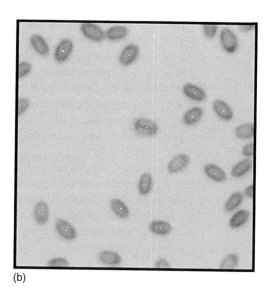

(a)   (b)

FIGURE 21.11

Image showing grain location using the sampling approach. (a) Sampling points. (b) Final
center locations.

Source: © EURASIP 1998

approaches is a viable one. In addition, the procedure is very fast, as the number of
pixels that are visited is a small proportion of the total number in each image.

Finally, we show why the triple bisection algorithm presented above is appropriate. First, note that it is correct for a circle, for reasons of symmetry. Second, note
that in orthographic projection, circles become ellipses, straight lines remain straight
lines, parallel lines remain parallel lines, chords remain chords, and midpoints
remain midpoints. Hence, choosing the right orthogonal projection to transform the
circle into a correctly orientated ellipse of appropriate eccentricity, the midpoints
and center location shown in the diagram of Figure 21.10 must be validly marked.
This proves the algorithm. (For a rigorous algebraic proof, see Davies, 1999b.)

21.4.3 Summary

This case study has studied sampling strategies for the rapid location of objects in
digital images. Motivated by the success of an earlier line-based sampling strategy
(Davies, 1987f), it has shown that point samples lead to the minimum computational effort when the 180°-rotated object shapes form a perfect tiling of the image
space. In practice imperfect tilings have to be used, but these can be extremely
efficient, especially when the image intensity patterns permit thresholding, the
images are sparsely populated with objects, and the latter are convex in shape. An
important feature of the approach is that detection speed is *improved* for larger
objects, although exact location involves some additional effort. In the case of
ellipses, the latter process is considerably aided by the triple bisection algorithm.

The method has been applied successfully to the location of well separated cereal grains, which can be modeled as ellipses with 2:1 aspect ratio, ready for scrutiny to assess damage, varietal purity or other relevant parameters.

In a more recent development (Davies, 2007c, 2008a), the line- and point-based sampling approaches have been analyzed to determine how much can be learnt by making the particular sampling tests. It turns out that the greatest gain in certainty (amount that can be learnt) is obtained by employing a logarithmic estimate of information and seeking a maximum of the entropy function:

$$E = - [P \log P + (1 - P) \log (1 - P)] \qquad (21.9)$$

where P is the chance of hitting an object at any point. The result of this theory is that a point-based solution will be optimal when object positions are totally unknown, whereas when their positions are already fairly well known—and *a fortiori* when knowledge of their positions is being refined—line-based solutions will be optimal.

21.5 OPTIMIZING THE OUTPUT FOR SETS OF DIRECTIONAL TEMPLATE MASKS

Several earlier sections of this chapter highlighted the value of low-level operations based on template masks, but did not enter into all the details of their design. In particular, they did not tackle the problem of how to calculate the signal from a set of n masks of various orientations—a situation exemplified by the case of corner detection, for which eight masks are frequently used in a 3×3 window (see Section 21.2). The standard approach is to take the maximum of the n responses, although clearly this represents a lower bound on the signal magnitude λ and gives only a crude indication of the orientation of the feature. Applying the geometry shown in Fig. 21.12, the true value clearly has components:

$$c_1 = \lambda \cos \alpha \qquad (21.10)$$

$$c_2 = \lambda \cos \beta \qquad (21.11)$$

Finding the ratio between the components yields the first important result:

$$\frac{c_2}{c_1} = \frac{\cos \beta}{\cos \alpha} = \cos (\gamma - \alpha) \sec \alpha \qquad (21.12)$$

$$= \cos \gamma + \sin \gamma \tan \alpha$$

and leads to a formula for $\tan \alpha$:

$$\tan \alpha = \frac{(c_2/c_1 - \cos \gamma)}{\sin \gamma} \qquad (21.13)$$

Hence:

$$\alpha = \arctan \left[\left(\frac{c_2}{c_1} \right) \operatorname{cosec} \gamma - \cot \gamma \right] \qquad (21.14)$$

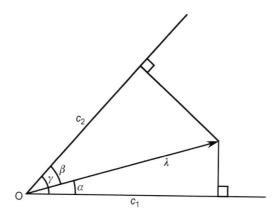

FIGURE 21.12

Geometry for vector calculation.

Source: © IEE 2000

Next, we find from Eq. (21.10):

$$\lambda = c_1 \sec \alpha \tag{21.15}$$

and elimination of α gives:

$$\lambda = c_1 \left[1 + \left(\frac{c_2/c_1 - \cos \gamma}{\sin \gamma} \right)^2 \right]^{1/2} \tag{21.16}$$

On expansion, we finally obtain the required symmetric formula:

$$\lambda = (c_1^2 + c_2^2 - 2c_1 c_2 \cos \gamma)^{1/2} \operatorname{cosec} \gamma \tag{21.17}$$

The closeness to the form of the cosine rule is striking and can be explained by simple geometry (Davies, 2000b).

Interestingly, Eqs. (21.17) and (21.14) are generalizations of the standard vector results for edge detection:

$$\lambda = (c_1^2 + c_2^2)^{1/2} \tag{21.18}$$

$$\alpha = \arctan \left(\frac{c_2}{c_1} \right) \tag{21.19}$$

which apply when $\gamma = 90°$. Specifically, the results apply for oblique axes where the relative orientation is γ.

21.5.1 Application of the Formulae

To proceed further, we define features that have $2\pi/m$ rotation invariance as m-vectors. It follows that edges are 1-vectors, and line segments (following the discussion in Section 21.3.1) are 2-vectors. In addition, it would appear that m-vectors with $m > 2$ are rarely important in practice.

So far we have established that the formulae given in Section 21.5 apply for 1-vectors such as edges. However, to apply the results to 2-vectors such as line segments, the 1:2 relation between the angles in real and quadrature manifolds has to be acknowledged. Thus, the proper formulae are obtained by replacing α, β, γ, respectively by 2α, 2β, 2γ in the earlier equations. This leads to the components:

$$c_1 = \lambda \cos 2\alpha \tag{21.20}$$

$$c_2 = \lambda \cos 2\beta \tag{21.21}$$

with $2\gamma = 90°$ replacing $\gamma = 45°$ in the formula giving the final orientation:

$$2\alpha = \arctan\left[\left(\frac{\cos 2\beta}{\cos 2\alpha}\right)\csc 90° - \cot 90°\right]$$

$$= \arctan\left[\left(\frac{\cos 2\beta}{\cos 2\alpha}\right)\right] \tag{21.22}$$

Hence:

$$\alpha = \frac{1}{2}\arctan\left(\frac{c_2}{c_1}\right) \tag{21.23}$$

as found in Section 21.3.1.

Finally, we apply the methods described above to obtain interpolation formulae for sets of 8 masks for both 1-vectors and 2-vectors. For 1-vectors (such as edges), 8-mask sets have $\gamma = 45°$, and Eqs. (21.17) and (21.14) take the form:

$$\lambda = \sqrt{2}(c_1^2 + c_2^2 - \sqrt{2}c_1c_2)^{1/2} \tag{21.24}$$

$$\alpha = \arctan\left[\sqrt{2}\left(\frac{c_2}{c_1}\right) - 1\right] \tag{21.25}$$

For 2-vectors (such as line segments), 8-mask sets have $\gamma = 22.5°$, so $2\gamma = 45°$, and the relevant equations are:

$$\lambda = \sqrt{2}(c_1^2 + c_2^2 - \sqrt{2}c_1c_2)^{1/2} \tag{21.26}$$

$$\alpha = \frac{1}{2}\arctan\left[\sqrt{2}\left(\frac{c_2}{c_1}\right) - 1\right] \tag{21.27}$$

21.5.2 Discussion

This section has examined how the responses to directional template matching masks can be interpolated to give optimum signals. It has arrived at formulae that cover the cases of 1-vector and 2-vector features, typified respectively by edge and line segments. While it appears that there are few likely instances of *m*-vectors with $m > 2$, features can easily be tested for rotation invariance, and it turns out that corners and most other features should be classed as 1-vectors. An exception to this is a symmetrical S-shape, which should be classed as a 2-vector. Support for classifying corners as 1-vectors arises as they form a subset of the edges in an image. These considerations mean that the solutions arrived at for 1-vectors and 2-vectors will solve all the foreseeable problems of signal interpolation between responses for masks of different orientations. Thus the main aim of this section is accomplished, to find a useful nonarbitrary solution to determining how to calculate combined responses for *n*-mask operators.

Oddly, these useful results (Davies, 2000b) do not seem to have been reported in the earlier literature of the subject.

21.6 CONCLUDING REMARKS

This chapter has covered three main case studies relating to cereal grain inspection. This topic is quite specialized: not only is it just one aspect of automated visual inspection, but also it covers just one sector of food processing. Yet its study has taken these topics to the limits of the capabilities of a number of algorithms and some useful theory has been developed and applied: this has made the chapter considerably more generic than might *a priori* have been thought. Indeed, this is a major reason for the rather strange inclusion of a section on template matching using sets of directional masks later on in the chapter: it happened to be needed to take the subject onward. This also emphasizes that there is often the need for more theory than is actually available in the literature. Another aspect of the situation is that there is a tendency for inspection and other algorithms to be developed *ad hoc*, whereas there is a need for solid theory to underpin this sort of work and above all to ensure that any techniques that are used are effective, robust, and close to optimal.

Food inspection is an important aspect of automated visual inspection. This chapter has shown how bulk cereals may be scrutinized for insects, rodent droppings, and other contaminants. The application is subject to highly challenging speed constraints and special sampling techniques had to be developed to attain the ultimate speeds of object location.

21.7 BIBLIOGRAPHICAL AND HISTORICAL NOTES

Food inspection is beset with problems of variability, the expression "Like as two peas in a pod" giving a totally erroneous indication of the situation. The author summarized the position in his book (Davies, 2000a) and later reviews (Davies,

2001a, 2003b, 2009). Graves and Batchelor (2003) edited a volume that collects much further data on the problems of variability.

The work on insect detection described in this chapter concentrates on the linear feature detector approach, which was developed in a series of papers already cited but founded on theory described in Davies (1997e, 1999e) and Davies et al. (2002, 2003a, 2003b). (For other relevant work on linear feature detection but not motivated by insect detection, see Chauduri et al. (1989), Spann et al. (1989), Koller et al. (1995), and Jang and Hong (1998).) Zayas and Flinn (1998) offer an alternative strategy based on discriminant analysis, but this targets the lesser grain borer beetle, which has a different shape, and the data and methods are not really comparable. That different methods are needed to detect different types and shapes of insect is obvious, and indeed, the author's own work shows that morphological methods are more useful for detecting large insects, as well as rodent droppings and ergot (Davies et al., 2003b; Ridgway et al., 2002). In fact, the issue is partly one of sensitivity of detection for the smaller insects *versus* functionality for the larger ones, coupled with speed of processing—optimization issues typical of those discussed in Chapter 27.

Work in this area also covers NIR detection of insect larvae growing within wheat kernels (Davies et al., 2003c), although the methodology is totally different, being centered on the location of bright (at NIR wavelengths) patches on the surfaces of the grains.

The fast processing issue arises again in respect of the inspection of wheat grains, not least because the 30 tonnes of wheat that constitute a typical lorry load contain some 6000 million grains, and the turnaround time for each lorry can be as little as 3 min. The theory for fast processing by image sampling and the subsequent rapid centering of elliptical shapes is developed in Davies (1997d, 1999b, 1999d, 2001b): see Davies (2007c, 2008a) for an integrated analysis in which entropy is used to optimize sampling when object locations vary from totally unknown to approximately known.

Sensitive feature matching is another aspect of the work described in this chapter. Relevant theory was developed by Davies (2000b), but there are other issues such as the "equal area rule" for designing template masks (Davies, 1999a) and the effect of foreground and background occlusion on feature matching (Davies, 1999c): these theories should all have been developed far earlier in the history of the subject and merely serve to show how little is still known about the basic design rules for image processing and analysis. A summary of much of this work appears in Davies (2000d).

21.7.1 More Recent Developments

Recent work in the cereals and grain industry has included investigations aimed at separating touching grain kernels (Zhang et al., 2005), classifying cereal grains (Choudhary et al., 2008), differentiating wheat classes (Mahesh et al., 2008), detecting sprouted wheat grains (Neethirajan et al., 2007), sorting grains by color

(Bayram and Öner, 2006), detecting the creases in wheat grains (Sun et al., 2007), and detecting insects inside wheat kernels (Manickavasagan et al., 2008). The different approaches have involved near infrared, thermal imaging, X-ray, and hyperspectral modalities. While the methods have employed a number of relatively standard image analysis techniques—ellipse fitting, wavelets, morphology, color analysis, and stereo vision—the work is far from trivial, because the various varieties of grain and their defects and contaminants involve high levels of variation: this means that getting the techniques to work well is anything but straightforward.

22

Surveillance

Surveillance is nowadays used widely in transport and civil centers for monitoring traffic and people, and is increasingly being carried out by computers. The motivation is largely to locate instances of undesirable behavior—theft, loitering with intent, speeding, and so on. What is special about surveillance is the rate at which pictorial information is delivered, and the fact that many of the objects being monitored are in motion. To cope with this, there is considerable emphasis on identification and elimination of the background and effective tracking of moving objects. At the same time algorithms need to be fast in operation, though some help can be obtained with fast dedicated hardware systems.

Look out for:

- the geometry of surveillance.
- the need to separate foreground from background.
- the basics of particle filters and their use for tracking.
- the use of color histograms for tracking.
- chamfer matching and its use for identification and tracking.
- how multiple cameras are used to obtain coverage over wide areas.
- systems for monitoring traffic flow.
- identification of the ground plane as an early stage in the analysis of many types of scene incorporating motion.
- the need for "occlusion reasoning" when objects repeatedly pass behind one another and then re-emerge.
- the importance of the Kalman filter in motion applications.
- license plate location.
- how studies of the motions of complex objects may have to take into account 3-D articulated models of linked parts.
- basic concepts of human gait analysis.
- animal tracking.

While this chapter covers the situation of static cameras being used to monitor moving objects, the following chapter covers the more complex case of in-vehicle vision systems, where moving cameras are used to monitor both stationary and moving objects.

"'T is Cinna; I do know him by his gait"

(William Shakespeare, 1599)

22.1 INTRODUCTION

Visual surveillance is a long-standing area of computer vision, and one of its main early uses was to obtain information on military activities—whether from high flying aircraft or from satellites. However, with the advent of ever cheaper video cameras, it subsequently became widely used for monitoring road traffic, and most recently it has become ubiquitous for monitoring pedestrians. In fact, its application has actually become much wider than this, the aim being to locate criminals or people acting suspiciously—for example, wandering around car-parks with the potential purpose of theft. However, by far the majority of visual surveillance cameras are connected to video recorders and gather miles of videotape, most of which will never be looked at—though, following criminal or other activity, some hours of videotape may be scanned for relevant events. Further cameras will be attached to closed circuit television monitors where human operators may be able to extract some fraction of the events displayed, though human attentiveness and reliability when overseeing a dozen or so screens will not be high. Clearly, it would be far better if video cameras could be connected to automatic computer vision monitoring systems, which would call human operators' attention to potential hazards or misdemeanors of various types. Even if this were not carried out in real time in specific applications, it would be useful if it could be achieved at high speed with selected videotapes: this could save huge amounts of police time in locating and identifying perpetrators of crime.

Surveillance can cover other useful activities, including riot control, monitoring of crowds on football pitches, checking for overcrowding on underground stations, and generally helping with safety as well as crime. To some extent, human privacy must suffer when surveillance is called into play, and there is clearly a tradeoff between privacy and security: suffice it to say here that many would be happier to have increased levels of security, a small loss of privacy being a welcome price to pay to achieve it.

In fact, there are many difficulties to be solved before the "people tracking" aspects of surveillance are fully solved. First, in comparison to cars, people are articulated objects that change shape markedly as they move: that their motion is often largely periodic can help visual analysis, though the irregularities in human motion may be considerable—especially if obstacles have to be avoided. Second, human motions are partly self-occluding, one leg regularly disappearing behind another, while arms can similarly disappear from view. Third, people vary in size and apparent shape, having a variety of clothes that can disguise their outlines. Fourth, when pedestrians are observed on a pavement, or on the underground, there is some possibility of losing track when one person passes behind another, as the two outlines tend to coalesce before re-emerging from the combined object shape.

It could be said that all these problems have been solved. However, many of the algorithms that have been applied to these tasks have limited intelligence: indeed, some employ rather simplified algorithms, as the need to operate continuously in real time generally overrides the need for absolute accuracy. In any case, given the visual data that the computer actually receives, it is doubtful whether a human operator could always guarantee making correct interpretations: for example, there are occasions when humans turn around in their tracks because they have forgotten something, and this could cause confusion when trying to track every person in a complex scene. Further complexities can be caused by varying illumination, fixed shadows from buildings, moving shadows from clouds or vehicles, and so on.

In the following sections we cover two main areas of surveillance—those in which people or pedestrians are the prime targets, and those in which vehicles are the prime targets. Of course, there are many transport scenarios where both would be observed by the same systems. In addition, similar techniques and considerations would apply in both cases. The next section, on the geometry underlying camera positioning, talks mainly about pedestrians: this is done for definiteness, though most of the considerations apply equally when vehicles are the prime targets, as happens on motorways, for example.

22.2 SURVEILLANCE—THE BASIC GEOMETRY

Perhaps the most obvious way of monitoring pedestrians is indicated in Fig. 22.1(a). As we have seen in Chapter 15, this leads to the following relations between real-world (X, Y, Z) and image coordinates (x, y):

$$x = \frac{fX}{Z} \tag{22.1}$$

$$y = \frac{fY}{Z} \tag{22.2}$$

Here, Z is the (horizontal) depth in the scene, X represents lateral position, Y represents vertical position (downward from the camera axis), and f is the focal length of the camera lens. This method of observation is useful in providing undistorted profiles of pedestrians from which they may be recognized. However, it provides virtually no information about depth in the scene beyond what can be deduced from knowledge of the pedestrian's size; and as size may be one of the key parameters to be determined by the vision system, this is an unsatisfactory situation. Note also that this view of the scene is subject to gross occlusion of one pedestrian by another.

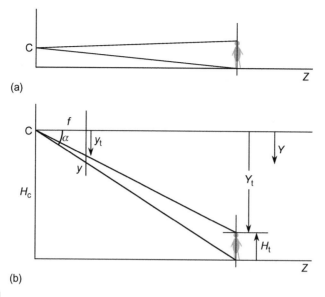

FIGURE 22.1

3-D monitoring: camera axis horizontal. (a) Camera axis mounted at eye-level. (b) Camera mounted higher up to obtain a less restricted view.

To overcome these problems, an overhead view would be better. However, it is difficult to obtain views from directly overhead; in any case, any one view would give a highly restricted range, and again pedestrian height could not be measured. An alternative approach is to place the camera in Fig. 22.1(a) higher up, as shown in Fig. 22.1(b), so that the positions of the feet of any pedestrian on the ground plane can be seen: this makes it possible to obtain a reasonable esti-mate of depth in the scene. In fact, if the camera is at a height H_c above the ground plane, Eq. (22.2) gives the depth Z as:

$$Z = \frac{fH_c}{y} \tag{22.3}$$

while the modified value of y at the top of the pedestrian is given by:

$$y_t = \frac{fY_t}{Z} = \frac{yY_t}{H_c} \tag{22.4}$$

The height of the pedestrian H_t can now be estimated from the following equation:

$$H_t = H_c - Y_t = H_c(1 - y_t/y) \tag{22.5}$$

Note that to achieve this, H_c must be known from prior on-site measurements, or alternatively by camera calibration using test objects.

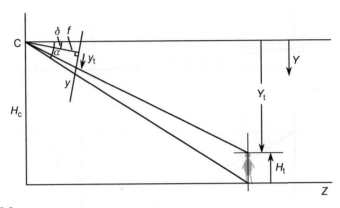

FIGURE 22.2

3-D monitoring: camera tilted downwards. δ is the angle of declination of the camera optical axis.

In practice, it is better to modify the above scheme by tilting the optical axis of the camera slightly downward (see Fig. 22.2), as this allows the range of observation to be increased, and particularly for nearby pedestrians to be kept in view. However, the geometry of the situation becomes somewhat more complicated, leading to the following basic formulae:

$$\tan \alpha = \frac{H_c}{Z} \qquad (22.6)$$

$$\tan (\alpha - \delta) = \frac{y}{f} \qquad (22.7)$$

where δ is the angle of declination of the camera. Substituting for $\tan(\alpha-\delta)$ using the formula:

$$\tan (\alpha - \delta) = \frac{(\tan \alpha - \tan \delta)}{(1 + \tan \alpha \tan \delta)} \qquad (22.8)$$

and using the above equations to eliminate α, we obtain the following formula for Z in terms of y:

$$Z = H_c \frac{(f - y \tan \delta)}{(y + f \tan \delta)} \qquad (22.9)$$

So far we have not allowed for the heights of any objects, but have only considered points on the ground plane. To estimate the heights of pedestrians, we need to bring in the additional equation:

$$Z = Y_t \frac{(f - y_t \tan \delta)}{(y_t + f \tan \delta)} \qquad (22.10)$$

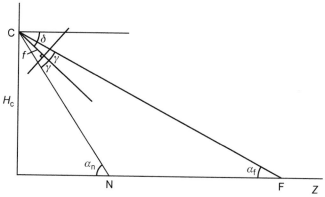

FIGURE 22.3

Geometry for considering optimum camera tilt. δ is the angle of declination of the camera optical axis. 2γ is the overall vertical field of view of the camera.

which is simply derived by substituting Y_t for H_c and y_t for y in Eq. (22.9). Eliminating Z between these two equations now allows us to find Y_t:

$$Y_t = H_c \frac{(f - y \tan \delta)(y_t + f \tan \delta)}{(y + f \tan \delta)(f - y_t \tan \delta)} \qquad (22.11)$$

thereby permitting $H_t = H_c - Y_t$ to be calculated in this case too.

Next, we consider the optimum value for the angle of declination δ of the camera optical axis. We assume that the viewing range of the camera has to vary from a near point given by Z_n to a far point given by Z_f, corresponding to respective values of α, α_n, and α_f (Fig. 22.3). We also assume that the overall vertical field of view of the camera is 2γ. This immediately results in the following formulae:

$$\frac{H_c}{Z_n} = \tan \alpha_n = \tan (\delta + \gamma) \qquad (22.12)$$

$$\frac{H_c}{Z_f} = \tan \alpha_f = \tan (\delta - \gamma) \qquad (22.13)$$

Taking the ratio between these equations now shows that:

$$\eta = \frac{Z_n}{Z_f} = \frac{\tan (\delta - \gamma)}{\tan (\delta + \gamma)} \qquad (22.14)$$

so specifying either Z_n or Z_f immediately gives the alternate value. In the case that Z_f is taken to be infinity, Eq. (22.13) shows that δ has to be equal to γ, in which case Eq. (22.12) leads to the relation $Z_n = H_c \cot 2\gamma$. Note that $\delta = \gamma = 45°$ is a limiting case that covers all points on the ground plane, i.e. $Z_n = 0$ and

$Z_f = \infty$. For smaller values of γ, values of Z_n and Z_f are determined by δ, e.g. for $\gamma = 30°$, the optimum value of η (namely, zero) occurs both at $\delta = 30°$ and at $\delta = 60°$, and the worst case ($\eta \approx 0.072$) occurs at $\delta = 45°$.

Finally, it is instructive to consider the minimum separation Z_s that is needed between pedestrians if they are not to occlude each other at all. Equating $\tan \alpha$ to both H_t/Z_s and H_c/Z (see Eq. (22.6)), we find:

$$Z_s = \frac{H_t Z}{H_c} \tag{22.15}$$

As might have been expected, this varies inversely with camera height; but note that it is also proportional to Z.

Overall, we have seen that placing the camera high up permits both depth and height to be estimated and the incidence of occlusion to be considerably reduced. In addition, tilting the camera downward permits the maximum range to be achieved. Importantly, two cameras placed at the far ends of a courtyard should be able to cover it quite well. Pedestrians can be identified as having a particular position on the ground plane, though they could then be recognized pictorially from knowledge of their size, shape, and coloring. The formulae that are involved reflect all the complications of perspective projection, and some are quite complex. Note that even in the simple case of Fig. 22.1(b), the inverse relation between y and Z is highly nonlinear (see Eq. (22.3)), and equal intervals in the Z direction by no means correspond to equal vertical intervals in the image plane: see Section 17.8 for further theory underpinning this point.

22.3 FOREGROUND—BACKGROUND SEPARATION

One of the first problems of surveillance is to locate the targets that are to be placed under observation. In principle, we could follow all the recognition methods of earlier chapters (see also Chapter 24), and just proceed to recognize the targets individually. However, there are two reasons why we should approach this differently. First, cars moving along a road, or pedestrians in a precinct, are highly variegated, unlike the situation for products appearing on a product line. Second, there is usually a significant real-time problem, especially when vehicles are moving at up to 100 mph on a highway, and cameras typically deliver 30 frames per second under highly variable conditions. Thus, it pays to capitalize on the motion of the targets and perform motion-based segmentation.

In these circumstances, it is natural to think of frame differencing and optical flow. Indeed, frame differencing has been applied to this task, but it is prone to noise problems and consequent unreliability. In any case, when applied between adjacent frames, it locates only limited sections of target outlines—in accordance with the $-\nabla I.\mathbf{v}$ formula of Chapter 19. The simplest way out of this difficulty is that of background modeling.

22.3.1 Background Modeling

The idea of background modeling is to create an idealized background image that can be subtracted from any frame to yield the target or foreground image. To achieve this, the simplest strategy is to take a frame when there are known to be no targets present and use that as the background model. In addition, to eliminate noise, it is useful to average a number of frames prior to making observations of targets. The problems with this strategy are: (a) How to know when there are no targets present, so that frames represent true background? (b) How to cope with the usual outdoor situation of illumination that varies with the weather and the time of day?

To solve the latter problem, only the most recent frames can reasonably be used, and if this path is followed, it is difficult to tackle the former problem (in any case, on highways with a lot of traffic, or precincts with a continuous mêlée of people, there may seldom be a chance of obtaining a clear background frame). One compromise solution is to take an average of many background frames over the most recent period Δt, whether or not targets are present. If targets are reasonably rare, most of the frames will be clear and a good approximation to an ideal background model will be built: of course, any targets will not be eliminated so much as averaged in, and the result will sometimes be visible "tails" in the model. To optimize the model, Δt can be increased, thereby minimizing problem (a), or decreased, thereby minimizing problem (b). Clearly, there is a tradeoff between the two difficulties: while it can be adjusted to suit the time of day and prevailing weather and illumination levels, this approach is limited.

Part of the problem is due to the "averaging in" mentioned above, and this can be partially eliminated by using a temporal median filter. Note that this means applying a median filter to the $\mathbf{I}$ (intensity and color) values of each pixel, over the sequence of frames arising during the most recent period Δt. This is a computation intensive process, but is considerably better than taking a raw average, as mentioned above. It is effective to take the median because it eliminates outliers, but ultimately it will still lead to biased estimates. In particular, if we suppose that vehicles are on the whole darker than the road, then the temporal median will also tend to end up darker than the road. To overcome this problem, a temporal mode filter can be used, and hopefully, the intensity distribution will have separate modes—one from the road and one from the vehicles, so the former can be used, and could probably be identified even if it became a minor mode when there were a lot of vehicles. However, there is no guarantee that there would only be one mode for vehicles, or even that any such modes would be clearly separated from the one corresponding to the road, and bias would again be the most likely result. Figures 22.4–22.6 illustrate some of the problems.

In fact, there are significant further problems with background modeling. In many situations the background objects are themselves subject to motion. In particular, shadows will move with time and their crispness will change with the weather; while leaves, branches of trees, and flags will flutter and sway in the wind, with highly variable frequencies; even the camera may sway, especially if it is mounted on a pole,

(a)   (b)

(c)   (d)

FIGURE 22.4

Background subtraction using a temporal median filter. The lines of black graphics dots demarcate the relevant road region: almost all of the fluttering vegetation lies outside this region (it is indicated by fainter boundaries than for the foreground objects: see (b)). Note the plethora of stationary shadows that are completely eliminated during the process of background subtraction. The stationary bus is progressively eaten away in (a) and (b), while in (c) and (d) the ghost of the bus appears and then starts to merge back into the background. These problems are largely eliminated in Fig. 22.5, which includes the same four frames.

but we defer that type of problem until Chapter 23. Motions of small animals and birds may also have to be considered. At this point we shall concentrate on fluttering vegetation, which is often prevalent in outdoor scenarios, even within cities.

The fluttering of vegetation can be more serious than might at first be imagined. It can result in the **I** values of some pixels oscillating between those of leaves, branches, and sky (or ground, buildings, etc.). Thus, the distributions of intensities and colors for any pixel may best be regarded as the superposition of several distributions corresponding to two or three component sources. Here what

(a)

(b)

(c)

(d)

(e)

(f)

FIGURE 22.5

Background subtraction using a restrained temporal median filter. This figure shows a much more comprehensive set of frames than Fig. 22.4 because the method is more accurate. In particular, its responses (d, e, g, h) to the bus problems of Fig. 22.4 are vastly improved. Fluttering vegetation problems are indicated by fainter boundaries than for vehicles, but are entirely absent from the road region. In all frames the stationary shadows are completely eliminated by background subtraction: even the prominent bridge shadow is ignored; neither does it have much effect on the integrity of foreground objects. Note the low false negative rate for vehicles, and the fact that they only tend to be joined together in the distance. Overall, foreground object fragmentation and false shapes (including the effects of moving shadows) are the worst problems.

FIGURE 22.5

(Continued)

(a)   (b)

FIGURE 22.6

Problems arising immediately after background subtraction. These two frames show clearly the noise problems that arise during background subtraction: the white pixels indicate where the current image fails to closely match the background model. Most of the noise effects occur for fluttering vegetation outside the road region. Morphological operations (see text) are used to largely eliminate the noise and to integrate the vehicle shapes as far as possible, as shown in Figs. 22.4 and 22.5.

is important is that each of the component distributions could be quite narrow and well defined. This means that if each is known from ongoing training, any current intensity **I** can be checked to determine whether it is likely to correspond to background. If not, it has to correspond to a new foreground object.

Models formed from multiple component distributions are commonly called mixture models: in practice, the component distributions are approximated by Gaussians, because the odd shape of the overall distributions is largely attributable to the existence of the separate components. Thus, we arrive at the terms Gaussian mixture models (GMMs) and mixtures of Gaussians (MoG). Note that the number of components at any pixel is initially unknown; indeed, a large proportion of pixels will have only a single component, and it may seem unlikely that the number would be much larger than three in practice. However, the fact that every pixel will have to be analyzed to determine its GMM is computationally burdensome, while the analysis can be unstable if the component distributions are not as tidy as suggested above. These factors mean that a computation intensive algorithm, the expectation maximization (EM) algorithm, has to be used to analyze the situation. In fact, while it is usual to use this rigorous approach to *initialize* the background generation process, many workers use simpler more efficient techniques for updating it, so that the ongoing process can proceed in real time. The GMM method determines for itself the number of component distributions to use, the judgment being based on a threshold value for the fraction of the total weight given to the background model.

Unfortunately, the GMM approach fails when the background has very high frequency variations. Essentially, this is because the algorithm has to cope with rapidly varying distributions that change dramatically over very short periods of time, so the statistics become too poorly defined. To tackle this problem, Elgammal et al. (2000) moved away from the parametric approach of the GMM (the latter essentially finds the weights and variances of the component distributions, and thus is parametric). Their nonparametric method involves taking a kernel smoothing function (typically a Gaussian) and for each pixel, applying it to the N samples of $\mathbf{I}$ for frames appearing during the period Δt prior to the current time t. This approach is able to rapidly adapt to jumps from one intensity value to another, while at the same time obtaining the local variances at each pixel. Thus, its value lies in its capability to quickly forget old intensities and to reflect local variances rather than random intensity jumps. In addition, it is a probabilistic approach, but has no need of the EM algorithm, and this enables it to run highly efficiently in real time. In addition, it is capable of sensitive detection of foreground objects coupled with low false alarm rates. To achieve all this, it incorporates two further features:

1. It assumes independence between three different color channels, each having its own kernel bandwidth (variance). Together with the adoption of a Gaussian kernel function, this leads to a probability estimate given by:

$$P(\mathbf{I}) = \frac{1}{N} \sum_{i=1}^{N} \prod_{j=1}^{C} \frac{1}{(2\pi\sigma_j^2)^{1/2}} \; e^{-(I_j - I_{j,i})^2 / 2\sigma_j^2} \qquad (22.16)$$

   where i runs over the N samples taken in the time period Δt and j runs over the C color channels; this function is simple to calculate, though computation is further speeded up by using pre-calculated kernel function lookup tables.
2. It uses chromaticity coordinates for suppressing shadows. As these coordinates are independent of the level of illumination, and shadows can be regarded as poorly lit background, this means that they should largely merge back into the background. Hence, the foreground is much less likely to have shadows accompanying it after background subtraction. The chromaticity coordinates r, g, b are derived from the usual R, G, B coordinates by equations $r = R/(R + G + B)$, and so on, with $r + g + b = 1$.

In fact, shadows can be particularly problematic: not only do they distort the apparent shapes of foreground objects after background subtraction, but also they can connect separate foreground objects, and thus cause under-segmentation. The problem is reviewed by Prati et al. (2003), while Xu et al. (2005) have proposed a hybrid shadow removal method that makes use of morphology; see also Guan (2010). Whatever method is used for background modeling leading to background subtraction and hence to foreground detection, the various blobs will need to be clustered and labeled using connected components analysis. Frame-to-frame tracking is then carried out by making correspondences between the blobs in the different frames. As in the case of Xu et al. (2005), morphology can be used to help with this process. Nevertheless, false positives tend to arise because of shadows and illumination effects, while false negatives can arise from color similarities between foreground and background.

Overall, failures arise in two categories: (1) *the stationary background problem*, in which the shape of the foreground object is not defined accurately enough; (2) *the transient background problem*, in which the start and stop of the foreground object aren't found quickly enough. If the accuracy or reactivity of the background model are inadequate, background subtraction will lead to the detection of false objects: these are called "ghosts" by Cucchiara et al. (2003). In addition, as indicated above, shadows tend to compound these problems.

22.3.2 Practical Examples of Background Modeling

To add concreteness to the above discussion, a traffic surveillance video was taken and submitted to some of the algorithms mentioned above. For illustrative purposes the algorithms were kept as simple as possible. The raw data consisted of an AVI video from a digital camera (Canon Ixus 850 IS), which was decompiled into individual JPG frames, and though the JPG artifacts were fairly severe, no specific attempt was made to eliminate them. The frame size was 320×240 pixels in RGB color, but only the 8-bit lightness component was used for the main tests. While the video was taken at 15 frames per second, only every tenth frame was used for the test, which comprised 113 frames. Of these, the first 10 can be regarded as initialization training material and these are not considered further. During the test, a bus arrived and was stationary at a bus stop for some time. The overall sequence is illustrated in Fig. 22.5; however, for reasons of space the only frames included in the figures are those that illustrate the problems well. Note that the video was taken on a sunny day, and that there are a great many shadows, which over the minute or so of the video did not change markedly. On the other hand, some camera motion is detectable, possibly due to movement of the bridge. Thus, there are many ways in which the raw data were not ideal; these therefore impose exacting conditions on the success of any algorithm.

Figure 22.4 shows some of the results obtained by applying a temporal median. Frames (a) and (b) show the bus stationary at the bus stop and being progressively eaten up as it starts to merge with the background. Frame (c) shows the bus moving away from the bus stop, leaving a large "ghost" behind it. Frame (d) shows that the ghost remains for some time and is a substantial factor to be taken into account by any foreground interpretation procedure.

To overcome these problems, the median was restrained so that it could only take into account pixel intensities within a limited number of gray levels of the current median value; in this way, it took on something of the characteristics of a mode filter (a temporal mode filter *per se* would lock on to the current value too inflexibly and not adapt well to the changing intensity distribution). The results are shown in Fig. 22.5. It is clear that the restrained median largely eliminates the two problems mentioned above (*viz.* the observed vehicle being eaten away while stationary and leaving a ghost behind it when moving on). For this reason the remainder of the tests used only the restrained median. Problems seen in Fig. 22.5 include:

1. Eating away of foreground objects, leaving unusual shapes (e.g. (d), (i)).
2. Fragmentation of foreground objects (e.g. (b), (f)).
3. Shadows accompanying the moving foreground objects (e.g. (c), (g), (j)).
4. Joining of foreground objects that should appear separated (e.g. (i), (j)).
5. Signals from fluttering vegetation (e.g. (a), (k)).

Item 2 can be considered as an extreme case of item 1. Item 3 is bound to arise as the shadows are moving at the same speed as the vehicles that give rise to them, and straightforward background suppression or alternatively moving object detection alone will not eliminate them. In general, unless color interpretation will help (we return to this possibility below), high-level interpretation is needed to achieve satisfactory elimination. Item 4 is due partly to the effects of vehicle shadows, which tend to connect vehicles, especially when seen in the distance. The morphological operations that were applied (see below) also tended to make vehicles become joined. Item 5 is never manifest in the road region, i.e. between the lines of black graphics dots shown in the frames. This is because, in this case, the vegetation is high up, away from the road region. In addition, it is largely eliminated by morphological operations. In fact, Fig. 22.6 shows the results obtained immediately after background subtraction. It is clear that there is a serious noise problem, caused largely by (a) camera noise, (b) the effects of JPG artifacts, (c) fluttering vegetation, and (d) the effects of slight camera motion. Interestingly, two applications of a single pixel erosion operation were sufficient to eliminate the noise almost completely, these being followed by four applications of a single dilation operation to help restore vehicle shapes. (Overall, this corresponds to a 2-pixel opening operation followed by a 2-pixel dilation operation.) These morphological operations were selected to give roughly optimal results—in particular, low probability of failing to capture foreground objects in any individual frame, coupled with pressure to maintain object shapes as far as reasonable, and not to join vehicles together unavoidably. The point is that background subtraction must aim to pass on sufficient useful information to subsequent foreground object identification, tracking, and interpretation stages.

One of the remarkable aspects of the results is the total elimination of stationary shadows and lack of problems arising from this. On the other hand, two other sorts of shadows are manifest—those arising from moving objects, and those falling on moving objects (the latter arise in the video both from the bridge and from the other causes of ground shadows): neither of these sorts of shadows are eliminated. Other problems are those of reflections, particularly from the windows of the bus (see the frame in Fig. 22.5(g)), and secondary illumination from moving vehicles.

Lastly, it was felt worthwhile to attempt utilizing the original color images and augmenting the background model by using the chromaticity coordinates as outlined in the previous section. In fact, while in some respects improvements occurred, these were more than canceled out by increased numbers of false negatives and fragmented shapes of the foreground objects. No results are shown here, but whereas Elgammal et al. (2000) were able to show excellent results obtained in this way, with the video used here no improvement seemed to be achievable by this approach. This requires some explanation. High up among the reasons is the effect of the highly variegated colors and intensities of the many different vehicles. In particular, some vehicles turned out to have body intensities close to those of shadows, whereas others had windows or transparent roofs of similar intensity. These had the effect of eliminating large portions of vehicles together with the shadows, thereby increasing the incidence of false negative and false shape information. However, even Elgammal et al. (2000) point out that intensities have to be used carefully in a way that will bolster up the shadow removing capabilities of the chromaticity information, and here there appeared to be no way this could be achieved. Overall, all the color and grayscale information has to be taken into account in a more considered and strategic way, and this demands a thoroughgoing statistical pattern recognition approach in which objects are identified one by one in a high-level schema rather than by relatively chancy *ad hoc* methods: the latter definitely have their place but their use must not be pushed beyond what is reasonable. One example of the use of object-by-object recognition is the identification of road markings, which need to be, and could easily be, identified whether or not vehicle shadows cover them. This could then lead to a much more viable strategy for identifying, tracking, and eliminating vehicle shadows. Meanwhile, in situations where the road has almost no color content, as in the traffic surveillance trials described above, it is difficult to remove shadows effectively merely by using chromaticity information.

22.3.3 Direct Detection of the Foreground

In the previous subsection it has been seen that background modeling followed by background subtraction constitutes a powerful strategy for the location of moving targets in image sequences. Nevertheless, for all sorts of reasons it is limited in what it can achieve. While these reasons devolve into problems such as changes in ambient illumination, effects of shadows, irrelevant motions such as fluttering leaves, and color similarities between foreground and background, there is one whole tranche of information that is absent; specifically, there is a total lack of information on the nature of the target objects, including size, shape, location, orientation, color, speed, and probability of occurrence. If this sort of information were obtainable, there would be some chance of incorporating it into a complete target detection system and achieving close to perfect detection capability. Indeed, it seems possible that in some cases ignoring the background and attempting direct detection of the foreground might be a better first approximation. Such a procedure might well be both effective and efficient for the case of face detection, for example. In what follows we consider how direct foreground detection might be achieved.

Direct foreground detection is only possible if a suitable foreground model is available or can be constructed. It would seem that this requires a specialization to each particular application, such as pedestrian detection or vehicle detection. However, some workers (e.g. Khan and Shah, 2000) have managed to achieve it more generically by a bootstrapping process. They start with background modeling and background subtraction, locate foreground objects by an "exception to background" procedure, and thus create initial foreground models. In subsequent frames these are enhanced, mostly using Gaussian-based models: GMMs and non-parametric models have been employed for this. However, a difference relative to background modeling is that the latter applies continuously (with updates) for the same camera, whereas each foreground object must have its own individual model that is learnt anew for that object. So, background modeling is only applied to initially locate the foreground object: thereafter, the foreground model is built and tracked, albeit in a similar way to what happens with background modeling.

More recently, in a new class of algorithm, Yu et al. (2007) use a GMM for simultaneously modeling both foreground and background. In this way, a tension is built between foreground and background that potentially leads to higher segmentation accuracy, and this does seem to have been achieved in practice. Against this, the algorithm has to be initialized by marking areas of definite foreground and background, and then it continues to track autonomously. However, there seems to be no reason why initialization should not also be carried out autonomously with the help of an initial stage of background modeling.

22.4 PARTICLE FILTERS

When trying to track foreground objects, independent detection in each frame, followed by appropriate linking, does not make best use of available information; neither does it achieve optimum sensitivity: this will be obvious when noting that averaging slowly moving objects over a number of frames can boost signal-to-noise ratio. In addition, over time—and sometimes over very few frames—objects can change radically in appearance, so tracking is needed in order to ensure continued capture. Nowhere is this more obvious than in the case of a guided missile approaching a target over several miles, as during the time of flight the size, scale, and the resolution will increase dramatically. But even in cases where a person is being tracked, rotation of the head will present if anything even more dramatic changes in appearance. With the radically changing backgrounds arising with moving and rotating objects, sensitive robust tracking is clearly of fundamental importance. To achieve this, optimal methods are needed. In particular, in the face of radical change, we need to know what is the most *likely* position of an object that is being tracked. Optimal estimation of likelihood implies the need for Bayesian filtering.

To achieve this, we start by considering the observations $\mathbf{z}_1$ to $\mathbf{z}_k$ of an object in successive frames, and the corresponding deduced states of the object $\mathbf{x}_0$ to $\mathbf{x}_k$ (there is no zeroth value of $\mathbf{z}$ because it takes at least two frames to estimate the velocity $\mathbf{v}_k$, which forms part of the state information). At each stage, we need to estimate the most probable state of the object, and Bayes rule gives us the *a posteriori* probability density:[1]

$$p(\mathbf{x}_{k+1}|\mathbf{z}_{1:k+1}) = \frac{p(\mathbf{z}_{k+1}|\mathbf{x}_{k+1})p(\mathbf{x}_{k+1}|\mathbf{z}_{1:k})}{p(\mathbf{z}_{k+1}|\mathbf{z}_{1:k})} \tag{22.17}$$

where the normalizing constant is:

$$p(\mathbf{z}_{k+1}|\mathbf{z}_{1:k}) = \int p(\mathbf{z}_{k+1}|\mathbf{x}_{k+1})p(\mathbf{x}_{k+1}|\mathbf{z}_{1:k})d\mathbf{x}_{k+1} \tag{22.18}$$

The prior density is obtained from the previous time-step:

$$p(\mathbf{x}_{k+1}|\mathbf{z}_{1:k}) = \int p(\mathbf{x}_{k+1}|\mathbf{x}_k)p(\mathbf{x}_k|\mathbf{z}_{1:k})d\mathbf{x}_k \tag{22.19}$$

but note that this is only valid because of the Markov process (of order one) assumption commonly taken to simplify Bayesian analysis, which leads to:

$$p(\mathbf{x}_{k+1}|\mathbf{x}_k, \mathbf{z}_{1:k}) = p(\mathbf{x}_{k+1}|\mathbf{x}_k) \tag{22.20}$$

In other words, the transition probability for the update $\mathbf{x}_k \rightarrow \mathbf{x}_{k+1}$ depends only indirectly on $\mathbf{z}_{1:k}$, via previous updates.

General solutions of this set of equations—in particular, Eqs. (22.17) and (22.19)—do not exist. However, restricted solutions are possible, as in the case of the Kalman filter (see Chapter 19), which assumes that all posterior densities are Gaussian. In addition, particle filters can be used to approximate the optimal Bayesian solution when Gaussian constraints are inapplicable.

The particle filter, also known as sequential importance sampling (SIS), the sequential Monte Carlo approach, bootstrap filtering and condensation, is a recursive (iteratively applied) Bayesian approach that at each stage employs a set of samples of the posterior density function. It is an attractive concept because in the limit of large numbers of samples (or "particles"), the filter is known to approach the optimal Bayesian estimate (Arulampalam et al., 2002).

To apply this method, the posterior density is reformulated as a sum of delta function samples:

$$p(\mathbf{x}_k|\mathbf{z}_{1:k}) \approx \sum_{i=1}^{N} w_k^i \delta(\mathbf{x}_k - \mathbf{x}_k^i) \tag{22.21}$$

---

[1]It is much easier to see the relation to Bayes rule if conditional dependence on $\mathbf{z}_{1:k}$ is eliminated; once this is done, all remaining subscripts are equal to $k+1$, and suppressing them, Eqs. (22.17) and (22.18) become standard Bayes rule. Reinstating dependence on $\mathbf{z}_{1:k}$ is of course necessary when dealing with tracking over $k+1$ frames involving previous observations $\mathbf{z}_{1:k}$.

where the weights are normalized by:

$$\sum_{i=1}^{N} w_k^i = 1 \qquad (22.22)$$

Substituting into Eqs. (22.17)–(22.19), we obtain the posterior:

$$p(\mathbf{x}_{k+1}|\mathbf{z}_{1:k+1}) \propto p(\mathbf{z}_{k+1}|\mathbf{x}_{k+1}) \sum_{i=1}^{N} w_k^i p(\mathbf{x}_{k+1}|\mathbf{x}_k^i) \qquad (22.23)$$

where the prior now takes the form of a mixture of N components.

In principle, this gives us a discrete weighted approximation to the true posterior density. In fact, it is often difficult to sample directly from the posterior density: this problem is normally solved by sequential importance sampling (SIS) from a suitable "proposal" density function $q(\mathbf{x}_{0:k}|\mathbf{z}_{1:k})$. It is useful to take an importance density function that can be factorized:

$$q(\mathbf{x}_{0:k+1}|\mathbf{z}_{1:k+1}) = q(\mathbf{x}_{k+1}|\mathbf{x}_{0:k}\mathbf{z}_{1:k+1})q(\mathbf{x}_{0:k}|\mathbf{z}_{1:k}) \qquad (22.24)$$

following which, the weight update equation can be obtained (Arulampalam et al., 2002) in the form:

$$
\begin{aligned}
w_{k+1}^i &= w_k^i \frac{p(\mathbf{z}_{k+1}|\mathbf{x}_{k+1}^i)p(\mathbf{x}_{k+1}^i|\mathbf{x}_k^i)}{q(\mathbf{x}_{k+1}^i|\mathbf{x}_{0:k}^i, z_{1:k+1})} \\
&= w_k^i \frac{p(\mathbf{z}_{k+1}|\mathbf{x}_{k+1}^i)p(\mathbf{x}_{k+1}^i|\mathbf{x}_k^i)}{q(\mathbf{x}_{k+1}^i|\mathbf{x}_k^i, \mathbf{z}_{k+1})}
\end{aligned}
\qquad (22.25)
$$

where the path $\mathbf{x}_{0:k}^i$ and history of observations $\mathbf{z}_{1:k}$ have been eliminated—as this is necessary if the particle filter is to be able to track recursively in a manageable way.

In fact, pure SIS has the largely unavoidable problem that all but one particle will have negligible weight after a few iterations. More precisely, the variance of the importance weights is only able to increase over time, leading ineluctably to this degeneracy problem. However, one simple means of limiting the problem is to resample particles so that those with small weights are eliminated, while those with large weights are enhanced by duplication. Duplication can be implemented relatively easily, but it also leads to so-called "sample impoverishment," i.e. it still results in some loss of diversity among the particles, which is itself a form of degeneracy. Nevertheless, if there is sufficient process noise, the result may prove to be adequate.

One basic algorithm for performing the resampling is "systematic resampling," and involves taking the cumulative discrete probability distribution (in which the original delta function samples are integrated into a series of steps) and subjecting it to uniform cuts over the range 0–1 to find appropriate indexes for the new samples. As seen in Fig. 22.7, this leads to small samples being eliminated and strong samples being duplicated, possibly several times. The result is called

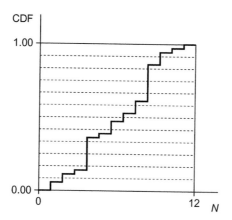

FIGURE 22.7

Use of the cumulative distribution function (CDF) to perform systematic resampling. Applying the regularly spaced horizontal sampling lines shows the cuts needed to find appropriate indexes (N) for the new samples. The cuts tend to ignore the small steps in the CDF and to accentuate the large steps by duplicating samples.

sampling importance resampling (SIR), and is a useful first step on the way to producing stable sets of samples. With this particular approach, the importance density is chosen to be the prior density:

$$q(\mathbf{x}_{k+1}|\mathbf{x}_k^i, \mathbf{z}_{k+1}) = p(\mathbf{x}_{k+1}|\mathbf{x}_k^i) \tag{22.26}$$

Appealing to Eq. (22.25) shows that the weight update equation becomes enormously simplified to:

$$w_{k+1}^i = w_k^i p(\mathbf{z}_{k+1}|\mathbf{x}_{k+1}^i) \tag{22.27}$$

Moreover, as resampling is applied at every time index, previous weights w_k^i are all given the value $1/N$, so we can simplify this equation to:

$$w_{k+1}^i \propto p(\mathbf{z}_{k+1}|\mathbf{x}_{k+1}^i) \tag{22.28}$$

As can be seen in Eq. (22.26), the importance density is taken to be independent of measurement $\mathbf{z}_{k+1}$, so the algorithm is restricted with regard to observational evidence, and this is one cause of the loss of particle diversity mentioned earlier.

The Condensation method of Isard and Blake (1996) goes some way to eliminating these problems by following the resampling with a prediction phase during which a diffusion process separates any duplicated samples, thereby helping to maintain sample diversity. This is achieved by applying a stochastic dynamical model that has been trained on sample object motions. Figure 22.8 gives an overall perspective on the approach, and includes all the sampling and other processes that have been discussed above.

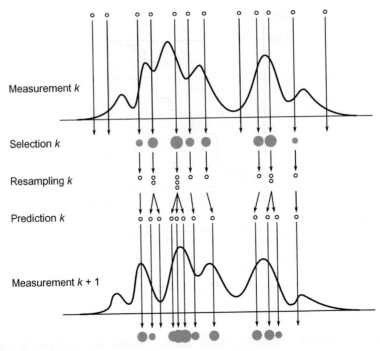

Measurement k

Selection k

Resampling k

Prediction k

Measurement $k + 1$

FIGURE 22.8

Perspective on the processes involved in particle filtering. Note how the filter cycles repeatedly through the same basic sequence.

The concept is taken further in the condensation approach (Isard and Blake, 1998) by using a mixture of samples, some using standard SIR and some using an importance function depending on the most recent measurement $\mathbf{z}_{k+1}$ but ignoring the dynamics. Thus, this complex method reflects the need to ensure continued sample diversity: it also aims to combine low and high-level approaches to tracking by noting that model switching may be necessary when handling real-world tasks such as tracking human hands.

Similar ideas and motivation were employed by Pitt and Shephard (1999) in their auxiliary particle filter (APF). This generates particles from an importance distribution depending on the most recent observations, and then samples the posterior using this importance density. The algorithm involves an additional likelihood computation for each particle, but overall the computational efficiency is improved because fewer particles are needed. Nevertheless, Nait-Charif and McKenna (2004) found that the method gave only limited improvement relative to SIR. They went on to make a comparison with the iterated likelihood weighting (ILW) scheme. In this approach, after an initial iteration of SIR, the sample set is split randomly into two sets of equal size; one of these is migrated to regions of high likelihood and the other is handled normally. The purpose is to cope on the one hand with situations where the prior is sound and on the other hand with situations where it is not and regions of high likelihood need to be explored. When tracking human heads, the method proved to be a significantly more robust tracker than either SIR or the APF. Perhaps oddly, the ILW is designed to reduce approximation error rather than to give unbiased estimates of a posterior. This means that it is not based completely on probabilistic methods. On the other hand, as for the Isard and Blake condensation approach mentioned above, it is intended to match a variety of scenarios that can be found when tracking under real-world conditions, where it is difficult to model all probabilities accurately.

Many more particle filter methods have been developed over the past decade or so. Several incorporate the Kalman filter and its extended and "unscented" versions, in attempts to optimize likelihoods when sample diversity turns out to be insufficient. More recently "regularized" and "kernel" particle filters (Schmidt et al., 2006) have been developed to tackle the problem of sample impoverishment. These perform resampling using a continuous approximation to the posterior density, typically using the Epanechnikov kernel (Comaniciu and Meer, 2002). The mean shift approach is in this category. Essentially, the mean shift algorithm is a means of climbing density gradients to identify underlying modes in sparse distributions, and involves moving a sampling sphere around the space being searched. This makes it a good iterative search technique, though it is only suitable for locating one mode at a time. It complements the particle filter formalism well, as it can be used to refine the accuracy with which objects may be found, and works well even if a limited number of particles are employed. It has recently been applied by Chang and Lin (2010) for tracking various parts of the moving human body.

At this stage it is starting to become apparent that different tracking applications will demand different types of particle filter. This will depend on a variety of factors including how jerky the motion is, whether rotations will be involved, whether occlusions will occur and if so for how long—and of course on the appearance and variability of the objects being tracked. It should be noted that all the theory and most of the ideas presented above reflect abstract situations and the concentration is on relatively small, well localized objects that are considered locally, i.e. the filters themselves will not have a global understanding of the situation. Thus, they must be categorized as low or intermediate level vision. In contrast, the human eye is an excellent tracker by virtue of its capability for thinking about what objects are present and which ones have moved where, including passing behind other objects or even temporarily out of the scene. Clearly, we must not expect too much from particle filters just because they are based on probabilistic models.

An important advantage of particle filters is that they can be used to track multiple objects in an image sequence. This is because there is no record of which object is being tracked by which particles. However, this possibility arises only because no restrictions are placed on the posterior densities; in particular, they are not assumed to be Gaussian as in the case of Kalman filters. Indeed, if Kalman filters are used for tracking, each object must be tracked by its own Kalman filter.

Once a suitable approach to particle filtering has been arrived at, it is necessary to determine how to implement it. The basic means of achieving this is via appearance models. In particular, color and shape models are frequently used for this purpose. However, before delving into this topic, it will be useful to find what can be achieved by color analysis using what is by now quite an old approach—that of color indexing via color histogram matching.

22.5 USE OF COLOR HISTOGRAMS FOR TRACKING

One of the most useful tools that is available for object tracking was developed as early as 1991 by Swain and Ballard (1991) in a paper called "Color indexing." The aim of that work was to index from a color image into a large database of models. In a sense that idea is the inverse of the tracking problem, as its purpose was to search for the model with the best match to a given image rather than to search for instances of a given model in frames from an image sequence.[2] However, apart from one important difference which will be discussed below, this is a rather minor point.

The main idea behind the color indexing approach is that of matching color histograms rather than the images themselves. There is an obvious validity in this approach, in that if the images match, so will their color histograms. What is more, as the histograms have no memory of where in the image a particular color originated, histograms are invariant to translation and rotation about the viewing axis (so-called "in-plane rotation"). Also relevant is the fact that a planar object that is subject to out of plane rotation will still have the same color histogram, although it will involve different numbers of pixels, so normalization will be required. The same applies to objects that are at different depths in the scene: the histogram profile will be unchanged, but it will have to be normalized to allow for the different numbers of pixels that are involved. Finally, a spherical or cylindrical object with the same set of colors distributed similarly over its surface will again have the same histogram. While exact adherence to this scenario might be relatively rare, it would apply almost perfectly for a ball of wool or a football, and with varying degrees of exactness for a shaven human head or torso. In fact, the main problem with use of histograms for recognition is the possible ambiguity it could bring, but when tracking a known object that has moved only a small distance between frames, this problem should be a minor one.

The above explanation demonstrates the potential power of the histogram approach, but raises an important question about what happens when the object moves in such a way as to be larger or smaller than the model, whether through

---

[2]Actually, database searches fall in the realm of classification, whereas the process of tracking assumes implicitly that the object in question has already been identified.

depth scaling or through out of plane rotation. In particular, if it becomes smaller, this will mean that the model will be matched partly against the object background. Swain and Ballard sought to minimize this effect by taking the following intersection measure rather than any sort of correlation between the image I and model M histograms:

$$\sum_{i=1}^{n} \min(I_i, M_i) \tag{22.29}$$

This would have the effect of discounting any pixels of a given color in excess of those expected in the model histogram (including both those whose colors are simply not represented in the model and those that have limited representation). The above expression was then normalized by the number of pixels in the model histogram. However, here we follow Birchfield (1998) in normalizing by the number of pixels in the image histogram, to reflect the point made earlier that we are searching for the best image match rather than the best model match:

$$H_N(I, M) = \frac{\sum_{i=1}^{n} \min(I_i, M_i)}{\sum_{i=1}^{n} I_i} \tag{22.30}$$

At first sight, this formula might appear wrong, in that a match over fewer pixels would be normalized out, still representing perfect agreement and giving a normalized intersection of unity. Note for example that in shape matching, it is common to use the formula $(A \cap B)/(A \cup B)$, where A and B are sets representing object areas, which would give a value less than unity in the case $A \supset B$. However, Eq. (22.30) is designed to cope well with partial occlusions in the image, which will lead to the intersection with M being reduced, yielding the I values, which would then cancel with the denominator, giving the answer 1.

The overall effect of using the normalized intersection of Eq. (22.30) is that the method has the twin advantages of minimizing the effects of background and canceling the effects of occlusion, while also coping well (and in some cases exactly) with varying viewpoints. The problem of varying scale remains, but this can be countered by preliminary segmentation of the object and scaling its histogram to the size of the model histogram.

There remains one further important consideration when matching an image against a model—that the model might have become out of date under varying levels of illumination. To a large extent, the latter can be considered as varying levels of *luminance*, with the *chrominance* parameters remaining more or less unchanged. This problem can be addressed by changing to different color representations. For example, we can move from the RGB representation to the HSI representation (see Chapter 20), and then use the hue (H) and saturation (S) parameters. However, more protection will be available by color normalization (dividing by I)—though it is far easier and less computation intensive to normalize the RGB parameters directly:

$$r = \frac{R}{(R+G+B)} \qquad (22.31)$$

$$g = \frac{G}{(R+G+B)} \qquad (22.32)$$

$$b = \frac{B}{(R+G+B)} \qquad (22.33)$$

but because $r + g + b = 1$, we should ignore one of the parameters, e.g. b.

While the above arguments suggest that luminance should be totally ignored, this is inadvisable, as colors that are close to the black–white line in color space (where saturation $S \approx 0$) would be indistinguishable. Indeed, Birchfield (1998) cites the "dangerous" case of dark brown hair looking similar to a white wall if luminance is ignored. For these reasons, most workers use different sizes and numbers of histogram bins for luminance and chrominance information. Here, we have to remember that a full-sized color histogram with 256 bins in each of the color dimensions would be both large and clumsy, and would not easily be searchable in real time—an especially important factor in tracking applications. In addition, such a histogram would not be well populated and would lead to very noisy statistics. For this reason, 16–40 bins per color dimension are much more typical. In particular, $16 \times 16 \times 8$ bins are widely used, 8 being the number in the luminance channel. Note that these numbers correspond, respectively, to 16, 16 and 32 levels per channel, and that a 512×512 image would lead to an average occupation number of 128 per bin; however, a 256×256 image would give an average occupation of just 32 per bin, which is distinctly low and liable to be inaccurate (though this would depend very much on the type of data).

Birchfield (1998) reported that when used for head tracking, the color histogram method was able to follow a head reliably, though it became "unstable" when the head was in front of a white board whose color was quite close to that of skin. This behavior is understandable, as it has already been noted that the histogram approach is invariant to translation (hence, as long as the head is somewhere within the image, the histogram tracker will be unlikely to lose it). These points show that ultimately the histogram tracker approach is limited, and needs to be enhanced by other means, in particular some means of detecting object outlines. To achieve this, Fieguth and Terzopoulos (1997) used (1) simple M-ary hypothesis testing of position around the previous position, the displacements merely being those at the $M = 9$ points in a 3×3 window; (2) a highly nonlinear velocity prediction scheme involving step incremental corrections for acceleration, deceleration, and damping to avoid oscillations; and (3) color histograms bins based exclusively on chrominance. The reason for these simplifications was to achieve real-time operation for full frames (640 × 480 pixels) at 30 frames per second—at which rate object displacements become much smaller and easier to track.

Birchfield (1998) developed a more sophisticated approach, based on approximating the shape of the human head by a vertical ellipse with a fixed aspect ratio of 1:2. Then, in common with previous contour trackers he measured the goodness of match by computing the normalized sum of gradient magnitudes around the boundary of the ellipse, though (a) he summed the gradient values at all points on the boundary rather than just at selected points,[3] and (b) he took the component of the gradient along the perpendicular to the boundary. This led to a shape model $s(x, y, \sigma)$ with three parameters—x, y denoting the ellipse location and σ denoting its semi-minor axis—and the following goodness of fit parameter:

$$\psi(\mathbf{s}) = \frac{1}{N_\sigma} \sum_{i=1}^{N_\sigma} \left| \mathbf{n}_\sigma(i) \cdot \mathbf{g}_s(i) \right| \qquad (22.34)$$

Here, N_σ is the number of pixels on the boundary of an ellipse with semi-minor axis σ, $\mathbf{n}_\sigma(i)$ is the unit vector normal to the ellipse at pixel i, and $\mathbf{g}_s(i)$ is the local intensity gradient vector. Normalized goodness of fit parameters for boundary shape (ψ_b) and color (ψ_c) are added and used to obtain an optimum fit:

$$\mathbf{s}_{opt} = \arg \max_{\mathbf{s}_i} \{ \psi_b(\mathbf{s}_i) + \psi_c(\mathbf{s}_i) \} \qquad (22.35)$$

As discussed earlier, when the color module was tested on its own, it performed well, but somewhat unstably when the background color was close to that of skin. All this was corrected by adding the gradient module. However, the gradient module on its own performed less well than the color module on its own. As time progressed the gradient model tended to became distracted by the background, not having any inbuilt design features to counteract this. Moreover, in a cluttered background it behaved even less well; and it was not able to handle large accelerations adequately because of its limited ability to probe for high gradient regions, and its consequent propensity for attaching itself to the wrong ones. Fortunately, the two modules were able to complement each other's capabilities; in particular, the color module helped the gradient module by its ability to ignore background clutter, and by providing a larger region of attraction. Finally, when the human subject turned around, so that only his hair was visible, the gradient module was able to take over and handle rescaling correctly as the subject moved; and it was able to prevent the color tracker from slipping down the subject's neck, which had a similar color histogram. All this signals that two or more strategies for tracking can be useful in real-world situations where enough

---

[3]In fact, this is unusual: most workers sample at a set of 100 or so points on the boundary.

information needs to be brought to bear to provide correct tracking interpretations on an ongoing basis. It also signals that the color histogram type of tracking module is exceptionally powerful, and tends to need only minor tweaks to keep it properly on lock. Overall, however, the outstanding factor that needs further detailed attention and development is the handling of occlusion: this needs to be arranged by design rather than by tweaks, as we shall see in a later section.

22.6 IMPLEMENTATION OF PARTICLE FILTERS

The particle filter formalism is a very powerful one, mediated by generic probability-based optimization, yet needing to be taken further to achieve its promise. To arrange this, it needs to be applied to real objects and thus appearance models have to be taken into account—though in the present context it will be more accurate and general to refer to them as observation models. Our particle filter formalism already embodies these in the form of conditional densities $p(\mathbf{z}_k|\mathbf{x}_k)$, see Eq. (22.28).

At this stage we need to specialize the observations. Here, we illustrate the process by considering the color and assumed elliptical shape of a human head: these can be thought of as region-based (r) and boundary-based (b) properties, each with their own likelihoods. Taking the latter to be conditionally independent, we can factorize $p(\mathbf{z}_k|\mathbf{x}_k)$ as follows:

$$p(\mathbf{z}_k|\mathbf{x}_k) = p(\mathbf{z}_k^r|\mathbf{x}_k)p(\mathbf{z}_k^b|\mathbf{x}_k) \tag{22.36}$$

Clearly, the region-based likelihood will depend not only on the color but also on the shape of the region it is in. Nevertheless, the conditional independence assumption will be valid, as we are really interested in the colors within the boundary and the gradient values along it.

To proceed further, we assume that color histograms I and M have been obtained for the image and the target model, within the current region r. Following Nummiaro et al. (2003) and many other workers—and at this point abandoning the Swain and Ballard (1991) normalized intersection formalism—we normalize them to unity as p^I, p^M, respectively. To compare these distributions, it is convenient to use the Bhattacharyya coefficient (here expressed as a sum rather than an integral) that expresses the similarity between the distributions:

$$\rho(p^I, p^M) = \sum_{i=1}^{m} \sqrt{p_i^I p_i^M} \tag{22.37}$$

To display the *distance* between the distributions, we simply apply the measure:

$$d = \sqrt{1 - \rho(p^I, p^M)} \tag{22.38}$$

Ideally, the color distribution will be close to the target distribution, so these should differ only as a Gaussian error function. Remembering that p^I is actually a function of $\mathbf{x}_k$, we now find the region (and color) conditional likelihood:

$$p(\mathbf{z}_k^r|\mathbf{x}_k) = \frac{1}{(2\pi\sigma_r^2)^{1/2}} e^{-(d^2/2\sigma_r^2)} = \frac{1}{(2\pi\sigma_r^2)^{1/2}} e^{-[(1-\rho(p^I(\mathbf{x}_k),p^M))/2\sigma_r^2]} \tag{22.39}$$

Making a similar assumption that the estimated gradient positions in the image I will differ from those in the target model M by a Gaussian error function, we find the boundary conditional likelihood:

$$p(\mathbf{z}_k^b|\mathbf{x}_k) = \frac{1}{(2\pi\sigma_b^2)^{1/2}} e^{-(G^2/2\sigma_b^2)} \tag{22.40}$$

where G represents the sum of the gradient magnitude values perpendicular to the local boundary positions.

Combining the last two equations, as specified by Eq. (22.36), now provides the required estimate of $p(\mathbf{z}_k|\mathbf{x}_k)$, which in turn leads via the particle filter formulation to an estimate of $p(\mathbf{x}_k|\mathbf{z}_{1:k})$. This essentially completes the long series of arguments and calculations comprising the particle filter scenario.

In fact, there are several further aspects to consider. The first is that it is natural to weight the contributions made by the various pixels to the color histograms. In particular, the pixels nearest to the centers of the ellipses should be weighted higher than those near their boundaries, so that any inaccuracies in the center locations will be minimized. For example, Nummiaro et al. (2003) used the weighting function:

$$k(r) = \begin{cases} 1 - \dfrac{r^2}{r_0^2} & : r < r_0 \\ 0 & : r \geq r_0 \end{cases} \tag{22.41}$$

with $r_0 = \sqrt{a^2 + b^2}$, a and b being the semi-major and semi-minor axes of the ellipses. In fact, Nummiaro et al. (2003) placed such reliance on this weighting that their particle filter did not use a separate boundary likelihood $p(\mathbf{z}_k^b|\mathbf{x}_k)$. In contrast, Zhang et al. (2006) used both, almost exactly as described above, albeit with an auxiliary particle filter incorporating mean shift filtering.

Another important aspect not so far mentioned is the need to adapt the target model M to keep it up to date, e.g. with regard to the size and orientation of the real-world target. Nummiaro et al. (2003) achieved this using the commonly applied "learning/forgetting" operation:

$$p_{k+1,i}^M = \alpha p_{k,i}^M + (1 - \alpha) p_{k,i}^I, \quad i = 1, 2, \ldots, m \tag{22.42}$$

which mixes in a little of the recent image data while forgetting a correspondingly small amount of the old model data. During this process care is taken to avoid mixing in outlier data, such as when an object is partly occluded. Even with this precaution, it should be borne in mind that use of an adaptive model is potentially dangerous; while it helps by valid adaptation to appearance changes, it gives an increased sensitivity to extended occlusions and loss of target.

While heads are typically tracked using 2-D position (x, y) and ellipse shape parameters (a, b), it can normally be assumed that the ellipse is vertically aligned. However, when viewed from overhead, in-plane orientation (θ) is also an important parameter. Sometimes, similar models are used for individual human limbs, though rectangles have also been employed. However, ellipses provide a simple, easily parametrized shape, and can be specified with as few as three parameters (x, y, b); these can even be used to track whole human figures using 3 or 4 parameters (Nummiaro et al., 2003). On the other hand, when torsos or hands are being tracked, closed curves may not be appropriate, and it is common to use parametric spline curves.

With the type of particle filter design outlined above, performance in the event of occlusions is a vexed question.[4] In principle, if a significant change such as a strong partial occlusion occurs, the simple artifice of putting the tracker on hold is often sufficient to allow it to recover and continue tracking. However, to ensure recovery, the tracker might have to wait for a background subtraction routine to signal that the object is again present (Nait-Charif and McKenna, 2006). In any case, a background subtraction module is useful for signaling when a totally new object has entered the scene. Finally, when objects leave the scene, some memory of their appearance and position is useful in case they re-enter the scene after a short time in the same or other location (when humans appear indoors, there are usually a limited number of entry and exit points, and re-entry via the same one will generally be the most likely possibility). However, there is a danger of instituting a rather *ad hoc* set of algorithms to solve such problems, when what is needed is a more absolute object recognition module to positively identify individuals, or at least to search for the most likely identifications, together with sets of probabilities. A particular example of this type of situation is when two pedestrians walking in opposite directions (a) pass each other without interacting, but with the one momentarily occluding the other, or (b) stop, shake hands, and then proceed, or (c) stop, shake hands, and then retrace their steps. Scenario (c) involves merging of profiles and can be as difficult to handle as occlusion: in any case temporary partial occlusion involves merging of the figures; only seldom does complete occlusion and total disappearance of one figure occur. It ought to be stressed that scenario (a) is handled well by a Kalman filter module that uses

---

[4]In this area, many claims and counter-claims about relative effectiveness of tracking and occlusion handling capabilities are made in various papers. As the claims are often made on different datasets, it is difficult to know the true position. However, the particle filter has quite a high level of intrinsic robustness. This is because "less likely object states have a chance to temporarily remain in the tracking process, [so] particle filters can deal with short-lived occlusions" (Nummiaro et al., 2003). Hence minor propping up in a judicious way using other modules can often boost performance significantly.

continuity of velocity to aid interpretation; scenario (b) is handled badly or not at all by such a module, depending on the time delay; while scenario (c) is not handled at all by such a module.[5] In general, when processing human interactions, the Kalman filter has to be used tentatively, to throw up *hypotheses* about motion. However, it is possible to incorporate Kalman filters usefully into a particle filter (van der Merwe et al., 2000); equally, they can be incorporated into supervisory programs that oversee the whole tracking process, as indicated above (see also Comaniciu et al., 2003).

22.7 CHAMFER MATCHING, TRACKING, AND OCCLUSION

As we have seen, one of the perennial problems of matching and tracking is that of occlusion of objects within the field of view. A variety of measures can be applied to make single camera systems as robust as possible against overlap. Leibe et al. (2005) have devised methods based on chamfer matching and segmentation, together with a minimum description length procedure for hypothesis verification. The latter evaluates hypotheses in terms of the savings that can be made by explaining part of the image by the hypotheses. Here, we concentrate on the concept of chamfer matching, as it has achieved considerable use for matching pedestrians, notably by Gavrila (1998, 2000).

The basic idea behind chamfer matching relates to the process of matching objects to templates via their boundaries—a strategy that should be much less computation intensive than matching via whole object regions. However, since this would not give much indication of a potential match until very close to the match position, some means is required of making the approach to a match far smoother. This should also permit substantial speedup of the process by employing a hierarchical coarse-to-fine search. To achieve a smoother transition, edge points in the image are first located, and then a distance function image is generated, starting with the edge points, which are initialized to zero distance values. Application of the template, also in the form of edge points, will ideally yield a zero sum (of image distance function values) along the template points: this will rise to a higher value when the template is misplaced or the shape of the object is distorted, corresponding to the sum of distances of each image point from the ideal position. Taking the distance function as $DF_I(i)$, we can express the degree of match by the average "chamfer" distance, i.e. the average distance from each edge point to the nearest edge point in the template T:

$$D_{chamfer}(T, I) = \frac{1}{N_T} \sum_{i=1}^{N_T} DF_I(i) \qquad (22.43)$$

where N_T is the number of edge points within the template. $D_{chamfer}(T, I)$ is actually a dissimilarity measure, having a value of zero for a perfect match.

---

[5]These points about use of Kalman filters are well illustrated by Nummiaro et al. (2003) in relation to a quite different scenario—that of a bouncing ball.

In fact, there is no necessity to take edge points for the image and the template: corner points or other feature points can be utilized, and the method is quite general. However, the method works best when the point set is sparse, so that (a) accurate location is achieved, and (b) computation is reduced. On the other hand, reducing the number of points too far will result in lack of sensitivity and robustness as parts of the image and template will not be adequately represented.

As it stands, this approach is limited because any outliers (caused by occlusion or segmentation errors, for example) will lead to substantial matching problems. To limit this problem Leibe et al. (2005) used a truncated distance for matching:

$$D_{\text{chamfer}}(T, I) = \frac{1}{N_T} \sum_{i=1}^{N_T} \min(DF_I(i), d) \tag{22.44}$$

with a suitable empirical value of d. On the other hand, Gavrila (1998) applied an order-based method for limiting the number of interfering distance function values, taking the kth of the ordered values (1 to N_T) as the solution value:

$$D_{\text{chamfer}}(T, I) = \arg \text{order}_k^{i=1:N_T} DF_I(i) \tag{22.45}$$

When applying this formula, it may seem attractive to use the median value, for which $k = \frac{1}{2}(N_T + 1)$. However, it can easily happen that a large proportion of the template area is obscured, so we usually need to take a smaller value of k (e.g. $0.25N_T$) that reflects this. In fact, this will reduce accuracy when none of the template is obscured, so, in the end, Eq. (22.44) might give a more useful result. We take this discussion no further here as a lot depends on the type of data that is involved. In passing, it is worth observing that when $k = N_T$, Eq. (22.45) gives the well-known Hausdorff distance (Huttenlocher et al., 1993):

$$D_{\text{chamfer}}(T, I) = \max_{i=1:N_T} DF_I(i) \tag{22.46}$$

(This formula for the Hausdorff distance may appear different from the usual one that involves a max−min operation; however, as computation of a distance function involves taking local minima of possible distances (see Chapter 9), there is concurrence in the two formulations.)

Note that, in the foregoing discussion, the distance function of the image is used rather than that of the template. This is because in practical situations many templates will have to be applied in order to cover expected variations in the objects being detected. For example, if the method is being applied for pedestrian detection, various sizes, poses, positions of limbs, and types of clothing will have to be allowed for, as well as variations in the background and possible overlaps. In these circumstances it is far more efficient to use DF_I than DF_T. Gavrila (1998) showed with considerable success how all the variations listed above can be dealt with and how the method can be made to work well to detect pedestrians.

Finally, returning to the work of Leibe et al. (2005), limitations of the chamfer matching technique were compensated by using segmentation information. This meant obtaining a similarity function from the chamfer distance (which is a dis-similarity measure), and then combining with a Bhattacharyya coefficient repre-senting overlap with the hypothesized segmentation $Seg_I(i)$ to produce an overall similarity measure:

$$S = a\left[1 - \frac{1}{b}D_{\text{chamfer}}(\mathbf{T}, \mathbf{I})\right] + (1 - a)\sum_{i} \sqrt{Seg_I(i)R_T(i)} \qquad (22.47)$$

Here, $R_T(i)$ is the region within $\mathbf{T}$, and the sum covers the pixels in this region. In addition, a somewhat arbitrary but nonetheless reasonable pair of weights is applied to balance the two similarity measures: a is the proportion of the overall similarity allotted to chamfer matching, and b is a weight expressing the fact that chamfer matching is applied over a significant boundary distance; in the work of Leibe et al. (2005), a and b were taken to be 0.45 and 50, respectively. The over-all effect was to produce much improved solutions in respect of placement accu-racy and elimination of false positives, relative to the chamfer distance method taken on its own (Eq. (22.44)).

22.8 COMBINING VIEWS FROM MULTIPLE CAMERAS

Over the past decade or so there has been a surge of interest in multi-camera surveil-lance systems. Multiple cameras are clearly necessary if, e.g. long stretches of motor-way are to be monitored, or if pedestrians are to be tracked around cities or shopping precincts. The field of view (FOV) of a single camera is quite restricted and the reso-lution available for viewing in the distance will almost certainly be inadequate for detailed observation. Another reason for the use of several cameras is that of viewing in stereo and obtaining sufficient depth information. A further reason is that pedes-trians in a precinct will frequently be partially or wholly occluded by architectural features such as statues or other pedestrians, but the chance of missing a pedestrian will be much less if the scene is viewed by multiple cameras; this sort of situation will also apply on roads, where many other possibilities for occlusion exist.

On roads, cameras are often mounted on overhead gantries, and maintaining observation over long distances will require many cameras. This raises the ques-tion of whether the observation should be unbroken, i.e. whether the cameras will have overlapping, contiguous, or nonoverlapping views. On motorways, cameras may be separated by several miles, and can usefully be sited at junctions, so it will be possible to keep track of all vehicles without too much expense, though breakdowns at intermediate locations may not be observed. On the other hand, in a shopping precinct, if pedestrians are to be monitored closely enough for attacks or terrorist activities to be detected, contiguous, or overlapping views will be mandatory. In fact, there will be a problem in ensuring that all pedestrians are positively identified as they progress from one FOV to the next: to facilitate this, and for ease of setting up the system, overlapping views are normally required.

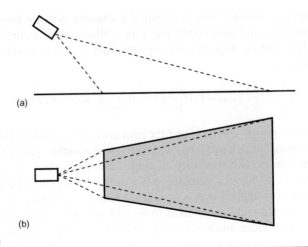

(a)

(b)

FIGURE 22.9

Area on the ground plane viewed by a camera. (a) Side view with the camera canted slightly downwards. (b) Plan view of the symmetrical trapezium seen by the camera on the ground plane.

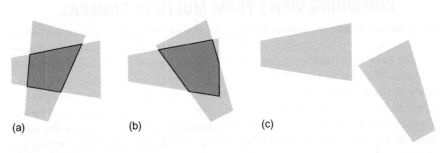

(a)   (b)   (c)

FIGURE 22.10

Areas on the ground plane viewed by multiple cameras. (a) Overlapping trapezia forming a quadrilateral. (b) Overlapping trapezia forming another shape—here a pentagon. (c) Trapezia that do not overlap, though tracking across the gap can in some cases be achieved by making spatial and temporal correspondences (see text).

Next we consider the layout of a multi-camera system. To do this we must examine the area of the ground plane that lies within the FOV of the camera. First, note that the optical axis of the camera passes through the center of the image plane and that the latter has a rectangular shape given by the minimum and maximum values of x and y, $\pm x_m$ and $\pm y_m$. The FOV is therefore limited by four planes, at horizontal and vertical angles $\pm \alpha$ and $\pm \beta$, where $\tan \alpha = x_m/f$ and $\tan \beta = y_m/f$, f being the focal length of the camera lens. Each plane will intersect with the ground plane in a line, and for a camera with a horizontal x-axis, the viewed area on the ground plane will be a symmetrical trapezium (Fig. 22.9). However, following on from the discussion in Section 22.2, if the camera is not inclined slightly downward, the distant side of the trapezium will not be visible. Since this would not make the most of the camera FOV, we will assume that it has been arranged for the distant side to fall on the ground plane.

When an adjacent camera views an adjacent section of the ground plane, there are two possibilities: (1) it will view the next stretch in the same direction, as on a motorway; (2) it will not be restricted to lie, or point, in the same direction, but just to overlap in some convenient way. For example, in a precinct or park a typical placement would be as shown in Fig. 22.10(a), where two opposite sides of the common viewing area would arise from the FOV of the first camera and the other two from that of the second camera—thereby forming a quadrilateral rather than a trapezium. However, other situations are possible, as shown in Fig. 22.10(b), where the trapezia of the two cameras overlap in a more complex way, and the common viewing area is not a quadrilateral.

No matter which of the reasons for using a multi-camera system apply, there is a need to relate the views from the separate cameras in order to obtain a consistent labeling of the objects passing between them. The obvious means of achieving this is by appearance, i.e. to apply recognition algorithms to establish that the same person or vehicle is being tracked across the various camera fields of view. Unfortunately, while this correspondence problem can normally be solved straightforwardly in binocular vision, when the two cameras are close together and pointing in a similar direction, this is by no means true for wide baseline cases such as those shown in Fig. 22.10. This is so for two reasons: (1) a person seen in two disparate views may have an altogether different appearance, e.g. the face may be visible in one and the back of the head in the other; or the back of a shirt may have a different design or color from the front; (2) the illumination may be quite different for each of the views, and this will make it even more difficult to confirm the person's identity from the other camera.

The obvious solution to this problem is to confirm identity not by appearance but by position and time. If we know that person P is at position $\mathbf{X}$ in the scene at time t, this must be the case in all views. So all that has to be done is to relate the common areas of the ground plane uniquely between cameras. Following the widely used and usually sufficiently accurate assumption that everything is happening on the same flat ground plane, we only need to set up a homography between the two cameras to arrange for the same correct interpretation from any view. Under perspective projection, it requires a minimum of four common feature points to set up a homography (the number is as small as this because of the planar constraint, as is made clear in Table 16.1), though more points can be used to improve accuracy; note also that at least one more point is needed to validate the homography.

In the work of Calderara et al. (2008) greater accuracy was achieved by finding the straight lines bounding the common quadrilateral and using its corners as highly accurate points by which to define the homography. While this might seem trivial, in fact the common quadrilateral has to be located by experiment. This can be achieved most easily when the scene is empty (e.g. overnight), and one individual can be sent to walk repeatedly around the site until a sufficient number of boundary points—as determined by the individual entering or leaving one of the fields of view—have been measured in both views. Note that to ensure that this gives sound results, temporal synchronization of the two camera systems is crucial. Once all this has been carried out, methods such as Hough transforms or RANSAC are applied to collate the boundary points into the straight lines bounding the quadrilateral: and because of the averaging inherent in this process, the straight lines will be known accurately; therefore, the corner positions will be known accurately, so there will be no need to use more points to establish an accurate homography.

Interestingly, Khan and Shah (2003) consider this approach an overkill to solve the consistent labeling problem. They assert that there is no need to determine the homography in this numerical sort of way: rather, it should be done by finding the FOV boundary lines and then merely noting when a pedestrian passes over one of these lines and making the identity at that point in time, i.e. if an individual crosses a line at time t, this will be detected at the same time t in each camera and the person's identity can be passed across at that moment. This process is commonly called camera "handoff" (whereas it might appear to be more natural to call it "handover," there is a subtlety in that the latter term would tend to imply that the fields of view are contiguous rather than overlapping). However, if a group of people all cross the line together, this could obviously give rise to difficulties. Indeed, the whole problem of tracking groups of individuals is a difficult one, and becomes almost insuperable in dense crowd situations.

While finding FOV boundary lines can be carried out when no crowds are present, and ideally when a single individual walks around, there are limits to the performance of the trained system. This is because a homography relates to a plane, and the simplest way of defining and using a plane is to use the foot locations to provide the plane contact points. (In principle, this is easily done by taking the lowest point on the individual.) However, when the calibrated system is used, the feet of one individual will often be obscured by another individual—a situation that will be virtually unavoidable in crowds. Consequently, there has been a fair amount of attention to recognizing and locating individuals from the tops of their heads (e.g. Eshel and Moses, 2008, 2010). Clearly, tops of heads are much less likely than feet to be occluded. Hence, even in crowd conditions, as long as cameras are quite high up and canted down at quite high angles (say 40°), all but the shortest individuals should be identifiable. Interestingly, apart from orientation, tops of heads may actually look similar in different views. As the camera cant angle will be known, altered head orientation can be allowed for and recognition and cross identification between cameras can proceed. With fully calibrated cameras (see Chapter 18), tops of heads can be located in 3-D space, and the positions of feet and heights of individuals can be deduced. Unfortunately, full camera calibration is a tedious process and may need frequent updating, so it is better not to rely on that approach in "informal" (and therefore changeable) surveillance situations such as shopping centers. Instead, camera views can be related using the fundamental matrix formulation (Chapter 18), which only requires that epipoles should be known so that epipolar lines can be determined; however, finding them requires considerable computation, though this can be done offline prior to actual use (Calderara et al., 2008).

An intriguing approach to top-of-head location is to try various homographies differing only in the parameter H signifying distance from the floor. When a homography is found that indicates the same value of H, the foot locations can be calculated for each camera view, even though the feet themselves are obscured. However, to achieve this a somewhat complex and subtle process is required (Eshel and Moses, 2008, 2010). Four vertical poles are set up at the corners of each viewing quadrilateral (or other convenient location), each pole having three bright lights along it (e.g. at the top, bottom and middle of the pole). Then standard homographies are set up for each of these, so that at any location in an image, three heights can be deduced. Finally, a height that is to be measured can be related to the three known ones for that location, a cross-ratio calculated along a vertical line, and the actual height deduced; at the same time the foot position in each camera view can be identified unambiguously.

Overall, the simplest and most powerful approach is that of prior training by getting someone to walk around the site and thus demarcate the boundaries of each common viewing zone. Then, applying the fundamental matrix for pairs of cameras will permit homographies to be set up relating all the mutually viewable regions of ground planes. The paper by Calderara et al. (2008) contains a number of other subtleties, but space prevents them from being described in detail here. Finally, if heights and exact locations of people are to be found from top-of-head positions, elegant though fairly complex methods using several homographies have to be used, but in some applications, such as observation of crowds, the additional complexity may well be justified. However, segmentation of crowd views and identification of all individuals remains a research topic, especially when the people are tightly packed—as can easily happen in metro stations and football matches.

22.8.1 The Case of Nonoverlapping Fields of View

Next we move on to the case of nonoverlapping fields of view. Here there seems to be no basis for homographies or for reliable camera handoff. However, some degree of similarity in appearance will still be detectable between views; in addition, there will be strong correlations between the time of leaving one FOV and arriving in another. The situation will often be helped if there is some restriction of access, such as would occur if there is a single adjoining door. (On a motorway, there is anyway such a restriction, and temporal correlations can be strong.)

Pflugfelder and Bischof (2008) have obtained significant success in this sort of situation, and made no assumptions about appearance. In particular, they have found how to relate the camera calibration matrices when overlapping views are not available. While this seems intrinsically impossible because no common image points can be found and hence no equations can be obtained linking the parameters (recall that the 8-point algorithm requires eight points in order to obtain a sufficient number of equations), they have found that if velocities are assumed to be more or less constant across the intervening space, this provides the continuity needed to permit enough equations to be found. Thus, a minimum of two positions per view for each trajectory is sufficient, these being immediately before and after camera handoff. Strict temporal correspondences are required, as is data on relative camera orientations, but a common ground plane is not assumed. Under these conditions, tracking across gaps of up to 4 m was achieved (Fig. 22.10(c)). The method works by making use of Rother and Carlsson's (2001) 2-point technique for determining the relative positions of two cameras with overlapping views: the new method simulates this situation by utilizing a separate pair of points in the second nonoverlapping view in order to emulate and replace the two points that would ideally have been present in an overlapping view.

For a differently motivated probabilistic strategy tackling this problem, based on transition probabilities between nonoverlapping views, see Makris et al. (2004): what is special about this approach is that it is quite general as it is entirely unsupervised and has no direct knowledge of camera placement or camera characteristics.

22.9 APPLICATIONS TO THE MONITORING OF TRAFFIC FLOW
22.9.1 The System of Bascle et al.

One important area of surveillance is the visual analysis of traffic flow. In an early study (Bascle et al., 1994) it was found that the complexity of the analysis was reduced because vehicles run on the roadway and because their motions are generally smooth. Nevertheless, the methods that had to be used to make scene interpretation reliable and robust were nontrivial.

First, motion-based segmentation is used to initialize the interpretation of the sequence of scenes. The motion image is used to obtain a rough mask of the object, and then the object outline is refined by classical edge detection and linking. B-splines are used to obtain a smoother version of the outline, which is fed to a snake-based tracking algorithm. The latter updates the fit of the object outline and proceeds to repeat this for each incoming image.

However, snake-based segmentation concentrates on isolation of the object boundary, and therefore ignores motion information from the main region of the object. It is therefore more reliable to perform motion-based segmentation of the entire region bounded by the snake, and to use this information to refine the description of the motion and to predict the position of the object in the next

image. The overall process is thus to feed the output of the snake boundary estimator into a motion-based segmenter and position predictor that re-initializes the
snake for the next image—so both constituent algorithms perform the operations
they are best adapted to. It is especially relevant that the snake has a good starting
approximation in each frame, both to help eliminate ambiguities and to save on
computation. The motion-based region segmenter operates principally by the analysis of optical flow, though in practice the increments between frames are not
especially small: this means that while true derivatives are not obtained, the result
is not as bedevilled by noise as it might otherwise be.

   Various refinements were incorporated into the basic procedure:

- B-splines are used to smooth the outlines.
- The motion predictions are carried out using an affine motion model that
  works on a point-by-point basis. (The affine model is sufficiently accurate for
  this purpose if perspective is weak so that motion can be approximated locally
  by a set of linear equations.)
- A multi-resolution procedure is invoked to perform a more reliable analysis of
  the motion parameters.
- Temporal filtering of the motion is performed over several image frames.
- The overall trajectories of the boundary points are smoothed by a Kalman
  filter.[6]

The affine motion model used in the algorithm involves six parameters:[7]

$$\begin{bmatrix} x(t+1) \\ y(t+1) \end{bmatrix} = \begin{bmatrix} a_{11}(t) & a_{12}(t) \\ a_{21}(t) & a_{22}(t) \end{bmatrix} \begin{bmatrix} x(t) \\ y(t) \end{bmatrix} + \begin{bmatrix} b_1(t) \\ b_2(t) \end{bmatrix} \tag{22.48}$$

This leads to an affine model of image velocities, also with six parameters:

$$\begin{bmatrix} u(t+1) \\ v(t+1) \end{bmatrix} = \begin{bmatrix} m_{11}(t) & m_{12}(t) \\ m_{21}(t) & m_{22}(t) \end{bmatrix} \begin{bmatrix} u(t) \\ v(t) \end{bmatrix} + \begin{bmatrix} c_1(t) \\ c_2(t) \end{bmatrix} \tag{22.49}$$

Once the motion parameters have been found from the optical flow field, it is
straightforward to estimate the following snake position.

   An important factor in the application of this type of algorithm is the degree
of robustness it permits. In this case, both the snake algorithm and the motion-
based region segmentation scheme are claimed to be relatively robust to partial
occlusions: the abundance of available motion information for each object, the
insistence on consistent motion, and the recursive application of smoothing procedures including a Kalman filter, all help to achieve this end. However, no specific

---

[6]A basic treatment of Kalman filters is given in Section 19.8.
[7]An affine transformation is one which is linear in the coordinates employed. This type of transformation includes the following geometric transformations: translation, rotation, scaling and skewing
(see Chapter 18). An affine motion model is one which takes the motion to lead to co-ordinate
changes describable by affine transformations.

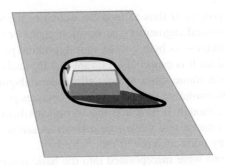

FIGURE 22.11

Vehicles located with their shadows. In many practical situations, shadows move with the objects that cause them, and simple motion segmentation procedures produce composite objects that include the shadows. Here a snake tracker envelops the car and its shadow.

nonlinear outlier rejection process is mentioned, which could help if two vehicles merged together and became separated later on or if total occlusion occurred.

Finally, the initial motion segmentation scheme locates the vehicles with their shadows since these are also moving (see Fig. 22.11); subsequent analysis seems able to eliminate the shadows and arrive at smooth vehicle boundaries.

22.9.2 The System of Koller et al.

Another scheme for automatic traffic scene analysis was described by Koller et al. (1994). This contrasts with the system described above by placing heavy reliance on high-level scene interpretation through the use of belief networks. The basic system incorporates a low-level vision system employing optical flow, intensity gradient, and temporal derivatives. These provide feature extraction, and lead to snake approximations to contours; since convex polygons would be difficult to track from image to image (because the control points would tend to move randomly), the boundaries are smoothed by closed cubic splines having 12 control points; tracking is then achieved using Kalman filters. The motion is again approximated by an affine model, though in this case only three parameters are used, one being a scale parameter and the other two being velocity parameters:

$$\Delta \mathbf{x} = s(\mathbf{x} - \mathbf{x}_m) + \Delta \mathbf{x}_m \tag{22.50}$$

Here the second term gives the basic velocity component of the center of a vehicle region, and the first term gives the relative velocity for other points in the region, s being the change in scale of the vehicle ($s = 0$ if there is no change in scale). The rationale for this is that vehicles are constrained to move on the roadway, and rotations will be small. In addition, motion with a component toward the camera will result in an increase in size of the object and a corresponding increase in its apparent speed of motion.

Occlusion reasoning is achieved by assuming that the vehicles are moving along the roadway, and are proceeding in a definite order, so that later vehicles (when viewed from behind) may partly or wholly obscure earlier ones. This depth ordering defines the order in which vehicles are able to occlude each other, and appears to be the minimum necessary to rigorously overcome problems of occlusion.

As stated above, belief networks are employed in this system to distinguish between various possible interpretations of the image sequence. Belief networks are directed acyclic graphs in which the nodes represent random variables and arcs between them represent causal connections. In fact, each node has an associated list of the conditional probabilities of its various states corresponding to assumed states of its parents (i.e. the previous nodes on the directed network). Thus, observed states for subsets of nodes permit deductions to be made about the probabilities of the states of other nodes. The reason for using such networks is to permit rigorous analysis of probabilities of different outcomes when a limited amount of knowledge is available about the system. Likewise, once various outcomes are known with certainty (e.g. a particular vehicle has passed beneath a bridge), parts of the network will become redundant and can be removed; however, before removal their influence must be "rolled up" by updating the probabilities for the remainder of the network. Clearly, when applied to traffic, the belief network has to be updated in a manner appropriate to the vehicles that are currently being observed; indeed, each vehicle will have its own belief network that will contribute a complete description of the entire traffic scene. However, one vehicle will have some influence on other vehicles, and special note will have to be taken of stalled vehicles or those making lane changes. In addition, one vehicle slowing down will have some influence on the decisions made by drivers in following vehicles. All these factors can be encoded into the belief network and can aid in arriving at globally correct interpretations. General road and weather conditions can also be taken into account.

Further work was planned to enable the vision part of the system to deal with shadows, brake lights and other signals, and a wide enough variety of weather conditions. Overall, the system was designed in a very similar manner to that of Bascle et al. (1994), though its use of belief networks made it rather more sophisticated.

In a later version of the system (Coifman et al., 1998), it was decided that a greater degree of robustness with regard to partial occlusion was required. Hence, the idea of tracking objects as a whole was abandoned and corner features were used for detection. This led to a different problem—that of grouping corner features to infer the presence of the vehicles, a process that was simplified by using a common motion constraint, so that features that were seen to be rigidly moving together were grouped together. The new version of the system also applied a homography between the image plane and the ground plane. The reason for this was to generate world parameters so that ground-based positions, trajectories, velocities, and densities could be established. Note for example that

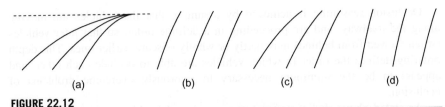

FIGURE 22.12

Adjusting the inverse perspective mapping of the roadway. (a) shows the roadway as observed by the camera. (b) shows an inverse perspective mapping with the roadway adjusted for constant width. (c) and (d) show cases of incorrect adjustment of the mapping.

a vehicle traveling at constant speed on the road would have variable speed when viewed in an image. In addition, the right information could more easily be brought to bear when problems of partial or total occlusion are being investigated.

When designing a much later system, Magee (2004) made several interesting observations: (a) corner features are unreliable because of the small size of the objects of interest; (b) connected components analysis is a poor tool for combining parts of vehicles because of fragmentation and similarity of some object foreground points to background; (c) particle filter trackers have high computational cost that does not scale linearly with the number of objects present—a serious matter when 30 or more vehicles in close proximity are being tracked simultaneously. He found that a sound way to track vehicles was to dynamically model vehicle invariants such as size, color, and speed: in other words, object appearance and recognition were important to systematic and accurate tracking; and the only way they could be achieved was by establishing a homography between the image and the ground plane. In that way vehicle parameters properly became invariants as required. The homography is expressible as a nonlinear perspective transformation (or "inverse perspective mapping"),[8] and some care is required in setting it up. However, if the camera x-axis is horizontal, the homography only requires a rotation through an angle θ about the image x-axis, together with a scaling, in order to relate the image coordinates to the ground plane coordinates. Ignoring the scaling, there is only one parameter (θ) to be determined. Magee adopted the simple strategy of estimating θ as the angle required to make the roadway appear to have constant width, a procedure that proved to be adequate in his particular application (Fig. 22.12). The calculation was made sufficiently accurate by approximating the road centerlines and outlines by three polynomials and performing a fit by iteratively adjusting θ. The reason for adopting this procedure is

---

[8]Note that such a mapping is mathematically valid only for points known to lie on the ground plane. When points not lying on the ground plane are back-projected to it, they give rise to weird, nonsensical effects, such as buildings that appear to lean backwards.

that the roadway has no absolute predefined shape so a heuristic approach seemed appropriate. Ideally, however, the ground truth for the road centerlines and outlines would be known and the value of θ could be adjusted to fit the ground truth without having to assume that the roadway has constant width.

22.10 LICENSE PLATE LOCATION

Over the past decade there has been intensive effort to identify vehicles automatically by their license plates. Although license plates were introduced many years ago for the purpose of checking ownership and detecting stolen vehicles, nowadays two other important reasons for automatically identifying vehicles are (1) for taxation within tolling zones and (2) for exacting fines in the case of parking offences—because considerable sums of money can be obtained in these ways with very little human intervention. Also, considering all the possible applications of computer vision in surveillance, identification of license plates represents a potentially straightforward application of current methodology. Nevertheless, there are many problems, not least because of the different styles of license plate from different countries.

Identification of license plates progresses through three main stages: (1) location and segmentation of the license plate; (2) segmentation of the individual characters; (3) recognition of the individual characters. Here, we concentrate on the first of these stages, as the other two are more specialized and less generic, considering the different styles, fonts, and character sets in use in different countries. In any case, the first stage is probably the most difficult to engineer.

A priori, it might be thought that the best way of locating license plates would be via their colors, which are generally well specified for each country. However, many problems arise from variations in ambient lighting, particularly with the seasons, the weather, and the time of day, while shadows are also a source of difficulty. In this milieu, one of the best starting points has been found to be use of a simple Sobel or other vertical edge detection operator, in conjunction with horizontal nonmaximum suppression and thresholding. This has been found to locate not only the vertical lines at the ends of the number plates but also the vertical lines at the sides of the characters (Zheng et al., 2005). This generally gives a relatively dense set of vertical edges within the region of the license plate. To proceed further, long background edges and short noise edges are eliminated. Finally, moving a rectangle of license plate size over the image and counting the edge pixels within it turns out to be a highly reliable way of locating license plates (in fact, this process is a form of correlation). The whole process is shown in Fig. 22.13, with the difference that in the case shown, the final stages are carried out solely using morphological operations (horizontal closing followed by horizontal opening, in each case by 16 pixels).

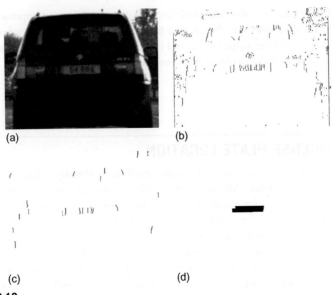

(a)   (b)

(c)   (d)

FIGURE 22.13

Simple procedure for locating license plates. (a) Original image with license plate pixellated to prevent identification. (b) Vertical edges of original image. (c) Vertical edges selected for length. (d) Region of license plate located by horizontal closing followed by horizontal opening, each over a substantial distance (in this case 16 pixels).

This method has been developed considerably further by Abolghasemi and Ahmadyfard (2009) using color and texture cues. They found that a particular advantage of color object analysis is robustness to viewpoint changes. They also used morphological closing to link all the vertical edge points, and followed this by opening to eliminate the effects of isolated noise points.

Before characters can be segmented and recognized, another stage is needed— that of license plate distortion correction. This arises because license plates may not be observed from the most ideal viewpoint. This is something that requires careful attention. If vehicles are too far away from the camera, the resolution will be too low to permit vertical edges to be found; likewise, accurate identification of the characters will not be possible. If the license plate is viewed obliquely it will appear misorientated and will not even appear rectangular. However, if license plates were always viewed at a particular distance and location, a standard perspective transformation could be applied to correct such distortions. While it is acknowledged in the literature (e.g. Chang et al., 2004) that adding such a step would improve the performance of license number recognition, few systems seem to incorporate such a step. The reason is probably that OCR systems are already very accurate even when characters are slightly sheared and rotated.

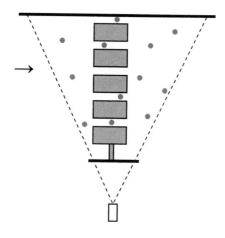

FIGURE 22.14

Typical situation of occlusion. This illustrates the case of turnstiles leading to underground trains, viewed from the side. The dots represent people (moving in the direction of the arrow) ranged at different distances from the camera.

22.11 OCCLUSION CLASSIFICATION FOR TRACKING

It will be clear from the many remarks made about occlusion on the preceding pages that this is a serious problem that needs in-depth analysis and careful algorithm design, particularly with regard to people tracking. To this end, Vezzani and Cucchiara (2008), and Vezzani et al. (2011) have made a careful analysis of the means by which occlusion can arise, starting with the definition of *nonvisible regions* as the parts of objects that are not visible in the current frame. They proceeded to classify these as "dynamic," "scene," or "apparent" occlusions:

1. *Dynamic occlusions* are due to moving objects that are readily identified.
2. *Scene occlusions* are due to static objects that are part of the background, but can nevertheless be in front of moving objects.
3. *Apparent occlusions* are sets of pixels that arise from shape variations of objects being tracked.

Here it is important to note the distinction between background and foreground. To the layman, "background" merely means a backdrop in front of which the actors perform: it is regarded as static, while the foreground is considered to consist of more interesting moving objects. However, in computer vision we have to consider the background as static wherever it is, with the moving "foreground" objects ranged at different distances from the camera and sometimes moving *behind* background objects (Fig. 22.14). Note that the background that is identified by background modeling algorithms is the static part of the scene. Of course, a complication that can disrupt this tidy situation is that the background may be

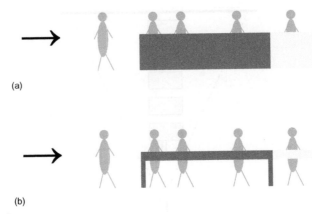

(a)

(b)

FIGURE 22.15

Further examples of occlusion. (a) Case of people walking behind a fence or barrier, potentially resulting in only the head and shoulders being tracked afterwards. (b) Case of people walking behind a table, potentially resulting in two parts of the body being tracked independently afterwards.

composed partly of objects that have come to rest, either permanently or temporarily, and it will be up to the vision algorithm to consider the available evidence from watching the scene and to assess various possibilities and probabilities.

Another factor to consider is whether occlusions are partial or total. For many static scenes and static situations, total occlusion is an eventuality that is normally disregarded; hence, all occlusions are taken to be partial, and they are simply referred to as "occlusions." However, when tracking objects, total occlusion is a possibility that has to be borne in mind indefinitely (though in practical situations a time-limit may have to be set).

So, when viewing image sequences containing motion, objects may temporarily be totally occluded, *or* they may be partially occluded—in which case they may be broken into several sections. And when objects re-emerge later on, these sections need to be reassembled into whole objects: this scenario arises when a person passes behind a table, for example. In addition, when a person passes behind a low fence, and the lower body is temporarily invisible, it is necessary for the model of the complete person to be remembered; for if the model becomes adapted to the changed situation, it may not be able to cope properly when the whole person re-emerges and it will continue to track only the top of the body. Clearly, to cope successfully with such situations (Fig. 22.15), the computer needs to have the means to deal with them holistically. Similarly, when two people merge into a larger blob when walking together, the computer will need to have the means to recall that two people were involved so that their identities will be preserved and reinstated when they separate again. Thus, we require substantial intelligence to be incorporated into tracking algorithms.

The various components that have to be incorporated into the algorithm would appear to be the following: (a) the usual background extraction capability, (b) the usual blob tracking capability, (c) full appearance and identity recall, (d) merge capability, (e) split capability, and (f) probabilistic analysis of interpretations. Indeed, (f) will probably have to be the unifying force that drives the whole algorithm.

All these aspects are included in the work of Vezzani and Cucchiara (2008) and Vezzani et al. (2011). In particular, they employ an appearance-based formalism that integrates the possible shape variations of each object and represents them by probabilistic maps. This means that when part of an object is obscured, its shape model hallucinates the whole of the object in the probabilistic shape it ought to have, so that when it re-emerges it is automatically re-integrated virtually instantaneously into its natural form.

So far we have not examined item 3 (see the beginning of this section) that mentioned apparent partial occlusions due to changes in shape. These arise because as a body rotates or bends slightly, or otherwise deforms, new parts will become visible though other parts will become invisible. While these could be regarded as arising through self-occlusion, this may not be the only possibility, e.g. if stretching is involved. We shall not delve further into this point, but merely underline that apparent partial occlusions are not caused by any other objects. This is quite an innovative observation relating to occlusion, and may be part of the reason why progress with occlusion has been drawn out over many years. Suffice it to say that the work of Vezzani et al. represents a sound and impressive advance through its cognizance of the many aspects of occlusion in tracking scenarios.

The overall system is very robust and fast and is well able to cope with more than 40 people in videos from the PETS2006 dataset. Nevertheless, it gives rise to some failures that devolve into the following categories: (a) identity change of one person, (b) split head/feet, (c) incorrect splitting of groups containing two or three people, (d) identity change of luggage. In fact, it appears that these are failures not of the *overall* system, including the handling of objects and occlusions, but of the part of the system handling appearance, which is arguably the part to which relatively little design effort has been devoted. Furthermore, it is not clear from the two papers whether a human observer could have performed better using the same video input. Nevertheless, the way forward, including the ability to handle problems (b) and (c) above, is probably to enhance the system using stick-figure based models of humans, which can take proper account of limb articulation constraints (see Section 22.13): the performance of a system that looks at the body as a whole and models it as a holistic probabilistic shape profile must in the end be limited without suitable enhancement.

22.12 DISTINGUISHING PEDESTRIANS BY THEIR GAIT

This section outlines a method for distinguishing pedestrians by their gait. Clearly, unlike many other moving objects such as vehicles, pedestrians have

FIGURE 22.16

Portions of frames extracted from video sequences by motion detector. Left: Three frames of moving vehicle. Right: Respective frames of pedestrian, runner and group of walkers.

Source: © *IET 2007*

cyclical motions, and it is actually possible to recognize individual people by their gait. However, here we consider only the methodology needed to locate pedestrians in image sequences.

The basis of the approach is to perform spatiotemporal differencing operations, in which spatiotemporal averaging is followed by temporal differencing. This "motion distillation" method (Sugrue and Davies, 2008) is implemented as a Haar wavelet and leads to a nonbinary motion map of the video at each time-step, according to the equation:

$$W = \sum_{t=t_0}^{t} \sum_i \sum_j x_{tij} - \sum_{t=t_1}^{t} \sum_i \sum_j x_{tij} \qquad (22.51)$$

where x_{tij} represents the video pixel data at the point (t, i, j) in spatiotemporal space.

In this method, undesirable contrast dependence is removed by normalizing W values across the detected object: the process involves taking the ratio R of positive (W_+) to negative (W_-) filter outputs:

$$R = \frac{\sum |W_+|}{\sum |W_-|} \qquad (22.52)$$

For a rigid object that retains its orientation relative to the camera, R will remain approximately constant over time. On the other hand, pedestrians deform as they move, and can quickly be detected by testing for changes, and particularly oscillations in the "rigidity parameter" R.

Figure 22.17(a) compares the motion signals R of a typical vehicle and a pedestrian (see Fig. 22.16 for typical frames from the original videos). While the vehicle signal changes only gradually as a result of slight rotation, changing perspective and noise, the pedestrian signal is highly variable and oscillatory because of gait motion. Note that over the period shown in Fig. 22.17(a), the area of the vehicle changes by a factor ~ 10, while the area of the pedestrian changes by only a few percent. This makes it all the more significant that R is so constant for the vehicle, demonstrating that it is a useful invariant of the motion.

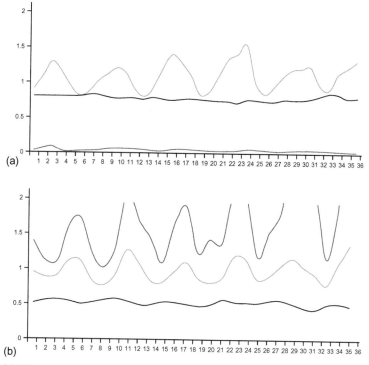

FIGURE 22.17

Motion analysis using rigidity and box parameters. (a) Top to bottom: result of applying the rigidity parameter R for the pedestrian; result of applying R for the vehicle; result of applying box parameter η for the pedestrian. The horizontal scales indicate video frames. (b) Top to bottom: result of applying the respective parameters R_{ex}, R, η for the runner. In all cases the originals are shown in Fig. 22.16.

Source: © IET 2007

After detection, a pedestrian's motion field can be further analyzed for a variety of behavior patterns. Normal behavior can be modeled by fitting a rectangular box to the subject's motion field. The rectangle is the full height of the figure with a width typically set at half the height (see below). The total motion area A is calculated as:

$$A = \sum |W_+| + \sum |W_-| \qquad (22.53)$$

where the sums are taken over the whole object; in addition, the corresponding area A_{ex} is calculated for the region outside the box. The box parameter η is then defined as the ratio of the two areas:

$$\eta = \frac{A_{ex}}{A} \qquad (22.54)$$

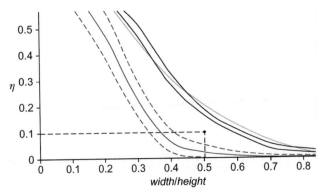

FIGURE 22.18

Discrimination permitted by box parameter η. The lower solid line represents the mean value of a sample of walkers (the broken lines indicate $\pm \sigma$ error bars), and the upper solid lines record runners. To discriminate between walkers and runners, the best operating point is close to (0.5, 0.1), as shown.

Source: © IET 2007

This parameter should also be an invariant both for rigid motion and for "compact" motion where A_{ex} is small, giving a measure of the type of behavior: this is because η is dimensionless and compares like with like but still contrasts two motions (*viz.* exterior to the box and overall). If the pedestrian is walking normally, the η value will be low at all times (see, for example, the bottom trace in Fig. 22.17(a)): the higher values typical of runners are demonstrated clearly in Fig. 22.17(b). In addition, individual sudden actions such as waving and jumping will result in spikes in η.

A third type of invariant R_{ex} has also been developed to help discriminate other more complex cases. This has the same definition as for R, except that it applies only to the part of the object external to the box. It provides additional useful information helping to discriminate runners from groups of walkers (see Fig. 22.16). Both of these categories have been found to have values of η around 0.5, so R_{ex} is useful in enabling them to be discriminated. (Specifically, $R_{ex} \approx 1.5R$ for runners and $R_{ex} \approx R$ for groups of walkers, though additionally R and R_{ex} are only well synchronized for runners.) While the R_{ex} information cannot be described as being specific to groups, it is nevertheless valuable, though ultimately only detailed analysis leading, e.g., to stick-figure models of people may provide the information that is required in a particular application (see Section 22.13).

Because of the importance of box size in determining the values of η and R_{ex}, a careful study was made to optimize discrimination between single walkers and runners. This gave the optimum box width/height ratio as close to 0.5, for which the walker–runner threshold was best set at about 0.1 (see Fig. 22.18).

Overall, the methods described here have been found to distinguish motions of rigid and nonrigid objects with ~97% accuracy. They are also able to classify single walkers with ~95% accuracy, and runners and groups of walkers with

~87% accuracy; in addition, they give useful indications of "extravagant" activities such as waving and jumping. Interestingly, all this was achieved via use of specially designed invariants, which save complexity and computation while being straightforward to set up and adjust.

22.13 HUMAN GAIT ANALYSIS

For several decades human motion has been studied using conventional cinematography. Often the aim of this work has been to analyze human movements in the context of various sports—in particular, tracking the swing of a golf club and thus helping the player to improve his game. To make the actions clearer, stroboscopic analysis coupled with bright markers attached to the body have been employed, and have resulted in highly effective action displays. In the 1990s, machine vision was applied to the same task. At this point the studies became much more serious and there was increased focus on accuracy. The reason for this was a widening of the area of application not only to other sports but also to medical diagnosis and to animation for modern types of film containing artificial sequences.

Because high accuracy is needed for many of these purposes—not least measuring limps or other imperfections of human gait—analysis of the motion of the whole human body in normally lit scenes proved insufficient, and body markers remained important. Typically, two are needed per limb, so that the 3-D orientation of each limb is deducible. Some work has been done to analyze human motions using single cameras, but the majority of the work employs two or more cameras: multiple cameras are valuable because of the occlusion that occurs when one limb passes behind another, or behind the body.

To proceed with the analysis, a kinematic model of the human body is required. In general, such models assume that limbs are rigid links between a limited number of ball-and-socket joints, which can be approximated as point junctions between stick limbs. For example, one such model (Ringer and Lazenby, 2000) employs two rotation parameters at the point where the hips join the backbone, three for the joint where the thigh bone joins the hips, plus one for the knee and another for the ankle. Thus, each leg has seven degrees of freedom, two of these being common (at the backbone): this leads to a total of 12 parameters covering leg movements (Fig. 22.19). Clearly, it is part of the nature of the skeleton that the joints are basically rotational, though there is some slack in the system, especially in the shoulders, while the knees have some lateral freedom. Finally, the whole situation is made more complex by constraints such as the inability of the knee to extend the lower leg too far forward.

Once a kinematic model has been established, tracking can be undertaken. It is relatively straightforward to identify the markers on the body with reasonable accuracy. The next problem is to distinguish one marker from another and to label them. Considering the huge number of combinations of labels that are possible,

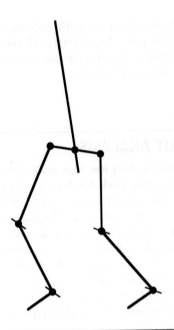

FIGURE 22.19

Stick skeleton model of the lower human body. This model takes the main joints on the skeleton as being universal ball-and-socket joints, which can be approximated by point junctions—albeit with additional constraints on the possible motions (see text). Here a thin line through a joint indicates the single rotational axis of that joint.

and the frequency with which occlusions of parts of a leg or arm are bound to take place, special association algorithms are required for the purpose. These include the Kalman filter that helps to predict how unseen markers will move until they come back into view. Such models can be improved by including acceleration parameters as well as position and velocity parameters (Dockstader and Tekalp, 2002). Their model is not merely theoretically deduced: it has to be trained, typically on sequences of 2500 images each separated by 1/30 s. In addition, the stick model of each human subject has to be initialized manually. Considerable training is necessary to overcome the slight inaccuracies of measurement and to build up the statistics sufficiently for practical application when testing. Errors are greatest when measuring hand and arm movements, because of the frequent occlusions they are subject to.

Overall, articulated motion analysis involves complex processing and a lot of training data. It is a key area of computer vision and the subject is evolving rapidly. It has already reached the stage of producing useful output, but accuracy will improve over the next few years and this will set the scene for practical medical monitoring and diagnosis, completely natural animation, detailed help with sports activities at affordable costs, not to mention recognition of criminals by their characteristic gaits. Certain requirements—such as multiple cameras—will probably remain, though the trend to markerless monitoring can be expected to continue. For further information, the reader could start by referring to the monograph by Nixon et al. (2006).

22.14 MODEL-BASED TRACKING OF ANIMALS

This section is concerned with the care of farm animals. Good stockmen noted many aspects of the behavior of the animals, and learnt to respond to them. Fighting, bullying, tail biting, activity, resting behavior, and posture are useful indicators of states of health, potential lameness, or heat stress, while group behavior may indicate the presence of predators or human intruders. In addition, feeding behavior is all-important, as is the incidence of animals giving birth or breaking away from the confinement of the pen. In all these aspects, automatic observation of animals by computer vision systems is potentially useful.

Some animals such as pigs and sheep are lighter than their usual backgrounds of soil and grass, and thus they can in principle be located by thresholding. However, the backgrounds may be cluttered with other objects such as fences, pen walls, drinking troughs, and so on—all of which can complicate interpretation. Thus, straightforward thresholding will rarely work well in normal farm scenes. McFarlane and Schofield (1995) tackled this problem by background subtraction. They used a background image obtained by temporal median filtering for a whole range of images taken over a fair period: during this process care was taken to mask out regions where piglets were known to be resting. Their algorithm modeled piglets as simple ellipses, and achieved fair success in its task of monitoring the animals.

We next examine the more rigorous modeling approach adopted by Marchant and Onyango (1995), and developed further by Onyango and Marchant (1996) and Tillett et al. (1997). These workers aimed to track movements of pigs within a pen by viewing them from overhead under not very uniform lighting conditions. The main aim of the work at this early stage was tracking the animals, though, as indicated above, it was intended to lead on to behavioral analysis in later work. To find the animals, some form of template matching is required. Shape matching is an attractive concept, but with live animals such as pigs, the shapes are highly variable: specifically, animals which are standing up or walking around will bend from side to side, and may also bend their necks sideways or up and down as they feed. It is insufficient to use a small number of template masks to match the shapes, as there is an infinity of shapes related by various values of the shape parameters mentioned. These parameters are additional to the obvious ones of position, orientation, and size.

Careful trials showed that matching with all these parameters is insufficient, as the model is quite likely to be shifted laterally by variations in illumination: if one side of a pig is closer to the source of illumination, it will be brighter, and hence the final template used for matching will also shift in that direction. The resulting fit could be so poor that many possible "goodness of fit" criteria will deny the presence of a pig. These factors mean that possible variations in lighting have to be taken into account in fitting the animal's intensity profile.

A rigorous approach involves principal components analysis (PCA). The deviation in position and intensity between the training objects and the model at a series of carefully chosen points is fed to a PCA system: the highest energy eigenvalues indicate the main modes of variation to be expected; then any specific test example is fitted to the model and amplitudes for each of these modes of variation are extracted, together with an overall parameter representing the goodness of fit. Unfortunately, this sort of approach is highly computation intensive because of the large number of free parameters; in addition, the position and intensity parameters are disparate measures that require quite different scale factors to be used to coax the schema into working. This means that some means is required for decoupling the position and intensity information. This is achieved by performing two independent principal components analyses in sequence—first on the position coordinates and then on the intensity values.

When this procedure is carried out, three significant shape parameters are found, the first being lateral bending of the pig's back, accounting for 78% of the variance from the mean; and the second being nodding of the pig's head: as the latter corresponded to only $\sim 20\%$ of the total variance, it was ignored in later analysis. In addition, the gray-level distribution model had three modes of variation amounting to a total of 77% of the intensity variance: the first two modes corresponded to (a) a general amplitude variation in which the distribution is symmetrical about the backbone, and (b) a more complex variation in which the intensity distribution is laterally shifted relative to the backbone (this arises largely from lateral illumination of the animal), see Fig. 22.20.

While principal components analyses yield the important modes of variation in shape and intensity, in any given case the animal's profile has still to be fitted using the requisite number of parameters—one for shape and two for intensity. The Simplex algorithm (Press et al., 1992) proved effective for this purpose. The objective function to be minimized to optimize the fit takes account of (a) the average difference in intensity between the rendered (gray-level) model and the image over the region of the model; and (b) the negative of the local intensity gradient in the image normal to the model boundary averaged along the model boundary (the local intensity gradient will be a maximum right around the animal if this is correctly outlined by the model).

One crucial factor has been skirted around in the preceding discussion: that the positioning and alignment of the model to the animal must be highly accurate (Cootes et al., 1992). This applies both for the initial PCA and later when fitting individual animals to the model is in progress. Here we concentrate on the PCA task. When using PCA, it should be borne in mind that it is a method of characterizing deviations: this means that the deviations must already be minimized by referring all variations to the mean of the distribution. Thus, it is very important when setting up the data to bring all objects to a common position, orientation, and scale

FIGURE 22.20

Effect of one mode of intensity variation found by PCA. This mode clearly arises from lateral illumination of the pig.

Source: ©World Scientific 2000

before attempting PCA. In the present context the PCA relates to shape analysis, and it is assumed that prior normalizations of position, orientation, and scale have already been carried out. (Note that in more general cases scaling may be included within PCA if required. However, PCA is a computation intensive task, and it is best to encumber it as little as possible with unnecessary parameters.)

Overall, the achievements outlined above are notable, particularly in the effective method for decoupling shape and intensity analysis. In addition, the work holds significant promise for application in animal husbandry, demonstrating that animal monitoring and ultimately behavioral analysis should be attainable with the aid of computer vision.

22.15 CONCLUDING REMARKS

This chapter has shown something of the purpose of surveillance, which is largely to do with monitoring the behavior patterns of people and vehicles on roads and precincts. It has also shown a number of the principles and methods by which surveillance may be implemented: these include identification and elimination of background; detection and tracking of moving objects; identification of the ground plane; occlusion reasoning; Kalman and particle filtering; capability for modeling complex motions including those of articulated objects; and use of multiple cameras for widening coverage in time and space.

Over time, some specialized application areas have appeared, such as location and identification of license plates, identification of vehicles exceeding the speed limit, human gait analysis, and even animal tracking: the chapter has aimed to indicate how all these can be achieved. Early methods included Kalman filters and chamfer matching, and later ones included particle filters, which rely on a probabilistic approach to tracking. Particle filters have come a long way, but it is doubtful whether they can go much further if based on probability assessment alone, as it is clear that humans bring to bear huge databases of relevant information when tracking moving objects.

An important lesson is that detection and tracking are distinct, complementary functions, and there is no reason why the same algorithms will be optimal for both. As indicated in Section 22.3.3, foreground detection requires the application of a suitable foreground model, or else a bootstrapping process involving an "exception to background" procedure. But once detection has been achieved, tracking can in principle proceed as a much simpler, more blinkered process. It remains to be seen whether future work will find ways of streamlining the detection + tracking model. Most likely it will be found lacking, because objects such as people radically alter in appearance as they walk by; hence, it is more natural to have both processes working in parallel (not necessarily at constant rates) all the time, rather than being applied serially. The same applies when a guided missile approaches a tank but can be misled into tracking a different object in the background as the scale of the target radically changes by several orders of magnitude: again the tracking algorithm needs to be monitored by a continuously acting detection algorithm. These points are labored because detection and tracking are at the core of surveillance, whatever the application area, and are thus very much the generic backcloth to this chapter.

While the chapter has covered the situation of static cameras being used to monitor moving objects, the following chapter covers the intrinsically more complex case of in-vehicle vision systems, where moving cameras are used to monitor both stationary and moving objects. This will call for a radical rethink of vision system strategy, because all parts of the scene will be eternally shifting and changing, and it will generally not be possible to rely on relatively trivial preliminary identification of a stationary background.

> This chapter has shown that surveillance is largely about the detection and tracking of moving objects, and that different types of algorithm will often be needed to achieve each of these functions. In most cases locating the ground plane is a necessary first step in the analysis, while occlusion reasoning, Kalman filtering, capability for modeling complex motions, and multiple cameras will often be needed to achieve the ultimate aim of analyzing developing behavior patterns.

22.16 BIBLIOGRAPHICAL AND HISTORICAL NOTES

As we have seen, surveillance involves many factors, from 3-D to motion, but paramount among these is the tracking of moving objects—in particular, vehicles and people. For some years tracking meant the use of Kalman filters, but the deficiencies of this approach led in the 1990s to the development of particle filters, including particularly the work of Isard and Blake (1996, 1998), Pitt and Shepherd (1999), van der Merwe et al. (2000), Nummiaro et al. (2003), Nait-Charif and McKenna (2004, 2006), Schmidt et al. (2006), and many others. Much of the early work is summarized in a tutorial paper by Arulampalam et al. (2002), though Doucet and Johansen (2011) have justifiably felt it necessary to produce another. The first of these and a 2008 preprint of the second might better be called reviews than tutorials, as the going is difficult—partly because of the lack of explanatory figures—and often it is easier to appeal to the original works than to them.

In parallel with these developments, much work took place on background modeling using both parametric and nonparametric methods, see for example Elgammal et al. (2000). Cucchiara et al. (2003) helped by defining the stationary and transient background problems and by clarifying the problem of "ghosts." Shadows have been a source of problems over the whole period, not least because they can be static or moving, and also because they can fall on static or moving objects: Elgammal et al. (2000) and Prati et al. (2003) carried out seminal work on this topic.

Khan and Shah (2000, 2003, 2009) were responsible for a thoroughgoing approach to the tracking of people both with single and with multiple cameras, this work being followed up by Eshel and Moses (2008, 2010) who found how to make good use of top-of-head tracking in crowd scenes. Pflugfelder and Bischof (2008, 2010) developed the approach to cover nonoverlapping views—a task that had previously (Makris et al., 2004) been solved with some degree of generality, but without knowledge of scene geometry, by learning transition probabilities for objects passing between views.

Vezzani and Cucchiara (2008) and Vezzani et al. (2011) made a careful analysis of the means by which occlusions can arise, and this enabled them to devise algorithms that cope better with temporary partial or full occlusions or temporary merging of moving objects—in the sense of not getting confused, and recovering faster from such events.

Work on traffic monitoring has stretched over many years (e.g. Fathy and Siyal, 1995; Kastrinaki et al., 2003). Early work on the application of snakes to tracking was carried out by Delagnes et al. (1995); on the use of Kalman filters for tracking by Marslin et al. (1991); and on the recognition of vehicles on the ground plane by Tan et al. (1994). For details of belief networks see Pearl (1988). Note that corner detectors (Chapter 6) have also been widely used for tracking, see Tissainayagam and Suter (2004) for an assessment of performance.

A huge amount of work has been carried out on the analysis of human motions (Aggarwal and Cai, 1999; Gavrila, 1999; Collins et al., 2000; Haritaoglu et al., 2000; Siebel and Maybank, 2002; Maybank and Tan, 2004). See Sugrue and Davies (2007) for a simple method of distinguishing pedestrians. However, note that rigorous analysis of human motion involves studies of articulated motion (Ringer and Lazenby, 2000; Dockstader and Tekalp, 2001), one of the earliest enabling techniques being that of Wolfson (1991). As a result, a number of workers have been able to characterize or even recognize human gait patterns (Foster et al., 2001; Dockstader and Tekalp, 2002; Vega and Sarkar, 2003): see Nixon et al. (2006) for a recent monograph on the subject. A particular purpose for this type of work has been the identification and avoidance of pedestrians from moving vehicles (Broggi et al., 2000; Gavrila, 2000). Much of this work has its roots in the early farsighted paper by Hogg (1983), which was later followed up by crucial work on eigenshape and deformable models (Cootes et al., 1992; Baumberg and Hogg, 1995; Shen and Hogg, 1995). Gavrila's work on pedestrian detection (Gavrila, 1998, 2000) used chamfer matching, while Leibe et al. (2005) developed the method further, albeit with the help of a minimum distance length (MDL) top-down segmentation scheme capable of handling multiple hypotheses.

While focussing on complex topics such as articulated motion and complications caused by occlusion, it is important not to lose sight of simple but elegant developments such as histograms of orientated gradients (HOGs), which have only appeared relatively recently (Dalal and Triggs, 2005). These were designed for, and are well-matched to, the detection of human shapes. Basically, they focus on the straight limbs of the human body, which have many edge points aligned along the same direction—though the latter will naturally change with walking or other motions. The basis of the method is to divide the image into "cells" (sets of pixels) and to produce orientation histograms for each of them. Voting into the orientation histogram bins takes place with weighting proportional to gradient magnitude. To provide strong illumination invariance, a robust normalization method is used. The cells are combined into larger overlapping blocks in several ways, with the result that some of the blocks end up with larger signals indicating the presence of human limbs. However, the result is that the HOG detectors cue mainly on silhouette contours and emphasize the head, shoulders, and feet. In a later paper, Dalal et al. (2006) combined the HOG detector with motion detectors and were able to achieve even better results (motion detection improved the false alarm rate by a factor of 10 relative to the best appearance-based detector). An interesting feature of the HOG approach is that it outperforms wavelet analysis because the latter eliminates vital abrupt edge information by prematurely blurring the image data.

Overall, work on surveillance has stretched over many years, but has vastly accelerated since the mid-1990s as workers have had access to more powerful computers that made it realistic to think of real-time implementation, both for experimentation and for on-road systems. Note that the past few years have seen developments in real-time systems involving FPGAs (a trend that was already present around 2000) and GPUs (a trend that is especially recent, and has arisen as a result of natural interaction between the video games industry and computer vision): for further details see Chapter 26.

22.16.1 More Recent Developments

Among the most recent works, Kim et al. (2010) have proposed a robust method for recognizing humans by their gait by using a hierarchical active shape model.

The approach is novel in that it is prediction-based and overcomes the drawbacks of existing methods by extracting a set of model parameters instead of directly analyzing the gait. Feature extraction proceeds by motion detection, object region detection, and Kalman prediction of the active shape model parameters. The method is able to alleviate tasks such as background generation, shadow removal, and obtaining high recognition rates. Ramanan (2006) has obtained good results by a new iterative parsing method for analyzing motions of articulated bodies ranging from humans playing games to horses frolicking and cantering. The approach has the advantage of being generic and does not depend on the location of skin or human faces. Lian et al. (2011) have obtained impressive performance when tracking pedestrians between camera view separations of more than 20 m— much greater than the separations ~ 4 m obtained by Pflugfelder and Bischof (2008, 2010).

Ulusoy and Yuruk (2011) have analyzed the problems of fusing data from visual and thermal images in order to make good use of their complementary properties to improve overall performance. They show that fusion should lead to a better recall rate (fewer false negatives), but at the same time result in a decrease in precision rate (more false positives); they also note that the infrared (thermal) domain always has higher precision (the underlying reasons for these observations are that thermal images effectively provide the *foreground* information containing the object pixels). In fact, it is only worth attempting fusion when an improved recall rate is required. This paper presents a more efficient method for fusing the data from the two domains and at the same time obtaining recall rates better than those previously obtained. The method was tested on outdoor images of human groups including those from a well-known database. The work in this paper leans heavily on the earlier work of Davis and Sharma (2007). Both papers refer to thermal imagery in spite of the title of the first which refers to "infrared" images.

Finally, three key papers have highlighted recent progress with the surveillance of road users: Buch et al. (2010) have employed 3-D wire-frame models for the classification of road users; Lazarevic-McManus et al. (2008) have demonstrated the value of the F-measure for optimising motion detection on the roads; and Xu et al. (2011) have obtained improved accuracy and robustness for tracking partially observable targets on the roads.

22.17 PROBLEM

1. When an inverse perspective mapping onto the ground plane is carried out, points on the ground plane are well represented in the new representation. Explain why this does not apply for buildings or people, and why they always appear to lean backward when presented in this representation.

23 In-Vehicle Vision Systems

This chapter considers the value of in-vehicle vision as part of the means for providing driver assistance systems. To achieve this, many objects have to be identified, including not only the roadway itself but also the lane and other markings on it, road signs, other vehicles, and pedestrians. The latter are particularly important as their actions are relatively unpredictable, and people who wander into the roadway are liable to cause accidents—unless the driver assistance system can help to avoid them.

Designing in-vehicle vision systems is anything but trivial, as they necessarily deploy moving cameras, which means that all objects in a scene are moving; hence it becomes quite difficult to eliminate the background from consideration. For these reasons, it becomes necessary to rely more on recognition of individual objects than on motion-based segmentation.

Look out for:

- how the roadway, road signs, and road markings may be located.
- the availability of several distinct methods for locating vehicles.
- what information can be obtained by viewing licence plates and wheels.
- how pedestrians may be located.
- how vanishing points can be used to provide a basic understanding of the scene.
- how the ground plane may be identified.
- how a plan view of the ground plane can be obtained and used to help with navigation.
- how vehicles can be guided using vision to compensate for roll, pitch, and yaw.

While it is easy to set out strategies for building in-vehicle vision systems that will work well in normal conditions on the roadway, it is far from simple to design them to operate on the less structured environments of farms or fields. Indeed, much additional reliance on GPS and other methodologies will often be needed for the purpose.

23.1 **INTRODUCTION**

This chapter provides an introduction to in-vehicle vision systems. The topic clearly overlaps with many of the ideas of the previous chapter, particularly regarding traffic surveillance, as here we are regarding the flow from inside a vehicle rather than from a stationary camera mounted (typically) on an overhead gantry. However, although the environment may be similar, the situation is essentially different, because the camera platform is in motion and almost nothing that is viewed appears stationary (Table 23.1). This means that it is extremely difficult to use methods such as background subtraction. Note that while it is theoretically possible to find a general perspective transformation that makes a sequence of frames exactly coincide so that background subtraction can be achieved, to do this would be to replace a technique that is intended to be a simple way of cueing into images into one that is highly complex; and the process of finding a sufficiently exact perspective transformation would itself require considerable computation, so this is unlikely to provide a useful strategy for analyzing image sequences.

Given the more difficult problem of analyzing scenes containing moving objects from a moving platform, we have to find ways of tackling the task equitably. Fortunately, with vehicles on a road, the range of types of scene is highly restricted. In particular, the roadway is always present in the image foreground, and thus is easily identifiable. Likewise, it normally has a characteristic dark intensity and thus, its recognition right into the distance need not be too problematic: the fact that it is moving relative to the camera is relatively immaterial. In fact, it may even be quite difficult to detect motion by looking downward toward the road surface. Next, there is a host of standard types of objects that are likely to be visible from within a vehicle—buildings, other vehicles, pedestrians, road markings, road signs, telegraph poles, lamp standards, bollards, and so on. The high frequency with which each of these can appear indicates that it will be necessary to have the capability of recognizing each of them independently, at any range and at any speed. This means that it is better, *as a first stage in the analysis*, to revert to ignoring speed of motion and to concentrate on pattern recognition. In fact, recognition can be helped by considering the range, which is readily deduced approximately at first from the lowest location on the object, which is where it meets the road (it is here assumed that the road has already been segmented from the remainder of the scene as an important preliminary stage of the analysis). Note that depending on the aim of the analysis (a point to which we shall return below), it is likely to be more

---

Table 23.1 Levels of Difficulty When Motions Can Occur

1. Locating stationary objects from a stationary platform
2. Locating moving objects from a stationary platform
3. Locating stationary objects from a moving platform
4. Locating moving objects from a moving platform

important to identify objects that lie within the road region, so segmentation of the latter is all the more important as a first stage. Then these objects—now restricted mainly to the subset, other vehicles, pedestrians, road markings, road signs, traffic lights—each needs to be identified in its own right. Later, the exact motion of the moving platform, and subsequently its location relative to all the other objects, will need to be ascertained.

We next consider the aims of implementing in-vehicle vision systems. Broadly, there are two: (1) navigation along the road, including staying in lane and finding out from road signs and traffic signals where to go, when to stop, and other such information (here, to simplify matters, we ignore use of GPS and other types of help, and how information from the various sources can be fused together reliably); (2) driver assistance, which can include a variety of matters, particularly informing the driver of all aspects included in (1), and alerting him or her to important factors, such as vehicles that are braking, or pedestrians who are moving onto the roadway. In fact, much of the information that is acquired by the vision system will need to be conveyed to the driver in one way or another. However, of particular interest is the fact that drivers will sometimes not be able to act rapidly enough to avoid pedestrians, vehicles that brake unpredictably, overtaking vehicles that suddenly cut in, and so on. There is also the problem that drivers may be drowsy or may for various reasons—e.g., because of distractions from other occupants or those caused by the simultaneous need to navigate—react too slowly, so that an accident could become imminent. In such cases, driver assistance that could automatically initiate breaking or swerving might be crucial. We can also envisage various situations where the vision system would be part of a fully automatic driving system: here there is bound to be a problem of legality, and who or what would be to blame for an accident (*viz.* the driver, car manufacturer, vision system designer, or whoever). We shall not delve into such problems here, but just consider the vision system as an enabling technology. However, once vision and driver assistance systems become sufficiently powerful, they will doubtless become part of other schemes such as those for driving in tight convoys—deemed by many to be the best way of achieving rapid safe transit along our motorways. In addition, there are other ways in which driver assistance can be valuable: these range from cruise control to automatic parking.

In this chapter, we focus generally on providing a vision system that can perceive all that might be needed for vehicle guidance and driver assistance, with emphasis on locating the roadway and road lanes, identifying other vehicles, and locating pedestrians close to or on the roadway. As indicated above, the whole process starts by locating the roadway, as discussed in the following section.

23.2 LOCATING THE ROADWAY

Chapter 4 described a technique that was capable of locating the roadway using a multilevel thresholding approach (see Fig. 4.9(b)). In fact, the roadway was identified by the third and fourth thresholds as that section of the image with gray levels in the approximate range 100–140. Similar results are obtained in other cases, e.g., Fig. 23.1(b), where the two threshold values demarcate an even greater grayscale range, approximately 60–160. While these can be construed as being reasonably ideal cases, thresholding is such a basic technique that it should be possible to extend it to cover less ideal situations. For example, if shadows appear on the roadway, the latter would in many cases appear as two contiguous sets of regions with two prominent intensity levels, and could indeed be identified by the

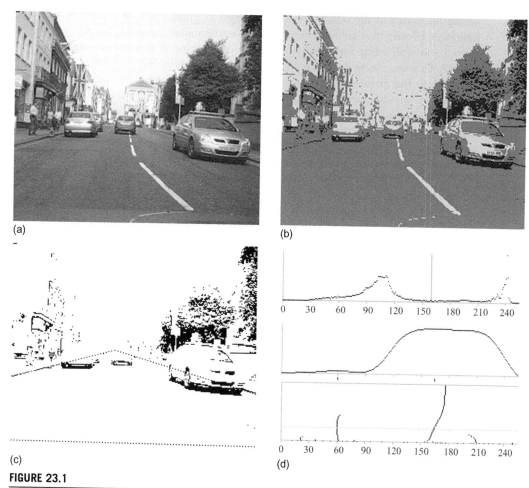

(a)   (b)

(c)   (d)

FIGURE 23.1

Frame of video taken from a moving vehicle. (a) Original image. (b) Doubly thresholded image. (c) Result of only applying the lower threshold. (d) Top: intensity histogram of the original scene. Middle: result of applying the global valley transformation and smoothing. Bottom: the dotted line shows the two thresholds used in (b) being located automatically. For further details, see text in Chapter 4. The graphics dots in (c) demonstrate that, within the road area, the lower threshold predominantly identifies under-vehicle shadows.

Source: © IET 2007

same method. Note also that varying illumination levels would be likely to make one intensity elide smoothly into another, and if a suitable range of intensities between thresholds (as in Figs. 4.9 and 23.1) were taken into account, the segmentation problem might still be solved in exactly the same way. However, ultimately the problem is one of pattern recognition, and can be solved by (a) eliminating other objects, such as road lane markings, (b) identifying the limits to the roadway, and (c) taking other features such as color or texture into account. Note that as the color of the roadway is often a bland gray, it may only be made to stand out by noting the colors of the other surroundings, such as grass, trees, or brickwork on buildings. Clearly, this would make the whole system more complex, but in a well known and well worn way—pattern recognition by now is a reasonably mature subject. To some extent the situation may be helped by bringing the motion of the vehicle into the picture (we have so far resisted this, to bring the discussion to the simplest possible base level). In that case, without calculating the exact motion of the vehicle, we can take account of the fact that the roadway stretches for a long distance ahead, so any part of it that is established to be roadway will remain so until the vehicle passes over it. Furthermore, on the road ahead, any vehicle that is located evidently runs on the roadway, so parts of it are continuously being identified. Thus, the camera vehicle merely needs to keep a record of all candidate regions that have been positively identified, so that any ambiguities from identification *via* intensities can be eliminated. Finally, this time taking motion parameters into account, keeping a tally on the road boundaries with the aid of Kalman filters will solve many of the remaining issues.

23.3 LOCATION OF ROAD MARKINGS

It has been noticed from Figs. 4.9 and 23.1 that the multilevel thresholding technique used to locate the gray surface of the road simultaneously segments white road markings. However, white road markings are seldom pure white and may be worn or even partly duplicated by older markings. In any case, segmenting them by thresholding is not the same as absolute identification. One way around this dilemma is that of fitting the road markings to suitable models. Often straight lines are adequate, although sometimes parabolas are used for the purpose. Figure 23.2 shows a case where continuous and broken road markings have been identified using the RANSAC technique, which helps to locate the vanishing point on the horizon to a reasonable approximation. The widths of the road lane markings can also be measured in this way. Figure 23.3 takes this even further. In this case, a greater degree of reliability and accuracy is obtained by locally bisecting each lane marking horizontally before feeding the data to RANSAC. In this way, extraneous signals can be eliminated—if necessary by filtering the horizontal widths. Note how RANSAC is able to find the best fit straight line section even when the road lane markings are curved. Likewise, it is able to eliminate lane markings that have been distorted by the presence of older lane

(a)   (b)

FIGURE 23.2

Application of RANSAC for locating road lane markings. (a) Original image of road scene with lane markings identified by RANSAC. (b) The edge point local maxima used by RANSAC for locating the road lane markings. While the lane markings converge to approximately the right point on the horizon line, the parallel sides of the individual lane markings do not converge quite so accurately, indicating the limits achievable with so few edge points. This is more a failure of the edge detector than of RANSAC itself.

markings (Fig. 23.3(a)). As described in Chapter 11, the version of RANSAC used for the tests successively eliminates the data points used to fit line segments, and the width delete threshold d_d is made larger than the fit threshold d_f so that no data points are retained that could mislead the algorithm while searching for subsequent line segments (see the algorithm flowchart in Fig. 23.4).

23.4 LOCATION OF ROAD SIGNS

We now continue with the process of analyzing the vehicle's environment and consider the most relevant remaining stationary parts that lie on or adjacent to the roadway. These include the traffic signs. It will not be possible to examine more than one or two cases, but among these are various relevant warnings, including those for road bumps and "GIVE WAY": note that many others appear in the same style—with the message in black on a white background and enclosed in a red triangle. To locate these signs, some tests were made without using the color aspect as this might represent too easy an approach (note also that in the wrong lighting conditions, color can be misleading): instead, an idealized small binary template of size 22×19 pixels was employed. While apparently crude, this small template had the advantage of requiring very little computation to locate the relevant objects. In fact, the chamfer matching technique (Section 22.7) was used for detecting the traffic signs shown in Fig. 23.5. While the template was primarily designed to detect

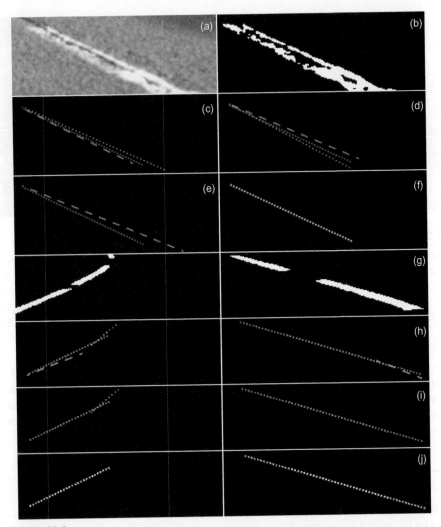

FIGURE 23.3

Further tests of RANSAC for locating road lane markings. (a) Original image 1: a distorted set of double line road markings. (b) Thresholded version of (a). (c) 3–3. (d) 3–6. (e) 3–10. (f) 3–11. (The notation "d_f–d_d" means that d_f is the "fit distance" and d_d is the "delete distance:" see text.) (g) Original image 2, already thresholded: the central section of the road containing no markings has been eliminated to save space. (h) 3–3, (i) 3–6, (j) 3–11. (f) and (j) show the final results as dense dotted lines: in other cases, dots and dashes are used to distinguish the different lines. Note that immediately after thresholding, the horizontal bisector algorithm finds the midpoints of white regions along horizontal lines, and feeds them to RANSAC for fitting.

Source: © IET 2007

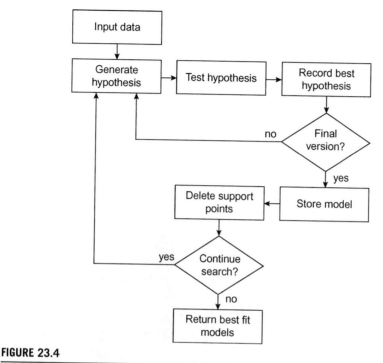

FIGURE 23.4

Flowchart of the lane detector algorithm used for the tests in Fig. 23.3.

Source: © *IET 2007*

the road bump sign, it also gave a sizeable signal for the GIVE WAY sign. Indeed, the two signals found using the template were both well above the signal-to-noise ratio elsewhere in the image, the closest possible false alarms being high up in the trees, which contain a plethora of random shapes. Note that the picture was taken under highly nonideal conditions on a wet day when there were a number of reflective areas on the road. Overall, the chamfer matching technique seems well suited to rapidly locating fixed road signs of various sorts.

There is some possibility of designing a single idealized template for locating all triangular signs. Note first that a blank white interior would be more suitable than the road bump structure in Fig. 23.5(e): this corresponds to disregarding the center of the template, taking it as being composed of "don't care" locations. In fact, the template should really be designed by a suitable training approach such as the one outlined by Davies (1992d). In this method, a matched filter approach is used in designing templates with local variability of training samples (represented by standard deviation $\sigma(\mathbf{x})$) being taken to correspond to noise, thereby necessitating reduced local weighting: the local matched filter weighting is thus (Davies, 1992d) taken as $\bar{S}(\mathbf{x})/\sigma(\mathbf{x})^2$ rather than $\bar{S}(\mathbf{x})$, where

FIGURE 23.5

Locating road signs using chamfer matching. (a) Original image showing two triangular road signs (indicating a road bump and "GIVE WAY"): each of the signs is marked with a white cross where it has been located by the chamfer matching algorithm. (b) Thresholded edge image after nonmaximum suppression. (c) Distance function image: note that with the display enhancement factor of 20 used here, the distance appears to saturate at 13 pixels. (d) The response obtained when moving the template (e) over the image.

$\overline{S}(\mathbf{x})$ is the mean local signal at $\mathbf{x}$ during training. For the types of road sign considered above, variable distributions of black within the central white area would be treated optimally by this method.

23.5 LOCATION OF VEHICLES

In recent years a number of algorithms have been designed for locating vehicles on the road, whether in surveillance applications or by in-vehicle vision systems. One notable means for achieving this has been by looking for the shadows induced by vehicles (Tzomakas and von Seelen, 1998; Lee and Park, 2006). Importantly, the strongest shadows are those appearing beneath the vehicle, not least because these are present even when the sky is overcast and no other shadows are visible. Such shadows are again identified by the multilevel thresholding approach of Chapter 4. Figure 23.1 shows a particular instance of this, where almost the only dark pixels appearing within the roadway region are the under-vehicle shadows. In fact, as under-vehicle shadows lie under vehicles, an excellent way of locating nearby vehicles is to move upward from the lowest part of the roadway until a dark entity appears: there is then a high probability that it will only locate vehicles. Note that in Fig. 23.1(c) the other main candidates are trees, but these are discounted as being well above the road region—as indicated by the dotted triangle.

As pointed out earlier when considering methods for locating the road region, it is useful to have a number of methods available for locating objects such as vehicles, in case of peculiar illumination conditions or other factors. Following this line of analysis we consider symmetry, which was first used for this purpose some years ago (e.g., Kuehnle, 1991; Zielke et al., 1993). Figure 23.6 shows a number of trials in which symmetry is applied to locate objects exhibiting a vertical axis of symmetry. The approach used is the 1-D Hough transform, taking the form of a histogram in which the bisector positions from pairs of edge points along horizontal lines through the image are accumulated. When applied to face detection, the technique is so sensitive that it will locate not only the centerlines of faces but also those of the eyes. In the case of Fig. 23.6(c), the algorithm was confused by the metal object at the bottom when locating the eye on the left, but when tested without that present it was found without difficulty. Note that some bias occurs there because the algorithm is averaging the contribution of the whole eye and the displacement between the iris and the rest of the eye becomes important. Similarly, the set of leaves in Fig. 23.6(e) is located without trouble, but the exact vertical axis that is located represents the combined peak signal from the lower two leaves and the uppermost leaf: in such a case it would be better to identify each one separately. These sorts of problem are less important in Fig. 23.6(g) where both vehicles are located quite accurately—in spite of the fact that the car on the right is not exactly horizontal. Interestingly, both vehicles would also be found using the under-vehicle shadow method. The fact that they both lie within their respective lanes also aids positive identification.

In spite of these successful applications of symmetry, note that the approach needs to be used with caution. In particular, the building on the left in Fig. 23.6(g) gives a plethora of signals because of the multiple symmetries between its windows. An interesting lesson is that three equally spaced vertical lines at locations $x = 1, 3, 5$ will have a symmetry not only at $x = 3$ but also at $x = 2$ and 4.

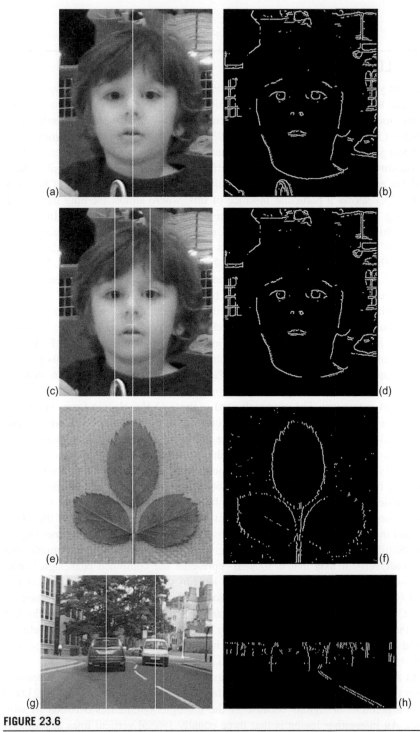

FIGURE 23.6

Searching for symmetry in images. (a) Original image of a face with a vertical axis of
symmetry. (b) Edge image used for determining the axis of symmetry in (a). (c) Original
image with symmetry axes of the eyes. (d) Slightly restricted edge image used

Finally, rotation symmetries and reflection symmetries about nonvertical axes are not especially useful in the present context. However, just as a 1-D HT can be used to locate symmetries about vertical axes, so 2-D HTs can be used to locate symmetries about lines of arbitrary direction. Thus, one can build a single 2-D parameter space, each horizontal line of which represents the symmetry in a different direction in the image. Such a parameter space might be expected to have a minor amount of coherence in the vertical direction, but we do not consider this further here.

23.6 INFORMATION OBTAINED BY VIEWING LICENCE PLATES AND OTHER STRUCTURAL FEATURES

Licence plate location has already been covered in Section 22.10. In this section, we consider what can be deduced from an oblique view of a licence plate of length R. We simplify the situation by assuming that both the image plane and the licence plate are vertical, and that they have their main axes aligned horizontally and vertically. Figure 23.7(a) and (b) shows respectively the oblique and plan views of the licence plate horizontal axis. The apparent horizontal projection (CQ) of the centerline of the licence plate is $R \cos \alpha$ when viewed in the direction PT. Following Fig. 23.7(c), its vertical projection (QT) is $R \sin \alpha \tan \beta$. However, when viewed in the more general direction PT', with lateral angle λ, its horizontal projection is CQ', which is equal to $R \cos \alpha - R \sin \alpha \tan \lambda$. From Fig. 23.7(d), we deduce that its apparent angle γ and length R' are given by the equations:

$$\tan \gamma = \frac{\tan \alpha \tan \beta}{1 - \tan \alpha \tan \lambda} \tag{23.1}$$

$$\begin{aligned} R' &= R \cos \alpha (1 - \tan \alpha \tan \lambda) \sec \gamma \\ &= R \sin \alpha \tan \beta \operatorname{cosec} \gamma \end{aligned} \tag{23.2}$$

These formulas seem intuitively correct, as for example, $\gamma = 0$ if $\alpha = 0$ or $\beta = 0$. In addition, under nonoblique viewing, $\beta = 0$, $\lambda = 0$, and $\gamma = 0$, so Eq. (23.2) reverts to the standard result for nonoblique viewing, $R' = R \cos \alpha$.

Perhaps a more important case is that of $\alpha = \pi/2$, leading to $\tan \gamma = -\tan \beta / \tan \lambda$. We can interpret this result by taking image plane coordinates (x, y) and 3-D coordinates (X, Y, Z). Noting that $\tan \beta = y/f$ and $\tan \lambda = x/f$, we deduce that

---

FIGURE 23.6 (Continued)

for determining the symmetry axes of the eyes. (e) Original image of a leaf triplet, with symmetry axis. (f) Vertical edge image used to determine the symmetry axis in (e). (g) Original image of a traffic scene, with symmetry axes marked. (h) Vertical edge image used for determining the symmetry axes in (f). The slight bias of the left-most symmetry axis in (c) is not surprising in view of the few pixels involved and the interfering effects of other edge pixels in the image.

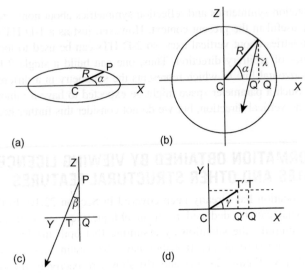

FIGURE 23.7

Horizontal line pose viewing geometry. (a) Oblique view of a horizontal straight line of length R, rotated through an angle α from the X-axis. (b) Plan view of the line. (c) Side view showing the viewing direction, along PT′, with a lateral angle λ; the angle of elevation β is that of T, not of T′. (d) Front view in the X–Y plane, which is parallel to the image plane x–y. Note that the horizontal line CP in (b) appears to lie at an angle γ in (d): it has an apparent length (CT′) of R′.

$\tan\gamma = -y/x = -Y/X$. This corresponds to viewing perspective lines on the roadway that are parallel to the optical axis of the camera. (Note that the minus sign in these equations corresponds to the fact that γ will be viewed in the range $\pi/2$ to π when $\alpha = \pi/2$.)

Finally, note that instead of obtaining the projection of the line as it would appear in the direction of viewing, we have determined its projection in the vertical plane X–Y, which is parallel to the image plane x–y. As a result, the equations correspond *exactly* to projective projection into the image plane, rather than merely to orthographic projection.

We now need to obtain an equation for α in terms of the other parameters. Solving Eq. (23.1) for α, we find:

$$\tan\alpha = \frac{\tan\gamma}{\tan\beta + \tan\gamma\tan\lambda} \tag{23.3}$$

Next, taking the projections of the centerline of the licence plate along the image x and y axes to be δx, δy, we find that the parameters β, γ, λ are all measurable, so α can be estimated:

$$\tan\alpha = \frac{\delta y/\delta x}{(y/f) + (\delta y/\delta x)(x/f)} = \frac{f\delta y}{y\delta x + x\delta y} \tag{23.4}$$

Thus, we now know the orientation in space of the licence plate. In principle, we can use Eq. (23.2) to estimate the range of the licence plate. To achieve this, we need to know the value of R. In fact, for standard UK licence plates, R is reasonably well defined (this assumes that the number of characters in the licence plate is known), and so Eq. (23.2) can be used to estimate R'. Next, the ratio of R' to the apparent length r of the licence plate gives the range Z:

$$Z = \frac{fR'}{r} = \frac{fR'}{[(\delta x)^2 + (\delta y)^2]^{1/2}} \qquad (23.5)$$

If we had also made use of the apparent lengths and orientations of the shorter sides of the licence plate, we could have eliminated dependence on the assumptions that the latter are vertical. However, it is unlikely that these short lines could be measured accurately enough to improve the situation significantly: instead we presume that the best that can be done is to use measurements on the longer sides to obtain preliminary estimates of the positions of vehicles, which can then be improved by other measurements.

Unfortunately, all the above theory is somewhat confounded by the variable camber of the road. But note that, while the camber will be considerably different on the opposite side of the road, its effects will tend to cancel when observing the licence plates of vehicles on the same side of the road. Next, the size of γ depends on y, and hence on the height of the camera above the target feature: this means that the observed value of γ will be smaller for the licence plate than for the rear wheels; hence, if the rear wheels are not occluded, it is likely that they will give a more accurate estimate of α than that from the licence plate. Nevertheless, licence plates are more satisfactory indicators than rear wheels both because they are less likely to be occluded and because they are uniquely recognizable: in fact, the rear wheels of one vehicle can sometimes be confused with those of other vehicles, and even the front wheels can cause confusion. Finally, another factor needs to be borne in mind—that we are attempting to estimate an often small quantity α from another small quantity γ when both are comparable to the interfering effect of the camber angle. Interestingly, this problem can be overcome more effectively by estimating $\tilde{\alpha} = \pi/2 - \alpha$ from $\tilde{\gamma} = \pi/2 - \gamma$ and applying these measures to views of the sides (particularly the sides of the wheels) of other vehicles. All this can be achieved by recalling that tan α and tan γ should respectively be replaced by cot $\tilde{\alpha}$ and cot $\tilde{\gamma}$ in Eqs. (23.1) and (23.2). Overall, it might be expected that side views of vehicles will be more valuable for estimating orientation than rear views, whether the latter use rear wheels or licence plates as indicators (although, obviously, only the rear view of a vehicle will be relevant when driving directly behind it). Consideration of Figs. 23.1, 23.6, 23.8 and 23.9 will provide adequate confirmation of these observations.

Finally, it might be asked why so much emphasis has been placed on measurement of angles *vis-à-vis* distances. This is basically because angles represent ratios of distances and thus they tend to provide scale-invariant information. In addition, they do not demand knowledge of absolute distances for interpretation.

(a)

(b)

FIGURE 23.8

Vehicles viewed obliquely. More accurate information about orientation is often obtained from the side of the vehicle than from its rear.

(a)

(b)

(c)

(d)

FIGURE 23.9

Chamfer matching to locate pedestrians from their lower legs. (a) and (b) show original images of road scenes containing pedestrians. The white dots are the peak signals after chamfer matching using an idealized binary U template. Note the plethora of false positives because of the number of vertical edges able to stimulate signals—as seen in (c) and (d).

23.7 LOCATING PEDESTRIANS

In principle, locating whole pedestrians would require many chamfer templates of varying shapes and sizes, to cover the many body profiles of moving people. The alternative chosen here is to look for specific sub-shapes that would be more general and invariant. Possibilities include leg, arm, head and body sections. Figure 23.9 shows lower legs being located using an idealized "U" template with parallel sides. However, a plethora of false positives arises because of the large number of vertical edges that are able to stimulate signals. Their presence means that the distance functions do not have the ideal maximum values that might be expected because the spurious edges reset the distance functions to zero in many places. This does not affect the sensitivity of the method in the sense that the templates are bound to locate instances of the profiles they represent. However, it does affect the numbers of false positives that are detected. In fact, in the examples shown, the result is not disastrous, because the lowest objects found, once road markings are eliminated, are the feet of the pedestrians. However, the fact that the method does not give ideal results makes it essential to back it up using alternative methods.

The Harris operator provides a useful alternative approach. As Fig. 23.10 shows, it is able to locate a range of features, including feet and heads, as well as road lane markings. Note that in the case shown in Fig. 23.10(a), the right foot has not been found as it is larger than the other foot and the particular Harris operator employed stretched over a range of only 7 pixels. Note that the Harris operator has no sense of polarity (preference for black or white): in the case of pedestrians this is useful as the clothing and shoes (or feet) are unpredictable and can appear dark on a light background or *vice versa*. (Lack of polarity also applies to chamfer matching, but for different reasons.)

(a)   (b)

FIGURE 23.10

Alternative approach to pedestrian location using the Harris operator. Here the operator has the effect of locating corners and interest points, some of which include pedestrian feet and heads: above all, road lane markings are also located with high probability. The operator has not been tuned in any way to recognize such features. In addition, it has no sense of polarity (preference for black or white).

FIGURE 23.11

Another approach to pedestrian location via skin color detection. (a) and (b) show that a lot can be achieved via skin color detection, detecting not only faces but also neck, chest, arms and feet: see also the detail in (c) and (d). With proper color classifier training, even more can be achieved, as shown in (e) and (f) (see also larger version of figure in color Plate 7).

Further approaches are useful to back up the two mentioned above and also to confirm detections that have already been made. In this respect unique identification of human skin color can be useful. That this is possible is shown in Fig. 23.11, one of the main problems clearly being the rather small numbers of pixels in the face regions. To carry out skin detection rigorously, it is necessary to train the color classifier on a set of training images. This was carried out for Fig. 23.11(e). While the method was highly successful (see Fig. 23.11(f)), it corresponded to supervised learning of skin color; in practice, with less tight control of the training images, this process could be compromised by the presence of sand, stone, cement, and a host of brown variants, which have colors close to those of darker or lighter people. Another important factor is that in-vehicle vision systems will not have sufficient time to gather enough training data, considering particularly that the whole point of a vehicle is to travel and thus adaptation from dark to light and other environmental factors are bound to be a source of serious problems. In this respect, in-vehicle systems are subject to far worse conditions than will be usual for surveillance systems.

Overall, we find that in-vehicle pedestrian detection systems involve a demanding set of pattern recognition problems. Earlier we emphasized the potential value of pattern recognition when moving objects are being detected from moving platforms: this approach to the subject was also useful for didactic reasons. However, we are now finding that there are limits to this. In fact, it would be an artificial restriction not to make use of motion by at least tracking features and grouping them according to velocity (a process that was already mentioned in Chapter 22). The problem with this approach is the large number of, e.g., interest point features that exist in an entire image, where almost all the features are moving. If each of them (say N) is to be compared with all others in a pair of adjacent frames, then $O(N^2)$ operations will have to be undertaken. However, by acknowledging the individuality and different characteristics of the various features, and their spatial arrangements, this vast number can be cut down to manageable proportions. In particular, feature points should only move a limited distance between frames, so there will only be a small number n of candidates that match a given feature as it moves from one frame to the next. This leaves us with $O(Nn)$ pairs of feature points to consider, a number that can be further minimized by examining the relative strengths and colors of the various pairs (ideally, the final result will be $O(N)$). Here, some of the ideas of Section 6.7, where features were characterized by a great many descriptors, may prove useful, even though wide baseline matching is not relevant for frame to frame tracking.

23.8 GUIDANCE AND EGOMOTION

An important aspect of driver assistance systems is that of vehicle guidance. In fact, this aspect is important both for vehicles with human drivers and for autonomous robot vehicles. In either case, vehicle egomotion is handled by a controlling computer that has to be fully aware of the situation. Incoming images contain complex information and reliable cues have to be found to key into them. Among the most widely used such cues are vanishing points (VPs), which are often very evident in city scenes (e.g., Fig. 17.11).

One of the ways in which VPs are most useful is in helping to identify the ground plane, and a lot of other information follows from this. In particular, local scale can be deduced: for example, objects on the ground plane have width that is referable to, and a known fraction of, the local width of the ground plane; in addition, VPs permit an estimate to be made of distance along the ground plane, by measuring the distance from the relevant image point to the VP, as we shall see below. Thus, they are useful for initiating the process of recognizing and measuring objects, determining their positions and orientations, and helping with the task of navigation.

Here, a lot will depend on the type of environment and the type of vehicle. There are many possibilities such as vacuum-cleaning robots, window-cleaning robots, lawn-mowing robots, invalid chair robots, weeding and spraying robots, maze-running robots, not to mention vehicles running autonomously on roads, or cars that park themselves automatically. In some cases, robots will have to undertake mapping, path planning, and navigational modeling and engage in detailed high-level analysis: this sort of situation has been explored by Kortenkamp et al. (1998). This approach will be important if a path has obstacles such as bollards or pillars

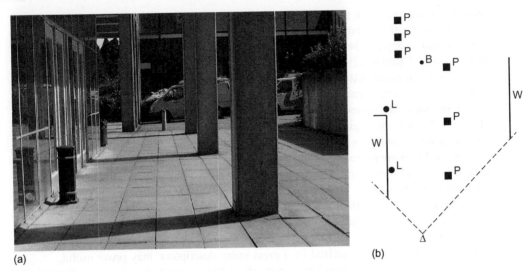

(a) (b)

FIGURE 23.12

Plan view obtained for navigation. (a) View of a scene showing the obstacles to be avoided. (b) Plan view of the ground plane showing what is visible from viewpoint Δ (for clarity, the full areas of the pillars P, bollards B, and litter-bins L are shown). The walls are marked W.

(Fig. 23.12); and it will be vital for a maze-running robot. In many such cases, vision or other sensors will provide only limited information about the working area, and knowledge will have to be augmented in a suitable representation: this makes a plan view model of the working area a natural solution. To proceed with this idea, we need to transfer the information from individual images into the plan view representation (see the algorithm of Table 23.2).

Basically, to construct a plan view of the ground plane, we start with a single view of a scene in which the vanishing point V has been determined and significant feature points on the ground plane (particularly regarding its boundaries) have been identified. Next, distance along the ground plane can be deduced as shown in Fig. 23.13. The angle of declination α of a general feature point $P(X, H, Z)$ on the ground plane, seen in the image as point (x, y), is given by:

$$\tan \alpha = \frac{H}{Z} = \frac{y}{f} \tag{23.6}$$

The value of Z is therefore given by:

$$Z = \frac{Hf}{y} \tag{23.7}$$

After obtaining a similar formula giving the lateral distance X, we deduce that:

$$X = \frac{Hx}{y} \tag{23.8}$$

Table 23.2 Computing Ongoing Plan Views of the Ground Plane

1. Detect all edges in the current frame.
2. Locate all straight lines in the current frame: e.g., use a Hough transform.
3. Locate all VPs: use a further HT, as described in Section 17.7.
4. Find the VP closest to the direction of motion: eliminate all other VPs.
5. Determine the closest section of G: this should be the part of the frame immediately in front of the robot.
6. Use this and other information to determine which lines through the primary VP lie on G: eliminate all other lines.
7. Segment objects on G.
8. Eliminate object boundaries on G that are unrelated to lines passing although the primary VP.
9. Tentatively identify as shadows any dark regions lying on G.
10. Take the remaining object and shadow boundaries and check for consistency between frames: e.g., use the 5-point cross-ratio values, as described in Section 17.3.
11. Label all remaining feature points on G with their (X, Z) coordinates: use Eqs. (23.7) and (23.8).
12. Check for consistency with previous frames.
13. Update list of objects with inconsistent boundaries as not lying on G, or as being otherwise unreliable: these could be due to moving shadows or noise.
14. Update history of feature point coordinates on G.

This table presents an algorithm showing how a plan view of the ground plane G may be computed. It is assumed that the robot sees a sequence of video frames and that it has to update its knowledge base as each frame comes along. The algorithm is set up assuming that it is best to analyze each frame ab initio, and then to look for consistency with previous frames.

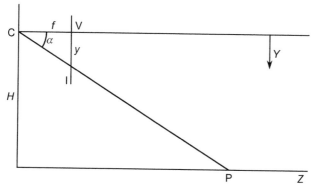

FIGURE 23.13

Geometry relating the image and the ground plane. C is the center of projection of the camera, I is the image plane, V is the vanishing point, and P is a general point on the ground plane. f is the focal length of the camera lens and H is the height of C above the ground plane. The optical axis of the camera is assumed to be parallel to the ground plane.

The world (plan view) coordinates (X, Z) have now been found in terms of the image coordinates (x, y). Note that y has to be measured from the vanishing point V rather than the top of the image. Note also that as X and Z vary inversely with y, they vary rapidly when y is small, so digitization and other errors will markedly affect the accuracy with which far away objects can be located from the plan view.

When the optical axis of the camera is not parallel to the ground plane, the calculations are best dealt with using homogeneous coordinates as shown in Chapter 18.

23.8.1 A Simple Path Planning Algorithm

In this subsection, we assume that a plan view of the environment has been built up using the methods of the previous section. While it is by no means clear that humans use an instantaneous plan view model to help them to walk or drive around an environment (an image-based representation seems more likely), it is clear that they use plan views for deductive, logical analysis of the situation and when reading maps. In any case, plan views probably constitute the most natural means for storing navigational knowledge and arriving at globally optimal routes. Here we leave aside conjecture of exactly how humans juggle the information between the two representations, and concentrate on how a robot might reasonably undertake path planning using a plan view it has built up. In fact, a maze-running robot would need to be provided with a suitable algorithm for this purpose.

Figure 23.14(a) shows a simple maze in which the robot has to proceed from the entrance E to the final goal G (respectively marked "↓" and "☺" in the figure). We assume that a plan view of the maze has been built up and that a systematic means is needed to find the optimum path to the goal G. The envisaged algorithm starts from G and propagates a distance function over the whole region, constrained only by the walls of the maze (Fig. 23.14). If a parallel algorithm is used, it is terminated when the distance function arrives at E; if a sequential algorithm is used, it must carry on until the whole maze has been covered—assuming that an optimal path is required. When the distance function has been completed, finding an optimum path necessitates proceeding downhill along the distance function until G is reached: at each point, the locally greatest gradient must be used (Kanesalingam et al., 1998). Connected components analysis could be used to confirm that a path exists, but a distance function has to be used to guarantee finding the shortest path. Note that the method will find only one of several paths of equal length: these arise because of the limitations of this type of method, which assigns integer values to distances between adjacent pixels.

23.9 VEHICLE GUIDANCE IN AGRICULTURE

In recent years, there has been increasing pressure on farmers to reduce the quantities of chemicals used for crop protection. This cry has come both from

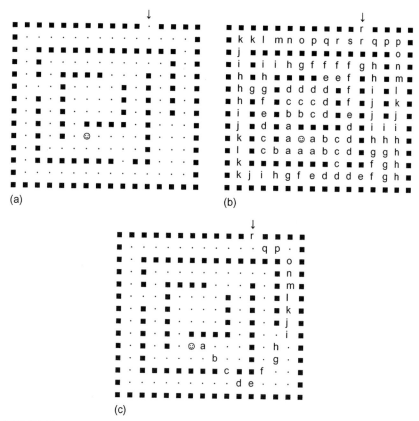

FIGURE 23.14

Method for finding an optimal path through a maze. (a) Plan view of maze. (b) Distance function of the maze, starting at the goal (marked ☺), and presenting distance values by successive letters, starting with a = 1. (c) Optimum path obtained by tracking from the maze entrance (marked ↓) along maximum gradient directions.

environmentalists and from the consumers themselves. The solution to this problem lies in more selective spraying of crops. For example, it would be useful to have a machine that would recognize and spray weeds with herbicides, leaving the vegetable crops themselves unharmed: alternatively, the individual plants could be sprayed with pesticides. This case study relates to the design of a vehicle that is capable of tracking plant rows and selecting individual plants for spraying (Marchant and Brivot, 1995; Marchant, 1996; Brivot and Marchant, 1996; Sanchiz et al., 1996; Marchant et al., 1998).[1]

---

[1]Many of the details of this work are remarkably similar to those for the totally independent project undertaken in Australia by Billingsley and Schoenfisch (1995).

FIGURE 23.15

Value of color in agricultural applications. In agricultural scenes such as this, color helps with segmentation and with recognition. It may be crucial in discriminating between weeds and crops if selective robot weedkilling is to be carried out (see also color Plate 1).

Source: © World Scientific 2000

The problem would be enormously simplified if plants grew in highly regular placement patterns, so that the machine could tell from their positions whether they were weeds or plants, and deal with them accordingly. However, the growth of biological systems is somewhat unpredictable and renders such a simplistic approach impracticable. Nevertheless, if plants are grown from seed in a greenhouse, and transplanted to the field when they are approaching 100 mm high, they can be placed in straight parallel rows, which will be approximately retained as they grow to full size. There is then the hope (as in the case shown in Fig. 23.15) that the straight rows can be extracted by relatively simple vision algorithms, and the plants themselves located and identified straightforwardly.

At this stage the main problems are: (1) the plants will have grown to one side or another, and will thus be out of line; (2) some will have died; (3) weeds will have appeared near some of the plants; (4) some plants will have grown too slowly, and will not be recognized as plants. Thus, a robust algorithm will be required to perform the initial search for the plant rows. The Hough transform (HT) approach is well adapted to this type of situation: specifically, it is well suited to looking for line structure in images.

The first step in the process is to locate the plants. This can be achieved with reasonable accuracy by thresholding the input images (this process is eased if

FIGURE 23.16

Perspective view of plant rows after thresholding. In this idealized sketch, no background clutter is shown.

Source: © World Scientific 2000

infrared wavelengths are used to enhance contrast). However, at this stage the plant images become shapeless blobs or clumps (Fig. 23.16). These contain holes and lobes (the leaves, in the case of cabbages or cauliflowers), but a certain amount of tidying up can be achieved either by placing a bounding box around the object shape or by performing a dilation of the shape that will regularize it and fill in the major concavities (the real-time solution employed the first of these methods). Then the position of the center of mass of the shape is determined, and it is this that is fed to the HT straight line (plant row) detector. In common with the usual HT approach, votes are accumulated in parameter space for all possible parameter combinations consistent with the input data. Here this means taking all possible line gradients and intercepts for lines passing through a given plant center and accumulating them in parameter space. To help find the most meaningful solution, it is useful to accumulate values in proportion to the plant area. In addition, note that if three rows of plants appear in any image, it will not initially be known which plant is in which row, and therefore each plant should be allowed to vote for all the row positions: this will naturally only be possible if the inter-row spacing is known and can be assumed in the analysis. However, if this procedure is followed, the method will be far more resistant to missing plants and to weeds that are initially mistakenly assumed to be plants.

The algorithm is improved by preferentially eliminating weeds from the images before applying the HT. Weed elimination is achieved by three techniques—hysteresis thresholding, dilation, and blob size filtering. Dilation refers to the standard shape expansion technique described in Chapter 7 and is used here to fill in the holes in the plant blobs. Filtering by blob area is reasonable since the weeds are seldom as strong as the plants, which were transplanted only when they had become well established.

Hysteresis thresholding is a widely used technique that involves use of two threshold levels. In this case, if the intensity is greater than the upper level t_u, the object is taken to be plant; if lower than the lower level t_1, it is taken to be weed; if at an intermediate level and next to a region classified as plant, it is taken to be plant; the plant region is allowed to extend sequentially as far as necessary, given only that there is a contiguous region of intensity between t_1 and t_u connecting a given point to a true plant ($\geq t_u$) region. Note that this application is unusual in that whole-object segmentation is achieved using hysteresis thresholding: more usually the technique is used to help create connected object boundaries (see Section 5.10).

Once the HT has been obtained, the parameter space has to be analyzed to find the most significant peak position. Normally, there will be no doubt as to the correct peak—even though the method of accumulation permits plants from adjacent rows to contribute to each peak. The reason for this is that with three rows each permitted to contribute to adjacent peaks, the resultant voting patterns in parameter space are: 1,1,1,0,0; 0,1,1,1,0; 0,0,1,1,1—totaling 1,2,3,2,1—thereby making the true center position the most prominent (actually, the position is more complicated than this as several plants will be visible in each row, thus augmenting the central position further). However, the situation could be erroneous if any plants are missing. It is therefore useful to help the HT arrive at the true central position. This can be achieved by applying a Kalman filter (Section 19.8) to keep track of the previous central positions, and anticipate where the next one will be—thereby eliminating false solutions. This concept is taken furthest in the paper by Sanchiz et al. (1996), where the individual plants are all identified on a reliable map of the crop field and errors from any random motions of the vehicle are systematically allowed for.

23.9.1 3-D Aspects of the Task

So far we have assumed that we are looking at simple 2-D images that represent the true 3-D situation in detail. In practice this is not so. The reason for this is that the rows of plants are being viewed obliquely and therefore appear as straight lines but with perspective distortions, which shift and rotate their positions. The full position can only be worked out if the vehicle motions are kept in mind. In practice, vehicles moving along the rows of plants exhibit variations in speed and are subject to roll, pitch, and yaw. The first two of these motions correspond respectively to rotations about horizontal axes along and perpendicular to the direction of motion: these are less relevant and are ignored here. The last is important as it corresponds to rotation about a vertical axis and affects the immediate direction of motion of the vehicle.

To proceed, we have to relate the position (X, Y, Z) of a plant in 3-D with its location (x, y) in an image. We can achieve this with a general translation:

$$T = (t_x, t_y, t_z)^T \tag{23.9}$$

together with a general rotation:

$$
R = \begin{bmatrix} r_1 & r_2 & r_3 \\ r_4 & r_5 & r_6 \\ r_7 & r_8 & r_9 \end{bmatrix} \tag{23.10}
$$

giving:

$$
\begin{bmatrix} X \\ Y \\ Z \end{bmatrix} = \begin{bmatrix} r_1 & r_2 & r_3 \\ r_4 & r_5 & r_6 \\ r_7 & r_8 & r_9 \end{bmatrix} \begin{bmatrix} x \\ y \\ z \end{bmatrix} + \begin{bmatrix} t_x \\ t_y \\ t_z \end{bmatrix} \tag{23.11}
$$

The lens projection formulae are also relevant:

$$
x = \frac{fX}{Z} \tag{23.12}
$$

$$
y = \frac{fY}{Z} \tag{23.13}
$$

We shall not give a full analysis here, but assuming that roll and pitch are zero, and that the heading angle (direction of motion relative to the rows of plants) is ψ, and that this is small, we obtain a quadratic equation for ψ in terms of t_x. This means that two sets of solutions are in general possible. However, it is soon found that only one solution matches the situation, as the wrong solution is not supported by the other feature point positions. This shows the complications introduced by perspective projection—even when highly restrictive assumptions can be made about the geometrical configuration (in particular, ψ being small).

23.9.2 Real-Time Implementation

Finally, it was found to be possible to implement the vehicle guidance system on a single processor augmented by two special hardware units—a color classifier and a chaincoder. The latter is useful for fast shape analysis following boundary tracking. The overall system was able to process the input images at a rate of 10 Hz, which is sufficient for reliable vehicle guidance. Perhaps more important, the claimed accuracy was in the region of 10 mm and 1° of angle, making the whole guidance system adequate to cope with the particular slightly constrained application considered. A later implementation (Marchant et al., 1998) did a more thorough job of segmenting the individual plants (although still not using the blob size filter), obtaining a final 5 Hz sampling rate—again fast enough for real-time application in the field. All in all, this case study demonstrates the possibility of highly accurate selective spraying of weeds, thereby very significantly cutting down the amount of herbicide needed for crops such as cabbages, cauliflowers, and wheat.

23.10 CONCLUDING REMARKS

This chapter has considered the value of in-vehicle vision as part of the means for providing driver assistance systems. It has also considered the design of such systems. This process is rendered far from trivial because the camera is necessarily moving, so all objects in a scene will appear to be in motion. Hence, it becomes quite difficult to eliminate the background from consideration and less easy to rely on motion-based segmentation. This makes it natural to adopt the alternative approach of placing reliance on recognition of individual objects. Sections 23.2 and 23.3 showed how this concept can be applied to the location not only of the roadway, but also of road markings and road signs. The principle also applied to location of vehicles, but as these vary in appearance, it proved necessary to have several distinct methods for locating them, including under-vehicle shadows, symmetry, wheels, and licence plates (the latter acting not merely as unique vehicle identifiers but also as characteristics of vehicles in general). Curiously, licence plates offered a possible means of finding the orientations of vehicles on roads as well as their locations, although the result was dependent on the relative heights of the camera and licence plate under observation. This meant that, when they are not occluded, tire and wheel location will probably be more accurate indicators of vehicle orientation.

Pedestrian location was also seen to be a challenge—particularly as people are articulated objects, and walk with bobbing motions, and also because they tend to have unique appearances and clothing. This makes it natural to use specific templates for leg, arm, head, and body detection rather than whole-body templates. Here, symmetry is also a possible cue as well as skin color. All these approaches were studied in Section 23.7 and tallied with findings in the literature.

The chapter also included aspects of path planning consequent on projecting vehicles and other obstructions onto a plan view of the ground plane: this has some consequence for robot egomotion and navigation. It is also relevant for guidance of agricultural vehicles that are being used for cultivation, selective spraying, and so on. Here, it is also important to consider the much greater degrees of roll, pitch, and yaw that will be experienced by a tractor or other vehicle moving over ploughed fields, and the visual compensation needed to cope with this. Some indication was given about how these factors have been coped with: because the principles are known, it seemed better for readers to refer to the original papers for further details.

Finally, we should remark on the almost explosive growth of interest in in-vehicle driver assistance systems, particularly since 2000. This is so important that the following section looks very closely at developments in this area and provides separate bibliographies relating to the various aspects. It was felt that it would be clearer presenting these separately once the principles of the subject had been dealt with, as has been done relatively didactically in the preceding sections.

> In-vehicle vision systems necessarily deploy moving cameras, so the usual surveillance strategy of eliminating the stationary background becomes difficult to apply. However, considerable success can be achieved using the alternative strategy of directly locating the most relevant objects, such as the roadway, road signs, road markings, vehicles (e.g., *via* their symmetry, shadows, wheel, and licence plates), and pedestrians (e.g., *via* their legs, arms, body, and head). Plan views of the ground plane form useful adjuncts to the information obtained in these ways.

23.11 MORE DETAILED DEVELOPMENTS AND BIBLIOGRAPHIES RELATING TO ADVANCED DRIVER ASSISTANCE SYSTEMS

As indicated earlier in the chapter, in recent years (and particularly since 2000) there has been an almost explosive growth of interest in in-vehicle vision systems. The prime although often unwritten underlying aim has been that of driver assistance—a general term that ultimately includes vehicle guidance. However, in 1998, it at first appeared that Bertozzi and Broggi (1998) had largely solved the problem. In fact, they had laid down many of the ground rules, including finding lane markings with the aid of morphological filters, locating obstacles without constraints on symmetry or shape, analyzing stereo images to find free space on the road ahead, removing the perspective effect, implementing the system on a rapidly operating software plus massively parallel hardware architecture, presenting feedback information to the driver *via* a TV monitor and control panel, testing the system on the road, and above all demonstrating robustness with respect to shadows, changing illumination conditions, varying road texture, and typical motions on the road. Nevertheless, the system was subject to basic assumptions such as the road being flat and road markings being visible; in addition, it placed a great deal of reliance on the stereo system, which had limited range; furthermore, it treated each pair of stereo images individually, and was unable to exploit temporal correlations. Finally, while it never failed to detect vehicles on the road ahead, it sometimes detected false obstacles because of noise arising from the various image remapping processes.

In the light of this work, other workers continued development with increased pace, pressing to eliminate deficiencies with the basic strategy; interestingly, many abandoned the stereo vision approach which brings with it many complications: in fact, appeal to the human vision system demonstrates all too clearly that stereo brings few real advantages for the restricted tasks involved in driving a vehicle (whatever is the case when assembling a gyroscope or other instrument on a workbench). We shall return to this point below.

First, it is worth outlining the findings of Connolly (2009) who has described in a general way the gains to be achieved by advanced driver assistance systems (ADASs). The main keys to success appear to be the provision of lane departure warnings, help for lane changing, collision avoidance, adaptive cruise control, and driver vigilance monitoring. However, it is important that the ADAS should not give too many warnings, or the driver may become annoyed and deactivate it: neither should it fail to act soon enough or give the driver too much confidence or too much freedom. In fact, it is vital for drowsiness to be detected because approximately 30% of motorway accidents are caused by drivers undergoing micro- or macro-sleeps. While much work has been carried out on blink-rate analysis for detecting these conditions, the method has limited effectiveness in probing the state of the brain itself. Nevertheless, it is clear that vision systems can do much to monitor the driver's behavior, and specifically to monitor his direction of gaze and state of *apparent* awareness. Overall, it is probably in the realm of lane departure warnings and of collision avoidance that an ADAS can do the most good, without annoying the driver. Indeed, in the event of the driver's unawareness of an impending collision, or incapability of acting soon enough, the ADAS should be permitted to act autonomously. While this could in principle be legally contentious, it is not without precedent, as anti-lock braking systems are in common use.

There are many causes of collision, and a large proportion of them are due to driver error, even when drowsiness is not a specific factor. Failure to see a vehicle or pedestrian because of preoccupation with other events on or off the road, failure to estimate speeds or trajectories of vehicles sufficiently accurately, failure to judge how rapidly braking can be performed in the prevailing conditions, and lack of awareness of what other drivers intend to do are all involved in causing accidents: this list does not include gross vehicle malfunctions, such as unpredictable tire bursts. In fact, all these factors arise from or are exacerbated by lack of the right information being available soon enough. Thus, it is obvious that vision has a large part to play in overcoming the problems. While radar, lidar, ultrasonics or other technologies may help, vision provides far more of the right sort of information with the right sort of response rates, and computer vision should be able to cope reliably and rapidly enough to make this possible. The main questions are: What will be the cost? Where will the cameras be placed? Can enough of them be used to ensure that relevant information is made available? Fortunately, cameras are by now so cheap that cost—relative to that of a vehicle or of the damage caused in a crash—is no longer a serious problem. On the other hand, the real problems are the sophistication and speed of the associated software (or in the latter case, how the system is to be implemented in hardware—a consideration that is largely postponed until Chapter 26). For the remainder of the chapter, we therefore concentrate mainly on the sophisticated software aspects and what has been achieved since the turn of the Millennium.

23.11.1 Developments in Vehicle Detection

One area of vital concern has been the detection of other vehicles, especially those overtaking (Zhu et al., 2004; Wang et al., 2005; Hilario et al., 2006; Cherng et al., 2009). The last of these papers considers patterns of driving, such as "cutting in" after overtaking, but more subtly how interactions between events involving more than two vehicles can cause distractions that prevent optimal actions being taken: this is because not all dynamic obstacles are predicable; in fact, multiple critical situations can occur simultaneously. The paper takes the line that the computer must follow attention patterns that emulate those of the human brain, and concentrate cyclically on eliminating the various critical phases that are being experienced. The necessary dynamic visual model is in this case tackled using a spatiotemporal attention (STA) neural network. The system of Kuo et al. (2011) concentrates on detecting vehicles on the road ahead, but is also able to assess longitudinal distance information and thus to provide adaptive cruise control (albeit no indications of accuracy are given in the paper). Note that this system uses a monocular camera and thus avoids the difficulties of stereo systems mentioned earlier.

Sun et al. (2004, 2006) reviewed the methods used by various workers to detect vehicles. They reported knowledge-based methods using symmetry, color, shadow, corners, horizontal and vertical edges, texture, and lights. In addition, stereo and motion approaches have been used. They also reported template matching and appearance-based methods, and noted that sensor fusion is needed to ensure that sufficient information is brought to bear to make vehicle detection reliable. They emphasized that hypothesis generation and verification are important for obtaining reliable solutions. Overall, they offered no silver bullet solution, apart from sensor fusion, although (looking at their conclusions as a whole) *method* fusion appears to be rather more important. Among the worst challenges they found were those of "all hours—all weathers" operation. In particular, bad illumination (especially at night) and the results of rain and snow will affect many well-known algorithms for vehicle detection, including those based on shadows. While in principle vehicle lights should provide an easy way of detecting vehicles, in the dark they can prove confusing, especially when rain-soaked roads cause reflections. Sun et al. therefore "believe that these cues have limited employability." However, there are bound to be conditions under which some methods will not work well, but by using method fusion in a dynamic way, giving different methods different weights in different conditions, viable solutions should in the end be obtainable. Whereas humans could be confused in dark situations where no information at all is available, it is difficult to imagine them not being able to solve vehicle detection problems because of rain, snow, or random reflections, and certainly not simply because no shadows are visible.

While the difficulties of dealing with the problems of driving at speed on a motorway can be hugely complicated, with vehicles overtaking on either side and sometimes cutting in, the solution is often to drive more slowly thereby minimizing risks and lowering the data rate to manageable levels. However, the problems of dealing with pedestrians are considerably more complicated. This is because, in contrast to the case of vehicles that travel at more or less constant speeds in constant directions for considerable periods of time—and also have a fair amount of free space immediately around them—pedestrians are unpredictable, sometimes running to get across roads between vehicles, sometimes jay-walking, and sometimes moving in groups having even more unpredictable behavior. A basic problem is that it is unknown when a stationary pedestrian might suddenly move into the roadway, and with a temporary acceleration that exceeds that of most vehicles. Hence, a great many workers have been, and are producing algorithms for pedestrian detection and tracking.

23.11.2 Developments in Pedestrian Detection

Geronimo et al. (2010) have recently reviewed pedestrian detection systems for ADASs. As their paper is very thorough and contains 146 references, the reader is recommended to work carefully through it. Nevertheless, some useful points can be made here. They emphasize that pedestrians exhibit high variability in size, pose, clothing, objects carried, and so on; they appear in cluttered scenes, can be partially occluded, and may be in poor contrast regions; they have to be identified in dynamically varying scenes when both they and the camera are moving; they often appear radically different when viewed from different directions. Geronimo et al. note that silhouette matching, e.g., using the chamfer matching technique, is widely used for detection, yet it needs to be augmented by an additional appearance-based step. (This is not an argument against silhouette matching, but one for using it as a cue in accordance with the idea expressed above that method fusion is required—i.e., method redundancy is needed to cope robustly with *real* scenes containing substantial clutter.) Geronimo et al. (2010) underline the need for verification and refinement. Interestingly, they note that the Kalman filter is (still) by far the most heavily used tracking algorithm—a surprising fact considering that pedestrian motions along pavements, in precincts or crossing the road exhibit far from steady motion (in fact their motions tend to be jerky and indecisive, as they find their way around obstacles and other people). Finally, Geronimo et al. emphasize the need for all hours—all weathers performance; here they note that NIR imaging gives pictures not dissimilar to visible light images, so similar algorithms can be used for analysis. This is less true for thermal (far infrared or FIR) images, which are commonly called "night vision." In any case the latter respond to relative temperature, which is useful for distinguishing hot targets, including pedestrians for vehicles, but inappropriate for examining most of the background or objects such as road signs. Thus, thermal cameras need to be backed up with visible light cameras in the day or NIR cameras in the night, and so would generally constitute an unnecessary expense.

Gavrila and Munder (2007) describe a multi-cue pedestrian detection system: after extensive field tests in difficult urban traffic conditions, they reasonably claim it to be at the (2007) leading edge. The four main detection modules are sparse stereo-based ROI generation, shape-based detection, texture-based classification, and verification using dense stereo, these being complemented by a tracking module. In fact, the paper builds on earlier work (Gavrila et al., 2004), and its main contributions are the method of integration into a multi-cue system for pedestrian detection and a systematic ROC-based procedure for parameter setting and system optimization. In part, the success of the system is due to the use of a novel mixture-of-experts architecture for shape and texture-based classification: here the idea is to take the known shape information and to use texture to partition the feature space into regions of reduced variability—a process that matches well the types of clothing worn by humans. Importantly, the approach using a texture-based mixture-of-experts weighted by the outcome of shape matching was found to outperform an approach based on single texture classifiers. Also notable is the (continued) use of chamfer matching for shape detection, prominent in much of Gavrila's earlier work.

It was remarked earlier that stereo adds considerable complication to a vision system, which may not be justified for an in-vehicle system when most of the objects being viewed will be many meters away. This makes it no surprise that the review article by Enzweiler and Gavrila (2009) concentrates on monocular pedestrian detection. The paper also included descriptions of a number of experimental comparisons of methods for pedestrian detection. Apart from temporal integration and tracking, methods that were tested included the following: (1) Haar wavelet-based cascades, (2) neural networks using local receptive fields, (3) histograms of orientated gradients (HOGs) together with linear SVM classifiers, and (4) combined shape and texture-based approaches. The fourth of these was subsequently disregarded as its main advantage was processing speed, which was not considered relevant to the comparison. The investigation found that the HOG approach outperformed the wavelet and neural network approaches (Section 22.16 contains a brief outline of the HOG approach and also explains why it outperforms the wavelet approach in this type of application). In particular, at a sensitivity of 70%,[2] the respective false positive rates were 0.045, 0.38, and 0.86, representing huge reduction factors for false positives. Similarly, at a sensitivity of 60%, the precision rates were vastly improved for the HOG approach, particularly relative to the neural network approach. It should be emphasized that these results apply for intermediate resolutions with pedestrian images $\sim 48 \times 96$ pixels, while earlier low-resolution work with pedestrian images $\sim 18 \times 36$ pixels led to Haar wavelets being the most viable option. Overall, there seemed to be slight doubt about what the critical factors actually are: in particular, the authors state "perhaps it is the data that matters most, after all," meaning that increased performance may be at least partly due to increases in the size of the training set. In addition, quite a bit depends on the processing constraints that are applied, and for tighter constraints the Haar wavelet approach comes back into its own. However, as ever, it is difficult to standardize or specify image data, or *a fortiori*,

---

[2]This assumes that the term "detection rate" used by the authors actually means "sensitivity" (or "recall"): see Chapter 24. In this paragraph, note that sensitivity gives a reverse measure of false negative rate, $1 - FN/(TP + FN)$, while precision gives a reverse measure of false positive rate, $1 - FP/(TP + FP)$.

image sequence data, so this paper is not able to tell the whole story. Finally, it should be noted that at this point in time, shape-based detection, and in particular the chamfer matching approach, has dropped out of sight because its main advantage was that of speed, and here recognition accuracy measures were the main performance criteria.

Looking back to the work of Curio et al. (2000)—who use Hausdorff distance rather than chamfer matching for template matching—the attention is very much on analyzing limb movements, modeling human walking, and observing human gait patterns. However, they note that the upper body shows a high degree of variation in its appearance, so it is better to restrict pedestrian detection to the lower body: in fact this strategy is both more reliable and more computationally efficient. They also point out that exact modeling is more complicated for women wearing skirts. (A similar situation must apply for men wearing robes or mackintoshes.) Overall, just as the driver is aware of motion and gait as well as the body models of pedestrians, these need to be incorporated into practical pedestrian detection algorithms in order to provide maximum reliability and robustness.

Zhang et al. (2007) performed tests on pedestrian detection in "IR images" (these were actually thermal images taken with a camera operating in the spectral range $7-14$ μm). Their motivation was to make a system that was capable of working at night time, although they also noted that many undesirable activities occur at night or in relative darkness, so the methodology should be useful in other applications as well. They found that IR images are by no means dissimilar to visible light images, so similar algorithms can be used for analyzing them: i.e., there is no need to invent radically different methods for the IR domain. In particular, they found that edgelet and histogram of orientated gradients (HOG) methods (see Dalal and Triggs, 2005) could be adapted to work with IR images, and similarly for boosting and SVM cascade classification methods (Viola and Jones, 2001). Hence they achieved detection performance for IR images comparable to state-of-the-art results for visible light. The underlying reason for this seemed to be that IR and visible light lead to similar silhouettes.

23.11.3 Developments in Road and Lane Detection

Zhou et al. (2006) developed a lane detection and tracking system using a monocular monochromatic camera. They used a deformable template model to initially locate the lane markings, with tabu search for optimal location; then they used a particle filter for tracking the markings. Their experimental results showed that the resulting system was robust against broken lane markings, curved lanes, shadows, distracting edges, and occlusions. Kim (2008) also used a particle filter for tracking lane markings, but employed RANSAC for initial detection. Similarly, Mastorakis and Davies (2011) used RANSAC for detection but modified it for increased reliability, as described in Sections 11.6 and 23.3: see also Borkar et al. (2009). Finally, Marzotto et al. (2010) showed how a RANSAC-based system could be implemented in real time using an FPGA platform: for more details, see Chapter 26.

While the above approaches are suitable for urban roads, which normally have well-defined lane markings, many roads, especially in rural regions, are unstructured and lacking in markings—and the road boundaries may be overgrown with vegetation. Cheng et al. (2010) devised a system with the ability to handle both structured and unstructured types of road using a monocular camera. To achieve this they devised a hierarchical lane detection strategy which was able to achieve high accuracy using quite simple algorithms. First, environment classification of pixels was carried out with high dimensional feature vectors using eigenvalue decomposition regularized discriminant analysis (EDRDA). For unstructured roads, mean-shift segmentation was used and then road boundary candidates were selected from the region boundaries: Bayes rule was used to select the most probable of these as actual boundaries. When the vehicle moved from one type of road to another, the environment classifier indicated that a different algorithm should be used so that accuracy could be maintained.

There is one way in which road and lane mapping schemes are restricted—namely, by the view available from the chosen camera. Typically, this will give an overall viewing angle of up to $\sim 45°$. In fact, ideally, a vehicle-borne camera should have a full 360° viewing angle, so that overtaking vehicles and pedestrians about to approach from the side can be seen clearly. Omnidirectional (catadioptric) cameras may be the best answer to this problem, and many workers are actively pursuing this possibility. Cheng and Trivedi (2007) tested a system which used an omnidirectional camera for the dual tasks of lane detection and monitoring the head pose of the driver (the reason for monitoring head pose is to check that the driver is aware of the situation on the road). Their tests showed that accuracy of lane detection is reduced by a factor of (only) 2−3 because of the reduced resolution available with this sort of camera. Thus, it should prove possible to make savings in the numbers of sensors employed in practical implementations.

23.11.4 Developments in Road Sign Detection

It is a sign of the seriousness with which ADASs are nowadays being taken that a good many papers describing research into the detection and recognition of road signs have been published since the turn of the Millennium. Fang et al. (2003) describe a system that uses neural networks for detecting and tracking road signs by their color and shape. The shapes considered are circles, triangles, octagons, diamonds, and rectangles. Initial detection takes place at some distance, where the road signs appear small and relatively undistorted, and tracking is carried out by a Kalman filter. At each distance, due account is taken of changes in size and shape due to increasing projective distortions, and when a potential sign has become large enough the system verifies that it is a road sign or discards it.

Actual recognition is not discussed in the paper, but detection and tracking are said to be accurate and robust: although speed was slow on a single PC, the neural networks could conveniently be run in parallel on other processors. A related paper by Fang et al. (2004) describes the types of neural network used in this sort of application. Kuo and Lin (2007) describe a similar system, again involving use of neural networks. The latter paper makes use of greater amounts of structural analysis of the images at the detection stage, e.g., using corner detection, Hough transforms and morphology. De la Escalera et al. (2003) describe a system which starts the analysis using color classification, uses genetic algorithms for narrowing down the search, and employs neural networks for sign classification.

McLoughlin et al. (2008) describe practically orientated work on road sign detection and also on the detection of "cats eyes." Their aim is to assess the road signage quality rather than to use it, and to this end they relate the signs to GPS information. They focus particularly on reflectivity aspects of the signs and are able to detect defective road studs and road signs. Their system is fully autonomous and thus the methodology is largely transferrable to ADASs.

Prieto and Allen (2009) describe a vision-based system for detecting and classifying traffic signs using self-organizing maps (SOM)—a type of neural network. A two-stage detection process is adopted—of first detecting potential road signs by analyzing the distribution of red pixels within the image, and then identifying the road signs from the distribution of dark pixels in their central pictograms. The HT approach and other structural analysis approaches were eschewed because they were felt to operate too slowly for (efficient) real-time operation, so the SOM approach was adopted. To achieve recognition of the pictogram, it was divided into 16 blocks arranged in the form of a triangle (or whatever shape the particular sign was found to possess). It was found necessary to normalize brightness over the region of the sign. The hardware of the embedded machine vision used for this application was a hybrid consisting of an FPGA together with a digital implementation of a SOM. Experiments showed that the system had good performance, being able to tolerate substantial changes in position, scale, orientation and partial occlusion of the road signs, and also being trainable, at least to within the model of colored surround and black on white pictograms. For further details of the SOM and the hybrid implementation, see the original paper and the references mentioned therein.

Ruta et al. (2010) have developed a system not based on neural networks (as for many of the above) but on color distance transforms, coupled with a nearest neighbor recognition system. The color distance transform is actually a set of three distance transforms, one for each color (RGB). If a particular color is absent during testing, it is accorded a maximum distance value of 10 pixels to avoid confusing the system. The color distance transform was tested for dependence on a variety of conditions, such as strong incident light, reflections, and deep shade, and was found to be robust to substantial illumination changes. Perhaps more important, it was found to be reasonably invariant to the effects of affine transformations, which a moving camera would be subject to. This is almost certainly because chamfer matching is subject to graceful degradation as distortions occur, so the distances at any template (edge) locations will *gradually* increase with the changing levels of distortion. When compared with other methods, the method performed well, the percentages of correct classifications being 22.3 for HOG/PCA, 62.6 for Haar/AdaBoost, 74.5 for HOG/AdaBoost, and 74.4 for the new method using the color distance transform. The main competitor to the new method, HOG/AdaBoost, offers an elegant solution but is much more complex than the new method, and did not outperform it in any real sense. Hence the new method seemed well adapted to the task it was set.

23.11.5 Developments in Path Planning, Navigation, and Egomotion

The subjects of vehicle guidance and egomotion date from as long ago as 1992 (Brady and Wang, 1992; Dickmanns and Mysliwetz, 1992), while automatic visual guidance in convoys dates from a similar period (Schneiderman et al., 1995; Stella et al., 1995). Mobile robots and the need for path planning were discussed by Kanesalingam et al. (1998) and by Kortenkamp et al. (1998), and later a survey was carried out by DeSouza and Kak (2002): see also Davison and Murray (2002). Guidance of outdoor vehicles, particularly on roads, has undergone increasingly rapid development: see for example Bertozzi and Broggi (1998), Guiducci (1999), Kang and Jung (2003), and Kastrinaki et al. (2003). Zhou et al. (2003) considered the situation for elderly pedestrians—although clearly such work could also be relevant for blind people or wheelchair users. Hofmann et al. (2003) showed that vision and radar can profitably be used together to combine the excellent spatial resolution of vision with the accurate range resolution of radar.

In spite of the evident successes, there is still only a limited number of fully automated visual vehicle guidance systems in everyday use. The main problem would appear to be *potential* lack of the robustness and reliability required to trust the system in "all hours—all weathers" situations—although there are also legal implications for a system that is to be used for control rather than merely for vehicle monitoring.

23.12 PROBLEM

1. Check that the path through the maze shown in Fig. 23.14(c) is optimal, (a) by a hand calculation and (b) by a computer calculation. Confirm that several other paths are also optimal. Obtain a more accurate result by taking the horizontal and vertical neighbors of any pixel as being 2 units away, and taking diagonal neighbors as being 3 units away.

24

Statistical Pattern Recognition

Pattern recognition is a task that humans are able to achieve "at a glance" with little apparent effort. Much of pattern recognition is structural, being achieved essentially by analyzing shape. In contrast, statistical pattern recognition (SPR) treats sets of extracted features as abstract entities that can be used to classify objects on a statistical basis, often by mathematical similarity to sets of features for objects with known classes. This chapter explores the subject, presenting relevant theory where appropriate, and shows how artificial neural networks (ANNs) are able to help with recognition tasks.

Look out for:

- the nearest neighbor algorithm—probably the most intuitive of all statistical pattern recognition techniques.
- Bayes' theory, which forms the ideal minimum error classification system.
- the relation linking the nearest neighbor method to Bayes' theory.
- the reason why the optimum number of features will always be finite.
- the receiver operating characteristic (ROC) curve, which allows an optimum balance between false positives and false negatives to be achieved.
- the distinction between supervised and unsupervised learning.
- the method of principal components analysis and its value.
- how artificial neural networks can be trained, avoiding problems of inadequate training and overfitting to training data.

SPR is a core methodology in the design of practical vision systems. As such it has to be used in conjunction with structural pattern recognition methods and many other relevant techniques—as already indicated in the title to Part 4.

24.1 INTRODUCTION

The earlier chapters of this book have tackled the task of interpreting images on the basis that when suitable cues have been found, the positions of various objects and their identities will emerge in a natural and straightforward way. When the objects that appear in an image have simple shapes, just one stage of processing may be required, as in the case of circular washers on a conveyor. For more complex objects, location requires at least two stages, as when graph matching methods are used. For situations where the full complexity of three dimensions occurs, more subtle procedures are usually required, as has already been seen in Part 3. Indeed, the very ambiguity involved in interpreting the 2-D images from a set of 3-D objects generally requires cues to be sought and hypotheses to be proposed before any serious attempt can be made with the task. Thus, cues are vital to keying into the complex data structures of many images. However, for simpler situations, concentration on small features is valuable in permitting image interpretation to be carried out efficiently and rapidly; neither must it be forgotten that in specific applications the problems tend to be much more restricted and simpler to solve; e.g., only widgets need to be inspected on a widget line, and only vehicles need to be scrutinized on a highway.

However, the fact that it is often expedient to start by searching for cues means that the vision system could lock on to erroneous interpretations. For example, HT-based methods lead to interpretations based on the most prominent peaks in parameter space, while the maximal clique approach supports interpretations based on the largest number of feature matches between parts of an image and the object model. Clearly, noise, background clutter, and other artifacts make it possible that an object will be seen (or its presence hypothesized) when none is there, whereas occlusions, shadows, etc. may cause an object to be missed altogether.

Hence, a rigorous interpretation strategy would ideally try to find all possible feature and object matches, and should have some means of distinguishing between them. An attractive idea is that of not being satisfied with an interpretation until the *whole* of any image has been understood. However, enough has been seen in previous chapters (e.g., Chapter 13) to demonstrate that the computational load of such a strategy will generally make it impracticable for real scenes. Yet there are more constrained types of situation in which the strategy is worth considering and in which the various possible solutions can be evaluated carefully before a final interpretation is reached. These situations arise practically where small relevant parts of images can be segmented and interpreted in isolation. One such case is that of optical character recognition (OCR); a commonly used approach for tackling it is SPR.

The following sections study some of the important principles of SPR. A description of all the work that has been carried out in this area would take several volumes to cover and cannot be attempted here. Fortunately, SPR has been researched for well over three decades and has settled down sufficiently so that an overview chapter can serve a useful purpose. The reader is referred to several existing texts

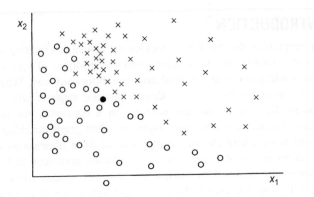

FIGURE 24.1

Principle of the nearest neighbor algorithm for a two-class problem. o, class 1 training set patterns; ×, class 2 training set patterns; •, test pattern.

for further details (Duda et al., 2001; Webb, 2002). We start by describing the nearest neighbor (NN) approach to SPR and then go on to consider Bayes' decision theory, which provides a more general model of the underlying process.

24.2 THE NEAREST NEIGHBOR ALGORITHM

The principle of the nearest neighbor algorithm is that of comparing input image[1] patterns against a number of paradigms and then classifying them according to the class of the paradigm that gives the closest match (Fig. 24.1). An instructive but rather trivial example is shown in Fig. 1.1. Here a number of binary patterns are presented to the computer in the training phase of the algorithm, then the test patterns are presented one at a time and compared bit by bit against each of the training patterns. It is clear that this gives a generally reasonable result, the main problems arising when (a) training patterns of different classes are close together in Hamming distance (i.e., they differ in too few bits to be readily distinguishable) and (b) minor translations, rotations, or noise cause variations that inhibit accurate recognition. More generally, problem (b) means that the training patterns are insufficiently representative of what will appear during the test phase. The latter statement encapsulates an exceptionally important principle and it implies that there must be sufficient patterns in the training set for the algorithm to be able to generalize over all possible patterns of each class. However, problem (a) implies that patterns of two different classes may in some cases be so close as to be

---

[1]Note that a number of the methods discussed in this chapter are very general and can be applied to the recognition of widely different datasets, including, e.g., speech and electrocardiograph waveforms.

indistinguishable by any algorithm, and then it is inevitable that erroneous classifications will be made. It is seen below that this is because the underlying distributions in feature space overlap.

The example of Fig. 1.1 is rather trivial but nevertheless carries important lessons. Note that general images have many more pixels than that in Fig. 1.1 and also are not merely binary. However, since it is pertinent to simplify the data as far as possible to save computation, it is usual to concentrate on various features of a typical image and classify on the basis of these. One example is provided by a more realistic version of the OCR problem, where characters have linear dimensions of at least 32 pixels (although we continue to assume that the characters have been located *reasonably* accurately so that it remains only to classify the subimages containing them). We can start by thinning the characters to their skeletons and making measurements on the skeleton nodes and limbs (see also Chapters 1 and 9). This gives (a) the numbers of nodes and limbs of various types, (b) the lengths and relative orientations of limbs, and perhaps (c) information on curvatures of limbs. Thus, we arrive at a set of numerical features that describe the character in the subimage.

The general technique is to plot the characters in the training set in a multidimensional feature space and to tag the plots with the classification index. Then test patterns are placed in turn in the feature space and classified according to the class of the nearest training set pattern. Clearly, this generalizes the method adopted in Fig. 24.1. In the general case, the distance in feature space is no longer Hamming distance but some more general measure such as Mahalanobis distance (Duda and Hart, 1973). In fact, a problem arises since there is no reason why the different dimensions in feature space should contribute equally to distance, rather they should each have different weights in order to match the physical problem more closely. The problem of weighting cannot be discussed in detail here and the reader is referred to other texts such as that by Duda and Hart (1973). Suffice it to say that with an appropriate definition of distance, the generalization of the method outlined above is adequate to cope with a variety of problems.

To achieve a suitably low error rate, large numbers of training set patterns are normally required. This then leads to significant storage and computation problems. Means have been found for reducing these problems by several important strategies. Notable among these is that of pruning the training set by eliminating patterns that are not near the boundaries of class regions in feature space, since such patterns do not materially help in reducing the misclassification rate.

An alternative strategy for obtaining equivalent performance at lower computational cost is to employ a piecewise linear or other functional classifier instead of the original training set. Clearly, the NN method itself can be replaced, with no change in performance, by a set of planar decision surfaces that are the perpendicular bisectors (or their analogs in multidimensional space) of the lines joining pairs of training patterns of different classes that are on the boundaries of class regions. If this system of planar surfaces is simplified by any convenient means, then the computational load may be reduced further (Fig. 24.2). This may be achieved either

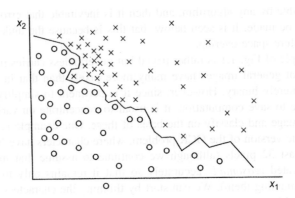

FIGURE 24.2

Use of planar decision surfaces for pattern classification: in this example the "planar decision surface" reduces to a piecewise linear decision boundary in two dimensions. Once the decision boundary is known, the training set patterns themselves need no longer be stored.

indirectly by some smoothing process, as implied above, or directly by finding training procedures that act to update the positions of decision surfaces immediately on receipt of each new training set pattern. The latter approach is in many ways more attractive, since it drastically cuts down storage requirements—although it must be confirmed that a training procedure is selected that converges sufficiently rapidly. Again, discussion of this well-researched topic is left to other texts (Nilsson, 1965; Duda and Hart, 1973; Devijver and Kittler, 1982).

We now turn to a more generalized approach—that of Bayes' decision theory—since this underpins all the possibilities thrown up by the NN method and its derivatives.

24.3 BAYES' DECISION THEORY

The basis of Bayes' decision theory is examined in this section. If we are trying to get a computer to classify objects, a sound approach is to get it to measure some prominent feature of each object such as its length and to use this feature as an aid to classification. Sometimes such a feature may give very little indication of the pattern class—perhaps because of the effects of manufacturing variation. For example, a hand-written character may be so ill-formed that its features are of little help in interpreting it; it then becomes much more reliable to make use of the known relative frequencies of letters, or to invoke context: in fact, either of these strategies can give a greatly increased probability of correct interpretation. In other words, when feature measurements are found to be giving an error rate above a certain threshold, it is more reliable to employ the *a priori* probability of a given pattern appearing.

The next step in improving recognition performance is to combine the information from feature measurements and from *a priori* probabilities. This is achieved by applying Bayes' rule. For a single feature x, this takes the form:

$$P(C_i|x) = \frac{p(x|C_i)P(C_i)}{p(x)} \tag{24.1}$$

where

$$p(x) = \sum_j p(x|C_j)P(C_j) \tag{24.2}$$

Mathematically, the variables here are (a) the *a priori* probability of class C_i, $P(C_i)$; (b) the probability density for feature x, $p(x)$; (c) the class-conditional probability density for feature x in class C_i, $p(x|C_i)$—i.e., the probability that feature x arises for objects known to be in class C_i; and (d) the *a posteriori* probability of class C_i when x is observed, $P(C_i|x)$.

The notation $P(C_i|x)$ is a standard one, being defined as the probability that the class is C_i when the feature is known to have the value x. Bayes' rule says that to find the class of an object we need to know two sets of information about the objects that might be viewed: the first is the basic probability $P(C_i)$ that a particular class might arise; the second is the distribution of values of the feature x for each class. Fortunately, each of these sets of information can be found straightforwardly by observing a sequence of objects as they move along a conveyor. As before, such a sequence of objects is called the training set.

Many common image analysis techniques give features that may be used to help identify or classify objects. These include the area of an object, its perimeter, the numbers of holes it possesses, and so on. Note that classification performance may be improved not only by making use of the *a priori* probability but also by employing a number of features simultaneously. Generally, increasing the number of features helps to resolve object classes and reduce classification errors (Fig. 24.3); however, the error rate is rarely reduced to zero merely by adding more and more features, and indeed the situation eventually deteriorates for reasons explained in Section 24.5.

Bayes' rule can be generalized to cover the case of a generalized feature $\mathbf{x}$, in multidimensional feature space, by using the modified formula:

$$P(C_i|\mathbf{x}) = \frac{p(\mathbf{x}|C_i)P(C_i)}{p(\mathbf{x})} \tag{24.3}$$

where $P(C_i)$ is the *a priori* probability of class C_i, and $p(\mathbf{x})$ is the overall probability density for feature vector $\mathbf{x}$:

$$p(\mathbf{x}) = \sum_j p(\mathbf{x}|C_j)P(C_j) \tag{24.4}$$

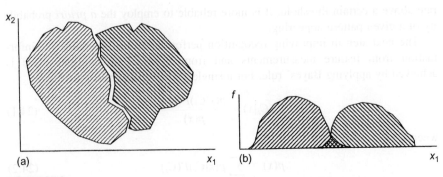

FIGURE 24.3

Use of several features to reduce classification errors: (a) the two regions to be separated in 2-D (x_1, x_2) feature space; (b) frequencies of occurrence of the two classes when the pattern vectors are projected onto the x_1-axis. Clearly, error rates will be high when either feature is used on its own, but will be reduced to a low level when both features are employed together.

The classification procedure is then to compare the values of all the $P(C_j|\mathbf{x})$ and to classify an object as class C_i if:

$$P(C_i|\mathbf{x}) > P(C_j|\mathbf{x}) \quad \text{for all } j \neq i \tag{24.5}$$

24.3.1 The Naive Bayes' Classifier

Many classification methods, including the nearest neighbor method and the Bayes' classifier, can involve substantial amounts of storage and computation if the amount of training is to be sufficient to achieve low error rates. Hence, there is considerable value in employing methods that minimize computation while retaining adequate classification accuracy. In fact, the naive Bayes' classifier is able to achieve this in many applications—particularly those where individual features can be selected that are approximately independent. Features in this category include roundness, size, and redness in the case of oranges.

To understand this, take the expression $p(\mathbf{x}|C_i)P(C_i) = p(x_1, x_2, \dots, x_N|C_i)P(C_i)$ in Eq. (24.4), and re-express it as appropriate for independent (uncorrelated) features $x_1, x_2, \dots, x_N$:

$$p(\mathbf{x}|C_i)P(C_i) = p(x_1|C_i)p(x_2|C_i)\dots p(x_N|C_i) \cdot P(C_i) = \prod_j p(x_j|C_i) \cdot P(C_i) \tag{24.6}$$

This is valid because the overall probability of a set of independent variables is the product of the individual probabilities. First, note that this is a significant simplification of the original general expression. Second, its computation involves only the means and variances of the N individual variables and not the whole $N \times N$ covariance matrix. Clearly, reducing the number of parameters makes the naive Bayes' classifier less powerful. However, this is counterbalanced by the fact that, if the same training set is used, the remaining parameters will be much more accurately determined. The result is that, given the right combination of features, the naive Bayes' classifier can indeed be highly effective in practice.

24.4 RELATION OF THE NEAREST NEIGHBOR AND BAYES' APPROACHES

When Bayes' theory is applied to simple pattern recognition tasks, it is immediately clear that *a priori* probabilities are important in determining the final classification of any pattern, since these probabilities arise explicitly in the calculation. However, this is not so for the NN type of classifier. Indeed, the whole idea of the NN classifier appears to be to get away from such considerations, instead classifying patterns on the basis of training set patterns that lie nearby in feature space. However, there must be a definite answer to the question of whether *a priori* probabilities are or are not taken into account *implicitly* in the NN formulation, and therefore of whether an adjustment needs to be made to the NN classifier to minimize the error rate. Since it is clearly important to have a categorical statement of the situation, Section 24.4.1 is devoted to providing such a statement together with necessary analysis.

24.4.1 Mathematical Statement of the Problem

This section considers in detail the relation between the NN algorithm and Bayes' theory. For simplicity (and with no ultimate loss of generality), we here take all dimensions in feature space to have equal weight, so that the measure of distance in feature space is not a complicating factor.

For greatest accuracy of classification, many training set patterns will be used and it will be possible to define a density of training set patterns in feature space, $D_i(\mathbf{x})$, for position $\mathbf{x}$ in feature space and class C_i. Clearly, if $D_k(\mathbf{x})$ is high at position $\mathbf{x}$ in class C_k, then training set patterns lie close together and a test pattern at $\mathbf{x}$ will be likely to fall in class C_k. More particularly, if:

$$D_k(\mathbf{x}) = \max_i D_i(\mathbf{x}) \tag{24.7}$$

then our basic statement of the NN rule implies that the class of a test pattern $\mathbf{x}$ will be C_k.

However, according to the outline given above, this analysis is flawed in not showing explicitly how the classification depends on the *a priori* probability of class C_k. To proceed, note that $D_i(\mathbf{x})$ is closely related to the conditional probability density $p(\mathbf{x}|C_i)$ that a training set pattern will appear at position $\mathbf{x}$ in feature space if it is in class C_i. Indeed, the $D_i(\mathbf{x})$ are merely non-normalized values of the $p(\mathbf{x}|C_i)$:

$$p(\mathbf{x}|C_i) = \frac{D_i(\mathbf{x})}{\int D_i(\mathbf{x})d\mathbf{x}} \tag{24.8}$$

The standard Bayes' formulae (Eqs. (24.3) and (24.4)) can now be used to calculate the *a posteriori* probability of class C_i.

So far it has been seen that the *a priori* probability should be combined with the training set density data before valid classifications can be made using the NN rule. As a result, it seems invalid merely to take the nearest training set pattern in feature space as an indicator of pattern class. However, note that when clusters of training set patterns and the underlying within-class distributions scarcely overlap, there is anyway a rather low probability of error in the overlap region, and the result of using $p(\mathbf{x}|C_i)$ rather than $P(\mathbf{x}|C_i)$ to indicate class often introduces only a very small bias in the decision surface. Hence, although invalid *mathematically*, the error introduced need not be disastrous.

We now consider the situation in more detail, finding how the need to multiply by the *a priori* probability affects the NN approach. In fact, multiplying by the *a priori* probability can be achieved either *directly*, by multiplying the densities of each class by the appropriate $P(\mathbf{x}|C_i)$, or *indirectly*, by providing a suitable amount of additional training for classes with high *a priori* probability. It may now be seen that the amount of additional training required is *precisely* the amount that would be obtained if the training set patterns were allowed to appear with their natural frequencies (see equations (24.9)–(24.13)). For example, if objects of different classes are moving along a conveyor, we should not first separate them and then train with equal numbers of patterns from each class; we should instead allow them to proceed normally and train on them all at their normal frequencies of occurrence in the training stream. Clearly, if training set patterns do not appear for a time with their proper natural frequencies, this will introduce a bias into the properties of the classifier. Thus, we must make every effort to permit the training set to be representative not only of the *types* of pattern of each class but also of the *frequencies* with which they are presented to the classifier during training.

The above ideas for *indirect* inclusion of *a priori* probabilities may be expressed as follows:

$$P(C_i) = \frac{\int D_i(\mathbf{x})\mathrm{d}\mathbf{x}}{\sum_j \int D_j(\mathbf{x})\mathrm{d}\mathbf{x}} \tag{24.9}$$

Hence

$$P(C_i|\mathbf{x}) = \frac{D_i(\mathbf{x})}{\left(\sum_j \int D_j(\mathbf{x})\mathrm{d}\mathbf{x}\right) p(\mathbf{x})} \tag{24.10}$$

where

$$p(\mathbf{x}) = \frac{\sum_k D_k(\mathbf{x})}{\sum_j \int D_j(\mathbf{x})\mathrm{d}\mathbf{x}} \tag{24.11}$$

Substituting for $p(\mathbf{x})$ now gives:

$$P(C_i|\mathbf{x}) = \frac{D_i(\mathbf{x})}{\sum_k D_k(\mathbf{x})} \tag{24.12}$$

so the decision rule to be applied is to classify an object as class C_i if:

$$D_i(\mathbf{x}) > D_j(\mathbf{x}) \quad \text{for all } j \neq i \tag{24.13}$$

The following conclusions have now been arrived at:

1. The NN classifier may well not include *a priori* probabilities and hence could give a classification bias.
2. It is in general wrong to train an NN classifier in such a way that an equal number of training set patterns of each class are applied.
3. The correct way to train an NN classifier is to apply training set patterns at the natural rates at which they arise in raw training set data.

The third conclusion is perhaps the most surprising and the most gratifying. Essentially, it adds further fire to the principle that training set patterns should be representative of the class distributions from which they are taken, although we now see that it should be generalized to the following: *training sets should be fully representative of the populations from which they are drawn*, where "fully representative" includes ensuring that the frequencies of occurrence of the various classes are representative of those in the whole population of patterns. Phrased in this way, the principle becomes a general one, which is relevant to many types of trainable classifier.

24.4.2 The Importance of the Nearest Neighbor Classifier

The NN classifier is important in being perhaps the simplest of all classifiers to implement on a computer. In addition, it has the advantage of being guaranteed to give an error rate within a factor of two of the ideal error rate (obtainable with a Bayes' classifier). By modifying the method to base classification of any test pattern on the most commonly occurring class among the k nearest training set patterns (giving the "k-NN" method), the error rate can be reduced further until it is arbitrarily close to that of a Bayes' classifier (note that Eq. (24.12) can be interpreted as covering this case too). However, both the NN and (*a fortiori*) the k-NN methods have the disadvantage that they often require enormous storage to record enough training set pattern vectors, and correspondingly large amounts of computation to search through them to find an optimal match for each test pattern, hence necessitating the pruning and other methods mentioned earlier for cutting down the load.

24.5 THE OPTIMUM NUMBER OF FEATURES

It has been stated in Section 24.3 that error rates can be reduced by increasing the number of features used by a classifier, but that there is a limit to this, after which performance actually deteriorates. We here consider why this should happen. Basically, the reason is similar to the situation where many parameters are used to fit a curve to a set of D data points. As the number of parameters P is increased, the fit of the curve becomes better and better, and in general becomes perfect when $P = D$. However, by that stage the significance of the fit is poor, since the parameters are no longer overdetermined and no averaging of their values is taking place. Essentially, all the noise in the raw input data is being transferred to the parameters. The same thing happens with training set patterns in feature space. Eventually, training set patterns are so sparsely packed in feature space that the test patterns have reduced probability of being nearest to a pattern of the same class, so error rates become very high. This situation can also be regarded as due to a proportion of the features having negligible statistical significance, i.e., they add little additional information and serve merely to add uncertainty to the system.

However, an important factor is that the optimum number of features depends on the amount of training a classifier receives. If the number of training set patterns is increased, more evidence is available to support the determination of a greater number of features and hence to provide more accurate classification of test patterns. Indeed, in the limit of very large numbers of training set patterns, performance continues to increase as the number of features is increased.

This situation was first clarified by Hughes (1968) and verified in the case of n-tuple pattern recognition (a variant of the NN classifier due to Bledsoe and Browning, 1959) by Ullmann (1969). Both workers produced clear curves showing the initial improvement in classifier performance as the number of features increased, this improvement being followed by a fall in performance for large numbers of features.

Before leaving this topic, note that the above arguments relate to the number of features that should be used but not to their selection. Clearly, some features are more significant than others, the situation being very data-dependent. It is left as a topic for experimental tests to determine in any case which subset of features will minimize classification errors (see also Chittineni, 1980).

24.6 COST FUNCTIONS AND ERROR—REJECT TRADEOFF

In the foregoing sections, it has been implied that the main criterion for correct classification is that of maximum *a posteriori* probability. However, although probability is always relevant, in a practical engineering environment it can be more important to minimize costs. Hence, it is necessary to compare the costs involved in making correct or wrong decisions. Such considerations can be expressed mathematically by invoking a loss function $L(C_i|C_j)$ that represents the cost involved in making a decision C_i when the true class for feature $\mathbf{x}$ is C_j.

To find a modified decision rule based on minimizing costs, we first define a function known as the conditional risk:

$$R(C_i|\mathbf{x}) = \sum_j L(C_i|C_j)P(C_j|\mathbf{x}) \tag{24.14}$$

This function expresses the expected cost of deciding on class C_i when $\mathbf{x}$ is observed. As it is wished to minimize this function, we decide on class C_i only if:

$$R(C_i|\mathbf{x}) < R(C_j|\mathbf{x}) \quad \text{for all } j \neq i \qquad (24.15)$$

If we were to choose a particularly simple cost function, of the form:

$$L(C_i|C_j) = \begin{cases} 0 & \text{for } i = j \\ 1 & \text{for } i \neq j \end{cases} \qquad (24.16)$$

then the result would turn out to be identical to the previous probability-based decision rule, relation (24.5). Clearly, it is only when certain errors lead to relatively large (or small) costs that it pays to deviate from the normal decision rule. Such cases arise when we are in a hostile environment and must, e.g., give precedence to the sound of an enemy tank over that of other vehicles—it is better to be oversensitive and risk a false alarm than to retain a small chance of not noticing the hostile agent. Similarly, on a production line it may in some circumstances be better to reject a small number of good products than to risk selling a defective product. Cost functions therefore permit classifications to be biased in favor of a safe decision in a rigorous, predetermined, and controlled manner, and the desired balance of properties obtained from the classifier.

Another way of minimizing costs is to arrange for the classifier to recognize when it is "doubtful" about a particular classification, because two or more classes are almost equally likely. Then one solution is to make a safe decision, the decision plane in feature space being biased away from its position for maximum probability classification. An alternative is to reject the pattern, i.e., place it into an "unknown" category: in that case some other means can be employed for making an appropriate classification. Such a classification could be made by going back to the original data and measuring further features, but in many cases it is more appropriate for a human operator to be available to make the final decision. Clearly, the latter approach is more expensive and so introducing a "reject" classification can incur a relatively large cost factor. A further problem is that the error rate is reduced only by a fraction of the amount that the rejection rate is increased.[2] Indeed, in a simple two-class system, the initial decrease in error rate is only one-half the initial increase in reject rate (i.e., a 1% decrease in error rate is obtained only at the expense of a 2% increase in reject rate), and the situation gets rapidly worse as progressively lower error rates are attempted (Fig. 24.4). Thus, very careful cost analysis of the error−reject tradeoff curve must be made before an optimal scheme can be developed. Finally, note that the overall error rate of the classification system depends on the error rate of the classifier that examines the rejects (e.g., the human operator), and this needs to be taken into account in determining the exact tradeoff to be used.

---

[2]All error and reject rates are assumed to be calculated as proportions of the total number of test patterns to be classified.

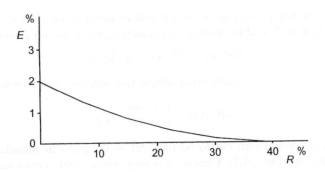

FIGURE 24.4

An error–reject tradeoff curve (E, error rate; R, reject rate). In this example, the error rate E drops substantially to zero for a reject rate R of 40%. More usually E cannot be reduced to zero until R is 100%.

24.7 THE RECEIVER OPERATING CHARACTERISTIC

In the early sections of this chapter, there has been an implicit understanding that classification error rates have to be reduced as far as possible, although in the last section it was acknowledged that it is cost rather than error that is the practically important parameter. It was also found that a tradeoff between error rate and reject rate allows a further refinement to be made to the analysis.

Here we consider another refinement that is required in many practical cases where binary decisions have to be made. Radar provides a good illustration of this, showing that there are two basic types of misclassification: first, radar displays may indicate an aircraft or missile when none is present, in which case the error is called a false positive (or in popular parlance, a false alarm); second, they may indicate that no aircraft or missile is present when there actually is one, in which case the error is called a false negative. Similarly, in automated industrial inspection, when searching for deficient products, a false positive corresponds to finding one when none is present, whereas a false negative corresponds to missing a deficient product when one is present.

In fact, there are four relevant categories: (1) true positives (positives that are correctly classified), (2) true negatives (negatives that are correctly classified), (3) false positives (positives that are incorrectly classified), and (4) false negatives (negatives that are incorrectly classified). If many experiments are carried out to determine the proportions of these four categories in a given application, we can obtain the four probabilities of occurrence. Using an obvious notation, these will be related by the following formulae:

$$P_{\text{TP}} + P_{\text{FN}} = 1 \qquad (24.17)$$

$$P_{TN} + P_{FP} = 1 \qquad (24.18)$$

(In case the reader finds the combinations of probabilities in these formulae confusing, note that an object that is actually a faulty product will be either *correctly* detected as such or *incorrectly* categorized as acceptable, in which case it is a false negative.)

It will be apparent that the probability of error P_E is the sum:

$$P_E = P_{FP} + P_{FN} \qquad (24.19)$$

In general, false positives and false negatives will have different costs. Thus, the loss function $L(C_1|C_2)$ will not be the same as the loss function $L(C_2|C_1)$. For example, missing an enemy missile or failing to find a glass splinter in baby food may be far more costly than the cost of a few false alarms (which in the case of food inspection merely means the rejection of a few good products). In fact, there are a good many applications where there is a prime need to reduce as far as possible the number of false negatives (the number of failures to detect the requisite targets).

But how far should we go in aiming to reduce the number of false negatives? This is an important question that should be answered by systematic analysis rather than by *ad hoc* means. The key to achieving this is to note that the proportions of false positives and false negatives will vary independently with the system setup parameters, although frequently only a single threshold parameter need be considered in any detail. In that case we can eliminate this parameter and determine how the numbers of false positives and false negatives depend on each other. The result is the receiver operating characteristic or "ROC" curve (Fig. 24.5).[3]

The ROC curve will often be approximately symmetrical, and, if expressed in terms of probabilities rather than numbers of items, will pass through the points (1, 0), (0, 1)—as shown in Fig. 24.5. It will generally be highly concave so it will pass well below the line $P_{FP} + P_{FN} = 1$, except at its two ends. The point closest to the origin will often be close to the line $P_{FP} = P_{FN}$. This means that if false positives and false negatives are assigned equal costs, the classifier can be optimized simply by minimizing P_E with the constraint $P_{FP} = P_{FN}$. Note, however, that in general the point closest to the origin is *not* the point that minimizes P_E: the point that minimizes total error is actually the point on the ROC curve where the gradient is -1 (Fig. 24.5).

Unfortunately, there is no general theory predicting the shape of the ROC curve. Furthermore, the number of samples in the training set may be limited (especially in inspection if rare contaminants are being sought), and then it may not prove possible to make an accurate assessment of the shape—especially in the extreme wings of the curve. In some cases this problem can be tackled by modeling, e.g., using exponential or other functions—as shown in Fig. 24.6, where the exponential functions lead to reasonably accurate descriptions. However, the

---

[3]While this text defines the ROC curve in terms of P_{FP} and P_{FN}, many other texts use alternative definitions based, e.g., on P_{TP} and P_{FP}, in which case the graph will appear inverted.

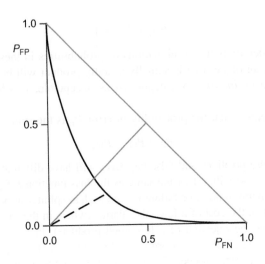

FIGURE 24.5

Idealized ROC curve. The gray line of gradient +1 indicates the position that *a priori* might be expected to lead to minimum error. In fact, the optimum working point is that indicated by the dotted line, where the gradient on the curve is −1. The gray line of gradient −1 indicates the limiting worst case scenario: all practical ROC curves will lie below this line.

underlying shape can hardly be exactly exponential, as this would suggest that the ROC curve tends to zero at infinity rather than at the points (1, 0), (0, 1). Also, there will in principle be a continuity problem at the join of two exponentials. Nevertheless, if the model is reasonably accurate over a good range of thresholds, the relative cost factors for false positives and false negatives can be adjusted appropriately, and an ideal working point determined systematically. Of course, there may be other considerations: e.g., it may not be permissible for the false negative rate to rise above a certain critical level. For examples of the use of the ROC analysis, see Keagy et al. (1995, 1996) and Davies et al. (2003c).

24.7.1 On the Variety of Performance Measures Relating to Error Rates

In signal detection theory (typified by the radar type of application), it is usual to work with error rates rather than the probabilities used in Section 24.7. Hence, we define the following:

$$\text{True positive rate: } tpr = \frac{TP}{P} = \frac{TP}{TP + FN} \tag{24.20}$$

$$\text{True negative rate: } tnr = \frac{TN}{N} = \frac{TN}{TN + FP} \tag{24.21}$$

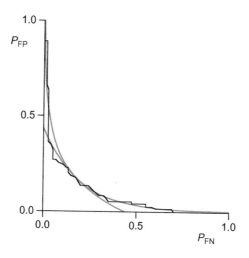

FIGURE 24.6

Fitting a ROC curve using exponential functions. Here the given ROC curve (see Davies et al., 2003c) has distinctive steps resulting from a limited set of data points. A pair of exponential curves fits the ROC curve quite well along the two axes, each having an obvious region that it models best. In this case, the crossover region is reasonably smooth, but there is no real theoretical reason for this. Furthermore, exponential functions will not pass through the limiting points (0, 1) and (1, 0).

$$\text{False positive rate: } fpr = \frac{FP}{N} = \frac{FP}{TN + FP} \tag{24.22}$$

$$\text{False negative rate: } fnr = \frac{FN}{P} = \frac{FN}{TP + FN} \tag{24.23}$$

where P and N are the actual numbers of objects in classes P and N, respectively. Following Eqs. (24.17) and (24.18), we have:

$$tpr + fnr = 1 \tag{24.24}$$

and

$$tnr + fpr = 1 \tag{24.25}$$

These two equations show the consistency of the four definitions presented above.

It is unfortunate and sometimes confusing that a plethora of names for these and related parameters have arisen in various fields of pattern recognition. Below we aim to make sense of these names as well as defining them:

Sensitivity	A parameter describing the success in finding a particular type of target. It is also a synonym for *hit-rate*. It is therefore equal to *tpr*.
Recall	A term used when describing the success in finding an item in a database. It is therefore also equal to *tpr*.

Specificity	A term that is important in medicine, and relates to the proportion of *well* patients who are accurately told after a test that they are not ill. It is therefore equal to $tnr = 1 - fpr$.
Discriminability	A term used when describing the success in differentiating a particular type of target from a similar type of target. It is therefore equal to $TP/(TP + FP)$.
Precision	A term describing the accuracy in picking out a particular type of target from any distractors, including noise and clutter. It is also equal to $TP/(TP + FP)$.
Accuracy	A term used to describe the overall success rate when distinguishing between foreground and background. We can deduce that it must be equal to $(TP + TN)/(P + N)$.
False alarm rate	Another synonym for *fpr*.
Positive predictive value	Another synonym for *precision*.
F-measure	A measure that is used as an overall performance indicator, combining the recall and precision measures (or alternatively the sensitivity and discriminability measures). Because it is the numbers of errors $(FP + FN)$ that have to be combined (rather than the error rates themselves), the formula for *F-measure* may initially appear overcomplex: $\dfrac{2}{1/recall + 1/precision}$.

A more general formula for F-measure is the following:

$$F_\gamma\text{-}measure = \frac{1}{\frac{\gamma}{recall} + \frac{(1-\gamma)}{precision}} \qquad (24.26)$$

where γ can be adjusted to give the most appropriate weighting between *recall* and *precision*: it is normally given a value close to 0.5.

Quite often, recall and precision are used together to form ROC-like graphs, although they are distinct from ROC curves which typically show *tpr* vs. *fpr*. (Note that, although *recall* = *tpr*, *precision* ≠ *fpr*.)

Finally, a valuable performance indicator for binary classifiers is obtained by finding the area under the curve (*AUC*) for a *tpr* vs. *fpr* ROC curve. It is largest when the ROC curve lies close to the *tpr* = 1, *fpr* = 0 axes. Using this measure, the best classifier is the one that has the largest *AUC*.

24.8 MULTIPLE CLASSIFIERS

In recent years, there have been moves to make the classification process more reliable by application of multiple classifiers working in cooperation. The basic concept is much like that of three magistrates coming together to make a more reliable judgement than any can make alone. Each is expert in a variety of things, but not in everything, so putting their knowledge together in an appropriate way should permit more reliable judgements to be made. A similar concept applies to expert AI systems: multiple expert systems should be able to make up for each other's shortcomings. In all these cases, some way should exist for getting the most out of the individual classifiers without confusion reigning.

Note that the idea is not just to take all the feature detectors that the classifiers use and to replace their output decision-making devices with a single more complex decision-making unit. Indeed, such a strategy could well run into the problem discussed in Section 24.5—of exceeding the optimum number of features: at best only a minor improvement would result from such a strategy, and at worst the system would be grossly failure-prone. On the contrary, the idea is to take the final classification of a number of complete but totally separate classifiers and to combine their outputs to obtain a substantially improved output. Furthermore, it could happen that the separate classifiers use totally different strategies to arrive at their decisions: one may be a nearest neighbor classifier; another may be a Bayes' classifier; and another may be a neural network classifier (see Sections 24.13–24.16). Likewise, one may employ structural pattern recognition, one might use SPR, and another might use syntactic pattern recognition. Each will be respectable in its own right: each will have its own strengths and weaknesses. Part of the idea is one of convenience: to make use of any soundly based classifier that is available, and to boost its effectiveness by using it in conjunction with other soundly based classifiers.

The next task is to see how to achieve this in practice. Perhaps the most obvious way forward is to get the individual classifiers to vote for the class of each input pattern. While this is a nice idea, it will often fail because the weaknesses of the individual classifiers may be worse than their strengths. Thus, the concept must be made more sophisticated.

Another strategy is again to allow the individual classifiers to vote, but this time to make them do so in an exclusive manner, so that as many classes as possible are eliminated for each input pattern. This is achievable with a simple intersection rule: a class is accepted as a possibility only if *all* the classifiers indicate that it is a possibility. The strategy is implemented by applying a threshold to each classifier in a special way, which will now be described.

A prerequisite for this strategy to work is that each classifier must not only give a class decision for each input pattern: it must also give the ranks of all possible classes for each pattern. In other words, it must give its first choice of class for any pattern, its second choice for that pattern, and so on. Then the classifier is labeled with the rank it assigned to the true class of that pattern. In fact, we apply each classifier to the whole training set and get a table of ranks (Table 24.1). Finally, we find the worst case (largest rank)[4] for each classifier, and take that as a threshold value that will be used in the final multiple classifier. When using this method for testing input patterns, only those classifiers that are not excluded by

---

[4]In everyday parlance, the worst case corresponds to the lowest rank, which is here the largest numerical rank; similarly, the highest rank is the smallest numerical rank. It is obviously necessary to be totally unambiguous about this nomenclature.

Table 24.1 Determining a Set of Classifiers for the Intersection Strategy

| | Classifier Ranks | | | | |
	C_1	C_2	C_3	C_4	C_5
D_1	5	3	7	1	8
D_2	4	9	6	4	2
D_3	5	6	7	1	4
D_4	4	7	5	3	5
D_5	3	5	6	5	4
D_6	6	5	4	3	2
D_7	2	6	1	3	8
thr	6	9	7	5	8

In the upper section of this table, the original classifier ranks are shown for each input pattern; in the bottom line of the table, only the worst-case rank is retained. When later applying test patterns, this can be used as the threshold (marked 'thr') to determine which classifiers should be employed.

the threshold have their outputs intersected to give the final list of classes for the input pattern.

The above "intersection strategy" focuses on the worst-case behavior of the individual classifiers, and the result could be that a number of classifiers will hardly reduce the list of possible classes for the input patterns. This tendency can be tackled by an alternative "union strategy," which focusses on the specialisms of the individual classifiers: the aim is then to find a classifier that recognizes each particular pattern well. To achieve this, we look for the classifier with the smallest rank (classifier rank being defined exactly as already defined above for the intersection strategy) for each individual pattern (Table 24.2). Having found the smallest rank for the individual input patterns, we determine the largest of these ranks that arises for each classifier as we go right through all the input patterns. Applying this value as a threshold now determines whether the output of the classifier should be used to help determine the class of a pattern. Note that the threshold is determined in this way using the training set and is later used to decide which classifiers to apply to individual test patterns. Thus, for any pattern a restricted set of classifiers is identified that can best judge its class.

To clarify the operation of the union strategy, let us examine how well it will work on the training set. In fact, it is guaranteed to retain enough classifiers to ensure that the true class of any pattern is not excluded (although naturally, this is not guaranteed for any member of the test set). Hence, the aim of employing a classifier that recognizes each particular pattern well is definitely achieved.

Unfortunately, this guarantee is not obtained without cost. Specifically, if a member of the training set is actually an outlier, the guarantee will still apply, and the overall performance may be compromised. This problem can be tackled

Table 24.2 Determining a Set of Classifiers for the Union Strategy

	Classifier Ranks					Best Classifiers				
	C_1	C_2	C_3	C_4	C_5	C_1	C_2	C_3	C_4	C_5
D_1	5	3	7	1	8	0	0	0	1	0
D_2	4	9	6	4	2	0	0	0	0	2
D_3	5	6	7	1	4	0	0	0	1	0
D_4	4	7	5	3	5	0	0	0	3	0
D_5	3	5	6	5	4	3	0	0	0	0
D_6	6	5	4	3	2	0	0	0	0	2
D_7	2	6	1	3	8	0	0	1	0	0
min–max threshold						3	0	1	3	2

In the left-hand section of this table, the original classifier ranks are shown for each input pattern; in the right-hand section of the table, only one rank is retained, namely that obtaining for the classifier that is best able to recognize that pattern. Note that to facilitate the next piece of analysis—finding the thresholds on the classifier ranks—all remaining places in the table are packed with zeros. A zero final threshold then indicates a classifier that is of no help in analyzing the input data.

in many ways, but a simple possibility is to eliminate excessively bad exemplars from the training set. Another way is to abandon the union strategy altogether and go for a more sophisticated voting strategy. Other approaches involve reordering the data to improve the rank of the correct class (Ho et al., 1994).

24.9 CLUSTER ANALYSIS

24.9.1 Supervised and Unsupervised Learning

In the earlier parts of this chapter, we made the implicit assumption that the classes of all the training set patterns are known, and in addition that they should be used in training the classifier. Indeed, this assumption might be thought of as inescapable. However, classifiers may actually use two approaches to learning—*supervised learning* (in which the classes are known and used in training) and *unsupervised learning* (in which they are either unknown or else known and not used in training). Unsupervised learning can frequently be advantageous in practical situations. For example, a human operator is not required to label all the products coming along a conveyor, as the computer can find out for itself both how many classes of product there are and which categories they fall into: in this way considerable operator effort is eliminated; in addition, it is not unlikely that a number of errors would thereby be circumvented. Unfortunately, unsupervised learning involves a number of difficulties, as will be seen in Subsection 24.9.2.

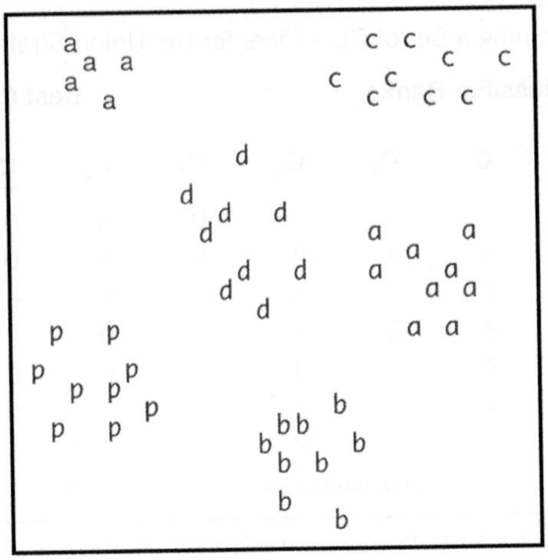

FIGURE 24.7

Location of clusters in feature space. Here the letters correspond to samples of characters taken from various fonts. The small cluster of a's with strokes bent over the top from right to left appear at a separate location in feature space: this type of deviation should be detectable by cluster analysis.

Before proceeding, we give two other reasons why unsupervised learning is useful. First, when the characteristics of objects vary with time—e.g., beans changing in size and color as the season develops—it will be necessary to track these characteristics within the classifier, and unsupervised learning provides an excellent means of approaching this task. Second, when setting up a recognition system, the characteristics of objects, and in particular their most important parameters (e.g., from the point of view of quality control), may well be unknown, and it will be useful to gain some insight into the nature of the data. Thus, types of fault will need to be logged, and permissible variants on objects will need to be noted. As an example, many OCR fonts (such as Times Roman) have a letter "a" with a stroke bent over the top from right to left, although other fonts (such as Monaco) do not have this feature. An unsupervised classifier will be able to flag this up by locating a cluster of training set patterns in a totally separate part of feature space (see Fig. 24.7). In general, unsupervised learning is about the location of clusters in feature space.

24.9.2 Clustering Procedures

As indicated above, an important reason for performing cluster analysis is characterization of the input data. However, the underlying motivation is normally to classify test data patterns reliably. To achieve these aims, it will be necessary both to partition feature space into regions corresponding to significant clusters and to label each region (and cluster) according to the type of data involved. In practice, this can happen in two ways:

1. By performing cluster analysis, and then labeling the clusters by specific queries to human operators on the classes of a small number of individual training set patterns
2. By performing supervised learning on a small number of training set patterns, and then performing unsupervised learning to expand the training set to realistic numbers of examples

In either case, there is ultimately no escape from the need for supervised classification. However, by placing the main emphasis on unsupervised learning we limit tedium and the possibility of preconceived ideas about possible classes from affecting the final recognition performance.

Before proceeding further, note that there are cases where we may have absolutely no idea in advance about the number of clusters in feature space: this occurs in classifying the various regions in satellite images. Such cases are in direct contrast with applications such as OCR or recognizing chocolates being placed in a chocolate box.

Cluster analysis involves a number of very significant problems. Not least is the visualization problem. First, in one, two, or even three dimensions, we can easily visualize and decide on the number and location of any clusters, but this capability is misleading: we cannot extend this capability to feature spaces of many dimensions. Second, computers do not visualize as we do, and special algorithms will have to be provided to enable them to do so. While computers could be made to emulate our capability in low-dimensional feature spaces, a combinatorial explosion would occur if we attempted this for high-dimensional spaces. This means that we will have to develop algorithms that operate on *lists* of feature vectors, if we are to produce automatic procedures for cluster location.

Available algorithms for cluster analysis fall into two main groups—*agglomerative* and *divisive*. Agglomerative algorithms start by taking the individual feature points (training set patterns, excluding class) and progressively grouping them together according to some similarity function until a suitable target criterion is reached. Divisive algorithms start by taking the whole set of feature points as a single large cluster, and progressively dividing it until some suitable target criterion is reached. Let us assume that there are P feature points. Then, in the worst case, the number of comparisons between pairs of individual feature point positions, which will be required to decide whether to combine a pair of clusters in an agglomerative algorithm, will be:

$$^{P}C_{2} = \tfrac{1}{2}P(P-1) \tag{24.27}$$

while the number of iterations required to complete the process will be of order $P-k$ (here we are assuming that the final number of clusters to be found is k,

Table 24.3 Basis of Forgy's Algorithm for Cluster Analysis

```
choose target number k of clusters;
set initial cluster centers;
calculate quality of clustering;
do {
    assign each data point to the closest cluster center;
    recalculate cluster centers;
    recalculate quality of clustering;
} until no further change in the clusters or the quality of the clusters;
```

where $k \le P$). On the other hand, for a divisive algorithm, the number of comparisons between pairs of individual feature point positions will be reduced to:

$$^{k}C_2 = \frac{1}{2}k(k-1) \tag{24.28}$$

while the number of iterations required to complete the process will be of order k.

Although it would appear that divisive algorithms require far less computation than agglomerative algorithms, this is not so. This is because any cluster containing p feature points will have to be examined for a huge number of potential splits into subclusters, the actual number being of order:

$$\sum_{q=1}^{p} {}^{p}C_q = \sum_{q=1}^{p} \frac{p!(p-q)!}{q!} \tag{24.29}$$

This means that in general the agglomerative approach will have to be adopted. In fact, the type of agglomerative approach outlined above is exhaustive and rigorous, and a less exacting, iterative approach can be used. First, a suitable number k of cluster centers are set (these can be decided from *a priori* considerations, or by making arbitrary choices. Second, each feature vector is assigned to the closest cluster center. Third, the cluster centers are recalculated. This process is repeated if any feature points have moved from one cluster to another during the iteration, although termination can also be instituted if the quality of clustering ceases to improve. The overall algorithm, which was originally due to Forgy (1965), is given in Table 24.3.

Clearly, the effectiveness of this algorithm will be highly data-dependent—in particular, with regard to the order in which the data points are presented. In addition, the result could be oscillatory or nonoptimal (in the sense of not arriving at the best solution). This could happen if at any stage a single cluster center arose near the center of a pair of small clusters. In addition, the method gives no indication of the most appropriate number of clusters. Accordingly, a number of variant and alternative algorithms have been devised. One such algorithm is the ISODATA algorithm (Ball and Hall, 1966). This is similar to Forgy's method, but is able to merge clusters that are close together and to split elongated clusters.

Table 24.4 Basis of MacQueen's *k-means* Algorithm

```
choose target number k of clusters;
set the k initial cluster centers at k data points;
for all other data points {//first pass
    assign data point to closest cluster center;
    recalculate relevant cluster center;
}
for all data points //second pass
    re-assign data point to closest cluster center;
```

Another disadvantage of iterative algorithms is that it may not be obvious when to get them to terminate: as a result, they are liable to be too computation intensive. Thus, there has been some support for noniterative algorithms. MacQueen's *k-means* algorithm (MacQueen, 1967) is one of the best known non-iterative clustering algorithms; it involves two runs over the data points, one being required to find the cluster centers and the other being required to finally classify the patterns (see Table 24.4). Again, the choice of which data points are to act as the initial cluster centers can be either arbitrary or on some more informed basis.

Noniterative algorithms are, as indicated earlier, very dependent on the order of presentation of the data points. With image data this is especially problematic, as the first few data points are quite likely to be similar (e.g., all derived from sky or other background pixels). A useful way of overcoming this problem is to randomize the choice of data points so that they can arise from anywhere in the image. In general, noniterative clustering algorithms are less effective than iterative algorithms because they are overinfluenced by the order of presentation of the data.

Overall, the main problem with the algorithms described above is the lack of indication they give of the most appropriate value of k. However, if a range of possible values for k is known, all of them can be tried, and the one giving the best performance in respect of some suitable target criterion can be taken as providing an optimal result. In that case, we will have found the set of clusters that, in some specified sense, gives the best overall description of the data. Alternatively, some method of analyzing the data to determine k can be used before final cluster analysis: the Zhang and Modestino (1990) approach falls into this category.

24.10 PRINCIPAL COMPONENTS ANALYSIS

Closely related to cluster analysis is the concept of data representation. One powerful way of approaching this task is that of principal components analysis. This involves finding the mean of a cluster of points in feature space and then finding the principal axes of the cluster in the following way. First an axis is found which passes through the mean position and which gives the maximum variance when

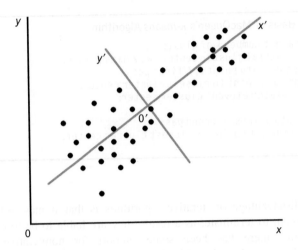

FIGURE 24.8

Illustration of principal components analysis. Here the dots represent patterns in feature space and are initially measured relative to the x- and y-axes. Then the sample mean is located at $0'$, and the direction $0'x'$ of the first principal component is found as the direction along which the variance is maximized. The direction $0'y'$ of the second principal component is normal to $0'x'$; in a higher dimensional space it would be found as the direction normal to $0'x'$ along which the variance is maximized.

the data is projected onto it. Then a second such axis is found which maximizes variance in a direction normal to the first. This process is carried out until a total of N principal axes have been found for an N-dimensional feature space. The process is illustrated in Fig. 24.8. In fact, the process is entirely mathematical and need not be undertaken in the strict sequence indicated above. It merely involves finding a set of orthogonal axes that diagonalizes the covariance matrix.

The covariance matrix for the input population is defined as:

$$C = E\{(\mathbf{x}_{(p)} - \mathbf{m})(\mathbf{x}_{(p)} - \mathbf{m})^{T}\} \qquad (24.30)$$

where $\mathbf{x}_{(p)}$ is the location of the pth data point, and $\mathbf{m}$ is the mean of the P data points; $E\{\ldots\}$ indicates expectation value for the underlying population. We can estimate C from the following equations:

$$C = \frac{1}{P}\sum_{p=1}^{P}\mathbf{x}_{(p)}\mathbf{x}_{(p)}^{T} - \mathbf{m}\mathbf{m}^{T} \qquad (24.31)$$

$$\mathbf{m} = \frac{1}{P}\sum_{p=1}^{P}\mathbf{x}_{(p)} \qquad (24.32)$$

Since C is real and symmetric, it is possible to diagonalize it using a suitable orthogonal transformation matrix A, obtaining a set of N orthonormal eigenvectors $\mathbf{u}_i$ with eigenvalues λ_i given by:

$$C\mathbf{u}_i = \lambda_i \mathbf{u}_i \quad (i = 1, 2, \ldots, N) \tag{24.33}$$

The vectors $\mathbf{u}_i$ are derived from the original vectors $\mathbf{x}_i$ by:

$$\mathbf{u}_i = A(\mathbf{x}_i - \mathbf{m}) \tag{24.34}$$

and the inverse transformation needed to recover the original data vectors is:

$$\mathbf{x}_i = \mathbf{m} + A^\mathsf{T} \mathbf{u}_i \tag{24.35}$$

Here we have recalled that, for an orthogonal matrix:

$$A^{-1} = A^\mathsf{T} \tag{24.36}$$

In fact it may be shown that A is the matrix whose rows are formed from the eigenvectors of C, and that the diagonalized covariance matrix C' is given by:

$$C' = ACA^\mathsf{T} \tag{24.37}$$

so that:

$$C' = \begin{bmatrix} \lambda_1 & 0 & \cdots & 0 \\ 0 & \lambda_2 & & 0 \\ \vdots & & \ddots & \vdots \\ 0 & 0 & \cdots & \lambda_N \end{bmatrix} \tag{24.38}$$

Note that in an orthogonal transformation, the trace of a matrix remains unchanged. Thus, the trace of the input data is given by:

$$\text{trace } C = \text{trace } C' = \sum_{i=1}^{N} \lambda_i = \sum_{i=1}^{N} s_i^2 \tag{24.39}$$

where we have interpreted the λ_i as the variances of the data in the directions of the principal component axes (note that for a real symmetric matrix, the eigenvalues are all real and positive).

In what follows, we shall assume that the eigenvalues have been placed in an ordered sequence, starting with the largest. In that case, λ_1 represents the most significant characteristic of the set of data points, with the later eigenvalues representing successively less significant characteristics. We could even go so far as to say that, in some sense, λ_1 represents the most interesting characteristic of the data, while λ_N would be largely devoid of "interest." More practically, if we ignored λ_N, we might not lose much useful information, and indeed the last few eigenvalues would frequently represent characteristics that are not statistically significant and are essentially noise. For these reasons, principal components analysis is commonly used for reduction in the dimensionality of the feature space from N to some lower value N'. In some applications, this would be taken as leading to a useful amount of data compression. In other applications, it would be taken as providing a reduction in the enormous redundancy present in the input data.

We can quantify these results by writing the variance of the data in the reduced dimensionality space as:

$$\text{trace}(C')_{\text{reduced}} = \sum_{i=1}^{N'} \lambda_i = \sum_{i=1}^{N'} s_i^2 \qquad (24.40)$$

Not only is it now clear why this leads to reduced variance in the data, but also we can see that the mean square error obtained by making the inverse transformation (Eq. (24.35)) will be:

$$\overline{e^2} = \sum_{i=1}^{N} s_i^2 - \sum_{i=1}^{N'} s_i^2 = \sum_{i=N'+1}^{N} s_i^2 \qquad (24.41)$$

One application in which principal components analysis has become especially important is the analysis of multispectral images, e.g., from earth-orbiting satellites. Typically, there will be six separate input channels (e.g., three color and three infra-red), each providing an image of the same ground region. If these images are 512×512 pixels in size, there will be about a quarter of a million data points and these will have to be inserted into a six-dimensional feature space. After finding the mean and covariance matrix for these data points, the latter is diagonalized and a total of six principal component images can be formed. Commonly, only two or three of these will contain immediately useful information, and the rest can be ignored. (For example, the first three of the six principal component images may well possess 95% of the variance of the input images.) Ideally, the first few principal component images in such a case will highlight such areas as fields, roads, and rivers, and this will be precisely the data that is required for map-making or other purposes. In general, the vital pattern recognition tasks can be aided and considerable savings in storage can be achieved on the incoming image data by attending to just the first few principal components.

Finally, it is as well to note that principal components analysis really provides a particular form of data representation. In itself it does not deal with pattern classification, and methods that are required to be useful for the latter type of task must possess useful discrimination. Thus, selection of features simply because they possess the highest variability does not mean that they will necessarily perform well in pattern classifiers. Another important factor that is relevant to the whole study of data analysis in feature space is the scales of the various features. Often, these will be an extremely variegated set, including length, weight, color, numbers of holes, and so on. Clearly, such a set of features will have no special comparability and are unlikely even to be measurable in the same units. This means that placing them in the same feature space and assuming that the scales on the various axes should have the same weighting factors must be invalid. One way of tackling this problem is to normalize the individual features to some standard scale given by measuring their variances. Such a procedure will naturally radically change the results of principal components calculations and further mitigates against principal components methodology being used thoughtlessly. On the other hand, there are some occasions when different features can be compatible, and where principal components analysis can be performed without such worries: one such situation is where all the features are pixel intensities in the same window (this case is discussed in Section 8.5).

24.11 THE RELEVANCE OF PROBABILITY IN IMAGE ANALYSIS

Having seen the success of Bayes' theory in pointing to apparently absolute answers in the interpretation of certain types of image, it is attractive to consider complex scenes in terms of the probabilities of various interpretations and the likelihood of a particular interpretation being the correct one. Given a sufficiently large number of such scenes and their interpretations, it seems that it ought to be possible to use them to train a suitable classifier. However, practical interpretation in real time is quite another matter. Next, note that the eye—brain system does not appear to operate in a manner corresponding to the algorithms we have studied. Instead, it appears to pay attention to various parts of an image in a nonpredetermined sequence of "fixations" of the eye, interrogating various parts of the scene in turn and using the newly acquired information to work out where the next piece of relevant information is to come from. Clearly, it is employing a process of *sequential pattern recognition*, which saves effort overall by progressively building up a store of knowledge about relevant parts of the scene and at the same time forming and testing hypotheses about its structure.

The above process can be considered as one of modifying and updating the *a priori* probabilities as analysis progresses. This is an inherently powerful process, since the eye is thereby not tied to "average" *a priori* probabilities for *all* scenes but is able to use information in a particular scene to improve on the average *a priori* probabilities. However, it will be difficult to estimate at all accurately the *a priori* probabilities for sequences of real, complex scenes. So while this is a tempting approach, its realization will be fraught with difficulty.

Nevertheless, the concept of probability is useful when it can validly be applied. This certainly covers cases where a restricted range of images can arise, such that the consequent image description contains relatively few bits of information—*viz.* those forming the various pattern class names. In addition, structured images containing several parts that can separately be recognized and then coupled together can also be dealt with under the SPR (probabilistic) formalism. Overall, there are limits to its application, but it can still be used in conjunction with structural, syntactic, and other forms of pattern recognition in the design of more powerful recognition systems.

24.12 ANOTHER LOOK AT STATISTICAL PATTERN RECOGNITION: THE SUPPORT VECTOR MACHINE

The support vector machine (SVM) is a new paradigm for SPR, and emerged during the 1990s as an important contender for practical applications. The basic concept relates to linearly separable feature spaces and is illustrated in Fig. 24.9(a). The idea is to find the pair of parallel hyperplanes that leads to the maximum separation between two classes of feature so as to provide the greatest protection against errors. In Fig. 24.9(a), the dashed set of hyperplanes has lower separation and thus represents a less ideal choice, with reduced protection against errors. Each pair of parallel hyperplanes is characterized by specific sets of feature points—the so-called "support vectors." In the feature space shown in Fig. 24.9 (a), the planes are fully defined by three support vectors, although clearly this particular value only applies for 2-D feature spaces: in N dimensions the number of support vectors required is $N+1$. This provides an important safeguard against overfitting; since however many data points exist in a feature space, the maximum number of vectors used to describe it is $N+1$.

For comparison, Fig. 24.9(b) shows the situation that would exist if the nearest neighbor method were employed. In this case the protection against errors would be higher, as each position on the separating surface is optimized to the highest local separation distance. However, this increase in accuracy comes at quite high cost in the much larger number of defining example patterns. Indeed, as indicated above, much of the gain of the SVM comes from its use of the smallest possible number of defining example patterns (the support vectors). The disadvantage is that the basic method only works when the dataset is linearly separable.

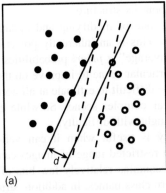 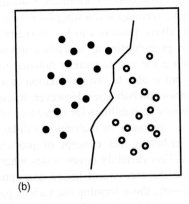

(a)   (b)

FIGURE 24.9

Principle of the support vector machine. Part (a) shows two sets of linearly separable feature points: the two parallel hyperplanes have the maximum possible separation d, and should be compared with alternatives such as the pair shown dashed. Part (b) shows the optimal piecewise linear solution that would be found by the nearest neighbor method.

To overcome this problem, it is possible to transform the training and test data to a feature space of higher dimension where the data does become linearly separable. In fact, this approach will tend to reduce or even eliminate the main advantage of the SVM and lead to overfitting of the data plus poor generalizing ability. However, if the transformation that is employed is nonlinear, the final (linearly separable) feature space could have a manageable number of dimensions, and the advantage of the SVM may not be eroded. Nevertheless, there comes a point where the basic restriction of linear separability has to be questioned. At that point, it has been found useful to build "slack" variables s_i into the optimization equations to represent the amount by which the separability constraint can be violated. This is engineered by adding a cost term $C\Sigma_i s_i$ to the normal error function: C is adjustable and acts as a regularizing parameter, which is optimized by monitoring the performance of the classifier on a range of training data.

For further information on this topic, the reader should consult the original papers by Vapnik, including Vapnik (1998), the specialized text by Cristianini and Shawe-Taylor (2000), or other texts on SPR, such as Webb (2002).

24.13 ARTIFICIAL NEURAL NETWORKS

The concept of an artificial neural network that could be useful for pattern recognition started in the 1950s and continued right through the 1960s. For example, Bledsoe and Browning (1959) developed the "n-tuple" type of classifier that involved bit-wise recording and lookup of binary feature data, leading to the "weightless" or "logical" type of ANN. Although the latter type of classifier maintained a continuous following for many years, it is probably no exaggeration to say that it is Rosenblatt's "perceptron" (1958, 1962), which has had the greatest influence on the subject.

The simple perceptron is a linear classifier that classifies patterns into two classes. It takes a feature vector $\mathbf{x} = (x_1, x_2,\ldots, x_N)$ as its input, and produces a single scalar output $\sum_{i=1}^{N} w_i x_i$, the classification process being completed by applying a threshold (Heaviside step) function at θ (see Fig. 24.10). The mathematics is simplified by writing $-\theta$ as w_0, and taking it to correspond to an input x_0 that is maintained at a constant value of unity. The output of the linear part of the classifier is then written in the form:

$$d = \sum_{i=1}^{N} w_i x_i - \theta = \sum_{i=1}^{N} w_i x_i + w_0 = \sum_{i=0}^{N} w_i x_i \tag{24.42}$$

and the final output of the classifier is given by:

$$y = f(d) = f\left(\sum_{i=0}^{N} w_i x_i\right) \tag{24.43}$$

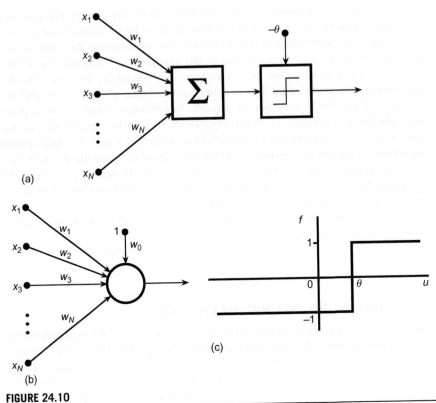

FIGURE 24.10

Simple perceptron. (a) shows the basic form of a simple perceptron: input feature values are weighted and summed, and the result fed via a threshold unit to the output connection. (b) gives a convenient shorthand notation for the perceptron; and (c) shows the activation function of the threshold unit.

This type of neuron can be trained using a variety of procedures, such as the *fixed increment rule* given in Table 24.5. (The original fixed increment rule used a learning rate coefficient η equal to unity.) The basic concept of this algorithm was to try to improve the overall error rate by moving the linear discriminant plane a fixed distance toward a position where no misclassification would occur—but only doing this when a classification error had occurred:

$$w_i(k+1) = w_i(k) \quad y(k) = \omega(k) \tag{24.44}$$

$$w_i(k+1) = w_i(k) + \eta[\omega(k) - y(k)]x_i(k) \quad y(k) \neq \omega(k) \tag{24.45}$$

In these equations, the parameter k represents the kth iteration of the classifier and $\omega(k)$ is the class of the kth training pattern. It is clearly important to know whether this training scheme is effective in practice. In fact, it is possible to show that if the algorithm is modified so that its main loop is applied sufficiently many

Table 24.5 Perceptron *Fixed Increment* Algorithm

```
initialize weights with small random numbers;
select suitable value of learning rate coefficient η in the range 0 − 1;
do {
    for all patterns in the training set {
        obtain feature vector x and class ω;
        compute perceptron output y;
        if (y != ω) adjust weights according to wᵢ = wᵢ + η(ω − y)xᵢ;
    }
} until no further change;
```

times, *and* if the feature vectors are linearly separable, then the algorithm will converge to a correct error-free solution.

Unfortunately, most sets of feature vectors are not linearly separable. Thus, it is necessary to find an alternative procedure for adjusting the weights. This is achieved by the Widrow–Hoff delta rule, which involves making changes in the weights in proportion to the error $\delta = \omega - d$ made by the classifier. (Note that the error is calculated *before* thresholding to determine the actual class, i.e., δ is calculated using d rather than $f(d)$.) Thus, we obtain the Widrow–Hoff delta rule in the form:

$$w_i(k + 1) = w_i(k) + \eta \delta x_i(k) = w_i(k) + \eta[\omega(k) - d(k)]x_i(k) \qquad (24.46)$$

There are two important ways in which the Widrow–Hoff rule differs from the fixed increment rule:

1. An adjustment is made to the weights whether or not the classifier makes an actual classification error.
2. The output function d used for training is different from the function $y = f(d)$ used for testing.

These differences underline the revised aim of being able to cope with nonlinearly separable feature data. Figure 24.11 clarifies the situation by appealing to a 2-D case. Figure 24.11(a) shows separable data, which is straightforwardly fitted by the fixed increment rule. However, the fixed increment rule is not designed to cope with nonseparable data of the type shown in Fig. 24.11(b) and results in instability during training and inability to arrive at an optimal solution. On the other hand, the Widrow–Hoff rule copes satisfactorily with this type of data. An interesting addendum to the case of Fig. 24.11(a) is that although the fixed increment rule apparently reaches an optimal solution, the rule becomes "complacent" once a zero error situation has occurred, whereas an ideal classifier would arrive at a solution that minimizes the probability of error. Clearly, the Widrow–Hoff rule goes some way to solving this problem.

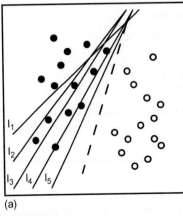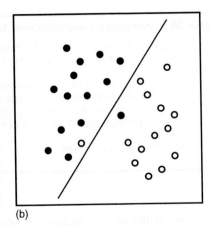

FIGURE 24.11

Separable and nonseparable data. Part (a) shows two sets of pattern data: lines l_1–l_5 indicate possible successive positions of a linear decision surface produced by the fixed increment rule. Note that the latter is satisfied by the final position l_5. The dotted line shows the final position that would have been produced by the Widrow–Hoff delta rule. Part (b) shows the stable position that would be produced by the Widrow–Hoff rule in the case of nonseparable data: in this case, the fixed increment rule would oscillate over a range of positions during training.

So far we have considered what can be achieved by a simple perceptron. Clearly, although it is only capable of dichotomizing feature data, a suitably trained array of simple perceptrons—the "single-layer perceptron" of Fig. 24.12—should be able to divide feature space into a large number of subregions bounded (in multidimensional space) by hyperplanes. However, in a multiclass application, this approach would require a very large number of simple perceptrons—up to $^cC_2 = \frac{1}{2}c(c-1)$ for a c-class system. Hence, there is a need to generalize the approach by other means. In particular, multilayer perceptron (MLP) networks (see Fig. 24.13)—which would emulate the neural networks in the brain—seem poised to provide a solution since they should be able to recode the outputs of the first layer of simple perceptrons.

Rosenblatt himself proposed such networks, but was unable to propose general means for training them systematically. In 1969, Minsky and Papert published their famous monograph, and in discussing the MLP raised the specter of "the monster of vacuous generality"; they drew attention to certain problems that apparently would never be solved using MLPs. For example, diameter-limited perceptrons (those that view only small regions of an image within a restricted diameter) would be unable to measure large-scale connectedness within images. These considerations discouraged effort in this area, and for many years attention was diverted to other areas such as expert systems. It was not until 1986 that Rumelhart et al. were successful in proposing a systematic approach to the training of MLPs. Their solution is known as the back-propagation algorithm.

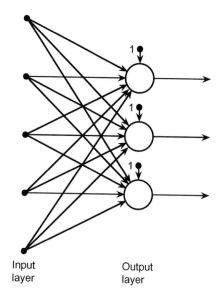

FIGURE 24.12

Single-layer perceptron. The single-layer perceptron employs a number of simple perceptrons in a single layer. Each output indicates a different class (or region of feature space). In more complex diagrams, the bias units (labeled "1") are generally omitted for clarity.

24.14 THE BACK-PROPAGATION ALGORITHM

The problem of training an MLP can be simply stated: a general layer of an MLP obtains its feature data from the lower layers and receives its class data from higher layers. Hence, if all the weights in the MLP are potentially changeable, the information reaching a particular layer cannot be relied upon: there is no reason why training a layer in isolation should lead to overall convergence of the MLP toward an ideal classifier (however defined). In addition, it is not evident what the optimal MLP architecture should be. While it might be thought that this is a rather minor difficulty, in fact this is not so: indeed, this is but one example of the so-called "credit assignment problem."[5]

One of the main difficulties in predicting the properties of MLPs and hence of training them reliably is the fact that neuron outputs swing suddenly from one state to another as their inputs change by infinitesimal amounts. Hence, we might

---

[5]This is not a good first example by which to define the credit assignment problem (in this case it would appear to be more of a deficit assignment problem). The credit assignment problem is the problem of correctly determining the local origins of global properties and making the right assignments of rewards, punishments, corrections, and so on, thereby permitting the whole system to be optimized systematically.

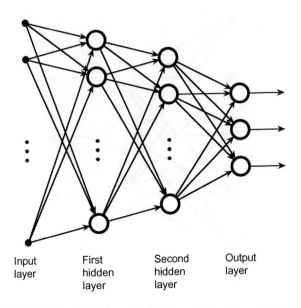

Input	First	Second	Output
layer	hidden	hidden	layer
	layer	layer	

FIGURE 24.13

Multilayer perceptron. The multilayer perceptron employs several layers of perceptrons. In principle, this topology permits the network to define more complex regions of feature space, and thus perform much more precise pattern recognition tasks. Finding systematic means of training the separate layers becomes the vital issue. For clarity, the bias units have been omitted from this and later diagrams.

consider removing the thresholding functions from the lower layers of MLP networks to make them easier to train. Unfortunately, this would result in these layers acting together as larger linear classifiers, with far less discriminatory power than the original classifier (in the limit we would have a single linear classifier with a single thresholded output connection, so the overall MLP would act as a single-layer perceptron).

The key to solving these problems was to modify the perceptrons composing the MLP by giving them a less "hard" activation function than the Heaviside function. As we have seen, a linear activation function would be of little use, but one of "sigmoid" shape, such as the tanh function (Fig. 24.14), is effective, and indeed is almost certainly the most widely used of the available functions.[6] Once these softer activation functions were used, it became possible for each layer of the MLP to

---

[6]We do not here make a marked distinction between symmetrical activation functions and alternatives that are related to them by shifts of axes, although the symmetrical formulation seems preferable as it emphasizes bidirectional functionality. In fact, the tanh function, which ranges from -1 to 1, can be expressed in the form: $\tanh u = (e^u - e^{-u})/(e^u + e^{-u}) = 1 - 2/(1 + e^{2u})$ and is thereby closely related to the commonly used function $(1 + e^{-v})^{-1}$. It can now be deduced that the latter function is symmetrical, although it ranges from 0 and 1 as v goes from $-\infty$ to ∞.

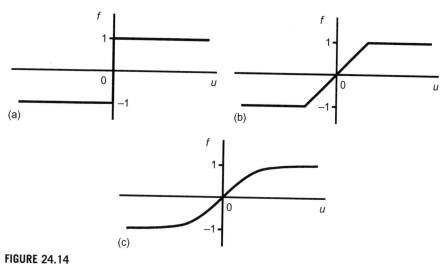

FIGURE 24.14

Symmetric activation functions. This figure shows a series of symmetric activation functions. (a) The Heaviside activation function used in the simple perceptron. (b) A linear activation function, which is, however, limited by saturation mechanisms. (c) A sigmoidal activation function that approximates to the hyperbolic tangent function.

"feel" the data more precisely and thus training procedures could be set up on a systematic basis. In particular, the rate of change of the data at each individual neuron could be communicated to other layers which could then be trained appropriately—though only on an incremental basis. We shall not go through the detailed mathematical procedure, or proof of convergence, beyond stating that it is equivalent to energy minimization and gradient descent on a (generalized) energy surface. Instead, we give an outline of the backpropagation algorithm (see Table 24.6). Nevertheless, some notes on the algorithm are in order:

1. The outputs of one node are the inputs of the next, and an arbitrary choice is made to label all variables as output (y) parameters rather than as input (x) variables; all output parameters are in the range 0 to 1.
2. The class parameter ω has been generalized as the target value t of the output variable y.
3. For all except the final outputs, the quantity δ_j has to be calculated using the formula $\delta_j = y_j(1 - y_j)(\Sigma_m \delta_m w_{jm})$, the summation having to be taken over all the nodes in the layer *above* node j.
4. The sequence for computing the node weights involves starting with the output nodes and then proceeding downward one layer at a time.
5. If there are no hidden nodes, the formula reverts to the Widrow–Hoff delta rule, except that the input parameters are now labeled y_i, as indicated above.

Table 24.6 The Back-Propagation Algorithm

```
initialize weights with small random numbers;
select suitable value of learning rate coefficient η in the range 0 − 1;
do {
    for all patterns in the training set
        for all nodes j in the MLP {
            obtain feature vector x and target output value t;
            compute MLP output y;
            if (node is in output layer)
                δⱼ = yⱼ(1 − yⱼ)(tⱼ − yⱼ);
            else δⱼ = yⱼ(1 − yⱼ)(Σₘδₘwⱼₘ);
            adjust weights i of node j according to wᵢⱼ = wᵢⱼ + ηδⱼyᵢ;
        }
} until changes are reduced to some predetermined level;
```

6. It is important to initialize the weights with random numbers to minimize the chance of the system becoming stuck in some symmetrical state from which it might be difficult to recover.

7. Choice of value for the learning rate coefficient η will be a balance between achieving a high rate of learning and avoidance of overshoot: normally a value of around 0.8 is selected.

When there are many hidden nodes, convergence of the weights can be very slow, and indeed this is one disadvantage of MLP networks. Many attempts have been made to speed convergence, and a method that is almost universally used is to add a "momentum" term to the weight update formula, it being assumed that weights will change in a similar manner during iteration k to the change during iteration $k − 1$:

$$w_{ij}(k + 1) = w_{ij}(k) + \eta\delta_j y_i + \alpha[w_{ij}(k) − w_{ij}(k − 1)] \qquad (24.47)$$

where α is the momentum factor. This technique is primarily intended to prevent networks becoming stuck at local minima of the energy surface.

24.15 MLP ARCHITECTURES

The preceding sections gave the motivation for designing an MLP and for finding a suitable training procedure, and then outlined a general MLP architecture and the widely used back-propagation training algorithm. However, having a general solution is only one part of the answer. The next question is how best to adapt the general architecture to specific types of problem. We shall not give a full answer to this question here. However, Lippmann attempted to answer this problem in 1987. He showed that a two-layer (single hidden layer) MLP can implement arbitrary convex decision boundaries, and indicated that a three-layer (two-hidden layer) network is

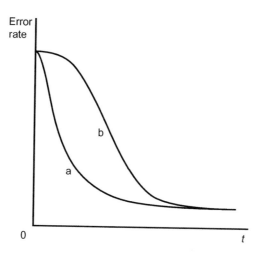

FIGURE 24.15

Learning curve for the multilayer perceptron. Here curve (a) shows the learning curve for a single-layer perceptron, and curve (b) shows that for a multilayer perceptron. Note that the multilayer perceptron takes considerable time to get going, since initially each layer receives relatively little useful training information from the other layers. Note also that the lower part of the diagram has been idealized to the case of identical asymptotic error rates, although this situation would seldom occur in practice.

required to implement more complex decision boundaries. It was subsequently found that it should never be necessary to exceed two hidden layers, as a three-layer network can tackle quite general situations if sufficient neurons are used (Cybenko, 1988). Subsequently, Cybenko (1989) and Hornik et al. (1989) showed that a two-layer MLP can approximate any continuous function, although nevertheless there may sometimes be advantages in using more than two layers.

Although the back-propagation algorithm can train MLPs of any number of layers, in practice, training one layer "through" several others introduces an element of uncertainty that is commonly reflected in increased training times (see Fig. 24.15). Thus, there is some advantage to be gained from using a minimal number of layers of neurons. In this context, the above findings on the necessary numbers of hidden layers are especially welcome.

24.16 OVERFITTING TO THE TRAINING DATA

When training MLPs and many other types of ANN, there is a problem of overfitting the network to the training data. One of the fundamental aims of SPR is for the learning machine to be able to generalize from the particular set of data it is trained on to other types of data it might meet during testing. In particular, the machine should be able to cope with noise, distortions, and fuzziness in the data,

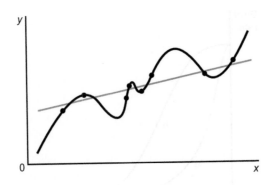

FIGURE 24.16

Overfitting of data. In this graph, the data points are rather too well fitted by the solid curve, which matches every nuance exactly. Unless there are strong theoretic reasons why the solid curve should be used, the gray line will give a higher confidence level.

although clearly not to the extent of being able to respond correctly to types of data different from that on which it has been trained. The main points to be made here are (1) that the machine should learn to respond to the underlying population from which the training data has been drawn and (2) that it must not be so well adapted to the specific training data that it responds less well to other data from the same population. Figure 24.16 shows in a 2-D case both a fairly ideal degree of fit and a situation where every nuance of the set of data has been fitted, thereby achieving a degree of overfit.

Typically, overfitting can arise if the learning machine has more adjustable parameters than are strictly necessary for modeling the training data: with too few parameters such a situation should not arise. However, if the learning machine has enough parameters to ensure that relevant details of the underlying population are fitted, there may be overmodeling of part of the training set; thus, the overall recognition performance will deteriorate. Ultimately, the reason for this is that recognition is a delicate balance between capability to discriminate and capability to generalize, and it is most unlikely that any complex learning machine will get the balance right for all the features it has to take account of.

Be this as it may, we clearly need to have some means of preventing over-adaptation to the training data. One way of achieving this is to curtail the training process before overadaptation can occur.[7] This is not difficult, since we merely need to test the system periodically during training to ensure that the point of

---

[7]It is often stated that this procedure aims to prevent overtraining. However, the term "overtraining" is ambiguous. On the one hand, it can mean recycling through the *same* set of training data until eventually the learning machine is overadapted to it. On the other hand, it can mean using more and more *totally new* data—a procedure that cannot produce overadaptation to the data, and on the contrary is almost certain to improve performance. In view of this ambiguity, it seems better not to use the term.

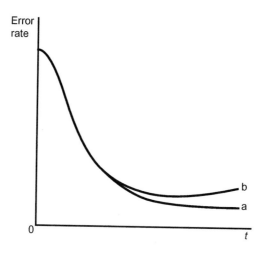

FIGURE 24.17

Cross-validation tests. This diagram shows the learning curve for a multilayer perceptron (a) when tested on the training data and (b) when tested on a special validation set. Curve (a) tends to go on improving even when overfitting is occurring. However, this situation is detected when curve (b) starts deteriorating. To offset the effects of noise (not shown on curves (a) and (b)), it is usual to allow 5–10% deterioration relative to the minimum in curve (b).

overadaptation has not been reached. Figure 24.17 shows what happens when testing is carried out simultaneously on a separate dataset: at first performance on the test data closely matches that on the training data, being slightly superior for the latter because a small degree of overadaptation is already occurring. But after a time, performance starts deteriorating on the test data while performance on the training data appears to go on improving. This is the point where serious overfitting is occurring, and the training process should be curtailed. The aim, then, is to make the whole training process far more rigorous by splitting the original training set into two parts—the first being retained as a normal training set and the second being called the *validation set*. Note that the latter is actually part of the training set in the sense that it is not part of the eventual test set.

The process of checking the degree of training by use of a validation set is called *cross-validation*, and is vitally important to proper use of an ANN. The training algorithm should include cross-validation as a fully integrated part of the whole training schedule; it should not be regarded as an optional extra.

It is useful to speculate how overadaptation could occur when the training procedure is completely determined by the back propagation (or other) provably correct algorithm. In fact, there are mechanisms by which overadaptation can occur. For example, when the training data do not control particular weights sufficiently closely, some could drift to large positive or negative values, while retaining a sufficient degree of cancellation so that no problems appear to arise with the training data; yet, when test or validation data are employed, the problems become all too clear. The fact that the form of the sigmoid function will permit some nodes to become "saturated out" does not help the situation, as it inactivates parameters and hides certain aspects of the incoming data. Yet it is intrinsic to the MLP architecture and the way it is trained that some nodes are *intended* to be saturated out in order to ignore irrelevant features of the training set. The problem is whether inactivation is inadvertent or designed. The answer probably lies in the quality of the training set and how well it covers the available or potential feature space.

Finally, let us suppose an MLP is being set up, and it is initially unknown how many hidden layers will be required or how many nodes there will have to be in each layer: it will also be unknown how many training set patterns will be required or how many training iterations will be needed—or what values of the momentum or learning parameters will be appropriate. A quite substantial number of tests will be required to decide all the relevant parameters. There is therefore a definite risk that the final system will be overadapted not only to the training set but also to the validation set. In such circumstances what we need is a second validation set that can be used after the whole network has been finalized and final training is being undertaken.

24.17 CONCLUDING REMARKS

The methods of this chapter make it rather surprising that so much of image processing and analysis is possible without any reference to *a priori* probabilities. It seems likely that this situation is due to several factors: (a) expediency and in particular the need for speed of interpretation; (b) the fact that algorithms are designed by humans who have knowledge of the types of input data and thereby incorporate *a priori* probabilities implicitly, e.g., via the application of suitable threshold values; and (c) tacit recognition of the situation outlined in Section 24.11, that probabilistic methods have limited applicability. In practice, it is at the stage of strong image structure and contextual analysis that probabilistic interpretations really come into their own.

Nonetheless, SPR is extremely valuable within its own range of utility. This includes identifying objects on conveyors and making value judgements of their quality, reading labels and codes, verifying signatures, checking fingerprints, and so on. Indeed, the number of distinct applications of SPR is huge and it forms an essential counterpart to the other methods described in this book.

This chapter has concentrated mainly on the supervised learning approach to SPR. However, unsupervised learning is also vitally important, particularly when training on huge numbers of samples (e.g., in a factory environment) is involved.

The section on this topic should therefore not be ignored as a minor and insignificant perturbation: much the same comments apply to the subject of principal components analysis which has had an increasing following in many areas of machine vision (see Section 24.10); nor should it go unnoticed that these topics link in strongly with ANNs, which are often able to play a powerful role.

> Vision is largely a recognition process with both structural and statistical aspects. This chapter has reviewed SPR, emphasizing fundamental classification error limits, and has shown the part played by Bayes' theory, the nearest neighbor algorithm, ROCs, PCA, and ANNs. Note that the last of these is subject to the same limitations as other SPR methods, particularly with regard to adequacy of training and the possibility of overfitting.

24.18 BIBLIOGRAPHICAL AND HISTORICAL NOTES

Although the subject of SPR tends not to be at the center of attention in image analysis work,[8] it provides an important background—especially in the area of automated visual inspection where decisions continually have to be made on the adequacy of products. Most of the relevant work on this topic was already in place by the early 1970s, including the work of Hughes (1968) and Ullmann (1969) relating to the optimum number of features to be used in a classifier. At that stage a number of important volumes appeared; see, e.g., Duda and Hart (1973) and Ullmann (1973), and these were followed a little later by Devijver and Kittler (1982).

In fact, the use of SPR for image interpretation dates from the 1950s. For example, in 1959 Bledsoe and Browning developed the n-tuple method of pattern recognition, which turned out (Ullmann, 1973) to be a form of NN classifier; however, it has been useful in leading to a range of simple hardware machines based on RAM (n-tuple) lookups (see, e.g., Aleksander et al., 1984), thereby demonstrating the importance of marrying algorithms and readily implementable architectures.

Many of the most important developments in this area have probably been those comparing the detailed performance of one classifier with another, particularly with respect to cutting down the amount of storage and computational effort. Papers in these categories include those by Hart (1968) and Devijver and Kittler (1980). Oddly, there appeared to be no overt mention in the literature of how *a priori* probabilities should be used with the NN algorithm, until the author's paper on this topic (Davies, 1988f); see Section 24.4.

On the unsupervised approach to SPR, Forgy's (1965) method for clustering data was soon followed by the famous ISODATA approach of Ball and Hall (1966), and then by MacQueen's (1967) k-means algorithm. Much related work

---

[8]Note, however, that it is vital to the analysis of multispectral data from satellite imagery.

ensued, and this was summarized by Jain and Dubes (1988), which became a classic text. However, cluster analysis is an exacting process and various workers have felt the need to push the subject further forward: e.g., Postaire and Touzani (1989) required more accurate cluster boundaries; Jolion and Rosenfeld (1989) wanted better detection of clusters in noise; Chauduri (1994) needed to cope with time-varying data; and Juan and Vidal (1994) required faster k-means clustering. Note that all this work can be described as conventional, and did not involve the use of robust statistics *per se*. However, elimination of outliers is central to the problem of reliable cluster analysis; for a discussion of this aspect of the problem, see Appendix A and the references cited therein.

While the field of pattern recognition has moved forward substantially since 1990, there are fortunately several quite recent texts that cover the subject relatively painlessly (Duda et al., 2001; Webb, 2002; Theodoridis and Koutroumbas, 2009). The reader can also appeal to the review article by Jain et al. (2000), which outlines new areas that appeared in the previous decade.

The multiple classifier approach is a relatively recent development, and is well reviewed by Duin (2002). Ho et al. (1994) dates from when the topic was rather younger, and lists an interesting set of options as seen at that point—some of these being covered in Section 24.8.

"Bagging" and "boosting" are further variants on the multiple classifier theme: they were developed by Breiman (1996) and Freund and Schapire (1996). Bagging (short for "bootstrap aggregating") means sampling the training set, with replacement, n times, generating b bootstrap sets to train b subclassifiers, and assigning any test pattern to the class most often predicted by the subclassifiers. The method is particularly useful for unstable situations (such as when classification trees are used), but is almost valueless when stable classification algorithms are used (such as the nearest neighbor algorithm). Boosting is useful for aiding the performance of weak classifiers. In contrast with bagging, which is a parallel procedure, boosting is a sequential deterministic procedure. It operates by assigning different weights to different training set patterns according to their intrinsic (estimated) accuracy. For further progress with these techniques, see Rätsch et al. (2002), Fischer and Buhmann (2003), and Lockton and Fitzgibbon (2002). Finally, Beiden et al. (2003) discuss a variety of factors involved in the training and testing of competing classifiers; in addition, much of the discussion relates to multivariate ROC analysis.

SVMs also came into prominence over the 1990s and have found an increasing number of applications: the concept was invented by Vapnik and the historical perspective is covered in Vapnik (1998). Cristianini and Shawe-Taylor (2000) provide a student-orientated text on the subject.

Next we digress to outline something of the history of ANNs. After a promising start in the 1950s and 1960s, they fell into disrepute (or at least, disregard) following the pronouncements of Minsky and Papert in 1969; they picked up again in the early 1980s; were subjected to an explosion in interest after the announcement of the back-propagation algorithm by Rumelhart et al. in 1986;

and in the mid-1990s settled into the role of normal tools for vision and other applications. Note that the back-propagation algorithm was invented several times (Werbos, 1974; Parker, 1985) before its relevance was finally recognized. In parallel with these MLP developments, Oja (1982) developed his Hebbian principal components network. Useful early references on ANNs include the volumes by Haykin (1999) and Bishop (1995), and papers on their application to segmentation and object location, such as Toulson and Boyce (1992) and Vaillant et al. (1994); for work on contextual image labeling, see Mackeown et al. (1994).

After the euphoria of the early 1990s, during which papers on ANNs applied to vision were ubiquitous, it was seen that the main value of ANNs lay in their unified approach to feature extraction and selection (even if this necessarily carries the disadvantage that the statistics are hidden from the user), and their intrinsic capability for finding moderately nonlinear solutions with relative ease. Later papers include the ANN face detection work of Rowley et al. (1998), among others (Fasel, 2002; Garcia and Delakis, 2002). For further general information on ANNs, see the book by Bishop (2006).

24.18.1 More Recent Developments

Returning to mainstream SPR, Jain (2010) presented a review of the subject of clustering, entitled "Data clustering: 50 years beyond k-means." He noted, "In spite of the fact that k-means was proposed over 50 years ago and thousands of clustering algorithms have been published since then, k-means is still widely used"—thereby reflecting the difficulty of designing a general purpose clustering algorithm and the ill-posed nature of the problem: emerging and useful research directions include semi-supervised clustering and ensemble clustering. The review presents the main challenges and issues facing the subject as of 2010: above all is the plea for a suite of benchmark data with ground truth to test and evaluate clustering methods.

Li and Zhang (2004) describe how a new boosting algorithm "FloatBoost" has been applied to produce the first real-time multiview face detection system reported. The method uses a backtrack mechanism after each iteration of AdaBoost learning to minimize the error rate directly; it also uses a novel statistical model for learning the best weak classifiers and a stagewise approximation to the posterior probability, thereby requiring fewer weak classifiers than AdaBoost. Gao et al. (2010) report on a modified version of AdaBoost to resolve the key problems of how to *select* the most discriminative weak learners and how to optimally *combine* the selected weak learners. Experiments confirm the utility of the algorithm including the capability to solve these two key problems; both synthetic and real scene data (car and non-car patterns) are used for the tests. Fumera et al. (2008) present a theoretical analysis of bagging as a linear combination of classifiers, thereby giving an analytical model of bagging misclassification probability as a function of ensemble size.

Youn and Jeong (2009) describe a class-dependent feature scaling method employing a naive Bayes' classifier for text data mining, including functions such as text categorization and search. While the reasons why the naive Bayes' independence assumption works well in many cases have not been well explained or understood until recently, this paper confirms that it is often a good choice for text analysis because the amount of data used is large (e.g., the number of features is about 100,000 for protein sequence data). In particular, the simplicity and the effectiveness of the naive Bayes' classifier maps well to text categorization. Rish (2001) provides an empirical study of naive Bayes, containing much useful information.

Decision trees provide a convenient fast-operating method of pattern recognition, and the methodology has developed quite rapidly in recent years. Chandra et al. (2010) describe a new node splitting procedure called the distinct class based splitting measure (DCSM) for decision tree construction. Node splitting measures are important as they help to produce compact decision trees with improved generalization abilities. Chandra et al. have shown that DCSM is well-behaved and produces decision trees that are more compact and provide better classification accuracy than trees constructed using other common node splitting measures. The DCSM measure also helps with pruning (which produces compact trees with better classification accuracy). Köktas et al. (2010) describe a multi-classifier for grading knee osteoarthritis using gait analysis. It employs a decision tree with MLPs at the leaves. In fact, three different MLPs (different "experts") with binary classifications are employed at different leaves of the tree. They showed that, for this type of data, this produced better results than a single multi-class classifier. Rodríguez et al. (2010) describe tests made on a large number of datasets using ensemble methods to generate more accurate classifiers. They show that, for multiclass problems, ensembles of decision trees ("forests") can be successfully combined with ensembles of nested dichotomies. The direct approach, using ensembles of nested dichotomies with a forest method as the base classifier, can be improved using ensemble methods with a nested dichotomy of decision trees as the base classifier.

Fawcett (2006) produced an excellent, largely tutorial summary of ROC analysis in which many descriptors employing true and false positives and negatives are used; a valuable feature of the paper is the unification of a subject in which many apparently different descriptors appear with different names according to the varying backgrounds of the workers. In particular, the recently much more widely used terms "precision" and "recall" are related to "sensitivity," "specificity," "accuracy," and others (for definitions and further discussion of these performance measures, see Section 24.7.1). In addition, measures such as "F-measure" are defined, and problems and pitfalls of using ROC graphs are pointed out. Ooms et al. (2010) underline the value of Fawcett's summary, but show that the ROC concept is limited and is not an optimal measure for *sorting* as distinct from cases where *misclassification costs* are the main concern. They propose a sorting optimization curve (SOC) to cope with sorting problems and help identify the best choice of operating point in that case. In contrast with the ROC curve, which plots *fpr* vs. *fnr* or *tpr* vs. *fpr*, the SOC curve plots yield rate (Y) vs. relative quality improvement rate (Q), where $Y = (TP + FP)/(P + N)$; this formula arises because no distinction is made between true and false positives when selling a product. Quality Q is defined in terms of the precision $Pr = TP/(TP + FP)$, *viz.* $Q = f(Pr)$, and uses whatever function f is needed to achieve this when sorting a particular commodity (such as apples). Typically, optimization involves moving up the Y vs. Q curve until reaching the lowest level of quality that is acceptable or legal.

Assessing the quality of the ROC curve has acquired some importance in the past decade, and the *AUC* (area under the curve) measure has been the main performance indicator for this (Fawcett, 2006). For example, Hu et al. (2008) have used it to advantage for optimal evaluation and selection of features.

24.19 PROBLEMS

1. Show that if the cost function of Eq. (24.16) is chosen, the decision rule (24.15) can be expressed in the form of relation (24.5).

2. Show that in a simple two-class system, introducing a reject classification to reduce the number of errors by R in fact requires $2R$ test patterns to be rejected, assuming that R is small. What is likely to happen as R increases?

3. Why is the point on a ROC curve closest to the origin *not* the point that minimizes total error? Prove that the point that minimizes the total error on a ROC curve is actually the point where the gradient is -1 (see Section 24.7).

4. Consider the four quantities TP, TN, FP, FN defined in Section 24.7. Arrange them in order of size for situations where positives are rare and recognition errors are likely to be low. If the rates *tpr*, *tnr*, *fpr*, *fnr* are also arranged in order of size, will the order be the same as for TP, TN, FP, FN?

5. Compare the shapes of ROC curves and precision–recall curves for identical classifiers. What mathematical relations link their shapes? Determine whether the ROC curves for two classifiers will cross each other the same number of times as the precision–recall curves.

6. Prove that Eq. (24.26) provides a mathematically sound way of combining precision and recall into a single measure.

25

Image Acquisition

In vision, everything depends on image acquisition, and in image acquisition, everything depends on illumination. Naturally, robust algorithms can be designed to largely overcome any problems of inadequacy on these fronts. On the other hand, care with acquisition often means that simpler, more reliable algorithms can be produced. This chapter considers these important aspects of vision system design.

Look out for:

- lighting effects, reflectance, and the appearance of highlights and shadows.
- the value of soft or diffuse lighting.
- how lighting can systematically be made uniform by use of several point or line sources.
- the types of camera and sensor that are commonly available.
- the sampling theorem and its implications.

The advent of solid-state cameras and widely available frame-grabbing devices has made one part of image acquisition straightforward: yet the other aspect—that of providing suitable illumination—is still rather a black art. However, the methods described here demonstrate that uniform illumination is subject to design rather than *ad hoc* experimentation.

This work described in this chapter necessarily provides underpinning for all practical vision systems, except perhaps those involving X-rays or other modalities such as ultrasonic imaging.

25.1 INTRODUCTION

When implementing a vision system, nothing is more important than image acquisition. Any deficiencies of the original images can cause great problems with image analysis and interpretation. An obvious example is that of lack of detail due to insufficient contrast or poor focusing of the camera: this can have the effect at best that the dimensions of objects will not be accurately measurable from the images, and at worst that the objects will not even be recognizable, so the purpose of vision cannot be fulfilled. This chapter examines the problems of image acquisition.

Before proceeding, it is as well to note that vision algorithms are of use in a variety of areas where visual pictures are not directly input. For example, vision techniques (image processing, image analysis, recognition, and so on) can be applied to seismographic maps, to pressure maps (whether these arise from handwriting on pressure pads or from weather data), infrared, ultraviolet, X-ray and radar images, and a variety of other cases. There is no space here to consider methods for acquisition in any of these instances and attention is concentrated on purely optical methods. In addition, space does not permit a detailed study of methods for obtaining range images using laser scanning and ranging techniques, while other methods that are specialized for 3-D work will also have to be passed by. Instead, we concentrate on (a) lighting systems for obtaining intensity images, (b) technology for receiving and digitizing intensity images, and (c) basic theory such as the Nyquist sampling theorem that underlies this type of work.

First we consider how to set up a basic system that might be suitable for the thresholding and feature detection work of Chapters 2–5.

25.2 ILLUMINATION SCHEMES

The simplest and most obvious arrangement for acquiring images is that shown in Fig. 25.1. A single source provides light over a cluster of objects on a worktable or conveyor, and this scene is viewed by a camera directly overhead. The source is typically a tungsten light that approximates to a point source. Assuming for now that the light and camera are some distance away from the objects, and are in different directions relative to them, it may be noted that:

1. different parts of the objects are lit differently, because of variations in the angle of incidence, and hence have different brightnesses as seen from the camera.
2. the brightness values also vary because of the differing absolute reflectivities[1] of the object surfaces.
3. the brightness values vary with the specularities of the surfaces in places where the incident, emergent, and phase angles are compatible with specular reflection (Chapter 15).
4. parts of the background and of various objects are in shadow and this again affects the brightness values in different regions of the image.

---

[1]Referring to Eq. (15.12), R_0 is the absolute surface reflectivity and R_1 is the specularity.

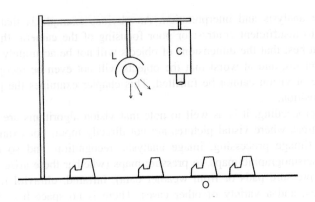

FIGURE 25.1

Simple arrangement for image acquisition: C, camera; L, light with simple reflector; O, objects on worktable or conveyor.

5. other more complex effects occur because light reflected from some objects will cast light over other objects—factors that can lead to complicated variations in brightness over the image.

Clearly, even in this apparently simple case—one point light source and one camera—the situation can become quite complex. However, effect 5 is normally reasonably marginal and is ignored in what follows. In addition, effect 3 can often be ignored except in one or two small regions of the image where sharply curved pieces of metal give rise to glints. This still leaves considerable scope for complication due to effects 1, 2, and 4.

There are two important reasons for viewing the surfaces of objects: the first is when we wish to locate objects and their facets, and the second is when we wish to scrutinize the surfaces themselves. In the first instance, it is important to try to highlight the facets by arranging that they are lit differently, so that their edges stand out clearly. In the second instance, it might be preferable to do the opposite—i.e., to arrange that the surfaces are lit very similarly, so that any variations in reflectivity caused by defects or blemishes stand out plainly. The existence of effects 1 and 2 implies that it is difficult to achieve both of these things at the same time: one set of lighting conditions is required for optimum segmentation and location, and another set for optimum surface scrutiny. In most of this book, object location has been regarded as the more difficult task and therefore the one that needs the most attention. Hence, we have imagined that the lighting scheme is set up for this purpose. In principle, a point source of light is well adapted to this situation. However, it is easy to see that if a very diffuse lighting source is employed, then angles of incidence will tend to average out and effect 2 will dominate over 1 so that, *to a first approximation*, the observed brightness values will represent variations in surface reflectance. In fact, "soft" or diffuse lighting also subdues specular reflections (effect 3), so that for the most part they can be ignored.

Returning to the case of a single point source, recall (effect 4) that shadows can become important. There is one special case when this is not so, and that is when the light is projected from exactly the same direction as the camera; we return to this case below. Shadows are a persistent cause of complications in image analysis. One problem is that it is not a trivial task to identify them, so they merely contribute to the overall complexity of any image and in particular add to the number of edges that have to be examined in order to find objects. They also make it much more difficult to use simple thresholding. (However, note that shadows can sometimes provide information that is of vital help in interpreting complex 3-D images; see, e.g., Section 15.6.)

25.2.1 Eliminating Shadows

The above considerations suggest that it would be highly convenient if shadows could be eliminated. A strategy for achieving this is to lower their contrast by using several light sources. Then the region of shadow from one source will be a region of illumination from another, and shadow contrast will be lowered dramatically. Indeed, if there are n lights, many positions of shadow will be illuminated by $n - 1$ lights and their contrast will be so low that they can be eliminated by straightforward thresholding operations. However, if objects have sharp corners or concavities, there may still be small regions of shadow that are illuminated by only one light or perhaps no light at all; these regions will be immediately around the objects, and if the objects appear dark on a light background, shadows could make the objects appear enlarged or cause shadow lines immediately around them. For light objects on a dark background this is normally less of a problem.

Clearly, it seems best to aim for large numbers of lights so as to make the shadows more diffuse and less contrasting, and in the limit it appears that we are heading for the situation of soft lighting discussed earlier. However, this is not quite so. What is often required is a form of diffuse lighting that is still directional—as in the case of a diffuse source of restricted extent directly overhead: this can be provided very conveniently by a continuous ring light around the camera. This technique is found to eliminate shadows highly effectively while retaining sufficient directionality to permit a good measure of segmentation of object facets to be achieved, i.e., it is an excellent compromise although it is certainly not ideal. For these reasons, we describe its effects in some detail. In fact, it is clear that it will lead to good segmentation of facets whose boundaries lie in horizontal planes, but to poor segmentation of those whose boundaries lie in vertical planes.

The situation just described is very useful for analyzing the shape profiles of objects with cylindrical symmetry. Note, e.g., the case shown in Fig. 25.2, which involves a special type of chocolate biscuit with jam underneath the chocolate. If this is illuminated by a continuous ring light fairly high overhead (the proper

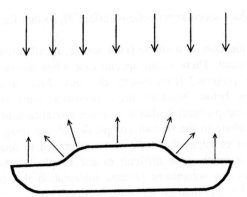

FIGURE 25.2

Illumination of a chocolate-and-jam biscuit. This figure shows the cross-section of a particular type of round chocolate biscuit with jam underneath the chocolate. The arrows show how light arriving from vertically overhead is scattered by the various parts of the biscuit.

working position), the region of chocolate above the edge of the jam reflects the light obliquely and appears darker than the remainder of the chocolate. On the contrary, if the ring light is lowered to near the worktable, the region above the edge of the jam appears *brighter* than the rest of the chocolate because it scatters light upward rather than sideways. Clearly there is also[2] a particular height at which the ring light can make the jam boundary disappear (Fig. 25.3), this height being dictated by the various angles of incidence and reflection and by the relative direction of the ring light. In comparison, if the lighting were made highly diffuse, these effects would tend to disappear and the jam boundary would always have very low contrast.

Suitable fluorescent ring lights are readily available and straightforward to use, and provide a solution that is more practicable than the alternative means of eliminating shadows mentioned above—that of illuminating objects directly from the camera direction, e.g., via a half-silvered mirror.

Earlier, the one case we did not completely solve arose when we were attempting to segment facets whose joining edges were in vertical planes. There appears to be no simple way of achieving a solution to this problem without recourse to switched lights (see Chapter 15); this option will not be discussed further here.

We have now identified various practical forms of lighting that can be used to highlight various object features and which eliminate complications as far as possible. These types of lighting are restricted in what they can achieve (as would clearly be expected from the shape-from-shading ideas of Chapter 15). However,

---

[2]The latter two situations are described for interest only.

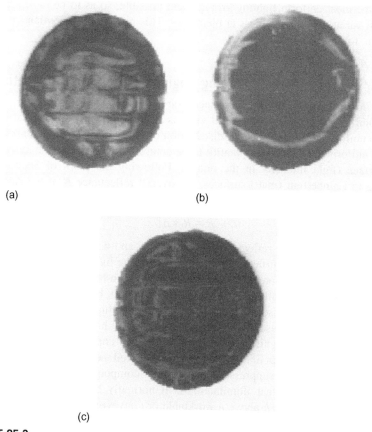

(a)   (b)

(c)

FIGURE 25.3

Appearance of the chocolate-and-jam biscuit of Fig. 25.2. (a) How the biscuit appears to a camera directly overhead when illuminated as in Fig. 25.2; (b) appearance when the lights are lowered to just above table level; (c) appearance when the lights are raised to an intermediate level making the presence of the jam scarcely detectable.

they are exceedingly useful in a variety of applications. A final problem is that two lighting schemes may have to be used in turn, the first for locating objects and the second for inspecting their surfaces. However, this problem can largely be overcome by *not* treating the latter case as a special one requiring its own lighting scheme, but rather noting the direction of lighting and *allowing for* the resulting variation in brightness values by taking account of the known shape of the object. The opposite approach is generally of little use unless other means are used for locating the object. However, the latter situation frequently arises in practice: imagine that a slab of concrete or a plate of steel is to be inspected for defects. In that case the position of the object is known and it is clearly best to set up the most uniform lighting arrangement possible, so as to be most sensitive to small variations in brightness at blemishes. This is, then, an important practical problem, to which we now turn.

25.2.2 Principles for Producing Regions of Uniform Illumination

While initially it may appear to be necessary to illuminate a worktable or conveyor uniformly, a more considered view is that a uniform flat material should *appear* uniform so that the spatial distribution of the light emanating from its surface is uniform. The relevant quantity to be controlled is therefore the radiance of the surface (light intensity in the image). Following the work of Section 15.4 relating to Lambertian (matt) surfaces, the overall reflectance R of the surface is given by:

$$R = R_0 \mathbf{s}.\mathbf{n} \tag{25.1}$$

where R_0 is the absolute reflectance of the surface and $\mathbf{n}$, $\mathbf{s}$ are unit vectors along the local normal to the surface and the direction of the light source, respectively.

Clearly, the assumption of a Lambertian surface can be questioned, since most materials will give a small degree of specular reflection, but in this section we are mainly interested in those nonshiny substances for which Eq. (25.1) is a good approximation. In any case, special provision normally has to be made for examining surfaces with a significant specular reflectance component. However, note that the continuous strip lighting systems considered below have the desirable property of very largely suppressing any specular components.

Next we recognize that illumination will normally be provided by a set of lights at a certain height h above a worktable or conveyor. We start by taking the case of a single point source at height h. Supposing that this is displaced laterally through a distance a, so that the actual distance from the source to the point of interest on the worktable is d, I will have the general form:

$$I = \frac{c \cos i}{d^2} = \frac{ch}{d^3} \tag{25.2}$$

where c is a constant factor (see Fig. 25.4).

Eq. (25.2) represents a distinctly nonuniform region of intensity over the surface. However, this problem may be tackled by providing a suitable distribution of lights. A neat solution is provided by a symmetrical arrangement of two strip lights that will clearly help to make the reflected intensity much more uniform (Fig. 25.5). We illustrate this idea by reference to the well-known arrangement of a pair of "Helmholtz" coils widely used for providing a uniform magnetic field, with the separation of the coils made equal to their radius so as to eliminate the second-order variation in field intensity.

In a similar way, the separation of the strip lights can be adjusted so that the second-order term vanishes (Fig. 25.5(b)). There is an immediate analogy also with the second-order Butterworth low-pass filter, which gives a maximally flat

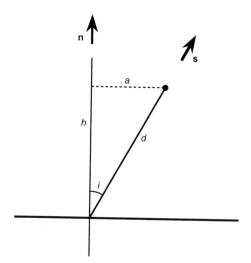

FIGURE 25.4

Geometry for a single point source illuminating a surface. Here a point light source at a height *h* above a surface illuminates a general point with angle of incidence *i*. **n** and **s** are respectively unit vectors along the local normal to the surface and the direction of the light source.

Source: © IEE 1997

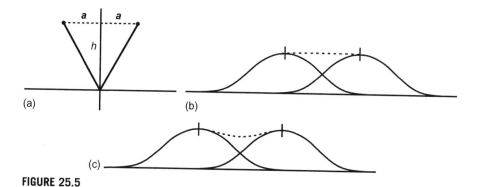

FIGURE 25.5

Effect of using two strip lights for illuminating a surface. Part (a) shows two strip lights at a height *h* above a surface, and part (b) shows the resulting intensity patterns for each of the lights; the dotted line shows the combined intensity pattern. Part (c) shows the corresponding patterns when the separation of the lights is increased slightly.

Source: © IEE 1997

response, the second-order term in the frequency response curve being made zero and the lowest order term then being the fourth-order term (Kuo, 1966). In fact, the latter example demonstrates how the method might be improved further—by aiming for a Chebychev type of response in which there is some ripple in the pass band, yet the *overall* pass-band response is flatter (Kuo, 1966). In a lighting application, we should aim to start with the strip lights not just far enough apart so that the second-order term vanishes, but slightly further apart so that the intensity is *almost* uniform over a rather larger region (Fig. 25.5(c)). This reflects the fact that in practice the prime aim will be to achieve a given degree of uniformity over the maximum possible size of region.

In principle it is easy to achieve a given degree of uniformity over a larger region by starting with a given response and increasing the linear dimensions of the *whole* lighting system proportionately. Although valid in principle, this approach will frequently be difficult to apply in practice: e.g., it will be limited by convenience and by availability of the strip lights; it must also be noted that as the size of the lighting system increases, so must the power of the lights. Hence, in the end we will have only one adjustable geometric parameter by which to optimize the response.

Finally, in many practical situations, it will be less useful to have a long narrow working region than one whose aspect ratio is close to unity. We shall consider two such cases—a circular ring light and a square ring light. The first of these is conveniently provided in diameters of up to at least 30 cm by commercially available fluorescent tubes, while the second can readily be constructed—if necessary on a very much larger scale—by assembling a set of four linear fluorescent tubes. In this case we take the tubes to be finite in length, and in contact at their ends: these are made into an assembly that can be raised or lowered to optimize the system. Thus, these two cases have fixed linear dimensions characterized in each case by the parameter a, and it is h that is adjusted rather than a. To make comparisons easier, we assume in all cases that a is the constant and h is the optimization parameter (Fig. 25.6).

25.2.3 Case of Two Infinite Parallel Strip Lights

First we take the case of two infinite parallel strip lights. In this case, the intensity I is given by the sum of the intensities I_1, I_2 for the two tubes:

$$I_1(x) = h \int_{-\infty}^{\infty} [(a-x)^2 + (v-y)^2 + h^2]^{-3/2} dv \qquad (25.3)$$

$$I_2(x) = I_1(-x) \qquad (25.4)$$

Suitable substitutions permit Eq. (25.3) to be integrated, and the final result is:

$$I = \frac{2h}{(a-x)^2 + h^2} + \frac{2h}{(a+x)^2 + h^2} \qquad (25.5)$$

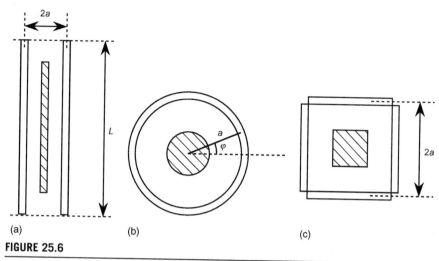

FIGURE 25.6

Lighting arrangements for obtaining uniform intensity. This diagram shows three arrangements of tubular lights for providing uniform intensity over fairly large regions, shown cross-hatched in each case. Part (a) shows two long parallel strip lights, part (b) shows a circular ring light, and part (c) shows four strip lights arranged to form a square "ring." In each case, height h above the worktable must also be specified.

Source: © *IEE 1997*

Differentiating I twice and setting $d^2I/dx^2 = 0$ at $x = 0$ eventually (Davies, 1997b) yields the maximally flat condition:

$$h = \sqrt{3}a \qquad (25.6)$$

However, as noted above, it should be better to aim for minimum *overall* ripple over a region $0 \le x \le x_1$. The situation is shown in Fig. 25.7. We take the ripple ΔI as the difference in height between the maximum intensity I_m and the minimum intensity I_0, at $x = 0$, and on this basis the maximum permissible deviation in x is the value of x where the curve again crosses the minimum value I_0.

A simple calculation shows that the intensity is again equal to I_0 for $x = x_1$, where:

$$x_1 = (3a^2 - h^2)^{1/2} \qquad (25.7)$$

the graph of h vs. x_1 being the circle $x_1^2 + h^2 = 3a^2$ (Fig. 25.8, top curve). Interestingly, the maximally flat condition is a special case of the new one, applying where $x_1 = 0$.

Further mathematical analysis of this case is difficult: numerical computation leads to the graphs presented in Fig. 25.8. The top curve in Fig. 25.8 has already been referred to, and shows the optimum height for selected ranges of values of x up to x_1. Taken on its own, this curve would be valueless as the accompanying nonuniformity in intensity would not be known. This information is provided by

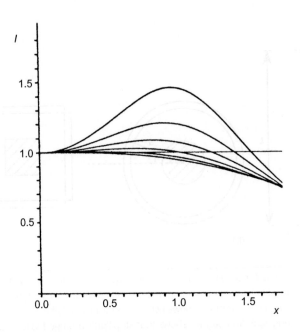

FIGURE 25.7

Intensity variation for two infinite parallel strip lights. This diagram shows the intensity variation I as a function of the distance x from the center of symmetry for six different values of h; h increases in steps of 0.2 from 0.8 for the top curve to 1.8 for the bottom curve. The value of h corresponding to the maximally flat condition is $h = 1.732$. x and h are expressed in units of a, while I is normalized to give a value of unity at $x = 0$.

Source: © IEE 1997

the left curve in Fig. 25.8. However, for design purposes, it is most important first to establish what range of intensities accompanies a given range of values of x, since this information (Fig. 25.8, bottom curve) will permit the necessary compromise between these variables to be made. Having decided on particular values of x_1 and ΔI, the value of the optimization parameter h can then be determined from one of the other two graphs: both are provided for convenience of reference (once two of the graphs are provided, the third gives no completely new information). Maximum acceptable variations in ΔI are assumed to be in the region of 20%, although the plotted variations are taken up to $\sim$50% to give a more complete picture; on the other hand, in most of the practical applications envisaged here, ΔI would be expected not to exceed 2–3% if accurate measurements of products are to be made.

The ΔI vs. x_1 variation varies faster than the fourth power of x_1, there being a very sharp rise in ΔI for higher values of x_1. This means that once ΔI has been specified for the particular application, there is little to be gained by trying to

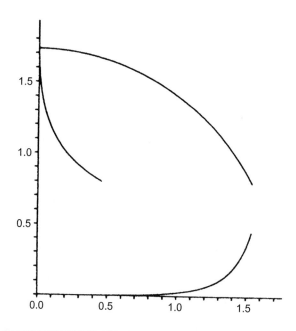

FIGURE 25.8

Design graphs for two parallel strip lights. Top, h vs. x_1. Left, h vs. ΔI. Bottom, ΔI vs. x_1. The information in these graphs has been extracted from Fig. 25.7. In design work, a suitable compromise working position would be selected on the bottom curve, and then h would be determined from one of the other two curves. In practice, ΔI is the controlling parameter, so the left and bottom curves are the important ones, the top curve containing no completely new information.

Source: © IEE 1997

squeeze extra functionality through going to higher values of x_1, i.e., in practice ΔI is the controlling parameter.

In the case of a circular ring light, the mathematics is more tedious (Davies, 1997b) and it is not profitable to examine it here. The final results are very similar to those for parallel strip lights. They would be used for design in the identical manner to that outlined earlier for the previous case.

In the case of a square ring light, the mathematics is again tedious (Davies, 1997b), but the results follow the same pattern and warrant no special comment.

25.2.4 Overview of the Uniform Illumination Scenario

Previous work on optical inspection systems has largely ignored the design of optimal lighting schemes. This section has tackled the problem in a particular case of interest—how to construct an optical system that makes a uniform matt surface appear uniformly bright, so that blemishes and defects can readily be detected with minimal additional computation. Three cases for which calculations have been carried out cover a good proportion of practical lighting schemes, and the design principles described here should be applicable to most other schemes that could be employed.

The results are best presented in the form of graphs. In any one case, the graph shows the tradeoff between variation in intensity and range of position on the working surface, from which a suitable working compromise can be selected. The other two graphs provide data for determining the optimization parameter (the height of the lights above the working surface).

Clearly, a wide variety of lighting arrangements are compatible with the general principles presented above: thus it is not worthwhile to give any detailed dimensional specifications here. However, the adjustment of just one parameter (the height) permits uniform illumination to be achieved over a reasonable region. Note that (as for a Chebyshev filter) it may be better to arrange a slightly less uniform brightness over a larger region than absolutely uniform brightness over a small region. It is left to empirical tests to finalize the details of the design.

Finally, it should be reiterated that such a lighting scheme is likely to be virtually useless for segmenting object facets from each other—or even for discerning relatively low curvatures on the surface of objects: its particular value lies in the scrutiny of surfaces via their absolute reflectivities, without the encumbrance of switched lights (see Chapter 15). It should also be emphasized that the aim of the discussion in the past few sections has been to achieve as much as possible with a simple static lighting scheme set up systematically. Naturally, such solutions are compromises and again no substitute for the full rigor of switched lighting schemes.

25.2.5 Use of Line-Scan Cameras

Throughout the above discussion it has been assumed implicitly that a conventional "area" camera is employed to view the objects on a worktable. However, when products are being manufactured in a factory they are very frequently moved from one stage to another on a conveyor. Stopping the conveyor to acquire an image for inspection would impose unwanted design problems: for this reason use is made of the fact that the speed of the conveyor is reasonably uniform, and an area image is built up by taking successive linear snapshots. This is achieved with a line-scan camera that consists of a row of photocells on a single integrated circuit sensor; the orientation of the line of photocells must of course be normal to the direction of motion. The internal design of line-scan and other cameras is discussed further below. However, we here concentrate on the lighting arrangement to be used with such a camera.

When using a line-scan camera, it is natural to select a lighting scheme that embodies the same symmetry as the camera: indeed, the most obvious such scheme is a pair of long fluorescent tubes parallel to the line of the camera (and perpendicular to the motion of the conveyor). We here caution against this "obvious" scheme, since a small round object, e.g., will not be lit symmetrically. Of course, there are difficulties in considering this problem in that different parts of the object are viewed by the line-scan camera at different moments, but note that for small objects a linear lighting scheme will not be isotropic: this could lead to small distortions being introduced in measurements of object dimensions. This means that in practice the ring and other symmetrical lighting schemes described above are likely to be more closely optimal even when a line-scan camera is used. For larger objects much the same situation applies, although the geometry is more complex to work out in detail.

Finally, the comment above that conveyor speeds are "reasonably uniform" should be qualified. The author has come across cases where this is true only as a first approximation. As with many mechanical systems, conveyor motion can be unreliable: e.g., it can be jerky, and in extreme cases not even purely longitudinal! Such circumstances frequently arise through a variety of problems that cause slippage relative to the driving rollers—the effects of wear or of an irregular join in the conveyor material, misalignment of the driving rollers, and so on. Furthermore, the motors controlling the rollers may not operate at constant speed, either in the short term (e.g., because of varying load) or in the longer term (e.g., because of varying mains frequency and voltage). While, therefore, it cannot be assumed that a conveyor will operate in an ideal way, careful mechanical design can minimize these problems. However, when high accuracy is required, it will be necessary to monitor the conveyor speed, perhaps by using optically coded disk devices, and feeding appropriate distance marker pulses to the controlling computer. Even with this method, it will be difficult to match in the longitudinal direction the extremely high accuracy[3] available from the line-scan camera in the lateral direction. However, images of 512×512 pixels that are within 1 pixel accuracy in each direction should normally be available.

25.2.6 Light Emitting Diode (LED) Sources

The past decade has seen a rapidly changing situation in the types of lighting available for various applications. The earlier tungsten lights coexisted for a long time with fluorescent tubes, and more recently compact fluorescents and halogen lights have moved forward. Low-power LEDs have been available for a significant time, but were initially only suitable as indicator lights rather than to provide illumination. In fact, there were problems in bringing them to higher power levels, and at the same time making their cost competitive. However, this position has been changing rapidly, and LED headlights and sidelights are now ubiquitous

---

[3]Line-scan cameras are available with 4096 or greater numbers of photocells in a single linear array. In addition, these arrays are fabricated using very high precision technology (see Section 25.3), so considerable reliance can be placed on the data they provide.

on road vehicles. To some extent the power problems have been solved by employing *arrays* of LEDs—a trend that has been very evident with vehicle lights. For inspection, LEDs now seem to provide the main route forward: for instance, they do not have the high-frequency firing problems of fluorescents, or the unreliability and short lives of fluorescents, tungstens and halogens. And the need for multiple LEDs to provide high illumination levels is synergistic with the need for uniform lighting, so that any illumination shape profile that would be useful for inspection, and could earlier have been provided by long or circular fluorescents or fiber-optic bundles, can now be achieved using arrangements of LEDs. While the home lighting market is still waiting for high-power LEDs to come down in price, they are easily within reach of important inspection applications. It should also be noted that LEDs are commonly guaranteed for 7 years, although their real lifetime is closer to 20 years or more (this figure assumes up to $\sim$50% usage per day). These long lifetimes are offset to some extent by gradually falling emission (a drop of 10% in $\sim$20,000 hours),[4] but this can be cancelled out by slightly derating and progressively increasing the dc supply current (the light intensity emitted by LEDs is directly proportional to the supply current). Overall, the relevant advantages of LEDs are easily controllable intensity, directly proportional to current; output that does not switch on and off at a rate determined by the mains frequency; exceptionally long life; high conversion efficiency; and an intense light from a small area that is readily focussed.

25.3 CAMERAS AND DIGITIZATION

For a good many years the camera that was normally used for image acquisition was the TV camera with a vidicon or related type of vacuum tube. The scanning arrangements of such cameras became standardized, first to 405 lines, then to 625 lines (or 525 lines in the United States). In addition, it is usual to interlace the image—i.e., to scan odd lines in one frame and even lines in the next frame, then repeat the process, each full scan taking 1/25 second (1/30 second in the United States). There are also standardized means for synchronizing cameras and monitors, using line and frame "sync" pulses. Thus, the vacuum TV camera left a legacy of scanning techniques that are in the process of being eliminated with the advent of digital TV.[5] However, as the result of the legacy is still present, we include a few more details here.

It is important to note that the output of these early cameras is inherently analog, consisting of a continuous voltage variation, although this applies only along

---

[4]A useful view of the situation appears on the following manufacturer's (Philips Colour Kinetics') website: http://www.colorkinetics.com/support/whitepapers/LEDLifetime.pdf (website accessed 23 July 2011).

[5]All the vestiges of the old system will not have been swept away until all TV receivers and monitors as well as the cameras are digital.

the line directions: the scanning action is discrete in that lines are used, making the output of the camera part analog and part digital. Hence, before the image is available as a set of discrete pixels, the analog waveform has to be sampled. Since some of the line scanning time is taken up with frame synchronization pulses, only about 550 lines are available for actual picture content. In addition, the aspect ratio of a standard TV image is 4:3 and it is common to digitize TV pictures as 512×768 or 512×512 pixels. Note also that after the analog waveform has been sampled and pixel intensity values have been established, it is still necessary to digitize the intensity values.

Modern solid-state cameras are much more compact and robust, and generate less noise; a very important additional advantage is that they are not susceptible to distortion,[6] because the pixel pattern is fabricated very accurately by standard integrated circuit photolithography techniques. They have thus replaced vacuum tube cameras in all except special situations.

Most solid-state cameras currently available are of the self-scanned charge-coupled device (CCD) type and attention is concentrated on these in what follows. In a solid-state CCD camera, the target is a piece of silicon semiconductor that possesses an array of photocells at the pixel positions. Hence, this type of camera digitizes the image from the outset, although in one respect—that signal amplitude represents light intensity—the image is still analog. The analog voltages (or more accurately the charges) representing the intensities are systematically passed along analog shift registers to the output of the instrument, where they may be digitized to give 6–8 bits of grayscale information (the main limitation here being lack of uniformity among the photosensors rather than noise *per se*). This architecture is important as it means that widely available CCD cameras can be triggered and read out at any desired rate by externally applied pulses.

Interestingly, the old vacuum tube TV cameras had a spectral response curve that peaked at much the same position as the spectral pattern of ("daylight") fluorescent lights—which itself matches the response of the human eye (Table 25.1). However, CCD cameras have significantly lower response to the spectral pattern of fluorescent tubes (Table 25.1). In general this may not matter too greatly, but when objects are moving, the integration time of the camera is limited and sensitivity can suffer. In such cases the spectral response is an important factor and may dictate against use of fluorescent lights (this is particularly relevant where CCD line-scan cameras are used with fast-moving conveyors).

An important factor in the choice of cameras is the delay lag that occurs before a signal disappears. This clearly causes problems with moving images. Fortunately, the effect is entirely eliminated with CCD camera, since the action of reading an image wipes the old image. However, moving images require frequent reading and this implies loss of integration time and therefore loss of sensitivity—a factor that normally has to be made up by increasing the power of

---

[6]However, this does not prevent distortions from being introduced by other mechanisms—poor optics, poor lighting arrangements, perspective effects, and so on.

Table 25.1 Spectral Responses

Device	Band (nm)	Peak (nm)
Vidicon	200–800	~550
CCD	400–1000	~800
Fluorescent tube	400–700	~600
Human eye	400–700	~550

In this table, the response of the human eye is included for reference. Note that the CCD response peaks at much higher wavelength than the vidicon or fluorescent tube, and therefore is often at a disadvantage when used in conjunction with the latter.

illuminating sources. Camera "burn-in" is another effect that is absent with CCD cameras but which causes severe problems with certain types of conventional camera: it is the long-term retention of picture highlights in the light-sensitive material, which makes it necessary to protect the camera against bright lights and to take care to make use of lens covers whenever possible. Finally, "blooming" is the continued generation of electron-hole pairs even when the light-sensitive material is locally saturated with carriers, with the result that the charge spreads and causes highlights to envelop adjacent regions of the image. Both CCD and conventional camera tubes are subject to this problem, although it is inherently worse for CCDs, and this has led to the production of antiblooming structures in these devices. Space precludes detailed discussion of the situation here.

25.3.1 Digitization

The remaining important item to be studied in this context is that of digitization, i.e., conversion of the original analog signals into digital form. There are many types of analog-to-digital converter (ADC) but the ones that are used for digitizing images have so much data to process—usually in a very short time if real-time analysis is called for—that special types have to be employed. The only one considered here is the "flash" ADC, so called because it digitizes all bits simultaneously, in a flash. In fact, it possesses $n - 1$ analog comparators to separate n gray levels, followed by a priority encoder to convert the $n - 1$ results into normal binary code. Such devices produce a result in a very few nanoseconds and their specifications are generally quoted in megasamples per second (typically in the range 50–200 megasamples/second). For some years these were available only in 6-bit versions (apart from some very expensive parts), but nowadays 8-bit versions are available at extremely low cost:[7] such 8-bit devices are probably sufficient for most needs considering that a certain amount of sensor noise, or

---

[7]Indeed, as is clear from the advent of cheap web cameras and digital cameras, it is becoming virtually impossible to get noncolor versions of such devices.

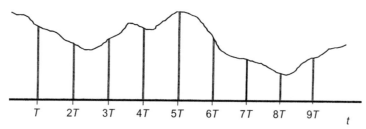

FIGURE 25.9

The process of sampling a time-varying signal: a continuous time-varying 1-D signal is sampled by narrow sampling pulses at a regular rate $f_r = 1/T$, which must be at least twice the bandwidth of the signal.

variability, is usually present below these levels and that it is difficult to engineer lighting to this accuracy. In fact, devices with even greater grayscale resolution can be obtained: these are useful for extending the overall dynamic range capability that is required when lighting levels are highly variable (which is the normal occurrence outdoors during the course of the day).

25.4 THE SAMPLING THEOREM

The Nyquist sampling theorem underlies all situations where continuous signals are sampled and is especially important where patterns are to be digitized and analyzed by computers. This makes it highly relevant both with visual patterns and with acoustic waveforms, hence it is described briefly in this section.

Consider the sampling theorem first in respect of a 1-D time-varying waveform. The theorem states that a sequence of samples (Fig. 25.9) of such a waveform contains all the original information and can be used to regenerate the original waveform exactly, but only if (a) the bandwidth W of the original waveform is restricted and (b) the rate of sampling f is at least twice the bandwidth of the original waveform—i.e., $f \geq 2W$. Assuming that samples are taken every T seconds, this means that $1/T \geq 2W$.

At first it may be somewhat surprising that the original waveform can be reconstructed exactly from a set of discrete samples. However, the two conditions for achieving this are very stringent. What they are demanding in effect is that the signal must not be permitted to change unpredictably (i.e., at too fast a rate), else accurate interpolation between the samples will not prove possible (the errors that arise from this source are called "aliasing" errors).

Unfortunately, the first condition is virtually unrealizable, since it is close to impossible to devise a low-pass filter with a perfect cut-off. Recall from Chapter 3 that a low-pass filter with a perfect cut-off will have infinite extent in the time domain, so any attempt at achieving the same effect by time domain operations must be doomed to failure. However, acceptable approximations can

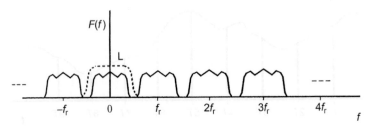

FIGURE 25.10

Effect of low-pass filtering to eliminate repeated spectra in the frequency domain (f_r sampling rate; L, low-pass filter characteristic). This diagram shows the repeated spectra of the frequency transform $F(f)$ of the original sampled waveform. It also demonstrates how a low-pass filter can be expected to eliminate the repeated spectra to recover the original waveform.

be achieved by allowing a "guard band" between the desired and actual cut-off frequencies. This means that the sampling rate must be higher than the Nyquist rate (in telecommunications, satisfactory operation can generally be achieved at sampling rates around 20% above the Nyquist rate—see Brown and Glazier, 1974).

One way of recovering the original waveform is by applying a low-pass filter. This approach is intuitively correct, since it acts in such a way as to broaden the narrow discrete samples until they coalesce and sum to give a continuous waveform. Indeed, this method acts in such a way as to eliminate the "repeated" spectra in the transform of the original sampled waveform (Fig. 25.10). This in itself shows why the original waveform has to be narrow-banded before sampling, so that the repeated and basic spectra of the waveform do not cross over each other and become impossible to separate with a low-pass filter. The idea may be taken further because the Fourier transform of a square cut-off filter is the sinc (sin u/u) function (Fig. 25.11). Hence, the original waveform may be recovered by convolving the samples with the sinc function (which in this case means replacing them by sinc functions of corresponding amplitudes). This has the effect of broadening out the samples as required, until the original waveform is recovered.

So far we have considered the situation only for 1-D time-varying signals. However, recalling that there is an exact mathematical correspondence between time and frequency domain signals on the one hand and spatial and spatial frequency signals on the other, the above ideas may all be applied immediately to each dimension of an image (although the condition for accurate sampling now becomes $1/X \geq 2W_X$, where X is the spatial sampling period and W_X is the spatial bandwidth). Here we accept this correspondence without further discussion and proceed to apply the sampling theorem to image acquisition.

Consider next how the signal from a camera may be sampled rigorously according to the sampling theorem. First, note that this has to be achieved both horizontally and vertically. Perhaps the most obvious solution to this problem is to perform the process optically, perhaps by defocusing the lens; however, the optical

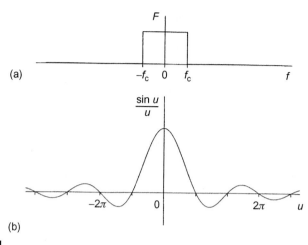

FIGURE 25.11

The sinc (sin u/u) function shown in (b) is the Fourier transform of a square pulse (a) corresponding to an ideal low-pass filter. In this case, $u = 2\pi f_c t$, f_c being the cut-off frequency.

FIGURE 25.12

Low-pass filtering carried out by averaging over the pixel region: an image with local high-frequency banding is to be averaged over the whole pixel region by the action of the sensing device.

transform function for this case is frequently (i.e., for extreme cases of defocusing) very odd, going negative for some spatial frequencies and causing contrast reversals; hence, this solution is far from ideal (Pratt, 2001). Alternatively, we could use a diffraction-limited optical system or perhaps pass the focussed beam through some sort of patterned or frosted glass to reduce the spatial bandwidth artificially. None of these techniques will be particularly easy to apply, nor (apart possibly from the second) will it give accurate solutions. However, this problem is not as serious as might be imagined. If the sensing region of the camera (per pixel) is reasonably large, and close to the size of a pixel, then the averaging inherent in obtaining the pixel intensities will in fact perform the necessary narrow-banding (Fig. 25.12). To analyze the situation in more detail, note that a pixel is essentially square with a sharp cut-off at its borders. Thus its spatial frequency pattern is a 2-D sinc function, which (taking the central positive peak) approximates to a low-pass spatial frequency filter. This approximation improves somewhat as the border between pixels becomes more fuzzy.

The point here is that the worst case from the point of view of the sampling theorem is that of extremely narrow discrete samples, but clearly this worst case is most unlikely to occur with most cameras. However, this does not mean that sampling is automatically ideal—and indeed it is not, since the spatial frequency pattern for a sharply defined pixel shape has (in principle) infinite extent in the spatial frequency domain. The review by Pratt (2001) clarifies the situation and shows that there is a tradeoff between aliasing and resolution error. Overall, quality of sampling will be one of the limiting factors if the greatest precision in image measurement is aimed for: if the bandwidth of the presampling filter is too low, resolution will be lost; if it is too high, aliasing distortions will creep in; and if its spatial frequency response curve is not suitably smooth, a guard band will have to be included and performance will again suffer.

25.5 HYPERSPECTRAL IMAGING

For a number of decades, multispectral imaging has been employed in remote sensing to obtain sufficient data to separate various land regions, such as soil, water, crops, roads, and buildings. In particular, multispectral data typically consists of four to six channels of color; infrared and microwave data from which relevant regions can be separated, e.g., with the aid of principal components analysis (PCA). However, for some time there has been a need in certain applications for even more image data, and this has led to the development of so-called "hyperspectral imaging." This is a generalization of multispectral data to cover a far larger number of spectral channels. In fact, hyperspectral imaging collects a whole *spectrum* of data for each pixel by application of suitable spectrometers. As a result, a "hyperspectral cube" of image data is formed. In contrast with multispectral images which contain at most tens of spectral bands at relatively isolated wavelengths, hyperspectral images contain hundreds or even thousands of contiguous channels. This is useful as it permits detailed selections of the required data to be made some time after the images have been obtained. Clearly, the technique is costly in storage and in the processing load needed both to select appropriate subsets of the data and to perform subsequent processing: the fundamental problem is that typical hyperspectral images usually contain hundreds of megabytes of data. Nevertheless, the technique is well adapted to remote sensing of crops from satellites, and as often happens, once a new technology develops, workers see the opportunity to apply it to their own particular types of data, and the technology is forced to develop further, for reliability, ease of use, and reduction of cost. As in the case of MRI, which is applied mainly but by no means exclusively to medical imaging, this new technology is poised to be applied extremely widely. Already it is being tested and applied to the inspection of fruit and vegetables, and to the sensitive detection of nematode worms in fish (Gómez et al., 2007; Heia et al., 2007).

In such applications, the objects under inspection are moved slowly, as would be normal for line-scan camera inspection, but here each line of pixels is scanned over the whole spectrum to image all the color channels. Thus, instead of a line of pixels being grabbed, a whole plane of pixels is grabbed in which the additional coordinate represents the color. This can be achieved in two ways: (1) the same line-scan camera is used and the spectrum is scanned sequentially over it using a spectrometer with a rotating mirror or rotating diffraction grating; (2) an area camera is used to view the whole spectral plane in parallel (this could in principle be carried out with a prism and an area camera). Clearly, use of standard area cameras applied sequentially in ~10 nm steps over the required spectral range is also a possibility. Typical applications employ visible and infrared measurements over a range 500–1000 nm, or infrared measurements over the range 960–1700 nm using an InGaAs NIR camera (Qin and Lu, 2008; Mahesh et al., 2008). Further details will not be considered here, because of the particularly rapid development of the subject.

25.6 CONCLUDING REMARKS

This chapter has aimed to give some background to the problems of acquiring images, particularly for inspection applications. Methods of illumination have been deemed to be worthy of considerable attention since they furnish means by which the practitioner can help to ensure that an inspection system operates successfully—and indeed that its vision algorithms are not unnecessarily complex, thereby necessitating excessive hardware expense for real-time implementation. Means of arranging reasonably uniform illumination and freedom from shadows have been taken to be of significant relevance and allotted fair attention (interestingly, these topics are scarcely mentioned in most books on this subject— a surprising fact that perhaps indicates that few authors ascribe much importance to this vital aspect of the work). For recent publications on illumination and shadow elimination, see the following section.

By contrast, camera systems and digitization techniques have been taken to be purely technical matters to which little space could be devoted (to be really useful—and considering that most workers in this area buy commercial cameras and associated frame-grabbing devices ready to plug into a variety of computers—whole chapters would have been required for each of these topics). Because of its theoretical importance, it was relevant to give some background to the sampling theorem and its implications although, considering the applications covered in this book, only limited space could be devoted to this topic (see Rosie, 1966 and Pratt, 2001 for further details).

Finally, the chapter included a section on hyperspectral imaging, which has recently been the subject of much research and development. In fact, the motivation for hyperspectral imaging, with the large amount of data it can bring to bear in fields such as agriculture and inspection, has only been made thinkable as a result of the vastly increased storage and processing capabilities of modern computers. There seems little doubt that progress in its technology and application will accelerate markedly in the next few years.

> Computer vision systems are commonly highly dependent on the quality of the incoming images. This chapter has shown that image acquisition can often be improved, particularly for inspection applications, by arranging regions of nearly uniform illumination so that shadows and glints are suppressed, and vision algorithms can be simplified and speeded up.

25.7 BIBLIOGRAPHICAL AND HISTORICAL NOTES

It is a regrettable fact that very few papers and books give details of lighting schemes that are used for image acquisition, and even fewer give any rationale or background theory for such schemes. Hence, some of the present chapters appear to have broken new ground in this area. However, Batchelor et al. (1985) and Browne and Norton-Wayne (1986) give much useful information on light sources, filters, lenses, light guides, and so on, thereby complementing the work of this chapter (indeed, the former of these books gives a wealth of detail on how unusual inspection tasks, such as those involving internal threads, may be carried out).

Details of various types of scanning system, camera tube, and solid-state (e.g., CCD) device are widely available—see, e.g., Batchelor et al. (1985), and also various manufacturers' catalogs. Note that much of the existing CCD imaging device technology dates from the mid-1970s (Barbe, 1975; Weimer, 1975) and is still undergoing development.

The sampling theorem is well covered in very many books on signal processing (see, e.g., Rosie, 1966). However, details of how band-limiting should be carried out prior to sampling are not so readily available. Only a brief treatment is given in Section 25.4; for further details the reader is referred to Pratt (2001) and references contained therein.

The work described in Section 25.2 arose from the author's work on food product inspection, which required carefully controlled lighting to facilitate measurement, improve accuracy, and simplify (and thereby speed up) the inspection algorithms (Davies, 1997b). Similar motivation drove Yoon et al. (2002) to attempt to remove shadows by switching different lights on and off, and then using logic to eliminate the shadows: under the right conditions, it was only necessary to find the maximum of the individual pixel intensities between the various images. However, in outdoor scenes, it is difficult to control the lighting: instead various rules must be worked out for minimizing the problems. Prati et al. (2001) have made a comparative evaluation of available methods. Other work on shadow location and elimination has been reported by Rosin and Ellis (1995), Mikić et al. (2000), and Cucchiara et al. (2003).

In addition, Koch et al. (2001) have presented results on the use of switched lights to maintain image intensity irrespective of changes of ambient illumination, and to limit the overall dynamic range of image intensities so that the risk of over- or underexposing the scene is drastically reduced. See also the bibliography in Section 15.13 for related solutions under the guise of photometric stereo.

25.7.1 More Recent Developments

Hyperspectral imaging has developed almost explosively since 2000. Not only have the techniques themselves developed in phase with the increasing power of the PC, but also significant numbers of applications have started to come forward. In particular, Gómez et al. (2007) and Gómez-Sanchis et al. (2008) have applied it to citrus fruit inspection; Qin and Lu (2008) have applied it more generally to fruit and vegetable inspection; Heia et al. (2007) have applied it to the detection of nematode worms in cod fillets (the parasites sometimes being detected well below the surface of the fillet by this means); and Mahesh et al. (2008) have applied it to differentiate Canadian wheat classes. Outside the food inspection arena, Yuen and Richardson (2010) have developed it for security, surveillance, and target acquisition. Here the potentialities are enormous, as can be seen from the fact that the method even provides the capability for assessing human stress by monitoring blood hemoglobin oxygenation levels in the face region. For further details, see Section 25.5 and the original papers mentioned above. Note that Yeen and Richardson's paper provides interesting details of the instrumentation and instrumentation schemes needed for hyperspectral imaging hardware, which range from dispersive spectrographs to narrow-band tunable filters.

26

Real-Time Hardware and Systems Design Considerations

In general, vision involves huge amounts of computation, as images are two-dimensional and in real-time applications are liable to arrive at rates of 10−20 per second. Although humans are able to cope easily with these data rates, they are often beyond the processing capabilities of conventional computers. This chapter explores the situation, and demonstrates how, using special computational hardware, the processing problems can be tackled and alleviated.

Look out for:

- how parallel processing can radically improve the speed at which vision algorithms run.
- the concept of a SIMD (single instruction stream, multiple data stream) computer, with one processor per pixel in a 2-D array.
- Flynn's classification of sequential and parallel computers.
- how a vision algorithm may optimally be partitioned between hardware and software.
- modern real-time hardware options.
- the increasingly important status of FPGAs and GPUs in real-time hardware design.
- vision system design considerations and the optimization process.

Much of this book has been devoted to the systematic design of vision algorithms, and of necessity has tended to focus on a great variety of sub-problems such as edge detection. However, when embarking upon design of a *complete* vision system, the situation is much less "clean," and indeed is subject to all sorts of financial and marketing constraints, as well as fiercely nonideal data. In this context, vision system design is as much an art as a science, and is more subject to cyclic improvement than in an ideal world—as the last few sections of this chapter aim to indicate. Suffice it to say here that the situation must be viewed realistically with an eye to improvement.

26.1 INTRODUCTION

In Chapter 1, we started by pointing out that of the five senses, vision has the advantage of providing enormous amounts of information at very fast rates: this was observed to be useful to humans and should also be of great value with robots and other machines. However, the input of large quantities of data necessarily implies large amounts of data processing, and this can create problems—especially in real-time applications. Hence, it is no wonder that speed of processing has been alluded to on numerous occasions in earlier chapters of this book.

It is now necessary to examine how serious this situation is and to suggest what can be done to alleviate the problem. Consider a simple situation where it is necessary to examine products of size 64×64 pixels moving at rates of 10–20 per second along a conveyor: this amounts to a requirement to process up to 100,000 pixels per second—or typically four times this rate if space between objects is taken into account. In fact, the situation can be significantly worse than indicated by these figures. First, even a basic process such as edge detection generally requires a neighborhood of at least 9 pixels to be examined before an output pixel value can be computed: thus, the number of pixel memory accesses is already 10 times that given by the basic pixel processing rate. Second, functions such as skeletonization or size filtering require a number of basic processes to be applied in turn: e.g., eliminating objects up to 20 pixels wide requires 10 erosion operations, whereas thinning similar objects using simple "north-south-east-west" algorithms requires at least 40 whole-image operations. Third, typical applications such as inspection require a number of tasks—edge detection, object location, surface scrutiny, and so on—to be carried out. All these factors mean that the overall processing task may involve anything between 1 and 100 million pixels or other memory accesses per second. Finally, this analysis ignores the complex computations required for some types of 3-D modeling, and for certain more abstract processing operations (see Chapters 14 and 15), although the expanding area of hyperspectral imaging (Chapter 25) also has extremely demanding computational requirements.

These formidable processing requirements imply a need for very carefully thought out algorithm strategies. This means that special hardware will normally be needed[1] for almost all real-time applications (exceptions might occur in those tasks where performance rates are governed by the speed of a robot, vehicle, or other slow mechanical device). Broadly speaking, there are two main strategies for improving processing rates. The first involves employing a single very fast

---

[1]Sometime in the future, this view will need to be reconsidered: see Section 26.7.

processor that is fabricated using advanced technology, e.g., with gallium arsenide semiconductor devices, Josephson junction devices or perhaps optical processing elements (PEs). Such techniques can be expected to yield speed increases by a factor of 10 or so, which might be sufficiently rapid for certain applications. However, it is more likely that a second strategy will have to be invoked—that of parallel processing: this involves employing N processors working in parallel, thereby giving the possibility of enhancing speed by a factor N. This second strategy is particularly attractive: to achieve a given processing speed, it should be necessary only to increase the number of processors appropriately—although it has to be accepted that cost will be increased by a factor of around N, as for the speed. Clearly, it is partly a matter of economics which of the two strategies will be the better choice but at any time the first strategy will be technology-limited, whereas parallel processing seems more flexible and capable of giving the required speed in any circumstances. Hence, for the most part, parallel processing is considered in what follows.

26.2 PARALLEL PROCESSING

There are two main approaches to parallel processing: in the first, the computational *task* is split into a number of functions, which are then implemented by different processors and in the second, the *data* are split into several parts and different processors handle the different parts. These two approaches are sometimes called *algorithmic parallelism* and *data parallelism*, respectively. Note that if the data are split, different parts of the data are likely to be nominally similar, so there is no reason to make the PEs different from each other. However, if the task is split functionally, the functions are liable to be very different and it is most unlikely that the PEs should be identical.

The example cited in Section 26.1 involves a fixed sequence of processes being applied in turn to the input images. On the whole, this type of task is well adapted to algorithmic parallelism and indeed to being implemented as a pipelined processing system, each stage in the pipeline performing one task such as edge detection or thinning (Fig. 26.1) (note that each stage of a thinning task will probably have to be implemented as a single stage in the pipeline). Clearly, such an approach lacks generality but it is cost-effective in a large number of applications, since it is capable of providing a speedup factor of around two orders of magnitude without undue complexity. Unfortunately, this approach is liable to be inefficient in practice. This is because the speed at which the pipeline operates is dictated by the speed of the slowest device on the pipeline—faster speeds of the other stages constitute wasted computational capability. Variations in the data passing along the pipeline add to this problem: e.g., a wide object would require many passes of a thinning operation, so either thinning would not proceed to completion (and the effect of this would have to be anticipated and allowed for), or

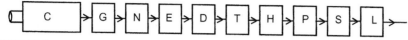

FIGURE 26.1

Typical pipelined processing system: C, input from camera; G, grab image; N, remove noise; E, enhance edge; D, detect edge; T, thin edge; H, generate Hough transform; P, detect peaks in parameter space; S, scrutinize products; L, log data and identify products to be rejected.

else the pipeline would have to be run at a slower rate. Obviously, it is necessary for such a system to be designed in accordance with worst-case rather than "average" conditions—although additional buffering between stages can help to reduce this latter problem.

Clearly, the design and control of a reliable pipelined processor is not trivial but, as mentioned above, it gives a generally cost-effective solution in many types of application. However, both with pipelined processors and with other machines that use algorithmic parallelism, there are significant difficulties in dividing tasks into functional partitions that match well the PEs on which they are to run. For this and other reasons, there have been many attempts at the alternative approach of data parallelism. Indeed, image data are, on the whole, reasonably homogeneous, so it is evidently worth searching for solutions incorporating data parallelism. Further consideration then leads to the SIMD type of machine in which each pixel is processed by its own PE. This method is described in Section 26.3.

26.3 SIMD SYSTEMS

In the SIMD (single instruction stream, multiple data stream) architecture, a 2-D array of PEs is constructed, which maps directly onto the image being processed; thus, each PE stores its own pixel value, processes it, and stores the processed pixel value. Furthermore, all PEs run the same program and indeed are subject to the same clock; this means that they execute the same instruction simultaneously (hence the existence of a single instruction stream). An additional feature of SIMD machines that are used for image processing is that each PE is connected to its immediate neighbors in the array, so that neighborhood operations can conveniently be carried out—the required input data are always available. This means that each PE is typically connected to eight others in a square array. Such machines therefore have the advantage not only of *image parallelism* but also of *neighborhood parallelism*—data from neighboring pixels are available immediately and several sequential memory accesses per pixel process are no longer required (for a useful review of these and other types of parallelism, see Danielsson and Levialdi (1981)).

The SIMD architecture is extremely attractive in principle since its processing structure seems closely matched to the requirements of many tasks, such as noise removal, edge detection, thinning, size analysis, and so on (although we return to this point below). However, in practice, it suffers from a number of disadvantages. Some of these are due to the compromises needed to keep costs at reasonable levels. For example, the PEs may not be powerful floating-point processors and may not contain much memory (this is because available cost is expended on including more PEs rather than making them more powerful); in addition, the processor array may be too small to handle the whole image at once, and problems of continuity and overlap arise when trying to process subimages separately (Davies et al., 1995); this can also lead to difficulties when global operations (such as finding an accurate convex hull) have to be performed on the whole image. Finally, getting the data in and out of the array can be a relatively slow process.

Although SIMD machines may appear to operate efficiently on image data, this is not always the case in practice, since many processors may be "ticking over" and not doing anything useful. For example, if a thinning algorithm is being implemented, much of the image may be bare of detail for most of the time, since most of the objects will have shrunk to a fraction of their original area. Thus, the PEs are not being kept *usefully* busy. Here the topology of the processing scheme is such that these inactive PEs are unable to get data they can act on, and efficiency drops off markedly. Hence, it is not obvious that an SIMD machine can always carry out the *overall* task any faster than a more modest MIMD machine (see definition and full explanation in Section 26.5), or a specially fast but significantly cheaper single processor (SISD) machine.

A more important characteristic is that although the SIMD machine is reasonably well adapted for image processing, it is quite restricted in its capabilities for image analysis. For example, it is virtually impossible to use *efficiently* for implementing Hough transforms, especially when these demand mapping image features into an abstract parameter space. In addition, most serial (SISD) computers are much more efficient at operations such as simple edge tracking, since their single processors are generally much faster than costs will permit for the many processors in an SIMD machine. Overall, these problems should be expected since the SIMD concept is designed for image-to-image transformations *via* local operators and does not map well to (a) image-to-image transformations that demand *nonlocal* operations, (b) image-to-abstract data transforms (intermediate-level processing), or (c) abstract-to-abstract data (high-level) processing (note that some would classify (a) as being a form of intermediate-level processing where *deductions* are made about what is happening in distant parts of an image—i.e., higher level interpretive data are being marked in the transformed image). This means that unaided SIMD machines are unlikely to be well suited for practical applications such as inspection.

Before leaving the topic of SIMD machines, recall that they incorporate two types of parallelism—image parallelism and neighborhood parallelism. Both of these contribute to high processing rates. Although it might at first appear that image parallelism contributes mainly through the high processing bandwidth[2] it offers, it also contributes through the high data accessing bandwidth: in contrast, neighborhood parallelism contributes only through the latter mechanism. However, what is important is that this type of parallel machine, in common with any successful parallel machine, incorporates both features. It is of little use to attend to the problem of achieving high processing bandwidth only to run into data bottlenecks through insufficient attention to data structures and data access rates: i.e., it is necessary to match the data access and processing bandwidths if full use is to be made of available processor parallelism.

26.4 THE GAIN IN SPEED ATTAINABLE WITH *N* PROCESSORS

It is interesting to speculate whether the gain in processing rate could ever be greater than N, say N^2. It could in principle be imagined that two robots used to make a bed would operate more efficiently than one, or four more efficiently than two, for a rectangular bed. Similarly, N robots welding N sections of a car body would operate more efficiently than a single one. The same idea should apply to N processors operating in parallel on an N-pixel image. At first sight, it does appear that a gain greater than N could result. However, closer study shows that any task is split between data organization and actual processing. Thus, the maximum gain that could result from the use of N processors is (exactly) N: any other factor is due to the difficulty, either for low or for high N, of getting the right data to the right processor at the right time. Thus, in the case of the bed-making robots, there is an overhead for $N = 1$ of having to run around the bed at various stages because the data (the sheets) are not presented correctly. More usually, it is at large N that the data are not available at the right place at the right moment. An immediate practical example of these ideas is that of accessing all eight neighbors in a 3×3 neighborhood where only four are directly connected, and the corner pixels have to be accessed *via* these four: then a *threefold* speedup in data access may be obtained by *doubling* the number of local links from four to eight.

There have been many attempts to model the utilization factor of both SIMD and pipelined machines when operating on branching and other algorithms. Minsky's conjecture (Minsky and Papert, 1971), that the gain in speed from a parallel processor is proportional to $\log_2 N$ rather than N, can be justified on this basis, and leads to an efficiency $\eta = \log_2 N/N$. Hwang and Briggs (1984) produced a more optimistic estimate of efficiency in parallel systems: $\eta = 1/\log_2 N$.

Following Chen (1971), the efficiency of a pipelined processor is usually estimated as $\eta = P/(N + P - 1)$, where there are on average P consecutive data

---

[2]In this context, it is conventional to use the term *bandwidth* to mean the maximum rate realizable *via* the stated mechanism.

points passing through a pipeline of N stages—the reasoning being based on the proportion of stages that are usefully busy at any one time. For imaging applications, such arguments are often somewhat irrelevant since the total delay through the pipelined processor is unimportant compared with the cycle time between successive input or output data values. This is because a machine that does not keep up with the input data stream will be unacceptable, whereas one that incorporates a fixed time delay may be acceptable in some cases (such as a conveyor belt inspection problem) although inadequate in others (such as a missile guidance system).

Broadly speaking, the situation that is being described here involves a speedup factor N coupled with an efficiency η, giving an overall speedup factor of $N' = \eta N$. Ultimately, the loss in efficiency is often due to frustrated algorithm branching processes but presents itself as underutilization of resources, which cannot be reduced because the incoming data are of variable complexity.

26.5 FLYNN'S CLASSIFICATION

Early in the development of parallel processing architectures, Flynn (1972) developed a now well-known classification, which has already been referred to above: architectures are either SISD, SIMD, or MIMD. Here SI (single instruction stream) means that a single program is employed for all the PEs in the system, whereas MI (multiple instruction stream) means that different programs can be used; SD (single data stream) means that a single stream of data is sent to all the PEs in the system, whereas MD (multiple data stream) means that the PEs are fed with data independently of each other.

The SISD machine is a single processor and is normally taken to refer to a conventional von Neumann computer. However, the definitions given above imply that SISD falls more naturally under the heading of a Harvard architecture, whose instructions and data are fed to it through separate channels: this gives it a degree of parallelism and makes it generally faster than a von Neumann architecture (in fact, there is almost invariably bit *parallelism* also, the data taking the form of words of data holding several bits of information, and the instructions being able to act on all bits simultaneously: however, this possibility is so universal that it will be accepted as standard in what follows).

The SIMD architecture has already been described reasonably thoroughly, although it is worth reiterating that the multiple data stream arises in imaging work through the separate pixels being processed by their PEs independently as separate, although similar, data streams. Note that the PEs of SIMD machines invariably embody the Harvard architecture.

The MISD architecture is notably absent from the above classification, although it is possible to envisage that pipelined processors fall into this category since a single stream of data is fed through all processors in turn, albeit being modified as it proceeds so the same *data* (as distinct from the same data stream) do not pass through each PE. However, many parties take the MISD category to be null (e.g., Hockney and Jesshope, 1981).

The MIMD category is a very wide one, containing all possible arrangements of separate PEs that get their data and their instructions independently: it even includes the case where none of the PEs are connected together in any way. However, such a wide interpretation does not solve practical problems. We can therefore envisage linking the PEs together by a common memory bus, or every PE being connected to every other one, or linkage by some other means. A common memory bus would tend to cause severe contention problems in a fast-operating parallel system, whereas maintaining separate links between all pairs of processors is clearly at the opposite extreme but would run into a combinatorial explosion as systems become larger. Hence, a variety of other arrangements is used in practice. Crossbar, star, ring, tree, pyramid, and hypercube structures have all been used (Fig. 26.2). In the crossbar arrangement, half of the processors can communicate directly with the other half *via* N links (and $N^2/4$ switches), although all processors can communicate with each other *indirectly*. In the star, there is one central PE so the maximum communication path has length 2. In the ring, all N PEs are placed symmetrically and the maximum communication path is of length $N - 1$ (note that this figure assumes unidirectional rings, which are easier to implement, and a number of notable examples of this type exist). In the

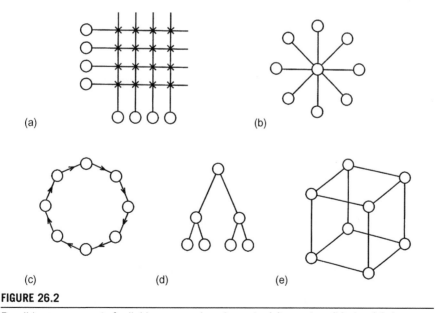

(a)   (b)

(c)   (d)   (e)

FIGURE 26.2

Possible arrangements for linking processing elements: (a) crossbar, (b) star, (c) ring, (d) tree and (e) hypercube. All links are bidirectional except where arrows indicate otherwise.

tree or pyramid, the maximum path length is of order $2(\log_2 N - 1)$, assuming that there are two branches for each node of the tree. Finally, for the hypercube of n dimensions (built in an n-dimensional space with two positions per dimension), the shortest path length between any two PEs is at most n; in this case, there are 2^n processors, so the shortest path length is at most $\log_2 N$. Overall, the basic principle must be to minimize path length while cutting down the possibility of data bottlenecks: this shows that the star configuration is very limited and explains the considerable attention that has been paid to the hypercube topology. In view of what was said earlier about the importance of matching the data bandwidth to the processing bandwidth, a very careful choice clearly needs to be made concerning a suitable architecture—and there is no lack of possible candidates (the above set of examples is by no means exhaustive).

Finally, note that factors other than speed enter into the choice of an architecture. For example, many have argued that the pyramid type of architecture closely matches the hierarchical data structures most appropriate for scene interpretation.

Much of the above discussion about architectures has been in the realm of the possible and the ideal (many of the existing systems are expensive experimental ones), and it is now necessary to consider more practical issues. How, e.g., are we to match the architecture to the data? More particularly, how do we lay down general guidelines for partitioning the tasks and implementing them on practical architectures? In the absence of general guidelines of this type, it is useful to look in detail at a given practical problem to see how to design an optimal hardware system to implement it: this is done in the next section.

26.6 OPTIMAL IMPLEMENTATION OF IMAGE ANALYSIS ALGORITHMS

The particular algorithm considered in this section was examined earlier by Davies and Johnstone (1986, 1989). It involves the inspection of round food products (see also Chapter 20). The purpose of the analysis is to show how to make a systematic selection between available hardware modules (including computers), so that it can be guaranteed that the final hardware configuration is optimal in specific ways and in particular with regard to relevant cost–speed tradeoffs.

In the particular food product application considered, biscuits are moving at rates of up to 20 per second along a conveyor. Since the conveyor is moving continuously, it is natural to use a line-scan camera to obtain images of the products. Before they can be scrutinized for defects and their sizes measured, they have to be located accurately within the images. Since the products are approximately circular, it is straightforward to employ the Hough transform technique for the purpose (see Chapter 12): it is also appropriate to use the radial intensity histogram approach to help with the task of product scrutiny (Chapter 20). In addition, simple thresholding can be used to measure the amount of chocolate cover and

Table 26.1 Breakdown of the Inspection Algorithm

	Function	Description	Time (s)	Cost (£)	c/t (£/ms)
1.	Acquire image	1 × 1	—	1000	—
2.	Clear parameter space	1 × 1	0.017	200	11.8
3.	Find edge points	3 × 3	4.265	3000	0.7
4.	Accumulate points in parameter space	1 × 1	0.086	2000	23.3
5.	Find averaged center	—	0.020	2000	100.0
6.	Find area of product	1 × 1	0.011	100	9.1
7.	Find light area (no chocolate cover)	1 × 1	0.019	200	10.5
8.	Find dark area (slant on product)	1 × 1	0.021	200	9.5
9.	Compute radial intensity histogram	1 × 1	0.007	400	57.1
10.	Compute radial histogram correlation	1-D	0.013	400	30.8
11.	Overheads for functions 6–10	—	0.415	1200	2.9
12.	Calculate product radius	1-D	0.047	4000	85.1
13.	Track parameters and log	—	0.037	4000	108.1
14.	Decide if rejection is warranted	—	0.002	4000	2000.0
	Time for whole algorithm		4.960		

Source: © IMechE 1986.

certain other product features. The main procedures in the algorithm are summarized in Table 26.1. Note that the Hough transform approach requires the rapid and accurate location of edge pixels, which is achieved using the Sobel operator and thresholding the resulting edge enhanced image. Edge detection is in fact the only 3 × 3 neighborhood operation in the algorithm and hence it is relatively time consuming; rather less processing is required by the 1 × 1 neighborhood operations (Table 26.1); then come various 1-D processes such as analysis of radial histograms. The fastest operations are those such as logging variables, which are neither 1-D nor 2-D processes.

26.6.1 Hardware Specification and Design

On finalization of the algorithm strategy, the overall execution time was found to be about 5 s per product (Davies and Johnstone, 1986).[3] With product flow rates of the order of 20 per second, and software optimization subject to severely

---

[3]Although the particular example is dated, the principles involved are still relevant and worth following here.

diminishing returns, a further gain in speed by a factor of around 100 could be obtained only by using special electronic hardware. In this application, a compromise was sought with a single CPU linked to a set of suitable hardware accelerators. In this case, the latter would have had to be designed specially for the purpose, and indeed some were produced in this way. However, in the discussion below, it is immaterial whether the hardware accelerators are made specially or purchased—the object of the discussion is to present rigorous means for deciding which software functions should be replaced by hardware modules.

As a prerequisite to the selection procedure, Table 26.1 lists execution times and hardware implementation costs of all the algorithm functions: the figures are somewhat notional since it is difficult to divide the algorithm rigorously into completely independent sections. However, they are sufficiently accurate to form the basis for useful decisions on the cost-effectiveness of hardware. Since the aim is to examine the principles underlying cost–speed tradeoffs, the figures presented in Table 26.1 are taken as providing a concrete example of the sort of situation that can arise in practice.

26.6.2 Basic Ideas on Optimal Hardware Implementation

The basic strategy that was adopted for deciding on a hardware implementation of the algorithm is as follows:

1. Prioritize the algorithm functions so that it is clear in which order they should be implemented.
2. Find some criterion for deciding when it is *not* worth proceeding to implement further functions in hardware.

From an intuitive point of view, function prioritization seems a simple process: basically functions should be placed in order of cost-effectiveness, i.e., those saving the most execution time per unit cost (when implemented in hardware) should be placed first, and those giving lesser savings should be placed later. Thus, we arrive at the *c/t* (*cost/time*) criterion function. Then with limited expenditure we achieve the maximum saving in execution time, i.e., the maximum speed of operation.

To decide at what stage it is not worth implementing further functions in hardware is arguably more difficult. Excluding here the practical possibility of strict cost or time limits, the ideal solution results in the optimal balance between total cost and total time. Since these parameters are not expressible in the same units, it is necessary to select a criterion function such as $C \times T$ (*total cost* $\times$ *total time*), which, when minimized, allows a suitable balance to be arrived at automatically.

The procedure outlined above is simple and does not take account of hardware that is common to several modules. This "overhead" hardware must be implemented for the first such module and is then available at zero cost for subsequent modules. In many cases, a speed advantage results from the use of overhead hardware. In the example system, it is found that significant economies are possible when implementing functions 6–10, since common pixel scanning circuitry may be used. In addition, note that any of these functions that are *not* implemented in hardware engender a time overhead in software. This, and the fact that the time overhead is much greater than the sum of the software times for functions 6–10, means that once the initial cost overhead has been paid it is proven to be best (in *this* case) to implement all these functions in hardware.

Trying out the design strategy outlined above gives the c/t ratio sequence shown in Table 26.2. A set of overall times and costs resulting from implementing in hardware all functions down to and including the one indicated is now deduced. Examination of the column of $C \times T$ products then shows where the tradeoff between hardware and software is optimized: this occurs here when the first 13 functions are implemented in hardware.

The analysis presented in this section clearly gives only a general indication of the required hardware–software tradeoff. Indeed, minimizing $C \times T$ indicates an overall "bargain package," whereas in practice the system might well have to meet certain cost or speed limits. In this food product application, it was necessary to aim at an overall cost of less than £10,000. By implementing functions 1, 3 and 6–11 in hardware, it was found possible to get to within a factor 3.6 of the optimal $C \times T$ product. Interestingly, using an upgraded host processor it proved possible to get within a much smaller factor (1.8) of the optimal tradeoff, with the same number of functions implemented in hardware (Table 26.2): indeed, it is a particular advantage of using this criterion function approach that the choice of which processor to use becomes automatic.

The paper by Davies and Johnstone (1989) goes into some depth concerning the choice of criterion function, showing that more general functions are available and there is a useful overriding geometrical interpretation—that global concavities are

Table 26.2 Speed–Cost Tradeoff Figures

Function (see Table 26.1)	c/t (£/ms)	t (s)	c (£)	T (s)	C (£)	C×T (£–s)	C′×T′ (£–s)
—		—	6000	4.990	6000	29,940	15,080
3	0.7	4.265	3000	0.725	9000	6530	3190
6–11	5.1	0.486	2500	0.239	11,500	2750	1400
2	11.8	0.017	200	0.222	11,700	2600	1350
4	23.3	0.086	2000	0.136	13,700	1860	1040
12	85.1	0.047	4000	0.089	17,700	1580	1010
5	100.0	0.020	2000	0.069	19,700	1360	860
13	108.1	0.037	4000	0.032	23,700	760	770
14	2000.0	0.002	4000	0.030	27,700	830	860

In this table, the first entry corresponds to the base system cost, including computer, camera, frame store, backplane, and power supply. The other entries are derived from Table 26.1 by ordering the c/t values (see text). The final column shows the figures obtained using an upgraded host processor. Source: © IMechE 1986.

being sought on the chosen curve in $f(C)$, $g(T)$ space. The paper also places the problem of overheads and relevant functional partitions on a more rigorous basis. Finally, a later paper (Davies et al., 1995) emphasizes the value of software solutions, achieved, e.g., with the aid of arrays of digital signal processing (DSP) chips.

26.7 SOME USEFUL REAL-TIME HARDWARE OPTIONS

In the 1980s, real-time inspection systems typically had many circuit boards containing hundreds of dedicated logic chips, although some of the functionality was often implemented in software on the host processor. This period was also a field-day for the Transputer type of microprocessor, which had the capability for straightforward coupling of processors to make parallel processing systems. In addition, the "bit-slice" type of microprocessor permitted easy expansion to larger word sizes, although the technology was also capable of use in multi-pixel parallel processors. Finally, the VLSI type of logic chip was felt by many to present a route forward, particularly for low-level image processing functions.

In spite of all these competing lines, what gradually emerged in the late 1990s as the predominant real-time implementation device was the digital signal processing (DSP) chip: this had evolved earlier in response to the need for fast 1-D signal processors, capable of performing such functions as Fast Fourier Transforms and processing of speech signals. The reason for its success lays in its convenience and programmability (and thus flexibility) and its high speed of operation. By coupling DSP chips together it was found that 2-D image processing operations could be performed both rapidly and flexibly, thereby to a large extent eliminating the need for dedicated random logic boards, and at the same time ousting Transputers and bit-slices.

However, over the same period, single-chip field-programmable logic arrays (FPGAs) were becoming more popular and considerably more powerful, whereas, in the 2000s, microprocessors are being embedded on the same chip. At this stage, the concept is altogether more serious and in the 2000s one has to think quite carefully to obtain the best balance between DSP and FPGA chips for implementing practical vision systems.

In fact, another contender in this race is the "ordinary" PC. Although in the 1990s, this had not normally been regarded as a suitable implementation vehicle for real-time vision, the possibility of implementing some of the slower real-time functions using a PC gradually arose though the relentless progression of Moore's Law, whereas other work on special software designs (see Chapter 21) showed how this line of development could be extended to faster running applications. In fact, we are now at the exciting stage that a single unaided PC with an embedded operating system is sufficient to run a proportion of machine vision applications—a proportion that is expected to grow substantially in the coming decade. Furthermore, we can anticipate that, over time, the whole emphasis of real-time vision will move away from speed being the dominating influence. At that stage effectiveness, accuracy, robustness, and reliability (what one might call "fitness for purpose") will be all that matters: for the first time we will be free to design ideal vision systems. The main word of caution is that this ideal will not necessarily apply for all possible applications—one can imagine exceptions, e.g., where ultrahigh speed aircraft have to be controlled or where huge image databases have to be searched rapidly.

Overall, we can see a progression amid all the hardware developments outlined above: this is a move from random logic design to the use of software-based PEs, and further, one where the software runs not on special devices with limited capability but on conventional computers for which (a) very long instruction words are not needed, (b) machine code or assembly language programming is not necessary to get the most out of the system (so standard languages can be used), and (c) overt parallel processing (beyond that available in the central processing chip of a standard PC) is no longer crucial.[4] The advantages in terms of flexibility are dramatic compared with the early days of machine vision.

The trends described above are underlined by the publications discussed in Section 26.12, and are summarized in Table 26.3.

26.8 SYSTEMS DESIGN CONSIDERATIONS

Having focussed on the problems of real-time hardware design, it is now necessary to get a clearer idea of the overall systems design process. In fact, one of the most important limitations on the rate at which machine vision systems can be produced is the lack of flexibility of existing design strategies: this applies especially to inspection systems. To some extent, this problem stems from lack of understanding of the basic principles of vision upon which inspection systems might optimally be based. In addition, there is the problem of lack of knowledge of what goes into the design process. It is difficult enough designing a complete inspection system, including all the effort that goes into producing a cost-effective real-time hardware implementation, without having to worry at the same time whether the schema used is generic or adaptable to other products. Yet this is a crucial factor that deserves a lot of attention.

An important factor impeding progress in this area is lack of detailed information on commercial systems, and lack of space in published papers: in the latter case what suffers is "know-how"—particularly on creativity aspects (journals see their role as promoting scientific methodology and results rather than subjective design notions). We explore the situation in more detail in the following section.

---

[4]Nevertheless, some functions such as image acquisition and control of mechanical devices will have to be carried out in parallel to prevent data bottlenecks and other holdups.

Table 26.3 Hardware Devices for the Implementation of Vision Algorithms

Device	Function	Summary of Properties
PC	*Personal computer or more powerful workstation*: complete computer with RAM, hard disk and other peripheral devices. Would need an embedded (restricted) operating system in a real-time application.	• Fast • Medium cost • Extremely flexible • Should be envisioned as a software device
MP	*Microprocessor*: single chip device containing CPU + cache RAM. The core element of a PC.	• Fast • Low cost • Extremely flexible • Should be envisioned as a software device
DSP	*Digital signal processor*: long instruction word MP chip designed specifically for signal processing—high processing speed on a restricted architecture.	• Very fast • Low cost • Highly flexible (some flexibility sacrificed for speed) • Should be envisioned as a software device
FPGA	*Field programmable gate array*: random logic gate array with programmable linkages; may even be dynamically reprogrammable within the application. The latest devices have flip-flops and higher level functions already made up on chip, ready for linking in; some such devices even have one or more MPs on board.	• Fast • Low-to-medium cost • Extremely flexible • Should be envisioned as a hardware device, commonly slaved to a DSP • Can be a software device if controlled by on-chip MPs
LUT	*Lookup table*: RAM or ROM. Useful for fast lookup of crucial functions. Normally slave to a MP or DSP.	• Very fast • Low cost • Extremely flexible, if built using RAM • Should be envisioned as a slave software device
ASIC	*Application-specific integrated circuit*: contains devices such as Fourier transforms, or a variety of specific SP or vision functions. Normally slave to a MP or DSP.	• Very fast • Medium cost • Inflexible (flexibility sacrificed for speed) • Should be envisioned as a slave software device
Vision chip	*Vision chips are ASICs that are devised specifically for vision*: they may contain several important vision functions, such as edge detectors, thinning algorithms, and	• Very fast • Medium cost • Inflexible (flexibility sacrificed for speed)

(Continued)

Table 26.3 (Continued)

Device	Function	Summary of Properties
VLSI	connected components analyzers. Normally slave to a MP or DSP. *Custom VLSI chip*: this commonly has many components from gate level upwards frozen into a fixed circuit with a particular functional application in mind. (Note, however, that the generic name includes MPs, DSPs, although we shall ignore this possibility here.)	• Should be envisioned as a slave software device • Fast • High cost • Inflexible • Should be envisioned as a hardware device • Normally slave to a MP or DSP
GPU	Graphics processing unit on a PC or other workstation. Although designed for computer games and other graphics applications, GPUs are able to provide valuable functionality for computer vision.	• Very fast • Substantial power requirements

The only high-cost item is the VLSI chip: *the cost of producing the masks is only justifiable for high-volume products, such as those used in digital TV. In general, high cost means more than £10,000, medium cost means around £2000, and low cost means less than £100.*

If there is a single winner in this table from the point of view of real-time applications, it is the FPGA, supposing only that it contains sufficient onboard raw computing power to make the optimum use of its available random logic. Its dynamic reprogrammability is potentially extremely powerful, but it needs to be known how to make best use of it. It has been usual to slave FPGAs to DSPs or MPs,[a] but the picture changes radically for FPGAs containing on-chip MPs.

For a discussion of GPUs, see Section 26.12.3.

[a]*In fact, the FPGA and the DSP complement each other exceptionally well, and it has been common practice to use them in tandem.*

26.9 DESIGN OF INSPECTION SYSTEMS—THE STATUS QUO

As practiced hitherto, the design of an inspection system has a number of stages, much as in the list presented in Table 26.4. While this list is clearly incomplete (e.g., it includes no mention of lighting systems), it is a useful start, and does reveal something about the creativity aspects—by admitting that reassessments of the efficacy of algorithms may be needed. In fact, there are necessarily one or two feedback loops, through which the efficacy can be improved systematically, again and again, until operation is adequate, or indeed until the process is abandoned.[5] The underlying "process" appears to be:

---

[5]It is in the nature of things that you can't be totally sure whether a venture will be a success without trying it. Furthermore, in the hard world of industrial survival, part of adequacy means producing a working system and part means making a profit out of it. This section must be read in this light.

Table 26.4 Stages in the Design of a Typical Inspection System

1. Hearing about the problem
2. Analyzing the situation
3. Looking at the data
4. Testing obvious algorithms
5. Realizing limitations
6. Developing algorithms further
7. Finding things are difficult and to some extent impossible
8. Doing theory to find the source of any limitations
9. Doing further tests
10. Getting an improved approach
11. Reassessing the specification
12. Deciding whether to go ahead
13. Completing a software system
14. Assessing the speed limitations
15. Starting again if necessary
16. Speeding up the software
17. Reaching a reasonable situation
18. Putting through 1000 images
19. Designing a hardware implementation
20. Revamping the software if necessary
21. Putting through another 100,000 images
22. Assessing difficulties regarding rare events
23. Assessing timing problems
24. Validating the final system

Source: © IEE 1997.

```
create a basic scheme;
do
        improve current scheme;
        if time up then stop;
        if no further ideas then stop;
until an adequate system is obtained;
```

The "if no further ideas" clause can be fulfilled if no way is found of making the system fast enough or low enough in cost, or of high enough specification. This would account for most contingencies that could arise.

The problem with the above process is that it represents *ad hoc* rather than scientific development, and there is no guarantee that the solution that is reached is optimal. Indeed, specification of the problem is not insisted upon (except to the level of "adequacy"), and if specification and aims are absent, it is impossible to

---

Table 26.5 Complexities of the Design Process

- It is not always evident that there is a solution, or at least a cost-effective one.
- Specifications cannot always be made in a nonfuzzy manner.
- There is often no rigorous scientific design procedure to get from specification to solution (there is certainly no guaranteed way of achieving this optimally).
- The optimization parameters are not obvious; nor are their relative priorities clear.
- It can be quite difficult to discern whether one solution is better than another.
- Some inspection environments make it difficult to tell whether a solution is valid or not.

Source: © IEE 1997.

---

judge whether the success that is obtained is optimal or not. In an engineering environment we ought to be insisting on problem specification first, solution second. However, things are more complex than this—as will be seen from Table 26.5.

In the last case in Table 26.5, there are some types of fault[6] that are particularly rare, so that only one may arise in many million cases, or, equivalently, every few weeks or months. Thus there is little statistical basis for making judgments of the risk of failure, and no proper means of training the system so that it can learn to discriminate these faults. For these sorts of reasons, making a rigorous specification and systematically trying to meet it are extremely difficult, though it is still worth trying to do so.

Let us return to the first of the complexities listed in Table 26.5. Although in principle it is difficult to know if there is a solution to a problem, nevertheless it is frequently possible to examine the computer image data that arise in the application, and see whether the eye can detect the faults or the foreign objects. If it can, this represents a significant step forward, as it means that it should be possible to devise a computer algorithm to do the same thing. What will then be in question will be whether we are creative enough to design such an algorithm, and to ensure that it is sufficiently rapid and cost-effective to be useful.

A key factor at this point is making an appropriate choice of design strategy: with structured types of image data, for example, we can ask the following:

1. Should boundary tracking be employed?
2. Should Hough transform line finding be used?
3. Should corner detection be used?

Which of these alternatives could, *with the particular dataset*, lead to algorithms of appropriate speed and robustness? On the other hand, for data that is fuzzy or

---

[6]An important class of rare faults is that of foreign objects, including the hard and sometimes soft contaminants targeted by X-ray inspection systems.

where objects are ill-defined, would artificial neural networks be more useful than conventional programming? Overall, the types of data, *and* the types of noise and background clutter that accompany it, are key to the final choice of algorithm.

26.10 SYSTEM OPTIMIZATION

To proceed further, we need to examine the optimization parameters that are relevant. In fact, there are arguably rather few of these:

1. Sensitivity
2. Accuracy
3. Robustness
4. Adaptability
5. Reliability
6. Speed
7. Cost

Of course, these one-word parameters are somewhat imprecise. For example, "reliability" can mean a multitude of things, including freedom from mechanical failures such as might arise with the camera being shaken loose, or the illumination failing; or where timing problems occur when additional image clutter results in excessive analysis time, so that the computer can no longer keep up with the real-time flow of product. Be this as it may, several of the parameters have been shown to depend on each other—e.g., sensitivity and accuracy, cost, and speed: more will be said about this in Chapter 27. Thus, we are working in a multidimensional space with various constraining curves and surfaces. In the worst case, all the parameters will be interlinked, corresponding to a single constraining surface, so that adjusting one parameter forces adjustment of at least one other. There is also the possibility that the constraining surface will impose hard limits on the values of some parameters.

In fact, each algorithm will have its own constraining surface which will in general be separate from that of others. Placing all such surfaces together will create some sort of envelope surface, corresponding to the limits of what is possible with *currently available* algorithms (Fig. 26.3). There will also be an envelope surface, corresponding to what is possible with *all possible* algorithms, i.e., including those that have not yet been developed. Thus there are limits imposed by creativity and ingenuity, and it is not known at what stage such limits might be overcome.

Returning to the constraining surface, this can be seen to provide an element of choice in the situation. Do we prefer a sensitive algorithm or a robust one, a reliable algorithm or a fast one? And so on. However, algorithms are rarely accurately describable as robust or nonrobust, reliable or nonreliable, and the multiparameter space concept with its envelope constraining surfaces clearly allows for variations in such parameters. But at the present stage of development of the

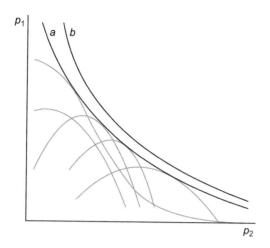

FIGURE 26.3

Optimization curves for a two-parameter system. The gray curves result from individual algorithms. (a) The envelope of the curves for all known algorithms. (b) The limiting curve for all possible algorithms.

Source: © World Scientific 2000

subject, the problem we have is that these constraining surfaces are rarely known; in addition, the algorithms that could provide access to their ideal forms are largely unknown: even the algorithms that are available are of largely unmapped capabilities. It is also the case that means are not generally known for selecting optimal working points (an interesting exception is that elucidated by Davies and Johnstone (1989) relating to optimization of cost/speed tradeoffs for image-processing hardware). Thus, there is a long way to go before the algorithm design and selection process can be made fully scientific.

26.11 CONCLUDING REMARKS

This chapter has studied the means available for implementing image analysis algorithms in real time. A number of sequential and parallel processing architectures have been considered, as well as other approaches involving use of DSP and FPGA chips or PCs with embedded operating systems. In addition, means of selecting between the various realizable schemes have been studied, these being based on criterion function optimizations.

The field is characterized on the one hand by elegant parallel architectures that would cost an excessive amount in most applications, and on the other by hardware solutions that are optimized to the application yet which are bound to lack generality. In fact, it is easy to gain too rosy a view of the elegant parallel architectures that are available: their power and generality could easily make them an overkill in dedicated applications, while reductions in generality could mean that their PEs spend much of their time waiting for data or performing null operations. Overall, the subject of image analysis is quite variegated and it is difficult to find a set of "typical" algorithms for which an "average" architecture can be designed whose PEs will always be kept usefully busy. Thus, the subject of mapping algorithms to architectures—or, better, of matching algorithms *and* architectures, i.e., designing them together as a system—is a truly complex one for which obvious solutions turn out to be less efficient in reality than when initially envisioned. Perhaps the main problem lies in envisaging the nature of image analysis, which turns out to be a far more abstract process than image processing (i.e., image-to-image transformation) ideas would indicate.

Interestingly, the problems of envisaging the nature of image analysis and of matching hardware to algorithms are gradually being bypassed by the relentless advance in the speed and power of everyday computers: we have seen that this means that there will be less need for special dedicated hardware for real-time implementation, especially if suitable algorithms such as those outlined in Section 21.4 can be developed: all this represents a very welcome factor for those developing vision systems for industry.

One topic which has been omitted for reasons of space is that of carrying out sufficiently rigorous timing analysis of image analysis and related tasks so that the overall process is *guaranteed* to run in real time. This is especially important in the case of inspection. In this context, a particular problem is that unusual image data conditions could arise which might engender additional processing, thereby setting the system behind schedule: this situation could arise because of excessive noise, because more than the usual number of products appear on a conveyor, because the heuristics in a search tree prove inadequate in some situation, or for a variety of other reasons. One way of eliminating this type of problem is to include a watchdog timer, so that the process is curtailed at a particular stage (with the product under inspection being rejected, or other suitable action being taken). However, this type of solution is crude, and sophisticated timing analysis methods now exist, based on Quirk analysis: Môtus and Rodd are protagonists of this type of rigorous approach. The reader is referred to Thomas et al. (1995) for further details and to Môtus and Rodd (1994) for an in-depth study.

Finally, the last few sections have aimed at highlighting certain weaknesses in the system design process and indicating how the subject can be developed further for the benefit of industry. Perusal of the earlier chapters will quickly show that we can now achieve a lot even though our global design base is limited. The key to our ability to solve the problems, and to know that they are solvable, lies in the capability of the human visual system for carrying out relevant visual tasks. Note, however, that there is sound reason for replacing the human eye for these visual tasks—so that (for example) 100% tireless inspection can be carried out and so that it can be achieved consistently and reliably to known standards. Clearly, the same sentiments apply for many other applications, such as driver assistance, surveillance for crime detection, and so on (see, for example, Chapters 22 and 23).

> The greatest proportion of the real-time vision system design effort used to be devoted to the development of dedicated hardware accelerators. This chapter has underlined that this problem is gradually receding, and that it is already possible to design reconfigurable FPGA systems that in many cases bypass the need for special parallel processor systems.

26.12 BIBLIOGRAPHICAL AND HISTORICAL NOTES[7]
26.12.1 General Background

The problem of implementing vision in real time presents exceptional difficulties, since in many cases megabytes of data have to be handled in times of the order of hundredths of a second, and often repeated scrutiny of the data is necessary. Hundreds of architectures have been considered for this purpose—most having some degree of parallelism. Of these, many have highly structured forms of parallelism, such as SIMD machines. The idea of a SIMD machine for image parallel operations goes back to Unger (1958), and strongly influenced later work both in the UK and in the USA. For an overview of early SIMD machines, see Fountain (1987).

The SIMD concept was generalized by Flynn (1972) in his well-known classification of computing machines (Section 26.5). Subsequently, there was a great proliferation of interconnection networks (Feng, 1981): Reeves (1984) reviewed a number that would be useful for image processing. An important problem is that of performing an optimal mapping between algorithms and architectures but for this to be possible a classification of parallel algorithms is needed to complement that of architectures: some such work was undertaken by Kung (1980), Cantoni and Levialdi (1983), and Chiang and Fu (1983).

For one period in the 1980s it became an important challenge to design VLSI chips for image processing and analysis. Offen (1985) and Fountain (1987) summarized many of the possibilities. The problems targeted in that era were defining the most useful image processing functions to encapsulate in VLSI and to understand how best to partition the algorithms taking account of chip limitations. At a

---

[7]Because of the rapid aging processes that computer and hardware development are subject to, it has been found necessary to divide this section into three parts. However, in spite of the obvious temporal development between these parts, the reader is recommended not to rush hastily past the first two of them, as they embody important principles, ideas and methods that are still highly relevant today.

more down-to-earth level, many image processing and analysis systems did not employ sophisticated parallel architectures but "merely" very fast serial processing techniques with hardwired functions—often capable of processing at video rates (i.e., giving results within one TV frame time).

Other solutions included the use of bit-slice types of device (Edmonds and Davies, 1991), and DSP chips (Davies et al., 1995), while special purpose multiprocessor designs were described for implementing multiresolution and other Hough transforms (Atiquzzaman, 1994).

26.12.2 Developments Since 2000

As indicated in Section 26.7, VLSI solutions to the production of rapid hardware for real-time applications have gradually given way to the much more flexible software solutions permitted by DSP chips and to the highly flexible FPGA type of system which permits random logic to be implemented with relative ease. In fact, FPGAs offer the possibility of dynamic reconfigurability, which may ultimately be very useful for space probes and the like, though it may prove to be an unnecessary design burden for most vision applications.[8] In this respect it is interesting to note the trend toward hybrid CPU/FPGA chips (Andrews et al., 2004) that will have optimal combinations of software and random logic available awaiting software and hardware programming (see also Batlle et al., 2002).

Be this as it may, in the early 2000s considerable attention was still focussed on VLSI solutions to vision applications (Tzionas, 2000; Mémin and Risset, 2001; Wiehler et al., 2001; Urriza et al., 2001), and it is clear that there are bound to be high-volume applications (such as digital television) where this will remain the best approach. Similarly, a lot of attention is still focussed on SIMD and linear array processor schemes, as there will always be facilities that offer ultrahigh-speed solutions of this type, and in any case small-scale implementations are likely to be cheap and usable for those (mainly low-level vision) applications that match this type of architecture. Examples of this appear in the papers by Hufnagl and Uhl (2000), Ouerhani and Hügli (2003), and Rabah et al. (2003).

Moving on to FPGA solutions, we find these embodying several types of parallelism and applied to underwater vision applications and robotics (Batlle et al., 2002), subpixel edge detection for inspection (Hussmann and Ho, 2003), and general low-level vision applications, including those implementable as morphological operators (Draper et al., 2003).

DSP solutions, some of which also involve FPGAs, include those by Meribout et al. (2002) and Aziz et al. (2003). It is useful to make comparisons with the Datacube MaxPCI (containing a pipeline of convolvers, histogrammers and other devices) and other commercially available boards and systems—see Broggi et al.

---

[8]However, see Kessal et al. (2003) for a highly interesting investigation of the possibilities—in particular showing that dynamic reconfigurability of a real-time vision system can already be achieved in milliseconds.

(2000a, 2000b), Yang et al. (2002), and Marino et al. (2001). This last paper employs a cleverly conceived architecture incorporating extensive use of lookup tables to perform high-speed matching functions that lead to road following: in fact, the matching functions involve location of interest points followed by high-speed search for matching them between corresponding blocks of adjacent frames. An interesting feature is the use of a residue number system to (effectively) factorize large lookup tables into several much smaller and more manageable lookup tables.

26.12.3 **More Recent Developments**

By 2011, hardware design for fast vision implementations has moved forward another stage, and various types of embedded real-time systems are vying with each other for supremacy—ASICs, DSPs, FPGAs, and GPUs (graphics processing units) being the main contenders. Ambrosch and Kubinger (2010) have developed stereo algorithms suitable for implementation in both ASIC and FPGA forms. However, as ASICs offer higher performance but also higher costs, these authors decided it would be more realistic to use FPGAs at least for prototyping and testing. Appiah et al. (2010) made a similar decision for their video object segmentation hardware, consisting of a single-chip FPGA together with four blocks of RAM. It embodied algorithms for foreground detection (using multimodal background modeling) and connected components labeling. Various parts of the algorithm run in parallel and the whole system is pipelined for maximum efficiency. While tackling nominally the same stereo imaging problem as Ambrosch and Kubinger (2010), Humenberger et al. (2010) produced three implementations using, respectively, a PC, a GPU, and a DSP. The latter two give real-time performance. However, the GPU is by far the fastest but has the highest power consumption. Interestingly, the DSP gives the most stable performance, with the processing times of successive frames being almost identical—in contrast to the other two implementations, which vary over several percent. In these cases, large data caches and high-level operating systems severely affect the predictability of the worst-case execution times.

The recent introduction of GPUs into the scheme of things is interesting: curiously, it was the games market that led to this way forward, as the demand for realistic computer games operating real-time 3-D HCI (human—computer interaction) necessitated exactly this sort of technology. Here, a relevant question is how dedicated GPUs are to games programs as distinct from operations that can be of value for vision. In fact, any fears on this score would appear to be groundless, as (to take one convenient example) May et al. (2010) have shown that (at minimum) *all* the SIFT modules (see Chapter 6) can be programmed on a general-purpose GPU.

Medeiros et al. (2010) developed algorithms for a parallel histogram-based particle filter for object tracking on SIMD-based smart cameras. The arrays of photoelectric elements in a camera map well to internal SIMD processors, though in practice cost and complexity considerations may, as here, limit the SIMD architecture to linear rather than area arrays. In addition, particle filters lend themselves well to parallel implementation since there are no data dependencies among particles. In fact, the research described in this paper makes much use of histograms, both color histograms and more complex histograms of oriented gradients (HOG), and again these map well to parallelism and to SIMD architectures. With these algorithms and this technology, the authors were able to achieve robust tracking of objects including humans at up to 30 frames per second.

Marzotto et al. (2010) developed a real-time roadway path extraction and tracking system for use on road vehicles, and implemented it on an FPGA platform. The ultimate purpose was that of driver assistance by providing a lane departure warning system. The proposed algorithm was designed to be completely embedded in FPGA hardware and to be capable of processing wide-VGA video sequences at 30 frames per second. The basic algorithm was targeted at locating road lane markings and made use of RANSAC for line and curve fitting. However, the overall algorithm also included a substantial amount of pre-processing in the form of noise reduction, histogram stretching, edge detection, edge thinning, automatic thresholding, and morphological filtering, together with post-processing using a Kalman filter. Nevertheless, all these and other functions were embedded within the FPGA, making full internal use of "DSPs" (programmable multiplier-accumulator units), "BRAMs" (blocks of RAM elements), and "slices" (configurable logic blocks). In fact, the system incorporated 34 DSPs, 32 BRAMs, and 8398 slices, but made use of only about 30% of the FPGA's hardware resources. While the main tests described in the paper involved simulations, preliminary tests on a real vehicle provided performances of up to 60 fps at normal VGA resolution.

Finally, to underline the role and importance of GPUs in this area, the success of the "Kinect" human motion capture system designed by engineers from Microsoft Research and provided in Xbox 360 should be noted: the unit sold 8 million devices within 2 months of its launch in November 2010, making it the fastest selling consumer electronics device in history.

Epilogue—*Perspectives in Vision* 27

27.1 INTRODUCTION

The preceding 26 chapters have covered many topics relating to vision: how images may be processed to remove noise, how features may be detected, how objects may be located from their features, how to set up lighting schemes, how to design hardware systems for automated visual inspection, and so on. The subject is one that has developed over a period of more than 40 years and has clearly come a long way. However, it has developed piecemeal rather than systematically. Often, development is motivated by the particular interests of small groups of workers and is relatively *ad hoc*. Coupled with this is the fact that algorithms, processes, and techniques are all limited by the creativity of the various researchers: the process of design tends to be intuitive rather than systematic, and so again some arbitrariness tends to creep in from time to time. As a result, sometimes no means has yet been devised for achieving particular aims, but more usually a number of imperfect methods are available and there is limited scientific basis for choosing between them.

All this poses the problem of how the subject may be placed on a firmer foundation. Time may help, but time can also have the effect of making things more difficult as more methods and results arise that have to be considered; in any case, there is no shortcut to intellectual analysis of the state of the art. This book has aimed to carry out a degree of analysis at every stage, but in this last chapter, it is worth trying to tie it all together, to make some general statements on methodology and to indicate the directions that might be taken in the future.

Computer vision is an engineering discipline and, like all such disciplines, it has to be based on science and understanding of fundamental processes.

However, as an engineering discipline, it should involve specification-based design. Once the specifications for a vision system have been laid down, it can be seen how they match up against the constraints provided by nature and technology. In what follows, we consider first the parameters of relevance for the specification of vision systems; then, we consider constraints and their origins. This leads to some clues as to how the subject could be developed further.

27.2 PARAMETERS OF IMPORTANCE IN MACHINE VISION

The first thing that can be demanded of any engineering design is that it should work. This applies as much to vision systems as to other parts of engineering. Clearly, there is no use in devising edge detectors that do not find edges, corner detectors that do not find corners, thinning algorithms that do not thin, 3-D object detection schemes that do not find objects, and so on. But in what way could such schemes fail? Even if we ignore the possibility of noise or artifacts preventing algorithms from operating properly, there remains the possibility that at any stage important fundamental factors have not been taken into account.

For example, a boundary-tracking algorithm can go wrong because it encounters a part of the boundary that is one pixel wide and crosses over instead of continuing. A thinning algorithm can go wrong because every possible local pattern has not been taken into account in the design and hence it disconnects a skeleton. A 3-D object detection scheme can go wrong because proper checks have not been made to confirm that a set of observed features is not coplanar. Of course, these types of problems may arise very rarely (i.e., only with highly specific types of input data), which is why the design error may not be noticed for a time. Often, mathematics or enumeration of possibilities can help to eliminate such errors, so problems can be removed systematically. However, being absolutely sure no error has been made is difficult—and it must not be forgotten that transcription errors in computer programs can contribute to the problems. These factors mean that algorithms should be put to extensive tests with large datasets in order to ensure that they are correct. There is no substitute for subjecting algorithms to variegated tests of this type to check out ideas that are "evidently" correct. This obvious fact is still worth stating, since silly errors continually arise in practice.

At this stage, imagine that we have a range of algorithms that all achieve the same results on ideal data, and that they really work. The next problem is to compare them critically and, in particular, to find how they react to *real* data and the nasty realities such as noise that accompany it. These nasty realities may be summed up as follows:

1. noise
2. background clutter
3. occlusions

4. object defects and breakages
5. optical and perspective distortions
6. nonuniform lighting and its consequences
7. effects of stray light, shadows, and glints

In general, algorithms need to be sufficiently robust to overcome these problems. However, things are not so simple in practice. For example, Hough transform (HT) and many other algorithms are capable of operating properly and detecting objects or features despite considerable degrees of occlusion. But how much occlusion is permissible? Or how much distortion, noise, or how much of any other of the nasty realities can be tolerated? In each specific case, we could state some figures that would cover the possibilities. For example, we may be able to state that a line detection algorithm must be able to tolerate 50% occlusion, and so a particular HT implementation is (or is not) able to achieve this. However, at this stage, we end with a lot of numbers that may mean very little on their own: in particular, they seem different and incompatible. In fact, this latter problem can largely be eliminated: each of the defects can be imagined to obliterate a definite proportion of the object (in the case of impulse noise, this is obvious; with Gaussian noise, the equivalence is not so clear but we suppose here that an equivalence can at least in principle be computed). Hence, we end up by establishing that artifacts in a particular dataset eliminate a certain proportion of the area and perimeter of all objects, or a certain proportion of all small objects.[1] This is a sufficiently clear statement to proceed with the next stage of analysis.

To go further, it is necessary to set up a complete specification for the design of a particular vision algorithm. The specification can be listed as follows (but generality is maintained by not stating any particular algorithmic function):

1. The algorithm must work on ideal data.
2. The algorithm must work on data that is $x\%$ corrupted by artifacts.
3. The algorithm must work to p pixels accuracy.
4. The algorithm must operate within s seconds.
5. The algorithm must be trainable.
6. The algorithm must be implemented with failure rate less than 1 per d days.
7. The hardware needed to implement the algorithm must cost less than £L.

(Note that the failure rate referred to in specification 6 will be taken to be a hardware characteristic and will be ignored in what follows.)

The set of specifications listed above may at any stage of technological (especially hardware) development be unachievable; this is because they are phrased in a particular way, so they are not compromisable. However, if a given specification is getting near to its limit of achievability, a switch to an alternative

---

[1]Clearly, certain of the nasty realities (such as optical distortions) tend to act in such a way as to cut down accuracy, but we concentrate here on robustness of object detection.

algorithm might be possible;[2] alternatively, an internal parameter might be adjusted, which keeps that specification within range, while pushing another specification closer to the limits of its range. In general, there will be some hard (nonnegotiable) specifications and others for which a degree of compromise is acceptable. As has been seen in various chapters of the book, this leads to the possibility of tradeoffs—a topic that is reviewed in the next section.

27.3 TRADEOFFS

Tradeoffs form one of the most important features of algorithms, since they permit a degree of flexibility subject only to what is possible in the nature of things. Ideally, the tradeoffs that are enunciated by theory provide absolute statements about what is possible so that if an algorithm approaches these limits, it is then probably as "good" as it can possibly be.

Next, there is the problem about where on a tradeoff curve an algorithm should be made to operate. The type of situation was examined carefully in Chapter 26 in a particular context—that of cost–speed tradeoffs of inspection hardware. Generally, the tradeoff curve (or surface) is bounded by hard limits. However, once it has been established that the optimum working point is somewhere within these limits, in a continuum, then it is appropriate to select a criterion function whereby an optimum can be located uniquely. Details will vary from case to case, but the crucial point is that an optimum must exist on a tradeoff curve, and that it can be found systematically once the curve is known. Clearly, all this implies that the science of the situation has been studied sufficiently so that relevant tradeoffs have been determined. We further illustrate this in the following subsections, which may be bypassed on a first reading.

27.3.1 Some Important Tradeoffs

Earlier chapters of this book including Chapters 5,6,12,14,24, and 26 have revealed some quite important tradeoffs that are more than just arbitrary relations between relevant parameters. Here, a few examples will have to suffice by way of summary.

First, in Chapter 5, the DG edge operators were found to have only one underlying design parameter—that of operator radius r. Ignoring here the important matter of the effect of a discrete lattice in giving preferred values of r, it was found that:

1. signal-to-noise ratio varies linearly with r because of underlying signal and noise averaging effects.

---

[2]But note that several, or all, relevant algorithms may be subject to almost identical limitations because of underlying technological or natural constraints.

2. resolution varies inversely with r, since relevant linear features in the image are averaged over the active area of the neighborhood: the scale at which edge positions are measured is given by the resolution.

3. the accuracy with which edge position (at the current scale) may be measured depends on the square root of the number of pixels in the neighborhood, and hence varies as r.

4. computational load, and associated hardware cost, is proportional to the number of pixels in the neighborhood, and hence varies as r^2.

Thus, operator radius carries with it four properties that are intimately related—signal-to-noise ratio, resolution (or scale), accuracy, and hardware/computational cost.

Another important problem was that of fast location of circle centers (Chapter 12); in this case, robustness was seen to be measurable as the amount of noise or signal distortion that can be tolerated. For HT-based schemes, noise, occlusions, distortions, etc., all reduce the peak height in parameter space, thereby reducing the signal-to-noise ratio and impairing accuracy. Furthermore, if a fraction β of the original signal is removed, leaving a fraction $\gamma = 1 - \beta$, either by such distortions or occlusions or else by deliberate sampling procedures, then the number of independent measurements of the center location drops to a fraction γ of the optimum. This means that the accuracy of estimation of the center location drops to a fraction around $\sqrt{\gamma}$ of the optimum.

What is important is that the effect of sampling is substantially the same as that of signal distortion so that the more distortion that must be tolerated, the higher α, the fraction of the total signal sampled, has to be. This means that as the level of distortion increases, the capability for withstanding sampling decreases, and therefore the gains in speed achievable from sampling are reduced—i.e., for fixed signal-to-noise ratio and accuracy, a definite robustness–speed tradeoff exists. Alternatively, the situation can be viewed as a three-way relation among accuracy, robustness, and speed of processing. This provides an interesting insight into how the edge operator tradeoff considered earlier might be generalized.

To underline the value of studying such tradeoffs, note that any given algorithm will have a particular set of adjustable parameters that are found to control—and hence lead to tradeoffs between—the important quantities such as speed of processing, signal-to-noise ratio, and attainable accuracy already mentioned. Ultimately, such *practically* realizable tradeoffs (i.e., arising from the given algorithm) should be considered against those that may be deduced on purely theoretical grounds. Such considerations would then indicate whether a better algorithm might exist than the one currently being examined.

27.3.2 Tradeoffs for Two-Stage Template Matching

Two-stage template matching has been mentioned a number of times in this book as a means whereby the normally slow and computationally intensive process of template matching may be speeded up. In general, it involves looking for easily distinguishable subfeatures so that locating the features that are ultimately being sought involves only the minor problem of eliminating false alarms. The reason this strategy is useful is that the first stage eliminates the bulk of the raw image data so that only a relatively trivial testing process remains. This latter process can then be made as rigorous as necessary. In contrast, the first "skimming" stage can be relatively crude, the main criterion being that it must not eliminate any of the desired features: false positives are permitted but not false negatives. However, the efficiency of the overall two-stage process is naturally limited by the number of false alarms thrown up by the first stage.

Suppose that the first stage is subject to a threshold h_1 and the second stage to a threshold h_2. If h_1 is set very low, then the process reverts to the normal template matching situation, since the first stage does not eliminate any part of the image. In fact, setting $h_1 = 0$ initially is useful so that h_2 may be adjusted to its normal working value. Then h_1 can be increased to improve efficiency (reduce overall computation); a natural limit arises when false negatives start to occur—i.e., some of the desired features are not being located. Further increases in h_1 now have the effect of cutting down available signal, although speed continues to increase. This clearly gives a tradeoff between signal-to-noise ratio, and hence accuracy of location, and speed.

In a particular application in which objects were being located by the HT, the numbers of edge points located were reduced as h_1 increased, so accuracy of object location was reduced (Davies, 1988g). A criterion function approach was then used to determine an optimum working condition. A suitable criterion function turned out to be $C = T/A$, where T is the total execution time and A the achievable accuracy. Although this approach gave a useful optimum, the optimum can be improved further if a mix of normal two-stage template matching and random sampling is used. This turns the problem into a 2-D optimization problem with adjustable parameters h_1 and u (the random sampling coefficient, equal to $1/\alpha$). However, in reality, these types of problems are even more complex than indicated so far: in general, this is a 3-D optimization problem, the relevant parameters being h_1, h_2, and u, although in fact a good approximation to the global optimum may be obtained by the procedure of adjusting h_2 first, and then optimizing h_1 and u together—or even of adjusting h_2 first, then h_1, and then u (Davies, 1988g). Further details are beyond the scope of the present discussion.

27.4 MOORE'S LAW IN ACTION

It has been indicated once or twice that the constraints and tradeoffs limiting algorithms are sometimes not accidental but rather the result of underlying technological or natural constraints. If so, it is important to determine this in as many cases as possible; otherwise, workers may spend much time on algorithm development only to find their efforts repeatedly being thwarted. Usually, this is more easily said than done but it underlines the necessity for scientific analysis of fundamentals.

The well-known law due to Moore (Noyce, 1977) relating to computer hardware states that the number of components that can be incorporated onto a single integrated circuit increases by a factor of about 2 per year. Certainly, this was so for the 20 years following 1959, although the rate subsequently decreased somewhat (not enough, however, to prevent the growth from remaining approximately exponential). It is not the purpose of this chapter to speculate on the accuracy of Moore's law. However, it is useful to suppose that computer memory and power will grow by a factor approaching 2 per year in the foreseeable future. Similarly, computer speeds may also grow at roughly this rate in the foreseeable future. When then of vision?

Unfortunately, many vision processes such as search are inherently NP-complete and hence demand computation that grows exponentially with some internal parameter such as the number of nodes in a match graph. This means that the advance of technology is able to give only a roughly linear improvement in this internal parameter (e.g., something like one extra node in a match graph every 2 years): it is therefore not solving the major search and other problems but only easing them.

NP-completeness apart, Chapter 26 was able to give an optimistic view that the relentless advance of computer power described by Moore's law is leading to an era when conventional PCs will be able to cope with a fair proportion of vision tasks. Certainly, when combined with specially designed algorithms (see Section 21.4), it should prove possible to implement many of the simpler tasks in this way, leading to a much less strenuous life for the vision systems designer.

27.5 HARDWARE, ALGORITHMS, AND PROCESSES

The previous section raised the hope that improvements in hardware systems will provide the key to the development of impressive vision capabilities. However, it seems likely that breakthroughs in vision algorithms will also be required before this can come about. My belief is that until robots can play with objects and materials in the way that tiny children do, they will not be able to build up sufficient information and the necessary databases for handling the complexities of real vision. The real world is too complex for all the rules to be written down overtly: these rules have to be internalized by training each brain individually. In some ways, this approach is better, since it is more flexible and adaptable and at the same time more likely to be able to correct for the errors that would arise in direct transference of huge databases or programs. Nor should it be forgotten that it is the underlying processes of vision and intelligence that are important: hardware merely provides a means of implementation. If an idea is devised for a hardware solution to a visual problem, it reflects an underlying algorithmic process that either is or is not effective. Once it is known to be effective, then the hardware implementation can be analyzed to confirm its utility. However, we must not segregate algorithms too much from hardware design: in the end, it is necessary to optimize the whole system, which means considering both together. Ideally at least, the underlying processes should be considered first before a hardware solution is frozen in. Hardware should not be the driving force, since there is a danger that some type of hardware implementation (especially one that is temporarily new and promising) will take over and make workers blind to underlying processes. And many readily designed hardware architectures (from SIMD to FPGA to GPU) are restricted and embody low-level vision capability rather than high-level functionality. Hardware should not be the tail that wags the vision dog.

27.6 THE IMPORTANCE OF CHOICE OF REPRESENTATION

This book has progressed steadily from low-level ideas, through intermediate-level methods to high-level processing, covering 3-D image analysis, the necessary technology, and so on—admittedly with its own type of detailed examples and emphasis. Many ideas have been covered and many strategies described. But where have we got to, and to what extent have we solved the problems of vision referred to in Chapter 1?

Among the worst of all the problems of vision is that of minimizing the amount of processing required to achieve particular image recognition and measurement tasks. Not only do images contain huge amounts of data, but often they need to be interpreted in frighteningly small amounts of time and the underlying search and other tasks tend to be subject to combinatorial explosions. Yet, in retrospect, we seem to have remarkably few *general* tools for coping with these problems. Indeed, the truly general tools available[3] appear to be:

1. reducing high-dimensional problems to lower dimensional problems that can be solved in turn.
2. the Hough transform and other indexing techniques, together with RANSAC and other hypothesis-based techniques.[4]
3. location of features that are in some sense sparse, and which can hence help to reduce redundancy quickly (obvious examples of such features are edges and corners).
4. two-stage and multistage template matching.
5. random sampling.

These are said to be general tools, since they appear in one guise or another in a number of situations, with totally different data. However, it is pertinent to ask to

---

[3]We here consider only intermediate-level processing, ignoring for example efficient AI tree-search methods relevant for purely abstract high-level processing.

[4]For simplicity, in this section it will be best for the reader to consider the synergies between the Hough transform and RANSAC rather than their differences.

what extent these are genuine tools rather than almost accidental means (or tricks) by which computation may be reduced. Further analysis yields interesting answers to this question, as will now be seen.

First, consider the Hough transform, which takes a variety of forms—the normal parameterization of a line in an abstract parameter space, the GHT that is parameterized in a space congruent to image space, the adaptive thresholding transform (Chapter 4) that is parameterized in an abstract 2-D parameter space, and so on. What is common about these forms is *the choice of a representation in which the data peak naturally at various points*, so that analysis can proceed with improved efficiency. The relation with item 3 above now becomes clear, making it less likely that either of these procedures is purely accidental in nature.

Next, item 1 appears in many guises—see, e.g., the approaches used to locate ellipses (Chapter 12). Thus, item 1 has much in common with item 4. Note also that item 5 can be considered a special case of item 4 (random sampling is a form of two-stage template matching with a "null" first stage, capable of eliminating large numbers of input patterns with particularly high efficiency: see Davies, 1988g). Finally, note that the example of so-called two-stage template matching covered in Section 27.3.2 was actually part of a larger problem that was really multistage: the edge detector was two-stage but this was incorporated in an HT that was itself two-stage, making the whole problem at least four-stage. It can now be seen that items 1–5 are all forms of multistage matching (or sequential pattern recognition), which are potentially more powerful and efficient than a single-stage approach. Similar conclusions are arrived at in Appendix A, which deals with robust statistics and their application to machine vision.

Hence, we are coming to a view that there is just one general tool for increasing efficiency. However, in practical terms, this may not itself be too useful a conclusion, since the subject of image analysis is also concerned with the ways in which this underlying idea may actually be realized—how are complex tasks to be broken down into the most appropriate multistage processes, and how then is the most suitable representation found for sparse feature location? Probably, the five-point list given above throws the right sort of light onto this particular problem—although it is clearly not the whole truth.

Finally, when looking at representations for vision algorithms, it needs to be noted that *all representations impose their own order on a system*: for a time, this may be a good imposition, but in the end it may turn into a dire restriction that is past its sell-by date. (This is what happened to the old chain code representation for boundary coding.)

27.7 PAST, PRESENT, AND FUTURE

In some sense, the contents of a book such as this have to be concentrated on subject matter that is definite: indeed, it is the duty of an author to provide information on the definite rather than the ephemeral, so there has to be some concentration on the past. Yet, a book must also concentrate on fundamental principles, and these necessarily continue from the past to the present and the future. The difference is that principles that will be known in the future cannot possibly be included, and here a sound framework together with the current difficulties and unsolved problems can at least provide readers with a readiness for any principles that are to come. In fact, this book has solved some of the problems it set itself—starting with low-level processing, concentrating on strategies, limitations, and optimizations of intermediate-level processing, going some way with higher level tasks, and attempting to create an awareness of the underlying processes of vision. At the same time, there are many interesting current developments that will prove even more interesting in the future. For the subject has passed the stage of overconcentration on hardware and absolute efficiency and has focused on the important need to extend effectiveness and capability. In addition, the developments of the past decade or so have taken the subject out of the era of the *ad hoc* into that of mathematical precision and probabilistic formulation, so that whatever vision is expected to achieve is written down in terms of estimators that are mathematically defined and turned into rigorous implementations. Nowhere is this clearer than for the new invariant feature detectors with their massive descriptors that arguably make 3-D interpretation and motion tracking almost trivial to implement. All this means that exotic yet direly needed applications such as vision-based driver assistance systems are able to come into being—and it is possible to predict that they will be with us in the cars of the immediate future, if only we and the legal system will allow this.

Only a fool would make rash predictions (and many predictions within AI have remained elusive for more than 30 years), but it is different if the principles are clear: and they are evident to many vision workers nowadays; in fact, there is an air of euphoria over the rapidly growing maturity of the latest vision algorithms and the capability of the newest computers to implement them, so the very momentum is starting to make it straightforward to estimate when various developments will happen—a situation that is advancing all types of video analytics, in areas ranging from transport to crime detection and prevention, not to mention face recognition, biometrics, and robotics. It is hoped that the present volume will be able to communicate some of the excitement underlying these present and future developments and also some means for understanding their basis.

While some of the concentration of this chapter has been on tradeoffs and optimization, deeper issues are involved, such as finding out how to make valid specifications for image data, what representations are needed within vision algorithms, and how the latter break down the overall process into viable subprocesses. There are also questions about the way in which vision algorithms are set up to rigorously estimate key parameters—a factor that relates directly to reliability, robustness, and fitness for purpose. Added to this are the exciting new applications of this rapidly maturing subject.

27.8 **BIBLIOGRAPHICAL AND HISTORICAL NOTES**

Much of this chapter has summarized the work of earlier chapters and attempted to give it some perspective. In particular, two-stage template matching has been highlighted in this chapter: the earliest work on this topic was carried out by Rosenfeld and VanderBrug (1977) and VanderBrug and Rosenfeld (1977), while the ideas of Section 27.3.2 were developed by Davies (1988g). Two-stage template matching harks back to the spatial matched filtering concept discussed in Chapter 13, though it also appears independently in an earlier chapter, Chapter 10. Ultimately, this concept is limited by the variability of the objects to be detected. However, it has been shown that some account can be taken of this problem, for example, in the design of filter masks (see Davies, 1992d). It ought also to be stated that this topic is highly formative, and although it is here developed in the context of template matching, it is possible to see shadows and reflections of it right through the whole subject: one has only to ask how any new algorithm breaks down visual analysis into an efficient set of subprocesses and what representations they are operating in, in order to see the ramifications of this concept.

Robust Statistics

At an early stage, science students learn that averaging is an effective way of eliminating noise and improving accuracy. However, Chapter 3 demonstrated unequivocally that median filtering of images is far better than mean filtering, both in retaining the form of the underlying signal and in suppressing impulse noise. Robust statistics is the subject of systematically eliminating outliers from visual or other data. This appendix aims to give useful insights into this important subject.

Look out for:

- the concepts "breakdown point" and "relative efficiency."
- M-, R-, and L-estimators.
- the idea of an influence function.
- the least median of squares (LMedS) approach.
- the RANSAC approach.
- the ways these methods can be applied in machine vision.

Although robust statistics is a relatively young discipline, dating largely from the 1980s, it has acquired a considerable following in machine vision, and is crucial, for example, in the development of robust 3-D vision algorithms. A basic problem to be tackled is the impossibility of knowing how much of the input data is in the form of outliers.

A.1 INTRODUCTION

We have found many times in this volume that noise can interfere with image signals and result in inaccurate measurements—e.g., of object shapes, sizes, and positions. Perhaps more important, however, is the fact that signals other than the particular one being focussed upon can lead to gross shape distortions and can thus prevent an object from being recognized or even being discerned at all. In many cases, this will render some obvious interpretation algorithm useless, although algorithms with intrinsic "intelligence" may be able to save the day. For this reason, the Hough transform has achieved some prominence: indeed, this approach to image interpretation has frequently been described as "robust," although no rigorous definition of robustness has been ventured so far in this volume. This appendix aims to throw further light on the problem.

Research into robustness did not originate in machine vision but evolved as the specialist area of statistics now known as robust statistics. Perhaps the paradigm problem in this area is that of fitting a straight line to a set of points. In the physics laboratory, least-squares analysis is commonly used to tackle the task. Figure A.1(a) shows a straightforward situation, where all the data points can be fitted with a reasonably uniform degree of exactness, in the sense that the residual errors[1] approximate to the expected Gaussian distribution. Figure A.1(b) shows a less straightforward case, where a particular data point seems not to fall within a Gaussian distribution. Intuition indicates that this particular point represents data that has become corrupted in some way, for example, by misreading an instrument or through a transcription error. Although the wings of a Gaussian distribution stretch out to infinity, the probability that a point will be more than five standard deviations from the center of the distribution is very small, and indeed, $\pm 3\sigma$ limits are commonly taken as demarcating practical limits of correctness: it is taken as reasonable to disregard data points lying outside this range.

Unfortunately, the situation can be much worse than this simple example suggests. Suppose that there is a rogue data point that is a very long way off. In least squares analysis, it will have such a large leverage that the correct solution may not be found. And if the correct solution is not found, there will be no basis for excluding the rogue data point. This situation is illustrated in Fig. A.1(c), where the obviously correct solution has been ignored by the numerical analysis procedure.

A worse case of line fitting occurs when there are many rogue points, and it is not clear which points lie on the straight line and which do not (Fig. A.1(d)). In fact, it may not be known whether there are several lines to be fitted, or whether there are *any* lines to be fitted. Although this circumstance would appear not to occur while data points are being plotted in physics experiments, it can arise when high energy particles are being tracked; it also occurs frequently in images of indoor and outdoor scenes where a myriad of straight lines of various lengths can appear in a great many orientations and positions. Thus, it is a real problem for which answers are required. An attempt at a full statement for this type of problem might be: devise a means for finding all the straight lines—of whatever length—in a generalized[2] image so as to obtain the best overall fit to the dataset. Unfortunately, there are likely to be many solutions to any line fitting task, particularly if the data points are not especially accurate (if they are highly accurate, then the number of solutions will be small, and it should be easy to decide intuitively or automatically what the best solution is). In fact, a rigorous

---

[1]The residual errors or "residuals" are the deviations between the observed values and the theoretical predictions of the current model or current iteration of that model.

[2]That is, an image that might correspond to off-camera images, or to situations such as data points being plotted on a graph.

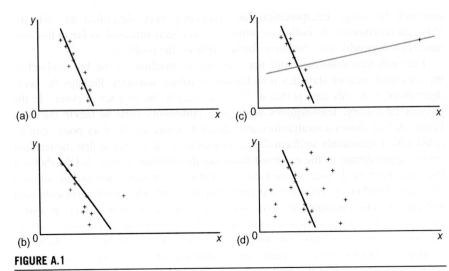

FIGURE A.1

Fitting of data points to straight lines. (a) Straightforward situation where all the data points can be fitted with reasonable precision; (b) a less straightforward case where a particular data point seems not to fall within a Gaussian distribution; (c) a situation where the correct solution has been ignored by the numerical analysis procedure; and (d) a situation where there are many rogue points, and it is not clear which points lie on the straight line and which do not: in such cases, it may not be known whether there are several, or any, lines to be fitted.

answer to the question of which solution provides the best fit requires the defini-tion of a criterion function that in some way takes account of the number of lines and the *a priori* length distribution. We shall not pursue this line of attack here, as the purpose of this appendix is to give a basic account of the subject of robust statistics, not one that is tied to a particular task. Hence, we shall focus mainly on to the simpler case where there is only one line present in the generalized image, and there are a substantial number of rogue data points or "outliers" present.

A.2 PRELIMINARY DEFINITIONS AND ANALYSIS

In the previous section, we saw that robustness is an important factor in deciding on a scheme for fitting experimental data to numerical models. It is clearly impor-tant to have an exact measure of robustness, and the concept of a "breakdown point" long ago emerged as such a measure. The breakdown point ε of a regres-sion scheme is defined as the smallest proportion of outlier contamination, which may force the value of the estimate to exceed an arbitrary range. As we have seen, even a single outlier in a set of plots can cause least-squares regression to give completely erroneous results. However, a much simpler example is to hand,

Table A.1 Breakdown Points for Means and Medians

n	Mean	Median
1	1	1
3	1/3	2/3
5	1/5	3/5
11	1/11	6/11
∞	0	0.5

The table shows how the respective breakdown points for the mean and median approach 0 and 0.5 as n tends to infinity, in the case of 1-D data.

namely, a 1-D distribution for which the mean is computed: here again, a single outlier can cause the mean to exceed any stated bound. This means that the breakdown point for the mean must be zero. On the other hand, the median of a distribution is well known to be highly robust to outliers, and remains unchanged if nearly half the data is corrupted. Specifically, for a set of n data points, the median will remain unchanged if the lowest $\lfloor n/2 \rfloor$ points[3] are moved to arbitrary lower values, or the highest $\lfloor n/2 \rfloor$ points are moved to arbitrary higher values, but in either case the median value will be changed to an arbitrary value if $\lfloor n/2 \rfloor + 1$ points are so moved. By definition (see above), this means that the breakdown point of the median is $(\lfloor n/2 \rfloor + 1)/n$; this value should be compared with the value $1/n$ for the mean. In the case of the median, the breakdown point approaches 0.5 as n tends to infinity (Table A.1). Thus, the median attains the apparently maximum achievable breakdown point of 0.5, and is therefore optimal—at least in the 1-D case described in this paragraph.

In fact, the breakdown point is not the only relevant parameter for characterizing regression schemes. For example, the "relative efficiency" is also important, and is defined as the ratio between the lowest achievable variance and the actual variance achieved by the regression method. In fact, the relative efficiency depends on the particular noise distribution that the data is subject to. It can be shown that the mean is optimal for elimination of Gaussian noise, having a relative efficiency of unity, while the median has a relative efficiency of $2/\pi = 0.637$. However, when dealing with impulse noise, the median has a higher relative efficiency than the mean, the exact values depending on the nature of the noise. This point is discussed in more detail in the following paragraphs.

Time complexity is a further parameter that is needed for characterizing regression methods. We shall not pursue this aspect further here, beyond making the observation that the time complexity of the mean is $O(n)$, while that for the median

---

[3]The function $\lfloor \cdot \rfloor$ denotes the "floor" (rounding down) operation and indicates the largest integer less than or equal to the enclosed value. In the present case, we have $\lfloor n/2 \rfloor \le n/2 \le \lfloor n/2 \rfloor + 1$.

varies with the method of computation (e.g., $O(n)$ for the histogram approach of Section 3.3 and $O(n^2)$ when using a bubble sort): in any case, the absolute time for computing a median normally far exceeds than that for the mean.

Of the parameters referred to above, the breakdown point has been at the forefront of workers' minds when devising new regression schemes. While it might appear that the median already provides an optimal approach for robust regression, its breakdown value of 0.5 only applies to 1-D data. It is therefore worth considering what breakdown point could be achieved for tasks such as line fitting, bearing in mind the poor performance of least-squares regression. Let us take the method of Theil (1950) in which the slope of each pair of a set of n data points is computed, and the median of the resulting set of ${}^nC_2 = \frac{1}{2}n(n-1)$ values is taken as the final slope; in fact, the intercept can be determined more simply because the problem has at that stage been reduced to one dimension. As the median is used in this procedure, at least half the slopes have to be correct in order to obtain a correct estimate of the actual slope. If we assume that the proportion of outliers in the data is η, the proportion of inliers[4] will be $1 - \eta$, and the proportion of correct slopes will be $(1 - \eta)^2$, and this has to be at least 0.5. This means that η has to lie in the range:

$$\eta \leq 1 - \frac{1}{\sqrt{2}} = 1 - 0.707 = 0.293 \tag{A.1}$$

Thus, the breakdown point for this approach to linear regression is less than 0.3. In a 3-D data space where a best-fit plane has to be found, the best breakdown point will be even smaller, with a value $1 - 2^{-1/3} \approx 0.2$. The general formula for p dimensions is:

$$\eta_p \leq 1 - 2^{-1/p} \tag{A.2}$$

Clearly, there is a need for more robust regression schemes, which becomes more urgent for larger values of p.

The development of robust multidimensional regression schemes took place relatively recently, in the 1970s. The basic estimators that were developed at that time, and classified by Huber in 1981, were the M-, R-, and L-estimators. The M-estimator is by far the most widely used, and appears in a variety of forms that encompass median and mean estimators and least-squares regression: we shall study this type of estimator in more detail below. The L-estimators employ linear combinations of order statistics, and include the alpha-trimmed mean, with the median and mean as special cases. However, it will be easier to consider the median and the mean under the heading of M-estimators, and in what follows we concentrate on this approach.

---

[4]Inliers are normal valid data points: the dataset is to be regarded as composed of inliers and outliers.

A.3 THE M-ESTIMATOR (INFLUENCE FUNCTION) APPROACH

M-estimators operate by minimizing the sum of a suitable function ρ of the residuals r_i. Normally, ρ is taken to be a positive definite function, and for least-squares (L_2) regression, it is the square of the residuals:

$$\rho(r_i) = r_i^2 \qquad (A.3)$$

In general, it is necessary to perform the M-estimation minimization operation iteratively until a stable solution is obtained (at each iteration, the new set of offsets has to be added to the previous set of parameter values).

To improve upon the poor robustness of L_2 regression, reflected by its zero breakdown point, an improved function ρ must be obtained that is well adapted to the particular noise[5] and outlier content of the data. To understand this process, it is easiest to analyze the situation for 1-D datasets, and to consider the influence of each data point. We represent the influence of a data point by an influence function $\psi(r_i)$, where:

$$\psi(r_i) = \frac{d\rho(r_i)}{d(r_i)} \qquad (A.4)$$

Note that minimizing $\sum_{i=0}^{n} \rho(r_i)$ is equivalent to reducing $\sum_{i=0}^{n} \psi(r_i)$ to zero, and in the case of L_2 regression:

$$\psi(r_i) = 2r_i \qquad (A.5)$$

In one dimension, this equation has a simple interpretation—moving the origin of coordinates to a position where $\sum_{i=0}^{n} r_i = 0$, that is, to the position of the mean. Now that we have shown the equivalence of L_2 regression to simple averaging, the source of the lack of robustness becomes all too clear—however far away from the mean a data point is, it still retains a weight proportional to its residual value r_i. Accordingly, a wide range of possible alternative influence functions have been devised to limit the problem by cutting down the weights of distant points that are potential outliers.

An obvious approach is to limit the influence of a distant point to some maximum value: another is to eliminate its influence altogether once its residual error exceeds a certain limiting value (Fig. A.2). We could achieve this by a variety of schemes, either cutting off the influence suddenly at this limiting distance (as in the case of the $\pm 3\sigma$ points), or letting it approach zero according to a linear

---

[5]At this point, a certain ambiguity creeps into the discussion. "Noise" tends to originate from electronic processes in the image source, and typically leads to a Gaussian distribution in the pixel intensity values. By the time positions of objects are being measured, it is strictly speaking errors rather than noise that are being considered, and the error distribution is not necessarily identical to the noise distribution that gave rise to it. However, in the remaining sections of this appendix, we usually refer to noise and noise distributions: the term "noise" will be taken to refer either to the original noise source or to the derived errors, as appropriate to the discussion.

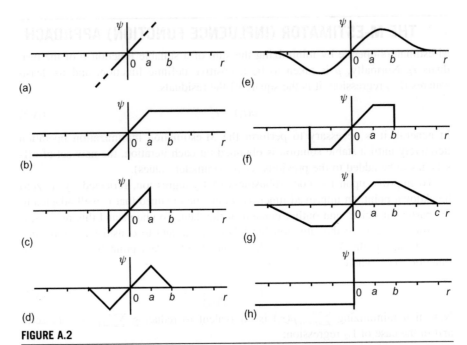

FIGURE A.2

Influence functions that limit the effects of outliers. (a) The case where no limit is placed on the influence of distant points; (b) how the influence is limited to some maximum value; (c) how the influence is eliminated altogether once the residual exceeds a certain maximum value; (d) a piecewise-linear profile that gives a less abrupt variation; (e) a mathematically more well-behaved influence function; (f) another possible piecewise-linear case; (g) a Hampel three-part redescending M-estimator that approximates the mathematically ideal case (e) with reasonable accuracy; and (h) the situation for a median estimator.

profile, or opting for a more mathematically ideal functional form with a smoother profile. In fact, there are other considerations, such as the amount of computation involved in dealing with large numbers of data points taken over a fair number of iterations. Thus, it is not surprising that a variety of piecewise linear profiles approximating to the smoother ideal profiles have been devised. In general, however, influence functions are linear near the origin, zero at large distances from the origin, and possess a region over which they give significant weight to the data points (Fig. A.2).

Prominent among these possibilities are the Hampel three-part redescending M-estimator, whose influence function is composed simply of convenient linear components, and the Tukey biweight estimator (Beaton and Tukey, 1974) that takes a form similar to that shown in Fig. A.2(e):

$$\psi(r_i) = r_i(\gamma^2 - r_i^2)^2 \quad |r_i| \leq \gamma$$
$$= 0 \quad\quad\quad\quad |r_i| > \gamma \tag{A.6}$$

It was remarked above that the median operation is a special case of the M-estimator: here all data points on one side of the origin have a unit positive weight, and all data points on the other side of the origin have unit negative weight:

$$\psi(r_i) = \text{sign}(r_i) \tag{A.7}$$

Thus, if more data points are on one side than the other, the solution will be pulled in that direction, iteration proceeding until the median is at the origin.

It is important to appreciate that while the median has exceptionally useful outlier suppression characteristics, it actually gives outliers significant weight: in fact, the median clearly ignores how far away an outlier is, but it still counts up how many outliers there are on either side of the current origin. As a result, the median is liable to produce a biased estimate. This is a good reason for considering other types of influence function for analyzing data. Finally, note that the median influence function leads to the value of ρ for L_1 regression:

$$\rho(r_i) = |r_i| \tag{A.8}$$

When selecting an influence function, it is important not only that the function must be appropriate but also that its scale must match that of the data. If the width of the influence function is too great, too few outliers will be rejected; if the width is too small, the estimator may be surrounded by a rather homogeneous sea of data points with no guarantee that it will do more than find a locally optimal fit to the data. These factors mean that preliminary measurements must be made to determine the optimal form of the influence function for any application.

It is now clear that we need a more scientific approach, which would permit the influence function to be calculated from the noise characteristics. Hence, if the expected noise distribution is given by $f(r_i)$, the optimal form of the influence function (Huber, 1964) has to be:

$$\psi(r_i) = -\frac{f'(r_i)}{f(r_i)} = -\frac{d}{dr_i}\ln[f(r_i)] \tag{A.9}$$

The logarithmic form of this solution is interesting and helpful, as it simplifies the situation for exponential-based noise distributions such as the Gaussian and double exponential functions. For the former, $\exp(-r_i^2/2\sigma^2)$, we find:

$$\psi(r_i) = \frac{r_i}{\sigma^2} \tag{A.10}$$

and for the latter, $\exp(-|r_i|/s)$:

$$\psi(r_i) = \frac{\text{sign}(r_i)}{s} \tag{A.11}$$

Since the constant multipliers may be ignored, we conclude that the mean and median are optimal estimators for signals in Gaussian and double exponential noise, respectively.

Gaussian noise may be expected to arise in many situations (most particularly because of the effects of the central limit theorem), demonstrating the intrinsic value of employing the mean or L_2 regression. On the other hand, the double exponential distribution has no obvious justification in practical situations. However, it represents situations where the wings of the noise distribution stretch out rather widely, and it is good to see under what conditions the widely used median would be optimal. Nevertheless, our purpose in wanting an explicit mathematical form for the influence function was to optimize the detection of signals in arbitrary noise conditions and specifically those where outliers may be present.

Let us suppose that the noise is basically Gaussian, but that outliers may also be present and that these would be drawn approximately from a uniform distribution: there might, for example, be a uniform (but low-level) distribution of outlier values over a limited range. An overall distribution of this type is shown in Fig. A.3. Near $r_i = 0$, the uniform distribution of outliers will have relatively little effect and $\psi(r_i)$ will approximate to r_i. For large $|r_i|$, the value of f' will be due mainly to the Gaussian noise contribution, whereas the value of f will arise mainly from the uniform distribution f_u, and the result will be:

$$\psi(r_i) \approx \frac{r_i}{s^2 f_u} \exp\left(-\frac{r_i^2}{2\sigma^2}\right) \tag{A.12}$$

a function that peaks at an intermediate value of r_i. This essentially proves that the form shown in Fig. A.2(e) is reasonable. However, there is a severe problem in that outliers are by definition unusual and rare, so it is almost impossible in most cases to be able to produce on optimum form of $\psi(r_i)$ as suggested above. Unfortunately, the situation is even worse than this discussion might indicate. Redescending M-estimators are even more limited in that they are sensitive to local densities of data points, and are therefore prone to finding false solutions—unique solutions are *not* guaranteed. Non-redescending M-estimators are guaranteed to arrive at unique solutions, although the accuracy of the latter depends on the accuracy of the preliminary scale estimate. In addition, the quality of the initial approximation tends to be of very great importance for M-estimators, particularly for redescending M-estimators.

Finally, we should point out that the above analysis has concentrated on optimization of accuracy and is ultimately based on maximum likelihood strategies

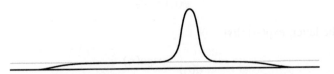

FIGURE A.3

Distribution resulting from Gaussian noise and outliers. Here, the usual Gaussian noise contribution is augmented by a distribution of outliers, which is nearly uniform over a limited range.

(Huber, 1964). It is really concerned with maximizing relative efficiency on the assumption that the underlying distribution is known. Robustness measured according to the breakdown point criterion is not optimized, and this factor will be of vital importance in any situation where the outliers form part of a totally unexpected distribution, or do not form part of a predictable distribution.[6] Clearly, methods that are intrinsically highly robust must be engineered according to the breakdown point criterion. This is what motivated the development of the least median of squares (LMedS) approach to regression during the 1980s.

A.4 THE LEAST MEDIAN OF SQUARES APPROACH TO REGRESSION

In Sections A.2 and A.3, we have seen that a variety of estimators exist, which can be used to suppress noise from numerical data, and to optimize the robustness and accuracy of the final result. The M-estimator (or influence function) approach is extremely widely used and is successful in eliminating the main problems associated with the use of least-squares regression (including, in 1-D, use of the mean). However, it does not in general achieve the ideal breakdown value of 0.5 and requires careful setting up to give optimal matching to the scale of the variation in the data. Accordingly, much attention has been devoted to a newer approach—LMedS regression.

The aim of LMedS regression is to capitalize on the known robustness of the median in a totally different way—by replacing the mean of the least (mean) squares averaging technique by the far more robust median. The effect of this is to ignore errors from the distant parts of the distribution and also from the central parts where the peak is often noisy and ill-defined, and to focus on the parts about halfway up and on either side of the distribution. Minimization then balances the contributions from the two sides of the distribution, thereby sensitively estimating the mode position, although clearly this is achieved rather indirectly. Perhaps the simplest view of the technique is that it determines the location of the narrowest width region that includes half the population of the distribution. In a 2-D straight-line location application, this interpretation amounts to locating the narrowest parallel-sided strip that includes half the population of the distribution (Fig. A.4). In principle, in such cases, the method operates just as effectively if the distribution is sparsely populated—as happens where the best-fit straight line for a set of experimental plots has to be determined.

The LMedS technique involves minimizing the median of the squares of the residuals r_j for all possible positions in the distribution that are potentially mode positions, that is, it is the position x_i that minimizes $M = \text{med}_j(r_j^2)$. While it might be thought that M is also equal to $M = \text{med}_j(|r_j|)$, this is not so if there are two

---

[6] It is perhaps a philosophical question whether an outlier distribution does not exist, cannot exist, or cannot be determined by any known experimental means, for example, because of rarity.

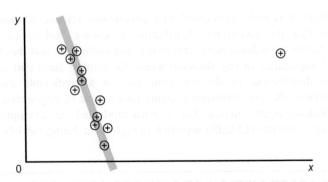

FIGURE A.4

Application of the least median of squares technique. Here, the narrowest parallel-sided strip is found that includes half the population of the distribution, in an attempt to determine the best-fit line. Note the effortless superiority in performance when compared with the situation in Fig. A.1(c).

adjacent central positions giving equal responses (as in Fig. A.5(a–c)); however, the form of M guarantees that a position midway between these two will give an appropriate minimum. For clarity, we shall temporarily ignore this technicality and concentrate on M: the reason for doing this is to take advantage of piecewise linear responses that considerably simplify theoretical analysis.

Figure A.5(a) shows the response M when the original distribution is approximately Gaussian. There is a clear minimum of M at the mode position, and the method works perfectly. Figure A.5(b) shows a case where there is a very untidy distribution, and there is a minimum of M at an appropriate position. Figure A.5(c) shows a more extreme situation in which there are two peaks, and again the response M is appropriate, except that it is now clear that the technique can only focus on one peak at a time. Nevertheless, it gets an appropriate and robust answer for the case it is focussing on. If the two peaks are identical, the method will still work, but will clearly not give a unique solution.

The LMedS approach to regression (Rousseeuw, 1984) has acquired considerable support, since it has the maximum possible breakdown point of 0.5. In particular, it has been used for pattern recognition and image analysis applications (see, e.g., Kim et al., 1989). In these areas, the method is useful for (a) location of straight lines in digital images, (b) location of Hough transform peaks in parameter space, and (c) location of clusters of points in feature space.

Unfortunately, the LMedS approach is liable to give a biased estimate of the modes if two distributions overlap, and in any case focusses on the main mode of a multimodal distribution. Thus, the LMedS technique has to be applied several times, alternating with necessary truncation processes, to find all the cluster centers, while weighted least-squares fitting is required to optimize accuracy. The result is a procedure of some complexity and considerable computational load. Indeed, the load is in general so large that it is normally approximated by taking

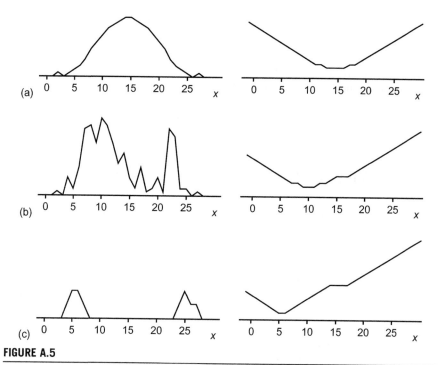

FIGURE A.5

Minimizing *M* for various distributions. The figure shows (left) the original distributions and (right) the resulting response functions *M*, in the following cases: (a) an approximately Gaussian distribution, (b) an "untidy" distribution, and (c) a distribution with two peaks.

subsets of the data points, although this aspect cannot be examined in detail here (see, e.g., Kim et al., 1989). Once this has been carried out, the method can give quite impressive results.

Ultimately, the value of the LMedS approach lies in its increased breakdown point in situations of multidimensional data. If we have n data points in p dimensions, the LMedS breakdown point is:

$$\varepsilon_{\mathrm{LMedS}} = \frac{(\lfloor n/2 \rfloor - p + 2)}{n} \tag{A.13}$$

which tends to 0.5 as n approaches infinity (Rousseeuw, 1984). This value must be compared with a maximum of

$$\varepsilon = \frac{1}{(p+1)} \tag{A.14}$$

for standard methods of robust regression such as the M-, R-, and L-estimators discussed earlier (Kim et al., 1989). (Eq. (A.2) represents the suboptimal solution achieved by the Theil approach to line estimation.) Thus, in these latter cases,

0.33 is the best breakdown point that can be achieved for $p = 2$, while the LMedS approach offers 0.5. However, the relative efficiency of LMedS is relatively low (ultimately because it is a median-based estimator); as stated above, this means that it has to be used with the weighted least-squares technique. We should also point out that the LMedS technique is intrinsically 1-D, so it has to be used in a "projection pursuit" manner (Huber, 1985), concentrating on one dimension at a time. For implementation details, the reader is referred to the literature (see Section A.8).

A.5 OVERVIEW OF THE ROBUSTNESS PROBLEM

For greatest success in solving the robustness and accuracy problems—represented, respectively, by the breakdown point and relative efficiency criteria—it has been found in the foregoing sections that the LMedS technique should be used for finding signals (whether peaks, clusters, lines, or hyperplanes, etc.), and weighted least squares regression should be used for refining accuracy, the whole process being iterated until satisfactory results are achieved. This is a complex and computationally intensive process, but reflects an overall strategy that has been outlined several times in earlier chapters (particularly Chapters 11–14)—namely, search for an approximate solution, and then refinement to optimize location accuracy. The major question to be considered at this stage is: what is the best method for performing an efficient and effective initial search? In fact, there is a further question that is of especial relevance: is there any means of achieving a breakdown point of greater than 0.5?

We now consider the extent to which the Hough transform tackles and solves these problems. First, it is a highly effective search procedure, although in some contexts its computational efficiency has been called into question (however, in the present context, it must be remembered that the LMedS technique is especially computationally intensive). Second, it seems able to yield breakdown points far higher than 0.5 and even approaching unity. Consider a parameter space where there are many peaks and also a considerable number of randomly placed votes. Then any individual peak includes perhaps only a small fraction of the votes, and the peak location proceeds without difficulty in spite of the presence of 90–99% contamination by outliers (the latter arising from noise and clutter). Thus, the strategy of searching for peaks appears to offer significant success at avoiding outliers. Yet this does not mean that the LMedS technique is valueless, since subsequent application of LMedS is essentially able to *verify* the identification of a peak, to locate it more accurately via its greater relative efficiency, and thus to feed reliable information to a subsequent least-squares regression stage. Overall, we can see that a staged progression is taking place from a high breakdown point, low relative efficiency procedure, to a procedure of intermediate breakdown point and moderate relative efficiency, and finally to a procedure of low breakdown point and high relative efficiency. We summarize the progression by giving possible figures for the relevant quantities in Table A.2.

Table A.2 Breakdown Points and Efficiency Values for Peak Finding

	HT	LMedS	LS	Overall
ε	0.98	0.50	0.2	0.98
η	0.2	0.4	0.95	0.95

The table gives possible breakdown points ε and relative efficiency values η for peak finding. A Hough transform is used to perform an initial search for peaks; then the LMedS technique is employed for validating the peaks and eliminating outliers; finally, least-squares regression is used to optimize location accuracy. The result is far higher overall effectiveness than that obtainable by any of the techniques applied alone; however, computational load is not taken into account, and is likely to be a major consideration.

A.6 THE RANSAC APPROACH

Over a good many years, RANSAC has become one of the most widely used outlier rejection and data fitting tools: it has achieved particular value in 3-D vision. RANSAC is an acronym for *random sample consensus*, and involves repeatedly trying to obtain a consensus (set of inliers) from the data until the degree of fit exceeds a given criterion.

To understand the process, let us first return to the LMedS approach, which is useful both in providing a graphic presentation of what it achieves and in requiring no parameters to be set in order to make it work. In fact, the latter feature is in many ways its undoing, because if the proportion of outliers in the data exceeds 50%, the resulting fit is liable to be heavily biassed. A simple modification of the method is to require a smaller number of inliers—indeed, whatever proportion would be expected in the incoming data. Thus, we may go for 20% inliers, 80% outliers if this seems appropriate. This naturally leads to problems, as ideally we will have to estimate the proportion of inliers in advance, or as part of the fitting process, and then apply the resulting value as part of the technique.

Once the "cleanness" of the LMedS method is lost, a variety of alternative solutions become possible. In fact, the RANSAC method involves not taking the proportion of inliers as fixed and finding how the residual distance (e.g., from a best-fit straight line) varies, but rather specifying a threshold residual distance t and finding how the proportion of inliers varies. Here, the word "inlier" is not a good term to use; as it implies, we already know that these data are acceptable points: instead they should be called consensus points—at least until the end of the process. In summary, we set a threshold residual distance t and ask how much consensus this gives. Note that in principle at least, t has to be iterated as part of the whole process of finding the best fit. However, it is possible to work on the basis that the experimental uncertainty is known in advance, and if, for example, t is made equal to three standard deviations, this should not lead to too much error in the final fit obtained.

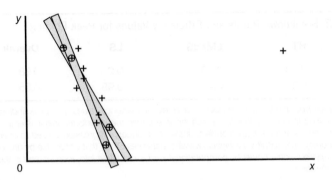

FIGURE A.6

The RANSAC technique. Here, the + signs indicate data points to be fitted, and two instances of pairs of data points (indicated by ⊕ signs) leading to hypothesized lines are also shown. Each hypothesized line has a region of influence of tolerance ± t within which the support of maximal numbers of data points is sought. The line with the most support indicates the best fit (although weighted least-squares analysis may subsequently be applied to improve it further).

Another aspect of RANSAC that must be brought out is the random extraction of n data points to specify each initial potential fit, following which the hypothesized solution is tested to find how much consensus there is; then out of k trials, the best solution is the one with the greatest consensus, and at this final stage, we can interpret the consensus as the set of inliers.

Finally, the number of data points n needed to specify a potential fit is made equal to the number of degrees of freedom of the data—two for a straight line in a plane, three for a circle, four for a sphere, and so on (Fig. A.6). All that remains to be specified is the number of iterations k of sets of n data points in order to reach the final best-fit solution. One way of estimating k is to calculate the risk that all the k sets of n data points chosen will contain only outliers so that no good data will be examined. Clearly, k must be sufficiently large to reduce the risk of this eventuality to a low enough level. Formulas to estimate k on this basis appear in several sources, for example, Hartley and Zisserman (2000).

Finally, note that, as happens with many other outlier identification processes, improved fits can be obtained by a final stage in which normal or weighted least-squares analysis is applied to the remaining (inlier) data.

A.7 CONCLUDING REMARKS

This appendix has aimed to place the discussion of robustness on a sounder basis than might have been thought possible in the earlier chapters of the book (particularly Chapters 11–14), where a more intuitive approach was presented. It has been necessary to delve quite deeply into the maturing and highly mathematical subject of robust statistics, and there are certain important lessons to be learnt. In particular, three relevant parameters have been found to form the basis for study in this area. The first is the breakdown point of an estimator, which shows the latter's resistance to outliers and provides the core meaning of robustness. The second is the relative efficiency of an estimator, which provides a measure of how efficiently it will use the inlier data at its disposal to arrive at accurate estimates. The third is the time complexity of the estimator when it is implemented as a computer algorithm. While the last parameter is a vital consideration in practical situations, available space has not permitted it to be covered in any depth here, although it is clear that the most robust techniques (especially LMedS) tend to be highly computationally intensive. It is also found that there is a definite tradeoff between the other two parameters—techniques that have high breakdown points have low relative efficiencies and vice versa.[7] These factors make it reasonable, and desirable, to use several techniques in sequence, or iteratively in cycle, in order to obtain the best overall performance. Thus, LMedS is frequently used in conjunction with least-squares regression (see, e.g., Kim et al., 1989).

Finally, it is worth pointing out that the basis of robust statistics is that of statistical analysis of the available data: there is thus a tendency to presume that outliers are rare events due typically to erroneous readings or transcriptions. Yet, in computer vision, the most difficult problems tend to arise from the clutter of irrelevant objects in the background, and only a tiny fraction of the incoming data may constitute the relevant inlier portion. This makes the problem of robustness all the more serious, and in principle *could* mean that until a whole image has been interpreted satisfactorily, no single object can finally be identified and its position and orientation measured accurately. However, it is rare that we need to take such an extreme view in practical applications of vision.

> Robust statistics is at the core of any practical vision system. This appendix has aimed to cover the intricacies of the subject in an accessible way, dealing with important concepts such as "breakdown point" and measurement "efficiency." What is really in question is *how* robust statistics will be incorporated into any practical vision system, not whether it needs to be.

A.8 BIBLIOGRAPHICAL AND HISTORICAL NOTES

This appendix has given a basic introduction to the rapidly maturing subject of robust statistics that has made a substantial impact on machine vision over the past 25 years or so. The most popular and successful approach to robust statistics must still be seen as the M-estimator (influence function) approach (which is

---

[7]The reason for this may be summarized as the aim of achieving high robustness requiring considerable potentially outlier data to be discarded, even when this could be accurate data that would contribute to the overall accuracy of the estimate.

broad enough to include least-squares regression and median filtering), although in high-dimensional spaces its robustness is called into question, and it is here that the newer LMedS approach has gathered a firm following. More recently, the value of using a sequence of estimators that can optimize the overall breakdown point and relative efficiency has been pointed out (Kim et al., 1989): in particular, the right combination of Hough transform (or other relevant technique), LMedS, and weighted least-squares regression would seem especially powerful.

Robust statistics has been applied in a number of areas of machine vision, including robust window operators (Besl et al., 1989), pose estimation (Haralick and Joo, 1988), motion studies (Bober and Kittler, 1993), camera location and calibration (Kumar and Hanson, 1989), and surface defect inspection (Koivo and Kim, 1989), to name but a few.

The original papers by Huber (1964) and Rousseeuw (1984) are still worth reading, and the books by Huber (1981), Hampel et al. (1986), and Rousseeuw and Leroy (1987) are valuable references, containing much insight and useful material. On the application of the LMedS technique, and for more reviews of robust regression in machine vision, see Meer et al. (1990, 1991).

Note that the RANSAC technique (Fischler and Bolles, 1981) was introduced before LMedS and presaged its possibilities: thus RANSAC was of great historical importance. The work of Siegel (1982) was also important historically in providing the background from which LMedS could take off, while the work of Steele and Steiger (1986) showed how LMedS might be implemented with attainable levels of computation.

While much of the work on robust statistics dates from the 1980s, one has only to look at the book by Hartley and Zisserman (2003) to see how deeply embedded it is in the current methodology and thinking on machine vision. An example of its application to 3-D correspondence matching is provided by Hasler et al. (2003): they consider exactly where the outlier data originates and model the whole process. Unexpected motion, occlusion of points in some views, and also viewing of convex boundaries from different positions all lead to mismatches and outliers; they arrive at a new way of calculating outliers in image pairs, which helps to put the subject area on a more secure footing.

A.8.1 More Recent Developments

In many applications, RANSAC requires an overly large number of hypotheses to be made before converging to an acceptable solution: this applies especially when searching in high-dimensional spaces. Many attempts have been made to overcome this problem. Myatt et al. (2002) tackled it by noting that in general inliers tend to be closer to one another than to outliers. Their algorithm, called NAPSAC, samples sets of adjacent points in a hypersphere: thereby the probability of selecting an inlying set is significantly increased—as demonstrated using wide baseline stereo matching data. Torr and Davidson (2003) also produced an improved version of RANSAC, which they called *importance sampling consensus*

(IMPSAC). It works in a hierarchical manner and is initialized at the coarsest level by RANSAC, but then goes on to sample at a finer level to refine relevant *a posteriori* estimates. While IMPSAC has been applied to 3-D matching tasks, it embodies statistical techniques that can be applied to a wide variety of statistical problems to eliminate outlier corrupted data.

Chum and Matas (2005) developed another idea for improving RANSAC. Instead of using randomly chosen hypotheses, they start by testing the most promising hypotheses and gradually revert to uniform sampling as diminishing returns set in. Their method, called PROSAC, achieves large computational savings and can be as much as 100 times faster than RANSAC, for example, with wide baseline stereo data. Effectively, PROSAC gains by ordering the hypotheses in an appropriate way. The worst-case performance essentially equals that of RANSAC, although no proof exists of this. Ni et al. (2009) developed another variant of RANSAC called GroupSAC. This relies on the assumption that there exists a grouping of the data in which some of the groups have a high inlier ratio while the others contain mostly outliers. When tested on wide baseline stereo data, GroupSAC was found to be much faster than PROSAC "most of the time," and RANSAC was always slower than either. Méler et al. (2010) devised yet another variant of RANSAC called BetaSAC. This was formulated as a general framework for including any relevant information for improving performance. BetaSAC offers a conditional sampling that is able to generate more suitable samples than pure random during the initial iterations. The only hypothesis required is that suitable samples can be built by successive data point selections. In the case of random ranking of samples, the method reverts to the same performance as RANSAC. When used for homography estimation, the method is always faster than RANSAC and typically 10–40 times faster than PROSAC.

A.9 PROBLEM

1. a. What is meant by the *breakdown point* of a data analysis method? Show how it is related to the concept of robustness. Consider also how accuracy of measurement is affected by the proportion of data points that are fully utilized by the data analysis method. Discuss the situation in relation to (i) the mean, (ii) the median, and (iii) the result of applying a Hampel three-part redescending M-estimator.

   b. A method for locating straight lines in digital images involves taking every pair of edge points and finding where a line through both points of a pair intercepts the x- and y-axes. Then medians for all such intercepts are found and the positions of any straight lines are deduced. Show that the effect of taking pairs is to reduce the breakdown point from 50% to around 30%, and give an exact answer for the breakdown point. (*Hint:* start by assuming that the fraction of outliers in the original set of edge points is ε and work out the probability of half the intercept values being correct.)

References

'Electronically published references without global page numbers are presented with numbers of pages in square brackets, e.g. [4 pp.]

Abdou, I.E., Pratt, W.K., 1979. Quantitative design and evaluation of enhancement/thresholding edge detectors. Proc. IEEE 67, 753–763.

Abolghasemi, V., Ahmadyfard, A., 2009. An edge-based color-aided method for license plate detection. Image Vis. Comput. 27, 1134–1142.

Abutaleb, A.S., 1989. Automatic thresholding of gray-level pictures using two-dimensional entropy. Computer Vis. Graph. Image Process. 47, 22–32.

Ade, F., 1983. Characterization of texture by "eigenfilters". Signal Process. 5 (5), 451–457.

Aggarwal, J.K., Cai, Q., 1999. Human motion analysis: a review. Comput. Vis. Image Underst. 73 (3), 428–440.

Agin, G.J., Binford, T.O., 1973. Computer description of curved objects. Proceedings of the Third International Joint Conference on Artificial Intelligence, Stanford, CA, pp. 629–640.

Agin, G.J., Binford, T.O., 1976. Computer description of curved objects. IEEE Trans. Comput. 25, 439–449.

Aguado, A.S., Montiel, M.E., Nixon, M.S., 2000. On the intimate relationship between the principle of duality and the Hough transform. Proc. R. Soc. 456 (1995), 503–526.

Aguilar, W., Frauel, Y., Escolano, F., Martinez-Perez, M.E., Espinosa-Romero, A., Lozano, M.A., 2009. A robust graph transformation matching for non-rigid registration. Image Vis. Comput. 27, 897–910.

Akey, M.L., Mitchell, O.R., 1984. Detection and sub-pixel location of objects in digitized aerial imagery. Proceedings of the Seventh International Conference on Pattern Recognition, Montreal, 30 July–2 August, pp. 411–414.

Aleksander, I., Thomas, W.V., Bowden, P.A., 1984. WISARD: a radical step forward in image recognition. Sensor Rev. 4, 120–124.

Alexandropoulos, T., Boutas, S., Loumos, V., Kayafas, E., 2008. Template-guided inspection of arbitrarily oriented targets. IET Comput. Vis. 2 (3), 150–163.

Ali, S.M., Burge, R.E., 1988. A new algorithm for extracting the interior of bounded regions based on chain coding. Computer Vis. Graph. Image Process. 43, 256–264.

Almansa, A., Desolneux, A., Vamech, S., 2003. Vanishing point detection without any a priori information. IEEE Trans. Pattern Anal. Machine Intell. 25 (4), 502–507.

Alter, T.D., 1994. 3-D pose from 3 points using weak-perspective. IEEE Trans. Pattern Anal. Machine Intell. 16 (8), 802–808.

Ambler, A.P., Barrow, H.G., Brown, C.M., Burstall, R.M., Popplestone, R.J., 1975. A versatile system for computer-controlled assembly. Artif. Intell. 6, 129–156.

Ambrosch, K., Kubinger, W., 2010. Accurate hardware-based stereo vision. Comput. Vis. Image Underst. 114, 1303–1316.

Amit, Y., 2002. 2D Object Detection and Recognition: Models, Algorithms and Networks. MIT Press, Cambridge, MA.

An, G., Wu, J., Ruan, Q., 2010. An illumination normalization model for face recognition under varied lighting conditions. Pattern Recogn. Lett. 31 (9), 1056–1067.

Andrews, D., Niehaus, D., Ashenden, P., 2004. Programming models for hybrid CPU/FPGA chips. IEEE Comput. 37 (1), 118−120.

Ansar, A., Daniilidis, K., 2003. Linear pose estimation from points or lines. IEEE Trans. Pattern Anal. Machine Intell. 25 (5), 578−589.

Appiah, K., Hunter, A., Dickinson, P., Meng, H., 2010. Accelerated hardware video object segmentation: from foreground detection to connected components labelling. Comput. Vis. Image Underst. 114, 1282−1291.

Aragon-Camarasa, G., Siebert, J.P., 2010. Unsupervised clustering in Hough space for recognition of multiple instances of the same object in a cluttered scene. Pattern Recogn. Lett 31, 1274−1284.

Arcelli, C., di Baja, G.S., 1985. A width-independent fast-thinning algorithm. IEEE Trans. Pattern Anal. Machine Intell. 7, 463−474.

Arcelli, C., Ramella, G., 1995. Finding grey-skeletons by iterated pixel removal. Image Vis. Comput. 13 (3), 159−167.

Arcelli, C., Cordella, L.P., Levialdi, S., 1975. Parallel thinning of binary pictures. Electron. Lett. 11, 148−149.

Arcelli, C., Cordella, L.P., Levialdi, S., 1981. From local maxima to connected skeletons. IEEE Trans. Pattern Anal. Machine Intell. 3, 134−143.

Arnold, R.D., 1978. Local context in matching edges for stereo vision. Proceedings of the DARPA Image Understanding Workshop, Cambridge, MA, USA, May, pp. 65−72.

Arulampalam, M.S., Maskell, S., Gordon, N., Clapp, T., 2002. A tutorial on particle filters for online nonlinear/non-Gaussian Bayesian tracking. IEEE Trans. Signal Process. 50 (2), 174−188.

Assheton, P., Hunter, A., 2011. A shape-based voting algorithm for pedestrian detection and tracking. Pattern Recogn. 44, 1106−1120.

Åström, K., 1995. Fundamental limitations on projective invariants of planar curves. IEEE Trans. Pattern Anal. Machine Intell. 17 (1), 77−81.

Atherton, T.J., Kerbyson, D.J., 1999. Size invariant circle detection. Image Vis. Comput. 17 (11), 795−803.

Atiquzzaman, M., 1994. Pipelined implementation of the multiresolution Hough transform in a pyramid multiprocessor. Pattern Recogn. Lett. 15 (9), 841−851.

Atiquzzaman, M., Akhtar, M.W., 1994. Complete line segment description using the Hough transform. Image Vis. Comput. 12 (5), 267−273.

Aziz, M., Boussakta, S., McLernon, D.C., 2003. High performance 2D parallel block-filtering system for real-time imaging applications using the Sharc ADSP21060. Real-Time Imaging 9 (2), 151−161.

Babaud, J., Witkin, A.P., Baudin, M., Duda, R.O., 1986. Uniqueness of the Gaussian kernel for scale-space filtering. IEEE Trans. Pattern Anal. Machine Intell. 8, 26−33.

Bai, X., Latecki, L.J., 2008. Path similarity skeleton graph matching. IEEE Trans. Pattern Anal. Machine Intell. 30 (7), 1282−1292.

Bai, X.Z., Zhou, F.G., 2010. Top-hat selection transformation for infrared dim small target enhancement. Imaging Sci. 58 (2), 112−117.

Bajcsy, R., 1973. Computer identification of visual surface. Comput. Graph Image Process. 2, 118−130.

Bajcsy, R., Liebermann, L., 1976. Texture gradient as a depth cue. Comput. Graph Image Process. 5 (1), 52−67.

Baker, S., Sim, T., Kanade, T., 2003. When is the shape of a scene unique given its light-field: a fundamental theorem of 3D vision? IEEE Trans. Pattern Anal. Machine Intell. 25 (1), 100−109.

Ball, G.H., Hall, D.J., 1966. ISODATA, an iterative method of multivariate data analysis and pattern classification. IEEE International Communications Conference Philadelphia, PA, Digest of Techn. Papers II, 15−17 June, pp. 116−117.

Ballard, D.H., 1981. Generalizing the Hough transform to detect arbitrary shapes. Pattern Recogn. 13, 111−122.

Ballard, D.H., Brown, C.M., 1982. Computer Vision. Prentice-Hall, Englewood Cliffs, NJ.

Ballard, D.H., Sabbah, D., 1983. Viewer independent shape recognition. IEEE Trans. Pattern Anal. Machine Intell. 5, 653−660.

Bangham, J.A., Marshall, S., 1998. Image and signal processing with mathematical morphology. IEE Electronics Commun. Eng. J. 10 (3), 117−128.

Barbe, D.F., 1975. Imaging devices using the charge-coupled principle. Proc. IEEE 63, 38−66.

Barker, A.J., Brook, R.A., 1978. A design study of an automatic system for on-line detection and classification of surface defects on cold-rolled steel strip. Opt. Acta 25, 1187−1196.

Barnard, S., 1983. Interpreting perspective images. Artif. Intell. 21, 435−462.

Barnard, S.T., Thompson, W.B., 1980. Disparity analysis of images. IEEE Trans. Pattern Anal. Machine Intell. 2 (4), 333−340.

Barnea, D.I., Silverman, H.F., 1972. A class of algorithms for fast digital image registration. IEEE Trans. Comput. 21, 179−186.

Barrett, E.B., Payton, P.M., Haag, N.N., Brill, M.H., 1991. General methods for determining projective invariants in imagery. Comput. Vis. Graph. Image Process. 53 (1), 46−65.

Barrow, H.G., Popplestone, R.J., 1971. Relational descriptions in picture processing. In: Meltzer, B., Michie, D. (Eds.), Machine Intelligence 6. Edinburgh University Press, Edinburgh, pp. 377−396.

Barrow, H.G., Tenenbaum, J.M., 1981. Computational vision. Proc. IEEE 69, 572−595.

Barrow, H.G., Ambler, A.P., Burstall, R.M., 1972. Some techniques for recognising structures in pictures. In: Watanabe, S. (Ed.), Frontiers of Pattern Recognition. Academic Press, New York, pp. 1−29.

Barsky, S., Petrou, M., 2003. The 4-source photometric stereo technique for three-dimensional surfaces in the presence of highlights and shadows. IEEE Trans. Pattern Anal. Machine Intell. 25 (10), 1239−1252.

Bartoli, A., Sturm, P., 2004. Nonlinear estimation of the fundamental matrix with minimal parameters. IEEE Trans. Pattern Anal. Machine Intell. 26 (3), 426−432.

Bartz, M.R., 1968. The IBM 1975 optical page reader. IBM J. Res. Dev. 12, 354−363.

Bascle, B., Bouthemy, P., Deriche, R., Meyer, F., 1994. Tracking complex primitives in an image sequence. Proceedings of the Twelfth International Conference on Pattern Recognition, Jerusalem, Israel, 9−13 October, vol. A, pp. 426−431.

Batchelor, B.G., 1979. Using concavity trees for shape description. Comput. Digit. Technol. 2, 157−165.

Batchelor, B.G., Davies, E.R., Graves, M., 2004. Using X-rays to detect foreign bodies in food materials and packs. In: Edwards, M. (Ed.), Detecting Foreign Bodies in Food. Woodhead Publishing Ltd, Cambridge, UK.

Batchelor, B.G., Hill, D.A., Hodgson, D.C., 1985. Automated Visual Inspection. IFS (Publications) Ltd, Bedford/North-Holland/Amsterdam.

Batlle, J., Marti, J., Ridao, P., Amat, J., 2002. A new FPGA/DSP-based parallel architecture for real-time image processing. Real-Time Imaging 8 (5), 345–356.

Baumberg, A., Hogg, D., 1995. An adaptive eigenshape model. Proceedings of the British Machine Vision Association Conference, September, pp. 87–96.

Bay, H., Ess, A., Tuytelaars, T., Van Gool, L., 2008. Speeded-up robust features (SURF). Comput. Vis. Image Underst. 110 (3), 346–359.

Bay, H., Tuytelaars, T., Van Gool, L., 2006. SURF: speeded up robust features. Proceedings of the Ninth European Conference on Computer Vision (ECCV), Springer, LNCS, Berlin, Heidelberg, vol. 3951, part 1, pp. 404–417.

Bayram, M., Öner, M.D., 2006. Determination of applicability and effects of colour sorting system in bulgur production line. J. Food Eng. 74, 232–239.

Beaton, A.E., Tukey, J.W., 1974. The fitting of power series, meaning polynomials, illustrated on band-spectroscopic data. Technometrics 16 (2), 147–185.

Beaudet, P.R. 1978. Rotationally invariant image operators. Proceedings of the Fourth International Conferences on Pattern Recognition, Kyoto, Japan, pp. 579–583.

Beckers, A.L.D., Smeulders, A.W.M., 1989. A comment on "A note on 'Distance transformations in digital images'". Computer Vis. Graph. Image Process. 47, 89–91.

Beiden, S.V., Maloof, M.A., Wagner, R.F., 2003. A general model for finite-sample effects in training and testing of competing classifiers. IEEE Trans. Pattern Anal. Machine Intell. 25 (12), 1561–1569.

Bergholm, F., 1986. Edge focusing. Proceedings of the Eighth International Conference on Pattern Recognition, Paris, 27–31 October, pp. 597–600.

Berman, S., Parikh, P., Lee, C.S.G., 1985. Computer recognition of two overlapping parts using a single camera. IEEE Comput. March, 70–80.

Bertozzi, M., Broggi, A., 1998. GOLD: a parallel real-time stereo vision system for generic obstacle and lane detection. IEEE Trans. Image Process. 7 (1), 62–81.

Besl, P.J., Birch, J.B., Watson, L.T., 1989. Robust window operators. Machine Vis. Appl. 2, 179–191.

Besl, P.J., Delp, E.J., Jain, R., 1985. Automatic visual solder joint inspection. IEEE J. Robot. Autom. 1, 42–56.

Beun, M., 1973. A flexible method for automatic reading of handwritten numerals. Philips Techn. Rev. 33, pp. 89–101, 131–137.

Billingsley, J., Schoenfisch, M., 1995. Vision-guidance of agricultural vehicles. Auton. Robot 2 (1), 65–76.

Birchfield, S., 1998. Elliptical head tracking using intensity gradients and color histograms. Proceedings of the IEEE Conference on Computer Vision and Pattern Recognition, Santa Barbara, CA, pp. 232–237.

Bishop, C., 1995. Neural Networks for Pattern Recognition. Oxford University Press, Oxford, UK.

Bishop, C.M., 2006. Pattern Recognition and Machine Learning. Springer-Verlag, Berlin, Heidelberg.

Blake, A., Zisserman, A., Knowles, G., 1985. Surface descriptions from stereo and shading. Image Vision Comput. 3, 183–191.

Bledsoe, W.W., Browning, I., 1959. Pattern recognition and reading by machine. Proceedings of the Eastern Joint Computer Conference, Boston, Massachusetts, USA, 1–3 December, pp. 225–232.

Blum, H., 1967. A transformation for extracting new descriptors of shape. In: Wathen-Dunn, W. (Ed.), Models for the Perception of Speech and Visual Form. MIT Press, Cambridge, MA, pp. 362−380.

Blum, H., Nagel, R.N., 1978. Shape description using weighted symmetric axis features. Pattern Recogn. 10, 167−180.

Bober, M., Kittler, J. 1993. Estimation of complex multimodal motion: an approach based on robust statistics and Hough transform. Proceedings of the Fourth British Machine Vision Association Conference, University of Surrey, 21−23 September, vol. 1, pp. 239−248.

Boerner, H., Strecker, H., 1988. Automated X-ray inspection of aluminium castings. IEEE Trans. Pattern Anal. Machine Intell. 10 (1), 79−91.

Bolles, R.C., 1979. Robust feature matching via maximal cliques. SPIE, 182. Proceedings of the Technical Symposium on Imaging Applications for Automated Industrial Inspection and Assembly, Washington, DC, April, pp. 140−149.

Bolles, R.C., Cain, R.A., 1982. Recognizing and locating partially visible objects: the local-feature-focus method. Int. J. Robot. Res. 1 (3), 57−82.

Bolles, R.C., Horaud, R., 1986. 3DPO: a three-dimensional part orientation system. Int. J. Robot. Res. 5 (3), 3−26.

Borkar, A., Hayes, M., Smith, M.T., 2009. Robust lane detection and tracking with RANSAC and Kalman filter. IEEE International Conference on Image Processing, Cairo, Egypt, 7−10 November, pp. 3261−3264.

Bors, A.G., Hancock, E.R., Wilson, R.C., 2003. Terrain analysis using radar shape-from-shading. IEEE Trans. Pattern Anal. Machine Intell. 25 (8), 974−992.

Boufama, B., Mohr, R., Morin, L., 1998. Using geometric properties for automatic object positioning. In Special Issue on Geometric Modelling and Invariants for Computer Vision. Image Vision Comput. 16 (1), 27−33.

Bovik, A.C., Huang, T.S., Munson, D.C., 1983. A generalization of median filtering using linear combinations of order statistics. IEEE Trans. Acoustics Speech Signal Process. 31 (6), 1342−1349.

Bovik, A.C., Huang, T.S., Munson, D.C., 1987. The effect of median filtering on edge estimation and detection. IEEE Trans. Pattern Anal. Machine Intell. 9, 181−194.

Boykov, Y., Funka-Lea, G., 2006. Graph cuts and efficient N-D image segmentation. Int. J. Comput. Vision 70 (2), 109−131.

Boykov, Y., Jolly, M.-P., 2001. Interactive graph cuts for optimal boundary and region segmentation of objects in N-D images. Proceedings of the International Conference on Computer Vision, Vancouver, British Columbia, Canada, 7−14 July, vol. I, pp. 105−112.

Boykov, Y., Kolmogorov, V., 2004. An experimental comparison of min-cut/max-flow algorithms for energy minimization in vision. IEEE Trans. Pattern Anal. Machine Intell 26 (9), 1124−1137.

Brady, J.M., Wang, H., 1992. Vision for mobile robots. Phil. Trans. Roy. Soc. (London) B337, 341−350.

Brady, J.M., Yuille, A., 1984. An extremum principle for shape from contour. IEEE Trans. Pattern Anal. Machine Intell 6, 288−301.

Brady, M., 1982. Computational approaches to image understanding. Comput. Surv. 14, 3−71.

Breiman, L., 1996. Bagging predictors. Machine Learn. 24 (2), 123−140.

Bretschi, J., 1981. Automated Inspection Systems for Industry. IFS Publications Ltd, Bedford, UK.

Brink, A.D., 1992. Thresholding of digital images using two-dimensional entropies. Pattern Recogn. 25, 803−808.

Brivot, R., Marchant, J.A., 1996. Segmentation of plants and weeds for a precision crop protection robot using infrared images. IEE Proc. Vis. Image Signal Process. 143 (2), 118–124.

Broggi, A., Bertozzi, M., Fascioli, A., 2000a. Architectural issues on vision-based automatic vehicle guidance: the experience of the ARGO project. Real-Time Imaging 6 (4), 313–324.

Broggi, A., Bertozzi, M., Fascioli, A., Sechi, M., 2000b. Shape-based pedestrian detection. Proceedings of the IEEE Intelligent Vehicles Symposium, Dearborn, MI, 3–5 October, pp. 215–220.

Bron, C., Kerbosch, J., 1973. Algorithm 457: finding all cliques in an undirected graph [H]. Comm. ACM 16, 575–577.

Brooks, M.J., 1976. Locating intensity changes in digitised visual scenes. Computer Science Memo-15 (from MSc Thesis), University of Essex, UK.

Brooks, M.J., 1978. Rationalising edge detectors. Computer Graph. Image Process. 8, 277–285.

Brown, C.M., 1984. Peak-finding with limited hierarchical memory. Proceedings of the Seventh International Conference on Pattern Recognition, Montreal, Canada, 30 July–2 August, pp. 246–249.

Brown, J., Glazier, E.V.D, 1974. Telecommunications. 2nd edn Chapman and Hall, London.

Brown, M.Z., Burschka, D., Hager, G.D., 2003. Advances in computational stereo. IEEE Trans. Pattern Anal. Machine Intell. 25 (8), 993–1008.

Browne, A., Norton-Wayne, L., 1986. Vision and Information Processing for Automation. Plenum Press, New York.

Bruckstein, A.M., 1988. On shape from shading. Computer Vision Graph. Image Process 44, 139–154.

Buch, N., Orwell, J., Velastin, S.A., 2010. Urban road user detection and classification using 3D wire frame models. IET Comput. Vis. 4 (2), 105–116.

Bunke, H., 1999. Error correcting graph matching: on the influence of the underlying cost function. IEEE Trans. Pattern Anal. Machine Intell. 21 (9), 917–922.

Bunke, H., Shearer, K., 1998. A graph distance metric based on the maximal common subgraph. Pattern Recogn. Lett 19, 255–259.

Burr, D.J., Chien, R.T., 1977. A system for stereo computer vision with geometric models. Proceedings of the Fifth International Joint Conference on Artificial Intelligence, Boston, MA, p. 583.

Cai, H., Mikolajczyk, K., Matas, J., 2011. Learning linear discriminant projections for dimensionality reduction of image descriptors. IEEE Trans. Pattern Anal. Machine Intell. 33 (2), 338–352.

Calderara, S., Prati, A., Cucchiara, R., 2008. HECOL: homography and epipolar-based consistent labeling for outdoor park surveillance. Comput. Vis. Image Underst. 111 (1), 21–42.

Califano, A., Mohan, R., 1994. Multidimensional indexing for recognizing visual shapes. IEEE Trans. Pattern Anal. Machine Intell. 16 (4), 373–392.

Canny, J., 1986. A computational approach to edge detection. IEEE Trans. Pattern Anal. Machine Intell. 8, 679–698.

Cantoni, V., Levialdi, S., 1983. Matching the task to an image processing architecture. Comput. Vis. Graph Image Process 22, 301–309.

Caselles, V., Kimmel, R., Sapiro, G., 1997. Geodesic active contours. Int. J. Comput. Vision 21 (1), 61−79.

Cauchie, J., Fiolet, V., Villers, D., 2008. Optimization of an Hough transform algorithm for the search of a center. Pattern Recogn. 41, 567−574.

Celebi, M.E., 2009. Real-time implementation of order-statistics-based directional filters. IET Image Process. 3 (1), 1−9.

Chakravarty, I., Freeman, H., 1982. Characteristic views as a basis for three-dimensional object recognition. Proc. Soc. Photo-opt. Instrum. Eng. Conf. Robot Vis. 336, 37−45.

Chakravarty, V.S., Kompella, B., 2003. The shape of handwritten characters. Pattern Recogn. Lett. 24 (12), 1901−1913.

Chan, J.P., Batchelor, B.G., Harris, I.P., Perry, S.J., 1990. Intelligent visual inspection of food products. Proc. SPIE Conf. on Machine Vis. Syst. in Ind. 1386, 171−179.

Chandra, B., Kothari, R., Paul, P., 2010. A new node splitting measure for decision tree construction. Pattern Recogn. 43, 2725−2731.

Chang, I.-C., Lin, S.-Y., 2010. 3D human motion tracking based on a progressive particle filter. Pattern Recogn. 43, 3621−3635.

Chang, S.-L., Chen, L.-S., Chung, Y.-C., Chen, S.-W., 2004. Automatic license plate recognition. IEEE Trans. Intell. Transport. Syst. 5 (1), 42−53.

Chao, S.-M., Tsai, D.-M., 2008. An anisotropic diffusion-based defect detection for low-contrast glass substrates. Image Vision Comput. 26, 187−200.

Charles, D., Davies, E.R., 2003a. Properties of the mode filter when applied to colour images. Proceedings of the IEE International Conference on Visual Information Engineering, VIE 2003, Surrey, 7−9 July, IEE Conference Publication 495, pp. 101−104.

Charles, D., Davies, E.R., 2003b. Distance-weighted median filters and their application to colour images. Proceedings of the IEE International Conference on Visual Information Engineering, VIE 2003, Surrey, 7−9 July, IEE Conference Publication 495, pp. 117−120.

Charles, D., Davies, E.R., 2004. Mode filters and their effectiveness for processing colour images. Imaging Sci. 52 (1), 3−25.

Chasles, M., 1855. Question no. 296. Nouv. Ann. Math. 14, 50.

Chauduri, B.B., 1994. Dynamic clustering for time incremental data. Pattern Recogn. Lett. 15 (1), 27−34.

Chauduri, S., Chatterjee, S., Katz, N., Nelson, M., Goldbaum, M., 1989. Detection of blood vessels in retinal images using two-dimensional matched filters. IEEE Trans. Med. Imaging 8 (3), 263−269.

Chen, S., Yang, X., Cao, G., 2009. Impulse noise suppression with an augmentation of ordered difference noise detector and an adaptive variational method. Pattern Recogn. Lett 30 (4), 460−467.

Chen, T.C., 1971. Parallelism, pipelining and computer efficiency. Comput. Design (Jan.), 69−74.

Chen, W., Zhang, M.-J., Xiong, Z.-H., 2011. Fast semi-global stereo matching via extracting disparity candidates from region boundaries. IET Comput. Vis. 5 (2), 143−150.

Chen, Y., Adjouadi, M., Han, C., Wang, J., Barreto, A., Rishe, N., Andrian, J., 2010. A highly accurate and computationally efficient approach for unconstrained iris segmentation. Image Vision Comput. 28, 261−269.

Cheng, H.-Y., Yu, C.-C., Tseng, C.-C., Fan, K.-C., Hwang, J.-N., Jeng, B.-S., 2010. Environment classification and hierarchical lane detection for structured and unstructured roads. IET Comput. Vis. 4 (1), 37−49.

Cheng, S.Y., Trivedi, M.M., 2007. Lane tracking with omnidirectional cameras: algorithms and evaluation. EURASIP J. Embedded Syst. 1−8, Article 46972.

Cherng, S., Fang, C.Y., Chen, C.P., Chen, S.W., 2009. Critical motion detection of nearby moving vehicles in a vision-based driver-assistance system. IEEE Trans. Intell. Transportation Syst. 10 (1), 70−82.

Chiang, Y.P., Fu, K.-S., 1983. Matching parallel algorithm and architecture. In: Proceedings of the International Conference on Parallel Processing. *IEEE* Computer Society Press, Columbus, Ohio, USA, August, pp. 374−380.

Chittineni, C.B., 1980. Efficient feature subset selection with probabilistic distance criteria. Inf. Sci. 22, 19−35.

Chiverton, J., Mirmehdi, M., Xie, X., 2008. Variational logistic maximum a posteriori model similarity and dissimilarity matching. Proceedings of the International Conference on Pattern Recognition, Tampa, FL, 8−11 December.

Chojnacki, W., Brooks, M.J., van den Hengel, A., Gawley, D., 2003. Revisiting Hartley's normalized eight-point algorithm. IEEE Trans. Pattern Anal. Machine Intell. 25 (9), 1172−1177.

Choudhary, R., Paliwal, J., Jayas, D.S., 2008. Classification of cereal grains using wavelet, morphological, colour, and textural features of non-touching kernel images. Biosyst. Eng. 99, 330−337.

Chow, C.K., Kaneko, T., 1972. Automatic boundary detection of the left ventricle from cineangiograms. Comput. Biomed. Res. 5, 388−410.

Choy, S.S.O., Choy, C.S.-T., Siu, W.-C., 1995. New single-pass algorithm for parallel thinning. Comput. Vis. Image Underst. 62 (1), 69−77.

Chum, O., Matas, J., 2005. Matching with PROSAC − progressive sample consensus. Proc. IEEE Conf. Comput. Vis. Pattern Recogn. 1, 220−226.

Chung, C.-H., Cheng, S.-C., Chang, C.-C., 2010. Adaptive image segmentation for region-based object retrieval using generalized Hough transform. Pattern Recogn. 43, 3219−3232.

Chung, K.-L., Lin, Z.-W., Huang, S.-T., Huang, Y.-H., Liao, H.-Y.M., 2010. New orientation-based elimination approach for accurate line-detection. Pattern Recogn. Lett. 31 (1), 11−19.

Clark, P., Mirmehdi, M., 2002. On the recovery of oriented documents from single images. Proceedings of the Advanced Concepts for Intelligent Vision Systems (ACIVS), Ghent, Belgium, 9−11 September, pp. 190−197.

Clark, P., Mirmehdi, M., 2003. Rectifying perspective views of text in 3D scenes using vanishing points. Pattern Recogn. 36, 2673−2686.

Clarke, J.C., Carlsson, S., Zisserman, A., 1996. Detecting and tracking linear features efficiently. Proceedings of the British Machine Vision Association Conference, Edinburgh, UK, 9−12 September, pp. 415−424.

Clerc, M., Mallat, S., 2002. The texture gradient equation for recovering shape from texture. IEEE Trans. Pattern Anal. Machine Intell. 24 (4), 536−549.

Coeurjoly, D., Klette, R., 2004. A comparative evaluation of length estimators of digital curves. IEEE Trans. Pattern Anal. Machine Intell. 26 (2), 252−258.

Coifman, B., Beymer, D., McLauchlan, P., Malik, J., 1998. A real-time computer vision system for vehicle tracking and traffic surveillance. Transport. Res. Part C 6, 271−288.

Coleman, G.B., Andrews, H.C., 1979. Image segmentation by clustering. Proc. IEEE 67, 773−785.

Collins, R.T., Lipton, A.J., Kanade, T. (Eds.), 2000. Special section on video surveillance. IEEE Trans. Pattern Anal. Machine Intell. 22, 8.

Comaniciu, D., Meer, P., 2002. Mean shift: a robust approach toward feature space analysis. IEEE Trans. Pattern Anal. Machine Intell. 24 (5), 603–619.

Comaniciu, D., Ramesh, V., Meer, P., 2003. Kernel-based object tracking. IEEE Trans. Pattern Anal. Machine Intell. 25 (5), 564–577.

Conners, R.W., Harlow, C.A., 1980a. A theoretical comparison of texture algorithms. IEEE Trans. Pattern Anal. Machine Intell. 2 (3), 204–222.

Conners, R.W., Harlow, C.A., 1980b. Toward a structural textural analyzer based on statistical methods. Comput. Graph Image Process. 12, 224–256.

Connolly, C., 2009. Driver assistance systems aim to halve traffic accidents. Sensor Rev. 29 (1), 13–19.

Cook, R.L., Torrance, K.E., 1982. A reflectance model for computer graphics. ACM Trans. Graphics 1, 7–24.

Cootes, T.F., Taylor, C.J., Cooper, D.H., Graham, J., 1992. Training models of shape from sets of examples. Proceedings of the Third British Machine Vision Association Conference, Leeds, UK, 22–24 September, pp. 9–18.

Corneil, D.G., Gottlieb, C.C., 1970. An efficient algorithm for graph isomorphism. J. ACM 17, 51–64.

Cosío, F.A., Flores, J.A.M., Castañeda, M.A.P., 2010. Use of simplex search in active shape models for improved boundary segmentation. Pattern Recogn. Lett. 31 (9), 806–817.

Costa, L., da, F., Cesar, R.M., 2000. Shape Analysis and Classification: Theory and Practice. CRC Press, Boca Raton, FL.

Coudray, N., Buessler, J.-L., Urban, J.-P., 2010. Robust threshold estimation for images with unimodal histograms. Pattern Recogn. Lett. 31 (9), 1010–1019.

Cowan, G., 1998. Statistical Data Analysis. Oxford University Press, Oxford.

Cremers, D., Rousson, M., Deriche, R., 2007. A review of statistical approaches to level set segmentation: integrating color, texture, motion and shape. Int. J. Comput. Vision 72 (2), 195–215.

Crimmins, T.R., Brown, W.R., 1985. Image algebra and automatic shape recognition. IEEE Trans. Aerosp. Electron. Syst. 21, 60–69.

Cristianini, N., Shawe-Taylor, J., 2000. An Introduction to Support Vector Machines. Cambridge University Press, Cambridge.

Cross, A.D.J., Wilson, R.C., Hancock, E.R., 1997. Inexact graph matching with genetic search. Pattern Recogn. 30 (6), 953–970.

Crowley, J.L., Bobet, P., Schmid, C., 1993. Auto-calibration by direct observation of objects. Image Vis. Comput. 11 (2), 67–81.

Cucchiara, R., Grana, C., Piccardi, M., Prati, A., 2003. Detecting moving objects, ghosts, and shadows in video streams. IEEE Trans. Pattern Anal. Machine Intell. 25 (10), 1337–1342.

Cumani, A., Guiducci, A., 1995. Geometric camera calibration: the virtual camera approach. Machine Vis. Appl. 8 (6), 375–384.

Curio, C., Edelbrunner, J., Kalinke, T., Tzomakas, C., von Seelen, W., 2000. Walking pedestrian recognition. IEEE Trans. Intell. Transport. Syst. 1 (3), 155–163.

Cybenko, G., 1988. Continuous valued neural networks with two hidden layers are sufficient. Technical Report, Department of Computer Science, Tufts University, Medford, MA.

Cybenko, G., 1989. Approximation by superpositions of a sigmoidal function. Math. Control, Signals Syst. 2 (4), 303–314.

da Gama Leitão, H.C., Stolfi, J., 2002. A multiscale method for the reassembly of two-dimensional fragmented objects. IEEE Trans. Pattern Anal. Mach. Intell. 24 (9), 1239–1251.

Dalal, N., Triggs, B., 2005. Histograms of oriented gradients for human detection. Proceedings of the Conference on Computer Vision and Pattern Recognition, San Diego, CA, pp. 886–893.

Dalal, N., Triggs, B., Schmid, C., 2006. Human detection using oriented histograms of flow and appearance. In: Leonardis, A., Bischof, H., Prinz, A. (Eds.), Proceedings of the European Conference on Computer Vision, Part II, LNCS 3952, Springer-Verlag, Berlin, Heidelberg, pp. 428–441.

Dance, C., Willamowski, J., Fan, L., Bray, C., Csurka, G., 2004. Visual categorization with bags of keypoints. Proceedings of the ECCV International Workshop on Statistical Learning in Computer Vision, Prague.

Danielsson, P.-E., 1981. Getting the median faster. Computer Graph. Image Process. 17, 71–78.

Danielsson, P.-E., Levialdi, S., 1981. Computer architectures for pictorial information systems. IEEE Comput. 14 (November), 53–67.

Daugman, J.G., 1993. High confidence visual recognition of persons by a test of statistical independence. IEEE Trans. Pattern Anal. Mach. Intell. 15, 1148–1161.

Daugman, J.G., 2003. Demodulation by complex-valued wavelets for stochastic pattern recognition. Int. J. Wavelets, Multiresol. Inform. Process. 1 (1), 1–17.

Davies, E.R., 1984a. The median filter: an appraisal and a new truncated version, Proceedings of the Seventh International Conference on Pattern Recognition, Montreal, Canada, 30 July–2 August, pp. 590–592.

Davies, E.R., 1984b. Circularity – a new principle underlying the design of accurate edge orientation operators. Image Vis. Comput. 2, 134–142.

Davies, E.R., 1984c. Design of cost-effective systems for the inspection of certain food products during manufacture. In: Pugh, A. (Ed.), Proceedings of the Fourth International Conference on Robot Vision and Sensory Controls, London, 9–11 October, IFS Publications Ltd, Bedford, UK, pp. 437–446.

Davies, E.R., 1985. Radial histograms as an aid in the inspection of circular objects. IEE Proc. D 132 (4, Special Issue on Robotics), 158–163.

Davies, E.R., 1986a. Constraints on the design of template masks for edge detection. Pattern Recogn. Lett. 4, 111–120.

Davies, E.R., 1986b. Image space transforms for detecting straight edges in industrial images. Pattern Recogn. Lett 4, 185–192.

Davies, E.R., 1987a. A new framework for analysing the properties of the generalised Hough transform. Pattern Recogn. Lett. 6, 1–7.

Davies, E.R., 1987b. A new parametrisation of the straight line and its application for the optimal detection of objects with straight edges. Pattern Recogn. Lett. 6, 9–14.

Davies, E.R., 1987c. Design of optimal Gaussian operators in small neighbourhoods. Image Vis. Comput. 5, 199–205.

Davies, E.R., 1987d. Improved localisation in a generalised Hough scheme for the detection of straight edges. Image Vis. Comput. 5, 279–286.

Davies, E.R., 1987e. The effect of noise on edge orientation computations. Pattern Recogn. Lett. 6, 315–322.

Davies, E.R., 1987f. A high speed algorithm for circular object location. Pattern Recogn. Lett. 6, 323–333.

Davies, E.R., 1988a. Application of the generalised Hough transform to corner detection. IEE Proc. E 135, 49–54.

Davies, E.R., 1988b. A modified Hough scheme for general circle location. Pattern Recogn. Lett. 7, 37−43.

Davies, E.R., 1988c. On the noise suppression and image enhancement characteristics of the median, truncated median and mode filters. Pattern Recogn. Lett. 7, 87−97.

Davies, E.R., 1988d. Median-based methods of corner detection. In: Kittler, J. (Ed.), Proceedings of the Fourth BPRA International Conference on Pattern Recognition, Cambridge, Lecture Notes in Computer Science, Springer-Verlag, Heidelberg, 28−30 March, vol. 301, pp. 360−369.

Davies, E.R., 1988e. A hybrid sequential-parallel approach to accurate circle centre location. Pattern Recogn. Lett. 7, 279−290.

Davies, E.R., 1988f. Training sets and *a priori* probabilities with the nearest neighbour method of pattern recognition. Pattern Recogn. Lett. 8, 11−13.

Davies, E.R., 1988g. Tradeoffs between speed and accuracy in two-stage template matching. Signal Process. 15, 351−363.

Davies, E.R., 1989a. Finding ellipses using the generalised Hough transform. Pattern Recogn. Lett. 9, 87−96.

Davies, E.R., 1989b. Edge location shifts produced by median filters: theoretical bounds and experimental results. Signal Process. 16, 83−96.

Davies, E.R., 1989c. Minimising the search space for polygon detection using the generalised Hough transform. Pattern Recogn. Lett. 9, 181−192.

Davies, E.R., 1989d. Occlusion analysis for object detection using the generalised Hough transform. Signal Process. 16, 267−277.

Davies, E.R., 1991a. The minimal match graph and its use to speed identification of maximal cliques. Signal Process. 22 (3), 329−343.

Davies, E.R., 1991b. Median and mean filters produce similar shifts on curved boundaries. Electron. Lett. 27 (10), 826−828.

Davies, E.R., 1991c. Insight into operation of Kulpa boundary distance measure. Electron. Lett. 27 (13), 1178−1180.

Davies, E.R., 1992a. Simple fast median filtering algorithm, with application to corner detection. Electron. Lett. 28 (2), 199−201.

Davies, E.R., 1992b. Modelling peak shapes obtained by Hough transform. IEE Proc. E 139 (1), 9−12.

Davies, E.R., 1992c. Locating objects from their point features using an optimised Hough-like accumulation technique. Pattern Recogn. Lett. 13 (2), 113−121.

Davies, E.R., 1992d. Procedure for generating template masks for detecting variable signals. Image Vis. Comput. 10 (4), 241−249.

Davies, E.R., 1992e. Accurate filter for removing impulse noise from one- or two-dimensional signals. IEE Proc. E 139 (2), 111−116.

Davies, E.R., 1992f. Simple two-stage method for the accurate location of Hough transform peaks. IEE Proc. E 139 (3), 242−248.

Davies, E.R., 1992g. A framework for designing optimal Hough transform implementations. Proceedings of the Eleventh IAPR International Conference on Pattern Recognition, The Hague, 30 August−3 September, vol. III, pp. 509−512.

Davies, E.R., 1993. Electronics, Noise and Signal Recovery. Academic Press, London.

Davies, E.R., 1995. Machine vision in manufacturing − what are the real problems? Proceedings of the Second International Conference on Mechatronics and Machine Vision in Practice, Hong Kong, Japan, 12−14 September, pp. 15−24.

Davies, E.R., 1997a. Shifts produced by mode filters on curved intensity contours. Electron. Lett. 33 (5), 381−382.

Davies, E.R., 1997b. Principles and design graphs for obtaining uniform illumination in automated visual inspection. Proceedings of the Sixth IEE International Conference on Image Processing and its Applications, Dublin, Ireland, 14−17 July, IEE Conference Publication 443, pp. 161−165.

Davies, E.R., 1997c. Designing efficient line segment detectors with high orientation accuracy. Proceedings of the Sixth IEE International Conference on Image Processing and its Applications, Dublin, Ireland, 14−17 July, IEE Conference Publication 443, pp. 636−640.

Davies, E.R., 1997d. Lower bound on the processing required to locate objects in digital images. Electron. Lett. 33 (21), 1773−1774.

Davies, E.R., 1997e. Vectorial strategy for designing line segment detectors with high orientation accuracy. Electron. Lett. 33 (21), 1775−1777.

Davies, E.R., 1998. Rapid location of convex objects in digital images. Proceedings of European Signal Processing Conference (EUSIPCO'98), Rhodes, Greece, 8−11 September, pp. 589−592.

Davies, E.R., 1999a. Designing optimal image feature detection masks: equal area rule. Electron. Lett. 35 (6), 463−465.

Davies, E.R., 1999b. Chord bisection strategy for fast ellipse location. Electron. Lett. 35 (9), 703−705.

Davies, E.R., 1999c. Effect of foreground and background occlusion on feature matching for target location. Electron. Lett. 35 (11), 887−889.

Davies, E.R., 1999d. Algorithms for ultra-fast location of ellipses in digital images. Proceedings of seventh IEE International Conference on Image Processing and its Applications, Manchester, 13−15 July, IEE Conf. Publication no. 465, pp. 542−546.

Davies, E.R., 1999e. Isotropic masks make efficient linear feature detectors. Electron. Lett. 35 (17), 1450−1451.

Davies, E.R., 1999f. Image distortions produced by mean, median and mode filters. IEE Proc. Vis. Image Signal Process. 146 (5), 279−285.

Davies, E.R., 2000a. Image Processing for the Food Industry. World Scientific, Singapore.

Davies, E.R., 2000b. Obtaining optimum signal from set of directional template masks. Electron. Lett. 36 (15), 1271−1272.

Davies, E.R., 2000c. Resolution of problem with use of closing for texture segmentation. Electron. Lett. 36 (20), 1694−1696.

Davies, E.R., 2000d. Low-level vision requirements. Electron. Commun. Eng. J. 12 (5), 197−210.

Davies, E.R., 2000e. Accuracy of multichannel median filter. Electron. Lett. 36 (25), 2068−2069.

Davies, E.R., 2000f. A generalized model of the geometric distortions produced by rank-order filters. Imag. Sci. 48 (3), 121−130.

Davies, E.R., 2001a. Some problems in food and cereals inspection and methods for their solution. Proceedings of International Conference on Quality Control by Artificial Vision − 2001, Le Creusot, France, 21−23 May, pp. 35−46.

Davies, E.R., 2001b. A sampling approach to ultra-fast object location. Real-Time Imaging 7 (4), 339−355.

Davies, E.R., 2003a. Formulation of an accurate discrete theory of median shifts. Signal Process. 83, 531−544.

Davies, E.R., 2003b. Design of real-time algorithms for food and cereals inspection. Imaging Sci. 51 (2), 63—78.

Davies, E.R., 2003c. An analysis of the geometric distortions produced by median and related image processing filters. Adv. Imaging Electron Phys. 126, 93—193.

Davies, E.R., 2005. Using an edge-based model of the Plessey operator to determine localisation properties. Proceedings of IET International Conference on Visual Information Engineering, University of Glasgow, Glasgow, 4—6 April, pp. 385—391.

Davies, E.R., 2007a. Efficient transformation for identifying global valley locations in 1D data. Electron. Lett. 43 (6), 332—333.

Davies, E.R., 2007b. Fast implementation of generalised median filter. Electron. Lett. 43 (9), 505—507.

Davies, E.R., 2007c. Towards an integrated theory of guided sampling for rapid object location. Proceedings of IET International Conference on Visual Information Engineering, Royal Statistical Society, London, 25—27 July, paper 67, pp. 1—6.

Davies, E.R., 2008a. A generalised approach to the use of sampling for rapid object location. Int. J. Appl. Math. Comput. Sci. 18 (1), 7—19.

Davies, E.R., 2008b. Stable bi-level and multi-level thresholding of images using a new global transformation Special Issue on Visual Information Engineering. IET Computer Vision 2, no. 2, In: Valestin, S. (Ed.), 60—74.

Davies, E.R., 2009. The application of machine vision to food and agriculture: a review. Imaging Sci. 57, 197—217.

Davies, E.R., Celano, D., 1993a. Analysis of skeleton junctions in 3×3 windows. Electron. Lett. 29 (16), 1440—1441.

Davies, E.R., Johnstone, A.I.C., 1986. Engineering trade-offs in the design of a real-time system for the visual inspection of small products. Proceedings of the Fourth Conference on UK Research in Advanced Manufacture, 10—11 December. IMechE Conference Publications, London, pp. 15—22.

Davies, E.R., Johnstone, A.I.C., 1989. Methodology for optimising cost/speed tradeoffs in real-time inspection hardware. IEE Proc. E 136, 62—69.

Davies, E.R., Plummer, A.P.N., 1981. Thinning algorithms: a critique and a new methodology. Pattern Recogn. 14, 53—63.

Davies, E.R., Ip, H.H.S. (Eds.), 1998. Special Issue on Real-Time Visual Monitoring and Inspection. Real-Time Imaging 4, 5.

Davies, E.R., Bateman, M., Chambers, J., Ridgway, C., 1998a. Hybrid non-linear filters for locating speckled contaminants in grain. IEE Digest no. 1998/284, Colloquium on Non-Linear Signal and Image Processing, IEE, 22 May, pp. 12/1—5.

Davies, E.R., Bateman, M., Mason, D.R., Chambers, J., Ridgway, C., 2003a. Design of efficient line segment detectors for cereal grain inspection. Pattern Recogn. Lett. 24 (1—3), 421—436.

Davies, E.R., Chambers, J., Ridgway, C., 2002. Combination linear feature detector for effective location of insects in grain images. Measurement Sci. Technol. 13 (12), 2053—2061.

Davies, E.R., Chambers, J., Ridgway, C., 2003b. Design of a real-time grain inspection system with high sensitivity for insect detection. In: Parkin, R.M., Al-Habaibeh, A., Jackson, M.R. (Eds.), Proceedings of the International Conference on Mechatronics (ICOM 2003), Loughborough, 18—20 June, Professional Engineering Publishing, pp. 377—382.

Davies, E.R., Mason, D.R., Bateman, M., Chambers, J., Ridgway, C., 1998b. Linear feature detectors and their application to cereal inspection. Proceedings of European Signal Processing Conference (EUSIPCO'98), Rhodes, Greece, 8—11 September, pp. 2561—2564.

Davies, E.R., Patel, D., Johnstone, A.I.C., 1995. Crucial issues in the design of a real-time contaminant detection system for food products. Real-Time Imaging 1 (6), 397−407.

Davies, E.R., Ridgway, C., Chambers, J., 2003c. NIR detection of grain weevils inside wheat kernels. Proc. IEE Int. Conf. on Visual Information Engineering, VIE 2003, Surrey, 7−9 July, IEE Conference Publication 495, pp. 173−176.

Davis, J.W., Sharma, V., 2007. Background-subtraction using contour-based fusion of thermal and visible imagery. Comput. Vis. Image Underst. 106, 162−182, nos. 2−3.

Davison, A.J., Murray, D.W., 2002. Simultaneous localization and map-building using active vision. IEEE Trans. Pattern Anal. Machine Intell. 24 (7), 865−880.

de la Escalera, A., Armingol, J.Ma, Mata, M., 2003. Traffic sign recognition and analysis for intelligent vehicles. Image Vision Comput. 21 (3), 247−258.

Deans, S.R., 1981. Hough transform from the Radon transform. IEEE Trans. Pattern Anal. Machine Intell. 3, 185−188.

Delagnes, P., Benois, J., Barba, D., 1995. Active contours approach to object tracking in image sequences with complex background. Pattern Recogn. Lett. 16 (2), 171−178.

Derganc, J., Likar, B., Bernard, R., Tomaževič, D., Pernuš, F., 2003. Real-time automated visual inspection of color tablets in pharmaceutical blisters. Real-Time Imaging 9 (2), 113−124.

DeSouza, G.N., Kak, A.C., 2002. Vision for mobile robot navigation: a survey. IEEE Trans. Pattern Anal. Machine Intell. 24 (2), 237−267.

Devijver, P.A., Kittler, J., 1980. On the edited nearest neighbour rule. Proceedings of fifth International Conference on Pattern Recognition, IEEE Computer Society, Miami Beach, FL, pp. 72−80.

Devijver, P.A., Kittler, J., 1982. Pattern Recognition: A Statistical Approach. Prentice-Hall, Englewood Cliffs, NJ.

Dewaele, P., Van Gool, L., Wambacq, P., Oosterlinck, A., 1988. Texture inspection with self-adaptive convolution filters. Proceedings of Ninth International Conference on Pattern Recognition, Rome, Italy, 14−17 November, pp. 56−60.

Dickinson, S., Pelillo, M., Zabih, R. (Eds.), 2001. Special Section on *Graph Algorithms and Computer Vision*. IEEE Trans. Pattern Anal. Machine Intell. 23 (10), 1049−1151.

Dickmanns, E.D., Mysliwetz, B.D., 1992. Recursive 3-D road and relative ego-state recognition. IEEE Trans. Pattern Anal. Machine Intell. 14 (2), 199−213.

Dinic, E.A., 1970. Algorithm for solution of a problem of maximum flow in networks with power estimation. Soviet Math. Dokl 11, 1277−1280.

Dockstader, S.L., Tekalp, A.M., 2001. On the tracking of articulated and occluded video object motion. Real-Time Imaging 7 (5), 415−432.

Dockstader, S.L., Tekalp, A.M., 2002. A kinematic model for human motion and gait analysis. Proc. Workshop on Statistical Methods in Video Processing (ECCV), Copenhagen, Denmark, 1−2 June, pp. 49−54.

Dodd, G.G., Rossol, L. (Eds.), 1979. Computer Vision and Sensor-Based Robots. Plenum Press, New York.

Dorst, L., Smeulders, A.W.M., 1987. Length estimators for digitized contours. Computer Vision Graph. Image Process. 40, 311−333.

Doucet, A., Johansen, A.M., 2011. A Tutorial on particle filtering and smoothing: Fifteen years later. In: Crisan, D., Rozovsky, B. (Eds.), Oxford Handbook of Nonlinear Filtering, Oxford University Press, Oxford, UK.

Dougherty, E.R., Giardina, C.R., 1988. Morphology on umbra matrices. Int. J. Pattern Recogn. Artif. Intell. 2, 367−385.

Dougherty, E.R., Sinha, D., 1995a. Computational gray-scale mathematical morphology on lattices (a comparator-based image algebra) Part I: Architecture. Real-Time Imaging 1 (1), 69−85.

Dougherty, E.R., Sinha, D., 1995b. Computational gray-scale mathematical morphology on lattices (a comparator-based image algebra) Part II: Image operators. Real-Time Imaging 1 (4), 283−295.

Doyle, W., 1962. Operations useful for similarity-invariant pattern recognition. J. ACM 9, 259−267.

Draper, B.A., Beveridge, J.R., Böhm, A.P.W., Ross, C., Chawathe, M., 2003. Accelerated image processing on FPGAs. IEEE Trans. Image Process. 12 (12), 1543−1551.

Dreschler, L., Nagel, H.-H., 1981. Volumetric model and 3D-trajectory of a moving car derived from monocular TV-frame sequences of a street scene. Proceedings of International Joint Conference on Artificial Intelligence, Vancouver, British Columbia, Canada, August, pp. 692−697.

du Buf, J.M.H., Kardan, M., Spann, M., 1990. Texture feature performance for image segmentation. Pattern Recogn. 23, 291−309.

Duda, R.O., Hart, P.E., 1972. Use of the Hough transformation to detect lines and curves in pictures. Commun. ACM 15, 11−15.

Duda, R.O., Hart, P.E., 1973. Pattern Classification and Scene Analysis. Wiley, New York.

Duda, R.O., Hart, P.E., Stork, D.G., 2001. Pattern Classification. Wiley, New York.

Dudani, S.A., Luk, A.L., 1978. Locating straight-line edge segments on outdoor scenes. Pattern Recogn. 10, 145−157.

Dudani, S.A., Breeding, K.J., McGhee, R.B., 1977. Aircraft identification by moment invariants. IEEE Trans. Comput. 26, 39−46.

Duin, R.P.W., 2002. The combining classifier: to train or not to train? Proceedings of 16th International Conference on Pattern Recognition, Québec, Canada, 11−15 August, vol. II, pp. 765−770.

Duin, R.P.W., Haringa, H., Zeelen, R., 1986. Fast percentile filtering. Pattern Recogn. Lett. 4, 269−272.

Edmonds, J.M., Davies, E.R., 1991. High-speed processor for realtime visual inspection. Microprocessors Microsyst. 15 (1), 11−19.

Ehsan, S., Kanwal, N., Clark, A.F., McDonald-Maier, K.D., 2010. Improved repeatability measures for evaluating performance of feature detectors. Electron. Lett 46 (14), 998−1000.

Ehsan, S., Kanwal, N., Clark, A.F. McDonald-Maier, K.D., 2011. Measuring the coverage of interest point detectors. Proceedings of Eighth International Conference on Image Analysis and Recognition (ICIAR), LNCS, British Columbia, Canada, 22−24 June, Volume, 6753, pp. 253−261.

Eisberg, R.M., 1961. Fundamentals of Modern Physics. Wiley, New York.

Elgammal, A., Harwood, D. and Davis, L., 2000. Non-parametric model for background subtraction. Proceedings of European Conference on Computer Vision LNCS, Dublin, Ireland, 26 June−1 July, 1843, 751−767.

Ellis, T.J., Abbood, A., Brillault, B., 1992. Ellipse detection and matching with uncertainty. Image Vis. Comput. 10 (5), 271−276.

Eng, H.-L., Ma, K.-K., 2001. Noise adaptive soft-switching median filter. IEEE Trans. Image Process. 10 (2), 242−251.

Enzweiler, M., Gavrila, D.M., 2009. Monocular pedestrian detection: survey and experiments. IEEE Trans. Pattern Anal. Machine Intell. 31 (12), 2179−2195.

Eshel, R., Moses, Y., 2008. Homography based multiple camera detection and tracking of people in a dense crowd. Proceedings of IEEE Conference on Computer Vision and Pattern Recognition, Anchorage, Alaska, USA, 23–28 June, pp. 1–8.

Eshel, R., Moses, Y., 2010. Tracking in a dense crowd using multiple cameras. Int. J. Computer Vis. 88 (1), 129–143.

Evans, A.N., Nixon, M.S., 1995. Mode filtering to reduce ultrasound speckle for feature extraction. IEE Proc. – Vis. Image Signal Process. 142 (2), 87–94.

Fang, C.Y., Chen, S.W., Fuh, C.S., 2003. Road-sign detection and tracking. IEEE Trans. Vehicular Technol. 52 (5), 1329–1341.

Fang, C.Y., Fuh, C.S., Yen, P.S., Cherng, S., Chen, S.W., 2004. An automatic road sign recognition system based on a computational model of human recognition processing. Comput. Vis. Image Underst. 96, 237–268.

Fang, X., Luo, B., Zhao, H., Tang, J., Zhai, S., 2010. New multi-resolution image stitching with local and global alignment. IET Comput. Vis. 4 (4), 231–246.

Fasel, B., 2002. Robust face analysis using convolutional neural networks. Proc. 16th Int. Conf. on Pattern Recognition, Québec, Canada, 11–15 August, vol. II, pp. 40–43.

Fathy, M., Siyal, M.Y., 1995. Real-time image processing approach to measure traffic queue parameters. IEE Proc. Vis. Image Signal Process. 142 (5), 297–303.

Fathy, M.E., Hussein, A.S., Tolba, M.F., 2011. Fundamental matrix estimation: a study of error criteria. Pattern Recogn. Lett. 32 (2), 383–391.

Faugeras, O., 1992. What can be seen in three dimensions with an uncalibrated stereo rig? In: Sandini, G. (Ed.), Proceedigs of the Second European Conference on Computer Vision: Lecture Notes in Computer Science, Springer-Verlag, Berlin, Heidelberg, vol. 588, pp. 563–578.

Faugeras, O., 1993. Three-Dimensional Computer Vision – A Geometric Viewpoint. MIT Press, Cambridge, MA.

Faugeras, O., Luong, Q.-T., 2001. The Geometry of Multiple Images. MIT press, Cambridge, MA.

Faugeras, O. Luong, Q.-T., Maybank, S.J., 1992. Camera self-calibration: theory and experiments. In: Sandini, G. (Ed.), Proceedings of the Second European Conference on Computer Vision: Lecture Notes in Computer Science, Springer-Verlag, Berlin, Heidelberg, vol. 588, pp. 321–334.

Faugeras, O., Quan, L., Sturm, P., 2000. Self-calibration of a 1D projective camera and its application to the self-calibration of a 2D projective camera. IEEE Trans. Pattern Anal. Machine Intell. 22 (10), 1179–1185.

Faugeras, O.D., 1978. Texture analysis and classification using a human visual model. Proceedings of the Fourth International Joint Conference on Pattern Recognition, Kyoto, 7–10 November, pp. 549–552.

Faugeras, O.D., Hebert, M., 1983. A 3-D recognition and positioning algorithm using geometrical matching between primitive surfaces. Proceedings of the Eighth International Joint Conference on Artificial Intelligence, Karlsruhe, Germany, August, pp. 996–1002.

Fawcett, T., 2006. An introduction to ROC analysis. Pattern Recogn. Lett. 27, 861–874.

Feng, T.-Y., 1981. A survey of interconnection networks. IEEE Computer 14, 12–17.

Ferrie, F.P., Levine, M.D., 1989. Where and why local shading analysis works. IEEE Trans. Pattern Anal. Machine Intell. 11, 198–206.

Fesenkov, V.P., 1929. Photometric investigations of the lunar surface. Astronomochhesk. Zh 5, 219–234.

Fieguth, P., Terzopoulos, D., 1997. Color-based tracking of heads and other mobile objects at video frame rates. Proceedings of IEEE Conference on Computer Vision and Pattern Recognition, San Juan, PR, USA, 17–19 June, pp. 21–27.

Fischer, B., Buhmann, J.M., 2003. Bagging for path-based clustering. IEEE Trans. Pattern Anal. Machine Intell 25 (11), 1411–1415.

Fischler, M.A., Bolles, R.C., 1981. Random sample consensus: a paradigm for model fitting with applications to image analysis and automated cartography. Commun. ACM 24 (6), 381–395.

Fish, R.K., Ostendorf, M., Bernard, G.D., Castanon, D.A., 2003. Multilevel classification of milling tool wear with confidence estimation. IEEE Trans. Pattern Anal. Machine Intell. 25 (1), 75–85.

Fitch, J.P., Coyle, E.J., Gallagher, N.C., 1985. Root properties and convergence rates of median filters. IEEE Trans. Acoustics Speech Signal Process. 33, 230–239.

Flynn, M.J., 1972. Some computer organizations and their effectiveness. IEEE Trans. Comput. 21, 948–960.

Föglein, J., 1983. On edge gradient approximations. Pattern Recogn. Lett. 1, 429–434.

Ford, L., Fulkerson, D., 1962. Flows in Networks. Princeton University Press, Princeton, NJ, USA.

Forgy, E.W., 1965. Cluster analysis of multivariate data: efficiency versus interpretability of classification. Biometrics 21, 768–769.

Förstner, W., Dickscheid, T., Schindler, F., 2009. Detecting Interpretable and Accurate Scale-Invariant Keypoints. Proceedings of International Conference on Computer Vision (ICCV), Kyoto, Japan, pp. 2256–2263.

Forsyth, D.A., Ponce, J., 2003. Computer Vision: A Modern Approach. Pearson Education International, Upper Saddle River, NJ.

Forsyth, D.A., Mundy, J.L., Zisserman, A., Coelho, C., Heller, A., Rothwell, C.A., 1991. Invariant descriptors for 3-D object recognition and pose. IEEE Trans. Pattern Anal. Machine Intell. 13 (10), 971–991.

Foster, J.P., Nixon, M.S., Prugel-Bennett, A., 2001. New area based metrics for automatic gait recognition. Proceedings of British Machine Vision Association Conference, Manchester, UK, 10–13 September, pp. 233–242.

Fountain, T, 1987. Processor Arrays: Architectures and Applications. Academic Press, London.

Frankot, R.T., Chellappa, R., 1990. Estimation of surface topography form SAR imagery using shape from shading techniques. Artif. Intell. 43, 271–310.

Freeman, H., 1961. On the encoding of arbitrary geometric configurations. IEEE Trans. Electron. Comput. 10, 260–268.

Freeman, H., 1974. Computer processing of line drawing images. Comput. Surveys 6, 57–97.

Freeman, H., 1978. Shape description via the use of critical points. Pattern Recogn. 10, 159–166.

Frei, W., Chen., C.-C., 1977. Fast boundary detection: a generalization and a new algorithm. IEEE Trans. Comput. 26, 988–998.

Freund, Y., Schapire, R., 1996. Experiments with a new boosting algorithm. Proceedings of Thirteenth International Conference on Machine Learning, Bari, Italy, 3–6 July, pp. 148–156.

Fu, K.-S., Mui, J.K., 1981. A survey on image segmentation. Pattern Recogn. 13, 3–16.

Fumera, G., Fabio, R., Alessandra, S., 2008. A theoretical analysis of bagging as a linear combination of classifiers. IEEE Trans. Pattern Anal. Machine Intell. 30 (7), 1293–1299.

Gallagher, N.C., Wise, G.L., 1981. A theoretical analysis of the properties of median filters. IEEE Trans. Acoustics Speech Signal Process. 29, 1136–1141.

Gallo, O., Manduchi, R., Rafii, A., 2011. CC-RANSAC: fitting planes in the presence of multiple surfaces in range data. Pattern Recogn. Lett. 32 (3), 403–410.

Gao, C., Sang, N., Tang, Q., 2010. On selection and combination of weak learners in AdaBoost. Pattern Recogn. Lett. 31 (9), 991–1001.

Garcia, C., Delakis, M., 2002. A neural architecture for fast and robust face detection. Proceedings of 16th International Conference on Pattern Recognition, Québec, Canada, 11–15 August, vol. II, pp. 44–47.

Gavrila, D., 1999. The visual analysis of human movement: a survey. Comput. Vis. Image Underst. 73 (1), 82–98.

Gavrila, D., 2000. Pedestrian detection from a moving vehicle. In: Vernon, D. (Ed.), Proceedings of the European Conference on Computer Vision, Dublin, Ireland, June, pp. 37–49.

Gavrila, D.M., 1998. Multi-feature hierarchical template matching using distance transforms. Proceedings of the IEEE International Conference on Pattern Recognition, Brisbane, Australia.

Gavrila, D.M., Groen, F.C.A., 1992. 3D object recognition from 2D images using geometric hashing. Pattern Recogn. Lett. 13 (4), 263–278.

Gavrila, D.M., Munder, S., 2007. Multi-cue pedestrian detection and tracking from a moving vehicle. Int. J. Computer Vis. 73 (1), 41–59.

Gavrila, D.M., Giebel, J., Munder, S., 2004. Vision-based pedestrian detection: The PROTECTOR + system. Proceedings of the IEEE Intelligent Vehicle Symposium, Parma, Italy.

Geiger, D., Liu, T.-L., Kohn, R.V., 2003. Representation and self-similarity of shapes. IEEE Trans. Pattern Anal. Machine Intell. 25 (1), 86–99.

Gerig, G., Klein, F., 1986. Fast contour identification through efficient Hough transform and simplified interpretation strategy. Proceedings of the Eighth International Conference on Pattern Recognition, Paris, 27–31 October, pp. 498–500.

Geronimo, D., Lopez, A.M., Sappa, A.D., Graf, T., 2010. Survey of pedestrian detection for advanced driver assistance systems. IEEE Trans. Pattern Anal. Machine Intell. 32 (7), 1239–1258.

Ghosh, A., Petkov, N., 2005. Robustness of shape descriptors to incomplete contour representations. IEEE Trans. Pattern Anal. Machine Intell. 27 (11), 1793–1804.

Gibbons, A., 1985. Algorithmic Graph Theory. Cambridge University Press, Cambridge.

Giblin, P.J., Kimia, B.B., 2003. On the intrinsic reconstruction of shape from its symmetries. IEEE Trans. Pattern Anal. Machine Intell. 25 (7), 895–911.

Gibson, J.J., 1950. The Perception of the Visual World. Houghton Mifflin, Boston, MA.

Gil, J., Kimmel, R., 2002. Efficient dilation, erosion, opening, and closing algorithms. IEEE Trans. Pattern Anal. Machine Intell. 24 (12), 1606–1617.

Goetcherian, V., 1980. From binary to grey tone image processing using fuzzy logic concepts. Pattern Recogn. 12, 7–15.

Goldberg, A.V., Tarjan, R.E., 1988. A new approach to the maximum-flow problem. J. Assoc. Comput. Machinery 35 (4), 921–940.

Goldman, D.B., Curless, B., Hertzmann, A., Seitz, S.M., 2010. Shape and spatially-varying BRDFs from photometric stereo. IEEE Trans. Pattern Anal. Machine Intell. 32 (6), 1060–1071.

Golightly, I., Jones, D., 2003. Corner detection and matching for visual tracking during power line inspection. Image Vis. Comput. 21 (9), 827–840.

Golub, G.H., van Loan, C.F., 1983. Matrix Computations. North Oxford, Oxford.

Gómez, J., Blasco, J., Moltó, E., Camps-Valls, G., 2007. Hyperspectral detection of citrus damage with a Mahalanobis kernel classifier. Electron. Lett 43 (20), 1082–1084.

Gómez-Sanchis, J., Moltó, E., Camps-Valls, G., Gómez-Chova, L., Aleixos, N., Blasco, J., 2008. Automatic correction of the effects of the light source on spherical objects. An application to the analysis of hyperspectral images of citrus fruits. J. Food Eng. 85, 191–200.

Gong, S., McKenna, S., Psarrou, A., 2000. Dynamic Vision: From Images to Face Recognition. Imperial College Press, London.

Gonnet, G.H., 1984. Handbook of Algorithms and Data Structures. Addison-Wesley, London.

Gonzalez, R.C., Woods, R.E., 1992. Digital Image Processing. Addison Wesley, Reading, MA.

Gonzalez, R.C., Woods, R.E., 2008. Digital Image Processing. 3rd ed. Prentice Hall, Upper Saddle River, NJ.

Gope, C., Kehtarnavaz, N., 2007. Affine invariant comparison of point-sets using convex hulls and Hausdorff distances. Pattern Recogn. 40, 309–320.

Goulermas, J.Y., Liatsis, P., 1998. Genetically fine-tuning the Hough transform feature space, for the detection of circular objects. In: Davies, E.R., Atiquzzaman, M. (Eds.), Special Issue on Projection-Based Transforms, Image Vision Computing, 16. pp. 615–626.

Granlund, G.H., 1980. Description of texture using the general operator approach. Proceedings of the Fifth International Conference on Pattern Recognition, Miami Beach, FL, 1–4 December, pp. 776–779.

Graves, M., Batchelor, B.G. (Eds.), 2003. Machine Vision Techniques for Inspecting Natural Products. Springer Verlag, Berlin, Heidelberg.

Graves, M., Batchelor, B.G., Palmer, S., 1994. 3D X-ray inspection of food products. Proceedings of SPIE Conference on Applications of Digital Image Processing 17, San Diego, California, July, vol. 2298, pp. 248–259.

Greenhill, D., Davies, E.R., 1993. Texture analysis using neural networks and mode filters. Proceedings of the Fourth British Machine Vision Association Conference, University of Surrey, 21–23 September, vol. 2, pp. 509–518.

Greenhill, D., Davies, E.R., 1994a. Relative effectiveness of neural networks for image noise suppression. In: Gelsema, E.S., Kanal, L.N. (Eds.), Pattern Recognition in Practice IV. Elsevier Science B.V, Amsterdam, The Netherlands, pp. 367–378.

Gregory, R.L., 1971. The Intelligent Eye. Weidenfeld and Nicolson, London.

Gregory, R.L., 1972. Eye and Brain. 2nd ed. Weidenfeld and Nicolson, London.

Griffin, L.D., 2000. Mean, median and mode filtering of images. Proc. R. Soc. A 456 (2004), 2995–3004.

Grimson, W.E.L., Huttenlocher, D.P., 1990. On the sensitivity of the Hough transform for object recognition. IEEE Trans. Pattern Anal. Machine Intell. 12 (3), 255–274.

Grimson, W.E.L., Lozano-Perez, T., 1984. Model-based recognition and localisation from sparse range or tactile data. Int. J. Robot. Res. 3 (3), 3–35.

Gruen, A., Huang, T.S. (Eds.), 2001. Calibration and Orientation of Cameras in Computer Vision. Springer-Verlag, Heidelberg.

Guan, Y.-P., 2010. Spatio-temporal motion-based foreground segmentation and shadow suppression. IET Comput. Vis. 4 (1), 50–60.

Guiducci, A., 1999. Parametric model of the perspective projection of a road with applications to lane keeping and 3d road reconstruction. Comput. Vis. Image Underst. 73, 414–427.

Guo, S., Pridmore, T., Kong, Y., Zhang, X., 2009. An improved Hough transform voting scheme utilizing surround suppression. Pattern Recogn. Lett. 30 (13), 1241–1252.

Guru, D.S., Shekar, B.H., Nagabhushan, P., 2004. A simple and robust line detection algorithm based on small eigenvalue analysis. Pattern Recogn. Lett. 25 (1), 1–13.

Hall, E.L., 1979. Computer Image Processing and Recognition. Academic Press, New York.

Hall, E.L., Tio, J.B.K., McPherson, C.A., Sadjadi, F.A., 1982. Measuring curved surfaces for robot vision. IEEE Computer 15 (12), 42–54.

Hampel, F.R., Ronchetti, E.M., Rousseeuw, P.J., Stahel, W.A., 1986. Robust Statistics, The Approach Based on Influence Functions. Wiley, New York.

Hannah, I., Patel, D., Davies, E.R., 1995. The use of variance and entropic thresholding methods for image segmentation. Pattern Recogn. 28 (8), 1135–1143.

Hansen, D.W., Ji, Q., 2010. In the eye of the beholder: a survey of models for eyes and gaze. IEEE Trans. Pattern Anal. Machine Intell. 32 (3), 478–500.

Hansen, F.R., Elliott, H., 1982. Image segmentation using simple Markov field models. Computer Graph. Image Process. 20, 101–132.

Haralick, R.M., 1979. Statistical and structural approaches to texture. Proc. IEEE 67 (5), 786–804.

Haralick, R.M., 1980. Edge and region analysis for digital image data. Computer Graph. Image Process. 12, 60–73.

Haralick, R.M., 1984. Digital step edges from zero crossing of second directional derivatives. IEEE Trans. Pattern Anal. Machine Intell. 6, 58–68.

Haralick, R.M., 1989. Determining camera parameters from the perspective projection of a rectangle. Pattern Recogn. 22, 225–230.

Haralick, R.M., Chu, Y.H., 1984. Solving camera parameters from the perspective projection of a parameterized curve. Pattern Recogn. 17 (6), 637–645.

Haralick, R.M., Joo, H., 1988. 2D-3D pose estimation. Proceedings of ninth International Conference on Pattern Recognition, Rome, Italy, 14–17 November, pp. 385–391.

Haralick, R.M., Shapiro, L.G., 1985. Image segmentation techniques. Computer Vis. Graph. Image Process. 29, 100–132.

Haralick, R.M., Shapiro, L.G., 1992. Computer and Robot Vision, Vol. I. Addison Wesley, Reading, MA.

Haralick, R.M., Shapiro, L.G., 1993. Computer and Robot Vision, Vol. II. Addison Wesley, Reading, MA.

Haralick, R.M., Chu, Y.H., Watson, L.T., Shapiro, L.G., 1984. Matching wire frame objects from their two dimensional perspective projections. Pattern Recogn. 17 (6), 607–619.

Haralick, R.M., Shanmugam, K., Dinstein, I., 1973. Textural features for image classification. IEEE Trans. Systems, Man Cybern. 3 (6), 610–621.

Haralick, R.M., Sternberg, S.R., Zhuang, X., 1987. Image analysis using mathematical morphology. IEEE Trans. Pattern Anal. Machine Intell. 9 (4), 532–550.

Haritaoglu, I., Harwood, D., Davis, L.S., 2000. W4: Real-time surveillance of people and their activities. In Special Section on Video Surveillance. IEEE Trans. Pattern Anal. Machine Intell. 22 (8), 809–830.

Harris, C., Stephens, M., 1988. A combined corner and edge detector. Proceedings of fourth Alvey Vision Conference, Manchester, 31 August–2 September, pp. 147–151.

Hart, P.E., 1968. The condensed nearest neighbour rule. IEEE Trans. Inf. Theory 14, 515–516.

Hartley, R., Zisserman, A., 2000. Multiple View Geometry in Computer Vision. Cambridge University Press, Cambridge.

Hartley, R., Zisserman, A., 2003. Multiple View Geometry in Computer Vision. 2nd ed. Cambridge University Press, Cambridge.

Hartley, R.I., 1992. Estimation of relative camera positions for uncalibrated cameras. In: Sandini, G. (Ed.), Proceedings of the Second European Conference on Computer Vision: Lecture Notes in Computer Science, Springer-Verlag, Berlin, Heidelberg, vol. 588, pp. 579−587.

Hartley, R.I., 1995. A linear method for reconstruction from lines and points. Proceedings of the International Conference on Computer Vision, Boston, MA, USA, 20−23 June, pp. 882−887.

Hartley, R.I., 1997. In defense of the eight-point algorithm. IEEE Trans. Pattern Anal. Machine Intell. 19 (6), 580−593.

Harvey, N.R., Marshall, S., 1994. Using genetic algorithms in the design of morphological filters. IEE Colloquium on Genetic Algorithms in Image Processing and Vision, IEE, 20 October, IEE Digest no. 1994/193, pp. 6/1−5.

Harvey, N.R., Marshall, S., 1995. Rank-order morphological filters: a new class of filters. Proceedings of the IEEE Workshop on Nonlinear Signal and Image Processing, Halkidiki, Greece, June, pp. 975−978.

Harwood, D., Subbarao, M., Davis, L.S., 1985. Texture classification by local rank correlation. Computer Vis. Graph. Image Process. 32, 404−411.

Hasler, D., Sbaiz, L., Süsstrunk, S., Vetterli, M., 2003. Outlier modelling in image matching. IEEE Trans. Pattern Anal. Machine Intell. 25 (3), 301−315.

Haykin, S., 1999. Neural Networks: A Comprehensive Introduction. Prentice-Hall, New Jersey, USA.

Heia, K., Sivertsen, A.H., Stormo, S.K., Elvevoll, E., Wold, J.P., Nilsen, H., 2007. Detection of nematodes in cod (Gadus morhua) fillets by imaging spectroscopy. J. Food Sci. 72 (1), 11−15.

Heijmans, H., 1991. Theoretical aspects of gray-level morphology. IEEE Trans. Pattern Anal. Machine Intell. 13, 568−582.

Heikkilä, J., 2000. Geometric camera calibration using circular control points. IEEE Trans. Pattern Anal. Machine Intell. 22 (10), 1066−1076.

Heikkonen, J., 1995. Recovering 3-D motion parameters from optical flow field using randomized Hough transform. Pattern Recogn. Lett. 16 (9), 971−978.

Heinemann, P.H., Varghese, Z.A., Morrow, C.T., Sommer III, H.J., Crassweller, R.M., 1995. Machine vision inspection of 'Golden Delicious' apples. Appl. Eng. Agric. 11 (6), 901−906.

Heinonen, P., Neuvo, Y., 1987. FIR-median hybrid filters. IEEE Trans. Acoust. Speech Signal Process. 35, 832−838.

Henderson, T.C., 1984. A note on discrete relaxation. Computer Vision Graph. Image Process. 28, 384−388.

Herault, L., Horaud, R., Veillon, F., Niez, J.J., 1990. Symbolic image matching by simulated annealing. Proceedings of the British Machine Vision Association Conference, Oxford, UK, 24−27 September, pp. 319−324.

Hernandez, C., Vogiatzis, G., Cipolla, R., 2011. Overcoming shadows in 3-source photometric stereo. IEEE Trans. Pattern Anal. Machine Intell. 33 (2), 419−426.

Hilario, C., Collado, J.M., Armingol, J.M., de la Escalera, A., 2006. Visual perception and tracking of vehicles for driver assistance systems. Proceedings of the Intelligent Vehicles Symposium, 13−15, June Tokyo, Japan, pp. 94−99.

Hildreth, E.C., 1984. Measurement of Visual Motion. MIT Press, Cambridge, MA.

Hlaoui, A., Wang, S., 2002. A new algorithm for inexact graph matching. Proceedings of the Sixteenth International Conference on Pattern Recognition, Québec, Canada, 11−15 August, vol. IV, pp. 180−183.

Ho, T.K., Hull, J.J., Srihari, S.N., 1994. Decision combination in multiple classifier systems. IEEE Trans. Pattern Anal. Machine Intell. 16 (1), 66−75.

Hockney, R.W., Jesshope, C.R., 1981. Parallel Computers. Adam Hilger Ltd, Bristol.

Hodgson, R.M., Bailey, D.G., Naylor, M.J., Ng, A.L.M., McNeil, S.J., 1985. Properties, implementations, and applications of rank filters. Image Vis. Comput. 3, 4−14.

Hofmann, U., Rieder, A., Dickmanns, E.D., 2003. Radar and vision data fusion for hybrid adaptive cruise control on highways. Machine Vis. Appl. 14 (1), 42−49.

Hogg, D., 1983. Model-based vision: a program to see a walking person. Image Vis. Comput. 1 (1), 5−20.

Horaud, R., 1987. New methods for matching 3-D objects with single perspective views. IEEE Trans. Pattern Anal. Machine Intell. 9, 401−412.

Horaud, R., Brady, M., 1988. On the geometric interpretation of image contours. Artif. Intell. 37, 333−353.

Horaud, R., Sossa, H., 1995. Polyhedral object recognition by indexing. Pattern Recogn. 28 (12), 1855−1870.

Horaud, R., Conio, B., Leboulleux, O., Lacolle, B., 1989. An analytic solution for the perspective 4-point problem. Computer Vis. Graph. Image Process. 47, 33−44.

Horn, B.K.P., 1975. Obtaining shape from shading information. In: Winston, P.H. (Ed.), The Psychology of Computer Vision, McGraw-Hill, New York, pp. 115−155.

Horn, B.K.P., 1977. Understanding image intensities. Artif. Intell. 8, 201−231.

Horn, B.K.P., 1986. Robot Vision. MIT Press, Cambridge, MA.

Horn, B.K.P., Brooks, M.J., 1986. The variational approach to shape from shading. Computer Vision Graph. Image Process. 33, 174−208.

Horn, B.K.P., Brooks, M.J. (Eds.), 1989. Shape from Shading. MIT Press, Cambridge, MA.

Horn, B.K.P., Schunck, B.G., 1981. Determining optical flow. Artif. Intell. 17 (1−3), 185−203.

Horng, J.-H., 2003. An adaptive smoothing approach for fitting digital planar curves with line segments and circular arcs. Pattern Recogn. Lett. 24 (1−3), 565−577.

Hornik, K., Stinchcombe, M., White, H., 1989. Multilayer feedforward networks are universal approximators. Neural Networks 2, 359−366.

Horowitz, S.L., Pavlidis, T., 1974. Picture segmentation by a directed split-and-merge procedure. Proceedings of the Second International Joint Conference on Pattern Recognition, Copenhagen, Denmark, August, pp. 424−433.

Hough, P.V.C., 1962. Method and means for recognising complex patterns. US Patent 3069654, Alexandria, Virginia, US.

Hsiao, J.Y., Sawchuk, A.A., 1989. Supervised textured image segmentation using feature smoothing and probabilistic relaxation techniques. IEEE Trans. Pattern Anal. Machine Intell. 11 (12), 1279−1292.

Hsiao, J.Y., Sawchuk, A.A., 1990. Unsupervised textured image segmentation using feature smoothing and probabilistic relaxation techniques. Computer Vis. Graph. Image Process. 48, 1−21.

Hu, M.K., 1961. Pattern recognition by moment invariants. Proc. IEEE 49, 1428.

Hu, M.K., 1962. Visual pattern recognition by moment invariants. IRE Trans. Inf. Theory 8, 179−187.

Hu, X.-P., Dempere-Marco, L., Davies, E.R., 2008. Bayesian feature evaluation for visual saliency estimation. Pattern Recogn. 41, 3302–3312.

Huang, C.T., Mitchell, O.R., 1994. A Euclidean distance transform using greyscale morphology decomposition. IEEE Trans. Pattern Anal. Machine Intell. 16 (4), 443–448.

Huang, T.S. (Ed.), 1983. Image Sequence Processing and Dynamic Scene Analysis. Springer-Verlag, New York.

Huang, T.S., Bruckstein, A.M., Holt, R.J., Netravali, A.N., 1995. Uniqueness of 3D pose under weak perspective: a geometrical proof. IEEE Trans. Pattern Anal. Machine Intell. 17 (12), 1220–1221.

Huang, T.S., Yang, G.J., Tang, G.Y., 1979. A fast two-dimensional median filtering algorithm. IEEE Trans. Acoust. Speech Signal Process. 27, 13–18.

Hubel, D.H., 1995. Eye, Brain and Vision. Scientific American Library, New York.

Huber, P.J., 1964. Robust estimation of a location parameter. Annals Math. Statist. 35, 73–101.

Huber, P.J., 1981. Robust Statistics. Wiley, New York.

Huber, P.J., 1985. Projection pursuit. Ann. Stat. 13 (2), 435–475.

Hufnagl, C., Uhl, A., 2000. Algorithms for fractal image compression on massively parallel SIMD arrays. Real-Time Imaging 6 (4), 267–281.

Hughes, G.F., 1968. On the mean accuracy of statistical pattern recognisers. IEEE Trans. Inf. Theory 14, 55–63.

Humenberger, M., Zinner, C., Weber, M., Kubinger, W., Vincze, M., 2010. A fast stereo matching algorithm suitable for embedded real-time systems. Comput. Vis. Image Underst. 114, 1180–1202.

Hummel, R.A., Zucker, S.W., 1983. On the foundations of relaxation labelling processes. IEEE Trans. Pattern Anal. Machine Intell. 5, 267–287.

Hussmann, S., Ho, T.H., 2003. A high-speed subpixel edge detector implementation inside a FPGA. Real-Time Imaging 9 (5), 361–368.

Huttenlocher, D.P., Klanderman, G.A., Rucklidge, W.J., 1993. Comparing images using the Hausdorff distance. IEEE Trans. Pattern Anal. Machine Intell. 15 (9), 850–863.

Hwang, K., Briggs, F.A., 1984. Computer Architecture and Parallel Processing. McGraw-Hill, New York.

Hwang, V.S., Davis, L.S., Matsuyama, T., 1986. Hypothesis integration in image understanding systems. Computer Vis. Graph. Image Process. 36, 321–371.

Ikeuchi, K., Horn, B.K.P., 1981. Numerical shape from shading and occluding boundaries. Artif. Intell. 17, 141–184.

Illingworth, J., Kittler, J., 1987. The adaptive Hough transform. IEEE Trans. Pattern Anal. Machine Intell. 9, 690–698.

Illingworth, J., Kittler, J., 1988. A survey of the Hough transform. Computer Vision Graph. Image Process 44, 87–116.

Isard, M., Blake, A., 1996. Contour tracking by stochastic propagation of conditional density. Proceedings of the European Conference on Computer Vision, Cambridge, UK, 15–18 April, vol. 1, pp. 343–356.

Isard, M., Blake, A., 1998. Icondensation: unifying low-level and high-level tracking in a stochastic framework. Proceedings of the European Conference on Computer Vision, Freiburg, Germany, vol. I, pp. 893–908.

Ito, M., Ishii, A., 1986. Three-view stereo analysis. IEEE Trans. Pattern Anal. Machine Intell. 8 (4), 524−532.

Jacinto, C.N., Arnaldo, J.A., George, S.M., 2003. Using middle level features for robust shape tracking. Pattern Recogn. Lett. 24, 295−307.

Jackway, P.T., Deriche, M., 1996. Scale-space properties of the multiscale morphological dilation-erosion. IEEE Trans. Pattern Anal. Machine Intell. 18 (1), 38−51.

Jain, A.K., 2010. Data clustering: 50 years beyond *k*-means. Pattern Recogn. Lett. 31 (8), 651−666.

Jain, A.K., Dubes, R.C., 1988. Algorithms for Clustering Data. Prentice-Hall, Englewood Cliffs, NJ.

Jain, A.K., Duin, R.P.W., Mao, J., 2000. Statistical pattern recognition: a review. IEEE Trans. Pattern Anal. Machine Intell. 22 (1), 4−37.

Jain, R., 1983. Direct computation of the focus of expansion. IEEE Trans. Pattern Anal. Machine Intell. 5, 58−63.

Jang, J.-H., Hong, K.-S., 1998. Detection of curvilinear structures using the Euclidean distance transform. Proceedings of the IAPR Workshop on Machine Vision Applications (MVA'98), Chiba, Japan, pp. 102−105.

Jang, Y.K., Kang, B.J., Park, K.R., 2008. A study on eyelid localization considering image focus for iris recognition. Pattern Recogn. Lett. 29, 1698−1704.

Janney, P., Geers, G., 2010. Texture classification using invariant features of local textures. IET Image Process. 4 (3), 158−171.

Jiang, J.-A., Chuang, C.-L., Lu, Y.-L., Fahn, C.-S., 2007. Mathematical-morphology-based edge detectors for detection of thin edges in low-contrast regions. IET Image Process. 1 (3), 269−277.

Jolion, J.-M., Rosenfeld, A., 1989. Cluster detection in background noise. Pattern Recogn. 22 (5), 603−607.

Juan, A., Vidal, E., 1994. Fast *K*-means-like clustering in metric spaces. Pattern Recogn. Lett. 15 (1), 19−25.

Kadir, T., Brady, M., 2001. Scale, saliency and image description. Int. J. Computer Vis. 45 (2), 83−105.

Kadir, T., Brady, M., Zisserman, A., 2004. An affine invariant method for selecting salient regions in images. Proceedings of the Eighth European Conference on Computer Vision (ECCV), Prague, Czech Republic, 11−14 May, pp. 345−457.

Kadyrov, A., Petrou, M., 2001. The trace transform and its applications. IEEE Trans. Pattern Anal. Machine Intell. 23, 811−828.

Kadyrov, A., Petrou, M., 2002. Affine parameter estimation from the trace transform. Proceeedings of the Sixteenth International Conference on Pattern Recognition, Québec, Canada, 11−15 August, vol. II, pp. 798−801.

Kaizer, H., 1955. A quantification of textures on aerial photographs. MS Thesis, Boston University, Boston, MA.

Kamat-Sadekar, V., Ganesan, S., 1998. Complete description of multiple line segments using the Hough transform. In: Davies, E.R., Atiquzzaman, M. (Eds.), Special Issue on Projection-Based Transforms, Image Vision Computing 16, nos. 9−10, 597−614.

Kamel, M.S., Shen, H.C., Wong, A.K.C., Hong, T.M., Campeanu, R.I., 1994. Face recognition using perspective invariant features. Pattern Recogn. Lett. 15 (9), 877−883.

Kanatani, K., Sugaya, Y., Niitsuma, H., 2008. Triangulation from two views revisited: Hartley-Sturm vs. optimal correction. Proceedings of the British Machine Vision Association Conference, Leeds, UK, 1–4 September, pp. 173–182.

Kanesalingam, C., Smith, M.C.B., Dodds, S.A., 1998. An efficient algorithm for environmental mapping and path planning for an autonomous mobile robot. Proceedings of the Twentieth International Symposium on Robotics, Birmingham, UK, pp. 133–136.

Kang, D.-J., Jung, M.-H., 2003. Road lane segmentation using dynamic programming for active safety vehicles. Pattern Recogn. Lett. 24, 3177–3185.

Kapur, J.N., Sahoo, P.K, Wong, A.K.C., 1985. A new method for gray-level picture thresholding using the entropy of the histogram. Computer Vis. Graph. Image Process. 29, 273–285.

Kasif, S., Kitchen, L., Rosenfeld, A., 1983. A Hough transform technique for subgraph isomorphism. Pattern Recogn. Lett. 2, 83–88.

Kass, M., Witkin, A., 1987. Analyzing oriented patterns. Computer Vis. Graph. Image Process. 37 (3), 362–385.

Kass, M., Witkin, A., Terzopoulos, D., 1988. Snakes: active contour models. Int. J. Computer Vis. 1, 321–331.

Kastrinaki, V., Zervakis, M., Kalaitzakis, K., 2003. A survey of video processing techniques for traffic applications. Image Vis. Comput. 21 (4), 359–381.

Keagy, P.M., Parvin, B., Schatzki, T.F., 1995. Machine recognition of navel orange worm damage in x-ray images of pistachio nuts. Opt. Agric. Forestry Biol. SPIE 2345, 192–203.

Keagy, P.M., Parvin, B., Schatzki, T.F., 1996. Machine recognition of navel orange worm damage in X-ray images of pistachio nuts. Lebens.-Wissu.-Technol. 29, 140–145.

Keefe, P.D., 1992. A dedicated wheat grain image analyser. Plant Varieties Seeds 5, 27–33.

Kégl, B., Krzyżak, A., 2002. Piecewise linear skeletonization using principal curves. IEEE Trans. Pattern Anal. Machine Intell. 24 (1), 59–74.

Kehtarnavaz, N., Mohan, S., 1989. A framework for estimation of motion parameters from range images. Computer Vis. Graph. Image Process. 45, 88–105.

Kelly, P., Beardsley, P., Cooke, E., O'Connor, N., Smeaton, A., 2005. Detecting shadows and low-lying objects in indoor and outdoor scenes using homographies. Proceedings of the IET Conference on Visual Information Engineering, Glasgow 4–6 April, pp. 393–400.

Kender, J.R., 1980. Shape from texture. Computer Science Technical Report CMU-CS-81-102. Carnegie-Mellon University.

Kender, J.R., 1983. Shape from texture. Technical Report CMU-CS-81-102. Carnegie-Mellon University.

Kenney, C.S., Manjunath, B.S., Zuliani, M., Hewer, G.A., van Nevel, A., 2003. A condition number for point matching with application to registration and postregistration error estimation. IEEE Trans. Pattern Anal. Machine Intell. 25 (11), 1437–1454.

Kerbyson, D.J., Atherton, T.J., 1995. Circle detection using Hough transform filters. IEE Conference Publication no. 410, pp. 370–374.

Kesidis, A.L., Papamarkos, N., 2000. On the grayscale inverse Hough transform. Image Vis. Comput. 18 (8), 607–618.

Kessal, L., Abel, N., Demigny, D., 2003. Real-time image processing with dynamically reconfigurable architecture. Real-Time Imaging 9 (5), 297–313.

Khan, S., Shah, M., 2000. Tracking people in presence of occlusion. Proceedings of the Asian Conference on Computer Vision, Taipei, Taiwan, 8−11 January, [6 pp.].

Khan, S., Shah, M., 2003. Consistent labeling of tracked objects in multiple cameras with overlapping fields of view. IEEE Trans. Pattern Anal. Machine Intell. 25 (10), 1355−1360.

Khan, S., Shah, M., 2009. Tracking multiple occluding people by localizing on multiple scene planes. IEEE Trans. Pattern Anal. Machine Intell. 31 (3), 505−519.

Kim, D., Kim, D., Paik, J., 2010. Gait recognition using active shape model and motion prediction. IET Comput. Vis. 4 (1), 25−36.

Kim, D.-S., Lee, W.-H., Kweon, I.-S., 2004. Automatic edge detection using 3×3 ideal binary pixel patterns and fuzzy-based edge thresholding. Pattern Recogn. Lett. 25 (1), 101−106.

Kim, D.Y., Kim, J.J., Meer, P., Mintz, D., Rosenfeld, A., 1989. Robust computer vision: a least median of squares based approach. Proceedings of the DARPA Image Understanding Workshop, Palo Alto, CA, 23−26 May, pp. 1117−1134.

Kim, Z.-W., 2008. Robust lane detection and tracking in challenging scenarios. IEEE Trans. Intell. Transport. Syst. 9 (1), 16−26.

Kimme, C., Ballard, D., Sklansky, J., 1975. Finding circles by an array of accumulators. Commun. ACM 18, 120−122.

Kimura, A., Watanabe, T., 2002. An extension of the generalized Hough transform to realize affine-invariant two-dimensional (2D) shape detection. Proceedings of the Sixteenth International Conference on Pattern Recognition, Québec, Canada, 11−15 August, vol. I, pp. 65−69.

King, T.G., Tao, L.G., 1995. An incremental real-time pattern tracking algorithm for line-scan camera applications. Mechatronics 4 (5), 503−516.

Kirsch, R.A., 1971. Computer determination of the constituent structure of biological images. Comput. Biomed. Res. 4, 315−328.

Kiryati, N., Bruckstein, A.M., 1991. Antialiasing the Hough transform. Computer Vis. Graph. Image Process.: Graph. Models Image Process. 53 (3), 213−222.

Kitchen, L., Rosenfeld, A., 1979. Discrete relaxation for matching relational structures. IEEE Trans. Systems Man Cybern. 9, 869−874.

Kitchen, L., Rosenfeld, A., 1982. Gray-level corner detection. Pattern Recogn. Lett. 1, 95−102.

Kittler, J., 1983. On the accuracy of the Sobel edge detector. Image Vis. Comput. 1, 37−42.

Kittler, J., Illingworth, J., Föglein, J., 1985. Threshold selection based on a simple image statistic. Computer Vis. Graph. Image Process. 30, 125−147.

Klassen, E., Srivistava, A., Mio, W., Joshi, S.H., 2004. Analysis of planar shapes using geodesic paths on shape spaces. IEEE Trans. Pattern Anal. Machine Intell. 26 (3), 372−383.

Koch, C., Park, S., Ellis, T.J., Georgiadis, A., 2001. Illumination technique for optical dynamic range compression and offset reduction. Proceedings of the British Machine Vision Association Conference, Manchester, UK, September, pp. 293−302.

Koenderink, J.J., van Doorn, A.J., 1979. The internal representation of solid shape with respect to vision. Biol. Cybern. 32, 211−216.

Koivo, A.J., Kim, C.W., 1989. Robust image modelling for classification of surface defects on wood boards. IEEE Trans. Syst. Man Cybern. 19 (6), 1659−1666.

Köktas, N.S., Yalabik, N., Yavuzer, G., Duin, R.P.W., 2010. A multi-classifier for grading knee osteoarthritis using gait analysis. Pattern Recogn. Lett. 31 (9), 898−904.

Koller, D., Weber, J., Huang, T., Malik, J., Ogasawara, G., Rao, B., Russell, S., 1994. Towards robust automatic traffic scene analysis in real-time. Proceedings of the Twelfth International Conference on Pattern Recognition, Jerusalem, Israel, 9–13 October, pp. 126–131.

Koller, Th.M., Gerig, G., Székely, G., Dettwiler, D., 1995. Multiscale detection of curvilinear structures in 2-D and 3-D image data. Proceedings of the Fifth International Conference on Computer Vision (ICCV'95), Boston, MA, USA, 20–23 June, pp. 864–869.

Koplowitz, J., Bruckstein, A.M., 1989. Design of perimeter estimators for digitized planar shapes. IEEE Trans. Pattern Anal. Machine Intell. 11, 611–622.

Kortenkamp, D., Bonasso, R.P., Murphy, R. (Eds.), 1998. Artificial Intelligence and Mobile Robots. AAAI Press and MIT Press, Menlo Park, CA/Cambridge, MA/London/England.

Kröger, C., Bartle, C.M., West, J.G., Purchas, R.W., Devine, C.E., 2006. Meat tenderness evaluation using dual energy X-ray absorptiometry (DEXA). Comput. Electron. Agric. 54, 93–100.

Kuehnle, A., 1991. Symmetry-based recognition of vehicle rears. Pattern Recogn. Lett. 12, 249–258.

Kulpa, Z., 1977. Area and perimeter measurement of blobs in discrete binary pictures. Computer Graph. Image Process. 6, 434–451.

Kumar, R., Hanson, A.R., 1989. Robust estimation of camera location and orientation from noisy data having outliers. Proceedings of the Workshop on Interpretation of 3D Scenes, Austin, TX, 27–29 November, pp. 52–60.

Kung, H.T., 1980. The structure of parallel algorithms. Adv. Comput. 19, 69–112.

Kuo, F.F., 1966. Network Analysis and Synthesis. 2nd ed. Wiley, New York.

Kuo, P., Makris, D., Nebel, J.-C., 2011. Integration of bottom-up/top-down approaches for 2D pose estimation using probabilistic Gaussian modelling. Comput. Vis. Image Underst. 115 (2), 242–255.

Kuo, W.-J., Lin, C.-C., 2007. Two-stage road sign detection and recognition. Proceedings of the IEEE International Conference on Multimedia and Expo, Beijing, China, 2–5 July, pp. 1427–1430.

Kuo, Y.-C., Pai, N.-S., Li, Y.-F., 2011. Vision-based vehicle detection for a driver assistance system. Computers Math. Appl. 61, 2096–2100.

Kwok, P.C.K., 1989. Customising thinning algorithms. Proceedings of the Third International Conference on Image Processing and its Applications, Warwick, 18–20 July, IEE Conf. Publ. 307, pp. 633–637.

Lacroix, V., 1988. A three-module strategy for edge detection. IEEE Trans. Pattern Anal. Machine Intell. 10, 803–810.

Lamdan, Y., Wolfson, H.J., 1988. Geometric hashing: a general and efficient model-based recognition scheme. Proceedings of the IEEE Second International Conference on Computer Vision, Tampa, FL, December, pp. 238–249.

Lane, R.A., Thacker, N.A., Seed, N.L., 1994. Stretch-correlation as a real-time alternative to feature-based stereo matching algorithms. Image Vis. Comput. 12 (4), 203–212.

Laurentini, A., 1994. The visual hull concept for silhouette-based image understanding. IEEE Trans. Pattern Anal. Machine Intell. 16 (2), 150–162.

Laws, K.I., 1979. Texture energy measures. Proceedings of Image Understanding Workshop, November, pp. 47–51.

Laws, K.I., 1980a. Rapid texture identification. Proceedings of the SPIE Conference on Image Processing for Missile Guidance, 238, San Diego, CA, 28 July–1 August, pp. 376–380.

Laws, K.I., 1980b. Textured Image Segmentation. PhD Thesis, University of Southern California, Los Angeles, CA.

Lazarevic-McManus, N., Renno, J.R., Makris, D., Jones, G.A., 2008. An object-based comparative methodology for motion detection based on the F-Measure. Comput. Vis. Image Underst. 111 (1), 74−85.

Leavers, V.F, 1993. Which Hough transform? Computer Vis. Graph. Image Process.: Image Understanding 58 (2), 250−264.

Leavers, V.F., Boyce, J.F., 1987. The Radon transform and its application to shape parametrization in machine vision. Image Vis. Comput. 5, 161−166.

Lebègue, X., Aggarwal, J.K., 1993. Significant line segments for an indoor mobile robot. IEEE Trans. Robot. Autom. 9 (6), 801−815.

Lee, D.H., Park, Y.T., 2006. Robust vehicle detection based on shadow classification. Proceedings of the Eighteenth International Conference on Pattern Recognition, Hong Kong, China, 20−24 August, vol. 3, pp. 1167−1170.

Lee, M.-S., Medioni, G., Mordohai, P., 2002. Inference of segmented overlapping surfaces from binocular stereo. IEEE Trans. Pattern Anal. Machine Intell. 24 (6), 824−837.

Lei, Y., Wong, K.C., 1999. Ellipse detection based on symmetry. Pattern Recogn. Lett. 20 (1), 41−47.

Leibe, B., Seemann, E., Schiele, B., 2005. Pedestrian detection in crowded scenes. Proceedings of the Conference on Computer Vision and Pattern Recognition, San Diego, USA, 20−25 June, pp. 878−885.

Lepetit, V., Moreno-Noguer, F., Fua, P., 2008. EPnP: An Accurate O(n) Solution to the PnP Problem. Int. J. Computer Vision 81 (2), 155−166.

Lev, A., Zucker, S.W., Rosenfeld, A., 1977. Iterative enhancement of noisy images. IEEE Trans. Systems Man. Cybern. 7, 435−442.

Levine, M.D., 1985. Vision in Man and Machine. McGraw-Hill, New York.

Lézoray, O., Charrier, C., 2009. Color image segmentation using morphological clustering and fusion with automatic scale selection. Pattern Recogn. Lett. 30 (4), 397−406.

Li, H., Lavin, M.A., 1986. Fast Hough transform based on bintree data structure. Proceedings of the Conference on Computer Vision and Pattern Recognition, Miami Beach, FL, pp. 640−642.

Li, H., Lavin, M.A., LeMaster, R.J., 1985. Fast Hough transform. Proceedings of the Third Workshop on Computer Vision: Representation and Control, Bellair, IL, pp. 75−83.

Li, L., Tan, C.L., 2010. Recognizing planar symbols with severe perspective deformation. IEEE Trans. Pattern Anal. Machine Intell. 32 (4), 755−762.

Li, P., Liu, X., Xiao, L., Song, Q., 2010. Robust and accurate iris segmentation in very noisy iris images. Image Vision Comput. 28, 246−253.

Li, S.Z., Zhang, Z.Q., 2004. FloatBoost learning and statistical face detection. IEEE Trans. Pattern Anal. Machine Intell. 26 (9), 1112−1123.

Li, Z., Yang, J., Liu, G., Cheng, Y., Liu, C., 2011. Unsupervised range-constrained thresholding. Pattern Recogn. Lett. 32 (2), 392−402.

Lian, G., Lai, J., Zheng, W.-S., 2011. Spatial−temporal consistent labeling of tracked pedestrians across non-overlapping camera views. Pattern Recogn. 44, 1121−1136.

Liao, K., Paulsen, M.R., Reid, J.F., 1994. Real-time detection of colour and surface defects of maize kernels using machine vision. J. Agric. Eng. Res 59, 263−271.

Liao, P.-S., Chen, T.-S., Chung, P.-C., 2001. A fast algorithm for multilevel thresholding. J. Information Science and Engineering 17, 713−727.

Lin, C.C., Chellappa, R., 1987. Classification of partial 2-D shapes using Fourier descriptors. IEEE Trans. Pattern Anal. Machine Intell. 9, 686–690.

Lindeberg, T., 1998. Feature detection with automatic scale selection. Int. J. Computer Vision 30 (2), 79–116.

Lippmann, R.P., 1987. An introduction to computing with neural nets. IEEE Acoustics, Speech, Signal Process. Magazine 4 (2), 4–22.

Liu, L., Sclaroff, S., 2004. Deformable model-guided region split and merge of image regions. Image Vision Comput. 22 (4), 343–354.

Liu, M.L., Wong, K.H., 1999. Pose estimation using four corresponding points. Pattern Recogn. Lett. 20 (1), 69–74.

Liu, W., Wen, X.-Z., Duan, B., Yuan, H., Wang, N., 2007. Rear Vehicle Detection and Tracking for Lane Change Assist. Proc. IEEE Intelligent Vehicles Symposium, Istanbul, 13–15 June, pp. 252–257.

Lladós, J., Martí, E., Villanueva, J.J., 2001. Symbol recognition by error-tolerant subgraph matching between region adjacency graphs. IEEE Trans. Pattern Anal. Machine Intell. 23 (10), 1137–1143.

Lockton, R., Fitzgibbon, A., 2002. Real-time gesture recognition using deterministic boosting. Proceedings of the British Machine Vision Association Conference, Cardiff, UK, 2–5 September, pp. 817–826.

Longuet-Higgins, H.C., 1981. A computer algorithm for reconstructing a scene from two projections. Nature 293, 133–135.

Longuet-Higgins, H.C., 1984. The visual ambiguity of a moving plane. Proc. Roy. Soc. (London) B233, 165–175.

Longuet-Higgins, H.C., Prazdny, K., 1980. The interpretation of a moving retinal image. Proc. Roy. Soc. (London) B208, 385–397.

Lowe, D., 2004. Distinctive image features from scale-invariant keypoints. Int. J. Computer Vision 60, 91–110.

Lowe, D.G., 1999. Object recognition from local scale-invariant features. Proceedings of the Seventh International Conference on Computer Vision (ICCV), Corfu, Greece, pp. 1150–1157.

Lüdtke, N., Luo, B., Hancock, E., Wilson, R.C., 2002. Corner detection using a mixture model of edge orientation. Proceedings of the Sixteenth International Conference on Pattern Recognition, Québec, Canada, 11–15 August, vol. II, pp. 574–577.

Lukac, R., 2003. Adaptive vector median filtering. Pattern Recogn. Lett. 24 (12), 1889–1899.

Luo, B., Hancock, E.R., 2001. Structural graph matching using the EM algorithm and singular value decomposition. IEEE Trans. Pattern Anal. Machine Intell. 23 (10), 1120–1136.

Luong, Q.-T., Faugeras, O., 1997. Self-calibration of a moving camera from point correspondences and fundamental matrices. Int. J. Computer Vision 22 (3), 261–289.

Lutton, E., Maître, H., Lopez-Krahe, J., 1994. Contribution to the determination of vanishing points using Hough transform. IEEE Trans. Pattern Anal. Machine Intell 16 (4), 430–438.

Lyvers, E.R., Mitchell, O.R., 1988. Precision edge contrast and orientation estimation. IEEE Trans. Pattern Anal. Machine Intell. 10, 927–937.

Ma, L., Tan, T., Wang, Y., Zhang, D., 2003. Personal identification based on iris texture analysis. IEEE Trans. Pattern Anal. Machine Intell. 25 (12), 1519–1533.

Mackeown, W.P.J., Greenway, P., Thomas, B.T., Wright, W.A., 1994. Contextual image labelling with a neural network. IEE Proc. Vis. Image Signal Process. 141 (4), 238–244.

MacQueen, J.B., 1967. Some methods for classification and analysis of multivariate observations. Proceedings of the Fifth Berkeley Symposium on Mathematical Statistics and Probability, Berkeley, CA, USA, vol. I, pp. 281−297.

Magee, D.R., 2004. Tracking multiple vehicles using foreground, background and motion models. Image Vision Comput. 22, 143−155.

Magee, M.J., Aggarwal, J.K., 1984. Determining vanishing points from perspective images. Computer Vision Graph. Image Process. 26 (2), 256−267.

Mahesh, S., Manickavasagan, A., Jayas, D.S., Paliwal, J., White, N.D.G., 2008. Feasibility of near-infrared hyperspectral imaging to differentiate Canadian wheat classes. Biosystems Engineering 101, 50−57.

Mak, K.L., Peng, P., Yiu, K.F.C., 2009. Fabric defect detection using morphological filters. Image Vision Comput. 27, 1585−1592.

Makris, D., Ellis, T., Black, J., 2004. Bridging the Gaps between Cameras. Proceedings of the IEEE Conference on Computer Vision and Pattern Recognition, Washington DC, pp. 205−210.

Manickavasagan, A., Jayas, D.S., White, N.D.G., 2008. Thermal imaging to detect infestation by Cryptolestes ferrugineus inside wheat kernels. J. Stored Products Research 44, 186−192.

Manthalkar, R., Biswas, P.K., Chatterji, B.N., 2003. Rotation invariant texture classification using even symmetric Gabor filters. Pattern Recogn. Lett. 24 (12), 2061−2068.

Marchant, J.A., 1996. Tracking of row structure in three crops using image analysis. Comput. Electron. Agric. 15, 161−179.

Marchant, J.A., Brivot, R., 1995. Real-time tracking of plant rows using a Hough transform. Real-Time Imaging 1 (5), 363−371.

Marchant, J.A., Onyango, C.M., 1995. Fitting grey level point distribution models to animals in scenes. Image Vision Comput. 13 (1), 3−12.

Marchant, J.A., Tillett, R.D., Brivot, R., 1998. Real-time segmentation of plants and weeds. Real-Time Imaging 4, 243−253.

Marino, F., Stella, E., Branca, A., Veneziani, N., Distante, A., 2001. Specialized hardware for real-time navigation. Real-Time Imaging 7 (1), 97−108.

Marr, D., 1976. Early processing of visual information. Phil. Trans. Roy. Soc. (London) B275, 483−524.

Marr, D., Hildreth, E., 1980. Theory of edge detection. Proc. Roy. Soc. (London) B207, 187−217.

Marr, D., Poggio, T., 1979. A computational theory of human stereo vision. Proc. Roy. Soc. (London) B204, 301−328.

Marshall, S., 2004. New direct design method for weighted order statistic filters. IEE Proc. Vis. Image Signal Process. 151 (1), 1−8.

Marshall, S., Harvey, N., Shah, D. (Eds.), 1998. Proc. Noblesse Workshop on Non-linear Model Based Image Analysis, Springer-Verlag, Glasgow, 1−3 July, London.

Marslin, R.F., Sullivan, G.D., Baker, K.D., 1991. Kalman filters in constrained model based tracking. Proceedings of the Second British Machine Vision Association Conference, Glasgow, UK, 23−26 September, pp. 371−374.

Marzotto, R., Zoratti, P., Bagni, D., Colombari, A., Murino, V., 2010. A real-time versatile roadway path extraction and tracking on an FPGA platform. Comput. Vis. Image Underst. 114, 1164−1179.

Mastorakis, G., Davies, E.R., 2011. Improved line detection algorithm for locating road lane markings. Electronics Lett. 47 (3), 183−184.

Matas, J., Chum, O., Urban, M., Pajdla, T., 2002. Robust wide baseline stereo from maximally stable extremal regions. Proceedings of the British Machine Vision Conference (BMVC), Cardiff University, UK, pp. 384−393.

May, M., Turner, M.J., Morris, T., 2010. Scale invariant feature transform: a graphical parameter analysis. Proceedings of the British Machine Vision Association Conference UK Postgraduate Workshop, pp. 5.1−5.11.

Maybank, S., 1992. Theory of reconstruction from image motion. Springer-Verlag, Berlin, Heidelberg.

Maybank, S., Tan, T. (Eds.), 2004. Special Issue: Visual Surveillance. Image Vision Comput. 22 (7), 515−582.

Maybank, S.J., 1986. Algorithm for analysing optical flow based on the least squares method. Image Vision Comput. 4, 38−42.

Maybank, S.J., 1996. Stochastic properties of the cross ratio. Pattern Recogn. Lett. 17 (3), 211−217.

Maybank, S.J., Faugeras, O., 1992. A theory of self-calibration of a moving camera. Int. J. Computer Vision 8 (2), 123−151.

Maybeck, P.S., 1979. Stochastic Models, Estimation, and Control, Volume 1. Academic Press, New York and London.

McFarlane, N.J.B., Schofield, C.P., 1995. Segmentation and tracking of piglets in images. Machine Vis. Appl. 8 (3), 187−193.

McGunnigle, G., Chantler, M., 2003. Resolving handwriting from background printing using photometric stereo. Pattern Recogn. 36, 1869−1879.

McGunnigle, G., Dong, J., 2011. Augmenting photometric stereo with coaxial illumination. IET Comput. Vis. 5 (1), 33−49.

McLoughlin, S., Deegan, C., Mulvihill, C., Fitzgerald, C., Markham, C., 2008. Mobile mapping for the automated analysis of road signage and delineation. IET Intelligent Transport Systems 2 (1), 61−73.

Medeiros, H., Holguín, G., Shin, P.J., Park, J., 2010. A parallel histogram-based particle filter for object tracking on SIMD-based smart cameras. Comput. Vis. Image Underst. 114, 1264−1272.

Medina-Carnicer, R., Muñoz-Salinas, R., Carmona-Poyato, A., Madrid-Cuevas, F.J., 2011. A novel histogram transformation to improve the performance of thresholding methods in edge detection. Pattern Recogn. Lett. 32 (5), 676−693.

Meer, P., Georgescu, B., 2001. Edge detection with embedded confidence. IEEE Trans. Pattern Anal. Machine Intell. 23 (12), 1351−1365.

Meer, P., Mintz, D., Rosenfeld, A., 1990. Least median of squares based robust analysis of image structure. Proceedings of the DARPA Image Understanding Workshop, Pittsburgh, PA, 11−13 September, pp. 231−254.

Meer, P., Mintz, D., Rosenfeld, A., Kim, D.Y., 1991. Robust regression methods for computer vision: a review. Int. J. Computer Vision 6 (1), 59−70.

Méler, A., Decrouez, M., Crowley, J., 2010. BetaSAC: A new conditional sampling for RANSAC. Proceedings of the British Machine Vision Association Conference, Aberystwyth, UK, 31 August−3 September, pp. 42.1−42.11.

Mémin, É., Risset, T., 2001. VLSI design methodology for edge-preserving image reconstruction. Real-Time Imaging 7 (1), 109−126.

Meribout, M., Nakanishi, M., Ogura, T., 2002. Accurate and real-time image processing on a new PC-compatible board. Real-Time Imaging 8 (1), 35−51.

Merlin, P.M., Farber, D.J., 1975. A parallel mechanism for detecting curves in pictures. IEEE Trans. Comput. 28, 96–98.

Mikić, I., Kogut, P.C.G., Trivedi, M., 2000. Moving shadow and object detection in traffic scenes. Proceedings of the International Conference on Pattern Recognition, Barcelona, Spain, 3–8 September, pp. 321–324.

Mikolajczyk, K., 2002. Interest point detection invariant to affine transformations. PhD Thesis, Institut National Polytechnique de Grenoble (INPG), France.

Mikolajczyk, K. and Schmid, C., 2002. An affine invariant interest point detector. Proceedings of the European Conference on Computer Vision (ECCV), Copenhagen, Denmark, pp. 128–142.

Mikolajczyk, K., Schmid, C., 2004. Scale and affine invariant interest point detectors. Int. J. Computer Vision 60 (1), 63–86.

Mikolajczyk, K., Schmid, C., 2005. A performance evaluation of local descriptors. IEEE Trans. Pattern Anal. Machine Intell. 27 (10), 1615–1630.

Mikolajczyk, K., Tuytelaars, T., Schmid, C., Zisserman, A., Matas, J., Schaffalitzky, F., Kadir, T., Van Gool, L., 2005. A comparison of affine region detectors. Int. J. Computer Vision 65, 43–72.

Min, T.-H., Park, R.-H., 2009. Eyelid and eyelash detection method in the normalized iris image using the parabolic Hough model and Otsu's thresholding method. Pattern Recogn. Lett 30, 1138–1143.

Minsky, M., Papert, S., 1971. On some associative, parallel and analog computations. In: Jacks, E.J. (Ed.), Associative Information Techniques. Elsevier, New York, pp. 27–47.

Minsky, M.L., Papert, S.A., 1969. Perceptrons. MIT Press, Cambridge, MA.

Mirmehdi, M., Petrou, M., 2000. Segmentation of colour textures. IEEE Trans. Pattern Anal. Machine Intell. 22 (2), 142–159.

Mirmehdi, M., Xie, X., Suri, J. (Eds.), 2008. Handbook of Texture Analysis. Imperial College Press, London.

Mishra, A.K., Fieguth, P.W., Clausi, D.A., 2011. Decoupled active contour (DAC) for boundary detection. IEEE Trans. Pattern Anal. Machine Intell. 33 (2), 310–324.

Mohr, R., Wu, C. (Eds.), 1998. Special Issue on Geometric Modelling and Invariants for Computer Vision, Image Vision Computing 16, no. 1.

Mokhtarian, F., Bober, M., 2003. Curvature Scale Space Representation: Theory, Applications and MPEG-7 Standardisation, Kluwer Academic Publishers, Dordrecht.

Mokhtarian, F., Abbasi, S., Kittler, J., 1996. Efficient and robust shape retrieval by shape content through curvature scale space. Proceedings of the First International Conference Image Database and Multi-Search, Amsterdam, August, pp. 35–42.

Montiel, E., Aguado, A.S., Nixon, M.S., 2001. Improving the Hough transform gathering process for affine transformations. Pattern Recogn. Lett. 22 (9), 959–969.

Moravec, H.P., 1977. Towards automatic visual obstacle avoidance. Proceedings of the Fifth International Joint Conference on Artificial Intelligence, Cambridge, MA, 22–25 August, pp. 584.

Moravec, H.P., 1980. Obstacle avoidance and navigation in the real world by a seeing robot rover. Stanford Artificial Intelligence Lab. Memo AIM-340.

Mori, G., Belongie, S., Malik, J., 2005. Efficient shape matching using shape contexts. IEEE Trans. Pattern Anal. Machine Intell. 27 (11), 1832–1837.

Môtus, L., Rodd, M.G., 1994. Timing Analysis of Real-Time Software. Elsevier, Oxford.

Mundy, J.L., Zisserman, A., 1992b. Appendix – projective geometry for machine vision. In: Mundy, J.L., Zisserman, A. (Eds.), op. cit., pp. 463–519.

Mundy, J.L., Zisserman, A. (Eds.), 1992a. Geometric Invariance in Computer Vision, MIT Press, Cambridge, MA.

Myatt, D.R., Torr, P.H.S., Nasuto, S.J., Bishop, J.M., Craddock, R., 2002. Napsac: High noise, high dimensional robust estimation − it's in the bag. Proceedings of the British Machine Vision Association Conference, Cardiff, UK, 2−5 September, pp. 458−467.

Nagao, M., Matsuyama, T., 1979. Edge preserving smoothing. Computer Graph. Image Process. 9, 394−407.

Nagel, H.-H., 1983. Displacement vectors derived from second-order intensity variations in image sequences. Computer Vision Graph. Image Process 21, 85−117.

Nagel, H.-H., 1986. Image sequences − ten (octal) years − from phenomenology towards a theoretical foundation. Proceedings of the Eighth International Conference on Pattern Recognition, Paris, 27−31 October, pp. 1174−1185.

Nagel, R.N., Rosenfeld, A., 1972. Ordered search techniques in template matching. Proc. IEEE 60, 242−244.

Nait-Charif, H., McKenna, S.J., 2004. Tracking poorly modelled motion using particle filters with iterated likelihood weighting. Proceedings of the Asian Conference on Computer Vision, Jeju Island, Korea, pp. 156−161.

Nait-Charif, H., McKenna, S.J., 2006. Tracking the activity of participants in a meeting. Machine Vis. Appl. 17 (2), 83−93.

Nakagawa, Y., 1982. Automatic visual inspection of solder joints on printed circuit boards. Proc. SPIE, Robot Vision 336, 121−127.

Nakagawa, Y., Rosenfeld, A., 1979. Some experiments on variable thresholding. Pattern Recogn. 11, 191−204.

Narendra, P.M., 1978. A separable median filter for image noise smoothing. Proceedings of the IEEE Computer Society Conference on Pattern Recognition and Image Process., Chicago, IL, 31 May−2 June, pp. 137−141.

Neethirajan, S., Jayas, D.S., White, N.D.G., 2007. Detection of sprouted wheat kernels using soft X-ray image analysis. J. Food Eng. 81, 509−513.

Nesi, P., Trucco, E. (Eds.), 1999. Special Issue on Real-Time Defect Detection. Real-Time Imaging 5 (1), 1−74.

Newman, T.S., Jain, A.K., 1995. A survey of automated visual inspection. Comput. Vis. Image Underst. 61 (2), 231−262.

Ng, H.-F., 2006. Automatic thresholding for defect detection. Pattern Recogn. Lett. 27 (14), 1644−1649.

Ni, K., Jin, H., Dellaert, F., 2009. GroupSAC: efficient consensus in the presence of groupings. Proceedings of the International Conference on Computer Vision, Kyoto, Japan, 27 September− 4 October, pp. 2193−2200.

Niblack, W., 1985. An Introduction to Digital Image Processing. Strandberg, Birkeroed.

Nieminen, A., Heinonen, P., Neuvo, Y., 1987. A new class of detail-preserving filters for image processing. IEEE Trans. Pattern Anal. Machine Intell. 9, 74−90.

Nilsson, N.J., 1965. Learning Machines—Foundations of Trainable Pattern-Classifying Systems, McGraw-Hill, New York.

Nitzan, D., Brain, A.E., Duda, R.O., 1977. The measurement and use of registered reflectance and range data in scene analysis. Proc. IEEE 65, 206−220.

Nixon, M., 1985. Application of the Hough transform to correct for linear variation of background illumination in images. Pattern Recogn. Lett. 3, 191−194.

Nixon, M., Aguado, A., 2008. Feature Extraction and Image Processing. 2nd ed. Academic Press, Oxford, UK.

Nixon, M.S., Tan, T.N., Chellappa, R., 2006. Human Identification Based on Gait. Springer, New York.

Noble, A., Hartley, R., Mundy, J., Farley, J., 1994. X-ray metrology for quality assurance. Proceedings of the IEEE International Conference on Robotics and Automation, vol. 2, San Diego, CA, May, pp. 1113–1119.

Noble, J.A., 1988. Finding corners. Image Vision Comput. 6, 121–128.

Noble, J.A., 1995. From inspection to process understanding and monitoring: a view on computer vision in manufacturing. Image Vision Comput. 13 (3), 197–214.

North, D.O., 1943. An analysis of the factors which determine signal/noise discrimination in pulsed-carrier systems. RCA Lab., Princeton, NJ, Rep. PTR-6C; reprinted in Proc. IEEE 51, 1016–1027 (1963).

Noyce, R.N., 1977. Microelectronics. Scientific Amer. 237 September, 62–69.

Nummiaro, K., Koller-Meier, E., Van Gool., L., 2003. An adaptive color-based particle filter. Image Vision Comput. 21 (1), 99–110.

O'Gorman, F., 1978. Edge detection using Walsh functions. Artif. Intell. 10, 215–223.

O'Gorman, F., Clowes, M.B., 1976. Finding picture edges through collinearity of feature points. IEEE Trans. Comput. 25, 449–456.

Offen, R.J. (Ed.), 1985. VLSI Image Processing. Collins, London.

Ohanian, P.P., Dubes, R.C., 1992. Performance evaluation for four classes of textural features. Pattern Recogn. 25, 819–833.

Ohta, Y., Maenobu, K., Sakai, T., 1981. Obtaining surface orientation from texels under perspective projection. Proceedings of the Seventh International Joint Conference on Artificial Intelligence, Vancouver, Canada, pp. 746–751.

Oja, E., 1982. A simplified neuron model as a principal component analyzer. Int. J. Neural Systems 1, 61–68.

Ojala, T., Pietikäinen, M., Mäenpää, T., 2002. Multiresolution gray-scale and rotation-invariant texture classificaiton with local binary patterns. IEEE Trans. Pattern Anal. Machine Intell. 24 (7), 971–987.

Olague, G., Hernández, B., 2002. Flexible model-based multi-corner detector for accurate measurements and recognition. Proceedings of the Sixteenth International Conference on Pattern Recognition, Québec, Canada, 11–15 August, vol. II, pp. 578–583.

Olson, C.F., 1998. Improving the generalized Hough transform through imperfect grouping. In: Davies, E.R., Atiquzzaman, M. (Eds.), Special Issue on Projection-Based Transforms, Image Vision Computing 16, nos. 9–10, 627–634.

Olson, C.F., 1999. Constrained Hough transforms for curve detection. Comput. Vis. Image Underst. 73 (3), 329–345.

Onyango, C.M., Marchant, J.A., 1996. Modelling grey level surfaces using three-dimensional point distribution models. Image Vision Comput. 14, 733–739.

Ooms, D., Palm, R., Leemans, V., Destain, M.-F., 2010. A sorting optimization curve with quality and yield requirements. Pattern Recogn. Lett. 31 (9), 983–990.

Osteen, R.E., Tou, J.T., 1973. A clique-detection algorithm based on neighbourhoods in graphs. Int. J. Comput. Inf. Sci. 2, 257–268.

Otsu, N., 1979. A threshold selection method from gray-level histograms. IEEE Trans. Systems Man. Cybern. 9 (1), 62–66.

Ouerhani, N., Hügli, H., 2003. Real-time visual attention on a massively parallel SIMD architecture. Real-Time Imaging 9 (3), 189–196.

Pal, N.R., Pal, S.K., 1989. Object-background segmentation using new definitions of entropy. IEE Proc. E 136 (4), 284–295.

Paler, K., Kittler, J., 1983. Greylevel edge thinning: a new method. Pattern Recogn. Lett. 1, 409–416.

Paler, K., Föglein, J., Illingworth, J., Kittler, J., 1984. Local ordered grey levels as an aid to corner detection. Pattern Recogn. 17, 535–543.

Pan, X.-B., Brady, M., Bowman, A.K., Crowther, C., Tomlin, R.S.O., 2004. Enhancement and feature extraction for images of incised and ink texts. Image Vision Comput. 22 (6), 443–451.

Pan, X.D., Ellis, T.J., Clarke, T.A., 1995. Robust tracking of circular features. Proceedings of the Sixth British Machine Vision Association Conference, Birmingham, UK, 11–14 September, pp. 553–562.

Panda, D.P., Rosenfeld, A., 1978. Image segmentation by pixel classification in (gray level, edge value) space. IEEE Trans. Comput. 27, 875–879.

Papadakis, N., Bugeau, A., 2011. Tracking with occlusions via graph cuts. IEEE Trans. Pattern Anal. Machine Intell. 33 (1), 144–157.

Paragios, N., Deriche, R., 2000. Geodesic active contours and level sets for the detection and tracking of moving objects. IEEE Trans. Pattern Anal. Machine Intell. 22 (3), 266–280.

Parker, D.B., 1985. Learning-logic: casting the cortex of the human brain in silicon. Technical Report TR-47, Center for Computer Research in Economics and Management Sci., MIT Press, Cambridge, MA.

Parker, J.R., 1994. Practical Computer Vision Using C, Wiley, New York.

Patel, D., Davies, E.R., Hannah, I., 1995. Towards a breakthrough in the detection of contaminants in food products. Sensor Review 15 (2), 27–28.

Patel, D., Davies, E.R., Hannah, I., 1996. The use of convolution operators for detecting contaminants in food images. Pattern Recogn. 29 (6), 1019–1029.

Patel, D., Hannah, I., Davies, E.R., 1994. Texture analysis for foreign object detection using a single layer neural network. Proceedings of the IEEE International Conference on Neural Networks, Florida, 28 June–2 July, vol. VII, pp. 4265–4268.

Pavlidis, T., 1980. Algorithms for shape analysis of contours and waveforms. IEEE Trans. Pattern Anal. Machine Intell. 2, 301–312.

Pearl, J., 1988. Probabilistic Reasoning in Intelligent Systems: Networks of Plausible Inference. Morgan Kaufmann, San Mateo, CA.

Pelillo, M., 1999. Replicator equations, maximal cliques and graph isomorphism. Neural Computation 11 (8), 1933–1955.

Penman, D., Olsson, O., Beach, D., 1992. Automatic X-ray inspection of canned products for foreign material. Machine Vision Applications, Architectures and Systems Integration. SPIE 1823, 342–347.

Pentland, A.P., 1984. Fractal-based description of natural scenes. IEEE Trans. Pattern Anal. Machine Intell. 6 (6), 661–674.

Perdoch, M., Matas, J., Obdrzalek, S., 2007. Stable affine frames on isophotes. Proceedings of the International Conference on Computer Vision (ICCV), Rio de Janeiro, Brazil, 14–20 October, [8 pp.].

Persoon, E., Fu, K.-S., 1977. Shape discrimination using Fourier descriptors. IEEE Trans. Systems Man Cybern 7, 170–179.

Petrou, M., Kittler, J., 1988. On the optimal edge detector. Proceedings of the Fourth Alvey Vision Conference, Manchester, UK, 31 August–2 September, pp. 191–196.

Petrou, M., Petrou, C., 2010. Image Processing: The Fundamentals. 2nd ed. Wiley, Ltd, Chichester, UK.

Petrou, M., Sevilla, P.G., 2006. Image Processing: Dealing With Texture. Wiley, Chichester, UK.

Pfaltz, J.L., Rosenfeld, A., 1967. Computer representation of planar regions by their skeletons. Comm. ACM 10, 119—125.

Pflugfelder, R., Bischof, H., 2008. Tracking across non-overlapping views via geometry. Proceedings of the IEEE International Conference on Pattern Recognition, Tampa, Florida, USA, 8—11 December, [4 pp.].

Pflugfelder, R., Bischof, H., 2010. Localization and Trajectory Reconstruction in Surveillance Cameras with Nonoverlapping Views. IEEE Trans. Pattern Anal. Machine Intell. 32 (4), 709—721.

Phong, B.-T., 1975. Illumination for computer-generated pictures. Comm. ACM 18, 311—317.

Pietikäinen, M., Rosenfeld, A., Davis, L.S., 1983. Experiments with texture classification using averages of local pattern matches. IEEE Trans. Systems Man. Cybern. 13 (3), 421—426.

Pitt, M.K., Shephard, N., 1999. Filtering via simulation: auxiliary particle filters. J. Amer. Statistical Assoc. 94 (446), 590—599.

Plummer, A.P.N., Dale, F., 1984. The Picture Processing Language Compiler Manual. National Physical Laboratory, Teddington.

Pollard, S.B., Porrill, J., Mayhew, J.E.W., Frisby, J.P., 1987. Matching geometrical descriptions in three-space. Image Vision Comput. 5 (2), 73—78.

Postaire, J.G., Touzani, A., 1989. Mode boundary detection by relaxation for cluster analysis. Pattern Recogn. 22 (5), 477—489.

Prati, A., Cucchiara, R., Mikić, I., Trivedi, M., 2001. Analysis and detection of shadows in video streams: a comparative evaluation. Proceedings of the IEEE International Conference on Computer Vision and Pattern Recognition, Kauai, Hawaii, USA, 11—13 December, vol. 2, pp. 571—576.

Prati, A., Mikić, I., Trivedi, M.M., Cucchiara, R., 2003. Detecting moving shadows: algorithms and evaluation. IEEE Trans. Pattern Anal. Machine Intell. 25 (7), 918—923.

Pratt, W.K., 2001. Digital Image Processing. 2nd ed. Wiley-Interscience.

Press, W.H., Teukolsky, S.A., Vetterling, W.T., Flannery, B.P., 1992. Numerical Recipes in C. The Art of Scientific Computing. 2nd ed., Cambridge Univ. Press, Cambridge.

Prewitt, J.M.S., 1970. Object enhancement and extraction. In: Lipkin, B.S., Rosenfeld, A. (Eds.), Picture Processing and Psychopictorics. Academic Press, New York, pp. 75—149.

Prieto, M.S., Allen, A.R., 2003. A similarity metric for edge images. IEEE Trans. Pattern Anal. Machine Intell. 25 (10), 1265—1273.

Prieto, M.S., Allen, A.R., 2009. Using self-organising maps in the detection and recognition of road signs. Image Vision Comput. 27, 673—683.

Princen, J., Illingworth, J., Kittler, J., 1989a. A hierarchical approach to line extraction. Proceedings of the IEEE Computer Vision and Pattern Recognition Conference, San Diego, CA, pp. 92—97.

Princen, J., Illingworth, J., Kittler, J., 1994. Hypothesis testing: a framework for analyzing and optimizing Hough transform performance. IEEE Trans. Pattern Anal. Machine Intell. 16 (4), 329—341.

Princen, J., Yuen, H.K., Illingworth, J., Kittler, J., 1989b. Properties of the adaptive Hough transform. Proceedings of the Sixth Scandinavian Conference on Image Analysis, Oulu, Finland, 19—22 June, pp. 613—620.

Pringle, K.K., 1969. Visual perception by a computer. In: Grasselli, A. (Ed.), Automatic Interpretation and Classification of Images. Academic Press, New York, pp. 277–284.

Pritchard, D., Heidrich, W., 2003. Cloth motion capture. Computer Graphics Forum (Eurographics 2003) 22 (3), 263–271.

Pugh, A. (Ed.), 1983. Robot Vision. IFS (Publications) Ltd., Bedford/Berlin/Springer-Verlag.

Pun, C.-M., Lee, M.-C., 2003. Log-polar wavelet energy signatures for rotation and scale invariant texture classification. IEEE Trans. Pattern Anal. Machine Intell. 25 (5), 590–603.

Pun, T., 1980. A new method for grey-level picture thresholding using the entropy of the histogram. Signal Processing 2, 223–237.

Pun, T., 1981. Entropic thresholding, a new approach. Computer Graph. Image Process. 16, 210–239.

Qin, J., Lu, R., 2008. Measurement of the optical properties of fruits and vegetables using spatially resolved hyperspectral diffuse reflectance imaging technique. Postharvest Biology and Technology 49, 355–365.

Rabah, H., Mathias, H., Weber, S., Mozef, E., Tanougast, C., 2003. Linear array processors with multiple access modes memory for real-time image processing. Real-Time Imaging 9 (3), 205–213.

Rabbani, H., Gazor, S., 2010. Image denoising employing local mixture models in sparse domains. IET Image Process. 4 (5), 413–428.

Rajashekhar, Chaudhuri, S., Namboodiri, V.P., 2007. Retrieval of images of man-made structures based on projective invariance. Pattern Recogn. 40, 296–308.

Ramanan, D., 2006. Learning to parse images of articulated bodies. Proceedings of the Neural Information Processing Systems Conference, Vancouver, British Columbia, Canada, 4–7 December, pp. 1129–1136.

Rätsch, G., Mika, S., Schölkopf, B., Müller, K.-R., 2002. Constructing boosting algorithms from SVMs: an application to one-class classification. IEEE Trans. Pattern Anal. Machine Intell. 24 (9), 1184–1199.

Reed, T.R., du Buf, J.M.H., 1993. A review of recent texture segmentation and feature extraction techniques. Computer Vision Graph. Image Process.: Image Underst. 57, 359–372.

Reeves, A.P., 1984. Parallel computer architectures for image processing. Computer Vision Graph. Image Process. 25, 68–88.

Reeves, A.P., Akey, M.L., Mitchell, O.R., 1983. A moment-based two-dimensional edge operator. Proceedings of the IEEE Computer Society Conference on Computer Vision and Pattern Recognition, Arlington, VA, 19–23 June, pp. 312–317.

Ren, J., Jiang, J., Wang, D., Ipson, S.S., 2010. Fusion of intensity and inter-component chromatic difference for effective and robust colour edge detection. IET Image Process. 4 (4), 294–301.

Ridgway, C., Chambers, J., 1996. Detection of external and internal insect infestation in wheat by near-infrared reflectance spectroscopy. J. Sci. Food Agric. 71, 251–264.

Ridgway, C., Davies, E.R., Chambers, J., Mason, D.R., Bateman, M., 2002. Rapid machine vision method for the detection of insects and other particulate bio-contaminants of bulk grain in transit. Biosyst. Eng. 83 (1), 21–30.

Rindfleisch, T., 1966. Photometric method for lunar topography. Photogrammetr. Eng 32, 262–276.

Ringer, M., Lazenby, J., 2000. Modelling and tracking articulated motion from multiple camera views. Proceedings of the Eleventh British Machine Vision Association Conference, Bristol, UK, 11–14 September, pp. 172–181.

Rish, I., 2001. An empirical study of the naive Bayes classifier. Proceedings of the IJCAI-01 Workshop on Empirical Methods in Artificial, Intelligence, International Joint Conference on Artificial Intelligence, Seattle, Washington, USA, 4–10 August, pp. 41–46.

Robert, L., 1996. Camera calibration without feature extraction. Comput. Vis. Image Underst. 63 (2), 314–325.

Roberts, L.G., 1965. Machine perception of three-dimensional solids. In: Tippett, J., et al., (Eds.), Optical and Electro-optical Information Processing, MIT Press, Cambridge, MA, pp. 159–197.

Robinson, G.S., 1977. Edge detection by compass gradient masks. Computer Graph. Image Process. 6, 492–501.

Robles-Kelly, A., Hancock, E.R., 2002. A graph-spectral approach to correspondence matching. Proceedings of the Sixteenth International Conference on Pattern Recognition, Québec, Canada, 11–15 August, vol. IV, pp. 176–179.

Rocket, P.I., 2003. Performance assessment of feature detection algorithms: a methodology and case study of corner detectors. IEEE Trans. Image Process. 12 (12), 1668–1676.

Rodríguez, J.J., García-Osorio, C., Maudes, J., 2010. Forests of nested dichotomies. Pattern Recogn. Lett. 31, 125–132.

Rogers, D.F., 1985. Procedural Elements for Computer Graphics. McGraw-Hill, New York.

Roh, Y.J., Park, W.S., Cho, H., 2003. Correcting image distortion in the X-ray digital tomosynthesis system for PCB solder joint inspection. Image Vis. Comput. 21 (12), 1063–1075.

Rosenblatt, F., 1958. The perceptron: a probabilistic model for information storage and organisation in the brain. Psychol. Rev. 65, 386–408.

Rosenblatt, F., 1962. Principles of Neurodynamics. Spartan, New York.

Rosenfeld, A., 1969. Picture Processing by Computer. Academic Press, New York.

Rosenfeld, A., 1970. Connectivity in digital pictures. J. ACM 17, 146–160.

Rosenfeld, A., Pfaltz, J.L., 1966. Sequential operations in digital picture processing. J. ACM 13, 471–494.

Rosenfeld, A., Pfaltz, J.L., 1968. Distance functions on digital pictures. Pattern Recogn. 1, 33–61.

Rosenfeld, A., Troy, E.B., 1970a. Visual texture analysis. Computer Science Center, University of Maryland Technical Report TR-116.

Rosenfeld, A., Troy, E.B., 1970b. Visual texture analysis. Conference Record for Symposium on Feature Extraction and Selection in Pattern Recognition, IEEE Publication 70C-51C, Argonne, IL, October, pp. 115–124.

Rosenfeld, A., VanderBrug, G.J., 1977. Coarse-fine template matching. IEEE Trans. Syst., Man Cybern. 7, 104–107.

Rosenfeld, A., de la Torre, P., 1983. Histogram concavity analysis as an aid in threshold selection. IEEE Trans. Syst., Man Cybern. 13 (3), 231–235.

Rosenfeld, A., Hummel, R.A., Zucker, S.W., 1976. Scene labelling by relaxation operations. IEEE Trans. Syst., Man Cybern. 6, 420–433.

Rosie, A.M., 1966. Information and Communication Theory. Blackie, London.

Rosin, P., 2000. Fitting superellipses. IEEE Trans. Pattern Anal. Machine Intell. 22 (7), 726–732.

Rosin, P., Ellis, T.J., 1995. Image difference threshold strategies and shadow detection. Proceedings of the British Machine Vision Association Conference, Birmingham, UK, 11–14 September, pp. 347–356.

Rosin, P.L., 2001. Unimodal thresholding. Pattern Recogn. 34 (11), 2083–2096.

Rosin, P.L., West, G.A.W., 1995. Curve segmentation and representation by superellipses. IEE Proc. Vis. Image Signal Process. 142 (5), 280–288.

Rosten, E., Porter, R., Drummond, T., 2010. Faster and better: a machine learning approach to corner detection. IEEE Trans. Pattern Anal. Machine Intell. 32 (1), 105–119.

Roth, G., Whitehead, A., 2002. Some improvements on two autocalibration algorithms based on the fundamental matrix. Proceedings of the Sixteenth International Conference on Pattern Recognition, Québec, Canada, 11–15 August, vol. II, pp. 312–315.

Rother, C., Carlsson, S., 2001. Linear multi view reconstruction and camera recovery. Proceedings of the Eighth IEEE International Conference on Computer Vision, Vancouver, Canada, 7–14 July, Vol. 1, pp. 42–50.

Rothwell, C.A., 1995. Object Recognition through Invariant Indexing. Oxford University Press, Oxford.

Rothwell, C.A., Zisserman, A., Forsyth, D.A., Mundy, J.L., 1992a. Canonical frames for planar object recognition. Proceedings of the Second European Conference on Computer Vision, Santa Margherita Ligure, Italy, 19–22 May, pp. 757–772.

Rothwell, C.A., Zisserman, A., Marinos, C.I., Forsyth, D.A., Mundy, J.L., 1992b. Relative motion and pose from arbitrary plane curves. Image Vis. Comput. 10 (4), 250–262.

Rousseeuw, P.J., 1984. Least median of squares regression. J. Am. Stat. Assoc. 79 (388), 871–880.

Rousseeuw, P.J., Leroy, A.M., 1987. Robust Regression and Outlier Detection. Wiley, New York.

Rowley, H., Baluja, S., Kanade, T., 1998. Neural network-based face detection. IEEE Trans. Pattern Anal. Machine Intell. 20 (1), 23–38.

Roy, P.P., Pal, U., Lladós, J., 2011. Document seal detection using GHT and character proximity graphs. Pattern Recogn. 44, 1282–1295.

Rumelhart, D.E., Hinton, G.E., Williams, R.J., 1986. Learning internal representations by error propagation. In: Rumelhart, D.E., McClelland, J.L. (Eds.), Parallel Distributed Processing: Explorations in the Microstructure of Cognition, MIT Press, Cambridge, MA, pp. 318–362.

Rummel, P., Beutel, W., 1984. Workpiece recognition and inspection by a model-based scene analysis system. Pattern Recogn. 17, 141–148.

Ruta, A., Li, Y., Liu, X., 2010. Real-time traffic sign recognition from video by class-specific discriminative features. Pattern Recogn. 43, 416–430.

Rutovitz, D., 1970. Centromere finding: some shape descriptors for small chromosome outlines. In: Meltzer, B., Michie, D. (Eds.), Machine Intelligence, Edinburgh University Press, Edinburgh, vol. 5, pp. 435–462.

Sahoo, P.K., Soltani, S., Wong, A.K.C., Chen, Y.C., 1988. A survey of thresholding techniques. Computer Vis. Graph. Image Process. 41, 233–260.

Sakarya, U., Erkmen, I., 2003. An improved method of photometric stereo using local shape from shading. Image Vis. Comput. 21 (11), 941–954.

Sanchiz, J.M., Pla, F., Marchant, J.A., Brivot, R., 1996. Structure from motion techniques applied to crop field mapping. Image Vis. Comput. 14, 353–363.

Sanfeliu, A., Fu, K.S., 1983. A distance measure between attributed relational graphs for pattern recognition. IEEE Trans. Syst., Man Cybern. 13 (3), 353−362.

Sangwine, S.J., Horne, R.E.N. (Eds.), 1998. The Colour Image Processing Handbook. Chapman and Hall, London.

Schaffalitsky, F., Zisserman, A., 2000. Planar grouping for automatic detection of vanishing lines and points. Image Vis. Comput. 18 (9), 647−658.

Schildt, H., 1995. C++, the Complete Reference. 3rd ed. McGraw-Hill, Osborne.

Schmid, C., Mohr, R., Bauckhage, C., 2000. Evaluation of interest point detectors. Int. J. Computer Vis. 37 (2), 151−172.

Schmidt, J., Fritsch, J., Kwolek, B., 2006. Kernel particle filter for real-time 3D body tracking in monocular color images, Proceedings of the IEEE International Conference on Automatic Face and Gesture Recognition, Southampton, UK, pp. 567−572.

Schneiderman, H., Nashman, M., Wavering, A.J., Lumia, R., 1995. Vision-based robotic convoy driving. Machine Vis. Appl. 8 (6), 359−364.

Scott, G.L., 1988. Local and Global Intepretation of Moving Images. Pitman and Morgan Kaufmann, London/San Mateo, CA.

Sebe, N., Lew, M.S., 2003. Comparing salient point detectors. Pattern Recogn. Lett. 24 (1−3), 89−96.

Sebe, N., Tian, Q., Loupias, E., Lew, M.S., Huang, T.S., 2003. Evaluation of salient point techniques. Image Vis. Comput. 21 (13−14), 1087−1095.

Semple, J.G., Kneebone, G.T., 1952. Algebraic Projective Geometry. Oxford University Press, Oxford.

Ser, P.-K., Siu, W.-C., 1995. Novel detection of conics using 2-D Hough planes. IEE Proc. Vis. Image Signal Process. 142 (5), 262−270.

Serra, J., 1982. Image Analysis and Mathematical Morphology. Academic Press, New York.

Sewisy, A.A., Leberl, F., 2001. Detection ellipses by finding lines of symmetry in the images via an hough transform applied to straight lines. Image Vis. Comput. 19 (12), 857−866.

Sezgin, M., Sankur, B., 2004. Survey over image thresholding techniques and quantitative performance evaluation. J. Electron. Imaging 13 (1), 146−168.

Shah, M.A., Jain, R., 1984. Detecting time-varying corners. Computer Vis. Graph. Image Process. 28, 345−355.

Shakespeare, W. (1599). The Tragedy of Julius Caesar. Act 1, Scene 3. Quotation confirmed from http://www.opensourceshakespeare.org/ (website accessed 14 December 2011).

Shapiro, L.G., Haralick, R.M., 1985. A metric for comparing relational descriptions. IEEE Trans. Pattern Anal. Machine Intell. 7 (1), 90−94.

Shen, F., Wang, H., 2002. Corner detection based on modified Hough transform. Pattern Recogn. Lett. 23 (8), 1039−1049.

Shen, X., Hogg, D., 1995. 3D shape recovery using a deformable model. Image Vis. Comput. 13 (5), 377−383.

Shima, T., Saito, S., Nakajima, M., 2010. Design and evaluation of more accurate gradient operators on hexagonal lattices. IEEE Trans. Pattern Anal. Machine Intell. 32 (6), 961−973.

Shioyama, T., Uddin, M.S., 2004. Detection of pedestrian crossings with projective invariants from image data. Measurement Sci. Technol. 15, 2400−2405.

Shirai, Y., 1972. Recognition of polyhedra with a range finder. Pattern Recogn. 4, 243–250.

Shirai, Y., 1987. Three-dimensional Computer Vision. Springer-Verlag, Berlin.

Shufelt, J.A., 1999. Performance evaluation and analysis of vanishing point detection techniques. IEEE Trans. Pattern Anal. Machine Intell. 21 (3), 282–288.

Shuster, R., Ansari, N., Bani-Hashemi, A., 1993. Steering a robot with vanishing points. IEEE Trans. Robot. Autom. 9 (4), 491–498.

Siebel, N.T., Maybank, S.J., 2002. Fusion of multiple tracking algorithms for robust people tracking. In Heyden, A., Sparr, G., Nielsen, M., Johansen, P. (Eds.) Proceedings of the Seventh European Conference on Computer Vision (ECCV), vol. IV, pp. 373–387.

Siegel, A.F., 1982. Robust regression using repeated medians. Biometrika 69 (1), 242–244.

Silberberg, T.M., Davies, L., Harwood, D., 1984. An iterative Hough procedure for three-dimensional object recognition. Pattern Recogn. 17, 621–629.

Silletti, A., Abate, A., Axelrod, J.D., Tomlin, C.J., 2011. Versatile spectral methods for point set matching. Pattern Recogn. Lett. 32 (5), 731–739.

Simard, P., Bottou, L., Haffner, P., LeCun, Y., 1999. Boxlets: a fast convolution algorithm for neural networks and signal processing, Advances in Neural Information Processing Systems, Vol. 11, MIT Press, Cambridge, MA.

Sjöberg, F., Bergholm, F., 1988. Extraction of diffuse edges by edge focussing. Pattern Recogn. Lett. 7, 181–190.

Sklansky, J., 1970. Recognition of convex blobs. Pattern Recogn. 2, 3–10.

Sklansky, J., 1978. On the Hough technique for curve detection. IEEE Trans. Comput. 27, 923–926.

Sklansky, J., Cordella, L.P., Levialdi, S., 1976. Parallel detection of concavities in cellular blobs. IEEE Trans. Comput. 25, 187–196.

Slama, C.C. (Ed.), 1980. Manual of photogrammetry. 4th ed. American Society of Photogrammetry, Falls Church, VA.

Smith, S., Brady, J.M., 1997. Susan − a new approach to low level image processing. Int. J. Computer Vis. 23 (1), 45–78.

Smolka, B., 2010. Peer group switching filter for impulse noise reduction in color images. Pattern Recogn. Lett. 31 (6), 484–495.

Sobey, P.J.M., 1989. The automated visual inspection and grading of timber. PhD Thesis, University of Adelaide.

Soille, P., 2003. Morphological Image Analysis: Principles and Applications. 2nd ed. Springer-Verlag, Heidelberg.

Soille, P., Vogt, P., 2009. Morphological segmentation of binary patterns. Pattern Recogn. Lett. 30 (4), 456–459.

Song, J., Cai, M., Lyu, M., Cai, S., 2002. A new approach for line recognition in large-size images using Hough transform. Proceedings of the Sixteenth International Conference on Pattern Recognition, Québec, Canada, 11–15 August, vol. I, pp. 33–36.

Sonka, M., Hlavac, V., Boyle, R., 2007. Image Processing, Analysis, and Machine Vision. 3rd ed. Thomson Engineering, Toronto.

Spann, M., Horne, C., du Buf, J.M.H., 1989. The detection of thin structures in images. Pattern Recogn. Lett. 10 (3), 175–179.

Spence, A., Robb, M., Timmins, M., Chantler, M., 2004. Real-time per-pixel rendering of textiles for virtual textile catalogues. Int. J. Clothing Sci. Technol. 16 (1/2).

Startchik, S., Milanese, R., Pun, T., 1998. Projective and illumination invariant representation of disjoint shapes. In Special Issue on Projection-Based Transforms. Image Vis. Comput. 16 (9–10), 713–723.

Steele, J.M., Steiger, W.L., 1986. Algorithms and complexity for least median of squares regression. Discrete Appl. Math. 14, 93–100.

Stella, E., Lovergine, F.P., D'Orazio, T., Distante, A., 1995. A visual tracking technique suitable for control of convoys. Pattern Recogn. Lett. 16 (9), 925–932.

Stephens, R.S., 1991. Probabilistic approach to the Hough transform. Image Vis. Comput. 9 (1), 66–71.

Stevens, K., 1980. Surface perception from local analysis of texture and contour. MIT Artificial Intelligence Laboratory Memo AI-TR-512.

Stockman, G.C., Agrawala, A.K., 1977. Equivalence of Hough curve detection to template matching. Commun. ACM 20, 820–822.

Straforini, M., Coelho, C., Campani, M., 1993. Extraction of vanishing points from images of indoor and outdoor scenes. Image Vis. Comput. 11 (2), 91–99.

Stroustrup, B., 1997. The C++ Programming Language, 3rd ed. Addison-Wesley, Reading, MA, USA.

Sturm, P., 2000. A case against Kruppa's equations for camera self-calibration. IEEE Trans. Pattern Anal. Machine Intell. 22 (10), 1199–1204.

Sugrue, M., Davies, E.R., 2007. Motion signals provide rapid discernment of pedestrians and pedestrian behaviour. Electron. Lett. 43 (23), 1267–1269.

Sugrue, M., Davies, E.R., 2008. Motion detection and tracking by mimicking neurological dorsal/ventral pathways. In: Bharath, A., Petrou, M. (Eds.), Reverse Engineering the Human Vision System: Next Generation Artificial Vision Systems, Artech House, Boston, USA; London, UK, pp. 217–247, Chapter 9

Sullivan, G.D., 1992. Visual interpretation of known objects in constrained scenes. Phil. Trans. R. Soc. (Lond.) B337, 361–370.

Sun, C., Berman, M., Coward, D., Osborne, B., 2007. Thickness measurement and crease detection of wheat grains using stereo vision. Pattern Recogn. Lett. 28 (12), 1501–1508.

Sun, T.-H., Tseng, C.-C., Chen, M.-S., 2010. Electric contacts inspection using machine vision. Image Vis. Comput. 28, 890–901.

Sun, Z., Bebis, G., Miller, R., 2004. On-road vehicle detection using optical sensors: a review. IEEE Int. Conf. Intell. Transport. Syst. 585–590.

Sun, Z., Bebis, G., Miller, R., 2006. On-road vehicle detection: a review. IEEE Trans. Pattern Anal. Machine Intell. 28 (5), 694–711.

Suzuki, K., Horiba, I., Sugie, N., 2003. Neural edge enhancer for supervised edge enhancement from noisy images. IEEE Trans. Pattern Anal. Machine Intell. 25 (12), 1582–1596.

Swain, M.J., Ballard, D.H., 1991. Color indexing. Int. J. Computer Vis. 7 (1), 11–32.

Tabandeh, A.S., Fallside, F., 1986. Artificial intelligence techniques and concepts for the integration of robot vision and 3D solid modellers. Proceedings of the International Conference on Intelligent Autonomous Systems, Amsterdam, 18–11 December, pp. 580–585.

Tan, T.N., 1995. Structure, pose and motion of bilateral symmetric objects. Proceedings of the Sixth British Machine Vision Association Conference, Birmingham, UK, 11–14 September, pp. 473–482.

Tan, T.N., Sullivan, G.D., Baker, K.D., 1994. Recognizing objects on the ground-plane. Image Vis. Comput. 12 (3), 164–172.

Tang, Y.Y., You, X., 2003. Skeletonization of ribbon-like shapes based on a new wavelet function. IEEE Trans. Pattern Anal. Machine Intell. 25 (9), 1118−1133.

Tao, W.-B., Tian, J.-W., Liu, J., 2003. Image segmentation by three-level thresholding based on maximum fuzzy entropy and genetic algorithm. Pattern Recogn. Lett. 24 (16), 3069−3078.

Teixeira, L.F., Corte-Real, L., 2009. Video object matching across multiple independent views using local descriptors and adaptive learning. Pattern Recogn. Lett. 30 (2), 157−167.

Theil, H., 1950. A rank-invariant method of linear and polynomial regression analysis (parts 1−3). Nederlandsche Akad. Wetenschappen Proc. A53, 386−392, 521−525 and 1397−1412.

Theodoridis, S., Koutroumbas, K., 1999. Pattern Recognition. Academic Press, London.

Thomas, A.D.H., Rodd, M.G., 1994. Knowledge-based inspection of electric lamp caps. Eng. Appl. Artif. Intell. 7 (1), 31−37.

Thomas, A.D.H., Rodd, M.G., Holt, J.D., Neill, C.J., 1995. Real-time industrial visual inspection: a review. Real-Time Imaging 1 (2), 139−158.

Tillett, R.D., Onyango, C.M., Marchant, J.A., 1997. Using model-based image processing to track animal movements. Comput. Electron. Agric. 17, 249−261.

Tissainayagam, P., Suter, D., 2004. Assessing the performance of corner detectors for point feature tracking applications. Image Vis. Comput. 22 (8), 663−679.

Toennies, K., Behrens, F., Aurnhammer, M., 2002. Feasibility of Hough-transform-based iris localisation for real-time application. Proceedings of the Sixteenth International Conference on Pattern Recognition, Québec, Canada, 11−15 August, vol. II, pp. 1053−1056.

Torr, P.H.S., Davidson, C., 2003. IMPSAC: synthesis of importance sampling and random sample consensus. IEEE Trans. Pattern Anal. Machine Intell. 25 (3), 354−364.

Torr, P.H.S., Fitzgibbon, A.W., 2003. Invariant fitting of two view geometry or In defiance of the 8 point algorithm. Proceedings of the British Machine Vision Association Conference, Norwich, UK, 9−11 September, pp. 83−92.

Torr, P.H.S., Fitzgibbon, A.W., 2004. Invariant fitting of two view geometry. IEEE Trans. Pattern Anal. Machine Intell. 26 (5), 648−650.

Torreão, J.R.A., 2001. A Green's function approach to shape from shading. Pattern Recogn. 34 (12), 2367−2382.

Torreão, J.R.A., 2003. Geometric−photometric approach to monocular shape estimation. Image Vision Comput. 21 (12), 1045−1062.

Toulson, D.L., Boyce, J.F., 1992. Segmentation of MR images using neural nets. Image Vis. Comput. 10 (5), 324−328.

Tsai, D.-M., 1995. A fast thresholding selection procedure for multimodal and unimodal histograms. Pattern Recogn. Lett. 16 (6), 653−666.

Tsai, D.-M., Huang, T.-Y., 2003. Automated surface inspection for statistical textures. Image Vis. Comput. 21 (4), 307−323.

Tsai, D.-M., Chang, C.-C., Chao, S.-M., 2010. Micro-crack inspection in heterogeneously textured solar wafers using anisotropic diffusion. Image Vis. Comput. 28, 491−501.

Tsai, F.C.D., 1996. A probabilistic approach to geometric hashing using line features. Comput. Vis. Image Underst. 63 (1), 182−195.

Tsai, R.Y., 1986. An efficient and accurate camera calibration technique for 3D machine vision. Proceedings of the Conference on Computer Vision Pattern Recognition, Miami, FL, pp. 364−374.

Tsai, R.Y., Huang, T.S., 1984. Uniqueness and estimation of three-dimensional motion parameters of rigid objects with curved surfaces. IEEE Trans. Pattern Anal. Machine Intell. 6, 13−27.

Tsuji, S., Matsumoto, F., 1978. Detection of ellipses by a modified Hough transform. IEEE Trans. Comput. 27, 777−781.

Tsukune, H., Goto, K., 1983. Extracting elliptical figures from an edge vector field. Proceedings of the IEEE Conference on Computer Vision and Pattern Recognition, Washington, DC, pp. 138−141.

Turin, G.L., 1960. An introduction to matched filters. IRE Trans. Inf. Theory 6, 311−329.

Turney, J.L., Mudge, T.N., Volz, R.A., 1985. Recognizing partially occluded parts. IEEE Trans. Pattern Anal. Machine Intell. 7, 410−421.

Tuytelaars, T., Mikolajczyk, K., 2008. Local invariant feature detectors: a survey. Foundations Trends Computer Graph. Vis. 3 (3), 177−280.

Tuytelaars, T., Van Gool, L., 2000. Wide baseline stereo matching based on local, affinely invariant regions. Proceedings of the British Machine Vision Conference (BMVC), Bristol University, UK, pp. 412−422.

Tuytelaars, T., Van Gool, L., 2004. Matching widely separated views based on affine invariant regions. Int. J. Computer Vis. 1 (59), 61−85.

Tuytelaars, T., Turina, A., Van Gool, L., 2003. Noncombinatorial detection of regular repetitions under perspective skew. IEEE Trans. Pattern Anal. Machine Intell. 25 (4), 418−432.

Tzionas, P., 2000. A cellular automaton processor for line and corner detection in grayscale images. Real-Time Imaging 6 (6), 461−470.

Tzomakas, C., von Seelen, W., 1998. Vehicle detection in traffic scenes using shadows. Internal report 98−06, Ruhr-Universität Bochum.

Ullman, S., 1979. The Interpretation of Visual Motion. MIT Press, Cambridge, MA.

Ullmann, J.R., 1969. Experiments with the n-tuple method of pattern recognition. IEEE Trans. Comput. 18, 1135−1137.

Ullmann, J.R., 1973. Pattern Recognition Techniques. Butterworth, London.

Ullmann, J.R., 1974. Binarisation using associative addressing. Pattern Recogn. 6, 127−135.

Ullmann, J.R., 1976. An algorithm for subgraph isomorphism. J. ACM 23, 31−42.

Ulusoy, I., Yuruk, H., 2011. New method for the fusion of complementary information from infrared and visual images for object detection. IET Image Process. 5 (1), 36−48.

Umeyama, S., 1988. An eigen decomposition approach to weighted graph matching problems. IEEE Trans. Pattern Anal. Machine Intell. 10 (5), 695−703.

Unger, S.H., 1958. A computer orientated towards spatial problems. Proc. IRE 46, 1744−1750.

Unser, M., 1986. Local linear transforms for texture measurements. Signal Process. 11, 61−79.

Unser, M., Eden, M., 1989. Multiresolution feature extraction and selection for texture segmentation. IEEE Trans. Pattern Anal. Machine Intell. 11 (7), 717−728.

Unser, M., Eden, M., 1990. Nonlinear operators for improving texture segmentation based on features extracted by spatial filtering. IEEE Trans. Syst., Man Cybern. 20 (4), 804−815.

Urriza, I., Artigas, J.I., Barragan, L.A., Garcia, J.I., Navarro, D., 2001. VLSI implementation of discrete wavelet transform for lossless compression of medical images. Real-Time Imaging 7 (2), 203−217.

Vaillant, R., Monrocq, C., Le Cun, Y., 1994. Original approach for the localisation of objects in images. IEE Proc. Vis. Image Signal Process. 141 (4), 245−250.

Valero, S., Chanussot, J., Benediktsson, J.A., Talbot, H., Waske, B., 2010. Advanced directional mathematical morphology for the detection of the road network in very high resolution remote sensing images. Pattern Recogn. Lett. 31 (10), 1120−1127.

van de Sande, K.E.A., Gevers, T., Snoek, C.G.M., 2010. Evaluating color descriptors for object and scene recognition. IEEE Trans. Pattern Anal. Machine Intell. 32 (9), 1582−1596.

van der Merwe, R., Doucet, A., De Freitas, N., Wan, E., 2000. The unscented particle filter. Proceedings of the Neural Information Processing Systems, Denver, Colorado, USA, 28−30 November, pp. 584−590.

van Digellen, J., 1951. Photometric investigations of the slopes and heights of the ranges of hills in the Maria of the moon. Bull. Astron. Inst. Neth. 11, 283−289.

van Dijck, H., van der Heijden, F., 2003. Object recognition with stereo vision and geometric hashing. Pattern Recogn. Lett. 24, 137−146, nos. 1−3.

Van Gool, L., Dewaele, P., Oosterlinck, A., 1985. Survey: texture analysis anno 1983. Computer Vis. Graph. Image Process. 29, 336−357.

Van Gool, L., Proesmans, M., Zisserman, A., 1998. Planar homologies as a basis for grouping and recognition. In Special Issue on Geometric Modelling and Invariants for Computer Vision. Image Vis. Comput. 16 (1), 21−26.

van Wyk, M.A., Durrani, T.S., van Wyk, B.J., 2002. A RKHS interpolator-based graph matching algorithm. IEEE Trans. Pattern Anal. Machine Intell. 24 (7), 988−995.

VanderBrug, G.J., Rosenfeld, A., 1977. Two-stage template matching. IEEE Trans. Comput. 26, 384−393.

Vapnik, V.N., 1998. Statistical Learning Theory. Wiley, New York.

Vega, I.R., Sarkar, S., 2003. Statistical motion model based on the change of feature relationships: human gait-based recognition. IEEE Trans. Pattern Anal. Machine Intell. 25 (10), 1323−1328.

Vezzani, R., Cucchiara, R., 2008. Ad-hoc: Appearance driven human tracking with occlusion handling. Proceedings of the First International Workshop on Tracking Humans for Evaluation of Motion in Image Sequences (THEMIS), Leeds, UK, pp. 9−18.

Vezzani, R., Grana, C., Cucchiara, R., 2011. Probabilistic people tracking with appearance models and occlusion classification: the AD-HOC system. Pattern Recogn. Lett. 32, 867−877.

Viola, P., Jones, M., 2001. Rapid object detection using a boosted cascade of simple features. Proceedings of the IEEE Conference on Computer Vision and Pattern Recognition, 8−14 December, Kauai, Hawaii, vol. 1, pp. 511−518.

Vistnes, R., 1989. Texture models and image measures for texture discrimination. Int. J. Computer Vis. 3, 313−336.

Wagner, G.G., 1988. Combining X-ray imaging and machine vision. Proc. SPIE 850, 43−53.

Wang, C., Sun, H., Yada, S., Rosenfeld, A., 1983. Some experiments in relaxation image matching using corner features. Pattern Recogn. 16, 167.

Wang, J., Bebis, G., Miller, R., 2005. Overtaking vehicle detection using dynamic and quasi-static background modeling. Proceedings of the IEEE Workshop on Machine Vision for Intelligent Vehicles, 21 June, San Diego, CA.

Wang, J.-G., Sung, E., 2001. Gaze determination via images of irises. Image Vis. Comput. 19 (12), 891−911.

Wang, J.-G., Sung, E., Venkateswarlu, R., 2003. Determining pose of a human face from a single monocular image. Proceedings of the British Machine Vision Association Conference, Norwich, UK, 9−11 September, pp. 103−112.

Wang, L., Bai, J., 2003. Threshold selection by clustering gray levels of boundary. Pattern Recogn. Lett. 24 (12), 1983–1999.

Wang, S., Siskind, J.M., 2003. Image segmentation with ratio cut. IEEE Trans. Pattern Anal. Machine Intell. 25 (6), 675–690.

Wang, T.W., Evans, J.P.O., 2003. Stereoscopic dual-energy X-ray imaging for target materials identification. IEE Proc. Vis. Image Signal Process. 150 (2), 122–130.

Ward, A.D., Hamarneh, G., 2010. The groupwise medial axis transform for fuzzy skeletonization and pruning. IEEE Trans. Pattern Anal. Machine Intell. 32 (6), 1084–1096.

Webb, A., 2002. Statistical Pattern Recognition. Wiley, Chichester.

Weiman, C.F.R., 1976. Highly parallel digitised geometric transformations without matrix multiplication. Proceedings of the International Joint Conference on Parallel Processing, Detroit, MI, USA, 24–27 August, pp. 1–10.

Weimer, P.K., 1975. From camera tubes to solid-state sensors. RCA Rev. 36, 385–405.

Werbos, P.J., 1974. Beyond Regression: New Tools for Prediction and Analysis in the Behavioral Sciences. PhD Thesis, Harvard University, Cambridge, MA.

Wermser, D., Liedtke, C.-E., 1982. Texture analysis using a model of the visual system. Proceedings of the Sixth International Conference on Pattern Recognition, Munich, 19–22 October, pp. 1078–1081.

Wermser, D., Haussmann, G., Liedtke, C.-E., 1984. Segmentation of blood smears by hierarchical thresholding. Computer Vis. Graph. Image Process. 25, 151–168.

Weska, J.S., 1978. A survey of threshold selection techniques. Computer Graph. Image Process. 7, 259–265.

Weska, J.S., Rosenfeld, A., 1976. An application of texture analysis to materials inspection. Pattern Recogn. 8, 195–199.

Weska, J.S., Nagel, R.N., Rosenfeld, A., 1974. A threshold selection technique. IEEE Trans. Comput. 23, 1322–1326.

Weszka, J.S., Dyer, C.R., Rosenfeld, A., 1976. A comparative study of texture measures for terrain classification. IEEE Trans. Syst., Man Cybern. 6 (4), 269–285.

Whelan, P.F., Molloy, D., 2001. Machine Vision Algorithms in Java. Springer, London.

White, J.M., Rohrer, G.D., 1983. Image thresholding for optical character recognition and other applications requiring character image extraction. IBM J. Res. Dev. 27, 400–411.

Wiehler, K., Heers, J., Schnörr, C., Stiehl, H.S., Grigat, R.-R., 2001. A one-dimensional analogue VLSI implementation for nonlinear real-time signal preprocessing. Real-Time Imaging 7 (1), 127–142.

Wiejak, J.S., Buxton, H., Buxton, B.F., 1985. Convolution with separable masks for early image processing. Computer Vis. Graph. Image Process. 32, 279–290.

Will, P.M., Pennington, K.S., 1971. Grid coding: a preprocessing technique for robot and machine vision. Artif. Intell. 2, 319–329.

Wilson, H.R., Giese, S.C., 1977. Threshold visibility of frequency gradient patterns. Vis. Res. 17, 1177–1190.

Witkin, A.P., 1981. Recovering surface shape and orientation from texture. Artif. Intell. 17, 17–45.

Witkin, A.P., 1983. Scale-space filtering. Proceedings of the Fourth International Joint Conference on Artificial Intelligence, Tbilisi, Georgi, USSR, pp. 1019–1022.

Wolfson, H.J., 1991. Generalizing the generalized Hough transform. Pattern Recogn. Lett. 12 (9), 565–573.

Wong, R.Y., Hall, E.L., 1978. Scene matching with invariant moments. Computer Graph. Image Process. 8, 16–24.

Woodham, R.J., 1978. Reflectance map techniques for analysing surface defects in metal castings. MIT Artificial Intelligence Laboratory Memo AI-TR-457.

Woodham, R.J., 1980. Photometric method for determining surface orientation from multiple images. Opt. Eng. 19, 139−144.

Woodham, R.J., 1981. Analysing images of curved surfaces. Artif. Intell. 17, 117−140.

Wu, A.Y., Hong, T.-H., Rosenfeld, A., 1982. Threshold selection using quadtrees. IEEE Trans. Pattern Anal. Machine Intell. 4, 90−94.

Wu, H., Yoshikawa, G., Shioyama, T., Lao, S., Kawade, M., 2002. Glasses frame detection with 3D Hough transform. Proceedings of the Sixteenth International Conference on Pattern Recognition, Québec, Canada, 11−15 August, vol. II, pp. 346−349.

Wu, T.-P., Tang, C.-K., 2010. Photometric stereo via expectation maximization. IEEE Trans. Pattern Anal. Machine Intell. 32 (3), 546−560.

Xie, Y., Ji, Q., 2002. A new efficient ellipse detection method. Proceedings of the Sixteenth International Conference on Pattern Recognition, Québec, Canada, 11−15 August, vol. II, pp. 957−960.

Xu, D., Li, Y.F., Tan, M., 2008. A general recursive linear method and unique solution pattern design for the perspective-n-point problem. Image Vis. Comput. 26, 740−750.

Xu, L., Oja, E., 1993. Randomized Hough transform (RHT): basic mechanisms, algorithms, and computational complexities. Computer Vis. Graph. Image Process.: Image Underst. 57 (2), 131−154.

Xu, L.-Q., Landabaso, J.L., Pardàs, M., 2005. Shadow removal with blob-based morphological reconstruction for error correction. Proceedings of the International Conference on Acoustics, Speech, Signal Process, Philadelphia, PA, USA, 18−23 March, pp. 729−732.

Xu, M., Ellis, T.J., Godsill, S.J., Jones, G.A., 2011. Visual tracking of partially observable targets with suboptimal filtering. IET Comput. Vis. 5 (1), 1−13.

Yan, C., Sang, N., Zhang, T., 2003. Local entropy-based transition region extraction and thresholding. Pattern Recogn. Lett. 24 (16), 2935−2941.

Yang, G.J., Huang, T.S., 1981. The effect of median filtering on edge location estimation. Computer Graph. Image Process. 15, 224−245.

Yang, M.-T., Gandhi, T., Kasturi, R., Coraor, L., Camps, O., Candless, J., 2002. Real-time implementation of obstacle detection algorithms on a Datacube MaxPCI architecture. Real-Time Imaging 8 (2), 157−172.

Yazdi, H.R., King, T.G., 1998. Application of 'Vision in the Loop' for inspection of lace fabric. Real-Time Imaging 4, 317−332.

Yitzhaky, Y., Peli, E., 2003. A method for objective edge detection evaluation and detector parameter selection. IEEE Trans. Pattern Anal. Machine Intell. 25 (8), 1027−1033.

Yoon, J.J., Koch, C., Ellis, T.J., 2002. Shadowflash: an approach for shadow removal in an active illumination environment. Proceedings of the British Machine Vision Association Conference, Cardiff, UK, 2−5 September, pp. 636−645.

Youn, E., Jeong, M.K., 2009. Class dependent feature scaling method using naive Bayes classifier for text datamining. Pattern Recogn. Lett. 30 (5), 477−485.

Yu, T., Zhang, C., Cohen, M., Rui, Y. and Wu, Y., 2007. Monocular video foreground/background segmentation by tracking spatial-color Gaussian mixture models. Proceedings of the IEEE Workshop on Motion and Video Computing, Austin, TX.

Yuen, H.K., Illingworth, J., Kittler, J., 1988. Ellipse detection using the Hough transform. Proceedings of the Fourth Alvey Vision Conference, Manchester, UK, 31 August−2 September, pp. 265−271.

Yuen, H.K., Princen, J., Illingworth, J., Kittler, J., 1989. A comparative study of Hough transform methods for circle finding. Proceedings of the Fifth Alvey Vision Conference, Manchester, UK, 31 August–2 September, pp. 169–174.

Yuen, P.W.T., Richardson, M., 2010. An introduction to hyperspectral imaging and its application for security, surveillance and target acquisition. Imaging Sci. 58 (5), 241–253.

Yuille, A., Poggio, T.A., 1986. Scaling theorems for zero crossings. IEEE Trans. Pattern Anal. Machine Intell. 8, 15–25.

Zahn, C.T., Roskies, R.Z., 1972. Fourier descriptors for plane closed curves. IEEE Trans. Comput. 21, 269–281.

Zayas, I.Y., Flinn, P.W., 1998. Detection of insects in bulk wheat samples with machine vision. Trans. Am. Soc. Agric. Eng. 41 (3), 883–888.

Zayas, I.Y., Steele, J.L., 1990. Image analysis applications for grain science. Optics in Agricultural SPIE 1379, 151–161.

Zhang, B., Tian, W., Jin, Z., 2006. Head tracking based on the integration of two different particle filters. Measurement Sci. Technol. 17 (11), 2877–2883.

Zhang, G., Wei, Z., 2003. A position-distortion model of ellipse centre for perspective projection. Measurement Sci. Technol. 14, 1420–1426.

Zhang, G., Jayas, D.S., White, N.D.G., 2005. Separation of touching grain kernels in an image by ellipse fitting algorithm. Biosyst. Eng. 92 (2), 135–142.

Zhang, J., Modestino, J.W., 1990. A model-fitting approach to cluster validation with application to stochastic model-based image segmentation. IEEE Trans. Pattern Anal. Machine Intell. 12 (10), 1009–1017.

Zhang, L., Wu, B., Nevatia, R., 2007. Pedestrian detection in infrared images based on local shape features. Proceedings of the Fourth Joint IEEE International Workshop on Object Tracking and Classification in and Beyond the Visible Spectrum, Minneapolis, MN, USA, Friday, 22 June, [8 pp.].

Zhang, Z., 1995. Motion and structure of four points from one motion of a stereo rig with unknown extrinsic parameters. IEEE Trans. Pattern Anal. Machine Intell. 17 (12), 1222–1227.

Zheng, D., Zhao, Y., Wang, J., 2005. An efficient method of license plate location. Pattern Recogn. Lett. 26, 2431–2438.

Zhou, H., Wallace, A.M., Green, P.R., 2003. A multistage filtering technique to detect hazards on the ground plane. Pattern Recogn. Lett. 24, 1453–1461.

Zhou, J., Fu, Z., Robles-Kelly, A., 2011. Structured learning approach to image descriptor combination. IET Comput. Vis. 5 (2), 134–142.

Zhou, Y., Xu, R., Hu, X., Ye, Q., 2006. A robust lane detection and tracking method based on computer vision. Measurement Sci. Technol. 17 (4), 736–745.

Zhu, Y., Comaniciu, D., Pellkofer, M., Koehler, T., 2004. Passing Vehicle Detection from Dynamic Background Using Robust information Fusion. Proceedings of the Conference on IEEE Intelligent Transportation Systems, Washington, DC, 3–6 October, pp. 564–569.

Zhuang, X., Haralick, R.M., 1986. Morphological structuring element decomposition. Comput. Vis., Graph, Image Process. 35, 370–382.

Zielke, T., Braukermann, M., von Seelen, W., 1993. Intensity and edge-based symmetry detection with an application to car-following. Computer Vis. Graph. Image Process.: Image Underst. 58 (2), 177–190.

Zisserman, A., Marinos, C., Forsyth, D.A., Mundy, J.L., Rothwell, C.A., 1990. Relative motion and pose from invariants. Proceedings of the First British Machine Vision Association Conference, Oxford, 24–27 September, pp. 7–12.

Zucker, S.W., 1976a. Toward a model of texture. Computer Graph. Image Process. 5, 190–202.

Zucker, S.W., 1976b. Region growing: childhood and adolescence. Computer Graph. Image Process. 5, 382–399.

Zuniga, O.A., Haralick, R.M., 1983. Corner detection using the facet model. Proc. IEEE Computer Vision Pattern Recognition Conference, pp. 30–37.

Zuniga, O.A., Haralick, R.M., 1987. Integrated directional derivative gradient operator. IEEE Trans. Pattern Anal. Machine Intell. 17, 508–517.

Author Index

845

Subject Index

PLATE 1 (FIG. 23.15)

Value of color in agricultural applications. In agricultural scenes such as this, color
helps with segmentation and with recognition. It may be crucial in discriminating between
weeds and crops if selective robot weedkilling is to be carried out.

Source: © World Scientific 2000

PLATE 2 (FIG. 2.2)

Value of color for segmentation and recognition. In natural outdoor scenes such as this,
color helps with segmentation and with recognition. While it may have been important to
the early human when discerning sources of food in the wild, robot drones may benefit by
using color to aid navigation.

PLATE 3 (FIG. 2.3)

Value of color in the built environment. Color plays an important role for the human in managing the built environment. In a vehicle, a plethora of bright lights, road signs and markings (such as yellow lines) are coded to help the driver: they may likewise help a robot to drive more safely by the provision of crucial information.

PLATE 4 (FIG. 2.4)

Value of color for food inspection. Much food is brightly colored, as with this Japanese meal. While this may be attractive to the human, it could also help the robot to check quickly for foreign bodies or toxic substances.

(a)   (b)

(c)   (d)

PLATE 5 (FIG. 3.12)

Color filtering of brightly colored objects. (a) Original color image of some sweets.
(b) Vector median filtered version. (c) Vector mode filtered version. (d) Version to which
a mode filter has been applied to each color channel separately. Note that (b) and (c)
show no evidence of color bleeding, though it is strongly evident in (d). It is most
noticeable as isolated pink pixels, plus a few green pixels, around the yellow sweets. For
further details on color bleeding, see Section 3.14.

(a)   (b)   (c)

PLATE 6 (FIG. 3.13)

Color filtering of images containing substantial impulse noise. (a) Version of the Lena
image containing 70% random color impulse noise. (b) Effect of applying a vector median
filter, and (c) effect of applying a vector mode filter. While the mode filter is designed more
for enhancement than for noise suppression, it has been found to perform remarkably
well at this task when the noise level is very high.

PLATE 7 (FIG. 23.11)

Another approach to pedestrian location via skin color detection. (a) and (b) show that a lot can be achieved via skin color detection, detecting not only faces but also neck, chest, arms and feet: see also the detail in (c) and (d). With proper color classifier training, even more can be achieved, as shown in (e) and (f).

Printed and bound by CPI Group (UK) Ltd, Croydon, CR0 4YY

03/10/2024

01040312-0003